Lecture Notes in Computer Science 4112

Commenced Publication in 1973
Founding and Former Series Editors:
Gerhard Goos, Juris Hartmanis, and Jan van Leeuwen

Lecture Notes in Computer Science 3102

Danny Z. Chen   D. T. Lee (Eds.)

Computing and Combinatorics

12th Annual International Conference, COCOON 2006
Taipei, Taiwan, August 15-18, 2006
Proceedings

 Springer

Volume Editors

Danny Z. Chen
University of Notre Dame
Department of Computer Science and Engineering
Notre Dame, IN 46556, USA
E-mail: chen@cse.nd.edu

D. T. Lee
Academia Sinica Taiwan
Institute of Information Science
No 128, Academia Road, Section 2, Nankang, Taipei 115, Taiwan
E-mail: dtlee@iis.sinica.edu.tw

Library of Congress Control Number: 2006929860

CR Subject Classification (1998): F.2, G.2, I.3.5, C.2.3-4, E.1, E.5, E.4

LNCS Sublibrary: SL 1 – Theoretical Computer Science and General Issues

ISSN   0302-9743
ISBN-10   3-540-36925-2 Springer Berlin Heidelberg New York
ISBN-13   978-3-540-36925-7 Springer Berlin Heidelberg New York

Springer is a part of Springer Science+Business Media

springer.com

© Springer-Verlag Berlin Heidelberg 2006
Printed in Germany

Typesetting: Camera-ready by author, data conversion by Scientific Publishing Services, Chennai, India
Printed on acid-free paper   SPIN: 11809678   06/3142   5 4 3 2 1 0

Preface

The papers in this volume were selected for presentation at the 12th Annual International Computing and Combinatorics Conference (COCOON 2006), held on August 15-18, 2006 in Taipei, Taiwan. Previous meetings of this conference were held in Xi'an (1995), Hong Kong (1996), Shanghai (1997), Taipei (1998), Tokyo (1999), Sydney (2000), Guilin (2001), Singapore (2002), Big Sky (2003), Jeju Island (2004), and Kunming (2005).

In response to the Call-for-Papers, 137 extended abstracts were submitted from 27 countries and regions, of which 52 were accepted (including a merged paper from two extended abstracts). The submitted papers were from Argentina (1), Australia (2), Bangladesh (1), Brazil (1), Canada (7), China (23), Germany (3), Denmark (1), France (2), UK (4), Greece (1), Hong Kong (1), Israel (4), India (11), Italy (1), Japan (12), North Korea (1), South Korea (23), Lebanon (1), Latvia (1), The Netherlands (2), Russian Federation (1), Sweden (2), Singapore (1), Taiwan (17), Ukraine (1), and USA (12).

The papers were evaluated by an international Program Committee consisting of Alberto Apostolico, Tetsuo Asano, Jin-Yi Cai, Amitabh Chaudhary, Bernard Chazelle, Danny Z. Chen, Siu-Wing Cheng, Francis Chin, Kyung-Yong Chwa, Peter Eades, Sandor Fekete, Rudolf Fleischer, Mordecai Golin, Michael Goodrich, Xiaodong Hu, Oscar H. Ibarra, Hiroshi Imai, Ming-Yang Kao, Naoki Katoh, Rolf Klein, Ming-Tat Ko, D.T. Lee, Chi-Jen Lu, Shuang Luan, Joseph S.B. Mitchell, Bernard Moret, Rajeev Motwani, David Mount, Kunsoo Park, Frank Ruskey, Michiel Smid, Chuan Yi Tang, Takeshi Tokuyama, Vijay V. Vazirani, Jie Wang, Lusheng Wang, Peter Widmayer, Xiaodong Wu, Jinhui Xu, Yinfeng Xu, Louxin Zhang, and Hong Zhu. Each paper was evaluated by at least three Program Committee members, with possible assistance of the external referees, as indicated by the referee list found in these proceedings. There were more acceptable papers than there was space available in the conference schedule, and the Program Committee's task was very difficult. In addition to the selected papers, the conference also included two invited presentations by Franco P. Preparata and Mikhail J. Atallah.

We thank all Program Committee members and the external referees for their excellent work, especially given the demanding time constraints. It was a wonderful experience to work with them. We also thank the two invited speakers and all the people who submitted papers for consideration: They all contributed to the high quality of the conference.

Finally, we thank all local organizers, colleagues, and the system team of the Institute of Information Science who worked tirelessly to put in place the logistical arrangements of the conference and to create and maintain the web system of the conference. It was their hard work that made the conference possible and enjoyable.

August 2006   Danny Z. Chen and D.T. Lee

Organization

COCOON 2006 was organized by the Institute of Information Science, Academia Sinica, Taipei, Taiwan. It was sponsored by National Science Council and Academia Sinica.

Program Committee Co-chairs

Danny Z. Chen (University of Notre Dame, USA)
D.T. Lee (Academia Sinica, Taiwan)

Program Committee

Alberto Apostolico (Georgia Institute of Technology, USA and University of Padova, Italy)
Tetsuo Asano (JAIST, Japan)
Jin-Yi Cai (University of Wisconsin at Madison, USA)
Amitabh Chaudhary (University of Notre Dame, USA)
Bernard Chazelle (Princeton University, USA)
Siu-Wing Cheng (Hong Kong University of Science and Technology, Hong Kong)
Francis Chin (University of Hong Kong, Hong Kong)
Kyung-Yong Chwa (KAIST, Korea)
Peter Eades (University of Sydney, Australia)
Sandor Fekete (Universitaet Braunschweig, Germany)
Rudolf Fleischer (Fudan University, China)
Mordecai Golin (Hong Kong University of Science and Technology, Hong Kong)
Michael Goodrich (University of California at Irvine, USA)
Xiaodong Hu (Chinese Academy of Sciences, China)
Oscar H. Ibarra (University of California at Santa Barbara, USA)
Hiroshi Imai (University of Tokyo, Japan)
Ming-Yang Kao (Northwestern University, USA)
Naoki Katoh (Kyoto University, Japan)
Rolf Klein (Universitaet Bonn, Germany)
Ming-Tat Ko (Academia Sinica, Taiwan)
Chi-Jen Lu (Academia Sinica, Taiwan)
Shuang Luan (University of New Mexico, USA)
Joseph S. B. Mitchell (State University of New York at Stony Brook, USA)
Bernard Moret (University of New Mexico, USA)
Rajeev Motwani (Stanford University, USA)
David Mount (University of Maryland at College Park, USA)
Kunsoo Park (Seoul National University, Korea)
Frank Ruskey (University of Victoria, Canada)
Michiel Smid (Carleton University, Canada)
Chuan Yi Tang (National Tsing Hua University, Taiwan)

Takeshi Tokuyama (Tohoku University, Japan)
Vijay V. Vazirani (Georgia Institute of Technology, USA)
Jie Wang (University of Massachusetts Lowell, USA)
Lusheng Wang (City University of Hong Kong, Hong Kong)
Peter Widmayer (Swiss Federal Institute of Technology Zurich, Switzerland)
Xiaodong Wu (University of Iowa, USA)
Jinhui Xu (State University of New York at Buffalo, USA)
Yinfeng Xu (Xi'an Jiaotong University, China)
Louxin Zhang (National University of Singapore, Singapore)
Hong Zhu (Fudan University, China)

Organizing Committee

Chi-Jen Lu (Academia Sinica, Taiwan; Chair)
Tsan-sheng Hsu (Academia Sinica, Taiwan)
Ming-Tat Ko (Academia Sinica, Taiwan)
Churn-Jung Liau (Academia Sinica, Taiwan)
Chris Tseng (Academia Sinica, Taiwan)
Da-Wei Wang (Academia Sinica, Taiwan)
Chiou-Feng Wang (Academia Sinica, Taiwan)

Referees

Richard Anstee	Ashish Gehani	William J. Proctor
Estie Arkin	Tero Harju	Kirk Pruhs
Abdullah Arslan	Thomas Hofmeister	B. Ravikumar
F. Betul Atalay	C. C. Huang	Nils Schweer
Martin Baca	Takehiro Itoh	Vikas Singh
Mihir Bellare	Volker Kaibel	Dilys Thomas
Paz Carmi	Hee-Chul Kim	Kenya Ueno
Maw-Shang Chang	Jae-Hoon Kim	Marc Uetz
Chandra Chekuri	Sven Krumke	Takeaki Uno
Zhide Chen	C. C. Li	Yang Wang
Minkyoung Cho	Guoliang Li	W. B. Wang
C. Y. Chong	Ming Li	Thomas Wolle
Stirling Chow	Mingen Lin	Jenni Woodcock
Jinhee Chun	Nargess Memarsadeghi	Yang Yang
Scott Craig	Sonoko Moriyama	Hsu-Chun Yen
Qi Duan	Shubha Nabar	S. M. Yiu
Richard Duke	Hiroyuki Nakayama	Maxwell R. Young
Adrian Dumitrescu	Hung Q. Ngo	K. L. Yu
Rusins Freivalds	C. Thach Nguyen	Huaming Zhang
Francois Le Gall	Chong-Dae Park	Lijuan Zhao

Table of Contents

Invited Talks

The Unpredictable Deviousness of Models
  Franco P. Preparata 1

Security Issues in Collaborative Computing
  Mikhail J. Atallah 2

Session A

Computational Economics, Finance, and Management

A Simplicial Approach for Discrete Fixed Point Theorems
  Xi Chen, Xiaotie Deng 3

On Incentive Compatible Competitive Selection Protocol
  Xi Chen, Xiaotie Deng, Becky Jie Liu 13

Edge Pricing of Multicommodity Networks for Selfish Users
with Elastic Demands
  George Karakostas, Stavros G. Kolliopoulos 23

Aggregating Strategy for Online Auctions
  Shigeaki Harada, Eiji Takimoto, Akira Maruoka 33

Graph Algorithms

On Indecomposability Preserving Elimination Sequences
  Chandan K. Dubey, Shashank K. Mehta 42

Improved Algorithms for the Minmax Regret 1-Median Problem
  Hung-I Yu, Tzu-Chin Lin, Biing-Feng Wang 52

Partitioning a Multi-weighted Graph to Connected Subgraphs
of Almost Uniform Size
  Takehiro Ito, Kazuya Goto, Xiao Zhou, Takao Nishizeki 63

Characterizations and Linear Time Recognition of Helly Circular-Arc
Graphs
  Min Chih Lin, Jayme L. Szwarcfiter   73

Computational Complexity and Computability

Varieties Generated by Certain Models of Reversible
Finite Automata
  Marats Golovkins, Jean-Eric Pin   83

Iterated TGR Languages: Membership Problem and Effective Closure
Properties
  Ian McQuillan, Kai Salomaa, Mark Daley   94

On the Negation-Limited Circuit Complexity of Sorting and Inverting
k-tonic Sequences
  Takayuki Sato, Kazuyuki Amano, Akira Maruoka 104

Quantum Computing

Robust Quantum Algorithms with ε-Biased Oracles
  Tomoya Suzuki, Shigeru Yamashita, Masaki Nakanishi,
  Katsumasa Watanabe 116

The Complexity of Black-Box Ring Problems
  V. Arvind, Bireswar Das, Partha Mukhopadhyay 126

Computational Biology and Medicine

Lower Bounds and Parameterized Approach for Longest Common
Subsequence
  Xiuzhen Huang 136

Finding Patterns with Variable Length Gaps
or Don't Cares
  M. Sohel Rahman, Costas S. Iliopoulos, Inbok Lee, Manal Mohamed,
  William F. Smyth 146

The Matrix Orthogonal Decomposition Problem in Intensity-Modulated
Radiation Therapy
  Xin Dou, Xiaodong Wu, John E. Bayouth,
  John M. Buatti 156

Computational Geometry

A Polynomial-Time Approximation Algorithm for a Geometric
Dispersion Problem
  Marc Benkert, Joachim Gudmundsson, Christian Knauer,
  Esther Moet, René van Oostrum, Alexander Wolff 166

A PTAS for Cutting Out Polygons with Lines
  Sergey Bereg, Ovidiu Daescu, Minghui Jiang 176

On Unfolding Lattice Polygons/Trees and Diameter-4 Trees
  Sheung-Hung Poon 186

Restricted Mesh Simplification Using Edge Contractions
  Mattias Andersson, Joachim Gudmundsson,
  Christos Levcopoulos 196

Graph Theory

Enumerating Non-crossing Minimally Rigid Frameworks
  David Avis, Naoki Katoh, Makoto Ohsaki, Ileana Streinu,
  Shin-ichi Tanigawa 205

Sequences Characterizing k-Trees
  Zvi Lotker, Debapriyo Majumdar, N.S. Narayanaswamy,
  Ingmar Weber 216

On the Threshold of Having a Linear Treewidth in Random
Graphs
  Yong Gao 226

Computational Biology

Reconciling Gene Trees with Apparent Polytomies
  Wen-Chieh Chang, Oliver Eulenstein 235

Lower Bounds on the Approximation of the Exemplar Conserved
Interval Distance Problem of Genomes
  Zhixiang Chen, Richard H. Fowler, Bin Fu, Binhai Zhu 245

Computing Maximum-Scoring Segments in Almost
Linear Time
  Fredrik Bengtsson, Jingsen Chen 255

Session B

Graph Algorithms and Applications

Enumerate and Expand: New Runtime Bounds for Vertex Cover
Variants
  Daniel Mölle, Stefan Richter, Peter Rossmanith 265

A Detachment Algorithm for Inferring a Graph from Path Frequency
  Hiroshi Nagamochi 274

The *d*-Identifying Codes Problem for Vertex Identification in Graphs:
Probabilistic Analysis and an Approximation Algorithm
  Ying Xiao, Christoforos Hadjicostis, Krishnaiyan Thulasiraman 284

Reconstructing Evolution of Natural Languages: Complexity and
Parameterized Algorithms
  Iyad A. Kanj, Luay Nakhleh, Ge Xia 299

On-Line Algorithms

On Dynamic Bin Packing: An Improved Lower Bound and Resource
Augmentation Analysis
  Wun-Tat Chan, Prudence W.H. Wong, Fencol C.C. Yung 309

Improved On-Line Broadcast Scheduling with Deadlines
  Feifeng Zheng, Stanley P.Y. Fung, Wun-Tat Chan,
  Francis Y.L. Chin, Chung Keung Poon, Prudence W.H. Wong 320

A Tight Analysis of Most-Requested-First for On-Demand Data
Broadcast
  Regant Y.S. Hung, H.F. Ting 330

On Lazy Bin Covering and Packing Problems
  Mingen Lin, Yang Yang, Jinhui Xu 340

Graph Theory

Creation and Growth of Components in a Random Hypergraph
Process
  Vlady Ravelomanana, Alphonse Laza Rijamamy 350

Optimal Acyclic Edge Colouring of Grid Like Graphs
  Rahul Muthu, N. Narayanan, C.R. Subramanian 360

An Edge Ordering Problem of Regular Hypergraphs
Hongbing Fan, Robert Kalbfleisch 368

Algorithms for Security and Systems

Efficient Partially Blind Signature Scheme with Provable
Security
Zheng Gong, Xiangxue Li, Kefei Chen 378

A Rigorous Analysis for Set-Up Time Models – A Metric
Perspective
Eitan Bachmat, Tao Kai Lam, Avner Magen 387

Discrete Geometry and Graph Theory

Geometric Representation of Graphs in Low Dimension
L. Sunil Chandran, Naveen Sivadasan 398

The On-Line Heilbronn's Triangle Problem in d Dimensions
Gill Barequet, Alina Shaikhet 408

Counting d-Dimensional Polycubes and Nonrectangular Planar
Polyominoes
Gadi Aleksandrowicz, Gill Barequet 418

Approximation Algorithms

Approximating Min-Max (Regret) Versions of Some Polynomial
Problems
Hassene Aissi, Cristina Bazgan, Daniel Vanderpooten 428

The Class Constrained Bin Packing Problem with Applications to
Video-on-Demand
E.C. Xavier, F.K. Miyazawa 439

MAX-SNP Hardness and Approximation of Selected-Internal Steiner
Trees
Sun-Yuan Hsieh, Shih-Cheng Yang 449

Minimum Clique Partition Problem with Constrained Weight
for Interval Graphs
Jianbo Li, Mingxia Chen, Jianping Li, Weidong Li 459

Computational Complexity and Computability

Overlap-Free Regular Languages
   Yo-Sub Han, Derick Wood 469

On the Combinatorial Representation of Information
   Joel Ratsaby 479

Finding Small OBDDs for Incompletely Specified Truth Tables Is Hard
   Jesper Torp Kristensen, Peter Bro Miltersen 489

Experimental Algorithms

Bimodal Crossing Minimization
   Christoph Buchheim, Michael Jünger, Annette Menze,
   Merijam Percan 497

Fixed Linear Crossing Minimization by Reduction to the Maximum
Cut Problem
   Christoph Buchheim, Lanbo Zheng 507

On the Effectiveness of the Linear Programming Relaxation of the 0-1
Multi-commodity Minimum Cost Network Flow Problem
   Dae-Sik Choi, In-Chan Choi 517

Author Index 527

The Unpredictable Deviousness of Models

Franco P. Preparata

Department of Computer Science, Brown University
franco@cs.brown.edu

The definition of computation models is central to algorithmic research. A model is designed to capture the essential features of the technology considered, dispensing with irrelevant and burdensome details. In other words, it is a judicious compromise between simplicity, for the ease of analysis, and fidelity (or reflectivity), for the value of the derived predictions. The achievement of this double objective has unleashed an enormous amount of valuable algorithmic research over the years.

However, the pursuit of simplicity may filter out details, once deemed irrelevant, which may later reassert their significance either under technological pressure or under more careful scrutiny. In such instances, the results may be invalidated by the inadequacy of the model, Examples of this situation, drawn from computational geometry, parallel computation, and computational biology, will be reviewed and examined in detail.

D.Z. Chen and D.T. Lee (Eds.): COCOON 2006, LNCS 4112, p. 1, 2006.
© Springer-Verlag Berlin Heidelberg 2006

Security Issues in Collaborative Computing
(Abstract of Keynote Talk)

Mikhail J. Atallah

Department of Computer Science
Purdue University
CERIAS, Recitation Bldg, West Lafayette, IN 47907, USA
mja@cs.purdue.edu

Even though collaborative computing can yield substantial economic, social, and scientific benefits, a serious impediment to fully achieving that potential is a reluctance to share data, for fear of losing control over its subsequent dissemination and usage. An organization's most valuable and useful data is often proprietary/confidential, or the law may forbid its disclosure or regulate the form of that disclosure. We survey security technologies that mitigate this problem, and discuss research directions towards enforcing the data owner's approved purposes on the data used in collaborative computing. These include techniques for cooperatively computing answers without revealing any private data, even though the computed answers depend on all the participants' private data. They also include computational outsourcing, where computationally weak entities use computationally powerful entities to carry out intensive computing tasks without revealing to them either their inputs or the computed outputs.

D.Z. Chen and D.T. Lee (Eds.): COCOON 2006, LNCS 4112, p. 2, 2006.

A Simplicial Approach for Discrete
Fixed Point Theorems
(Extended Abstract)

Xi Chen[1] and Xiaotie Deng[2,*]

[1] Department of Computer Science, Tsinghua University
`xichen00@mails.tsinghua.edu.cn`
[2] Department of Computer Science, City University of Hong Kong
`deng@cs.cityu.edu.hk`

Abstract. We present a new discrete fixed point theorem based on a
novel definition of direction-preserving maps over simplicial structures.
We show that the result is more general and simpler than the two recent discrete fixed point theorems by deriving both of them from ours.
The simplicial approach applied in the development of the new theorem
reveals a clear structural comparison with the classical approach for the
continuous case.

1 Introduction

There has recently been a sequence of works related to fixed point theorems in a
discrete disguise, started with the seminal work of Iimura [13] which introduced
a crucial concept of direction-preserving maps. Iimura, Murota and Tamura [14]
corrected the proof of Iimura for the definition domains of the maps. With a
different technique, Chen and Deng introduced another discrete fixed point theorem in order to achieve the optimal algorithmic bound for finding a discrete
fixed point for all finite dimensions [2]. In [15], Laan, Talman and Yang designed
an iterative algorithm for the discrete zero point problem. Based on Sperner's
lemma which is fundamental for deriving Brouwer's fixed point theorem, Friedl,
Ivanyosy, Santha and Verhoeven defined the black-box Sperner problems. They
also obtained a $\sqrt{n}$ upper bound for the two-dimensional case [11], which is also
a matching bound when combined with the lower bound of Crescenzi and Silvestri [8] (mirroring an early result of Hirsch, Papadimitriou and Vavasis on the
computation of 2D approximate fixed points [12]). On the other hand, Chen and
Deng [6] showed that the two theorems, that of Iimura, Murota and Tamura [14],
as well as that of Chen and Deng [2], cannot directly derive each other.

In this article, we derive a new discrete fixed point theorem based on simplicial
structures and a novel definition of direction-preserving maps. We show that both
previous discrete fixed point theorems can be derived from this simpler one.

The simplicial structure, together with Sperner's Lemma, has played an important role in establishing various continuous fixed point theorems. Our focus

---

* Work supported by an SRG grant (No. 7001838) of City University of Hong Kong.

D.Z. Chen and D.T. Lee (Eds.): COCOON 2006, LNCS 4112, pp. 3–12, 2006.

on the simplicial structure in the study of the discrete version will help us gain a full and clear understanding of the mathematical structures and properties related to discrete fixed point theorems. Furthermore, even for continuous fixed point theorems, discrete structural propositions are needed to derive them. Our study would provide a unified view of the fixed point theorem, both discrete and continuous, instead of treating them with ad hoc techniques. Our simplicial approach unveils the mystery behind the recent results on discrete fixed points and settles them under the same mathematical foundation as the classical continuous fixed point theorems.

The discrete nature of the fixed point theorem has been noticed previously, mainly due to the proof techniques of Sperner's lemma [16]. The recent effort in direct formulation of the discrete version of the fixed point theorem would be especially useful in the complexity analysis of related problems. The recent work in characterizing the complexity of Nash Equilibria, by Daskalakis, Goldberg, Papadimitriou [9], Chen and Deng [3], Daskalakis and Papadimitriou [10], Chen and Deng [4], has been based on another innovative formulation of the 2D (or 3D) discrete fixed point problem, where a fixed point is a collection of four [7] (or eight [9]) corners of a unit square (or cube). It's difficult to generalize such a formulation to high dimensional spaces, since a hypercube has an exponential number of corners, which is computationally infeasible. Instead, a simplicial definition has been necessary in extending those results to a non-approximability work obtained recently [10].

We first introduce notations and definitions with a review of previous works of Murota, Iimura and Tamura [14], as well as Chen and Deng [2]. The simplicial model is then introduced in section 3 and the fundamental discrete fixed point theorem is proved in section 4. In section 5, we present the discrete Brouwer's fixed point theorem for simplicial direction-preserving maps, with the theorem of Murota, Iimura and Tamura [14] derived as a simple corollary. In Section 6, we give an explicit explanation for the definition of bad cubes in [2] and show that, the theorem of Chen and Deng is a special case of the fundamental fixed point theorem. Finally, we conclude in section 7.

2   Preliminaries

2.1   Notations and Definitions

Informally speaking, map $\mathcal{F}$ (or function f) is hypercubic direction-preserving on a finite set $X \subset \mathbb{Z}^d$ if for every two neighboring points in X, their directions given by $\mathcal{F}$ (or f) are not opposite. The neighborhood relation considered here is defined by the infinity norm.

Definition 1 (Hypercubic Direction-Preserving Maps). *Let X be a finite subset of $\mathbb{Z}^d$. Map $\mathcal{F}$ from X to $\mathbb{R}^d$ is said to be hypercubic direction-preserving* on X if for every two points $r^1, r^2 \in X$ with $||r^1 - r^2||_\infty \leq 1$, we have $(\mathcal{F}_i(r^1) - r_i^1)(\mathcal{F}_i(r^2) - r_i^2) \geq 0$, for all $i : 1 \leq i \leq d$.

Definition 2 (Hypercubic Direction-Preserving Functions). *Let X be a finite subset of $\mathbb{Z}^d$. Function f from set X to $\{0, \pm e^1, \pm e^2 ... \pm e^{d-1}, \pm e^d\}$ is said to be hypercubic direction-preserving if for every two points r^1, $r^2 \in X$ such that $||r^1 - r^2||_\infty \leq 1$, we have $||f(r^1) - f(r^2)||_\infty \leq 1$.*

Point $r \in X$ is called a fixed point of $\mathcal{F}$ (or f) if $\mathcal{F}(r) = r$ (or $f(r) = 0$).

2.2 The Fixed Point Theorem of Murota, Iimura and Tamura

Murota, Iimura and Tamura proved in [14] that every hypercubic direction-preserving map from an integrally convex set X to $\overline{X}$ must have a fixed point. Here we use $\overline{X}$ to denote the convex hull of finite set $X \subset \mathbb{Z}^d$.

Definition 3 (Integrally Convex Sets). *Finite set $X \subset \mathbb{Z}^d$ is integrally convex if for all $x \in \overline{X}$, $x \in \overline{X \cap N(x)}$ where $N(x) = \{r \in \mathbb{Z}^d \mid ||r - x||_\infty < 1\}$.*

Theorem 1 ([14]). *Let X be an integrally convex set in $\mathbb{Z}^d$, then every hypercubic direction-preserving map $\mathcal{F}$ from X to $\overline{X}$ has a fixed point in X.*

2.3 The Fixed Point Theorem of Chen and Deng

Given a hypercubic direction-preserving function f on a lattice set $C_{a,b} \subset \mathbb{Z}^d$, Chen and Deng proved in [2] that if the number of bad $(d-1)$-cubes on the boundary of $C_{a,b}$ is odd, then f must have a fixed point in $C_{a,b}$.

Definition 4. *Lattice set $C_{a,b} \subset \mathbb{Z}^d$ is defined as $C_{a,b} = \{r \in \mathbb{Z}^d \mid \forall 1 \leq i \leq d, a \leq r_i \leq b\}$. For every $r \in \mathbb{Z}^d$ and $S \subset \{1, 2 ... d\}$ with $|S| = d - t$, the t-cube $C^t \subset \mathbb{Z}^d$ which is centered at r and perpendicular to S is defined as $C^t = \{p \in \mathbb{Z}^d \mid \forall 1 \leq i \leq d, \text{ if } i \in S, \text{ then } p_i = r_i. \text{ Otherwise, } p_i = r_i \text{ or } r_i + 1\}$.*

Definition 5 (Bad Cubes). *A 0-cube $C^0 \subset \mathbb{Z}^d$ is bad relative to function f if $f(C^0) = \{e^1\}$. For $1 \leq t \leq d - 1$, a t-cube $C^t \subset \mathbb{Z}^d$ is bad relative to f if:*

1. *$f(C^t) = \{e^1, e^2 ... e^{t+1}\}$;*
2. *the number of bad $(t-1)$-cubes in C^t is odd.*

Theorem 2 ([2]). *Let f be a hypercubic direction-preserving function on $C_{a,b} \subset \mathbb{Z}^d$, if N_B, i.e. the number of bad $(d-1)$-cubes on the boundary of $C_{a,b}$ is odd, then f must have a fixed point r in $C_{a,b}$ such that $f(r) = 0$.*

Although the theorem itself is succinct, the definition of bad cubes seems a little mysterious and lacks a satisfactory explanation. In section 6, we will use the fundamental discrete fixed point theorem for the simplicial model to resolve this puzzle.

3   Simplicial Direction-Preserving Maps and Functions

In this section, we introduce simplicial direction-preserving maps and functions based on simplicial structures. Let X be a finite set in $\mathbb{R}^d$. Here we only consider nondegenerate cases where $\overline{X} \subset \mathbb{R}^d$ is a convex d-polytope. For standard definitions concerning polytopes, readers are referred to [17] for details.

Definition 6. *A simplicial decomposition* $\mathcal{S}$ *of* $C \subset \mathbb{R}^d$ *is a collection of simplices satisfying:* **1**). $C = \cup_{S \in \mathcal{S}} S$; **2**). *For any* $S \in \mathcal{S}$, *if* S' *is a face of* S, *then* $S' \in \mathcal{S}$; **3**). *For every two simplices* $S_1, S_2 \in \mathcal{S}$, *if* $S_1 \cap S_2 \neq \emptyset$, *then* $S_1 \cap S_2$ *is a face of both* S_1 *and* S_2.

Definition 7. *Let* X *be a finite set in* $\mathbb{R}^d$. *A simplicial decomposition* $\mathcal{S}$ *of set* X *is a simplicial decomposition of* $\overline{X}$ *such that for every* $S \in \mathcal{S}$, $V_S \subset X$, *where* V_S *is the vertex set of simplex* S.

Given a simplicial decomposition $\mathcal{S}$ *of* X, *we use* $F_{\mathcal{S}}$ *to denote the set of* $(d-1)$-*simplices on the boundary of* $\overline{X}$, *and* B_X *to denote the set of points on the boundary of* $\overline{X}$: $F_{\mathcal{S}} = \{ (d-1)$-*simplex* $S \in \mathcal{S} \mid S \subset F$ *and* F *is a facet of* $\overline{X} \}$, *and* $B_X = \{ r \in X \mid r \in F$ *and* F *is a facet of* $\overline{X} \}$.

Definition 8 (Simplicial Direction-Preserving Maps). *A simplicial direction-preserving map is a triple* $M = (\mathcal{F}, X, \mathcal{S})$. *Here* X *is a finite set in* $\mathbb{R}^d$ *and* $\mathcal{S}$ *is a simplicial decomposition of* X. *Map* $\mathcal{F}$ *from* X *to* $\mathbb{R}^d$ *satisfies for every two points* $r^1, r^2 \in X$, *if there exists a simplex* $S \in \mathcal{S}$ *such that* $r^1, r^2 \in V_S$, *then* $(\mathcal{F}_i(r^1) - r_i^1)(\mathcal{F}_i(r^2) - r_i^2) \geq 0$, *for all* $i : 1 \leq i \leq d$.

Definition 9 (Simplicial Direction-Preserving Functions). *A triple* $G = (f, X, \mathcal{S})$ *is said to be a simplicial direction-preserving function if* X *is a finite set in* $\mathbb{R}^d$, $\mathcal{S}$ *is a simplicial decomposition of* X, *and function* f *from set* X *to* $\{ 0, \pm e^1, \ldots \pm e^d \}$ *satisfies for every two points* $r^1, r^2 \in X$, *if there exists* $S \in \mathcal{S}$ *such that* $r^1, r^2 \in V_S$, *then* $\| f(r^1) - f(r^2) \|_\infty \leq 1$.

In other words, for every two neighboring points in X, their directions given by map $\mathcal{F}$ (or function f) can't be opposite. The only difference with the hypercubic model is that the neighborhood relation is now defined by simplices in the simplicial decomposition $\mathcal{S}$ instead of unit d-cubes in $\mathbb{Z}^d$.

4   The Fundamental Discrete Fixed Point Theorem

In this section, we present the fundamental discrete fixed point theorem which is both simple and powerful. Any simplicial direction-preserving function which satisfies the boundary condition of the theorem must have a fixed point.

Definition 10 (Bad Simplices). *Let* $G = (f, X, \mathcal{S})$ *be a simplicial direction-preserving function, where* $X \subset \mathbb{Z}^d$. *A* t-*simplex* $S \in \mathcal{S}$ *where* $0 \leq t \leq d$ *is said to be bad (relative to function* G) *if* $f(V_S) = \{ e^1, e^2, \ldots e^{t+1} \}$, *where* V_S *is the vertex set of* S. *We use* N_G *to denote the number of bad* $(d-1)$-*simplices in* $F_{\mathcal{S}}$.

Lemma 1. *For any simplicial direction-preserving function $G = (f, X, \mathcal{S})$, if there exists no fixed point in X, then N_G is even.*

Proof. Firstly, one can show that for every $(d-1)$-simplex $S \in \mathcal{S}$, if $S \in F_{\mathcal{S}}$, then there exists exactly one d-simplex in $\mathcal{S}$ containing S. Otherwise, there are exactly two such simplices. Using this property, the parity of N_G is same as the one of the following summation:

$$\sum_{d\text{-simplex } S^d \in \mathcal{S}} \Big| \{ \text{ bad } (d-1)\text{-simplices in } S^d \} \Big|.$$

As G is direction-preserving and has no fixed point, the number of bad $(d-1)$-simplices in S^d is either 0 or 2. Therefore, the summation above must be even.

We now get the fundamental theorem as a simple corollary of Lemma 1.

Theorem 3 (The Fundamental Discrete Fixed Point Theorem). *Let $G = (f, X, \mathcal{S})$ be a simplicial direction-preserving function. If N_G, i.e. the number of bad $(d-1)$-simplices on the boundary is odd, then G must have a fixed point $r \in X$ such that $f(r) = 0$.*

5   The Discrete Brouwer's Fixed Point Theorem

In this section, the fundamental discrete fixed point theorem will be employed to prove a fixed point theorem concerning simplicial direction-preserving maps. It can be recognized as a discrete version of Brouwer's fixed point theorem. It states that for any simplicial direction-preserving map from some finite set to its convex hull, there must exist a fixed point in the definition domain.

We will also derive the theorem of Murota, Iimura and Tamura as a simple corollary. Actually, the one derived here is much stronger than theirs.

5.1   Preliminaries

We use e^k to denote the kth unit vector of $\mathbb{Z}^d$ where $e_k^k = 1$ and $e_i^k = 0$ for all $i : 1 \leq i \neq k \leq d$.

Definition 11. *For every $(d-1)$-simplex $S \in F_{\mathcal{S}}$, we let e_S be the unit vector which is outgoing and perpendicular to S. For all $r \in \overline{X}$ and $r_S \in S$, we have $e_S \cdot (r - r_S) \leq 0$.*
$S \in F_{\mathcal{S}}$ is visible from point $r \notin \overline{X}$ if $e_S \cdot (r - r_S) > 0$ for some $r_S \in S$.

Construction 1 (Extension of Simplicial Decomposition). *Let $X \subset \mathbb{R}^d$ be a finite set and $\mathcal{S}$ be a simplicial decomposition of X. For every point $r \notin \overline{X}$, we can add new simplices into $\mathcal{S}$ and build a simplicial decomposition $\mathcal{S}'$ of set $X' = X \cup \{r\}$ as follows. For every $(d-1)$-simplex $S \in F_{\mathcal{S}}$ visible from r, we add d-simplex $\mathrm{conv}(S, r)$ and all its faces into $\mathcal{S}$. One can check that $\mathcal{S}'$ is a simplicial decomposition of X', and $\mathcal{S} \subset \mathcal{S}'$.*

Given a simplicial direction-preserving map $M = (\mathcal{F}, X, \mathcal{S})$, we can convert it into a direction-preserving function $G = (f, X, \mathcal{S})$ as follows.

Construction 2. *Given a simplicial direction-preserving map $M = (\mathcal{F}, X, \mathcal{S})$, we can build a simplicial direction-preserving function $G = (f, X, \mathcal{S})$ as follows. For every $r \in X$, if $\mathcal{F}(r) = r$, then $f(r) = 0$. Otherwise, let $i : 1 \leq i \leq d$ be the smallest integer such that $\mathcal{F}_i(r) - r_i \neq 0$, then $f(r) = \mathrm{sign}\,(\mathcal{F}_i(r) - r_i)e^i$.*

5.2   The Key Lemma

Lemma 2. *Let $M = (\mathcal{F}, X, \mathcal{S})$ be a simplicial direction-preserving map where $\mathcal{F}$ is from X to $\overline{X}$, and $G = (f, X, \mathcal{S})$ be the function constructed above, then either f has a fixed point in B_X or N_G is odd.*

Proof (Proof Sketch). Let $n = \max_{r \in X, 1 \leq i \leq d} |r_i|$, then we can scale down X to be $X' \subset (-1, 1)^d$ where $X' = \{\, r/(n+1),\ r \in X \,\}$. We also get a simplicial decomposition $\mathcal{S}'$ of X' from $\mathcal{S}$ using the one-to-one correspondence between X and X', and a map $\mathcal{F}'$ from X' to $\overline{X'}$ where $\mathcal{F}'(r) = \mathcal{F}((n+1)r)/n + 1$.

Let G' be the function constructed from map $M' = (\mathcal{F}', X', \mathcal{S}')$, then it is easy to check that $N_G = N_{G'}$. Therefore, we only need to prove the lemma for maps $M = (\mathcal{F}, X, \mathcal{S})$ with $X \subset (-1, 1)^d$. From now on, we always assume that $X \subset (-1, 1)^d$.

If f has a fixed point in set B_X, then the lemma is proven. Otherwise, we extend (by applying Construction 1 for d times) $G = (f, X, \mathcal{S})$ to be a new function $G^* = (f^*, X^*, \mathcal{S}^*)$ such that $X \subset X^*$, $\overline{X^*} = [-1, 1]^d$ and $\mathcal{S} \subset \mathcal{S}^*$. After proving that G^* is simplicial direction-preserving, we show the following two properties of G and G^*:

Property 1. N_{G^*} is odd;

Property 2. $N_G \equiv N_{G^*} \pmod 2$,

and the lemma is proven.

Details of the proof can be found in the full version [1].

5.3   The Discrete Brouwer's Fixed Point Theorem

From Construction 2, every fixed point of function f is also a fixed point of map $\mathcal{F}$. By Theorem 3 and Lemma 2, we get the following theorem immediately.

Theorem 4 (The Discrete Brouwer's Fixed Point Theorem). *For every simplicial direction-preserving map $M = (\mathcal{F}, X, \mathcal{S})$ such that $\mathcal{F}$ maps X to $\overline{X}$, there must exist a fixed point in X.*

Now we prove the fixed point theorem of Murota, Iimura and Tamura [14] as a direct corollary of Theorem 4.

Lemma 3 (Property of Integrally Convex Sets [14]). *For every integrally convex set X, there exists a simplicial decomposition $\mathcal{S}$ of $\overline{X}$, which satisfies for every $x \in \overline{X}$, letting $S_x \in \mathcal{S}$ be the smallest simplex containing x, then all of its vertices belong to $N(x) = \{\, r \in \mathbb{Z}^d \mid ||r - x||_\infty < 1 \,\}$.*

Let $\mathcal{F}$ be a hypercubic direction-preserving map from integrally convex set $X \subset \mathbb{Z}^d$ to $\overline{X}$, and $\mathcal{S}$ be a simplicial decomposition of X which satisfies the condition in Lemma 3, then one can check that $M = (\mathcal{F}, X, \mathcal{S})$ is a simplicial direction-preserving map from X to $\overline{X}$. By Theorem 4, we know that there is a fixed point of $\mathcal{F}$ in X.

Moreover, the argument above shows that the theorem of Murota, Iimura and Tamura can be greatly strengthened. Actually, map $\mathcal{F}$ is not necessary to be hypercubic direction-preserving. Being simplicial direction-preserving relative to some simplicial decomposition of X is sufficient to ensure the existence of a fixed point in X.

6   An Explanation for the Definition of Bad Cubes

Chen and Deng [2] defined the badness of $(d-1)$-cubes relative to hypercubic direction-preserving functions in d-dimensional space, and showed that for any hypercubic direction-preserving function f on $C_{a,b} \subset \mathbb{Z}^d$, if the number of bad $(d-1)$-cubes on the boundary is odd, then f must have a fixed point in $C_{a,b}$. While the theorem itself is succinct, the definition of bad cubes seems a little mysterious and lacks a satisfactory explanation. In this section, we will use the simplicial model developed in section 3 and 4 to resolve this puzzle.

First, we add extra points into the lattice set $C_{a,b} \subset \mathbb{Z}^d$ and construct a simplicial decomposition for the new set $D_{a,b}$, where $\overline{C_{a,b}} = \overline{D_{a,b}}$. Then, we extend the hypercubic direction-preserving function f on $C_{a,b}$ to be a simplicial direction-preserving function on $D_{a,b}$. Finally, we prove that the parity of N_B is same as N_G, where N_B is the number of bad $(d-1)$-cubes and N_G is the number of bad $(d-1)$-simplices on the boundary. In this way, we show that Chen and Deng's theorem [2] is a special case of the fundamental discrete fixed point theorem.

6.1   Preliminaries

Definition 12. *A convex subdivision $\mathcal{P}$ of a finite set $X \subset \mathbb{R}^d$ is a collection of convex d-polytopes such that: 1). $\overline{X} = \cup_{P \in \mathcal{P}} P$, and for every polytope $P \in \mathcal{P}$, all of its vertices are drawn from X; 2). For every two polytopes $P_1, P_2 \in \mathcal{P}$, if $P_1 \cap P_2 \neq \emptyset$, then $P_1 \cap P_2$ is a face of both P_1 and P_2.*

Definition 13. *Let P be a convex t-polytope in $\mathbb{R}^d$ and V_P be its vertex set. The center point c_P of polytope P is defined as $c_P = \sum_{r \in V_P} (r/|V_P|)$. Obviously, we have $c_P \in P$ and $c_P \notin V_P$.*

For example, let $C \subset \mathbb{Z}^d$ be a t-cube centered at $r \in \mathbb{Z}^d$ and perpendicular to T, then the center point c of $\overline{C}$ satisfies that $c_k = r_k$ for every $k \in T$, and $c_k = r_k + 1/2$ for every $k \notin T$.

Let $\mathcal{P}$ be a convex subdivision of set X in $\mathbb{R}^d$. We now add extra points $r \in \overline{X}$ into X and construct a simplicial decomposition $\mathcal{S}'$ for the new set X'. Details of the construction are described by the algorithm in Figure 1.

1: $\mathcal{S}' = \{ \{r\} \mid r \in X \}$ and $X' = X$
2: **for any** t **from** 1 **to** d **do**
3:      **for any** F that is a t-face of some d-polytope in $\mathcal{P}$ **do**
4:          add the center point c_F of F into X'
5:          **for any** $(t-1)$-simplex $S \in \mathcal{S}'$ and $S \subset F$ **do**
6:              add every face of t-simplex $\mathrm{conv}(S, c_F)$ into $\mathcal{S}'$

Fig. 1. The Construction of $\mathcal{S}'$ and X'

Every lattice set $C_{a,b} \subset \mathbb{R}^d$ has a natural convex subdivision $\mathcal{P}$ where $\mathcal{P} = \{ \overline{C} \mid d\text{-cube } C \subset C_{a,b} \}$. Using Figure 1, we get a simplicial decomposition $\mathcal{S}$ of

$$ D_{a,b} = \left\{ r \in \mathbb{R}^d \mid \forall\, 1 \le i \le d,\ a \le r_i \le b \text{ and } \exists\, r' \in \mathbb{Z}^d,\ r = r'/2 \right\}. $$

6.2   Extension of Hypercubic Direction-Preserving Functions

Let f be a hypercubic direction-preserving function on $C_{a,b}$, we now extend it onto set $D_{a,b}$ as follows. For every $r \in D_{a,b} - C_{a,b}$, assume it is the center point of t-cube $C \subset C_{a,b}$. If $0 \in f(C)$, then $f(r) = 0$. Otherwise, let $1 \le t \le d$ be the largest integer such that $f(C) \cap \{\pm e^t\} \ne \emptyset$, then $f(r) = e^t$ if $e^t \in f(C)$ and $f(r) = -e^t$ if $-e^t \in f(C)$. One can prove the following two properties.

Property 1. *Let f be a hypercubic direction-preserving function on $C_{a,b} \subset \mathbb{Z}^d$, then $G = (f, D_{a,b}, \mathcal{S})$ is a simplicial direction-preserving function.*

Property 2. *If the extended function $G = (f, D_{a,b}, \mathcal{S})$ has a fixed point in $D_{a,b}$, then the original function must have a fixed point in $C_{a,b}$.*

6.3   The Nature of Bad Cubes

We are ready to give an explicit explanation for the definition of bad cubes.

Lemma 4. *Let f be a hypercubic direction-preserving function on $C_{a,b} \subset \mathbb{Z}^d$ and $G = (f, D_{a,b}, \mathcal{S})$ be the extend function. For every t-cube C^t in $C_{a,b}$ where $0 \le t \le d-1$, it is bad relative to f iff the cardinality of the following set is odd:*

$$ S_{C^t} = \left\{ t\text{-simplex } S \in \mathcal{S} \text{ is bad relative to } G \mid S \subset \overline{C^t} \right\}. $$

Proof. We use induction on t. The base case for $t = 0$ is trivial.

For $t > 0$, we assume the lemma is true for case $t - 1$. Let c be the center point of $\overline{C^t}$, then the way we build simplicial decomposition $\mathcal{S}$ implies that

$$ \left\{ t\text{-simplex } S^t \subset \overline{C^t} \right\} = \left\{ \mathrm{conv}(S^{t-1}, c),\ S^{t-1} \in \mathcal{S} \text{ is on the boundary of } \overline{C^t} \right\} $$

where S^{t-1} is used to denote $(t-1)$-simplices in $\mathcal{S}$.

Firstly, we prove that, if t-cube $C^t \subset \mathbb{Z}^d$ is not bad, then $|S_{C^t}|$ is even. If $0 \in f(C^t)$, then $f(c) = 0$. As each t-simplex in $\overline{C^t}$ has c as one of its vertices,

$S_{C^t} = \emptyset$ and we are done. Similarly, we can prove if $f(C^t) \cap \{\pm e^k\} \neq \emptyset$ where $k > t + 1$, then $S_{C^t} = \emptyset$. If $e^k \notin f(C^t)$ where $1 \leq k \leq t + 1$, then for every t-simplex $S \subset \overline{C^t}$, $e^k \notin f(V_S)$, and thus, $S_{C^t} = \emptyset$. Otherwise, we have $f(C^t) = \{e^1, ..., e^{t+1}\}$, and thus $f(c) = e^{t+1}$. Because C^t is not bad, the number of bad $(t-1)$-cubes on the boundary of C^t is even. Using the induction hypothesis on $t - 1$, a $(t-1)$-cube is bad iff the number of bad $(t-1)$-simplices in it is odd. As a result, the number of bad $(t-1)$-simplices on the boundary of C^t is even. Using the equation in the first paragraph, we know $|S_{C^t}|$ is even too.

On the other hand, we prove if C^t is bad, then $|S_{C^t}|$ is odd. Since $f(C^t) = \{e^1, ..., e^{t+1}\}$, we have $f(c) = e^{t+1}$. As the number of bad $(t-1)$-cubes on the boundary of C^t is odd, the number of bad $(t-1)$-simplices on the boundary of C^t is also odd, according to the induction hypothesis on case $t - 1$. Using the equation in the first paragraph again, we know $|S_{C^t}|$ is odd.

We now get Lemma 5 as a direct corollary of Lemma 4.

Lemma 5. *The parity of N_B (the number of bad $(d-1)$-cubes on the boundary of $C_{a,b}$) is same as the one of N_G (the number of bad $(d-1)$-simplices on the boundary of $D_{a,b}$).*

With Property 1, 2 and Lemma 5 above, Chen and Deng's theorem can be immediately derived from the fundamental discrete fixed point theorem.

7   Concluding Remarks

In this paper, we generalize the concept of direction-preserving maps and characterize a new class of discrete maps over simplicial structures. The fundamental discrete fixed point theorem is then proposed, which is based on the counting of bad $(d-1)$-simplices on the boundary. The power of this theorem is demonstrated in two ways. First, it is applied to prove the discrete Brouwer's fixed point theorem which is much more general than the one of Murota, Iimura and Tamura. Second, we resolve the puzzle of bad cubes, and show that the boundary condition of Chen and Deng's theorem is exactly equivalent to the one of the fundamental theorem.

Our work would immediately imply the corresponding discrete concept of *degree*. It would be an especially interesting problem to study the case when the fixed point is defined in the recent model of a set of points. An immediate follow-up research direction is to understand other concepts and theorems related to *degree*. A clear understanding would definitely advance the state of art of the numerical computation of related problems, such as the case of discrete fixed points versus approximate fixed points [2].

References

1. X. Chen and X. Deng. A Simplicial Approach for Discrete Fixed Point Theorems. *full version, available at* http://www.cs.cityu.edu.hk/~deng/.
2. X. Chen and X. Deng. On Algorithms for Discrete and Approximate Brouwer Fixed Points. In *STOC 2005*, pages 323–330.

3. X. Chen and X. Deng. 3-Nash is PPAD-complete. *ECCC, TR05-134*, 2005.
4. X. Chen and X. Deng. Settling the Complexity of 2-Player Nash-Equilibrium. *ECCC, TR05-140*, 2005.
5. X. Chen, X. Deng, and S-H Teng. Computing Nash Equilibria: Approximation and Smoothed Complexity. *ECCC, TR06-023*, 2006.
6. X. Chen and X. Deng. Lattice Embedding of Direction-Preserving Correspondence Over Integrally Convex Set. *Accepted by AAIM*, 2006.
7. X. Chen and X. Deng. On the Complexity of 2D Discrete Fixed Point Problem. *Accepted by ICALP*, 2006.
8. P. Crescenzi and R. Silvestri. Sperner's lemma and robust machines. *Comput. Complexity*, 7(2):163–173, 1998.
9. C. Daskalakis, P.W. Goldberg, and C.H. Papadimitriou. The Complexity of Computing a Nash Equilibrium. *ECCC, TR05-115*, 2005.
10. C. Daskalakis and C.H. Papadimitriou. Three-player games are hard. *ECCC, TR05-139*.
11. K. Friedl, G. Ivanyos, M. Santha, and F. Verhoeven. On the black-box complexity of Sperner's Lemma. In *FCT 2005*.
12. M.D. Hirsch, C.H. Papadimitriou, and S. Vavasis. Exponential lower bounds for finding Brouwer fixed points. *J.Complexity*, 5:379–416, 1989.
13. T. Iimura. A discrete fixed point theorem and its applications. *Journal of Mathematical Economics*, 39(7):725–742, 2003.
14. T. Iimura, K. Murota, and A. Tamura. Discrete Fixed Point Theorem Reconsidered. *Journal of Mathematical Economics*, to appear.
15. G. Laan, D. Talman, and Z. Yang. Solving discrete zero point problems. *Tinbergen Institute Discussion Papers*, 2004.
16. E. Sperner. Neuer Beweis fur die Invarianz der Dimensionszahl und des Gebietes.
17. G.M. Ziegler. *Lectures on Polytopes*. Springer-Verlag, New York, 1982.

On Incentive Compatible Competitive Selection Protocol

(Extended Abstract)

Xi Chen[1], Xiaotie Deng[2,*], and Becky Jie Liu[2]

[1] Department of Computer Science, Tsinghua University
xichen00@mails.tsinghua.edu.cn
[2] Department of Computer Science, City University of Hong Kong
csdeng@cityu.edu.hk, jliu@cs.cityu.edu.hk

Abstract. The selection problem of m highest ranked out of n candidates is considered for a model while the relative ranking of two candidates is obtained through their pairwise comparison. Deviating from the standard model, it is assumed in this article that the outcome of a pairwise comparison may be manipulated by the two participants. The higher ranked party may intentionally lose to the lower ranked party in order to gain group benefit. We discuss incentive compatible mechanism design issues for such scenarios and develop both possibility and impossibility results.

1 Introduction

Ensuring truthful evaluation of alternatives in human activities has always been an important issue throughout the history. In sport, in particular, such an issue is vital and the practice of the fair play principle has been consistently put forth at the foremost priority. In addition to reliance on the code of ethics and professional responsibility of players and coaches, the design of game rules is an important measure to make fair play enforced. The problem of tournament design consists of issues such as ranking, round-robin scheduling, timetabling, home-away assignment, etc. Ranking alternatives through pairwise comparisons is the most common approach in sports tournaments. Its goal is to find out the 'true' ordering among alternatives through complete or partial pairwise comparisons, and it has been widely studied in the decision theory.

In [4], Harary and Moser gave an extensive review of the properties of round-robin tournaments, and introduced the concept of 'consistency'. In [7], Rubinstein proved that counting the number of winning matches is a good scheme to rank among alternatives in round-robin tournaments; it is also the only scheme that satisfies all the nice rationality properties of ranking. Jech [5] proposed a ranking procedure for incomplete tournaments, which mainly depended on transitivity. He proved that if all players are comparable, i.e. there exists a beating

---

* Research supported by a CERG grant (CityU 1156/04E) of Research Grants Council of Hong Kong SAR, PR China.

D.Z. Chen and D.T. Lee (Eds.): COCOON 2006, LNCS 4112, pp. 13–22, 2006.

chain between each pair of players, then the ranking of players under a specific scheme uniquely exists. Chang et al. [1] investigated the ability of methods in revealing the true ranking in multiple incomplete round-robin tournaments. Works have also been done on evaluating the efficiency and efficacy of ranking methods. Steinhaus [8] proposed an upper bound for the number of matches required to reveal the overall ranking of all players. Mendonca et al. [6] developed a methodology for comparing the efficacy of ranking methods, and investigated their abilities of revealing the true ranking.

Such studies have been mainly based on the assumption that all the players play truthfully, i.e. with their maximal effort. It is, however, possible that some players cheat and seek for group benefit. For example, in the problem of choosing m winners out of n candidates, if the number of winning matches is the only parameter considered in selecting winners, some top players could intentionally lose some matches when confronting their 'friends', so the friends could earn a better ranking while the top players remain highly ranked. Such problems will be the focus of our study: Is there an ideal protocol which allows no cheating strategy under any circumstances, even when a majority of players, possibly many with high ranks, form a coalition to help lower ranked players in it?

The problem, that is, choosing m winners out of n players, is studied under two models. Under both models, a coalition will try to have more of its members be selected as winners than that under the true ranking. For the *collective incentive compatible model*, its only goal is to have more members be selected as winners, even by sacrificing some highly ranked players who ought to be winners. For the *alliance incentive compatible* model, it succeeds not only by having more winners, but also by ensuring the ones who ought to win remain winners, i.e. no players sacrifice their winning positions in order to bring in extra winners. Under both models, our objective is to find an incentive compatible protocol if it exists, or to prove the non-existence of such protocols.

We will formally introduce the models, notations and definitions in Section 2. In Section 3, we discuss the collective incentive compatible model and prove the non-existence of incentive compatible protocols under it. In Section 4, we present an incentive compatible selection protocol under the alliance incentive compatible model. Finally, we conclude with remarks and open problems.

2   Issues and Definitions

Firstly, we describe a protocol which is widely used in bridge tournaments, the *Swiss Team* Protocol. Using it as an example, we show collaboration is possible to improve the outcome of a subgroup of players, if the protocol is not properly designed.

2.1   Existence of Cheating Strategy Under the Swiss Team Protocol

The Swiss Team protocol chooses two winners out of four players. Let the four players $P_4 = \{p_1, p_2, p_3, p_4\}$ play according to the following arrangements. After all the three rounds, two of them will be selected as winners.

- Assign a distinct ID in $N_4 = \{1, 2, 3, 4\}$ to each player in P_4 by a randomly selected indexing function.
- In **round** 1, player (with ID) 1 vs. player 2, and player 3 vs. player 4.
- In **round** 2, two winners of the first round play against each other, and so as the two losers. The player continuously wins twice will be selected as the first winner of the whole game; the player continuously loses twice will be out. Therefore, there are only two players left.
- In **round** 3, the two remaining players play against each other. The winner will be selected as the second winner of the whole game.

Suppose the true capacity of the four players in P_4 is $p_1 > p_2 > p_3 > p_4$ and we consider the case in which p_1 and p_3 form a group. Their purpose is to get both winning positions by applying a cheating strategy, while the winners should be p_1 and p_2 according to the true ranking. Under the settings of the Swiss Team Protocol described above, the probability of this group $\{p_1, p_3\}$ having effective cheating strategies is non-negligible. Following is their strategy.

- Luckily, the IDs assigned to p_1, p_2, p_3 and p_4 are $1, 2, 3$ and 4 respectively.
- In **round** 1, p_1 plays against p_2 and p_3 plays against p_4. p_1 and p_3 win.
- In **round** 2, p_1 plays against p_3 and p_2 plays against p_4. In order to let p_3 be one of the winners, p_1 loses the match to p_3 intentionally. p_3 will then be selected as the first winner for winning twice. In the other match, both p_2 and p_4 play truthfully and p_2 wins.
- In **round** 3, p_1 and p_2 play against each other, and p_1 wins. Therefore, p_1 is selected as the second winner.

By applying the cheating strategy above, the group of bad players $\{p_1, p_3\}$ can break the Swiss Team protocol by letting p_1 confront p_2 twice, and earn an extra winning position.

2.2   Problem Description

Suppose a tournament is held among n players $P_n = \{p_1...p_n\}$ and m winners are expected to be selected by a selection protocol. Here a protocol $f_{n,m}$ is a predefined function to choose winners through pairwise competitions, with the intention of finding m players of highest capacity. When the tournament starts, a distinct ID in $N_n = \{1...n\}$ is assigned to each player in P_n by a randomly picked indexing function I. Then a match is played between each pair of players. The competition outcomes will form a tournament graph [2], whose vertex set is N_n and edges represent results of all the matches. Finally, the graph will be treated as input to $f_{n,m}$, and it will output a set of m winners.

　　Assume there exists a group of bad players play dishonestly, i.e. they might lose a match on purpose to gain overall benefit of the whole group, while all the other players always play truthfully, i.e. they try their best to win matches. We say that the group of bad players gains benefit if they are able to have more winning positions than that according to the true ranking. Given knowledge of the selection protocol $f_{n,m}$, the indexing function I and the true ranking of all players, the group of bad players tries to find a cheating strategy that can fool the selection protocol and gains benefit.

The problem is considered under two models in which the characterizations of bad players are different. Under the *collective incentive compatible model*, bad players are willing to sacrifice themselves to win group benefit; while the ones under the *alliance incentive compatible model* only cooperate if their individual interests are well maintained in the cheating strategy.

Our goal is to find an incentive compatible selection protocol, under which players or group of players maximize their benefits only by strictly following the fair play principle, i.e. always play with maximal effort. Otherwise, we prove the inexistence of such protocols.

2.3   Formal Definitions

When the tournament begins, an indexing function I is randomly picked and a distinct ID $I(p) \in N_n$ is assigned to each player $p \in P_n$. Then a match is played between each pair of players, and results are represented as a directed graph G. Finally, G is fed to the predefined selection protocol $f_{n,m}$, to produce a set of m winners $W = f_{n,m}(G) \subset N_n$.

Definition 1 (Indexing Function). *An indexing function I for a tournament attended by n players $P_n = \{p_1, p_2, \ldots p_n\}$ is a one-to-one correspondence from P_n to the set of IDs: $N_n = \{1, 2, \ldots n\}$.*

Definition 2. *A tournament graph of size n is is a directed graph $G = (N_n, E)$ such that, for any $i \neq j \in N_n$, either edge $ij \in E$ (player with ID i beats player with ID j) or edge $ji \in E_n$. We use K_n to denote the set of all such graphs.*

A selection protocol $f_{n,m}$ which chooses m winners out of n candidates is a function from K_n to $\{S \subset N_n \text{ and } |S| = m\}$.

The group of bad players not only know the selection protocol, but also the true ranking of players. We say a bad player group gains benefit if it has more members be selected as winners than that according to the true ranking.

Definition 3 (Ranking Function). *A ranking function R of is a one-to-one correspondence from P_n to N_n. $R(p) \in N_n$ represents the underlying true ranking of player p among the n players. The smaller, the stronger.*

Definition 4 (Tournament). *A tournament T_n among n players P_n is a pair $T_n = (R, B)$, where R is a ranking function from P_n to N_n and $B \subset P_n$ is the group of bad players.*

Definition 5 (Benefit). *Given a protocol $f_{n,m}$, a tournament $T_n = (R, B)$, an indexing function I and a tournament graph $G \in K_n$, the benefit of the group of bad players is*

$$\textbf{Ben}(f_{n,m}, T_n, I, G) = \left| \left\{ i \in f_{n,m}(G), \ I^{-1}(i) \in B \right\} \right| - \left| \left\{ p \in B, \ R(p) \leq m \right\} \right|.$$

Given $f_{n,m}$, T_n and I, not every graph $G \in K_n$ is a feasible strategy for the group of bad players. First, it depends on the tournament $T_n = (R, B)$, e.g. a player $p_b \in B$ cannot win player $p_g \notin B$ if $R(p_b) > R(p_g)$. Second, it depends on the property of bad players which is specified by the model considered.

We now, for each model, characterize tournament graphs which are recognized as feasible strategies. The key difference is that a bad player in alliance incentive compatible model is not willing to sacrifice his own winning position, while a player in the other model fights for group benefit at all costs.

Definition 6. *Given $f_{n,m}$, $T_n = (R, B)$ and I, a graph $G \in K_n$ is c-feasible if*

1. For every two players $p_i, p_j \notin B$, if $R(p_i) < R(p_j)$, then $I(p_i)I(p_j) \in E$;

2. For all $p_g \notin B$ and $p_b \in B$, if $R(p_g) < R(p_b)$, then edge $I(p_g)I(p_b) \in E$.

Graph $G \in K_n$ is a-feasible if it is c-feasible and also satisfies

3. For every bad player $p \in B$, if $R(p) \leq m$, then $I(p) \in f_{n,m}(G)$.

A cheating strategy is then a feasible tournament graph G that can be employed by the group of bad players to gain positive benefit.

Definition 7 (Cheating Strategy). *Given $f_{n,m}$, T_n and I, a cheating strategy for the group of bad players under the collective incentive compatible (alliance incentive compatible) model is a graph $G \in K_n$ which is c-feasible (a-feasible) and satisfies $\textbf{Ben}(f_{n,m}, T_n, I, G) > 0$.*

We ask the following two natural questions.

$\textbf{Q}_1$: Is there a protocol $f_{n,m}$ such that for all T_n and I, no cheating strategy exists under the collective incentive compatible model?

$\textbf{Q}_2$: Is there a protocol $f_{n,m}$ such that for all T_n and I, no cheating strategy exists under the alliance incentive compatible model?

In the following sections, we will present an impossibility proof for the first question, and design an incentive compatible protocol for the second model.

3 Incentive Compatible Protocol Under the Collective Incentive Compatible Model

In this section, we prove the inexistence of incentive compatible protocol under the collective incentive compatible model. For every $f_{n,m}$, we are able to find a large number of tournaments T_n where cheating strategy exists.

Definition 8. *For all integers n and m such that $2 \leq m \leq n - 2$, we define a graph $G_{n,m} = (N_n, E) \in K_n$ which consists of 3 parts, T_1, T_2 and T_3.*

1. $T_1 = \{1, 2, ... m - 2\}$. For all $i < j \in T_1$, edge $ij \in E$;
2. $T_2 = \{m - 1, m, m + 1\}$. $(m - 1)m$, $m(m + 1)$, $(m + 1)(m - 1) \in E$;

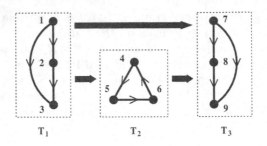

Fig. 1. Tournament Graph $G_{9,5}$

3. $T_3 = \{m+2, m+3, \ldots n\}$. For all $i < j \in T_3$, edge $ij \in E$;
4. For all $i' \in T_i$ and $j' \in T_j$ such that $i < j$, edge $i'j' \in E$.

Players in T_1 and T_3 are well ordered among themselves, but the ones in T_2 are not due to the existence of a cycle. All players in T_1 beat the ones in T_2 and T_3, and all players in T_2 beat the ones in T_3. Sample graph $G_{9,5}$ is shown in Figure 1. Proof of Lemma 1 can be found in the full version [3].

Lemma 1. *For every $f_{n,m}$ where $2 \leq m \leq n - 2$, if $T_n = (R, B)$ satisfies that $B = \{p_{m-r+1}\cdots p_{m+1}, p_{m+2}\}$ where $r \geq 2$ and $R(p_i) = i$ for all $1 \leq i \leq n$, then there exists an indexing function I such that $G_{n,m}$ is a cheating strategy.*

Corollary 1. *For every $f_{n,m}$ where $2 \leq m \leq n-2$, if $T_n = (R, B)$ satisfies that $B = R^{-1}(m - r + 1 \ldots m + 1, m + 2)$ where $r \geq 2$, then there exists an indexing function I such that $G_{n,m}$ is a cheating strategy.*

Corollary 2 can be derived from Lemma 1 immediately. Figure 2 shows the true ranking of a tournament T_n in which a cheating strategy exists.

By Lemma 2, one can extend Corollary 2 to Theorem 1 below.

Lemma 2. *Given $f_{n,m}$ and I, if $G \in K_n$ is a cheating strategy for tournament $T_n = (R, B)$, and there exist players $p_b \in B$ and $p_g \notin B$ such that $R(p_b) = R(p_g)+1 \leq m$, then graph G remains a cheating strategy of $T'_n = (R', B)$ where $R'(p_b) = R(p_g)$, $R'(p_g) = R(p_b)$ and $R'(p) = R(p)$ for every other player p.*

Theorem 1. *For every $f_{n,m}$ where $2 \leq m \leq n - 2$, if $T_n = (R, B)$ satisfies: **1**). at least one bad player ranks as high as $m - 1$; **2**). the ones ranked $m + 1$ and $m + 2$ are both bad players; **3**). the one ranked m is a good player, then there always exists an indexing function I such that $G_{n,m}$ is a cheating strategy.*

Theorem 1 describes a much larger class of tournaments in which cheating strategy exists. An example of such tournaments is shown in Figure 3.

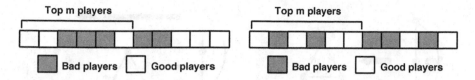

Fig. 2. An Example of Tournaments   **Fig. 3.** An Example of Tournaments

4 Incentive Compatible Protocol Under the Alliance Incentive Compatible Model

In this section, we answer question $\mathbf{Q_2}$ for arbitrary n and m. We prove that whether a successful protocol exists is completely determined by the value of $n - m$. When $n - m \leq 2$, cheating strategies can always be constructed, and thus we prove the inexistence of ideal protocol. When $n - m \geq 3$, we present a selection protocol $f^*_{n,m}$ under which no cheating strategy exists.

4.1 Inexistence of Selection Protocol When $n - m \leq 2$

Definition 9. *We define two classes of tournament graphs, graph G^*_n for any $n \geq 3$ and graph G'_n for any $n \geq 4$. Their structures are similar to $G_{n,m}$.*

- *For G^*_n, $T_1 = \{1, 2, \dots n - 3\}$, $T_2 = \{n - 2, n - 1, n\}$ and $T_3 = \emptyset$ with edges $(n - 2)(n - 1)$, $(n - 1)n$, $n(n - 2) \in G^*_n$. Graph G^*_6 is shown in Figure 4.*

- *For G'_n, $T_1 = \{1, 2, \dots n - 4\}$, $T_2 = \{n - 3, n - 2, n - 1\}$ and $T_3 = \{n\}$ with edges $(n - 3)(n - 2)$, $(n - 2)(n - 1)$, $(n - 1)(n - 3) \in G'_n$. Sample graph G'_7 is shown in Figure 5.*

By the following two lemmas, no ideal protocol exists when $n - m \leq 2$. The proofs can be found in the full version [3].

Lemma 3. *For every $f_{n,m}$ where $n - m = 1$ and $m \geq 2$, if $T_n = (R, B)$ satisfies $B = \{p_1, p_2, \dots p_{n-2}, p_n\}$ and $R(p_i) = i$ for all $1 \leq i \leq n$, then there exists an indexing function I such that graph G^*_n is a cheating strategy for the group of bad players under the alliance incentive compatible model.*

Lemma 4. *For every $f_{n,m}$ where $n - m = 2$ and $m \geq 2$, if $T_n = (R, B)$ satisfies $B = \{p_1, p_2, \dots p_{n-3}, p_{n-1}, p_n\}$ and $R(p_i) = i$ for all $1 \leq i \leq n$, then there exists an indexing function I such that graph G'_n is a cheating strategy for the group of bad players under the alliance incentive compatible model.*

4.2 Selection Protocol $f^*_{n,m}$ for Case $n - m \geq 3$

In this section, we'll first introduce some important properties of tournament graphs. Then a selection protocol $f^*_{n,m}$ will be described for case $n - m \geq 3$. Finally, we prove that for any tournament T_n and indexing function I, no cheating strategy exists for the group of bad players.

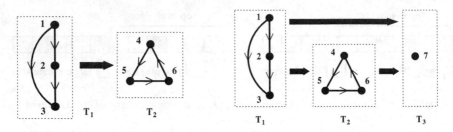

Fig. 4. Tournament graph G_6^* **Fig. 5.** Tournament graph G_7'

Definition 10. *A directed graph G is said to be strongly connected if there's a directed path between every pair of vertices. Any maximal subgraph of G that is strongly connected is called a strongly connected component of graph G.*

Let $G \in K_n$ be a tournament graph. We use $G_1 ... G_k$ to denote its strongly connected components which satisfy that for all $u \in G_i$ and $v \in G_j$ such that $i < j$, edge $uv \in G$. The proof of Lemma 5 below can be found in [2].

Definition 11. *A directed graph G of order $n \geq 3$ is pancyclic if it contains a cycle of length l for each $l = 3, 4, ... n$, and is vertex-pancyclic if each vertex v of G lies on a cycle of length l for each $l = 3, 4, ... n$.*

Lemma 5. *Every strongly connected tournament graph is vertex-pancyclic.*

Corollary 2. *Let G be a tournament graph with strongly connected components $G_1 ... G_k$. If there is no cycle of length l in G, then $|G_i| < l$ for all $1 \leq i \leq k$.*

Our protocol $f_{n,m}^*$ described in Figure 6 is an algorithm working on tournament graphs. The algorithm checks whether $3 \mid n - m$.

- When $n - m \equiv 1 \pmod 3$, if there exists a cycle of 4 vertices, delete all the vertices in the cycle; otherwise, delete the lowest ranked vertex in G. As a result, we have $n' - m \equiv 0 \pmod 3$ where n' is the number of remaining candidates after deletion.
- When $n - m \equiv 2 \pmod 3$, if there exists a cycle of 5 vertices in G, delete all the vertices in the cycle; otherwise, delete the two lowest ranked vertices. Similarly, it can also be reduced to the case of $n' - m \equiv 0 \pmod 3$.
- When $n - m \equiv 0 \pmod 3$, if there exist cycles of 3 vertices, continuously delete them until either **1)** no such cycle exists, then choose the m highest ranked ones as winners; or **2)** there're m vertices left, then choose all of the remaining candidates as winners.

The proof of the following theorem can be found in the full version [3].

Theorem 2. *For all T_n, I and a-feasible graph G, **Ben**$(f_{n,m}, T_n, I, G) \leq 0$.*

1: **Ensure** $n - m \geq 3$ and graph $G \in K_n$
2: let $G_1, G_2, \ldots G_k$ be the strongly connected components of graph $G = (N_n, E)$
3: **if** $n - m \equiv 1 \pmod 3$ **then**
4:     **if** there exists a cycle C of length 4 in G **then**
5:         delete all the 4 vertices in C from graph G
6:     **else**
7:         let t be the smallest vertex (integer) in G_k, and delete vertex t from G
8:     **endif**
9: **else if** $n - m \equiv 2 \pmod 3$ **then**
10:     **if** there exists a cycle C of length 5 in G **then**
11:         delete all the 5 vertices in C from graph G
12:     **else if** $|G_k| = 1$
13:         let $t_1 \in G_k$ and t_2 be the smallest vertex (integer) in G_{k-1}, delete t_1, t_2
14:     **else**
15:         let t_1 and t_2 be the two smallest vertices (integers) in G_k, delete t_1, t_2
16:     **end if**
17: **end if**
18: **while** the number of vertices in G is larger than m **do**
19:     **if** there exists a cycle C of length 3 in G **then**
20:         delete all the 3 vertices in C from graph G
21:     **else**
22:         vertices can be sorted as $k_1 \ldots k_{m'}$ such that $k_i k_j \in E, \forall\, 1 \leq i < j \leq m'$
23:         output set $\{ k_1, k_2, \ldots k_m \}$ and return
24:     **end if**
25: **end while**
26: output all the remaining vertices in G and return

Fig. 6. Details of Selection Protocol $f_{n,m}^*$

5 Conclusion Remarks

In this article, we discussed the possibility of an incentive compatible selection protocol to exist, by which the benefits of either individual players or a group of players are maximized by playing truthfully. Under the collective incentive compatible model, our result indicates that cheating strategies are available in at least 1/8 tournaments, if we assume the probability for each player to be in the bad group is 1/2. On the other hand, we showed that there does exist an incentive compatible selection protocol under the alliance incentive compatible model, by presenting a deterministic algorithm.

Many problems remain and require further analysis. Under the first model, could the general bound of 1/8 be improved? Could we find good selection protocols in the sense that the number of tournaments with cheating strategies is close to this bound? Though we have proved the inexistence of ideal protocol under this model, does there exist any probabilistic protocol, under which the probability of having cheating strategies is negligible?

Finally, we'd like to raise the issue of output truthful mechanism design. In our model, an output truthful mechanism would output a list of k players, each of which is among the top k players in the true ranking. It would be interesting to know whether there is such a mechanism or not. For a related problem we are going to describe next, this is possible. Consider a committee of 2n+1 to select one out of candidates. The expected output is the one favored by the majority of the committee. The following protocol will return the true outcome but not everyone will vote truthfully: After the voting, a fixed amount of bonus will be distributed to the voters who voted for the winner. Using this mechanism, every committee member will vote for the candidate favored by the majority though not everyone likes him or her.

Acknowledgement

We would like to thank professor Frances Yao for her contribution of both crucial ideas and many research discussions with us.

We would also like to thank Hung Chim and Xiang-Yang Li for a discussion in a research seminar about a year ago during which the idea of output truthful mechanism popped up, and the above example of voting committee was shaped.

References

1. P. Chang, D. Mendonca, X. Yao, and M. Raghavachari. An evaluation of ranking methods for multiple incomplete round-robin tournaments. In *Decision Sciences Institute conference 2004*.
2. G. Chartrand and L. Lesniak. *Graphs and Digraphs*. Chapman and Hall, London.
3. X. Chen, X. Deng, and B.J. Liu. On Incentive Compatible Competitive Selection Protocol. *full version, available at* http://www.cs.cityu.edu.hk/~deng/.
4. F. Harary and L.Moser. The theory of round robin tournaments. *The American Mathematical Monthly*, 73(3):231–246, Mar. 1966.
5. T. Jech. The ranking of incomplete tournaments: A mathematician's guide to popular sports. *The American Mathematical Monthly*, 90(4):246–266, Apr. 1983.
6. D. Mendonca and M. Raghavachari. Comparing the efficacy of ranking methods for multiple round-robin tournaments. *European Journal of Operational Research*, 123(2000):593–605, Jan. 1999.
7. A. Rubinstein. Ranking the participants in a tournament. *SIAM Journal of Applied Mathematics*, 38(1):108–111, 1980.
8. H. Steinhaus. *Mathematical Snapshots*. Oxford University Press, New York, 1950.

Edge Pricing of Multicommodity Networks for Selfish Users with Elastic Demands

George Karakostas[1],* and Stavros G. Kolliopoulos[2],**

[1] Department of Computing and Software, McMaster University
[2] Department of Informatics and Telecommunications, University of Athens

Abstract. We examine how to induce selfish heterogeneous users in a multicommodity network to reach an equilibrium that minimizes the social cost. In the absence of centralized coordination, we use the classical method of imposing appropriate taxes (tolls) on the edges of the network. We significantly generalize previous work [20,13,9] by allowing user demands to be *elastic*. In this setting the demand of a user is not fixed a priori but it is a function of the routing cost experienced, a most natural assumption in traffic and data networks.

1   Introduction

We examine a network environment where uncoordinated users, each with a specified origin-destination pair, select a path to route an amount of their respective commodity. Let f be a flow vector defined on the paths of the network, which describes a given routing according to the standard multicommodity flow conventions. The users are selfish: each wants to choose a path P that minimizes the cost $T_P(f)$. The quantity $T_P(f)$ depends typically on the latency induced on P by the aggregated flow of all users using some edge of the path.

We model the interaction of the selfish users by studying the system in the steady state captured by the classic notion of a *Wardrop equilibrium* [19]. This state is characterized by the following principle: in equilibrium, for every origin-destination pair (s_i, t_i), the cost on every used $s_i - t_i$, path is equal and less than or equal to the cost on any unused path between s_i and t_i. The Wardrop principle states that in equilibrium the users have no incentive to change their chosen route; under some minor technical assumptions the Wardrop equilibrium concept is equivalent to the Nash equilibrium in the underlying game. The literature on traffic equilibria is very large (see, e.g., [2,6,5,1]). The framework is in principle applicable both to transportation and decentralized data networks. In recent years, starting with the work of Roughgarden and Tardos [17], the latter area motivated a fruitful treatment of the topic from a computer science perspective.

---

* (www.cas.mcmaster.ca/~gk). Research supported by an NSERC Discovery grant and MITACS.
** (www.di.uoa.gr/~sgk). This work was partially supported by the EU-GME (Greek Ministry of Education) EPEAEK II project no. 2.2.3 Archimedes 2: Enhancing Research Groups in Technological Institutes.

D.Z. Chen and D.T. Lee (Eds.): COCOON 2006, LNCS 4112, pp. 23–32, 2006.

The behavior of uncoordinated selfish users can incur undesirable conse-quences from the point of view of the system as a whole. The *social cost* function, usually defined as the total user latency, expresses this societal point of view. Since for several function families [17] one cannot hope that the uncoordinated users will reach a traffic pattern which minimizes the social cost, the system designer looks for ways to induce them to do so. A classic approach, which we follow in this paper, is to impose economic disincentives, namely put nonneg-ative per-unit-of-flow *taxes* (tolls) on the network edges [2,12]. The tax-related monetary cost will be, together with the load-dependent latency, a component of the cost function $T_P(f)$ experienced by the users. As in [3,20] we consider the users to be *heterogeneous*, i.e., belonging to classes that have different sensitivi-ties towards the monetary cost. This is expressed by multiplying the monetary cost with a factor $a(i)$ for user class i. We call *optimal* the taxes inducing a user equilibrium flow which minimizes the social cost.

The existence of a vector of optimal edge taxes for heterogeneous users in multicommodity networks is not a priori obvious. It has been established for fixed demands in [20,13,9]. In this paper we significantly generalize this previous work by allowing user demands to be *elastic*. Elastic demands have been studied extensively in the traffic community (see, e.g., [10,1,12]). In this setting the demand d_i of a user class i is not fixed a priori but it is a function $D_i(u)$ of the vector u of routing costs experienced by the various user classes. Demand elasticity is natural in traffic and data networks. People may decide whether to travel based on traffic conditions. Users requesting data from a web server may stop doing so if the server is slow. Even more elaborate scenarios, such as multi-modal traffic, can be implemented via a judicious choice of the demand functions. E.g., suppose that origin-destination pairs 1 and 2 correspond to the same physical origin and destination points but to different modes of transit, such as subway and bus. There is a total amount d of traffic to be split among the two modes. The modeler could prescribe the modal split by following, e.g., the well-studied logit model [1]:

$$D_1(u) = d\frac{e^{\theta u_1 + A_1}}{e^{\theta u_1 + A_1} + e^{\theta u_2 + A_2}}, \quad D_2(u) = d - D_1(u)$$

for given negative constant θ and nonnegative constants A_1 and A_2. Here u_1 (resp. u_2) denotes the routing cost on all used paths of mode 1 (resp. 2).

For the elastic demand setting we show in Section 3 the existence of taxes that induce the selfish users to reach an equilibrium that minimizes the total latency. Note that for this result we only require that the vector $D(u)$ of the demand functions is monotone according to Definition 1. The functions $D_i(u)$ do not have to be strictly monotone (and therefore invertible) individually, and for some $i \neq j$, $D_i(u)$ can be increasing while $D_j(u)$ can be decreasing on a particular variable (as for example in the logit model mentioned above). The result is stated in Theorem 1 and constitutes the main contribution of this paper. The existence results for fixed demands in [20,13,9] follow as corollaries. Our proof is developed over several steps but its overall structure is explained at the the beginning of Section 3.1.

We emphasize that the equilibrium flow in the elastic demand setting satisfies the demand values that materialize in the same equilibrium, values that are not known a priori. This indeterminacy makes the analysis particularly challenging. On the other hand, one might argue that with high taxes, which increase the routing cost, the actual demand routed (which being elastic depends also on the taxes) will be unnaturally low. This argument does not take fully into account the generality of the demand functions $D_i(u)$ which do not even have to be decreasing; even if they do they do not have to vanish as u increases. Still it is true that the model is indifferent to potential lost benefit due to users who do not participate. Nevertheless, there are settings where users may decide not to participate without incurring any loss to either the system or themselves and these are settings we model in Section 3. Moreover in many cases the system designer chooses explicitly to regulate the effective use of a resource instead of heeding the individual welfare of selfish users. Charging drivers in order to discourage them from entering historic city cores is an example, among many others, of a social policy of this type. ˙

A more user-friendly agenda is served by the study of a different social cost function which sums total latency and the lost benefit due to the user demand that was not routed [10,11]. This setting was recently considered in [4] from a price of anarchy [14] perspective. In this case the elasticity of the demands is specified implicitly through a function $\Gamma_i(x)$ (which is assumed nonincreasing in [4]) for every user class i. $\Gamma_i(d_i)$ determines the minimum per-user benefit extracted if d_i users from the class decide to make the trip. Hence $\Gamma_i(d_i)$ also denotes the maximum travel cost that each of the first d_i users (sorted in order of nonincreasing benefit) from class i is willing to tolerate, in order to travel. In the full version of the paper we show the existence of optimal taxes for this model. We demonstrate however that for these optimal taxes to exist, participating users must tolerate, in the worst-case, higher travel costs than those specified by their $\Gamma(\cdot)$ function.

In this extended abstract we omit many technical details. A full version of the paper is available as AdvOL-Report 2006/02 at http://optlab.mcmaster.ca/

2 Preliminaries

The model: Let $G = (V, E)$ be a directed network (possibly with parallel edges but with no self-loops), and a set of *users*, each with an infinitesimal amount of traffic (flow) to be routed from an origin node to a destination node of G. Moreover, each user α has a positive *tax-sensitivity* factor $a(\alpha) > 0$. We will assume that the tax-sensitivity factors for all users come from a finite set of possible positive values. We can bunch together into a single *user class* all the users with the same origin-destination pair and with the same tax-sensitivity factor; let k be the number of different such classes. We denote by $\mathcal{P}_i, a(i)$ the the flow paths that can be used by class i, and the tax-sensitivity of class i, for all $i = 1, \ldots, k$ respectively. We will also use the term 'commodity i' for class i. Set $\mathcal{P} \doteq \cup_{i=1,\ldots,k} \mathcal{P}_i$. Each edge $e \in E$ is assigned a *latency function* $l_e(f_e)$ which gives the latency experienced by any user that uses e due to congestion caused by the total flow f_e that passes through e. In other words, as in [3], we assume the additive model in which for any path $P \in \mathcal{P}$ the latency is $l_P(f) = \sum_{e \in P} l_e(f_e)$, where $f_e = \sum_{e \ni P} f_P$ and f_P is the flow through path P. If every edge is assigned a per-unit-of-flow tax $b_e \geq 0$, a selfish user in class i that uses a path $P \in \mathcal{P}_i$ experiences total cost $T_P(f)$ equal to $\sum_{e \in P} l_e(f_e) + a(i) \sum_{e \in P} b_e$ hence the name 'tax-sensitivity' for the $a(i)$'s: they quantify the importance each user assigns to the taxation of a path.

A function $g : \mathbb{R}^n \to \mathbb{R}^m$ is *positive* if $g(x) > 0$ when $x > 0$. We assume that the functions l_e are strictly increasing, i.e., $x > y \geq 0$ implies $l_e(x) > l_e(y)$, and that $l_e(0) \geq 0$. This implies that $l_e(f_e) > 0$ when $f_e > 0$, i.e., the function l_e is positive.

Definition 1. *Let $f : K \to \mathbb{R}^n$, $K \subseteq \mathbb{R}^n$. The function f is* monotone *on K if $(x - y)^T (f(x) - f(y)) \geq 0$, $\forall x \in K, y \in K$. The function f is strictly monotone if the previous inequality is strict when $x \neq y$.*

In what follows we will use heavily the notion of a nonlinear complementarity problem. Let $F(x) = (F_1(x), F_2(x), \ldots, F_n(x))$ be a vector-valued function from the n-dimensional space $\mathbb{R}^n$ into itself. Then the nonlinear complementarity problem of mathematical programming is to find a vector x that satisfies the following system:

$$x^T F(x) = 0, \quad x \geq 0, \quad F(x) \geq 0.$$

3   The Elastic Demand Problem

In this section the social cost function is defined as the total latency $\sum_e f_e l_e(f_e)$. We set up the problem in the appropriate mathematical programming framework and formulate the main result for this model in Theorem 1.

The traffic (or Wardrop) equilibria for a network can be described as the solutions of the following mathematical program (see [1] p. 216):

$$(T_P(f) - u_i)f_P = 0 \quad \forall P \in \mathcal{P}_i, i = 1 \ldots k$$
$$T_P(f) - u_i \geq 0 \quad \forall P \in \mathcal{P}_i, i = 1 \ldots k$$
$$\sum_{P \in \mathcal{P}_i} f_P - D_i(u) = 0 \quad \forall i = 1 \ldots k$$
$$f, u \geq 0$$

where T_P is the cost of a user that uses path P, f_P is the flow through path P, and $u = (u_1, \ldots, u_k)$ is the vector of shortest travel times (or generalized costs) for the commodities. The first two equations model Wardrop's principle by requiring that for any origin-destination pair i the travel cost for all paths in $\mathcal{P}_i$ with nonzero flow is the same and equal to u_i. The remaining equations ensure that the demands are met and that the variables are nonnegative. Note that the formulation above is very general: every path $P \in \mathcal{P}_i$ for every commodity i has its own T_P (even if two commodities share the same path P, each may have its own T_P).

If the path cost functions T_P are positive and the $D_i(\cdot)$ functions take non-negative values, [1] shows that the system above is equivalent to the following nonlinear complementarity problem (Proposition 4.1 in [1]):

$$(T_P(f) - u_i)f_P = 0 \quad \forall i,\ \forall P \in \mathcal{P}_i \qquad \text{(CPE)}$$
$$T_P(f) - u_i \geq 0 \quad \forall i,\ \forall P \in \mathcal{P}_i$$

$$u_i\Big(\sum_{P\in\mathcal{P}_i} f_P - D_i(u)\Big) = 0 \quad \forall i$$

$$\sum_{P\in\mathcal{P}_i} f_P - D_i(u) \geq 0 \quad \forall i$$

$$f, u \geq 0$$

In our case the costs T_P are defined as $\sum_{e\in P} l_e(f_e) + a(i)\sum_{e\in P} b_e,\ \forall i,\ \forall P \in \mathcal{P}_i$, where b_e is the per-unit-of-flow tax for edge e, and $a(i)$ is the tax sensitivity of commodity i. In fact, it will be more convenient for us to define T_P slightly differently:

$$T_P(f) := \frac{l_P(f)}{a(i)} + \sum_{e\in P} b_e, \quad \forall i,\ \forall P \in \mathcal{P}_i.$$

The special case where $D_i(u)$ is constant for all i, was treated in [20,13,9]. The main complication in the general setting is that the minimum-latency flow $\hat{f}$ cannot be considered a priori given before some selfish routing game starts. At an equilibrium the u_i achieve some concrete value which in turn fixes the demands. These demands will then determine the corresponding minimum-latency flow $\hat{f}$. At the same time, the corresponding minimum-latency flow affects the taxes we impose and this, in turn, affects the demands. The outlined sequence of events serves only to ease the description. In fact the equilibrium parameters materialize simultaneously. We should not model the two flows (optimal and equilibrium) as a two-level mathematical program, since there is no the notion of leader-follower here, but as a complementarity problem as done in [1].

Suppose that we are given a vector u^* of generalized costs. Then the social optimum $\hat{f}^*$ for the particular demands $D_i(u^*)$ is the solution of the following mathematical program:

$$\min \sum_{e\in E} l_e(\hat{f}_e)\hat{f}_e \quad \text{s.t.} \qquad \text{(MP)}$$

$$\sum_{P\in\mathcal{P}_i} \hat{f}_P \geq D_i(u^*) \quad \forall i$$

$$\hat{f}_e = \sum_{P\in\mathcal{P}:e\in P} \hat{f}_P \quad \forall e \in E$$

$$\hat{f}_P \geq 0 \qquad \forall P$$

Under the assumption that the functions $xl_e(x)$ are continuously differentiable and convex, it is well-known that $\hat{f}^*$ solves (MP) iff $(\hat{f}^*, \mu^*)$ solves the following pair of primal-dual linear programs (see, e.g., [8, pp. 9–13]):

$$\min \sum_{e \in E} \left(l_e(\hat{f}_e^*) + \hat{f}_e^* \frac{\partial l_e}{\partial f_e}(\hat{f}_e^*) \right) \hat{f}_e \text{ s.t.}$$

$$\text{(LP2)}$$

$$\sum_{P \in \mathcal{P}_i} \hat{f}_P \geq D_i(u^*), \qquad \forall i$$

$$\hat{f}_e = \sum_{P \in \mathcal{P}: e \in P} \hat{f}_P, \qquad \forall e \in E$$

$$\hat{f}_P \geq 0, \qquad \forall P$$

$$\max \sum_i D_i(u^*) \mu_i \qquad \text{s.t.}$$

$$\text{(DP2)}$$

$$\mu_i \leq \sum_{e \in P} \left(l_e(\hat{f}_e^*) + \hat{f}_e^* \frac{\partial l_e}{\partial f_e}(\hat{f}_e^*) \right) \forall i, P \in \mathcal{P}_i$$

$$\mu_i \geq 0 \qquad \forall i$$

Let the functions $D_i(u)$ be bounded and set $K_1 := \max_i \max_{u \geq 0} \{D_i(u)\} + 1$. Then if n denotes $|V|$ the solutions $\hat{f}^*$, μ^* of (LP2), (DP2) are upper bounded as follows $\hat{f}_P^* \leq D_i(u^*) < K_1$, $\forall P \in \mathcal{P}_i$ $\mu_i \leq \sum_{e \in P} \left(l_e(\hat{f}_e^*) + \hat{f}_e^* \frac{\partial l_e}{\partial f_e}(\hat{f}_e^*) \right) <$ $n \cdot \max_{e \in E} \max_{0 \leq x \leq k \cdot K_1} \{l_e(x) + x \frac{\partial l_e}{\partial f_e}(x)\}$, $\forall i$. It is important to note that these upper bounds are *independent of* u^*.

We wish to find a tax vector b that will steer the edge flow solution of (CPE) towards $\hat{f}$. Similarly to [13] we add this requirement as a constraint to (CPE): for every edge e we require that $f_e \leq \hat{f}_e$. By adding also the Karush-Kuhn-Tucker conditions for (MP) we obtain the following complementarity problem:

$$f_P(T_P(f) - u_i) = 0, \ \forall i, P \qquad\qquad T_P(f) \geq u_i, \ \forall i, P$$

$$u_i \left(\sum_{P \in \mathcal{P}_i} f_P - D_i(u) \right) = 0, \ \forall i \qquad\qquad \sum_{P \in \mathcal{P}_i} f_P \geq D_i(u), \ \forall i$$

$$\text{(GENERAL CP)}$$

$$b_e(f_e - \hat{f}_e) = 0, \ \forall e \qquad\qquad f_e \leq \hat{f}_e, \ \forall e$$

$$\left(\sum_{e \in P} (l_e(\hat{f}_e) + \hat{f}_e \frac{\partial l_e}{\partial f_e}(\hat{f}_e)) - \mu_i \right) \hat{f}_P = 0, \forall i, P \qquad \sum_{e \in P} (l_e(\hat{f}_e) + \hat{f}_e \frac{\partial l_e}{\partial f_e}(\hat{f}_e)) \geq \mu_i, \forall i, P$$

$$\mu_i \left(\sum_{P \in \mathcal{P}_i} \hat{f}_P - D_i(u) \right) = 0, \ \forall i \qquad\qquad \sum_{P \in \mathcal{P}_i} \hat{f}_P \geq D_i(u), \ \forall i$$

$$f_P, b_e, u_i, \hat{f}_P, \mu_i \geq 0, \ \forall P, e, i$$

where $f_e = \sum_{P \ni e} f_P, \hat{f}_e = \sum_{P \ni e} \hat{f}_P$.

The users should be steered towards $\hat{f}$ without being conscious of the constraints $f_e \leq \hat{f}_e$; the latter should be felt only implicitly, i.e., through the corresponding tax b_e. Our main result is expressed in the following theorem. For convenience, we view $D_i(u)$ as the ith coordinate of a vector-valued function $D : \mathbb{R}^k \to \mathbb{R}^k$.

Theorem 1. *Consider the selfish routing game with the latency function seen by the users in class i being $T_P(f) := \sum_{e \in P} l_e(f_e) + a(i) \sum_{e \in P} b_e$, $\forall i$, $\forall P \in \mathcal{P}_i$. If (i) for every edge $e \in E$, $l_e(\cdot)$ is a strictly increasing continuous function with $l_e(0) \geq 0$ such that $xl_e(x)$ is convex and continuously differentiable and (ii) D_i*

are continuous functions bounded from above for all i such that $D(\cdot)$ is positive and $-D(\cdot)$ is monotone then there is a vector of per-unit taxes $b \in \mathbb{R}_+^{|E|}$ such that, if $\bar{f}$ is a traffic equilibrium for this game, $\bar{f}_e = \hat{f}_e$, $\forall e \in E$. Therefore $\bar{f}$ minimizes the social cost $\sum_{e \in E} f_e l_e(f_e)$.

3.1   Proof of the Main Theorem

The structure of our proof for Theorem 1 is as follows. First we give two basic Lemmata 1 and 2. We then argue that the two lemmata together with a proof that a solution to (GENERAL CP) exists imply Theorem 1. We establish that such a solution for (GENERAL CP) exists in Theorem 2. The proof of the latter theorem uses the fixed-point method of [18] and arguments from linear programming duality.

The following result of [1], can be easily extended to our case:

Lemma 1 (Theorem 6.2 in [1]). *Assume that the $l_e(\cdot)$ functions are strictly increasing for all $e \in E$, $D(\cdot)$ is positive and $-D(\cdot)$ is monotone. Then if more than one solutions (f, u) exist for (CPE), u is unique and f induces a unique edge flow.*

Lemma 2. *Let $(f^*, b^*, u^*, \hat{f}^*, \mu^*)$ be any solution of (GENERAL CP). Then $\sum_{P \in \mathcal{P}_i} f_P^* = D_i(u^*)$, $\forall i$ and $f_e^* = \hat{f}_e^*$, $\forall e \in E$.*

Let $(f^*, b^*, u^*, \hat{f}^*, \mu^*)$ be a hypothetical solution to (GENERAL CP). Then $\hat{f}^*$ is a minimum latency flow solution for the demand vector $D(u^*)$. Moreover $f_e^* \le \hat{f}_e^*$, $\forall e \in E$. After setting $b = b^*$ in (CPE), Lemma 1 implies that any solution $(\bar{f}, \bar{u})$ to (CPE) would satisfy $\bar{f}_e = f_e^*$ and $\bar{u} = u^*$. Therefore $\bar{f}_e \le \hat{f}_e^*$, $\forall e \in E$. Under the existing assumptions on $l_e(\cdot)$, We can show (proof omitted) that any *equilibrium flow* $\bar{f}$ for the selfish routing game where the users are conscious of the modified latency $T_P(f) := \frac{l_P(f)}{a(i)} + \sum_{e \in P} b_e^*$, $\forall i$, $\forall P \in \mathcal{P}_i$, is a minimum-latency solution for *the demand vector reached in the same equilibrium*. Therefore the b^* vector would be the vector of the optimal taxes. To complete the proof of Theorem 1 we will now show the existence of (at least) one solution to (GENERAL CP):

Theorem 2. *If $f_e l_e(f_e)$ are continuous, convex, strictly monotone functions for all $e \in E$, and $D_i(\cdot)$ are nonnegative continuous functions bounded from above for all i, then (GENERAL CP) has a solution.*

Proof. We provide only a sketch of the proof. See the full paper for details. (GENERAL CP) is equivalent in terms of solutions to the complementarity problem (GENERAL CP′) (proof omitted). The only difference between (GENERAL CP) and (GENERAL CP′) is that $T_P(f) = \sum_{e \in P} (\frac{l_e(f_e)}{a(i)} + b_e)$ is replaced by $T_P(\hat{f}) = \sum_{e \in P} (\frac{l_e(\hat{f}_e)}{a(i)} + b_e)$ in the first two constraints.

To show that (GENERAL CP′) has a solution, we will follow a classic proof method by Todd [18] that reduces the solution of a complementarity problem to a Brouwer fixed-point problem. In what follows, let $[x]^+ := \max\{0, x\}$. If $\phi :$ $\mathbb{R}^n \to \mathbb{R}^n$ with $\phi(x) = (\phi_1(x), \phi_2(x), \ldots, \phi_n(x))$ is a function with components $\phi_1, \ldots, \phi_n$ defined as

$$\phi_i(x) = [x_i - F_i(x)]^+,$$

then $\hat{x}$ is a fixed point to ϕ iff $\hat{x}$ solves the complementarity problem $x^T F(x) = 0, F(x) \geq 0, x \geq 0$. Following [1], we will restrict ϕ to a large cube with an artificial boundary, and show that the fixed points of this restricted version of ϕ are fixed points of the original ϕ by showing that no such fixed point falls on the boundary of the cube.

Note that for (GENERAL CP) $x = (f, u, b, \hat{f}, \mu)$. We start by defining the cube which will contain x. Let $K_{\hat{f}} := \max_i \max_{u \geq 0}\{D_i(u)\}+1$, $K_f := K_{\hat{f}}$, $K_\mu :=$ $n \cdot \max_{e \in E} \max_{0 \leq x \leq k \cdot K_{\hat{f}}}\{l_e(x) + x\frac{\partial l_e}{\partial f_e}(x)\}$. Let S be the maximum possible entry of the inverse of any ± 1 matrix of dimension at most $(k+m) \times (k+m)$, where m denotes $|E|$ (note that S depends only on $(k+m)$.) Also, let $a_{max} = \max_i\{1/a(i)\}$ and $l_{max} = \max_e\{l_e(k \cdot K_f)\}$. Then define $K_b := (k + m)Sm a_{max}l_{max} + 1$, $K_u := n \cdot \left(\max_{e \in E, i \in \{1, \ldots, k\}} \left\{\frac{l_e(k \cdot K_f)}{a(i)}\right\} + K_b\right) + 1$.

We allow x to take values from the cube $\{0 \leq f_P \leq K_f, P \in \mathcal{P}\}$, $\{0 \leq u_i \leq K_u, i = 1, \ldots k\}$, $\{0 \leq b_e \leq K_b, e \in E\}$, $\{0 \leq \hat{f}_P \leq K_{\hat{f}}, P \in \mathcal{P}\}$, $\{0 \leq \mu_i \leq K_\mu, i = 1, \ldots k\}$. We define $\phi = (\{\phi_P : P \in \mathcal{P}\}, \{\phi_i : i = 1, \ldots, k\}, \{\phi_e : e \in E\}, \{\phi_{\hat{P}} : P \in \mathcal{P}\}, \{\phi_{\hat{i}} : i = 1, \ldots k\})$ with $|\mathcal{P}| + k + m + |\mathcal{P}| + k$ components as follows:

$$\phi_P(f, u, b, \hat{f}, \mu) = \min\{K_f, [f_P + u_i - T_P(\hat{f})]^+\} \qquad \forall i, \forall P \in \mathcal{P}_i$$

$$\phi_i(f, u, b, \hat{f}, \mu) = \min\{K_u, [u_i + D_i(u) - \sum_{P \in \mathcal{P}_i} f_P]^+\} \qquad i = 1, \ldots, k$$

$$\phi_e(f, u, b, \hat{f}, \mu) = \min\{K_b, [b_e + f_e - \hat{f}_e]^+\} \qquad \forall e \in E$$

$$\phi_{\hat{P}}(f, u, b, \hat{f}, \mu) = \min\{K_{\hat{f}}, [\hat{f}_P + \mu_i - \sum_{e \in P} \frac{\partial l_e}{\partial f_e}(\hat{f}_e)]^+\} \qquad \forall i, \forall P \in \mathcal{P}_i$$

$$\phi_{\hat{i}}(f, u, b, \hat{f}, \mu) = \min\{K_{\hat{i}}, [\mu_i + D_i(u) - \sum_{P \in \mathcal{P}_i} \hat{f}_P]^+\} \qquad i = 1, \ldots, k$$

where $f_e = \sum_{P \ni e} f_P, \hat{f}_e = \sum_{P \ni e} \hat{f}_P$. By Brouwer's fixed-point theorem, there is a fixed point x^* in the cube defined above, i.e., $x^* = \phi(x^*)$. In particular we have that $f_P^* = \phi_P(x^*), u_i^* = \phi_i(x^*), b_e^* = \phi_e(x^*), \hat{f}_P^* = \phi_{\hat{P}}(x^*), \mu_i^* = \phi_{\hat{i}}(x^*)$ for all $P, \hat{P} \in \mathcal{P}, i = 1, \ldots, k, e \in E$.

Following the proof of Theorem 5.3 of [1] we can show that

$$\hat{f}_P^* = [\hat{f}_P^* + \mu_i^* - \sum_{e \in P}(l_e(\hat{f}_e^*) + \hat{f}_e^*\frac{\partial l_e}{\partial f_e}(\hat{f}_e^*))]^+, \forall P \quad \mu_i^* = [\mu_i^* + D_i(u^*) - \sum_{P \in \mathcal{P}_i} \hat{f}_P^*]^+, \forall i$$

$$f_P^* = [f_P^* + u_i^* - T_P(\hat{f}^*)]^+, \forall P. \tag{1}$$

Note that this implies that $(\hat{f}^*, \mu^*)$ satisfy the KKT conditions of (MP) for u^*. Here we prove only (1) (the other two are proven in a similar way). Let $f_P^* = K_f$

for some $i, P \in \mathcal{P}_i$ (if $f_P^* < K_f$ then (1) holds). Then $\sum_{P \in \mathcal{P}_i} f_P^* > D_i(u^*)$, which implies that $u_i^* + D_i(u^*) - \sum_{P \in \mathcal{P}_i} f_P^* < u_i^*$, and therefore by the definition of ϕ_i we have that $u_i^* = 0$. Since $T_P(\hat{f}^*) \geq 0$, this implies that $f_P^* \geq f_P^* + u_i^* - T_P(\hat{f}^*)$. If $T_P(\hat{f}^*) > 0$, the definition of ϕ_P implies that $f_P^* = 0$, a contradiction. Hence it must be the case that $T_P(\hat{f}^*) = 0$, which in turn implies (1).

If there are $i, P \in \mathcal{P}_i$ such that $f_P^* > 0$, then (1) implies that $u_i^* = T_P(\hat{f}^*) = \sum_{e \in P} \frac{l_e(\hat{f}_e^*)}{a(i)} + \sum_{e \in P} b_e^*$. In this case we have that $u_i^* < K_u$, because $u_i^* = K_u \Rightarrow \sum_{e \in P} \frac{l_e(\hat{f}_e^*)}{a(i)} + \sum_{e \in P} b_e^* = n \cdot \left(\max_{e \in E, i \in \{1,\dots,k\}} \left\{ \frac{l_e(K_f)}{a(i)} \right\} + K_b \right) + 1$ which is a contradiction since $b_e^* \leq K_b$. On the other hand, if there are $i, P \in \mathcal{P}_i$ such that $f_P^* = 0$, then (1) implies that $u_i^* \leq T_P(\hat{f}^*)$. Again $u_i^* < K_u$, because if $u_i^* = K_u$ we arrive at the same contradiction. Hence we have that

$$u_i^* = [u_i^* + D_i(u^*) - \sum_{P \in \mathcal{P}_i} f_P^*]^+, \ \forall i. \tag{2}$$

Next, we consider the following primal-dual pair of linear programs:

$$\min \sum_i \sum_{P \in \mathcal{P}_i} f_P \frac{l_P(\hat{f}^*)}{a(i)} \quad \text{s.t.} \quad (\text{LP*}) \qquad \max \sum_i D_i(u^*) u_i - \sum_{e \in E} \hat{f}_e^* b_e \ \text{s.t.}$$

$$(\text{DP*})$$

$$\sum_{P \in \mathcal{P}_i} f_P \geq D_i(u^*) \qquad i = 1, \dots, k \qquad u_i \leq \frac{l_P(\hat{f}^*)}{a(i)} + \sum_{e \in P} b_e \qquad \forall i, \forall P \in \mathcal{P}_i$$

$$f_e = \sum_{P \in \mathcal{P}: e \in P} f_P \qquad \forall e \in E \qquad b_e, u_i \geq 0 \qquad \forall e \in E, \forall i$$

$$f_e \leq \hat{f}_e^* \qquad \forall e \in E$$

$$f_P \geq 0 \qquad \forall P$$

From the above, it is clear that $\hat{f}^*$ is a feasible solution for (LP*), and (u^*, b^*) is a feasible solution for (DP*). Moreover, since the objective function of (LP*) is bounded from below by 0, (DP*) has at least one bounded optimal solution as well. There is an optimal solution $(\hat{u}, \hat{b})$ of (DP*) such that all the $\hat{b}_e$'s are suitably upper bounded:

Lemma 3 (folklore). *There is an optimal solution $(\hat{u}, \hat{b})$ of (DP*) such that $\hat{b}_e \leq K_b - 1, \ \forall e \in E$.*

Let $\hat{f}$ be the optimal primal solution of (LP*) that corresponds to the optimal dual solution $(\hat{u}, \hat{b})$ of (DP*). Exploiting the fact that $(\hat{f}, \hat{u}, \hat{b})$ is a *saddle point* for the Lagrangian (see e.g. [16]) of (LP*)-(DP*) we can show (derivation omitted) that

$$b_e^* = [b_e^* + f_e^* - \hat{f}_e^*]^+, \ \forall e \in E. \tag{3}$$

Equations (1),(2),(3) imply that $(f^*, u^*, b^*, \hat{f}^*, \mu^*)$ is indeed a solution of (GENERAL CP'), and therefore a solution to (GENERAL CP). The proof of Theorem 2 is complete.

References

1. H. Z. Aashtiani and T. L. Magnanti. Equilibria on a congested transportation network. *SIAM J. Algebraic and Discrete Methods*, 2:213–226, 1981.
2. M. Beckmann, C. B. McGuire, and C. B. Winsten. *Studies in the Economics of Transportation*. Yale University Press, 1956.
3. R. Cole, Y. Dodis, and T. Roughgarden. Pricing network edges for heterogeneous selfish users. In *Proc. 35th ACM STOC*, pp. 521–530, 2003.
4. R. Cole, Y. Dodis, and T. Roughgarden. Bottleneck links, variable demand and the tragedy of the commons. In *Proc. 17th ACM-SIAM SODA*, 668 - 677, 2006.
5. S. C. Dafermos. Traffic equilibria and variational inequalities. *Transportation Science* 14, pp. 42–54, 1980.
6. S. Dafermos and F. T. Sparrow. The traffic assignment problem for a general network. *J. Research National Bureau of Standards, Series B*, 73B:91–118, 1969.
7. S. Dafermos. Toll patterns for multiclass-user transportation networks. *Transportation Science*, 7:211–223, 1973.
8. F. Facchinei and J.-S. Pang. *Finite-Dimensional Variational Inequalities and Complementarity Problems, Vol 1*. Springer-Verlag, Berlin, 2003.
9. L. Fleischer, K. Jain, and M. Mahdian. Tolls for heterogeneous selfish users in multicommodity networks and generalized congestion games. In *Proc. 45th IEEE Symposium on Foundations of Computer Science*, 277–285, 2004.
10. N. H. Gartner. Optimal traffic assignment with elastic demands: a review. Part I: analysis framework. *Transportation Science*, 14:174–191, 1980.
11. N. H. Gartner. Optimal traffic assignment with elastic demands: a review. Part II: algorithmic approaches. *Transportation Science*, 14:192–208, 1980.
12. D. W. Hearn and M. B. Yildirim. A toll pricing framework for traffic assignment problems with elastic demand. In M. Gendreau and P. Marcotte, editors, *Transportation and Network Analysis: Current Trends. Miscellanea in honor of Michael Florian*. Kluwer Academic Publishers, 2002.
13. G. Karakostas and S. G. Kolliopoulos. Edge pricing of multicommodity networks for heterogeneous selfish users. In *Proc. 45th IEEE Symposium on Foundations of Computer Science*, 268–276, 2004.
14. E. Koutsoupias and C. Papadimitriou. Worst-case equilibria. In *Proc. 16th Symposium on Theoretical Aspects of Computer Science*, pages 404–413, 1999.
15. A. Nagurney. A multiclass, multicriteria traffic network equilibrium model. *Mathematical and Computer Modelling*, 32:393–411, 2000.
16. R. T. Rockafellar. *Convex Analysis*. Princeton University Press, 1970.
17. T. Roughgarden and É. Tardos. How bad is selfish routing? *J. ACM*, 49:236–259, 2002.
18. M. J. Todd. The computation of fixed points and applications. *Lecture Notes in Economics and Mathematical Systems*, 124, Springer-Verlag, 1976.
19. J. G. Wardrop. Some theoretical aspects of road traffic research. *Proc. Inst. Civil Engineers, Part II*, 1:325–378, 1952.
20. H. Yang and H.-J. Huang. The multi-class, multi-criteria traffic network equilibrium and systems optimum problem. *Transportation Research Part B*, 38:1–15, 2004.

Aggregating Strategy for Online Auctions

Shigeaki Harada[1,*], Eiji Takimoto[2], and Akira Maruoka[2]

[1] NTT Service Integration Laboratories
harada.shigeaki@lab.ntt.co.jp
[2] Graduate School of Information Sciences, Tohoku University
Aoba 6-6-05, Aramaki, Sendai 980-8579, Japan
{t2, maruoka}@maruoka.ecei.tohoku.ac.jp

Abstract. We consider the online auction problem in which an auction-eer is selling an identical item each time when a new bidder arrives. It is known that results from online prediction can be applied and achieve a constant competitive ratio with respect to the best fixed price profit. These algorithms work on a predetermined set of price levels. We take into account the property that the rewards for the price levels are not independent and cast the problem as a more refined model of online prediction. We then use Vovk's Aggregating Strategy to derive a new algorithm. We give a general form of competitive ratio in terms of the price levels. The optimality of the Aggregating Strategy gives an evidence that our algorithm performs at least as well as the previously proposed ones.

1 Introduction

We consider the online auction problem proposed by Bar-Yossef, Hildrum, and Wu [3]. This models the situation where an auctioneer is selling single items in unlimited supply to bidders who arrive one at a time and each desires one copy. A particularly interesting case is for a digital good, of which infinitely many copies can be generated at no cost. Precisely, when each bidder t arrives with bid m_t, the auctioneer puts a price r_t on the item and sells a copy to the bidder at price r_t if $r_t \leq m_t$ and rejects the bidder otherwise. The auctioneer is required to compute the price r_t prior to knowing the values $m_t, m_{t+1}, \ldots$. Below we give a formal description.

Definition 1 (Online Auction A). *For each bidder $t = 1, 2, \ldots, T$,*

1. *Compute (randomly) a price r_t.*
2. *Observe the bid $m_t > 0$.*
3. *If $r_t \leq m_t$, then sell to bidder t at price $g_{A,t} = r_t$.*
4. *Otherwise, reject bidder t and $g_{A,t} = 0$.*

The total profit of the auction A is $G_{A,T} = \sum_{t=1}^{T} g_{A,t}$.

The goal of the auction is to make the total expected profit $E[G_{A,T}]$ as much as the best fixed price profit, denoted OPT, no matter what the bidding sequence is. Note that $\text{OPT} = \max_{1 \leq k \leq T} k m_{(k)}$, where $m_{(k)}$ is the kth largest bid.

---

* This work was done while the author was at Tohoku University.

D.Z. Chen and D.T. Lee (Eds.): COCOON 2006, LNCS 4112, pp. 33–41, 2006.
© Springer-Verlag Berlin Heidelberg 2006

We first assume that the smallest value l and the largest values h of the bids in the auction are known. Discretizing the range $[l, h]$ with a finite set of price levels $h \geq b(1) > b(2) > \cdots > b(N) = l$, we have the problem reduced to an online prediction game with expert advice [4,5,8,9]. We use $b(i) = l\rho^{N-i}$ for some $\rho > 1$ with $N = O(\ln(h/l))$ so that $b(1) \geq h/\rho$. The idea is to introduce an expert for each price level $b(i)$ who always recommends the price $b(i)$. We can now use a number of expert-advice algorithms to achieve the total profit as much as that of the best expert, which is larger than OPT$/\rho$ by the choice of the set of price levels. Blum, Kumar, Rudra, and Wu employ the Hedge or the Randomized Weighted Majority algorithm [9,5] and give a lower bound of

$$E[G_{\text{Hedge},T}] \geq \frac{\ln \alpha}{\alpha - 1} \left(\frac{\text{OPT}}{\rho} - \frac{h}{\ln \alpha} \ln(\log_\rho(h/l) + 1) \right)$$

on the total profit [1], where $\alpha > 1$ is a parameter of the Hedge algorithm. Blum and Hartline improve the additional loss term to $O(h)$ by using the Following Perturbed Leader (FPL) approach with a slight modification [2]. They call the modified version the Hallucinated-Gain (HG) algorithm and give the following bound

$$E[G_{\text{HG},T}] \geq (1 - \delta) \left(\frac{\text{OPT}}{\rho} - 2h \left(\frac{2}{\delta} \ln \nu(\rho) + \frac{\nu(\rho)}{\delta^2}(1 - \delta)^{\nu(\rho)} + 1 \right) \right),$$

where $\nu(\rho) = \lfloor \log_\rho 2 \rfloor + 1$ and $\delta \in [0, 1]$ is a parameter of the HG algorithm. Moreover, the HG algorithm can be further improved so that it does not need to know l and h at a cost of only $O(h)$ additional loss.

In this paper, we first observe that, unlike the typical expert-advice setting, the rewards for the experts are not uniformly bounded. That is, the reward for expert i is either 0 or $b(i)$. So we could improve the algorithms using non-uniform risk information as in [7]. Furthermore, we have a further advantage in that the rewards for the experts are not independent. More precisely, when the bid m_t lies in $(b(i + 1), b(i)]$, then all experts j with $j \geq i$ get rewards $b(j)$ and others get no rewards. In other words, there are only N possible outcomes to be considered. Taking this advantage into account, we give a more refined model of online prediction and apply Vovk's Aggregating Strategy [10] to derive a new algorithm called the Aggregating Algorithm for Auction (AAA). We give its profit bound[1] given by

$$E[G_{\text{AAA},T}] \geq c(\alpha, B) \left(\frac{\text{OPT}}{\rho} - \frac{1}{\ln \alpha} \ln(\log_\rho(h/l) + 1) \right),$$

where $\alpha > 1$ is a parameter of the AAA and $c(\alpha, B)$ is a complicated function of α and $B = \{b(1), \ldots, b(N)\}$. It seems that the bound is somewhat better since it has only an $O(\log \log(h/l))$ additional loss term, but in order to make $c(\alpha, B)$ a constant, we need to choose α that depends on h so that it quickly converges to

---

[1] Actually we obtain a tighter form of bound.

1 as h is large. Unfortunately, we have not succeeded to give a useful expression for $c(\alpha, B)$ to compare the profit bound with that of the HG algorithm, but it is better than the Hedge bound by the optimality of the Aggregating Strategy. We conjecture that the AAA performs as well as the HG algorithm. Numerical computation shows that the bound of the AAA outperforms others for sufficiently large ranges $[l, h]$ with $l = 1$ and $h \leq 10^{14}$.

2   Online Prediction Game and the Aggregating Strategy

We will show that online auction can be modeled as an online prediction game to which the Aggregating Strategy can be applied. The Aggregating Strategy is a very general method for designing algorithms that perform optimally for various games. In this section, we describe the strategy with its performance bound in a generic form.

First we describe a game that involves the learner (an algorithm), N experts, and the environment. A game is specified by a triple $(\Gamma, \Omega, \lambda)$, where Γ is a fixed prediction space, Ω is a fixed outcome space, and $\lambda : \Omega \times \Gamma \to [0, \infty]$ is a fixed reward function. (Note that the game is often described in terms of a loss function in the literature.) At each trial $t = 1, 2, \ldots, T$, the following happens.

1. Each expert i makes a prediction $x_{i,t} \in \Gamma$.
2. The learner combines $x_{i,t}$ and makes its own prediction $\gamma_t \in \Gamma$.
3. The environment chooses some outcome $\omega_t \in \Omega$.
4. The learner gets reward $\lambda(\omega_t, \gamma_t)$ and experts i get reward $\lambda(\omega_t, x_{i,t})$.

The total reward of the learner A is $R_{A,T} = \sum_{t=1}^{T} \lambda(\omega_t, \gamma_t)$ and that of expert i is $R_{i,T} = \sum_{t=1}^{T} \lambda(\omega_t, x_{i,t})$. The goal of the learner A is to make predictions so that its total reward $R_{A,T}$ is not much less than the total reward of the best expert $\max_{1 \leq i \leq N} R_{i,T}$.

Now we give the Aggregating Strategy that derives an algorithm called the Aggregating Algorithm (AA) for each specific game. The AA uses a parameter $\alpha > 1$. For each trial t, the AA assigns to each expert i a weight $v_{i,t}$ given by

$$v_{i,t} = \frac{v_{i,1} \alpha^{R_{i,t-1}}}{\sum_{j=1}^{N} v_{j,1} \alpha^{R_{j,t-1}}}, \tag{1}$$

where $R_{i,t-1} = \sum_{q=1}^{t-1} \lambda(\omega_q, x_{i,q})$ is the sum of the rewards that expert i has received up to the previous trial. Initial weights $v_{i,1}$ can be set based on a prior confidence on the experts. Typically the uniform prior ($v_{i,1} = 1/N$) is used. When given predictions $x_{i,t}$ from experts, the AA predicts a $\gamma_t \in \Gamma$ given by

$$\gamma_t = \arg \sup_{\gamma \in \Gamma} \inf_{\omega \in \Omega} \frac{\lambda(\omega, \gamma)}{\log_\alpha \sum_{i=1}^{N} v_{i,t} \alpha^{\lambda(\omega, x_{i,t})}}. \tag{2}$$

The next theorem gives a performance bound of the AA.

Theorem 1 ([10]). *For any outcome sequence* $(\omega_1, \ldots, \omega_T) \in \Omega^*$,

$$R_{AA,T} \geq c(\alpha) \log_\alpha \sum_{i=1}^{N} v_{i,1} \alpha^{R_{i,T}} \geq c(\alpha) \max_{1 \leq i \leq N} \left(R_{i,T} - \frac{\ln(1/v_{1,i})}{\ln \alpha} \right),$$

where

$$c(\alpha) = \inf_{\boldsymbol{v},\boldsymbol{x}} \sup_{\gamma \in \Gamma} \inf_{\omega \in \Omega} \frac{\lambda(\omega, \gamma)}{\log_\alpha \sum_{i=1}^{N} v_i \alpha^{\lambda(\omega, x_i)}}, \tag{3}$$

where $\boldsymbol{v} = (v_1, \ldots, v_N)$ *ranges over all probability vectors of dimension* N *and* $\boldsymbol{x} = (x_1, \ldots, x_N)$ *ranges over all possible predictions of experts.*

3   The Game for Online Auction

Now we give the game $(\Gamma, \Omega, \lambda)$ reduced from the online auction problem. We first fix a finite set of price levels $B = \{b(1), \ldots, b(N)\}$ with $h \geq b(1) > \cdots > b(N) = l$ as options to choose from.

The prediction space Γ is the set of probability vectors of dimension N. The prediction $\gamma_t = \boldsymbol{p}_t = (p_t(1), \ldots, p_t(N)) \in \Gamma$ in the tth trial is interpreted as the way of choosing price r_t in the auction, i.e., letting $r_t = b(i)$ with probability $p_t(i)$. For each $1 \leq i \leq N$, we define an expert who always recommends the option $b(i)$. Formally, we let $x_{i,t} = \boldsymbol{e}_i (\in \Gamma)$, where $\boldsymbol{e}_i$ is the unit vector whose ith component is 1.

The outcome space Ω is the set of vectors whose ith component represents a reward for the ith option, which is either 0 (for the case where $m_t < b(i)$) or $b(i)$ (for the case where $m_t \geq b(i)$). Moreover, if the option $b(i)$ gets a positive reward, then all the options $b(j)$ with $j \geq i$ get positive rewards as well. Thus, we have only N possible reward vectors and

$$\Omega = \left\{ \begin{array}{c} (b(1), b(2), \ldots, b(N-1), b(N)), \\ (0, b(2), \ldots, b(N-1), b(N)), \\ \vdots \\ (0, 0, \ldots, 0, b(N)) \end{array} \right\}.$$

Let $\boldsymbol{b}_i = (0, \ldots, 0, b(i), \ldots, b(N))$ so that $\Omega = \{\boldsymbol{b}_1, \ldots, \boldsymbol{b}_N\}$. If the bid m_t lies in the interval $(b(i+1), b(i)]$ in the auction, then let the tth outcome be $\omega_t = \boldsymbol{b}_i$ in the reduced game.

Finally, our reward function is $\lambda(\boldsymbol{b}_i, \boldsymbol{p}) = \boldsymbol{b}_i \cdot \boldsymbol{p} = \sum_{j=i}^{N} b(j)p(j)$. Under the reduction just described, it is easy to see that $E[g_{A,t}] = \lambda(\boldsymbol{b}_i, \boldsymbol{p}_t)$ if the bid m_t is in $(b(i+1), b(i)]$, and so we have $E[G_{A,T}] = R_{A,T}$. Similarly, the total profit of a single sales price $b(i)$ equals $R_{i,T}$. So, Theorem 1 implies that the AA for the auction achieves profit nearly as large as the best single price sales $\max_i R_{i,T}$ in the set B. Moreover, if we choose $b(i) = l\rho^{N-i}$, then, no matter what the optimal price $r^* \in [l, h]$ is, there exists a $b(j)$ with $b(j) \le r^* < b(j+1) = \rho b(j)$ and so we have $\text{OPT}/\rho \le R_{j,T} \le \max_i R_{i,T}$. Therefore, the AA achieves profit nearly as large as OPT. We call this algorithm the Aggregating Algorithm for Auction (AAA).

4   The Aggregating Algorithm for Auction

In this section we show how the AAA works by giving the weights v_t it maintains and the prediction p_t in a closed form. First we rewrite (1) and (2) in terms of the notations used in our auction game as

$$v_{i,t} = \frac{v_{i,1} \alpha^{b(i)\tau_{i,t-1}}}{\sum_{j=1}^{N} v_{j,1} \alpha^{b(j)\tau_{j,t-1}}}, \tag{4}$$

where $\tau_{i,t} = \#\{1 \le q \le t \mid m_t \le b(i)\}$ is the number of trials up to t in which the price $b(i)$ receives reward, and

$$p_t = \arg\sup_{p \in \Gamma} \min_{1 \le k \le N} \frac{b_k \cdot p}{\log_\alpha \left(1 + \sum_{i=k}^{N} v_{i,t}(\alpha^{b(i)} - 1)\right)}. \tag{5}$$

Note that the Hedge algorithm predicts with $q_t(i) = v_{i,t}$ for determining the price at trial t. (More precisely, the normalized parameter $\alpha^{1/b(1)}$ is used instead of α in (4) [3].) The rest is to show the prediction of the AAA.

Theorem 2. *Let*

$$d_{k,t} = \log_\alpha \left(1 + \sum_{i=k}^{N} v_{i,t}(\alpha^{b(i)} - 1)\right)$$

for $1 \le k \le N$ with the convention $d_{N+1,t} = 0$. Then,

$$p_t(i) = \frac{\frac{1}{b(i)}(d_{i,t} - d_{i+1,t})}{\sum_{k=1}^{N} \frac{1}{b(k)}(d_{k,t} - d_{k+1,t})}$$

attains the supremum of (5).

Proof. Note that we want to solve

$$p_t = \arg\sup_{p \in \Gamma} \min_{1 \le k \le N} \frac{\sum_{i=k}^{N} b(i)p(i)}{d_{k,t}}. \tag{6}$$

We first claim that for any $p \in \Gamma$,

$$\min_{1 \le k \le N} \frac{\sum_{i=k}^{N} b(i)p(i)}{d_{k,t}} \le \frac{1}{\sum_{k=1}^{N} \frac{1}{b(k)}(d_{k,t} - d_{k+1,t})}. \tag{7}$$

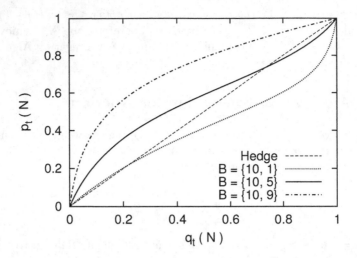

Fig. 1. The Hedge prediction $q_t(N)$ and the AAA prediction $p_t(N)$ for the lower price

Let M denote the r.h.s. of the above inequality. We prove the claim by contradiction. Assume on the contrary that the claim does not hold, i.e., for any $1 \leq k \leq N$, there exists a positive $\Delta_k > 0$ such that

$$\frac{\sum_{i=k}^{N} b(i)p(i)}{d_{k,t}} = M + \Delta_k.$$

Then we have

$$p(k) = \frac{1}{b(k)}\big((M + \Delta_k)d_{k,t} - (M + \Delta_{k+1})d_{k+1,t}\big)$$

$$= \frac{M(d_{k,t} - d_{k+1,t})}{b(k)} + \frac{\Delta_k d_{k,t}}{b(k)} - \frac{\Delta_{k+1}d_{k+1,t}}{b(k)}$$

$$> \frac{M(d_{k,t} - d_{k+1,t})}{b(k)} + \frac{\Delta_k d_{k,t}}{b(k)} - \frac{\Delta_{k+1}d_{k+1,t}}{b(k+1)}$$

since $b(k) > b(k+1)$. Summing up the both sides over all $1 \leq k \leq N$, we get

$$\sum_{k=1}^{N} p(k) > 1 + \frac{\Delta_1 d_{1,t}}{b(1)} > 1,$$

which contradicts the fact that p is a probability vector. So (7) holds.

On the other hand, the prediction $p \in \Gamma$ with

$$p(i) = \frac{\frac{1}{b(i)}(d_{i,t} - d_{i+1,t})}{\sum_{k=1}^{N} \frac{1}{b(k)}(d_{k,t} - d_{k+1,t})}$$

clearly satisfies the equality of (7). This implies that this prediction p attains the supremum.   □

The prediction of the AAA can be viewed as a nonlinear transformation of the Hedge prediction $q_t(i)$. Figure 1 illustrates the transformation for $N = 2$, $\alpha = 1.5$ and various sets of price levels $B = \{b(1), b(2)\}$. We fix $b(1) = 10$.

From the figure we can see that the AAA puts more weight on the lower price $b(N)$ when $q_t(N)$ is small. This is reasonable since the lower price is more likely to get reward. Curiously the weight on $b(N)$ gets larger when $b(N)$ gets closer to $b(1)$.

5   The Performance Bound of the AAA

In this section, we give the performance bound of the AAA by showing $c(\alpha)$ in terms of the set B of price levels. In what follows, we write $c(\alpha, B)$ to explicitly specify B. From the proof in Theorem 2, we can rewrite $c(\alpha)$ of (3) as

$$c(\alpha, B) = \inf_{v \in \Gamma} \frac{1}{\sum_{k=1}^{N} \frac{1}{b(k)}(d_k - d_{k+1})}, \tag{8}$$

where

$$d_k = \log_\alpha \left(1 + \sum_{i=k}^{N} v(i)(\alpha^{b(i)} - 1) \right)$$

Theorem 3. *Let $(r_1, \ldots, r_N)$ and $(s_1, \ldots, s_N)$ be the probability vectors in Γ defined as*

$$r_i = \left(\frac{1}{b(i)} - \frac{1}{b(i-1)} \right) b(N),$$

$$s_i = \left(\frac{1}{\alpha^{b(i)} - 1} - \frac{1}{\alpha^{b(i-1)} - 1} \right) (\alpha^{b(N)} - 1)$$

with the convention that $b(0) = \infty$. Then

$$c(\alpha, B) = \frac{b(N) \ln \alpha}{D(r\|s) + b(N) \ln \alpha}, \tag{9}$$

where $D(r\|s) = \sum_{i=1}^{N} r_i \ln(r_i/s_i)$ is the Kullback-Leibler divergence.

Proof. The problem is to maximize the denominator of (8)

$$f(v) = \sum_{k=1}^{N} \frac{1}{b(k)}(d_k - d_{k+1})$$

subject to $v \in \Gamma$. First we relax the constraint and find the maximum of $f(v)$ subject to $\sum_{i=1}^{N} v(i) = 1$. Then we will show that the maximizer v^* lies in the feasible solution, i.e., $v^*(i) \geq 0$ for all i. Since f is concave, the set of equations

$$\frac{\partial}{\partial v(j)} \left(f(v) + t \left(\sum_{i=1}^{N} v(i) - 1 \right) \right)$$

$$= -(\alpha^{b(j)} - 1) \sum_{k=1}^{j} \left(\frac{1}{b(k)} - \frac{1}{b(k-1)} \right) \frac{1}{1 + \sum_{i=k}^{N}(\alpha^{b(i)} - 1)v(i)} + t = 0$$

for $1 \leq j \leq N$ and $\sum_{i=1}^{N} v(i) = 1$ give the maximizer. It is straightforward to show that the solution is

$$v^*(j) = \frac{F(b(j), b(j-1)) - F(b(j+1), b(j))}{t(\alpha^{b(j)} - 1)}$$

and

$$t = F(b(N+1), b(N)) = \frac{1/b(N)}{1 + 1/(\alpha^{b(N)} - 1)},$$

where $b(N+1) = -\infty$ and

$$F(x, y) = \frac{\frac{1}{x} - \frac{1}{y}}{\frac{1}{\alpha^x - 1} - \frac{1}{\alpha^y - 1}}.$$

We can show that $F(a, b) < F(b, c)$ for any $a < b < c$ with $b > 0$. This gives $v^*(j) > 0$.

Plugging v^* into $f(v)$, we have the theorem. $\qquad\square$

6   Numerical Comparisons of the Performance Bounds

To compare the bound of the AAA with those of the Hedge and the HG algorithms, we need to give a useful form of $c(\alpha, B)$ with $b(i) = l\rho^{N-i}$ for $N = \lfloor \log_\rho(h/l) \rfloor + 1$. We have not succeeded to derive such an expression. So we show numerical experiments to compare the performance bounds. Recall that

$$E[G_{\text{Hedge},T}] \geq \frac{\ln \alpha}{\alpha - 1} \frac{\text{OPT}}{\rho} - \frac{h}{\alpha - 1} \ln \left(\log_\rho(h/l) + 1 \right),$$

$$E[G_{\text{HG},T}] \geq (1 - \delta) \frac{\text{OPT}}{\rho} - 2h(1 - \delta) \left(\frac{2}{\delta} \ln \nu(\rho) + \frac{\nu(\rho)}{\delta^2}(1 - \delta)^{\nu(\rho)} + 1 \right),$$

$$E[G_{\text{AAA},T}] \geq c(\alpha, B) \frac{\text{OPT}}{\rho} - \frac{c(\alpha, B)}{\ln \alpha} \ln \left(\lfloor \log_\rho h/l \rfloor + 1 \right).$$

We fix $l = 1$ and adjust the parameters of the algorithms so that the first terms of the bounds are all equal to $(1/(2\rho))\text{OPT}$. Thus, the bounds are all of the form of

$$E[G_{A,T}] \geq \frac{1}{2\rho}\text{OPT} - g_A(h)h$$

for some functions g_A. Note that $g_{\text{HG}}(h) = O(1)$ and $g_{\text{Hedge}}(h) = O(\log \log h)$ by definition. Figure 2 shows how fast the functions $g_A(h)$ grow for the three algorithms.

Although g_{AAA} seems to be slightly increasing, the value is much smaller than g_{Hedge} and g_{HG} for a reasonable range of h. In fact, for a typical choice of $\rho = 1.01$, g_{HG} is a large constant (17.97) while $g_{\text{AAA}} \leq 0.5$ for $\log \log h \leq 3.5$. It is interesting to note that the Hedge has a better bound than the HG bound in typical cases. We may improve the bound by using a tighter bound of Theorem 1 and choosing carefully the initial weights $v_{1,i}$.

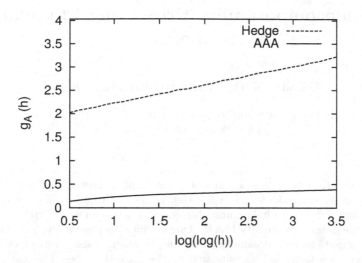

Fig. 2. The second term functions $g_A(h)$ for the three algorithms. We set $\rho = 1.01$. $g_{HG}(h) = 17.97$.

References

1. A. Blum, V. Kumar, A. Rudra, and F. Wu. Online Learning in Online Auctions. *Proc. 14th SODA*, 2003.
2. A. Blum and J. D. Hartline. Near-Optimal Online Auctions. *Proc. 16th SODA*, 1156–1163, 2005.
3. Z. Bar-Yossef, K. Hildrum and F. Wu. Incentive-Compatible Online Auctions for Digital Goods. *Proc. SODA 2002*, 964–970, 2002.
4. N. Cesa-Bianchi, Y. Freund, D. Haussler, D. P. Helmbold. R. E. Schapire, and M. K. Warmuth. How to use expert advice. *J. ACM*, 44(3):427–485, 1997.
5. Y. Freund and R. E. Schapire. A decision-theoretic generalization of on-line learning and an application to boosting. *JCSS*, 55(1):119–139, 1997.
6. M. Hutter and J. Poland. Prediction with expert advice by following the perturbed leader for general weights. *LNAI*, 3244, 279–293, 2004.
7. S. Harada, E. Takimoto and A. Maruoka. Online Allocation with Risk Information. *LNAI*, 3734, 343–355, 2005.
8. A. Kalai and S. Vempala. Efficient algorithms for online decision problems. *LNAI*, 2777, 26–40, 2003.
9. N. Littlestone and M. K. Warmuth. The weighted majority algorithm. *Inform. Comput.*, 108(2):212–261, 1994.
10. V. Vovk. A game of prediction with expert advice. *JCSS*, 56(2):153–173, 1998.

On Indecomposability Preserving Elimination Sequences

Chandan K. Dubey and Shashank K. Mehta*

Indian Institute of Technology, Kanpur - 208016, India
{cdubey, skmehta}@cse.iitk.ac.in

Abstract. A module of a graph is a non-empty subset of vertices such that every non-module vertex is either connected to all or none of the module vertices. An indecomposable graph contains no non-trivial module (modules of cardinality 1 and $|V|$ are trivial). We present an algorithm to compute indecomposability preserving elimination sequence, which is faster by a factor of $|V|$ compared to the algorithms based on earlier published work. The algorithm is based on a constructive proof of Ille's theorem [9]. The proof uses the properties of X-critical graphs, a generalization of critical indecomposable graphs.

Keywords: Module, indecomposable graph, critically indecomposable graph, elimination sequence.

1 Introduction

A non-trivial module (Fraïssé [7]) of an undirected graph is a proper subset with 2 or more vertices such that each vertex outside the subset is either connected to all vertices in the subset or to none. If each maximal module is replaced by a single vertex, then we get an *indecomposable* or *prime* or *base-level* graph. There are many graph algorithmic problems whose solution on indecomposable graphs will imply a solution on general graphs. These include the problems in domination, matching, coloring, optimal spanning tree, graph isomorphism etc. Therefore the study of indecomposable graph is very significant (see [11, 8, 16, 15, 12]).

Schmerl and Trotter [14] have studied critical indecomposable graphs in which deletion of any vertex transforms the graph into a decomposable graph. In [6] Dubey et.al. generalized this concept to X-critical graphs for any vertex subset X, where both G and $G(X)$ are indecomposable but $G(V - \{v\})$ is decomposable for all $v \in V - X$. So the critically indecomposable graphs are $\emptyset$-critical.

An ordered sequence of vertices of a graph, $v_1, v_2, v_3, \ldots$, is said to be an elimination sequence preserving a property $\mathcal{P}$ if that property remains valid in the graph after deletion of each vertex of the sequence in that order. Such

---

* Partly supported by Ministry of Human Resource Development, Government of India under grant no. MHRD/CD/20030320.

D.Z. Chen and D.T. Lee (Eds.): COCOON 2006, LNCS 4112, pp. 42–51, 2006.

sequences are useful in induction based proofs and iterative algorithms (see [14, 5, 4, 1]). In general an indecomposability preserving elimination sequence does not exist, since there are indecomposable graphs from which no vertex can be deleted without making the graph decomposable (i.e. critically indecomposable graphs). A generalization of elimination sequence considers a sequence of sets of vertices instead of single vertices. A result by Schmerl and Trotter [14] says that every indecomposable graph contains a pair of vertices which can be deleted without losing indecomposability. A further generalization of this result is due to Ille [9] which states that if the graph has an indecomposable induced-subgraph then such a pair of vertices can be extracted from the outside of that subgraph. Therefore, we can always construct an indecomposability preserving elimination sequence in which up to two vertices are eliminated at each step. In general, we can define an elimination sequence in which an initial subsequence has single vertices and the remainder has pairs. Such a sequence can also be viewed as a sequence of pairs, except at most one singleton element.

Based on Ille's result one can search a pair of vertices which leaves the graph indecomposable after elimination. Many papers [3, 10] discuss an $O(n + m)$ algorithm to test indecomposability of a graph, where n denote the number of vertices and m denotes the number of edges. Therefore by searching a pair to eliminate we can construct an elimination sequence which preserves indecomposability in $O(n^3(n + m))$ time. In this paper we present an $O(n^2(n + m))$ algorithm which uses the structure of X-critical graphs.

The most efficient algorithm to test indecomposability uses an approach based on modular decomposition and works in $O(n + m)$ time ([3, 10, 2]). These algorithms try to construct modular decomposition of graphs in linear time. The $O(n + m \log n)$ algorithm presented in Cournier and Habib [2] based on constructing induced indecomposable subgraphs for the test of indecomposability, gives an elimination sequence as a byproduct. It starts from an indecomposable subgraph $G(P_4)$ and at each step adds one or two vertices while preserving indecomposability, till entire graph is constructed. Here we present an elimination sequence which contains at most one single-vertex (at first) and remaining are all pairs. Such a sequence can be used in fast algorithms for problems such as perfect-matching, 3-coloring etc, see [5]. The sequence generated by Cournier and Habib cannot be used in these algorithms because it does not guarantee elimination sequence containing pairs.

Contribution of this paper. We describe an algorithm to compute a commutative elimination sequence which has $O(n^2)$ time complexity. The existance of such sequence was established in [6]. Following it we present a proof of the uniqueness of this sequence. Then we give an alternative (constructive) proof of Ille's theorem which leads to an $O(n + m)$ algorithm to compute a pair of vertices which can be eliminated while preserving X-criticality. Using these two algorithms we have an $O(n^2(n + m))$ algorithm to compute an elimination sequence for arbitrary indecomposable graph, which is the following computation. Given an indecomposable graph G with indecomposable subgraph $G(X)$, it computes 2-sets $D_1, \ldots, D_k$ which are mutually exclusive, $D_i \subset V - X$, and

$G(V - D_1 - \ldots - D_j)$ are indecomposable for all j and $|V - X - D_1 - \ldots - D_k| \leq 5$. The similar computation based on the original theorem by Ille involves explicit search and costs $O(n^3(n + m))$.

1.1 Basic Definitions and Results

In this paper we shall deal with undirected graphs only. If $G = (V, E)$ is a graph and $W \subseteq V$, then $G(W)$ will denote the induced subgraph on W. A vertex set $M \subseteq V$ is called a module if for all $x, y \in M$ and $z \in V - M$ $(x, z) \in E$ iff $(y, z) \in E$. A module M is said to be non-trivial if $2 \leq |M| \leq |V| - 1$. A graph is said to be prime or indecomposable if it has no non-trivial module. Note that a set M is a module in graph G, then it is also a module in $\overline{G}$, the complement of G.

Let $G = (V, E)$ be a graph and $X \subset V$ such that $G(X)$ is indecomposable. Then for any vertex $y \in V - X$, only one of the following three cases are possible: (i) $G(X \cup \{y\})$ is indecomposable, (ii) $G(X \cup \{y\})$ is decomposable with the unique module $\{y, z\}$ for some $z \in X$, and (iii) $G(X \cup \{y\})$ is decomposable with the unique module X. We partition the vertices of $V - X$ based on these cases. If it is case (i), then y belongs to a class denoted by $extn(X)$, in case of (ii) y belongs to a class denoted by $eq_X(z)$, finally in the third case y belongs to a class denoted by $[X]$. We denote this partition by $\mathcal{C}(V - X, X)$.

If a graph $G = (V, E)$ is indecomposable and its induced subgraphs $G(V - \{x\})$ are decomposable, for all $x \in V$, then G is said to be *critically indecomposable*. Schmerl and Trotter [14] have shown that the class of bipartite graphs given by $(\{a_i : 1 \leq i \leq k\}, \{b_i : 1 \leq i \leq k\}, \{(a_i, b_j) : i \leq j\})$ with $k > 1$, and their complements are the only critically indecomposable graphs. It is easy to see that if G is indecomposable or critically-indecomposable, then so is $\overline{G}$.

The concept of critically indecomposable graphs is generalized in [6]. Let $G = (V, E)$ be an indecomposable graph and $X \subset V$ such that $G(X)$ is also indecomposable. If $G(V - \{x\})$ is decomposable for all $x \in V - X$, then G is said to be *X-critical*. Therefore critically indecomposable graphs are $\emptyset$-critical.

1.2 Indecomposable Subgraphs

A basic theorem proved in [13, 14] states that every indecomposable graph with at least 4 vertices has a set of 4 vertices such that induced subgraph on it is also indecomposable. They also show that an induced indecomposable subgraph can be expanded to a larger indecomposable induced subgraph.

Theorem 1. *[14] Let $G = (V, E)$ be an indecomposable graph. If X is a subset of V such that $G(X)$ is indecomposable and $3 \leq |X| \leq |V| - 2$, then there exists a pair of vertices $a, b \in V - X$ such that $G(X \cup \{a, b\})$ is indecomposable.*

Let an indecomposable graph G have an indecomposable induced subgraph $G(X)$. Using the above result we can see there exists either a single or a pair of vertices in $V - X$ which can be eliminated while preserving indecomposability.

Formally, there is a vertex set Y, $X \subset Y \subset V$, such that $G(Y)$ is also indecomposable and $1 \leq |V - Y| \leq 2$. A stronger result proved by Ille [9] follows which states that there always exists a pair which can be eliminated.

Theorem 2. *[9] If G and an induced subgraph $G(X)$ are indecomposable and $|V - X| \geq 6$ then there exists a pair $\{a,b\} \in V - X$ s.t. $G(V - \{a,b\})$ is indecomposable.*

This result establishes the existence of an indecomposability preserving elimination sequence $D_1, D_2, \ldots, D_k$, where for all i, j $D_i \cap D_j = \emptyset$ for $i \neq j$, $|D_i| = 2$, $D_i \subset V - X$, each graph $G(V - D_1 - \ldots - D_i)$ is indecomposable, and $|V - D_1 - \ldots - D_k - X| \leq 5$.

The proof of the theorem is non-constructive. Therefore an indecomposable subgraph $G(V - \{a,b\})$ can be discovered by trying each pair $\{a,b\}$ and using the linear indecomposability test, in $O(n^2(n+m))$ time. The cost of computing an elimination sequence by this method would be $O(n^3(n+m))$. We shall present some results on X-critical graphs which will allow us to construct this elimination sequence in $O(n^2(n+m))$ time.

2 Computing Elimination Sequences

2.1 Computing Elimination Sequences in X-Critical Graphs

Let $G = (V, E)$ be an X-critical graph and $x \in V - X$. Then it is shown in [6] that $G(V - \{x\})$ has unique module, either of size $|V| - 2$ or of size 2. If $G = (V, E)$ is an X-critical graph and x, y is a pair of vertices in $V - X$ such that $G(V - \{x,y\})$ is also X-critical, then $\{x,y\}$ is called a *locked pair* in G. It is easy to verify, see [6], that at least one of the vertices of the pair is from class $eq_{V-\{x,y\}}(z)$ for some $z \in V-\{x,y\}$. The other will be either from $eq_{V-\{x,y\}}(z')$ for some $z' \in V-\{x,y\}$ or from $(V-\{x,y\})$. Following result from [6] establishes the existence of *commutative elimination sequence* in X-critical graphs.

Theorem 3. *[6] If $G = (V, E)$ is an X-critical graph, then vertices of $V - X$ can be partitioned into pairs $\{x_1, y_1\}, \ldots, \{x_k, y_k\}$ such that $G(V - \{x_{j_1}, y_{j_1}, \ldots, x_{j_s}, y_{j_s}\})$ is also X-critical for any subset $\{j_1, \ldots, j_s\}$ of $\{1, 2, \ldots, k\}$.*

In this section we present an algorithm to compute such a commutative elimination sequence for X-critical graphs. This algorithm is similar to the $O(n + m \log n)$ algorithm by Cournier and Habib [2].

Lemma 1. *A commutative elimination sequence for an X-critical graph can be computed in $O(n^2)$.*

Algorithm 1 computes the commutative elimination sequence of an X-critical graph. Starting from $Y = X$, iteratively expand Y till it becomes equal to V, identifying one locked pair in each step. This results into an elimination sequence but not necessarily commutative. Therefore in the last step the newly constructed pair, in some cases, exchanges elements with one of the previously constructed pair. The basic technique is based on computing $\mathcal{C}(Y, V - Y)$, introduced in [2]. If $\mathcal{C}$ denotes $\mathcal{C}(V - Y, Y)$ and $x \in V - Y$, then $update(\mathcal{C}, a)$ computes $\mathcal{C}(V - Y - \{a\}, Y \cup \{a\})$ from $\mathcal{C}(V - Y, Y)$.

Data: X-critical graph G, set X
Result: Commutative elimination sequence
1 $\mathcal{C} = \mathcal{C}(V - X, X)$;
   /* $extn(X)$ will be empty.   */
2 $Y = X$;
3 **for** $i = 1$ *to* $(|V| - |X|)/2$ **do**
4     $\mathcal{C}' = \mathcal{C}$;
5   Select any vertex a_i from some class $eq(u)$ of $\mathcal{C}$;
6     $\mathcal{C} = update(\mathcal{C}, a_i)$;
7     $b_i =$ An arbitrary vertex from $extn(Y \cup \{a_i\})$;
8     $\mathcal{C} = update(\mathcal{C}, b_i)$;
9   **if** $a_i \in eq(u)$ *and* $b_i \in eq(v)$ *in* $\mathcal{C}'$ *and* $(u, v) = (a_j, b_j)$ *for some* $j < i$ **then**
10 | $(a_j, b_j) = (a_i, v)$ and $(a_i, b_i) = (u, b_i)$;
11   **end**
12 **end**

Algorithm 1: Computation of a commutative elimination sequence

---

Let $G(Y)$ be an X-critical subgraph of G which is itself X-critical. Let a_i be a vertex in $V - Y$. Then due to Theorem 1 and X-criticality of G, $G(Y \cup \{a_i\})$ is not indecomposable so $\mathcal{C}(V - Y - \{a_i\}, Y \cup \{a_i\})$ is not defined. In the algorithm it denotes a generalization of the original classification. In addition to the original three types of classes, one more class is defined to be $eq(a_i, u)$. Suppose graph $G(Y \cup \{a_i\})$ is decomposable with the unique module $\{a_i, u\}$. Further if any vertex $x \in V - Y - \{a_i\}$ is such that $\{a_i, u, x\}$ is a module in $G(Y \cup \{a_i, x\})$, then x belongs to $eq(a_i, u)$.

The update in step 6 is performed as follows. Case $x \in eq_Y(y)$ with $y \neq u$: if $(x, a_i) \in E$ iff $(y, a_i) \in E$ then $x \in eq_{Y \cup \{a_i\}}(y)$ else $x \in extn(Y \cup \{a_i\})$; Case $x \in eq_Y(u)$: in this case $x \in eq_{Y \cup \{a_i\}}(a_i, u)$; Case $x \in [Y]$: if $(x, a_i) \in E$ iff $(x, z) \in E$ for any $z \in Y$, then $x \in [Y \cup \{a_i\}]$ else $x \in extn(Y \cup \{a_i\})$.

In step 7 b_i is chosen from $extn$-class so $G(Y \cup \{a_i, b_i\})$ is indecomposable, consequently class $eq_{Y \cup \{a_i.b_i\}}(z)$ becomes empty for all z. Existence of commutative elimination sequence ensures that such a b_i exists. Update in step 8 is similar to that in step 6 but for two exceptions. Case $x \in eq_{Y \cup \{a_i\}}(a_i, u)$: if $(b_i, a_i) \in E$ iff $(b_i, x) \in E$ then $x \in eq_{Y \cup \{a_i, b_i\}}(a_i)$ else $x \in eq_{Y \cup \{a_i, b_i\}}(u)$. Case $x \in eq_{Y \cup \{a_i\}}(v)$ where $b_i \in eq_Y(v)$: if $(x, a_i) \in E$ iff $(v, a_i) \in E$ then $x \in eq_{Y \cup \{a_i, b_i\}}(v)$ else $x \in eq_{Y \cup \{a_i, b_i\}}(b_i)$.

The correctness of the algorithm is based on few simple observations. In step 7 we select a vertex from $extn$-class so $G(Y \cup \{a_i, b_i\})$ is X-critical. Therefore we see that each successive graph $G(X), G(X \cup \{a_1, b_1\}), G(X \cup \{a_1, b_1, a_2, b_2\}), \ldots$ X-critical. In general $\{a_1, b_1\}, \ldots, \{a_k, b_k\}$ is not a commutative elimination sequence. Lemma 8 in [6] shows that the operation in Step 9 gives a commutative elimination sequence. The update steps take $O(n)$ time. The first step takes $O(n^2 + m)$ time since for each $x \in V - X$ it needs to be found out whether X is a module of $G(X \cup \{x\})$ or if there is $u \in X$ such that $\{x, u\}$ is a module of $G(X \cup \{x\})$ or neither. Therefore the entire process costs $O(n^2)$.

We prove here another interesting result for commutative elimination sequences.

Theorem 4. *The commutative elimination sequence in a X-critical graph is unique upto a permutation.*

Proof. Let S_1 and S_2 be two distinct commutative elimination sequences. Now consider $G(X \cup P)$, where P is the set of pairs which are in both S_1 and S_2. Call $Y = X \cup P$. If $\{a_1, b_1\}$ is a pair in S_1 which is not in S_2 then there must be pairs $\{a_1, b_2\}$ and $\{b_1, a_2\}$ in S_2. Let $Z = Y \cup \{a_1, a_2, b_1, b_2\}$.

Note that $\{a_2, b_2\}$ is a locked pair because $G(Z)$ is X-critical and $\{a_1, b_1\}$ is part of a commutative elimination sequence. For each $i, j \in \{1, 2\}$ consider the classes of $\mathcal{C}(\{a_i, b_j\}, Z - \{a_i, b_j\})$.

Claim. a_1 belongs to neither $[Z - p_{11}]$ nor $[Z - p_{12}]$.

Proof. If $Z - \{b_1\}$ has a module of size $|Z - \{b_1\}| - 1$ then that module must be $Z - \{a_1, b_1\}$ since it is a X-critical graph. But $Z - \{a_2, b_1\}$ is also X-critical therefore that module must be $Z - \{a_2, b_1\}$. Uniqueness of the module requires that $a_1 = a_2$ but that is not true. Therefore $a_1 \notin [Z - \{a_1, b_1\}]$. Similar argument shows that $a_1 \notin [Z - \{a_1, b_2\}]$. This leads to the conclusion that $a_1 \in eq_{Z - \{a_1, b_1\}}(p_1)$ and $a_1 \in eq_{Z - \{a_1, b_2\}}(p_2)$ for some p_1 and p_2. ♡

Claim. $p_1 = p_2 = a_2$.

Proof. By the definition $Z - p_{11} \cup \{a_1\} = Z - \{b_1\}$ has the module $\{a_1, p_1\}$. If $a_2 \neq p_1$, then $\{a_1, p_1\}$ is also a module of $Z - \{a_2, b_1\}$. But that is absurd since the latter is X-critical. Similarly, $p_2 = a_2$. ♡

We have shown that $\{a_1, a_2\}$ is a module of $G(Z - \{b_1\})$ as well as of $G(Z - \{b_2\})$. Therefore $\{a_1, a_2\}$ must also be a module of $G(Z)$ which is known to be X-critical. Therefore we conclude that S_1 and S_2 cannot be distinct. ∎

2.2 Indecomposability Preserving Elimination Sequence

In this section we shall show that if an indecomposable graph $G = (V, E)$ has an indecomposable subgraph $G(X)$ with $|V - X| > 5$, then a pair of vertices $a, b \in V - X$ can be computed in $O(n(n + m))$ time such that $G(V - \{a, b\})$ is also indecomposable. This gives an algorithm to compute an indecomposability preserving elimination sequence for any indecomposable graph with time complexity $O(n^2(n + m))$.

To find a pair of vertices from $V - X$ such that the reduced graph after deleting the pair remains indecomposable, we may arbitrarily delete a vertex and test the resulting graph for indecomposability. If this test fails for every vertex in $V - X$, then the graph is X-critical and we have already seen how to find a locked pair. If it succeeds for some vertex a, then we repeat this step on $G(V - \{a\})$. If this succeeds again, then we have the desired pair. The difficult case is when after deleting one vertex the graph reduces to X-critical. Following result addresses the problem of locating such a pair in these graphs.

Theorem 5. *Let G be an indecomposable graph on (V, E), X is a subset of V and $V - X = \{a, a_1, b_1, a_2, b_2, a_3, b_3\}$ where $G(V - \{a\})$ is X-critical and $p_1 = \{a_1, b_2\}, p_2 = \{a_2, b_2\}, p_3 = \{a_3, b_3\}$ is a commuting elimination sequence in $G(V - \{a\})$. Then for at least one locked pair, $p_i = \{a_i, b_i\}$, $G(V - \{a_i, b_i\})$ is indecomposable.*

Proof. Assume the contrary. Denote $V - p_i$ by Z_i. From the assumption $G(Z_i)$ is decomposable but $G(Z_i - \{a\})$ is indecomposable (actually X-critical) from the definition of commuting elimination sequence. It is known that if a subgraph $G(A)$ is indecomposable and $G(A \cup \{a\})$ is decomposable, then the latter has a unique module and it is either A or $\{a, b\}$ for some $b \in A$. Therefore either $a \in [Z_i - \{a\}]$ or $a \in eq_{Z_i - \{a\}}(u_i)$ for each i where u_i is some vertex in $Z_i - \{a\}$.

Assume that $a \in [Z_1 - \{a\}]$ and $a \in [Z_2 - \{a\}]$. Since $(Z_1 - \{a\}) \cap (Z_2 - \{a\})$ is non-empty, $(Z_1 - \{a\}) \cup (Z_2 - \{a\}) = V - \{a\}$ is a module of G which is absurd as G is indecomposable. Therefore $a \in [Z_i - \{a\}]$ for no more than one i. Without loss of generality, either a belongs to $[Z_1 - \{a\}]$, $eq_{Z_2 - \{a\}}(u_2)$, and $eq_{Z_3 - \{a\}}(u_3)$; or a belongs to $eq_{Z_i - \{a\}}(u_i)$ for all i. In the following discussions we show that these possibilities also lead to conflicts.

If $u_j = u_k = u$ for some $j \neq k$, then $\{a, u\}$ is a module of G, which is not possible as G is indecomposable. Thus $u_j \neq u_k$ for $j \neq k$.

Further if u_j and u_k both belong to $V - p_j - p_k$, then $\{a, u_j\}$ and $\{a, u_k\}$ are both modules in $G(V - p_2 - p_3)$ therefore $\{u_j, u_k\}$ must be a module in $G(V - \{a\} - p_j - p_k)$. This is impossible since the definition of commutative elimination sequence requires that $G(V - \{a\} - p_j - p_k)$ is X-critical. So we conclude that either $u_j \in p_k$ or $u_k \in p_j$. These observations lead to only two possibilities.

Case 1: Assume that $V - p_1 - \{a\}$, $\{a, u_2\}$ and $\{a, u_3\}$ be the modules of $G(Z_1)$, $G(Z_2)$ and $G(Z_3)$ respectively. From the previous paragraph we know that $u_3 \in p_2$ or $u_2 \in p_3$. Without loss of generality assume the latter. The facts that $V - p_1 - \{a\}$ is a module in $G(V - p_1)$ and $\{a, u_2\}$ is a module in $G(V - p_2)$ imply that $V - p_1 - p_2 - \{a, u_2\}$ is a module in $G(V - p_1 - p_2 - \{a\})$. This is absurd because $G(V - p_1 - p_2 - \{a\})$ is X-critical.

Case 2: Assume that $\{a, u_i\}$ is the module in $G(V - p_i)$ for all i. From the earlier observation all u_i are distinct and the following are true:

(i) $u_1 \in p_2$ or $u_2 \in p_1$, (ii) $u_2 \in p_3$ or $u_3 \in p_2$, and (iii) $u_3 \in p_1$ or $u_1 \in p_3$.

These condition require that $u_1 \in p_2, u_2 \in p_3, u_3 \in p_1$ or $u_1 \in p_3, u_2 \in p_1, u_3 \in p_2$. Without loss of generality assume the first with $u_1 = a_2, u_2 = a_3, u_3 = a_1$, as there is nothing here to distinguish between a_i from b_i.

Here, $\{a, u_1\} = \{a, a_2\}$ is a module in $G(V - p_1)$ and $\{a, a_3\}$ is a module in $G(V - p_2)$. Combining the two we have $\{a, a_2, a_3\}$ is a module in $G(V - p_1 - \{b_2\})$. Similarly $\{a, a_3, a_1\}$ is a module in $G(V - p_2 - \{b_3\})$ and $\{a, a_1, a_2\}$ is a module in $G(V - p_3 - \{b_1\})$. Together they imply that $\{a, a_1, a_2, a_3\}$ is a module in $G(V - \{b_1, b_2, b_3\})$. We can derive another fact from these three modules. $\{a, a_2, a_3\}$ is a module in $G(V - p_1 - \{b_2\})$ so $\{a_2, a_3\}$ is a module in $G(V - \{a\} - p_1 - \{b_2\})$. While $\{a_2, a_3\}$ cannot be a module of $G(V - \{a\} - p_1)$ because the latter is X-critical, it is necessary that $(a_2, b_2) \in E$ iff $(a_3, b_2) \notin E$. Since $\{a, a_1, a_2\}$ is a module in $G(V - p_3 - \{b_1\})$, $(a, b_2) \in E$ iff $(a_1, b_2) \in E$ iff $(a_2, b_2) \in E$. These relations and similar other relations are stated below:

(i) $(a, b_2) \in E$ iff $(a_1, b_2) \in E$ iff $(a_2, b_2) \in E$ iff $(a_3, b_2) \notin E$

(ii) $(a, b_1) \in E$ iff $(a_3, b_1) \in E$ iff $(a_1, b_1) \in E$ iff $(a_2, b_1) \notin E$ (1)

(iii) $(a, b_3) \in E$ iff $(a_2, b_3) \in E$ iff $(a_3, b_3) \in E$ iff $(a_1, b_3) \notin E$

As $\{a_1, b_1\}$ is a locked pair in $G(V - \{a\})$, either $a_1 \in (V - \{a\} - p_1)$ or $\{a_1, v_1\}$ is a module in $G(V - \{a, b_1\})$ for some $v_1 \in V - \{a, a_1, b_1\}$. Assume that $a_1 \in [V - \{a\} - p_1]$. We know that $\{a, a_1, a_2\}$ is a module in $G(V - p_3 - \{b_1\})$ so a_2 must be in $[V - p_1 - p_3 - \{a, a_2\}]$. This implies that $G(V - \{a\} - p_1 - p_3)$ is decomposable which is not true as it is X-critical. So $\{a_1, v_1\}$ must be the module in $G(V - \{a, b_1\})$. Similarly there exist v_2, v_3 such that $\{a_2, v_2\}$ is the module in $G(V - \{a, b_2\})$ and $\{a_3, v_3\}$ is the module in $G(V - \{a, b_3\})$.

Next we will show that v_i is b_j for some $j \neq i$. Firstly, $\{a_1, v_1\}$ is a module of $G(V - \{a, b_1\})$ so $(a_1, b_2) \in E$ iff $(v_1, b_2) \in E$. From relations (1) we find that $v_1 \neq a_3$. Similarly $(a_1, b_3) \in E$ iff $(v_1, b_3) \in E$ implies that $v_1 \neq a_2$. Similar arguments establishes that $\{v_1, v_2, v_3\} \cap \{a_1, a_2, a_3\} = \emptyset$. Secondly, suppose $v_1 \in V - p_1 - p_2 - p_3 - \{a\}$. Using the fact that $\{a, a_1, a_2\}$ is a module of $G(V - p_3 - \{b_1\})$ we can deduce that $\{v_1, a_2\}$ is a module of $G(V - \{a\} - p_1 - p_3)$ which is not possible for an X-critical graph. As $\{a_1, v_1\}$ is a module in $V - \{a, b_1\}$, $v_1 \neq b_1$. Thus we find that $v_1 \in \{b_2, b_3\}$. Similarly $v_2 \in \{b_3, b_1\}$ and $v_3 \in \{b_1, b_2\}$. We further show that all v_i are distinct. Let $v_1 = v_3 = b_2$. Now $\{a_1, v_1\}$ is a module in $G(V - \{a, b_1\})$ so $(a_3, a_1) \in E$ iff $(a_3, v_1) \in E$ iff $(a_3, b_2) \in E$. Also, $\{a_3, b_3\}$ is a module of $G(V - \{a, b_3\})$ so $(a_1, a_3) \in E$ iff $(a_1, v_3) \in E$ iff $(a_1, b_2) \in E$. This means $(a_1, b_2) \in E$ iff $(a_3, b_2) \in E$, which contradicts first of relations 1. Thus $\{v_1, v_2, v_3\} = \{b_1, b_2, b_3\}$.

Finally we put together the facts that $\{a_i, v_i\}$ is a module of $G(V - \{a, b_i\})$, $\{a, a_1, a_2, a_3\}$ is a module of $G(V - \{b_1, b_2, b_3\})$, and $\{v_1, v_2, v_3\} = \{b_1, b_2, b_3\}$. Consequently $\{a, a_1, a_2, a_3, b_1, b_2, b_3\}$ is a module of G, which is absurd as we had started with the assumption that G is indecomposable. So case 2 is also impossible. This completes the proof.   ∎

Now we return to our discussion of computing a pair of vertices which can be deleted while preserving indecomposability. The indecomposability of a graph can be computed in $O(n + m)$ time. Therefore in $O(n(n + m))$ either a pair can be determined or we can conclude that the graph is X-critical or that it turns X-critical after deleting one vertex. We have seen in the previous section that an entire commutative elimination sequence of an X-critical graph can be computed in $O(n^2)$ time. This not only solves the problem for the second case but in light of the Theorem 5 we can find an eliminatable pair for the third case too by considering any three locked pairs and testing the graph for indecomposability by eliminating one pair at a time.

Corollary 1. *Let $G = (V, E)$ be an indecomposable graph containing an indecomposable subgraph $G(X)$ and $|V - X| \geq 6$. Then a pair of vertices $a, b \in V - X$ can be computed in $O(n(n + m))$ time such that $G(V - \{a, b\})$ is also indecomposable.*

Proof. For each vertex $a \in V - X$ check if $G(V - \{a\})$ is indecomposable until one such vertex is located. If no such vertex exists, then G is X critical and from Lemma 1 we can compute a complete elimination sequence in $O(n^2)$ time. So total cost of the computation is $O(n(n+m)+n^2)$ since indecomposability can be tested in $(n + m)$. If a vertex a is located, then locate a vertex b in $V - X - \{a\}$ such that $G(V - \{a, b\})$ is indecomposable. If one such vertex is located then a, b is the desired pair and the cost of the computation is $O(n(n + m))$. Otherwise $G(V - \{a\})$ is X critical. Since $|V - \{a\}| \geq 5$, there are at least three locked pairs in the elimination sequence of $G(V - \{a\})$. Let $(a_1, b_1), (a_2, b_2), (a_3, b_3)$ are three of the pairs in the sequence. From the theorem we know that at least one of these pairs can be removed from G while preserving indecomposability, since $G(V')$ is an X-critical graph for $V' = V - \{a, a_1, b_1, a_2, b_2, a_3, b_3\}$. Therefore we compute the commutative elimination sequence of $G(V - \{a\})$ in $O(n^2)$ time and check the indecomposability of $G(V - \{a_i, b_i\})$, for $i = 1, 2, 3$. Then the desired pair is a_i, b_i if $G(V - \{a_i, b_i\})$ is indecomposable. The testing of indecomposability of the three subgraphs costs $O(n + m)$, so total cost in this case is $(n(n+m)+n^2)$.

An obvious consequence of this result is that an indecomposability preserving elimination sequence can be computed in $O(n^2(n + m))$.

Corollary 2. *Let $G = (V, E)$ be an indecomposable graph containing an indecomposable subgraph $G(X)$ with $|V - X| \geq 6$. Then a sequence of vertex pairs $D_1, D_2, \ldots, D_k$ can be computed in $O(n^2(n + m))$ such that for all i, j $D_i \subset V - X$; $D_i \cap D_j = \emptyset$ for $i \neq j$; $G(V - D_1 - D_2 \ldots - D_i)$ are indecomposable; and $|V - X - D_1 - \ldots - D_k| \leq 5$.*

We state another consequence of the results of this section without proof.

Theorem 6. *There exists an algorithm to compute a maximal X-critical subgraph of a given graph in $O(n^3)$ time.*

3   Conclusion

We have presented a constructive proof of Ille's theorem which states that every indecomposable graph, subject to some size conditions, has a pair of vertices which can be removed from the graph while preserving the indecomposability. Our proof gives an algorithm to compute this pair in $O(n(n + m))$ which is $O(n)$

better than a brute-force method of deleting each pair and checking for indecomposability. This leads to an $O(n^2(n + m))$ algorithm to compute an elimination sequence of pairs, which preserves indecomposability. In the proof we have used an earlier result that every X-critical graph has a commutative elimination sequence. We show that this sequence is unique and can be computed in $O(n^2)$ time.

References

[1] A. Cournier. Search in indecomposable graphs. In L. Kucera, editor, *WG*, volume 2573 of *Lecture Notes in Computer Science*, pages 80–91. Springer, 2002.

[2] A. Cournier and M. Habib. An efficient algorithm to recognize prime undirected graphs. In E. W. Mayr, editor, *WG*, volume 657 of *Lecture Notes in Computer Science*, pages 212–224. Springer, 1992.

[3] A. Cournier and M. Habib. A new linear algorithm for modular decomposition. In S. Tison, editor, *CAAP*, volume 787 of *Lecture Notes in Computer Science*, pages 68–84. Springer, 1994.

[4] A. Cournier and P. Ille. Minimal indecomposable graphs. *Discrete Mathematics*, 183:61–80, 1998.

[5] C. K. Dubey and S. K. Mehta. Some algorithms on conditionally critical indecomposable graphs. *ENDM*, 22:315–319, 2005.

[6] C. K. Dubey, S. K. Mehta, and J. S. Deogun. Conditionally critical indecomposable graphs. In L. Wang, editor, *COCOON*, volume 3595 of *Lecture Notes in Computer Science*, pages 690–700. Springer, 2005.

[7] R. Fraïssé. L'intervalle en théorie des relations, ses généralizations, filtre intervallire et clôture d'une relation. *Orders, Description and Roles*, 6:313–342, North-Holland, Amsterdam, 1984.

[8] M. Habib. Substitution des structures combinatoires, théorie et algorithmes. *Ph.D. Thesis, Université Pierre et Marie Curie, Paris VI*, 1981.

[9] P. Ille. Indecomposable graphs. *Discrete Mathematics*, 173:71–78, 1997.

[10] R. M. McConnell and J. P. Spinrad. Linear-time modular decomposition and efficient transitive orientation of comparability graphs. *SODA*, pages 535–545, 1994.

[11] R. H. Mohring and H. A. Buer. A fast algorithm for the decomposition of graphs and posets. *Math. Oper. Res.*, 8:170–184, 1983.

[12] R. H. Mohring and F. J. Radermacher. Substitution decomposition for discrete structures and connections with combinatorial optimization. *Ann. Discrete Mathematics*, 19:257–356, 1984.

[13] J. H. Schmerl. Arborescent structures, ii: interpretability in the theory of trees. *Transactions of the American Mathematical Scoiety*, 266:629–643, 1981.

[14] J. H. Schmerl and W. T. Trotter. Critically indecomposable partially ordered sets, graphs, tournaments and other binary relational structures. *Discrete Mathematics*, 113:191–205, 1993.

[15] J. Spinrad. p4-trees and substitution decomposition. *Discrete Applied Math.*, 39:263–291, 1992.

[16] D. P. Sumner. Graphs indecomposable with respect to the x-join. *Discrete Mathematics*, 6:281–298, 1973.

Improved Algorithms for the Minmax Regret 1-Median Problem

Hung-I Yu, Tzu-Chin Lin, and Biing-Feng Wang

Department of Computer Science, National Tsing Hua University Hsinchu, Taiwan
30043, Republic of China
herbert@cs.nthu.edu.tw, rems@cs.nthu.edu.tw, bfwang@cs.nthu.edu.tw

Abstract. This paper studies the problem of finding the 1-median on a graph where vertex weights are uncertain and the uncertainty is characterized by given intervals. It is required to find a minmax regret solution, which minimizes the worst-case loss in the objective function. Averbakh and Berman had an $O(mn^2\log n)$-time algorithm for the problem on a general graph, and had an $O(n\log^2 n)$-time algorithm on a tree. In this paper, we improve these two bounds to $O(mn^2 + n^3\log n)$ and $O(n\log n)$, respectively.

Keywords: Location theory, minmax regret optimization, medians.

1   Introduction

Over three decades, location problems on networks have received much attention from researchers in the fields of transportation and communication [9,10,11,12,17]. Traditionally, network location theory has been concerned with networks in which the vertex weights and edge lengths are known precisely. However, in practice, it is often impossible to make an accurate estimate of all these parameters [13,14]. Real-life data often involve a significant portion of uncertainty, and these parameters may change with time. Thus, location models involving uncertainty have attracted increasing research efforts in recent years [1,2,3,4,5,6,7,8,13,14,16,18,19,20,21].

Several ways for modeling network uncertainty have been defined and studied [13,16,18]. One of the most important models is the minmax regret approach, introduced by Kouvelis [13]. In the model, uncertainty of network parameters is characterized by given intervals, and it is required to minimize the worst-case loss in the objective function that may occur because of the uncertain parameters. During the last ten years, many important location problems have been studied on the minmax regret model. The 1-center problem was studied in [2,3,6], the p-center problem was studied in [2], and the 1-median problem was studied in [4,5,7,13].

The minmax regret 1-median problem is the focus of this paper. For a general graph with uncertain edge lengths, the problem is strongly NP-hard [1]. For a general graph with uncertain vertex weights, Averbakh and Berman [4] gave an $O(mn^2\log n)$-time algorithm, where n is the number of vertices and m is the number of edges. As to trees, it was proved in [7] that uncertainty in edge lengths

D.Z. Chen and D.T. Lee (Eds.): COCOON 2006, LNCS 4112, pp. 52–62, 2006.

can be ignored by setting the length of each edge to its upper bound. For a tree with uncertain vertex weights, Kouvelis et al. [13] proposed an $O(n^4)$-time algorithm. Chen and Lin [7] improved the bound to $O(n^3)$. Averbakh and Berman presented an $O(n^2)$-time algorithm in [4] and then improved it to $O(n\log^2 n)$ in [5]. In this paper, improved algorithms are presented for the minmax regret 1-median problem on a general graph and a tree with uncertain vertex weights. For general graphs, we improve the bound from $O(mn^2\log n)$ to $O(mn^2 + n^3\log n)$. For trees, we improve the bound from $O(n\log^2 n)$ to $O(n\log n)$.

The remainder of this paper is organized as follows. In Section 2, notation and definitions are introduced. In Sections 3 and 4, improved algorithms for the minmax regret 1-median problem on a general graph and a tree are proposed, respectively. Finally, in Section 5, we conclude this paper.

2   Notation and Definitions

Let $G = (V, E)$ be an undirected connected graph, where V is the vertex set and E is the edge set. Let $n = |V|$ and $m = |E|$. In this paper, G also denotes the set of all points of the graph. Thus, the notation $x \in G$ means that x is a point along any edge of G which may or may not be a vertex of G. Each edge $e \in E$ has a nonnegative length. For any two points $a, b \in G$, let $d(a, b)$ be the distance of the shortest path between a and b. Suppose that the matrix of shortest distances between vertices of G is given. Each vertex $v \in V$ is associated with two positive values w_v^- and w_v^+, where $w_v^- \leq w_v^+$. The weight of each vertex $v \in V$ can take any value randomly from the interval $[w_v^-, w_v^+]$. Let Σ be the Cartesian product of intervals $[w_v^-, w_v^+], v \in V$. Any element $S \in \Sigma$ is called a scenario and represents a feasible assignment of weights to the vertices of G. For any scenario $S \in \Sigma$ and any vertex $v \in V$, let w_v^S be the weight of v under the scenario S.

For any scenario $S \in \Sigma$, and a point $x \in G$, we define

$$F(S, x) = \sum_{v \in V}(w_v^S \times d(v, x)),$$

which is the total weighted distance from all the vertices to x according to S. Given a specific scenario $S \in \Sigma$, the classical 1-median problem is to find a point $x^* \in G$ that minimizes $F(S, x^*)$. The point x^* is called a 1-median of G under the scenario S. For any point $x \in G$, the regret of x with respect to a scenario $S \in \Sigma$ is $max_{y \in G}F(S, x) - F(S, y)$ and the maximum regret of x is

$$Z(x) = max_{S \in \Sigma}max_{y \in G}F(S, x) - F(S, y).$$

The minmax regret 1-median problem is to find a point $x \in G$ minimizing $Z(x)$.

3   Minmax Regret 1-Median on a General Graph

Averbakh and Berman [4] had an $O(mn^2\log n)$-time algorithm to find a minmax regret 1-median of a graph $G = (V, E)$. In this section, we give a new implementation of their algorithm, which requires $O(mn^2 + n^3\log n)$ time.

3.1   Averbakh and Berman's Algorithm

It is well-known that there is always a vertex that is a solution to the classical 1-median problem [10]. Thus, for any scenario $S \in \Sigma$, $min_{y \in G}F(S, y) = min_{y \in V}F(S, y)$. Therefore, the regret of x with respect to a scenario $S \in \Sigma$ can also be expressed as $max_{y \in V}F(S, x) - F(S, y)$. Consequently, we have

$$Z(x) = max_{S \in \Sigma}max_{y \in V}F(S, x) - F(S, y)$$
$$= max_{y \in V}max_{S \in \Sigma}F(S, x) - F(S, y).$$

For any point $x \in G$ and vertex $y \in V$, define $R(x, y) = max_{S \in \Sigma}F(S, x) - F(S, y)$. Then, we have

$$Z(x) = max_{y \in V}R(x, y).$$

The problem is to find a point $x \in G$ that minimizes $Z(x)$. For ease of presentation, only the computation of $min_{x \in G}Z(x)$ is described. For each edge $e \in E$, let $Z_e^* = min_{x \in e}Z(x)$. Averbakh and Berman solved the minmax regret 1-median problem by firstly determining Z_e^* for every $e \in E$ and then computing $min_{x \in G}Z(x)$ as the minimum among all Z_e^*. Their computation of each Z_e^* takes $O(n^2\log n)$ time and thus their solution to the minmax regret 1-median problem requires $O(mn^2\log n)$ time. Let $e = (a, b)$ be an edge in G. In the remainder of this section, Averbakh and Berman's algorithm for computing Z_e^* is described. For ease of description, e is regarded as an interval $[a, b]$ on the real line so that any point on e corresponds to a real number $x \in [a, b]$.

Consider the function $R(\cdot, y)$ for a fixed $y \in V$. For a point $x \in e$, any scenario $S \in \Sigma$ that maximizes $F(S, x) - F(S, y)$ is called a worst-case scenario of x according to y. For each point $x \in e$ and each vertex $v \in V$, let

$$w_v^{x,y} = \begin{cases} w_v^+ & \text{if } d(v, x) > d(v, y), \text{ and} \\ w_v^- & \text{if } d(v, x) \le d(v, y). \end{cases}$$

For each point $x \in e$, let $S_{(x,y)}$ be the scenario in which the weight of each $v \in V$ is $w_v^{x,y}$. It is easy to see that $S_{(x,y)}$ is a worst-case scenario of x according to y. Thus, we have the following lemma.

Lemma 3.1 [4]. $R(x, y) = F(S_{(x,y)}, x) - F(S_{(x,y)}, y)$.

For each $v \in V$, let x_v be the point on e that is farthest from v. For convenience, each x_v is called a *pseudo-node* of e. There are at most two points $x \in [a, b]$ with $d(v, x) = d(v, y)$. For ease of presentation, assume that there are two such points y_1^v and y_2^v, where $y_1^v < y_2^v$. For convenience, y_1^v and y_2^v are called *y-critical points* of e. In the interval $[a, b]$, $R(\cdot, y)$ is a piecewise linear function having $O(n)$ breakpoints, including at most n pseudo-nodes and at most $2n$ y-critical points. These breakpoints can be easily determined in $O(n)$ time. Let $B(y)$ be the non-decreasing sequence of the pseudo-nodes and y-critical points on e. Averbakh and Berman showed the following.

Lemma 3.2 [4]. Let $y \in V$ be a vertex. Given $B(y)$, the function $R(\cdot, y)$ on e can be constructed in $O(n)$ time.

Averbakh and Berman computed Z_e^* as follows. First, by sorting, $B(y)$ is obtained in $O(n^2\log n)$ time for every $y \in V$. Next, the function $R(\cdot, y)$ on e is constructed for every $y \in V$, which takes $O(n^2)$ time. Finally, $Z_e^* = min_{x \in e}Z(x) = min_{x \in e}\{max_{y \in V}R(x, y)\}$ is computed in $O(n^2)$ time, by using Megiddo's linear-time algorithm for two-variable linear programming [17].

3.2   The Improved Algorithm

Averbakh and Berman's algorithm for computing each Z_e^* requires $O(n^2\log n)$ time. The bottleneck is the computation of $B(y)$ for each $y \in V$. In this subsection, we show that with a simple $O(n^3\log n)$-time preprocessing, each $B(y)$ can be computed in $O(n)$ time.

The preprocessing computes a list for each edge $e \in E$ and computes n lists for each vertex $a \in V$. The list computed for each $e \in E$ is $P(e)$, which is the non-decreasing sequence of all pseudo-nodes of e. The n lists computed for each $a \in V$ are, respectively, denoted by $Y(a, y)$, $y \in V$. Each $Y(a, y)$, $y \in V$, stores the non-decreasing sequence of the values $d(v, y) - d(v, a)$ of all $v \in V$. By using sorting and with the help of the distance matrix, the computation of each $P(e)$ and each $Y(a, y)$ takes $O(n\log n)$ time. Thus, the preprocessing requires $O((m + n^2)n\log n) = O(n^3\log n)$ time.

Let $e = (a, b)$ be an edge. Consider the function $R(\cdot, y)$ for a fixed $y \in V$ on e. In the following, we show how to compute $B(y)$ in $O(n)$ time. The function $R(\cdot, y)$ has two kinds of breakpoints: pseudo-nodes and y-critical points. As mentioned, for each vertex $v \in V$ there are at most two y-critical points $x \in [a, b]$. To be more specific, if $d(v, a) < d(v, y) < d(v, x_v)$, there is a point x in the open interval (a, x_v) with $d(v, x) = d(v, y)$. In such a case, we say that v generates a *left y-critical point* and denote the point as y_1^v. On the other hand, if $d(v, b) < d(v, y) < d(v, x_v)$, we say that v generates a *right y-critical point* and denote the point as y_2^v. We obtain $B(y)$ as follows. First, we determine the y-critical points generated by all vertices $v \in V$. Then, we compute Y_1 as the non-decreasing sequence of the left y-critical points. Since the slope of $d(v, x)$ in the interval $[a, x_v]$ is $+1$, it is easy to conclude that each left y-critical point y_1^v is at a distance of $d(v, y) - d(v, a)$ from a. The sequence $Y(a, y)$ stores the non-decreasing sequence of the values $d(v, y) - d(v, a)$ of all $v \in V$. Thus, the order of the left y-critical points in Y_1 can be obtained from $Y(a, y)$ in $O(n)$ time by a simple scan. Next, we compute Y_2 as the non-decreasing sequence of the right y-critical points. Since the slope of $d(v, x)$ in the interval $[x_v, b]$ is -1, each right y-critical point y_2^v is at a distance of $d(v, y) - d(v, b)$ from b. The sequence $Y(b, y)$ stores the non-decreasing sequence of the values $d(v, y) - d(v, b)$ of all $v \in V$. Thus, the order of the right y-critical points in Y_2 can be obtained by scanning the reverse sequence of $Y(b, y)$ in $O(n)$ time. Finally, $B(y)$ is computed in $O(n)$ time by merging Y_1, Y_2, and $P(e)$.

With the above computation of $B(y)$, each Z_e^* can be computed in $O(n^2)$ time. Thus, we obtain the following.

Theorem 3.3. *The minmax regret 1-median problem on a general graph can be solved in $O(mn^2 + n^3 \log n)$ time.*

We remark that Averbakh and Berman's algorithm uses $O(n^2)$ space, whereas ours uses $O(n^3)$ space.

4   Minmax Regret 1-Median on a Tree

In Subsection 4.1, Averbakh and Berman's $O(n\log^2 n)$-time algorithm for the minmax regret 1-median problem on a tree $T = (V, E)$ is described. In Subsection 4.2, an $O(n\log n)$-time improved algorithm is presented.

4.1   Averbakh and Berman's Algorithm

An edge that contains a minmax regret 1-median is called an optimal edge. Averbakh and Berman's algorithm consists of two stages. The first stage finds an optimal edge e^*, which requires $O(n\log^2 n)$ time. The second stage finds the exact position of a minmax regret 1-median on e^* in $O(n)$ time. Our improvement is obtained by giving a more efficient implementation of the first stage. Thus, only the first stage is described in this subsection.

Let $v \in V$ be a vertex. By removing v and its incident edges, T is broken into several subtrees, each of which is called an *open v-branch*. For each open v-branch X, the union of v, X, and the edge connecting v and X is called a *v-branch*. For any vertex $p \in v$ in T, let $B(v, p)$ be the v-branch containing p. Let S^- be the scenario in which the weight of every $v \in V$ is w_v^-. Let S^+ be the scenario in which the weight of every $v \in V$ is w_v^+. For any subtree X of T, let $V(X)$ be the set of vertices in X. For any vertex $a \in V$ and any open a-branch X, the following auxiliary values are defined:

$$W^+(X) = \Sigma_{v \in V(X)} w_v^+ \qquad\qquad W^-(X) = \Sigma_{v \in V(X)} w_v^-$$
$$D^+(X, a) = \Sigma_{v \in V(X)} w_v^+ \times d(v, a) \quad D^-(X, a) = \Sigma_{v \in V(X)} w_v^- \times d(v, a)$$
$$F^+(a) = \Sigma_{v \in V} w_v^+ \times d(v, a) \qquad\quad F^-(a) = \Sigma_{v \in V} w_v^- \times d(v, a)$$

By using dynamic programming, these auxiliary values of all vertices a and all open a-branches X are pre-computed, which takes $O(n)$ time [5].

Let $R(x, y)$ and $S_{(x,y)}$ be defined the same as in Subsection 3.1. Consider the computation of $R(x, y)$ for a pair of vertices $x, y \in V$. An edge is a bisector of a simple path if it contains the middle point of the path. In case the middle point is located at a vertex, both the edges connecting the vertex are bisectors. Let $(i, j) \in E$ be a bisector of the path from x to y, where i is closer to x than j. Let X be the open j-branch containing x and Y be the open i-branch containing y. By Lemma 3.1, $R(x, y) = F(S_{(x,y)}, x) - F(S_{(x,y)}, y)$. Since T is a tree, it is easy to see that under the scenario $S_{(x,y)}$ the weights of all vertices in Y are equal to their upper bounds w_v^+ and the weights of all vertices in X are equal to their lower bounds w_v^-. Thus, $F(S_{(x,y)}, x)$ and $F(S_{(x,y)}, y)$ can be computed in $O(1)$ time according to the following equations [5]:

$$F(S_{(x,y)}, x) = F^-(x) - D^-(Y, i) - d(i, x) \times W^-(Y) + D^+(Y, i) + d(i, x) \times W^+(Y)$$
$$F(S_{(x,y)}, y) = F^+(y) - D^+(X, j) - d(j, y) \times W^+(X) + D^-(X, j) + d(j, y) \times W^-(X)$$

Based upon the above discussion, the following lemma is obtained.

Lemma 4.1 [5]. *Let $x \in V$ be a vertex. Given the bisectors of the paths from x to all the other vertices, $R(x, y)$ can be computed in $O(n)$ time for all $y \in V$.*

For each $x \in V$, let $\hat{y}(x)$ be a vertex in T such that $R(x, \hat{y}(x)) = max_{y \in V} R(x, y)$. We have the following.

Lemma 4.2 [5]. *Let $x \in V$ be a vertex. If $x = \hat{y}(x)$, x is a minmax regret 1-median of T; otherwise, $B(x, \hat{y}(x))$ contains a minmax regret 1-median.*

A *centroid* of T is a vertex $c \in V$ such that every open c-branch has at most $n/2$ vertices. It is easy to find a centroid of T in $O(n)$ time [12]. Averbakh and Berman's algorithm for finding an optimal edge is as follows.

Algorithm 1. OPTIMAL_EDGE(T)
begin
   1   $\hat{T} \leftarrow T$ // $\hat{T}$ is the range for searching an optimal edge
   2 **while** ($\hat{T}$ is not a single edge) **do**
   3 **begin**
   4     $x \leftarrow$ a centroid of $\hat{T}$
   5   **for** each $y \in V$ **do** compute a bisector of the path from x to y
   6   **for** each $y \in V$ **do** $R(x, y) \leftarrow F(S_{(x,y)}, x) - F(S_{(x,y)}, y)$
   7     $\hat{y}(x) \leftarrow$ the vertex that maximizes $R(x, y)$ over all $y \in V$
   8   **if** $x = \hat{y}(x)$ **then return** (x) // x is a minmax regret 1-median
   9   **else** $(\hat{T}) \leftarrow B(x, \hat{y}(x)) \cap \hat{T}$ // reduce the search range to a subtree
  10 **end**
  11 **return** $\hat{T}$ // $\hat{T}$ is an optimal edge
end

The while-loop in Lines 2-10 performs $O(\log n)$ iterations, in which Line 5 requires $O(n \log n)$ time [5] and is the bottleneck. Therefore, Algorithm 1 finds an optimal edge in $O(n \log^2 n)$ time.

4.2 An Improved Algorithm

For each $e = (i, j) \in E$, define S_e as the scenario in which the weight of every vertex v in the open i-branch containing j is w_v^+ and the weight of every vertex v in the open j-branch containing i is w_v^-. A key procedure of the new algorithm is to compute for every $e \in E$ a vertex that is a classical 1-median under the scenario S_e. The computation is described firstly in Subsection 4.2.1. Then, the new algorithm is proposed in Subsection 4.2.2.

4.2.1 Computing Medians

For ease of discussion, throughout this subsection, we assume that T is rooted at an arbitrary vertex $r \in V$. Let $v \in V$ be a vertex. Denote $P(v)$ as the path from r to v, $p(v)$ as the parent of v, and T_v as the subtree of T rooted at v. For convenience, the subtrees rooted at the children of v are called *subtrees of v*. For any subtree X of T, define the *weight* of X under a scenario $S \in \Sigma$ as $W(S, X) = \sum_{v \in V(X)} w_v^S$. For any scenario $S \in \Sigma$ and $v \in V$, define

$$\delta(S, v) = 2W(S, T_v) - W(S, T).$$

Note that $2W(S, T_v) - W(S, T) = W(S, T_v) - W(S, T \backslash T_v)$ and thus $\delta(S, v)$ is equal to the weight difference between T_v and its complement under the scenario S. By the definition of $\delta(S, v)$, it is easy to see the following.

Lemma 4.3. *For any scenario $S \in \Sigma$, the function value of $\delta(S, \cdot)$ is decreasing along any downward path in T.*

Under a scenario $S \in \Sigma$, a vertex v is a classical 1-median of T if and only if $W(S, X) \le 1/2 \times W(S, T)$ for all open v-branches X [12]. By using this property, the following lemma can be obtained.

Lemma 4.4 [15]. *For any scenario $S \in \Sigma$, the vertex $v \in V$ with the smallest $\delta(S, v) > 0$ is a classical 1-median.*

Under a scenario $S \in \Sigma$, T may have more than one median, but it has a unique vertex v with the smallest $\delta(S, v) > 0$, which is called *the median*, denoted as $m(S)$. For any $v \in V$, define the *heavy path* of v under a scenario $S \in \Sigma$, denoted by $h(S, v)$, as the path starting at v and at each time moving down to the heaviest subtree, with a tie broken arbitrarily, until a leaf is reached. With some efforts, the following can be proved.

Lemma 4.5. *Let $S \in \Sigma$ be a scenario and $v \in V$ be a vertex with $\delta(S, v) > 0$. Then, $h(S, v)$ contains the median $m(S)$.*

Lemma 4.3, 4.4 and 4.5 are useful for finding a classical 1-median for a fixed scenario $S \in \Sigma$. In our problem, all scenarios S_e, $e \in E$, need to be considered. Let $e = (i, j)$ be an edge in T such that i is the parent of j. For ease of presentation, denote $S_{\bar{e}}$ as the scenario $S_{(j,i)}$. For each $v \in V$, the weight of T_v and the heavy path of v only depend on the weight assignment to the vertices in T_v. Therefore, the following two lemmas can be obtained.

Lemma 4.6. *Let $e = (i, j)$ be an edge in T such that i is the parent of j. For any $v \in V$,*

$$W(S_e, T_v) = \begin{cases} W(S^+, T_v) & \text{if } v \in T_j, \\ W(S^-, T_v) & \text{if } v \notin T_j \text{ and } v \notin P(i), \\ W(S^-, T_v) - W(S^-, T_j) + W(S^+, T_j) & \text{if } v \in P(i); \text{ and} \end{cases}$$

$$W(S_{\bar{e}}, T_v) = \begin{cases} W(S^-, T_v) & \text{if } v \in T_j, \\ W(S^+, T_v) & \text{if } v \notin T_j \text{ and } v \notin P(i), \\ W(S^+, T_v) - W(S^+, T_j) + W(S^-, T_j) & \text{if } v \in P(i); \text{ and} \end{cases}$$

Lemma 4.7. *Let $e = (i, j)$ be an edge in T such that i is the parent of j. For any $v \notin P(i)$,*

$$h(S_e, v) = \begin{cases} h(S^+, v) & \text{if } v \in T_j, \\ h(S^-, v) & \text{if } v \notin T_j; \end{cases} \quad \text{and } h(S_{\bar{e}}, v) = \begin{cases} h(S^-, v) & \text{if } v \in T_j, \\ h(S^+, v) & \text{if } v \notin T_j. \end{cases}$$

According to Lemmas 4.6 and 4.7, maintaining information of T under the scenarios S^+ and S^- is very useful to the computation of $m(S_e)$ for every $e \in E$. With a bottom-up computation, we pre-compute the values $W(S^+, T_v)$ and $W(S^-, T_v)$ for all $v \in V$ in $O(n)$ time. With the pre-computed values, for any $e \in E$ and $v \in V$, $\delta(S_e, v) = 2W(S_e, T_v) - W(S_e, T_r)$ can be determined in $O(1)$ time by Lemmas 4.6. Besides, we preprocess T to construct the heavy paths $h(S^+, v)$ and $h(S^-, v)$ for every $v \in V$. The total length of the heavy paths of all $v \in V$ is $O(n^2)$. However, they can be constructed in $O(n)$ time and stored in $O(n)$ space [22].

Let $H^+ = \{h(S^+, v)|v \in V\}$ and $H^- = \{h(S^-, v)|v \in V\}$. Let $e \in E$ be an edge. Since the value of $\delta(S_e, \cdot)$ decreases along any downward path in T, we can compute the median $m(S_e)$ by firstly finding a path in $H^+ \cup H^-$ that contains the median and then locating the median by performing binary search on the path. The following two lemmas help in the finding. Due to the page limit, the proofs are omitted.

Lemma 4.8. *Let $e = (i, j)$ be an edge in T such that i is the parent of j. Let z be the lowest vertex on $P(j)$ with $\delta(S_e, z) > 0$. If $z = j$, $h(S^+, z)$ contains $m(S_e)$; otherwise, $h(S^-, z)$ contains $m(S_e)$.*

Lemma 4.9. *Let $e = (i, j)$ be an edge in T such that i is the parent of j. Let z be the lowest vertex on $P(j)$ with $\delta(S_{\bar{e}}, z) > 0$. If $z = j$, $h(S^-, z)$ contains $m(S_{\bar{e}})$; otherwise, either $h(S^+, z)$ or $h(S^+, c')$ contains $m(S_{\bar{e}})$, where c' is the child of z such that $T_{c'}$ is the second heaviest under the scenario S^+.*

Now, we are ready to present our algorithm for computing $m(S_e)$ for all $e \in E$. It is as follows.

Algorithm 2. COMPUTE_MEDIANS(T)
begin
  1   Orient T into a rooted tree with an arbitrary root $r \in V$
  2   Compute $W(S^+, T_v)$, $W(S^-, T_v)$, $h(S^+, v)$, and $h(S^-, v)$ for all $v \in V$
  3   **for** each $v \in V$ **do**
  4       $c'(v) \leftarrow$ the child of v such that $T_{c'(v)}$ is the second heaviest under S^+
  5   **for** each $j \in V - \{r\}$ **do** (in depth-first search order)
  6   **begin**
  7       $e \leftarrow (p(j), j)$
  8       $P(j) \leftarrow$ the path from r to j
  9       $m(S_e) \leftarrow$ the classical 1-median under S_e
  10      $m(S_{\bar{e}}) \leftarrow$ the classical 1-median under $S_{\bar{e}}$
  11 **end**
  12 **return** $(\{m(S_e)|e \in E\})$
end

The time complexity of Algorithm 2 is analyzed as follows. Lines 1-4 requires $O(n)$ time. Consider the for-loop in Lines 5-11 for a fixed $j \in V - r$. Line 7 takes

$O(1)$ time. Since the vertices are processed in depth-first search order, Line 8 also takes $O(1)$ time. Based upon Lemma 4.8, Line 9 is implemented in $O(\log n)$ time as follows. First, we compute z as the lowest vertex on $P(j)$ with $\delta(S_e, z) > 0$. Then, if $z = j$, we find $m(S_e)$ on $h(S^+, z)$; otherwise, we find $m(S_e)$ on $h(S^-, z)$. The computation of z and the finding of $m(S_e)$ are done in $O(\log n)$ time by using binary search. Similarly, based upon Lemma 4.9, Line 10 is implemented in $O(\log n)$ time. Therefore, each iteration of the for-loop takes $O(\log n)$ time. There are $n - 1$ iterations. Thus, the for-loop requires $O(n\log n)$ time in total. We obtain the following.

Theorem 4.10. *The computation of $m(S_e)$ of all $e \in E$ can be done in $O(n\log n)$ time.*

4.2.2   The New Approach

We pre-compute the medians $m(S_e)$ for all $e \in E$. Also, as in Subsection 4.1, the auxiliary values $W^+(X)$, $W^-(X)$, $D^+(X, a)$, $D^-(X, a)$, $F^+(a)$, and $F^-(a)$ of all vertices $a \in V$ and all open a-branches X are pre-computed.

Averbakh and Berman's algorithm finds an optimal edge by repeatedly reducing the search range $\hat{T}$ into a smaller subtree until only one edge is left. Let x be the centroid of the current search range $\hat{T}$. The reduction is done by determining a vertex $\hat{y}(x)$ with $R(x, \hat{y}(x)) = max_{y \in V} R(x, y)$ and then reducing $\hat{T}$ into $B(x, \hat{y}(x)) \cap \hat{T}$. A solution to the finding of bisectors is required for their determination of $\hat{y}(x)$. The new algorithm is obtained by using a different approach for the determination of $\hat{y}(x)$, which is based upon the following lemma. We omit the proof due to the page limit.

Lemma 4.11. *For any $x \in V$, there is an edge $e \in E$ such that $R(x, m(S_e)) = F(S_e, x) - F(S_e, m(S_e)) = max_{y \in V} R(x, y)$.*

According to Lemma 4.11, we modify Algorithm 1 by replacing Lines 5-7 with the following.

> 5   **for each** $e \in E$ **do** $R'(e) \leftarrow F(S_e, x) - F(S_e, m(S_e))$
> 6 $\hat{e} \leftarrow$ the edge that maximizes $R'(e)$ over all $e \in E$
> 7 $\hat{y}(x) \leftarrow m(S_{\hat{e}})$

Let $e = (i, j)$ be an edge in T. Let X be the open j-branch containing i and Y be the open i-branch containing j. Since T is a tree, for every $v \in V(X)$, we have $F(S_e, v) = F^-(v) - D^-(Y, i) - d(i, v) \times W^-(Y) + D^+(Y, i) + d(i, v) \times W^+(Y)$; and for every $v \in V(Y)$, we have $F(S_e, v) = F^+(v) - D^+(X, j) - d(j, v) \times W^+(X) + D^-(X, j) + d(j, v) \times W^-(X)$. By using these two equations, it is easy to implement the new Line 5 in $O(n)$ time. The new Lines 6 and 7 take $O(n)$ time. Therefore, Algorithm 2 requires $O(n\log n)$ time after the replacement.

Theorem 4.12. *The minmax regret 1-median problem on a tree can be solved in $O(n\log n)$ time.*

5   Concluding Remarks

During the last decade, minmax regret optimization problems have attracted significant research efforts. In [22], an $O(mn\log n)$-time algorithm is obtained for the minmax regret 1-center problem on a general graph, and an $O(n\log^2 n)$-time algorithm is obtained for the problem on a tree. For many location problems, however, there are still gaps between the time complexities of the solutions to their classical versions and those to their minmax regret versions. It would be a great challenge to bridge these gaps.

References

1. Averbakh, I.: On the complexity of a class of robust location problems. Working Paper. Western Washington University. Bellingham, WA. (1997)
2. Averbakh, I., Berman, O.: Minimax regret p-center location on a network with demand uncertainty. *Location Science* **5** (1997) 247–254
3. Averbakh, I., Berman, O.: Algorithms for the robust 1-center problem on a tree. *European Journal of Operational Research* **123** (2000) 292–302
4. Averbakh, I., Berman, O.: Minmax regret median location on a network under uncertainty. *Informs Journal on Computing* **12** (2000) 104–110
5. Averbakh, I., Berman, O.: An improved algorithm for the minmax regret median problem on a tree. *Networks* **41** (2003) 97–103
6. Burkard, R. E., Dollani, H.: A note on the robust 1-center problem on trees. *Annals of Operations Research* **110** (2002) 69–82
7. Chen, B. T., Lin, C. S.: Minmax-regret robust 1-median location on a tree. *Networks* **31** (1998) 93–103
8. Drezner, Z.: Sensitivity analysis of the optimal location of a facility. *Naval Research Logistics Quarterly* **33** (1980) 209–224
9. Goldman, A. J.: Optimal center location in simple networks. *Transportation Science* **5** (1971) 212–221
10. Hakimi, S. L.: Optimal locations of switching centers and the absolute centers and medians of a graph. *Operations Research* **12** (1964) 450–459
11. Kariv, O., Hakimi, S. L.: An algorithmic approach to network location problems. I: The p-centers. *SIAM Journal on Applied Mathematics* **37** (1979) 513–538
12. Kariv, O., Hakimi, S. L.: An algorithmic approach to network location problems. II: The p-medians. *SIAM Journal on Applied Mathematics* **37** (1979) 539–560
13. Kouvelis, P., Vairaktarakis, G., Yu, G.: Robust 1-median location on a tree in the presence of demand and transportation cost uncertainty. Working Paper 93/94-3-4. Department of Management Science and Information Systems, Graduate School of Business, The University of Texas at Austin (1994)
14. Kouvelis, P., Yu, G.: Robust discrete optimization and its applications. Kluwer Academic Publishers, Dordrecht (1997)
15. Ku, S. C., Lu, C. J., Wang, B. F., Lin, T. C.: Efficient algorithms for two generalized 2-median problems on trees. in *Proceedings of the 12th International Symposium on Algorithms and Computation* (2001) 768–778
16. Labbe, M., Thisse, J.-F., Wendell, R.: Sensitivity analysis in minisum facility location problems. *Operations Research* **38** (1991) 961–969
17. Megiddo, N.: Linear-time algorithms for linear-programming in R^3 and related problems. *SIAM Journal on Computing* **12** (1983) 759–776

18. Mirchandani, P. B., Odoni, A. R.: Location of medians on stochastic networks. *Transportation Science* **13** (1979) 85–97
19. Mirchandani, P. B., Oudjit, A., Wong, R. T.: Multidimensional extensions and a nested dual approach for the *M*-median problem. *European Journal of Operational Research* **21** (1985) 121–137
20. Oudjit, A.: Median locations on deterministic and probabilistic multidimensional networks. PhD Dissertation. Rennselaer Polytechnic Institute, Troy (1981)
21. Weaver, J. R., Church, R. L.: Computational procedures of location problems on stochastic networks. *Transportation Science* **17** (1983) 168–180
22. Yu, H. I, Lin, T. C., Wang, B. F.: Improved Algorithms for the Minmax Regret Single-Facilty Problems. Manuscript (2005)

Partitioning a Multi-weighted Graph to Connected Subgraphs of Almost Uniform Size

Takehiro Ito, Kazuya Goto, Xiao Zhou, and Takao Nishizeki

Graduate School of Information Sciences, Tohoku University,
Aoba-yama 6-6-05, Sendai, 980-8579, Japan
{take, kazzg}@nishizeki.ecei.tohoku.ac.jp, {zhou, nishi}@ecei.tohoku.ac.jp

Abstract. Assume that each vertex of a graph G is assigned a constant number q of nonnegative integer weights, and that q pairs of nonnegative integers l_i and u_i, $1 \leq i \leq q$, are given. One wishes to partition G into connected components by deleting edges from G so that the total i-th weights of all vertices in each component is at least l_i and at most u_i for each index i, $1 \leq i \leq q$. The problem of finding such a "uniform" partition is NP-hard for series-parallel graphs, and is strongly NP-hard for general graphs even for $q = 1$. In this paper we show that the problem and many variants can be solved in pseudo-polynomial time for series-parallel graphs. Our algorithms for series-parallel graphs can be extended for partial k-trees, that is, graphs with bounded tree-width.

1 Introduction

Let $G = (V, E)$ be an undirected graph with vertex set V and edge set E. Assume that each vertex $v \in V$ is assigned a constant number q of nonnegative integer weights $\omega_1(v), \omega_2(v), \cdots, \omega_q(v)$, and that q pairs of nonnegative integers l_i and u_i, $1 \leq i \leq q$, are given. We call $\omega_i(v)$ the i-th weight of vertex v, and call l_i and u_i the i-th lower bound and upper bound on component size, respectively. We wish to partition G into connected components by deleting edges from G so that the total i-th weights of all components are almost uniform for each index i, $1 \leq i \leq q$, that is, the sum of i-th weights $\omega_i(v)$ of all vertices v in each component is at least l_i and at most u_i for some bounds l_i and u_i with small $u_i - l_i$. We call such a partition a uniform partition of G. Figure 1(a) illustrates a uniform partition of a graph, where $q = 2$, $(l_1, u_1) = (10, 15)$, $(l_2, u_2) = (10, 20)$, each vertex v is drawn as a circle, the two weights $\omega_1(v)$ and $\omega_2(v)$ of v are written inside the circle, and the deleted edges are drawn by dotted lines.

The problem of finding a uniform partition often appear in many practical situations such as image processing [4,6], paging systems of operation systems [8], and political districting [3,9]. Consider, for example, political districting. Let M be a map of a country, which is divided into several regions, as illustrated in Fig. 1(b). Let G be a dual-like graph of the map M, as illustrated in Fig. 1(a). Each vertex v of G represents a region, the first weight $\omega_1(v)$ represents the number of voters in the region v, and the second weight $\omega_2(v)$ represents the

D.Z. Chen and D.T. Lee (Eds.): COCOON 2006, LNCS 4112, pp. 63–72, 2006.

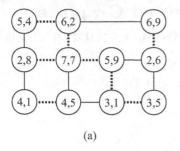
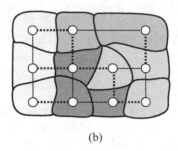

        (a)   (b)

Fig. 1. (a) A uniform partition of a graph into $p = 4$ components, and (b) electoral zoning of a map corresponding to the partition

area of the region. Each edge (u, v) of G represents the adjacency of the two regions u and v. For the political districting, one wishes to divide the country into electoral zones. Each zone must consist of connected regions, that is, the regions in each zone must induce a connected subgraph of G. Each zone must have an almost equal number of voters, and must be almost equal in area. Such electoral zoning corresponds to a uniform partition of the plane graph G for two appropriate pairs (l_1, u_1) and (l_2, u_2) of bounds.

In the paper we deal with the following three problems to find a uniform partition of a given graph G: the *minimum partition problem* is to find a uniform partition of G with the minimum number of components; the *maximum partition problem* is defined similarly; and the *p-partition problem* is to find a uniform partition of G with a given number p of components. All the problems are NP-hard for series-parallel graphs even when $q = 1$ [5]. Therefore, it is very unlikely that the three partition problems can be solved in polynomial time even for series-parallel graphs. Moreover, all the three partition problems are strongly NP-hard for general graphs even if $q = 1$ [5], and hence there is no pseudo-polynomial-time algorithm for any of the three problems on general graphs unless P = NP. Furthermore, for any $\varepsilon > 0$, there is no ε-approximation algorithm for the minimum partition problem or the maximum partition problem on series-parallel graphs unless P = NP [5], and the problems for the case $q = 1$ can be solved in pseudo-polynomial time for series-parallel graphs [5]; the minimum and maximum partition problems can be solved in time $O(u_1^4 n)$ and the p-partition problem can be solved in time $O(p^2 u_1^4 n)$ for series-parallel graphs G, where n is the number of vertices in G. However, it has not been known whether the problems can be solved in pseudo-polynomial time for the case $q \geq 2$.

In this paper, we obtain pseudo-polynomial-time algorithms to solve the three problems on series-parallel graphs for an arbitrary constant number q. More precisely, we show that the minimum and maximum partition problems can be solved in time $O(u^{4q} n)$ and hence in time $O(n)$ for any fixed constant u, and that the p-partition problem can be solved in time $O(p^2 u^{4q} n)$, where u is the maximum upper bound, that is, $u = \max\{u_i \mid 1 \leq i \leq q\}$. Our algorithms for series-parallel graphs can be extended for partial k-trees, that is, graphs with bounded tree-width [1,2].

2   Terminology and Definitions

In this section we give some definitions.

A (*two-terminal*) *series-parallel graph* is defined recursively as follows [7]:

(1) A graph G with a single edge is a series-parallel graph. The end vertices of the edge are called the *terminals* of G and denoted by $s(G)$ and $t(G)$. (See Fig. 2(a).)

(2) Let G' be a series-parallel graph with terminals $s(G')$ and $t(G')$, and let G'' be a series-parallel graph with terminals $s(G'')$ and $t(G'')$.

   (a) A graph G obtained from G' and G'' by identifying vertex $t(G')$ with vertex $s(G'')$ is a series-parallel graph, whose terminals are $s(G) = s(G')$ and $t(G) = t(G'')$. Such a connection is called a *series connection*, and G is denoted by $G = G' \bullet G''$. (See Fig. 2(b).)

   (b) A graph G obtained from G' and G'' by identifying $s(G')$ with $s(G'')$ and identifying $t(G')$ with $t(G'')$ is a series-parallel graph, whose terminals are $s(G) = s(G') = s(G'')$ and $t(G) = t(G') = t(G'')$. Such a connection is called a *parallel connection*, and G is denoted by $G = G' \parallel G''$. (See Fig. 2(c).)

The terminals $s(G)$ and $t(G)$ of G are often denoted simply by s and t, respectively. Since we deal with partition problems, we may assume without loss of generality that G is a simple graph and hence G has no multiple edges.

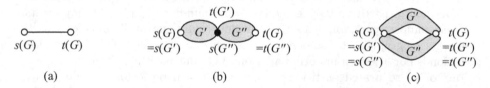

Fig. 2. (a) A series-parallel graph with a single edge, (b) series connection, and (c) parallel connection

A series-parallel graph G can be represented by a "binary decomposition tree" [7]. Figure 3(a) illustrates a series-parallel graph G, and Figure 3(b) depicts a binary decomposition tree T of G. Labels s and p attached to internal nodes in T indicate series and parallel connections, respectively. Nodes labeled s and p are called *s- and p-nodes*, respectively. Every leaf of T represents a subgraph of G induced by a single edge. Each node v of T corresponds to a subgraph G_v of G induced by all edges represented by the leaves that are descendants of v in T. Thus G_v is a series-parallel graph for each node v of T, and $G = G_r$ for the root r of T. Figure 3(c) depicts G_v for the left child v of the root r of T. Since a binary decomposition tree of a given series-parallel graph G can be found in linear time [7], we may assume that a series-parallel graph G and its binary decomposition tree T are given. We solve the three partition problems by a dynamic programming approach based on a decomposition tree T.

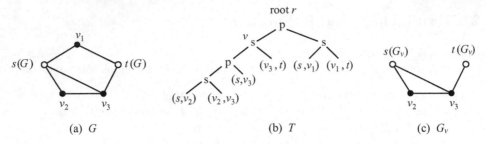

Fig. 3. (a) A series-parallel graph G, (b) its binary decomposition tree T, and (c) a subgraph G_v

3   Minimum and Maximum Partition Problems

In this section we have the following theorem.

Theorem 1. *Both the minimum partition problem and the maximum partition problem can be solved for any series-parallel graph G in time $O(u^{4q}n)$, where n is the number of vertices in G, q is a fixed constant number of weights, and u is the maximum upper bound on component size.*

In the remainder of this section we give an algorithm to solve the minimum partition problem as a proof of Theorem 1; the maximum partition problem can be similarly solved. We indeed show only how to compute the minimum number $p_{\min}(G)$ of components. It is easy to modify our algorithm so that it actually finds a uniform partition having the minimum number $p_{\min}(G)$ of components.

Every uniform partition of a series-parallel graph G naturally induces a partition of its subgraph G_v for a node v of a decomposition tree T of G. The induced partition is not always a uniform partition of G_v but is either a "connected partition" or a "separated partition" of G_v, which will be formally defined later and are illustrated in Fig. 4 where s and t represent the terminals of G_v. We denote by $\mathbb{X}$ a q-tuple $(x_1, x_2, \cdots, x_q)$ of integers with $0 \le x_i \le u_i$, $1 \le i \le q$. We introduce two functions f and h; for a series-parallel graph G_v and a q-tuple $\mathbb{X} = (x_1, x_2, \cdots, x_q)$, the value $f(G_v, \mathbb{X})$ represents the minimum number of components in some particular connected partitions of G_v; for a series-parallel graph G_v and a pair of q-tuples $\mathbb{X} = (x_1, x_2, \cdots, x_q)$ and $\mathbb{Y} = (y_1, y_2, \cdots, y_q)$, the value $h(G_v, \mathbb{X}, \mathbb{Y})$ represents the minimum number of components in some particular separated partitions of G_v. Our idea is to compute $f(G_v, \mathbb{X})$ and $h(G_v, \mathbb{X}, \mathbb{Y})$ from leaves of T to the root r of T by means of dynamic programming.

Fig. 4. (a) A connected partition, and (b) a separated partition

We now formally define the notion of connected and separated partitions of a series-parallel graph $G = (V, E)$. Let $\mathcal{P} = \{P_1, P_2, \ldots, P_m\}$ be a partition of the vertex set V of G into m nonempty subsets $P_1, P_2, \cdots, P_m$ for some integer $m \geq 1$. Thus $|\mathcal{P}| = m$. The partition $\mathcal{P}$ of V is called a *partition of* G if P_j induces a connected subgraph of G for each index j, $1 \leq j \leq m$. For a set $P \subseteq V$ and an index i, $1 \leq i \leq q$, we denote by $\omega_i(P)$ the sum of i-th weights of vertices in P, that is, $\omega_i(P) = \sum_{v \in P} \omega_i(v)$. Let $\omega_{st}(G, i) = \omega_i(s) + \omega_i(t)$. We call a partition $\mathcal{P}$ of G a *connected partition* if $\mathcal{P}$ satisfies the following two conditions (see Fig. 4(a)):

(a) there exists a set $P_{st} \in \mathcal{P}$ such that $s, t \in P_{st}$; and
(b) for each index i, $1 \leq i \leq q$, the inequality $\omega_i(P_{st}) \leq u_i$ holds, and the inequalities $l_i \leq \omega_i(P) \leq u_i$ hold for each set $P \in \mathcal{P} - \{P_{st}\}$.

Note that the inequality $l_i \leq \omega_i(P_{st})$, $1 \leq i \leq q$, does not necessarily hold for P_{st}. For a connected partition $\mathcal{P}$, we always denote by P_{st} the set in $\mathcal{P}$ containing both s and t. A partition $\mathcal{P}$ of G is called a *separated partition* if $\mathcal{P}$ satisfies the following two conditions (see Fig. 4(b)):

(a) there exist two distinct sets $P_s, P_t \in \mathcal{P}$ such that $s \in P_s$ and $t \in P_t$; and
(b) for each index i, $1 \leq i \leq q$, the two inequalities $\omega_i(P_s) \leq u_i$ and $\omega_i(P_t) \leq u_i$ hold, and the inequalities $l_i \leq \omega_i(P) \leq u_i$ hold for each set $P \in \mathcal{P} - \{P_s, P_t\}$.

Note that the inequalities $l_i \leq \omega_i(P_s)$ and $l_i \leq \omega_i(P_t)$, $1 \leq i \leq q$, do not always hold for P_s and P_t. For a separated partition $\mathcal{P}$, we always denote by P_s the set in $\mathcal{P}$ containing s and by P_t the set in $\mathcal{P}$ containing t.

We then formally define $f(G, \mathbb{X})$ for a series-parallel graph G and a q-tuple $\mathbb{X} = (x_1, x_2, \cdots, x_q)$ of integers with $0 \leq x_i \leq u_i$, $1 \leq i \leq q$, as follows:

$$f(G, \mathbb{X}) = \min\{p^* \geq 0 \mid G \text{ has a connected partition } \mathcal{P} \text{ such that}$$
$$x_i = \omega_i(P_{st}) - \omega_{st}(G, i) \text{ for each } i, \text{ and } p^* = |\mathcal{P}| - 1\}. \quad (1)$$

If G has no connected partition $\mathcal{P}$ such that $\omega_i(P_{st}) - \omega_{st}(G, i) = x_i$ for each i, then let $f(G, \mathbb{X}) = +\infty$.

We now formally define $h(G, \mathbb{X}, \mathbb{Y})$ for a series-parallel graph G and a pair of q-tuples $\mathbb{X} = (x_1, x_2, \cdots, x_q)$ and $\mathbb{Y} = (y_1, y_2, \cdots, y_q)$ of integers with $0 \leq x_i, y_i \leq u_i$, $1 \leq i \leq q$, as follows:

$$h(G, \mathbb{X}, \mathbb{Y}) = \min\{p^* \geq 0 \mid G \text{ has a separated partition } \mathcal{P} \text{ such that}$$
$$x_i = \omega_i(P_s) - \omega_i(s) \text{ and } y_i = \omega_i(P_t) - \omega_i(t) \text{ for each } i,$$
$$\text{and } p^* = |\mathcal{P}| - 2\}. \quad (2)$$

If G has no separated partition $\mathcal{P}$ such that $\omega_i(P_s) - \omega_i(s) = x_i$ and $\omega_i(P_t) - \omega_i(t) = y_i$ for each i, then let $h(G, \mathbb{X}, \mathbb{Y}) = +\infty$.

Our algorithm computes $f(G_v, \mathbb{X})$ and $h(G_v, \mathbb{X}, \mathbb{Y})$ for each node v of a binary decomposition tree T of a given series-parallel graph G from leaves to the root r of T by means of dynamic programming. Since $G = G_r$, one can compute the minimum number $p_{\min}(G)$ of components from $f(G, \mathbb{X})$ and $h(G, \mathbb{X}, \mathbb{Y})$ as follows:

$$p_{\min}(G) = \min\Big\{\min\{f(G, \mathbb{X}) + 1 \mid l_i \le x_i + w_{st}(G, i) \le u_i \text{ for each } i\},$$
$$\min\{h(G, \mathbb{X}, \mathbb{Y}) + 2 \mid l_i \le x_i + w_i(s) \le u_i \text{ and}$$
$$l_i \le y_i + w_i(t) \le u_i \text{ for each } i\}\Big\}. \quad (3)$$

Note that $p_{\min}(G) = +\infty$ if G has no uniform partition.

We first compute $f(G_v, \mathbb{X})$ and $h(G_v, \mathbb{X}, \mathbb{Y})$ for each leaf v of T, for which the subgraph G_v contains exactly one edge. We thus have

$$f(G_v, \mathbb{X}) = \begin{cases} 0 & \text{if } \mathbb{X} = (0, 0, \cdots, 0); \\ +\infty & \text{otherwise,} \end{cases} \quad (4)$$

and

$$h(G_v, \mathbb{X}, \mathbb{Y}) = \begin{cases} 0 & \text{if } \mathbb{X} = \mathbb{Y} = (0, 0, \cdots, 0); \\ +\infty & \text{otherwise.} \end{cases} \quad (5)$$

By Eq. (4) one can compute $f(G_v, \mathbb{X})$ in time $O(u^q)$ for each leaf v of T and all q-tuples $\mathbb{X} = (x_1, x_2, \cdots, x_q)$, where u is the maximum upper bound on component size, that is, $u = \max\{u_i \mid 1 \le i \le q\}$. Similarly, by Eq. (5) one can compute $h(G_v, \mathbb{X}, \mathbb{Y})$ in time $O(u^{2q})$ for each leaf v and all pairs of q-tuples $\mathbb{X} = (x_1, x_2, \cdots, x_q)$ and $\mathbb{Y} = (y_1, y_2, \cdots, y_q)$. Since G is a simple series-parallel graph, the number of edges in G is at most $2n - 3$ and hence the number of leaves in T is at most $2n - 3$. Thus one can compute $f(G_v, \mathbb{X})$ and $h(G_v, \mathbb{X}, \mathbb{Y})$ for all leaves v of T in time $O(u^{2q}n)$.

We next compute $f(G_v, \mathbb{X})$ and $h(G_v, \mathbb{X}, \mathbb{Y})$ for each internal node v of T from the counterparts of the two children of v in T.

We first consider a parallel connection.

[Parallel connection]

Let $G_v = G' \parallel G''$, and let $s = s(G_v)$ and $t = t(G_v)$. (See Figs. 2(c) and 5.)

We first explain how to compute $h(G_v, \mathbb{X}, \mathbb{Y})$ from $h(G', \mathbb{X}', \mathbb{Y}')$ and $h(G'', \mathbb{X}'', \mathbb{Y}'')$. The definitions of a separated partition and $h(G, \mathbb{X}, \mathbb{Y})$ imply that if $w_i(P_s) = x_i + w_i(s) > u_i$ or $w_i(P_t) = y_i + w_i(t) > u_i$ for some index i, then $h(G_v, \mathbb{X}, \mathbb{Y}) = +\infty$. One may thus assume that $x_i + w_i(s) \le u_i$ and $y_i + w_i(t) \le u_i$ for each index i, $1 \le i \le q$. Then every separated partition $\mathcal{P}$ of G_v can be obtained by combining a separated partition $\mathcal{P}'$ of G' with a separated partition $\mathcal{P}''$ of G'', as illustrated in Fig. 5(a). We thus have

$$h(G_v, \mathbb{X}, \mathbb{Y}) = \min\{h(G', \mathbb{X}', \mathbb{Y}') + h(G'', \mathbb{X} - \mathbb{X}', \mathbb{Y} - \mathbb{Y}') \mid$$
$$\mathbb{X}' = (x_1', x_2', \cdots, x_q') \text{ and } \mathbb{Y}' = (y_1', y_2', \cdots, y_q')$$
$$\text{such that } 0 \le x_i', y_i' \le u_i \text{ for each } i\}, \quad (6)$$

where $\mathbb{X} - \mathbb{X}' = (x_1 - x_1', x_2 - x_2', \cdots, x_q - x_q')$ and $\mathbb{Y} - \mathbb{Y}' = (y_1 - y_1', y_2 - y_2', \cdots, y_q - y_q')$.

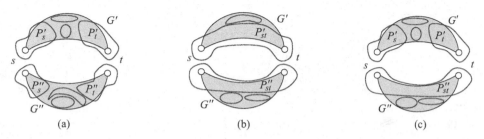

Fig. 5. The combinations of a partition $\mathcal{P}'$ of G' and a partition $\mathcal{P}''$ of G'' for a partition $\mathcal{P}$ of $G_v = G' \parallel G''$

We next explain how to compute $f(G_v, \mathbb{X})$ from $f(G', \mathbb{X}')$, $f(G'', \mathbb{X}'')$, $h(G', \mathbb{X}', \mathbb{Y}')$ and $h(G'', \mathbb{X}'', \mathbb{Y}'')$. If $\omega_i(P_{st}) = x_i + \omega_{st}(G_v, i) > u_i$ for some index i, then $f(G_v, \mathbb{X}) = +\infty$. One may thus assume that $x_i + \omega_{st}(G_v, i) \le u_i$ for each index i, $1 \le i \le q$. Then every connected partition $\mathcal{P}$ of G_v can be obtained by combining a partition $\mathcal{P}'$ of G' with a partition $\mathcal{P}''$ of G'', as illustrated in Figs. 5(b) and (c). There are the following two Cases (a) and (b), and we define two functions f^a and f^b for the two cases, respectively.

Case (a): both $\mathcal{P}'$ and $\mathcal{P}''$ are connected partitions. (See Fig. 5(b).)

Let

$$f^a(G_v, \mathbb{X}) = \min\{f(G', \mathbb{X}') + f(G'', \mathbb{X} - \mathbb{X}') \mid \mathbb{X}' = (x_1', x_2', \cdots, x_q')$$
$$\text{such that } 0 \le x_i' \le u_i \text{ for each } i\}. \quad (7)$$

Case (b): one of $\mathcal{P}'$ and $\mathcal{P}''$ is a separated partition and the other is a connected partition.

One may assume without loss of generality that $\mathcal{P}'$ is a separated partition and $\mathcal{P}''$ is a connected partition. (See Fig. 5(c).) Let

$$f^b(G_v, \mathbb{X}) = \min\{h(G', \mathbb{X}', \mathbb{Y}') + f(G'', \mathbb{X} - \mathbb{X}' - \mathbb{Y}') \mid$$
$$\mathbb{X}' = (x_1', x_2', \cdots, x_q') \text{ and } \mathbb{Y}' = (y_1', y_2', \cdots, y_q')$$
$$\text{such that } 0 \le x_i', y_i' \le u_i \text{ for each } i\}. \quad (8)$$

From f^a and f^b above, one can compute $f(G_v, \mathbb{X})$ as follows:

$$f(G_v, \mathbb{X}) = \min\{f^a(G_v, \mathbb{X}), f^b(G_v, \mathbb{X})\}. \quad (9)$$

By Eq. (6) one can compute $h(G_v, \mathbb{X}, \mathbb{Y})$ in time $O(u^{4q})$ for all pairs of q-tuples $\mathbb{X} = (x_1, x_2, \cdots, x_q)$ and $\mathbb{Y} = (y_1, y_2, \cdots, y_q)$ with $0 \le x_i, y_i \le u_i$, $1 \le i \le q$. By Eqs. (7)–(9) one can compute $f(G_v, \mathbb{X})$ in time $O(u^{3q})$ for all q-tuples $\mathbb{X} = (x_1, x_2, \cdots, x_q)$ with $0 \le x_i \le u_i$, $1 \le i \le q$. Thus one can compute $f(G_v, \mathbb{X})$ and $h(G_v, \mathbb{X}, \mathbb{Y})$ for each p-node v of T in time $O(u^{4q})$.

We next consider a series connection.

[Series connection]

Let $G_v = G' \bullet G''$, and let w be the vertex of G identified by the series connection, that is, $w = t(G') = s(G'')$. (See Figs. 2(b) and 6.)

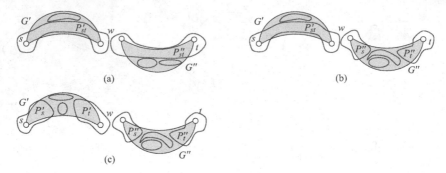

Fig. 6. The combinations of a partition $\mathcal{P}'$ of G' and a partition $\mathcal{P}''$ of G'' for a partition $\mathcal{P}$ of $G_v = G' \bullet G''$

We first explain how to compute $f(G_v, \mathbb{X})$. If $x_i + \omega_{st}(G_v, i) > u_i$ for some index i, then $f(G_v, \mathbb{X}) = +\infty$. One may thus assume that $x_i + \omega_{st}(G_v, i) \le u_i$ for each index i, $1 \le i \le q$. Then every connected partition $\mathcal{P}$ of G_v can be obtained by combining a connected partition $\mathcal{P}'$ of G' with a connected partition $\mathcal{P}''$ of G'', as illustrated in Fig. 6(a). We thus have

$$f(G_v, \mathbb{X}) = \min\{f(G', \mathbb{X}') + f(G'', \mathbb{X}'') \mid \mathbb{X}' = (x_1', x_2', \cdots, x_q') \text{ and}$$
$$\mathbb{X}'' = (x_1'', x_2'', \cdots, x_q'') \text{ such that } 0 \le x_i', x_i'' \le u_i \text{ and}$$
$$x_i' + x_i'' + \omega_i(w) = x_i \text{ for each } i\}. \quad (10)$$

We next explain how to compute $h(G_v, \mathbb{X}, \mathbb{Y})$. If $x_i + \omega_i(s) > u_i$ or $y_i + \omega_i(t) > u_i$ for some index i, then $h(G_v, \mathbb{X}, \mathbb{Y}) = +\infty$. One may thus assume that $x_i + \omega_i(s) \le u_i$ and $y_i + \omega_i(t) \le u_i$ for each index i, $1 \le i \le q$. Then every separated partition $\mathcal{P}$ of G_v can be obtained by combining a partition $\mathcal{P}'$ of G' with a partition $\mathcal{P}''$ of G'', as illustrated in Figs. 6(b) and (c). There are the following two Cases (a) and (b), and we define two functions h^a and h^b for the two cases, respectively.

Case (a): one of $\mathcal{P}'$ and $\mathcal{P}''$ is a connected partition and the other is a separated partition.

One may assume without loss of generality that $\mathcal{P}'$ is a connected partition and $\mathcal{P}''$ is a separated partition. (See Fig. 6(b).) Let

$$h^a(G_v, \mathbb{X}, \mathbb{Y}) = \min\{f(G', \mathbb{X}') + h(G'', \mathbb{X}'', \mathbb{Y}) \mid \mathbb{X}' = (x_1', x_2', \cdots, x_q') \text{ and}$$
$$\mathbb{X}'' = (x_1'', x_2'', \cdots, x_q'') \text{ such that } 0 \le x_i', x_i'' \le u_i \text{ and}$$
$$x_i' + x_i'' + \omega_i(w) = x_i \text{ for each } i\}. \quad (11)$$

Case (b): both $\mathcal{P}'$ and $\mathcal{P}''$ are separated partitions. (See Fig. 6(c).)
Let

$$h^b(G_v, \mathbb{X}, \mathbb{Y}) = \min\{h(G', \mathbb{X}, \mathbb{Y}') + h(G'', \mathbb{X}'', \mathbb{Y}) + 1 \mid$$
$$\mathbb{Y}' = (y_1', y_2', \cdots, y_q') \text{ and } \mathbb{X}'' = (x_1'', x_2'', \cdots, x_q'')$$
$$\text{such that } 0 \le y_i', x_i'' \le u_i \text{ and}$$
$$l_i \le y_i' + x_i'' + \omega_i(w) \le u_i \text{ for each } i\}. \quad (12)$$

From h^a and h^b above one can compute $h(G_v, \mathbb{X}, \mathbb{Y})$ as follows:

$$h(G_v, \mathbb{X}, \mathbb{Y}) = \min\{h^a(G_v, \mathbb{X}, \mathbb{Y}), h^b(G_v, \mathbb{X}, \mathbb{Y})\}. \tag{13}$$

By Eq. (10) one can compute $f(G_v, \mathbb{X})$ in time $O(u^{2q})$ for all q-tuples $\mathbb{X} = (x_1, x_2, \cdots, x_q)$ with $0 \le x_i \le u_i$, $1 \le i \le q$. By Eqs. (11)–(13) one can compute $h(G_v, \mathbb{X}, \mathbb{Y})$ in time $O(u^{4q})$ for all pairs of q-tuples $\mathbb{X} = (x_1, x_2, \cdots, x_q)$ and $\mathbb{Y} = (y_1, y_2, \cdots, y_q)$ with $0 \le x_i, y_i \le u_i$, $1 \le i \le q$. Thus one can compute $f(G_v, \mathbb{X})$ and $h(G_v, \mathbb{X}, \mathbb{Y})$ for each s-node v of T in time $O(u^{4q})$.

In this way one can compute $f(G_v, \mathbb{X})$ and $h(G_v, \mathbb{X}, \mathbb{Y})$ for each internal node v of T in time $O(u^{4q})$ regardless of whether v is a p-node or an s-node. Since T is a binary tree and has at most $2n - 3$ leaves, T has at most $2n - 4$ internal nodes. Since $G = G_r$ for the root r of T, one can compute $f(G, \mathbb{X})$ and $h(G, \mathbb{X}, \mathbb{Y})$ in time $O(u^{4q}n)$. By Eq. (3) one can compute the minimum number $p_{\min}(G)$ of components in a uniform partition of G from $f(G, \mathbb{X})$ and $h(G, \mathbb{X}, \mathbb{Y})$ in time $O(u^{2q})$. Thus the minimum partition problem can be solved in time $O(u^{4q}n)$. This completes our proof of Theorem 1.

4 p-Partition Problem

In this section we have the following theorem.

Theorem 2. *The p-partition problem can be solved for any series-parallel graph G in time $O(p^2 u^{4q} n)$, where n is the number of vertices in G, q is a fixed constant number of weights, u is the maximum upper bound on component size, and p is a given number of components.*

The algorithm for the p-partition problem is similar to the algorithm for the minimum partition problem in the previous section. So we present only an outline.

For a series-parallel graph G and an integer p^*, $0 \le p^* \le p - 1$, we define a set $F(G, p^*)$ of q-tuples $\mathbb{X} = (x_1, x_2, \cdots, x_q)$ as follows:

$$F(G, p^*) = \{\mathbb{X} \mid G \text{ has a connected partition } \mathcal{P} \text{ such that}$$
$$x_i = \omega_i(P_{st}) - \omega_{st}(G, i) \text{ for each } i, \text{ and } p^* = |\mathcal{P}| - 1\}.$$

For a series-parallel graph G and an integer p^*, $0 \le p^* \le p - 2$, we define a set $H(G, p^*)$ of pairs of q-tuples $\mathbb{X} = (x_1, x_2, \cdots, x_q)$ and $\mathbb{Y} = (y_1, y_2, \cdots, y_q)$ as follows:

$$H(G, p^*) = \{(\mathbb{X}, \mathbb{Y}) \mid G \text{ has a separated partition } \mathcal{P} \text{ such that}$$
$$x_i = \omega_i(P_s) - \omega_i(s) \text{ and } y_i = \omega_i(P_s) - \omega_i(t) \text{ for each } i,$$
$$\text{and } p^* = |\mathcal{P}| - 2\}.$$

Clearly $|F(G, p^*)| \le (u+1)^q$ and $|H(G, p^*)| \le (u+1)^{2q}$.

We compute $F(G_v, p^*)$ and $H(G_v, p^*)$ for each node v of a binary decomposition tree T of a given series-parallel graph G from leaves to the root r of T by means of dynamic programming. Since $G = G_r$, the following lemma clearly holds.

Lemma 1. *A series-parallel graph G has a uniform partition with p components if and only if the following condition* (a) *or* (b) *holds*:

(a) $F(G, p - 1)$ *contains at least one q-tuple* $\mathbb{X} = (x_1, x_2, \cdots, x_q)$ *such that* $l_i \leq x_i + \omega_{st}(G, i) \leq u_i$ *for each index i, $1 \leq i \leq q$; and*

(b) $H(G, p - 2)$ *contains at least one pair of q-tuples* $\mathbb{X} = (x_1, x_2, \cdots, x_q)$ *and* $\mathbb{Y} = (y_1, y_2, \cdots, y_q)$ *such that* $l_i \leq x_i + \omega_i(s) \leq u_i$ *and* $l_i \leq y_i + \omega_i(t) \leq u_i$ *for each index i, $1 \leq i \leq q$.*

One can compute in time $O(p)$ the sets $F(G_v, p^*)$ and $H(G_v, p^*)$ for each leaf v of T and all integers p^* ($\leq p - 1$), and compute in time $O(p^2 u^{4q})$ the sets $F(G_v, p^*)$ and $H(G_v, p^*)$ for each internal node v of T and all integers p^* ($\leq p - 1$) from the counterparts of the two children of v in T. Since $G = G_r$ for the root r of T, one can compute the sets $F(G, p - 1)$ and $H(G, p - 2)$ in time $O(p^2 u^{4q} n)$. By Lemma 1 one can know from the sets in time $O(u^{2q})$ whether G has a uniform partition with p components. Thus the p-partition problem can be solved in time $O(p^2 u^{4q} n)$.

5   Conclusions

In this paper we obtained pseudo-polynomial-time algorithms to solve the three uniform partition problems for series-parallel graphs. Both the minimum partition problem and the maximum partition problem can be solved in time $O(u^{4q} n)$. On the other hand, the p-partition problem can be solved in time $O(p^2 u^{4q} n)$.

One can observe that the algorithms for series-parallel graphs can be extended for partial k-trees, that is, graphs with bounded tree-width [1,2].

References

1. S. Arnborg, J. Lagergren and D. Seese, Easy problems for tree-decomposable graphs, *J. Algorithms*, Vol. 12, pp. 308–340, 1991.
2. H. L. Bodlaender, Polynomial algorithms for graph isomorphism and chromatic index on partial k-trees, *J. Algorithms*, Vol. 11, pp. 631–643, 1990.
3. B. Bozkaya, E. Erkut and G. Laporte, A tabu search heuristic and adaptive memory procedure for political districting, *European J. Operational Research*, Vol. 144, pp. 12–26, 2003.
4. R. C. Gonzalez and P. Wintz, Digital Image Processing, *Addison-Wesley, Reading, MA*, 1977.
5. T. Ito, X. Zhou and T. Nishizeki, Partitioning a graph of bounded tree-width to connected subgraphs of almost uniform size, *J. Discrete Algorithms*, Vol. 4, pp. 142–154, 2006.
6. M. Lucertini, Y. Perl and B. Simeone, Most uniform path partitioning and its use in image processing, *Discrete Applied Mathematics*, Vol. 42, pp. 227–256, 1993.
7. K. Takamizawa, T. Nishizeki and N. Saito, Linear-time computability of combinatorial problems on series-parallel graphs, *J. ACM*, Vol. 29, pp. 623–641, 1982.
8. D. C. Tsichritzis and P. A. Bernstein, Operating Systems, *Academic Press, New York*, 1974.
9. J. C. Williams Jr., Political redistricting: a review, *Papers in Regional Science*, Vol. 74, pp. 13–40, 1995.

Characterizations and Linear Time Recognition of Helly Circular-Arc Graphs

Min Chih Lin[1,*] and Jayme L. Szwarcfiter[2,**]

[1] Universidad de Buenos Aires, Facultad de Ciencias Exactas y Naturales,
Departamento de Computación, Buenos Aires, Argentina
oscarlin@dc.uba.ar
[2] Universidade Federal do Rio de Janeiro, Instituto de Matemática, NCE
and COPPE, Caixa Postal 2324, 20001-970 Rio de Janeiro, RJ, Brasil
jayme@nce.ufrj.br

Abstract. A circular-arc model $(C, \mathcal{A})$ is a circle C together with a collection $\mathcal{A}$ of arcs of C. If $\mathcal{A}$ satisfies the Helly Property then $(C, \mathcal{A})$ is a Helly circular-arc model. A (Helly) circular-arc graph is the intersection graph of a (Helly) circular-arc model. Circular-arc graphs and their subclasses have been the object of a great deal of attention, in the literature. Linear time recognition algorithm have been described both for the general class and for some of its subclasses. However, for Helly circular-arc graphs, the best recognition algorithm is that by Gavril, whose complexity is $O(n^3)$. In this article, we describe different characterizations for Helly circular-arc graphs, including a characterization by forbidden induced subgraphs for the class. The characterizations lead to a linear time recognition algorithm for recognizing graphs of this class. The algorithm also produces certificates for a negative answer, by exhibiting a forbidden subgraph of it, within this same bound.

Keywords: algorithms, circular-arc graphs, forbidden subgraphs, Helly circular-arc graphs.

1   Introduction

Circular-arc graphs form a class of graphs which has attracted much interest, since its first characterization by Tucker, almost fourty years ago [9]. There is a particular interest in the study of subclasses of it. The most common of these subclasses are the proper circular-arc graphs, unit circular-arc graphs and Helly circular-arc graphs (Golumbic [3]). Linear time recognition and representation algorithms have been already formulated for general circular-arc graphs (McConnell [7], Kaplan and Nussbaum [5]), proper circular-arc graphs (Deng,

---

* Partially supported by UBACyT Grants X184 and X212, PICT ANPCyT Grant 11-09112, CNPq under PROSUL project Proc. 490333/2004-4.
** Partially supported by the Conselho Nacional de Desenvolvimento Científico e Tecnológico, CNPq, and Fundação de Amparo à Pesquisa do Estado do Rio de Janeiro, FAPERJ, Brasil.

D.Z. Chen and D.T. Lee (Eds.): COCOON 2006, LNCS 4112, pp. 73–82, 2006.

Hell and Huang [1]) and unit circular-arc arc graphs (Lin and Szwarcfiter [6]). For Helly circular-arc graphs, the best recognition algorithm is by Gavril [2], which requires $O(n^3)$ time. Such an algorithm is based on characterizing Helly circular-arc graphs, as being exactly those graphs whose clique matrices admit the circular 1's property on their columns [2]. The book by Spinrad [8] contains an appraisal of circular-arc graph algorithms.

In the present article, we propose new characterizations for Helly circular-arc graphs, including a characterization by forbidden induced subgraphs for the class. The characterizations lead to a linear time algorithm for recognizing graphs of the class and constructing the corresponding Helly circular-arc models. In case a graph does not belong to the class, the method exhibits a certificate, namely a forbbiden induced subgraph of it, also in linear time.

Let G be a graph, V_G, E_G its sets of vertices and edges, respectively, $|V_G| = n$ and $|E_G| = m$. Write $e = v_i v_j$, for an edge $e \in E_G$, incident to $v_i, v_j \in V_G$. A *clique* of G is a maximal subset of pairwise adjacent vertices. Denote $N(v_i) = \{v_j \in V_G | v_i v_j \in E_G\}$, call $v_j \in N(v_i)$ a *neighbour* of v_i and write and $d(v_i) = |N(v_i)|$.

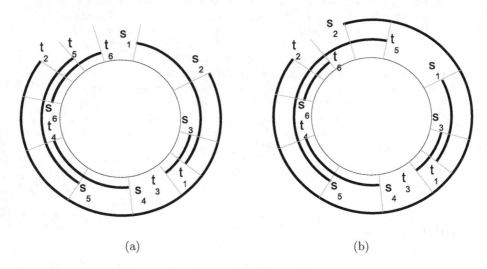

(a)   (b)

Fig. 1. Two circular-arc models

A *circular-arc (CA) model* $(C, \mathcal{A})$ is a circle C together with a collection $\mathcal{A}$ of arcs of C. Unless otherwise stated, we always traverse C in the clockwise direction. Each arc $A_i \in \mathcal{A}$ is written as $A_i = (s_i, t_i)$, where $s_i, t_i \in C$ are the *extreme points* of A_i, with s_i the *start point* and t_i the *end point* of the arc, respectively, in the clockwise direction. The *extremes* of $\mathcal{A}$ are those of all arcs $A_i \in \mathcal{A}$. As usual, we assume that no single arc of $\mathcal{A}$ covers C, that no two extremes of $\mathcal{A}$ coincide and that all arcs of $\mathcal{A}$ are open. When traversing C, we obtain a circular ordering of the extreme points of $\mathcal{A}$. Furthermore, we also consider a circular ordering $A_1, \ldots, A_n$ of the arcs of $\mathcal{A}$, defined by the corresponding circular ordering $s_1, \ldots, s_n$ of their respective start points. In general, when dealing with a sequence $x_1, \ldots, x_t$ of t objects circularly ordered, we assume that all the additions and subtractions of the indices i of the objects x_i are modulo t. Figure 1 illustrates two CA models, with the orderings of their arcs.

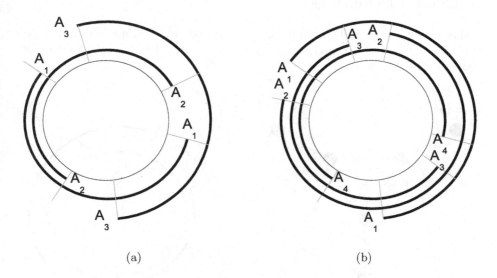

(a)   (b)

Fig. 2. Two minimally non Helly models

In a model $(C, \mathcal{A})$, the *complement* of an arc $A_i = (s_i, t_i)$ is the arc $\overline{A_i} = (t_i, s_i)$. Complements of arcs have been employed before by McConnell [7], under the name *arc flippings*. The *complement* of $(C, \mathcal{A})$ is the model $(C, \overline{\mathcal{A}})$, where $\overline{\mathcal{A}} = \{\overline{A_i} \mid A_i \in \mathcal{A}\}$.

In the model $(C, \mathcal{A})$, a subfamily of arcs of $\mathcal{A}$ is *intersecting* when they pairwise intersect. Say that $\mathcal{A}$ is *Helly*, when every intersecting subfamily of it contains a common point of C. In this case, $(C, \mathcal{A})$ is a *Helly circular-arc (HCA) model*. When $\mathcal{A}$ is not Helly, it contains a minimal non Helly subfamily $\mathcal{A}'$, that is $\mathcal{A}'$ is not Helly, but $\mathcal{A}' \setminus A_i$ is so, for any $A_i \in \mathcal{A}'$. The model $(C, \mathcal{A}')$ is then *minimally non HCA*. Figure 2 depicts two minimally non Helly models.

A *circular-arc (CA) graph* G is the intersection graph of some CA model $(C, \mathcal{A})$. Denote by $v_i \in V_G$ the vertex of G corresponding to $A_i \in \mathcal{A}$. Similarly, a *Helly circular-arc (HCA) graph* is the intersection graph of some HCA model. In a HCA graph, each clique $Q \subseteq V_G$ can be represented by a point $q \in C$, which is common to all those arcs of $\mathcal{A}$, which correspond to the vertices of Q. Clearly, two distinct cliques must be represented by distinct points. Finally, two CA models are *equivalent* when they share the same intersection graph.

In the next section, we present the main basic concepts, in which the proposed characterizations are based. In Section 3, we characterize HCA models, while HCA graphs are characterized in Section 4. In Section 5, we describe the construction of a special CA model, which is employed in the recognition algorithm. Finally, Section 6 describes the recognition algorithm, together with its certificates. Withou loss of generality, we consider all given graphs to be connected.

2   Central Definitions

In this section, we describe usefull concepts for the proposed method. Let G be a graph and $(C, \mathcal{A})$ a CA model of it. First, define special sequences of extremes of the arcs of $\mathcal{A}$.

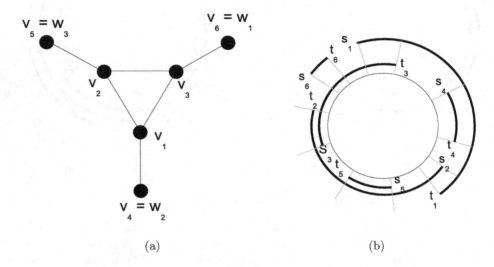

(a)   (b)

Fig. 3. An obstacle and its non Helly stable model

An *s-sequence (t-sequence)* is a maximal sequence of start points (end points) of $\mathcal{A}$, in the circular ordering of C. Write *extreme sequence* to mean an s-sequence or t-sequence. The $2n$ start points and end points are then partitioned into s-sequences and t-sequences, which alternate in C. For an extreme sequence E, denote by $FIRST(E)$ the first element of E, while the notations $NEXT(E)$ and $NEXT^{-1}(E)$ represent the extreme sequences which succeeds and preceeds E in C, respectively. For an extreme point $p \in C$, denote $SEQUENCE(p)$ the extreme sequence which contains p, while $NEXT(p)$ means the sequence $NEXT(SEQUENCE(p))$. Through the paper, we employ operations on the CA models, which possibly modify them, while preserving equivalence. A simple example of such operations is to permute the extremes of the arcs, whitin a same extreme sequence.

Next, we define a special model of interest.

Definition 1. *Let s_i be a start point of $\mathcal{A}$ and $S = SEQUENCE(s_i)$. Say that s_i is stable when $i = j$ or $A_i \cap A_j = \emptyset$, for every $t_j \in NEXT^{-1}(S)$.*

Definition 2. *A model $(C, \mathcal{A})$ is stable when all its start points are stable.*

As examples, the models of Figures 1(a) and 1(b) are not stable, while that of Figure 3(b) is.

We will employ stable models in the recognition process of HCA graphs.

Finally, define a special family of graphs.

Definition 3. *An obstacle is a graph H containing a clique $K_t \subseteq V_H$, $t \geq 3$, whose vertices admit a circular ordering $v_1, \ldots, v_t$, such that each edge $v_i v_{i+1}$, $i = 1, \ldots, t$, satisfies:*

(i) $N(w_i) \cap K_t = K_t \setminus \{v_i, v_{i+1}\}$, for some $w_i \in V_H \setminus K_t$, or
(ii) $N(u_i) \cap K_t = K_t \setminus \{v_i\}$ and $N(z_i) \cap K_t = K_t \setminus \{v_{i+1}\}$, for some adjacent vertices $u_i, z_i \in V_H \setminus K_t$.

As example, the graph of Figure 3(a) is obstacle.

We will show that the obstacles form a family of forbidden induced subgraphs for a CA graph to be HCA.

3   Characterizing HCA Models

In this section, we describe a characterization and a recognition algorithm for HCA models. The characterization is as follows:

Theorem 1. *A CA model $(C, \mathcal{A})$ is HCA if and only if*

(i) if three arcs of $\mathcal{A}$ cover C then two of these three arcs also cover it, and
(ii) the intersection graph of $(C, \overline{\mathcal{A}})$ is chordal.

Proof. By hypothesis, $(C, \mathcal{A})$ is a HCA model. Condition (i) is clear, otherwise $(C, \mathcal{A})$ can not be HCA. Suppose Condition (ii) fails. Then the intersection graph G^c of $(C, \overline{\mathcal{A}})$ contains an induced cycle C^c, with length $k > 3$. Let $\overline{\mathcal{A}'} \subseteq \overline{\mathcal{A}}$ be the set of arcs of $\overline{\mathcal{A}}$, corresponding to the vertices of C^c, and $\mathcal{A}' \subseteq \mathcal{A}$ the sets of the complements of the arcs $\overline{A_i} \in \overline{\mathcal{A}'}$. First, observe that no two arcs of $\overline{\mathcal{A}'}$ cover the circle, otherwise C^c would contain a chord. Consequently, $\overline{\mathcal{A}'}$ consists of k arcs circularly ordered as $\overline{A_1}, \ldots, \overline{A_k}$ and satisfying: $\overline{A_i} \cap \overline{A_j} \neq \emptyset$ if and only if $\overline{A_i}, \overline{A_j}$ are consecutive in the circular ordering. In general, comparing a model $(C, \mathcal{A})$ to its complement model $(C, \overline{\mathcal{A}})$, we conclude that two arcs of $\mathcal{A}$ intersect if and only if their complements in $\overline{\mathcal{A}}$ are either disjoint or intersect without covering the circle. Consequently, $\mathcal{A}'$ must be an intersecting family. On the other hand, the arcs of $\mathcal{A}'$ can not have a common point $p \in C$. Because, otherwise $p \notin \overline{A_i}$, for all $\overline{A_i}$, meaning that the arcs of $\overline{\mathcal{A}'}$ do not cover the circle, contradicting C^c to be an induced cycle. The inexistence of a common point in $\mathcal{A}'$ implies that $\mathcal{A}$ is not a Helly family, a contradiction. Then (ii) holds. The converse is similar.△

The following characterizes minimally non Hely models.

Corollary 1. *A model* $(C, \mathcal{A})$ *is minimaly non HCA if and only if*

(i) $\mathcal{A}$ *is intersecting and covers* C, *and*
(ii) two arcs of $\mathcal{A}$ *cover* C *precisely when they are not consecutive in the circular ordering of* $\mathcal{A}$.

Theorem 1 leads directly to a simple algorithm for recognizing Helly models, as follows. Given a model $(C, \mathcal{A})$ of some graph G, verify if $(C, \mathcal{A})$ satisfies Condition (i) and then if it satisfies Condition (ii). Clearly, $(C, \mathcal{A})$ is HCA if and only if both conditions are satisfied. Next, we describe methods for checking them.

For Condition (i), we seek directly for the existence of three arcs $A_i, A_j, A_k \in \mathcal{A}$ that cover C, two of them not covering it. Observe that there exist such arcs if and only if the circular ordering of their extremes is $s_i, t_k, s_j, t_i, s_k, t_j$. For each $A_i \in \mathcal{A}$, we repeat the following procedure, which looks for the other two arcs A_j, A_k whose extreme points satisfy this ordering. Let L_1 be the list of extreme points of the arcs contained in (s_i, t_i), in the ordering of C. First, remove from L_1 all pairs of extremes s_q, t_q of a same arc, which may possibly occur. Let L_2 be the list formed by the other extremes of the arcs represented in L_1. That is, $s_q \in L_1$ if and only if $t_q \in L_2$, and $t_q \in L_1$ if and only if $s_q \in L_2$, for any $A_q \in \mathcal{A}$. Clearly, the extremes points which form L_2 are all contained in (t_i, s_i), and we consider them in the circular ordering of C. Denote by $FIRST(L_1)$ and $LAST(L_2)$ the first and last extreme points of L_1 and L_2, in the considered orderings, respectively. Finally, iteratively perform the steps below, until either $L_1 = \emptyset$, or $FIRST(L_1) = t_k$ and $LAST(L_2) = t_j$, for some j, k.

if $FIRST(L_1)$ is a start point s_q then remove s_q from L_1 and t_q from L_2
if $LAST(L_2)$ is a start point s_q then remove s_q from L_2 and t_q from L_1

If the iterations terminate because $L_1 = \emptyset$ then there are no two arcs which together with A_i satisfy the above requirements, completing the computations relative to A_i. Otherwise, the arcs A_k and A_j, whose end points are $FIRST(L_1)$ and $LAST(L_2)$, form together with A_i a certificate for the failure of Condition (i). Each of the n lists L_2 needs to be sorted. There is no difficulty to sort them all together in time $O(m)$, at the beginning of the process. The computations relative to A_i require $O(d(v_i))$ steps. That is, the overall complexity of checking Condition (i) is $O(m)$.

For Condition (ii), the direct approach would be to construct the model $(C, \overline{\mathcal{A}})$, its intersection graph G^c and apply a chordal graph recognition algorithm to decide if G^c is chordal. However, the number of edges of G^c could be $O(n^2)$, breaking the linearity of the proposed method. Alternatively, we check whether the complement $\overline{G^c}$ of G^c is co-chordal. Observe that two vertices of $\overline{G^c}$ are adjacent if and only if their corresponding arcs in $\mathcal{A}$ cover the circle. Consequently, the number of edges of $\overline{G^c}$ is at most that of G, i.e. $\leq m$. Since co-chordal graphs can be recognized in linear time (Habib, McConnell, Paul and Viennot [4]), the complexity of the method for verifying Condition (ii) is $O(m)$.

Consequently, HCA models can be recognized in linear time.

4 Characterizing HCA Graphs

In this section, we describe the proposed characterizations for HCA graphs.

Theorem 2. *The following affirmative are equivalent for a CA graph G.*

(a) G is HCA.
(b) G does not contain obstacles, as induced subgraphs.
(c) All stable models of G are HCA.
(d) One stable model of G is HCA.

Proof. (a) $\Rightarrow$ (b): By hypothesis, G is HCA. Since HCA graphs are hereditary, it is sufficient to prove that no obstacle H is a HCA graph. By contrary, suppose H admits a HCA model $(C, \mathcal{A})$. Let K_t be the core of H. By Definition 3, there is a circular ordering $v_1, \ldots, v_t$ of the vertices of K_t which satisfies Conditions (i) or (ii) of it. Denote by $\mathcal{A}' = \{A_1, \ldots, A_t\} \subseteq \mathcal{A}$ the family of arcs corresponding to K_t. Define a clique C_i of H, for each $i = 1, \ldots, t$, as follows. If (i) of Definition 3 is satisfied then $C_i \supseteq \{w_i\} \cup K_t \setminus \{v_i, v_{i+1}\}$, otherwise (ii) is satisfied and $C_i \supseteq \{u_i, z_i\} \cup K_t \{v_i, v_{i+1}\}$. Clearly, all cliques $C_1, \ldots, C_t$ are distinct, because any two of them contain distinct subsets of K_t. Since H is HCA, there are distinct points $p_1, \ldots, p_t \in C$, representing $C_1, \ldots, C_t$, respectively. We know that $v_i \in C_j$ if and only if $i \neq j - 1, j$. Consequently, $p_j \in A_i$ if and only if $i \neq j - 1, j$. The latter implies that $p_1, \ldots, p_t$ are also in the circular ordering of C. On the other hand, because K_t is a clique distinct from any C_i, there is also a point $p \in C$ representing K_t. Try to locate p in C. Clearly, p lies between two consecutive points p_{i-1}, p_i. Examine the vertex $v_i \in K_t$ and its corresponding arc $A_i \in \mathcal{A}'$. We already know that $p \in A_i$, while $p_{i-1}, p_i \notin A_i$. Furthermore, because $t \geq 3$, there is $j \neq i - 1, i$ such that $p_j \in A_i$. Such situation can not be realized by arc A_i. Then $(C, \mathcal{A})$ is not HCA, a contradiction.

(b) $\Rightarrow$ (c): By hypothesis, G does not contain obstacles. By contrary, suppose that there exists a stable model $(C, \mathcal{A})$ of G, which is not HCA. Let $\mathcal{A}' \subseteq \mathcal{A}$ be a minimally non Helly subfamily of $\mathcal{A}$. Denote by $A_1, \ldots, A_t$ the arcs of $\mathcal{A}'$ in the circular ordering. Their corresponding vertices in G are $v_1, \ldots, v_t$, forming a clique $K_t \subseteq V_G$. Let A_i, A_{i+1} be two consecutive arcs of $\mathcal{A}'$, in the circular ordering. By Corollary 1, A_i, A_{i+1} do not cover C. Denote $T = SEQUENCE(t_{i+1})$ and $S = SEQUENCE(s_i)$. Because $(C, \mathcal{A})$ is stable, $S \neq NEXT(T)$. Let $S' = NEXT(T)$ and $T' = NEXT^{-1}(S)$. Choose $s_z \in S$ and $t_u \in T$. We know that A_z does not intersect A_{i+1}, nor does A_u intersect A_i, again because the model is stable. Since $s_z, t_u \in (t_{i+1}, s_i)$, Corollary 1 implies that $s_z, t_u \in A_j$, for any $A_j \in \mathcal{A}'$, $A_j \neq A_i, A_{i+1}$. Denote by z_i and u_i the vertices of G corresponding to A_z and A_u, respectively. Examine the following alternatives.

If z_i and v_i are not adjacent, rename z_i as w_i. Similarly, if u_i and v_{+1} are not adjacent, let w_i be the vertex u_i. In any of these two alternatives, it follows that $N(w_i) \cap K_t = K_t \setminus \{v_i, v_{i+1}\}$. The latter means that Condition (i) of Definition 3 holds. When none of the above alternatives occurs, the arcs A_z and A_u intersect, because s_z preceeds t_u in (t_{i+1}, s_i). That is, z_i and w_i are adjacent vertices satisfying $N(z_i) \cap K_t = K_t \setminus \{v_{i+1}\}$ and $N(u_i) \cap K_t = K_t \setminus \{v_i\}$. This corresponds to Condition (ii) of Definition 3. Consequently, for any pair of vertices $v_i, v_{i+1} \in K_t$ it is always possible to select a vertex $w_i \notin K_t$, or a pair of vertices $z_i, u_i \notin K_t$, so that Definition 3 is satisfied. That is, G contains an obstacle as an induced subgraph. This contradiction means all stable models of G are HCA.

The implications (c) $\Rightarrow$ (d) and (d) $\Rightarrow$ (a) are trivial, meaning that the proof is complete. $\triangle$

5   Constructing Stable Models

Motivated by the characterizations of HCA graphs in terms of stable models, described in the previous section, we present below an algorithm for constructing a stable model of a CA graph. Let $(C, \mathcal{A})$ be a CA model of some graph G, and $A_1, \ldots, A_n$ the circular ordering of the arcs of $\mathcal{A}$. Define the following expansion operations on the end points t_j and start points s_i of $\mathcal{A}$.

$expansion(t_j)$:
   Examine the extremes points of $\mathcal{A}$, starting from t_j, in the clockwise direction, and choosing the closest start point s_i satisfying $i = j$ or $A_i \cap A_j = \emptyset$. Then move t_j so as to become the extreme point preceeding s_i in the model.
$expansion(s_i)$:
   First, examine the extreme points of $\mathcal{A}$, starting from s_i, in the counterclockwise direction, and choosing the closest end point t_j satisfying $i = j$ or $A_i \cap A_j = \emptyset$. Let $T = SEQUENCE(t_j)$. Then move s_i counterclocwise towards T, transforming T into the sequences $T's_iT''$, where $T' = \{t_j \in T | i = j$ or $A_i \cap A_j = \emptyset\}$ and $T'' = T \setminus T'$.

The following lemma asserts that the intersections of the arcs are preserved by these operations.

Lemma 1. *The operations $expansion(t_j)$ or $expansion(s_i)$ applied to a model $(C, \mathcal{A})$ construct models equivalent to $(C, \mathcal{A})$.*

We describe the following algorithm for finding a stable model of a given CA model, with end points t_j and start points s_i:

1. Perform $expansion(t_j)$, for $j = 1, \ldots, n$.
2. Perform $expansion(s_i)$, for $i = 1, \ldots, n$.

The correctness of this algorithm then follows from Lemma 1 and from the following theorem.

Theorem 3. *The model constructed by the above algorithm is stable.*

Proof. Let $(C, \mathcal{A})$ be a given CA model, input to the algorithm. We show that all its start points are stable, at the end of the process. After the completion of Step 1, we know that $s_i = FIRST(S)$ is already stable, for any s-sequence S. Otherwise, there would exist some end point $t_j \in NEXT^{-1}(S)$ satisfying

$i \neq j$ and $A_i \cap A_j \neq \emptyset$, meaning that t_j would have been moved after s_i in the clockwise direction, by $expansion(t_j)$.

Next, examine Step 2. Choose a start point s_i and follow the operation $expansion(s_i)$. If s_i is already stable, the algorithm does nothing. Suppose s_i is not stable. Let $S^* = SEQUENCE(s_i)$ and S the s-sequence closest to S^* in the counterclockwise direction, where $T = NEXT^{-1}(S)$ contains some t_j satisfying $i = j$ or $A_i \cap A_j = \emptyset$. Then $expansion(s_i)$ transforms T into the sequences $T' s_i T''$, where $T' = \{t_j \in T \mid i = j \text{ or } A_i \cap A_j = \emptyset\}$ and $T'' = T \setminus T'$. Analyze the new sequences that have been formed. Clearly, $T' \neq \emptyset$, otherwise s_i would have been moved further from S^*. On the other hand, T'' could possibly be empty. However, the latter would only imply that T remains unchanged and that s_i has been incorporated to S. In any case, T' is the t-sequence which preceeds s_i. By the construction of T', it follows that s_i is now stable. In addition, previously stable start points of S remain so, because $T'' \subset T$. Furthermore, observe that $s_i \neq FIRST(S^*)$, because $FIRST(S^*)$ was before stable, whereas s_i was not. Consequently, S^* does not become empty by moving s_i out of it, implying that no parts of distinct t-sequences can be merged during the process. The latter assertion preserves the stability of the stable vertices belonging to the s-sequence which follows S^* in C. The remaining start points are not affected by $expansion(s_i)$. Consequently, s_i becomes now stable and all previousloy stable start points remain so. The algorithm is correct. △

Corollary 2. *Every CA model admits an equivalent stable model.*

Next, we discuss the complexity of the algorithm. The number of extreme points examined during the operation $expansion(t_j)$ is at most $d(v_j) + 1$, since the operation stops at the first extreme t_i, such that either $i = j$ or $A_i \cap A_j = \emptyset$. Consequently, Step 1 requires $O(m)$ time. As for the operation $expansion(s_i)$, we divide it into two parts. First, for finding the required s-sequence S, the above argument applies, that is, $O(m)$ time suffices for all s_i. As for the determination of the sequences T' and T'', a straightforward implementation of it would consist of examining the entire t-sequence $T = T' \cup T''$, for each corresponding s_i, meaning $O(n^2)$ time, overall. However, a more elaborate implementation is possible, as follows.

To start, after the completion of Step 1, order the end points of each t-sequence T, in reverse ordering of their corresponding start points. That is, if $t_j, t_k \in T$ then in the clockwise direction, the extreme points of A_j and A_k appear as $\ldots t_j \ldots t_k \ldots s_k \ldots s_j \ldots$. Such an ordering can be obtained in overall $O(n)$ time. With the end points so ordered, when traversing $T = NEXT^{-1}(S)$, for completing the operation $expansion(s_i)$, we can stop at the first $t_j \in T$ satisfying $i = j$ or $A_i \cap A_j = \emptyset$. In case the condition $i = j$ holds, we exchange in T, the positions of t_j and $FIRST(T)$. Afterwards, in any of the two alternatives, move s_i to the position just before t_j in the counterclockwise direction. We would need no more than additional $d(v_i) + 1$ steps for it, in the worst case. Consequently, $expansion(s_i)$ can be completed in $O(m)$ time, for all start points. Therefore the complexity of the algorithms is $O(m)$.

6   Recognition Algorithm for HCA Graphs

We are now ready to formulate the algorithm for recognizing HCA graphs. Let G be a graph.

1. Apply the algorithm [7] to recognize whether G is a CA graph. In the affirmative case, let $(C, \mathcal{A})$ be the model constructed by [7]. Otherwise terminate the algorithm (G is not HCA).
2. Transform $(C, \mathcal{A})$ into a stable model, applying the algorithm of Section 5.
3. Verify if $(C, \mathcal{A})$ is a HCA model, applying the algorithm of Section 3. Then terminate the algorithm (G is HCA if $(C, \mathcal{A})$ is HCA, and otherwise G is not HCA).

The correctness of the algorithm follows directly from Theorems 1, 2 and 3.

Each of the above steps can be implemented in $O(m)$ time. The complexity of the algorithm is $O(m)$.

The algorithm constructs a HCA model of the input graph G, in case G is HCA. If G is CA but not HCA, we can exhibit a certificate of this fact, by showing a forbidden subgraph of G, that is, an obstacle. In order to construct the obstacle, we may need certificates of non co-chordality. There is no difficulty to modify the algorithm [4] so as to produce such certificates. The entire process can also be implemented in linear time.

References

1. X. Deng and P. Hell and J. Huang, Linear time representation algorithms for proper circular-arc graphs and proper interval graphs, *SIAM J. Computing*, **25** (1996), pp. 390-403.
2. F. Gavril, Algorithms on circular-arc graphs, *Networks* **4** (1974), pp. 357-369.
3. M. C. Golumbic, *Algorithmic Graph Theory and Perfect Graphs*, Academic Press, 1980, 2nd ed. 2004.
4. M. Habib, R. M. McConnell, C. Paul, and L. Viennot, Lex-bfs and partition refinement, with applications to transitive orientation, interval graph recognition and consecutive ones testing, *Theoretical Computer Science*, **234** (2000), pp. 59-84.
5. H. Kaplan and Y. Nussbaum, A Simpler Linear-Time Recognition of Circular-Arc Graphs, accepted for publication in *10th Scandinavian Workshop on Algorithm Theory* (2006).
6. M. C. Lin and J. L. Szwarcfiter, Efficient Construction of Unit Circular-Arc Models, *Proceedings of the 17th Annual ACM-SIAM Symposium on Discrete Algorithms* (2006), pp. 309-315.
7. R. M. McConnell, Linear-time recognition of circular-arc graphs, *Algorithmica* **37** (**2**) (2003), pp. 93-147.
8. J. Spinrad, Efficient Graph Representations, *American Mathematical Society* (2003).
9. A. Tucker, Characterizing circular-arc graphs, *Bull. American Mathematical Society* **76** (1970), pp. 1257-1260.

Varieties Generated by Certain Models of Reversible Finite Automata

Marats Golovkins[1,*] and Jean-Eric Pin[2]

[1] Institute of Mathematics and Computer Science, University of Latvia,
Raiņa bulv. 29, Riga, Latvia
[2] LIAFA, Université Paris VII and CNRS, Case 7014, 2 Place Jussieu,
75251 Paris Cedex 05, France
marats@latnet.lv, Jean-Eric.Pin@liafa.jussieu.fr

Abstract. Reversible finite automata with halting states (RFA) were first considered by Ambainis and Freivalds to facilitate the research of Kondacs-Watrous quantum finite automata. In this paper we consider some of the algebraic properties of RFA, namely the varieties these automata generate. Consequently, we obtain a characterization of the boolean closure of the classes of languages recognized by these models.

1 Introduction

In this paper we study reversible finite automata (RFA). Being entirely classical, the model is however a special case of Kondacs-Watrous quantum finite automata and was introduced in [5]. Quantum finite automata (QFA) are of a specific interest, since the family of these models represent finite memory real-time quantum mechanical devices. On the other hand, recently it has been demonstrated [3] that these models are worth studying also from the point of view of classical algebraic automata theory. The first models of QFA are due to [11] and [13]. Other models are proposed and studied, for example, in [9,14,6,8,3,10,4], etc. In principle, the different types of QFA reflect the different ways how the results of computation can be interpreted, i.e., quantum measurements. By applying various restrictions, it is even possible to get deterministic and probabilistic special cases of QFA. Such models sometimes prove to be extremely useful in the research of the properties of QFA.

In Section 2 we introduce the finite automata models discussed further in the paper. Section 3 recalls the notations of the varieties used in this paper. Section 4 deals with injective finite automata (IFA), which are in turn a special case of RFA. IFA are closely related to a deterministic special case of Brodsky-Pippenger QFA [9]. We give an exact characterization of languages which are recognized by IFA and conclude that the syntactic monoids of this class generates the variety of commuting idempotent monoids, **ECom**. In Section 5 we show

---

* Supported by the Latvian Council of Science, grant No. 05.1528 and by the European Social Fund, contract No. 2004/0001/VPD1/ESF/PIAA/04/NP/3.2.3.1/0001/0063. The paper was prepared while visiting LIAFA, Université Paris VII and CNRS, and Electronics Research Laboratory, University of California, Berkeley.

D.Z. Chen and D.T. Lee (Eds.): COCOON 2006, LNCS 4112, pp. 83–93, 2006.

that the syntactic monoids of languages recognized by RFA generate the variety defined by the identity $x^\omega y^\omega x^\omega = x^\omega y^\omega$. Section 6 specifies algebraic conditions for a language to be recognized by RFA or IFA.

2   Preliminaries

In this paper, by *minimal automaton* of a regular language we understand a complete minimal deterministic finite automaton recognizing the language (the transition function is defined for any state and any input letter). Two automata (deterministic or not) are said to be *equivalent* if they accept the same language. We denote by L^c the complement of a language L. We do not recall the general definition for Kondacs-Watrous QFA, which can be found in [11]. The definition of RFA is obtained from Kondacs-Watrous QFA by adding the restriction that any transition is deterministic:

Definition 2.1. *A reversible finite automaton* $\mathcal{A} = (Q, \Sigma \cup \{\$\}, q_0, Q_a, Q_r, \cdot)$ *is specified by a finite set of states* Q, *a finite input alphabet* Σ, *an end-marker* $\$ \notin \Sigma$ *and an initial state* $q_0 \in Q$. *The set* Q *is the union of two disjoint subsets* Q_h *and* Q_n, *called the set of halting and non-halting states, respectively. Further, the set* Q_h *is the union of two disjoint subsets* Q_a *and* Q_r *of* Q, *called the set of accepting and rejecting states, respectively. The transition function* $(q, \sigma) \rightarrow q \cdot \sigma$ *from* $Q \times (\Sigma \cup \{\$\})$ *into* Q *satisfies the following conditions:*

$$\text{for all } \sigma \in \Sigma \cup \{\$\}, \quad q_1 \cdot \sigma = q_2 \cdot \sigma \text{ implies } q_1 = q_2; \tag{1}$$

$$\text{if } q \text{ is non-halting, then } q \cdot \$ \text{ is halting.} \tag{2}$$

The first condition is equivalent to each letter $\sigma \in \Sigma \cup \{\$\}$ inducing a bijection on Q. A RFA reads any input word starting with the first letter. As soon as the automaton enters a halting state, the computation is halted and the word is either accepted or rejected, depending on whether the state is accepting or rejecting. The end-marker $\$$ insures that any word is either accepted or rejected.

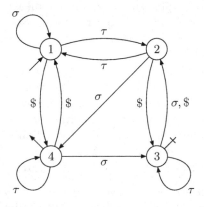

Fig. 1. A reversible finite automaton

In the example of Figure 1, state 4 is accepting and state 3 is rejecting. States 1 and 2 are non-halting.

A reversible finite automaton is called *end-decisive* [9], if it accepts a word only after reading the end-marker $. Dually, if the automaton rejects a word only after reading $, it is called *co-end-decisive*. If a reversible finite automaton is either end-decisive or co-end-decisive, it will be called a *deterministic Brodsky-Pippenger automaton* (DBPA).

It can be noticed that any RFA $\mathcal{A} = (Q, \Sigma \cup \{\$\}, q_0, Q_a, Q_r, \cdot)$ can be transformed into a classical finite automaton $\mathcal{B} = (Q, \Sigma, q_0, F, \cdot_\mathcal{B})$, where $F = Q_a \cup \{q \in Q_n \mid q \cdot \$ \in Q_a\}$ and the new transition function is defined in the following way: for all $\sigma \in \Sigma$ and $q \in Q$,

$$q \cdot_\mathcal{B} \sigma = \begin{cases} q \cdot \sigma & \text{if } q \text{ is non-halting,} \\ q & \text{if } q \text{ is halting.} \end{cases} \tag{3}$$

By eliminating in $\mathcal{B}$ the states which are not accessible from the initial state, we obtain an automaton $\mathcal{A}' = (Q', \Sigma, q_0, F', \cdot)$, where $F' = Q' \cap F$, which recognizes the same language as $\mathcal{A}$. For instance, if $\mathcal{A}$ is the automaton represented in Figure 1, the automata $\mathcal{B}$ and $\mathcal{A}'$ are represented in Figure 2.

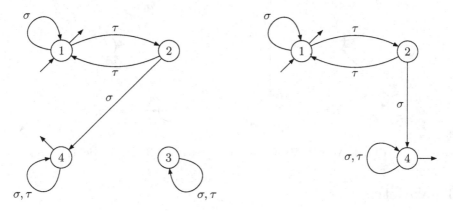

Fig. 2. The automata $\mathcal{B}$ and $\mathcal{A}'$

A state q such that, for every $\sigma \in \Sigma$, $q \cdot_\mathcal{B} \sigma = q$, will be called *absorbing*.

Proposition 2.2. *If $\mathcal{A}'$ is non-trivial, a state of Q' is absorbing if and only if it is halting.*

Consider the non-absorbing states of $\mathcal{A}'$, which are also, by Proposition 2.2, the non-halting states. It follows from (3) that each letter of Σ acts on these states as a partial injective function. All the absorbing states in F' are equivalent, so they can be merged. The same applies to non-final absorbing states.

The resulting deterministic automaton is equivalent to $\mathcal{A}$. It has at most two absorbing states and each letter defines a partial injective function on the set of non-absorbing states. An automaton with these properties will be called a *classical reversible finite automaton* (CRFA). Conversely, it is possible to show that any CRFA can be transformed into an equivalent RFA. Thus we have established the following result.

Proposition 2.3. *Any RFA is equivalent to some CRFA. Conversely, any CRFA is equivalent to some RFA.*

If a CRFA has no absorbing states, it is a *group automaton* (all letters define permutations on the set of states) and it recognizes a *group language*. If it has at most one absorbing state, it will be called an injective finite automaton (IFA), to illustrate the connection of this model to partial injective functions, as discussed in the next section. Similarly as RFA are equivalent to CRFA, IFA are equivalent to DBPA. We call IFA-A (resp. IFA-R) an injective automaton whose absorbing state (if it exists) is final (resp. nonfinal). IFA-A are equivalent to co-end-decisive automata and IFA-R to end-decisive automata. As we shall later see, the closure of IFA-R under finite union is equivalent to Pin's reversible automata [16,17].

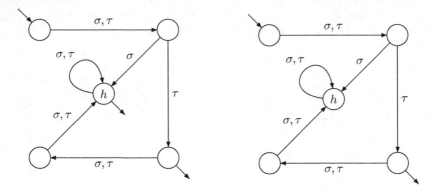

Fig. 3. An IFA-A (on the left) and an IFA-R (on the right)

3   Varieties

If x is an element of a monoid M, we denote by x^ω the unique idempotent of the subsemigroup of M generated by x.

An *ordered monoid* $(M, \leq)$ is a monoid M equipped with a *stable* order relation $\leq$ on M which means that, for every $u, v, x \in M$, $u \leq v$ implies $ux \leq vx$ and $xu \leq xv$.

Let M be a monoid and let s be an element of M. An *inverse* of s is an element $\bar{s}$ such that $s\bar{s}s = s$ and $\bar{s}s\bar{s} = \bar{s}$. An *inverse monoid* is a monoid in which every element has exactly one inverse. It is well known that the relation $\leq$ on M defined by

$$x \leq y \text{ if and only if } x = ye \text{ for some idempotent } e \text{ of } M$$

is a stable partial order, called the *natural order* of M.

Following [19], we call *ordered inverse monoid* an inverse monoid M, equipped with its natural order. We also call *dually ordered inverse monoid* an inverse monoid ordered by the dual order of its natural order.

A general overview on varieties of finite semigroups and monoids is given in [15], whereas introduction to varieties of ordered semigroups and monoids can be found in [18]. Given two varieties of ordered monoids $\mathbf{V}$ and $\mathbf{W}$, their semidirect product $\mathbf{V} * \mathbf{W}$ and Malcev product $\mathbf{V} \circledM \mathbf{W}$ are defined as in [19]. Theorems in [19, Section 3] imply that the semidirect product is an associative operation on varieties of ordered monoids.

In this paper, we shall use the following varieties of ordered monoids, which are defined by some simple identities:

(1) $\mathbf{G} = [\![x^\omega = 1]\!]$, the variety of groups;

(2) $\mathbf{J_1} = [\![x^2 = x, \ xy = yx]\!]$, the variety of commutative and idempotent monoids;

(3) $\mathbf{J_1^+} = [\![x^2 = x, \ x \leq 1]\!]$, the variety of ordered idempotent monoids in which the identity is the maximum element. Order implies $xy \leq y$, $xy \leq x$, and since monoids are idempotent, $xy \leq yx$. Hence $xy = yx$, and $\mathbf{J_1^+} \subset \mathbf{J_1}$;

(4) $\mathbf{J_1^-} = [\![x^2 = x, \ 1 \leq x]\!]$, the variety of ordered idempotent monoids in which the identity is the minimal element. Similarly, $\mathbf{J_1^-} \subset \mathbf{J_1}$;

(5) $\mathbf{R_1} = [\![xyx = xy]\!]$, the variety of idempotent and $\mathcal{R}$-trivial monoids;

(6) $\mathbf{ECom} = [\![x^\omega y^\omega = y^\omega x^\omega]\!]$, the variety of monoids with commuting idempotents: the set of idempotents form a submonoid which belongs to the variety $\mathbf{J_1}$. This variety is known [7] to be equal to $\mathbf{Inv}$, the variety of monoids generated by inverse monoids. Further, by [12], $\mathbf{Inv} = \mathbf{J_1} * \mathbf{G} = \mathbf{J_1} \circledM \mathbf{G} = \mathbf{ECom}$;

(7) $\mathbf{ECom^+} = [\![x^\omega y^\omega = y^\omega x^\omega, \ x^\omega \leq 1]\!]$, the variety of ordered monoids whose idempotents form an ordered submonoid which belongs to the variety $\mathbf{J_1^+}$. This variety is known [19] to be equal to $\mathbf{Inv^+}$, the variety of ordered monoids generated by ordered inverse monoids, and also to $\mathbf{J_1^+} * \mathbf{G}$;

(8) $\mathbf{ECom^-} = [\![x^\omega y^\omega = y^\omega x^\omega, \ 1 \leq x^\omega]\!]$ the variety of ordered monoids whose idempotents form an ordered submonoid which belongs to the variety $\mathbf{J_1^-}$. One can show that this variety is equal to $\mathbf{Inv^-}$, the variety of ordered monoids generated by dually ordered inverse monoids, and also to $\mathbf{J_1^-} * \mathbf{G}$.

By Vagner-Preston theorem [23,22], transition monoids of IFA, IFA-A, IFA-R generate the varieties $\mathbf{Inv}$, $\mathbf{Inv^+}$, $\mathbf{Inv^-}$, respectively. We elaborate this fact in the next section.

4   Injective Finite Automata

In this section we shall describe the languages recognized by IFA, as well as an algebraic characterization of the boolean closure of this class of languages. The transition monoid generated by an injective automaton is isomorphic to a submonoid of the monoid of injective partial functions from a finite set into itself, which justifies the name chosen for the model.

The classes of languages recognized by IFA-A and IFA-R will be denoted by $\mathbf{L}$ and $\mathbf{L^c}$, respectively. The intersection of $\mathbf{L}$ and $\mathbf{L^c}$ is the class of group languages. Recall that a class of languages is closed under inverse morphism if for any monoid morphism $\varphi : \Sigma^* \to \Gamma^*$ and for any language L in the class, the language $\varphi^{-1}(L)$ is also in the class. Given a word u and a language L of Σ^*, recall that the quotient of L by u on the left (resp. right) is the language $u^{-1}L = \{v \in \Sigma^* \mid uv \in L\}$ (resp. $Lu^{-1} = \{v \in \Sigma^* \mid vu \in L\}$).

Theorem 4.1. *The classes $\mathbf{L}$ and $\mathbf{L^c}$ are closed under inverse morphisms and word quotients. Furthermore, the class $\mathbf{L}$ is closed under finite union and the class $\mathbf{L^c}$ under finite intersection.*

Theorem 4.2. *A language of Σ^* is in $\mathbf{L}$ if and only if it is of the form $L_0 \cup \left(\bigcup_{\sigma \in \Sigma} L_\sigma \sigma \Sigma^* \right)$, where L_0 and the L_σ are group languages.*

Proof. First, if $L \subset \Sigma^*$ is a group-language and $\sigma \in \Sigma$, the languages L and $L\sigma\Sigma^*$ are recognized by IFA-A and therefore are in $\mathbf{L}$. Since by Theorem 4.1, $\mathbf{L}$ is closed under finite union, the languages described in the statement are in $\mathbf{L}$.

Consider now a language L recognized by an IFA-A $\mathcal{A} = (Q, \Sigma, q_0, F, \cdot)$ having an absorbing state h. Let $P = Q \setminus \{h\}$. Each letter of Σ induces an injective partial map on P. Completing these partial maps to bijections in an arbitrary way, we obtain a bijective automaton $\mathcal{B} = (Q, \Sigma, \cdot_\mathcal{B})$. Let L_0 be the language recognized by the automaton $\mathcal{A}_0 = (Q, \Sigma, q_0, F \setminus \{h\}, \cdot_\mathcal{B})$ and, for each letter $\sigma \in \Sigma$, let L_σ be the language recognized by the automaton $\mathcal{A}_\sigma = (Q, \Sigma, q_0, F_\sigma, \cdot_\mathcal{B})$, where $F_\sigma = \{q \in P \mid q \cdot \sigma = h\}$. If L is the language recognized by the IFA-A represented in Figure 3, the three automata $\mathcal{A}_0$, $\mathcal{A}_\sigma$ and $\mathcal{A}_\tau$ are pictured in Figure 4. Then by construction, $L = L_0 \cup \bigcup_{\sigma \in \Sigma^*} L_\sigma \sigma \Sigma^*$.   □

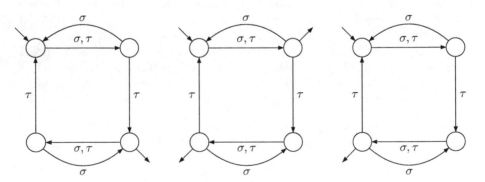

Fig. 4. The automata $\mathcal{A}_0$, $\mathcal{A}_\sigma$ and $\mathcal{A}_\tau$, respectively

Corollary 4.3. *A language of Σ^* is recognized by an IFA-R if and only if it can be written as $L_0 \cap \left(\bigcap_{\sigma \in \Sigma} (L_\sigma \sigma \Sigma^*)^c \right)$, where L_0 and the L_σ are group languages.*

So the class of languages recognized by IFA is characterized by Theorem 4.2 and Corollary 4.3.

By Theorem 4.1, $\mathbf{L}$ ($\mathbf{L^c}$, respectively) is closed under finite union (finite intersection), inverse morphisms and word quotients. Nevertheless, one can show that $\mathbf{L}$ ($\mathbf{L^c}$, respectively) does not form a disjunctive (conjunctive) variety in the sense of Polák [21], since it is not closed under inverse free semiring morphisms $\psi^{(-1)}$ ($\psi^{[-1]}$) defined there.

Consider the closure of $\mathbf{L}$ under finite intersection. The resulting class of languages is a positive variety of languages. By [20, Theorem 4.4], the corresponding variety of ordered monoids is $\mathbf{J_1^+} * \mathbf{G} = \mathbf{ECom^+}$. Combining this result with the description of the languages of $\mathbf{L}$ given by Theorem 4.2, we obtain the following result:

Proposition 4.4. *Let Z be a language of Σ^*. The following conditions are equivalent:*

(1) *Z belongs to the closure of $\mathbf{L}$ under finite intersection,*

(2) *Z is a positive boolean combination of languages of the form L or $L\sigma\Sigma^*$, where L is a group language,*

(3) *The syntactic ordered monoid of Z belongs to the variety $\mathbf{ECom^+}$.*

Similarly, the closure of $\mathbf{L^c}$ under finite union is exactly the class of languages recognized by Pin's reversible automata and the corresponding variety of ordered monoids is $\mathbf{ECom^-} = [\![x^\omega y^\omega = y^\omega x^\omega, \ x^\omega \geq 1]\!]$ [16,17].

Finally, by [12], the closure of $\mathbf{L}$ or $\mathbf{L^c}$ under boolean operations corresponds to the monoid variety $\mathbf{ECom}$, defined by the identity $x^\omega y^\omega = y^\omega x^\omega$.

5   Reversible Finite Automata

The class of languages recognized by CRFA (which, by Proposition 2.3, is also the class of languages recognized by RFA) will be denoted by $\mathbf{K}$.

In this section give a necessary condition for membership in $\mathbf{K}$, as well as an algebraic characterization of the boolean closure $\overline{\mathbf{K}}$ of this class of languages.

Theorem 5.1. *Any language of Σ^* recognized by a CRFA can be written as $K_0 \cup K_1\sigma_1\Sigma^* \cup \cdots \cup K_k\sigma_k\Sigma^*$, where $K_0, \ldots, K_k \in \mathbf{L^c}$ and $\sigma_1, \ldots, \sigma_k$ are letters.*

Proof. Consider a language Z recognized by a CRFA $\mathcal{A} = (Q, \Sigma, q_0, F, \cdot)$. If $\mathcal{A}$ has less than two absorbing states, the result follows from Theorem 4.2. Hence assume that $\mathcal{A}$ has two absorbing states: a non-final state g and a final state h. Let $J = Q \setminus \{h\}$. We first decompose Z as the union of two languages K_0 and Z_1. The language K_0 is recognized by the automaton $\mathcal{A}_0 = (J, \Sigma, q_0, F \setminus \{h\}, \cdot')$, where

$$q \cdot' \sigma = \begin{cases} q \cdot \sigma & \text{if } q \cdot \sigma \in J, \\ g & \text{otherwise.} \end{cases}$$

Then $\mathcal{A}_0$ is an IFA-R and thus $K \in \mathbf{L^c}$. The language Z_1 is recognized by the automaton $\mathcal{A}_1 = (Q, \Sigma, q_0, \{h\}, \cdot)$. For each transition in

$$T = \{(q, \sigma) \in J \times \Sigma \mid q \cdot \sigma = h\}$$

create an automaton $\mathcal{A}_{q,\sigma} = (Q, \Sigma, q_0, \{h\}, \cdot_{q,\sigma})$, where

$$p \cdot_{q,\sigma} \tau = \begin{cases} p \cdot \tau & \text{if } (p, \tau) \notin T \text{ or } (p, \tau) = (q, \sigma) \\ g & \text{otherwise.} \end{cases}$$

Denoting by $Z_{(q,\sigma)}$ the language recognized by $\mathcal{A}_{(q,\sigma)}$, we obtain $Z = \bigcup_{(q,\sigma) \in T} Z_{(q,\sigma)}$. Further, $Z_{(q,\sigma)} = K_{q,\sigma} \sigma \Sigma^*$, where $K_{q,\sigma}$ is the language in $\mathbf{L^c}$ that is recognized by the automaton $(J, \Sigma, q_0, \{q\}, \cdot'_{q,\sigma})$, where $\cdot'_{q,\sigma}$ is the restriction of $\cdot_{q,\sigma}$ to J, completed by the transition $q \cdot'_{q,\sigma} \sigma = g$. Hence $Z = K_0 \cup \left(\bigcup_{(q,\sigma) \in T} K_{q,\sigma} \sigma \Sigma^* \right)$. $\qquad \square$

Note that given a language $K \subseteq \Sigma^*$ of $\mathbf{L^c}$ and $\sigma \in \Sigma$, the language $K\sigma\Sigma^*$ is recognized by a CRFA.

Theorem 5.2. *The class $\mathbf{K}$ is closed under complement, inverse of morphisms between free monoids and word quotients.*

Corollary 5.3. *If a language of Σ^* is recognized by a CRFA, then it can be written as $K_0^c \cap (K_1 \sigma_1 \Sigma^*)^c \cap \cdots \cap (K_k \sigma_k \Sigma^*)^c$, where $k \geq 0$, $K_0, \ldots, K_k \in \mathbf{L^c}$ and $\sigma_1, \ldots, \sigma_k \in \Sigma$.*

Since $\mathbf{K}$ is closed under complement, its closure under positive boolean operations (finite unions and intersections) is equal to its boolean closure $\overline{\mathbf{K}}$.

Theorem 5.4. *A language belongs to $\overline{\mathbf{K}}$ if and only if its syntactic ordered monoid belongs to $\mathbf{J_1^+} * (\mathbf{J_1^-} * \mathbf{G})$.*

Proof. Let L be a regular language and let $M(L)$ be its syntactic ordered monoid. If $L \in \overline{\mathbf{K}}$, then it is by Theorem 5.1 a positive boolean combination of languages of the form K or $K\sigma\Sigma^*$, where $K \in \mathbf{L^c}$. Thus by the [16,17], $M(K) \in \mathbf{ECom^-} = \mathbf{J_1^-} * \mathbf{G}$. Therefore by [20, Theorem 4.4], $M(L) \in \mathbf{J_1^+} * (\mathbf{J_1^-} * \mathbf{G})$.

Suppose now that $M(L) \in \mathbf{J_1^+} * (\mathbf{J_1^-} * \mathbf{G})$. Then by [20, Theorem 4.4], L is a positive boolean combination of languages of the form Z or $Z\sigma\Sigma^*$, where $M(Z) \in \mathbf{J_1^-} * \mathbf{G}$. Further, Z is a positive boolean combination of languages of the form Y_i and $(Y_j \sigma \Sigma^*)^c$, where Y_i, Y_j are group languages. So $Z = \bigcup_i K_i$, where $K_i \in \mathbf{L^c}$. Now $Z\sigma\Sigma^* = (\bigcup_i K_i)\sigma\Sigma^* = \bigcup_i (K_i \sigma \Sigma^*)$. Hence $L \in \overline{\mathbf{K}}$. $\qquad \square$

By associativity, $\mathbf{J_1^+} * (\mathbf{J_1^-} * \mathbf{G}) = (\mathbf{J_1^+} * \mathbf{J_1^-}) * \mathbf{G}$, hence it is of interest to describe the variety $\mathbf{J_1^+} * \mathbf{J_1^-}$. Due to the lack of space, we omit the proof of this semigroup theoretic result.

Theorem 5.5. *The following equality holds: $\mathbf{J_1^+} * \mathbf{J_1^-} = \mathbf{J_1^-} * \mathbf{J_1^+} = \mathbf{R_1}$.*

The variety of monoids $\mathbf{R_1}$ is defined by the identity $xyx = xy$. Hence by [2], Corollary 4.3 and [1], p. 276, $\mathbf{R_1} * \mathbf{G} = [\![x^\omega y^\omega x^\omega = x^\omega y^\omega]\!]$.

The facts exposed above yield the following theorem, which essentially says that the languages recognized by RFA generate the variety $\mathbf{R_1} * \mathbf{G}$:

Theorem 5.6. *A language is in* **K** *if and only if its syntactic monoid belongs to the variety* $\mathbf{R}_1 * \mathbf{G} = [\![x^\omega y^\omega x^\omega = x^\omega y^\omega]\!]$.

6   Algebraic Conditions

Let us note that Ambainis and Freivalds have proved ([5], theorems 2 and 3) the following characterization for the class of languages recognized by RFA:

Theorem 6.1. [5] *Let $\mathcal{A}$ be the minimal automaton of a regular language L. Then L is recognized by a reversible finite automaton if and only if for any states q_1, q_2, q_3 of $\mathcal{A}$, $q_1 \neq q_2$, $q_2 \neq q_3$, and for any input words x, y, $\mathcal{A}$ does not contain the following configuration: $q_1 \cdot x = q_2$, $q_2 \cdot x = q_2$, $q_2 \cdot y = q_3$.*

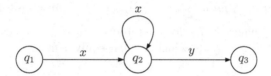

Fig. 5. The forbidden configuration in a RFA

The Ambainis-Freivalds condition can be translated into an algebraic condition. Let L a regular language of Σ^*. We denote by $M(L)$ its syntactic monoid, by $\varphi : \Sigma^* \longrightarrow M(L)$ its syntactic morphism and by $P = \varphi(L)$ the syntactic image of L. Let $\sim_r$ be the right congruence on $M(L)$ defined by $s \sim_r t$ if and only if, for all $u \in M(L)$, $su \in P$ is equivalent to $tu \in P$.

Corollary 6.2. *L is recognized by a reversible finite automaton if and only if for all $s, t, u \in M(L)$, $st^\omega \sim_r s$ or $st^\omega u \sim_r st^\omega$.*

Proof. Consider the minimal automaton $(Q, \Sigma, q_0, F, \cdot)$ of a language L. Due to Ambainis-Freivalds condition, a language is recognized by a reversible finite automaton if and only if for all $q_1, q_2, q_3 \in Q$ and $x, y \in \Sigma^*$,
$$q_1 \cdot x = q_2,\ q_2 \cdot x = q_2 \text{ and } q_2 \cdot y = q_3 \text{ imply } q_1 = q_2 \text{ or } q_2 = q_3$$
or, equivalently, for all $q \in Q$, for all $x, y \in \Sigma^*$,
$$q \cdot x = q \cdot x^2 \text{ implies } q = q \cdot x \text{ or } q \cdot x = q \cdot xy.$$
Now, choose $v \in \Sigma^*$ such that $q = q_0 \cdot v$ and let $s = \varphi(v)$ and $t = \varphi(x)$. We claim that the condition $q \cdot x = q \cdot x^2$ is equivalent to $st \sim_r st^2$. Indeed, by the definition of the Nerode equivalence, the first condition means that, for every $y \in \Sigma^*$, $q_0 \cdot vxy \in F$ if and only if $q_0 \cdot vx^2y \in F$, or, equivalently, for all $u \in M(L)$, $stu \in P$ if and only if $st^2u \in P$.

Therefore, Formula (6) can be rewritten as follows: for all $s, t, u \in M(L)$,
$$st \sim_r st^2 \text{ implies } s \sim_r st \text{ or } st \sim_r stu,$$
which is in turn equivalent to: for all $s, t, u \in M(L)$,
$$s \sim_r st^\omega \text{ or } st^\omega \sim_r st^\omega u. \qquad \square$$

Consider an injective automaton $\mathcal{A}$, which is not a group automaton, i.e., has one absorbing state. We assume that $\mathcal{A}$ is accessible. Then for any state q and any word w, exists $k > 0$ such that $q \cdot w^k = q$ or $q \cdot w^k = h$, where h is the absorbing state. Therefore we deduce that the absorbing state is accessible from any state. So the transition monoid $M(\mathcal{A})$ has a zero element ([15, Exercise 2.7]). Since $M(L)$ divides $M(\mathcal{A})$, $M(L)$ also has a zero element. One can view the syntactic monoid $M(L)$ as an automaton $(M(L), \Sigma, 1, P, \cdot)$, which recognizes L. Any of its states is accessible from the initial state 1. The right equivalence class containing 0 corresponds to the absorbing state in the minimal automaton of L. All the absorbing states of $M(L)$ are in this class. Hence if for every u $st^\omega \sim_r st^\omega u$, then $st^\omega \sim_r 0$. So in the case of DBPA, Corollary 6.2 may be rewritten as follows:

Corollary 6.3. *A language L is recognized by a deterministic Brodsky-Pippenger automaton if and only if, for all $s, t \in M(L)$, $st^\omega \sim_r s$ or $st^\omega \sim_r 0$.*

If L is a group language, $M(L)$ does not have a zero, so this condition reduces to: for all $s, t \in M(L)$, $st^\omega \sim_r s$, which is turn equivalent to $t^\omega = 1$.

References

1. J. Almeida. Finite Semigroups and Universal Algebra. *World Scientific*, Singapore, 1994.
2. J. Almeida, J.E. Pin, P. Weil. Semigroups whose Idempotents Form a Subsemigroup. *Math. Proc. Camb. Phil. Soc.*, Vol. 111, pp. 241-253, 1992.
3. A. Ambainis, M. Beaudry, M. Golovkins, A. Ķikusts, M. Mercer, D. Thérien. Algebraic Results on Quantum Automata. *STACS 2004, LNCS*, Vol. 2996, pp. 93-104, 2004.
4. A. Ambainis, R.F. Bonner, R. Freivalds, A. Ķikusts. Probabilities to Accept Languages by Quantum Finite Automata. *COCOON 1999, LNCS*, Vol. 1627, pp. 174-183, 1999.
5. A. Ambainis, R. Freivalds. 1-Way Quantum Finite Automata: Strengths, Weaknesses and Generalizations. *Proc. 39th FOCS*, pp. 332-341, 1998.
6. A. Ambainis, A. Nayak, A. Ta-Shma, U. Vazirani. Dense Quantum Coding and Quantum Finite Automata. *Journal of the ACM*, Vol. 49(4), pp. 496-511, 2002.
7. C.J. Ash. Finite Semigroups with Commuting Idempotents. *J. Austral. Math. Soc. (Series A)*, Vol. 43, pp. 81-90, 1987.
8. A. Bertoni, C. Mereghetti, B. Palano. Quantum Computing: 1-Way Quantum Finite Automata. *DLT 2003, LNCS*, Vol. 2710, pp. 1-20, 2003.
9. A. Brodsky, N. Pippenger. Characterizations of 1-Way Quantum Finite Automata. *SIAM Journal on Computing*, Vol. 31(5), pp. 1456-1478, 2002.
10. M. Golovkins, M. Kravtsev. Probabilistic Reversible Automata and Quantum Automata. *COCOON 2002, LNCS*, Vol. 2387, pp. 574-583, 2002.
11. A. Kondacs, J. Watrous. On The Power of Quantum Finite State Automata. *Proc. 38th FOCS*, pp. 66-75, 1997.
12. S.W. Margolis, J.E. Pin. Inverse Semigroups and Varieties of Finite Semigroups. *Journal of Algebra*, Vol. 110, pp. 306-323, 1987.
13. C. Moore, J.P. Crutchfield. Quantum Automata and Quantum Grammars. *Theoretical Computer Science*, Vol. 237(1-2), pp. 275-306, 2000.

14. A. Nayak. Optimal Lower Bounds for Quantum Automata and Random Access Codes. *Proc. 40th FOCS*, pp. 369-377, 1999.
15. J.E. Pin. Varieties of Formal Langages, North Oxford, London and Plenum, New-York, 1986.
16. J.E. Pin. On the Languages Accepted by Finite Reversible Automata. *ICALP 1987, LNCS*, Vol. 267, pp. 237-249, 1987.
17. J.E. Pin. On Reversible Automata. *LATIN 1992, LNCS*, Vol. 583, pp. 401-416, 1992.
18. J.E. Pin. Eilenberg's Theorem for Positive Varieties of Languages. *Russian Mathematics (Iz. VUZ)*, Vol. 39(1), pp. 80-90, 1995.
19. J.E. Pin, P. Weil. Semidirect Products of Ordered Semigroups. *Communications in Algebra*, Vol. 30(1), pp. 149-169, 2002.
20. J.E. Pin, P. Weil. The Wreath Product Principle for Ordered Semigroups. *Communications in Algebra*, Vol. 30(12), pp. 5677-5713, 2002.
21. L. Polák. Syntactic Semiring of a Language. *MFCS 2001, LNCS*, Vol. 2136, pp. 611-620, 2001.
22. G.B. Preston. Inverse Semi-groups with Minimal Right Ideals. *J. London Math. Soc.*, Vol. 29, pp. 404-411, 1954.
23. V.V. Vagner. Generalized Groups. *Dokl. Akad. Nauk SSSR*, Vol. 84(6), pp. 1119-1122, 1952.

Iterated TGR Languages: Membership Problem and Effective Closure Properties

(Extended Abstract)

Ian McQuillan[1], Kai Salomaa[2], and Mark Daley[3]

[1] Department of Computer Science, University of Saskatchewan, Saskatoon, Saskatchewan, Canada S7N 5A9
mcquillan@cs.usask.ca
[2] School of Computing, Queen's University, Kingston, Ontario, Canada K7L 3N6
ksalomaa@cs.queensu.ca
[3] Department of Computer Science and Department of Biology, University of Western Ontario, London, Ontario, Canada N6A 5B7
daley@csd.uwo.ca

Abstract. We show that membership is decidable for languages defined by iterated template-guided recombination systems when the set of templates is regular and the initial language is context-free. Using this result we show that when the set of templates is regular and the initial language is context-free (respectively, regular) we can effectively construct a pushdown automaton (respectively, finite automaton) for the corresponding iterated template-guided recombination language.

1 Introduction

The spirotrichous ciliates are a type of unicellular protozoa which possess a unique and fascinating genetic behaviour. Each ciliate cell contains two types of nuclei, *macronuclei* which are responsible for the day-to-day "genetic housekeeping" of the cell, and *micronuclei* which are functionally inert, but used in reproduction. This is in contrast to, e.g., mammalian cells which have only one micronucleus. Although they reproduce asexually, ciliates are also capable of sexual activity in which they exchange haploid micronuclear genomes. This results in each ciliate getting a "genetic facelift" by combining its own genes with those of a mate. After creating a new, hybrid, micronucelus, each ciliate will then regenerate its macronucleus. It is this process of macronuclear regeneration that is of principle interest to us here.

In the spirotrichous ciliates in particular, this macronuclear regeneration involves an intricate process of genetic gymnastics. Suppose that a functional gene in the macronucleus can be divided into 5 sections and written as follows: 1-2-3-4-5. In many cases, the micronuclear form of the same gene may have the segments in a completely different order and include additional segments not found in the macronucleus. For the example given above, a micronuclear gene

D.Z. Chen and D.T. Lee (Eds.): COCOON 2006, LNCS 4112, pp. 94–103, 2006.

may appear as: 3-x-5-y-1-z-4-2. For the ciliate to produce a functional macronucleus and continue living, it must *descramble* these micronuclear genes. (See, e.g., [10] for further detail).

A biological model for this descrambling process, based on template-guided DNA recombination, was proposed in [11]. This model was formalized as an operation on words and languages in [3] which also introduced the notion of a template-guided recombination system (TGR system). It was then shown in [4] that a TGR system with a regular set of templates preserves regularity, that is, for a regular initial language, the language resulting from iterated application of the TGR system is always regular. This is in striking contrast to splicing systems since the splicing language generated by a regular set of rules and a finite initial language need not be recursive [8]. In fact, [4] shows much more generally that the operation defined by a TGR system with a regular set of templates preserves any language family that is a full AFL [7,12].

However, the above results are non-constructive and, in particular, do not give an algorithm to decide the membership problem for the language defined by a TGR system, even in the case where the initial language is finite and the set of templates is regular. Here we show that the uniform membership problem for the language defined by a TGR system is decidable when the initial language is context-free and the set of templates is regular. The nonuniform membership problem (where the TGR system is fixed) can be decided in polynomial time. The decidability result is extended for languages that are extensions of the context-free languages, such as the indexed languages, or, more generally, for languages that belong to a full AFL satisfying certain natural effectiveness conditions.

Moreover, we use this result to positively solve the main open problem from [4]. That is, given a context-free (respectively, regular) initial language and a regular set of templates, we can effectively construct a pushdown automaton (respectively, a finite automaton) for the language defined by the TGR system. Using a variant of the decision algorithm for the membership problem, we effectively find a deterministic finite automaton (DFA) for the subset of templates that can be used in some recombination operation and this, together with the results of [4], enables us to construct the pushdown automaton (respectively, the finite automaton) for the language defined by the TGR system. This result also holds for regular sets of templates and initial languages from an arbitrary full AFL that satisfies certain effectiveness conditions.

Both the algorithm for the membership problem and the method for finding the set of useful templates use expensive brute-force techniques. It remains an open question, whether it is possible to find a more efficient algorithm, at least in the case where both the initial language and the set of templates are regular.

2 Preliminaries

Here we recall some basic definitions needed in the next section. For all unexplained notions related to formal languages we refer the reader e.g. to [12]. Recent work on language classes and bio-operations can be found e.g. in [2].

In the following Σ is a finite alphabet and the set of all words over Σ is Σ^*. The length of a word $w \in \Sigma^*$ is $|w|$. The ith symbol of a word $w \in \Sigma^*$ is denoted $w[i]$, $i = 1, \ldots |w|$. A language is a subset of Σ^*. The sets of all prefixes, all suffixes and all subwords of words in L are denoted, respectively, $\mathrm{pref}(L)$, $\mathrm{suf}(L)$, $\mathrm{subw}(L)$.

A family of languages is said to be a *full abstract family of languages* (full AFL) [7,12] if it contains a nonempty language and is closed under the following operations: union, Kleene plus, homomorphism, inverse homomorphism, and intersection with regular languages.

Definition 2.1. [3,4] *A template-guided recombination system (TGR system) is a tuple $\varrho = (T, \Sigma, n_1, n_2)$, where Σ is a finite alphabet, $T \subseteq \Sigma^*$ is the template language, and $n_1, n_2 \in \mathbb{N}$.*

Let $x, y \in \Sigma^*$ and $t \in T$. The recombination operation defined by ϱ is given by: $(x, y) \vdash_t^\varrho w$ if and only if we can write

$$x = u\alpha\beta d, \quad y = e\beta\gamma v, \quad t = \alpha\beta\gamma \quad and \quad w = u\alpha\beta\gamma v$$

for some $u, v, d, e \in \Sigma^*$, $\alpha, \gamma \in \Sigma^{\geq n_1}$ and $\beta \in \Sigma^{n_2}$. For $L \subseteq \Sigma^*$ we define $\varrho(L) = \{w \in \Sigma^* \mid (x, y) \vdash_t^\varrho w \text{ for some } x, y \in L, t \in T\}$.

Let $\varrho = (T, \Sigma, n_1, n_2)$ be a TGR system and let $L \subseteq \Sigma^*$. We define the iteration $\varrho^{(*)}$ of the operation ϱ by setting $\varrho^{(0)}(L) = L$, and defining

$$\varrho^{(i+1)}(L) = \varrho^{(i)}(L) \cup \varrho(\varrho^{(i)}(L)) \text{ for all } i \geq 0. \tag{1}$$

Denote $\varrho^{(*)}(L) = \bigcup_{i=0}^{\infty} \varrho^{(i)}(L)$.

Let $\varrho = (T, \Sigma, n_1, n_2)$ be a TGR system and let $L \subseteq \Sigma^*$. A word $t \in T$ is said to be *useful* on (L, ϱ) if t can be used in iterated application of ϱ on the initial language L. It is shown in [4] that $t \in T$ is useful on (L, ϱ) if and only if $|t| \geq 2n_1 + n_2$ and t is a subword of some word in $\varrho^{(*)}(L)$. The TGR system ϱ is said to be *useful on L* if every word of T is useful on (L, ϱ). The *useful subset of ϱ on L* is the set of all words in T which are useful on (L, ϱ).

3   Membership Problem

Here we show that for a context-free language L and a TGR system $\varrho = (T, \Sigma, n_1, n_2)$ where T is regular, the uniform membership problem for the language $\varrho^{(*)}(L)$ is decidable.

We want to establish properties concerning how many recombination operations are required to produce some subword of a word w when it is known that w requires a given number of recombination operations. For this purpose it turns out to be useful to consider "marked variants" of words over Σ. The marked variants associate states of a DFA recognizing the set of templates T and length information with certain positions in the word. This additional control information is used to keep track of the templates (or strictly speaking equivalence classes of templates) that can be used in the recombination operations.

For the above purpose we next introduce some technical notation. Let $\varrho = (T, \Sigma, n_1, n_2)$ be a TGR system and

$$A = (\Sigma, Q, q_0, F, \delta) \tag{2}$$

be a DFA that recognizes T. Denote $\overrightarrow{Q} = \{\overrightarrow{q} \mid q \in Q\}$, $\overleftarrow{Q} = \{\overleftarrow{q} \mid q \in Q\}$. For $n \in \mathbb{N}$ let $[n] = \{0, 1, \ldots, n\}$. We define the extended alphabet $\Sigma[\varrho]$ as

$$\Sigma[\varrho] = \Sigma \times \mathcal{P}((\overrightarrow{Q} \cup \overleftarrow{Q}) \times [n_1]). \tag{3}$$

The first component of elements of $\Sigma[\varrho]$ is an element of Σ and the second component consists of a set of states of Q each marked with a "right arrow" or a "left arrow". Additionally, each state is associated with an index from $\{0, 1, \ldots, n_1\}$.

The projections from $\Sigma[\varrho]$ to Σ and to $\mathcal{P}((\overrightarrow{Q} \cup \overleftarrow{Q}) \times [n_1])$ are denoted, respectively, π_1^ϱ and π_2^ϱ. When ϱ is clear from the context, we denote the projections simply as π_1 and π_2. The projection π_1 is in the natural way extended to a morphism $\Sigma[\varrho]^* \longrightarrow \Sigma^*$.

Let $L \subseteq \Sigma^*$. The T-controlled marked variant of L is the largest language $C_T(L) \subseteq \Sigma[\varrho]^*$ such that the below conditions (i) and (ii) hold[1]. The notations refer to (2) that gives a DFA for the language T.

(i) For every $w \in C_T(L)$, $\pi_1(w) \in L$.

(ii) Assume that $w \in C_T(L)$ and $(p, j) \in \pi_2(w[i])$, $1 \leq i \leq |w|$, $p \in \overrightarrow{Q} \cup \overleftarrow{Q}$, $j \in [n_1]$.

    (a) If $p \in \overrightarrow{Q}$, then $\pi_1(w)$ has a subword u starting at the $(i+1)$th position such that $|u| \geq j$ and $\delta(p, u) \in F$.

    (b) If $p \in \overleftarrow{Q}$, then $\pi_1(w)$ has a subword u ending at the $(i-1)$th position such that $|u| \geq j$ and $\delta(q_0, u) = p$.

Note that for any $w \in L$, the word w' is in $C_T(L)$ where w' is obtained from w by replacing each symbol $c \in \Sigma$ by $(c, \emptyset) \in \Sigma[\varrho]$. We identify words w and w' and in this way we can view L to be a subset of $C_T(L)$.

According to (i) and (ii) above, the elements (p, j), $p \in \overrightarrow{Q} \cup \overleftarrow{Q}$ occurring in symbols of a word $w \in C_T(L)$ place conditions on what kind of subwords w must have starting directly after or ending directly before that position. If $p \in \overrightarrow{Q}$, this means that $\pi_1(w)$ must have a subword u starting from the next position that is a suffix of a word in T, u is of length at least j, and the state p corresponds to this suffix (that is, $\delta(p, u) \in F$). If $p \in \overleftarrow{Q}$, this means that $\pi_1(w)$ must have a subword u ending at the previous position that is a prefix of a word in T, u has length at least j, and the state p corresponds to this prefix.

---

[1] Note that the union of languages satisfying this property also satisfies this property, and so the largest language must exist.

We still need the following notation to manipulate words over the alphabet $\Sigma[\varrho]$. Let $w \in \Sigma[\varrho]^*$, $1 \leq i \leq w$, $p \in (\overrightarrow{Q} \cup \overleftarrow{Q})$ and $j \in [n_1]$. Then $w[i \leftarrow (p,j)]$ denotes the word obtained from w by adding (p,j) to the second component of the ith symbol, that is, the second component of the ith symbol is changed to be $\pi_2(w[i]) \cup \{(p,j)\}$.

We say that a word $w \in \Sigma[\varrho]^*$ is *well formed* if $|w| \geq 2$ and the following three conditions hold: (i) $\pi_2(w[1]) \subseteq \overleftarrow{Q} \times [n_1]$, (ii) $\pi_2(w[|w|]) \subseteq \overrightarrow{Q} \times [n_1]$, and (iii) $\pi_2(w[j]) = \emptyset$ when $1 < j < |w|$.

In a well formed marked word the first symbol contains only elements of the type $(\overleftarrow{p},j)$ as markers, and the last symbol contains only elements of the type $(\overrightarrow{p},j)$ as markers, $p \in Q$, $j \in [n_1]$. Symbols of w other than the first or the last symbol have $\emptyset$ as the second component.

The set of all well formed words over $\Sigma[\varrho]$ is denoted by $\mathcal{WF}(\Sigma[\varrho])$

The following lemma says, very roughly speaking, that if w is a subword of $\varrho^{(k+1)}(L)$ but w is not a subword of $\varrho^{(k)}(L)$, then w has a proper subword that is a subword of $\varrho^{(k)}(L)$ but not a subword of $\varrho^{(k-1)}(L)$. The statement in the previous sentence is oversimplified and does not hold as such. To be precise, in order to be able to establish the required property we need to add to the subwords information on the states of the DFA for T associated with the templates used in the recombination operations, that is, we need to consider subwords of the T-controlled marked variant of $\varrho^{(k)}(L)$, $k \geq 1$.

For $m,n \in \mathbb{N}$ we define the non-negative difference of m and n, $m \ominus n$, as $m - n$ if $m \geq n$ and $m \ominus n = 0$ otherwise.

Lemma 3.1. *Let $\varrho = (T, \Sigma, n_1, n_2)$ where T is regular and let A as in (2) be a DFA that recognizes T. Let $k \geq 1$ and $L \subseteq \Sigma^*$.*

We claim that if $w \in \mathcal{WF}(\Sigma[\varrho])$ and

$$w \in \mathrm{subw}(C_T(\varrho^{(k+1)}(L))) - \mathrm{subw}(C_T(\varrho^{(k)}(L))) \tag{4}$$

then one of the below cases (P1)–(P4) holds:

(P1) $w = u\alpha\beta\gamma v$, $\pi_1(\alpha\beta\gamma) \in T$, $|\beta| = n_2$, $|\alpha|, |\gamma| \geq n_1$, $u\alpha\beta \in$ $\mathrm{subw}(C_T(\varrho^{(k)}(L))) \cap \mathcal{WF}(\Sigma[\varrho])$, $\beta\gamma v \in \mathrm{subw}(C_T(\varrho^{(k)}(L))) \cap \mathcal{WF}(\Sigma[\varrho])$,

(P2) $w = u\alpha\beta\gamma'$, $|\beta| = n_2$, $|\alpha| \geq n_1$, $|\gamma'| \geq 1$, $u\alpha\beta \in$ $\mathrm{subw}(C_T(\varrho^{(k)}(L))) \cap \mathcal{WF}(\Sigma[\varrho])$, $\beta\gamma'[|\beta\gamma'| \leftarrow (\overrightarrow{p}, n_1 \ominus |\gamma'|)] \in$ $\mathrm{subw}(C_T(\varrho^{(k)}(L))) \cap \mathcal{WF}(\Sigma[\varrho])$, *where* $p = \delta(q_0, \alpha\beta\gamma')$.

(P3) $w = \alpha'\beta\gamma v$, $|\beta| = n_2$, $|\gamma| \geq n_1$, $|\alpha'| \geq 1$, $\alpha'\beta[1 \leftarrow (\overleftarrow{p}, n_1 \ominus |\alpha'|)] \in$ $\mathrm{subw}(C_T(\varrho^{(k)}(L))) \cap \mathcal{WF}(\Sigma[\varrho])$, $p \in Q$, $\beta\gamma v \in$ $\mathrm{subw}(C_T(\varrho^{(k)}(L))) \cap \mathcal{WF}(\Sigma[\varrho])$, *where* $\delta(p, \alpha'\beta\gamma) \in F$.

(P4) $w = \alpha'\beta\gamma'$, $|\beta| = n_2$, $|\alpha'|, |\gamma'| \geq 1$, $\alpha'\beta[1 \leftarrow (\overleftarrow{p}, n_1 \ominus |\alpha'|)] \in$ $\mathrm{subw}(C_T(\varrho^{(k)}(L))) \cap \mathcal{WF}(\Sigma[\varrho])$, $p \in Q$, $\beta\gamma'[|\beta\gamma'| \leftarrow (\overrightarrow{p_1}, n_1 \ominus |\gamma'|)] \in$ $\mathrm{subw}(C_T(\varrho^{(k)}(L))) \cap \mathcal{WF}(\Sigma[\varrho])$, *where* $\delta(p, \alpha'\beta\gamma') = p_1$.

Furthermore, in any decomposition of w as in (P1)–(P4) at most one of the two mentioned marked words of $\mathrm{subw}(C_T(\varrho^{(k)}(L)))$ can be in $\mathrm{subw}(C_T(\varrho^{(k-1)}(L)))$.

We should note that in (P2), (P3) and (P4) in Lemma 3.1 it is essential that we add the new marker states to the resulting subwords. For example, using the notations of (P4), it is quite possible that $\pi_1(\alpha'\beta) \in \mathrm{subw}(\varrho^{(k-1)}(L))$ and $\pi_1(\beta\gamma') \in \mathrm{subw}(\varrho^{(k-1)}(L))$ because $\alpha'\beta$ could be part of a word that does not allow recombination using any template of T with the words where $\beta\gamma'$ occurs as a subword. The marked variants of the words prevent this possibility by storing the appropriate states and length information in the first symbol of α' and in the last symbol of γ'. The marker information forces that $\alpha'\beta$ (respectively, $\beta\gamma'$) must occur in a position where the immediately preceding (respectively, immediately following) subword contains a suffix (respectively, a prefix) that allows us to complete $\alpha'\beta\gamma'$ into a template of T.

Due to length restrictions the technical proof of Lemma 3.1 is omitted. We refer the reader to [9] for the proof of Lemma 3.1.

Using Lemma 3.1 we get the following property that will be essential for deciding the membership problem. Also we note that Lemma 3.2 (i) is not a special case of (ii) (although their proofs are similar) and hence we include both statements. The proof of Lemma 3.2 is available in [9].

Lemma 3.2. *Let* $\varrho = (T, \Sigma, n_1, n_2)$ *be a TGR-system where* T *is regular and* $L \subseteq \Sigma^*$.

(i) *If* $w \in \varrho^{(k)}(L) - \varrho^{(k-1)}(L)$, $k \geq 1$, *then* $|w| - n_2 - 1 \geq k$.
(ii) *If* $w \in \mathrm{subw}(\varrho^{(k)}(L)) - \mathrm{subw}(\varrho^{(k-1)}(L))$, *then* $|w| - n_2 - 1 \geq k$.

Theorem 3.1. *Given a TGR system* $\varrho = (T, \Sigma, n_1, n_2)$ *with* T *regular, a context-free language* L *and a word* $w \in \Sigma^*$, *it is decidable whether or not* $w \in \varrho^{(*)}(L)$.

Furthermore, it is decidable whether or not $w \in \mathrm{subw}(\varrho^{(*)}(L))$.

Proof. Let $A = (\Sigma, Q, q_0, F, \delta)$ be a DFA that recognizes T. Given a pushdown automaton B_i for $\varrho^{(i)}(L)$, $i \geq 0$, we can construct a pushdown automaton B_{i+1} for $\varrho^{(i+1)}(L)$ as follows. Let $\beta \in \Sigma^{n_2}$ and $q \in Q$. We define $L_1(B_i, \beta, q) = \{ w \in \mathrm{pref}(L(B_i)) \mid w = u\alpha\beta, |\alpha| \geq n_1, \delta(q_0, \alpha\beta) = q \}$, $L_2(B_i, \beta, q) = \{ w \in \beta^{-1}\mathrm{suf}(L(B_i)) \mid w = \gamma v, |\gamma| \geq n_1, \delta(q, \gamma) \in F \}$. Now it is clear that

$$\varrho^{(i+1)}(L) = \varrho^{(i)}(L) \cup \bigcup_{\beta \in \Sigma^{n_2}, q \in Q} L_1(B_i, \beta, q) \cdot L_2(B_i, \beta, q). \qquad (5)$$

Since context-free languages are effectively closed under prefix, suffix, union, and quotient and intersection with a regular language, using (5) we can construct a pushdown automaton B_{i+1} for $\varrho^{(i+1)}(L)$.

By Lemma 3.2, it is sufficient to construct the pushdown automaton $B_{|w|-n_2-1}$ and decide whether or not $B_{|w|-n_2-1}$ accepts w. The latter can be done effectively since membership is decidable for context-free languages.

Also, context-free languages are effectively closed under subword. Thus, we can test whether $w \in \mathrm{subw}(\varrho^{(|w|-n_2-1)}(L))$ and, by Lemma 3.2 (ii), this holds if and only if $w \in \mathrm{subw}(\varrho^{(*)}(L))$. ∎

The operation (5) uses union indexed over all words of length n_2 and consequently the algorithm given by Theorem 3.1 for the uniform membership problem requires exponential time. However, if ϱ is fixed, i.e., if we consider the non-uniform membership problem then the algorithm given by Theorem 3.1 uses polynomial time. The same is true even if only the value of n_2 is fixed. Note that the number of iterations of (5) is upper bounded by the length of w, i.e., the number of iterations is given in unary notation.

Corollary 3.1. *Let n_2 be fixed. Given a TGR system $\varrho = (T, \Sigma, n_1, n_2)$ with T regular, a context-free language L and a word $w \in \Sigma^*$, it is decidable in polynomial time whether or not $w \in \varrho^{(*)}(L)$.*

Lemma 3.2 does not make any assumptions on the initial language. The proof of Theorem 3.1 uses certain closure and decidability properties of context-free languages. A full AFL satisfies the required conditions, assuming that membership is decidable and closure under the AFL operations is effective, and a corresponding extended result is stated below in Corollary 3.2. Before that we introduce some terminology dealing with AFL's consisting of recursive languages. The terminology will be useful also in the next section in order to be able to rely in a uniform way on results from [4] that are formulated in terms of AFL's.

Definition 3.1. *We say that a property P of Turing machines is* syntactic *if given a Turing machine M it is decidable whether or not M has property P. The class of Turing machines satisfying a property P is denoted $\mathrm{TM}[P]$.*

A language family $\mathcal{L}$ is said to be a constructive full AFL *if $\mathcal{L}$ contains a nonempty language and there exists a syntactic property of Turing machines $P_{\mathcal{L}}$ such that*

(i) *a language L is in $\mathcal{L}$ if and only if L is recognized by some Turing machine in $\mathrm{TM}[P_{\mathcal{L}}]$,*

(ii) *given $M \in \mathrm{TM}[P_{\mathcal{L}}]$ and an input word w, it is decidable whether or not $w \in L(M)$, and*

(iii) *languages recognized by machines in $\mathrm{TM}[P_{\mathcal{L}}]$ are effectively closed under the AFL operations. That is, there is an algorithm that for given $M_1, M_2 \in \mathrm{TM}[P_{\mathcal{L}}]$ constructs $M_{\mathrm{union}} \in \mathrm{TM}[P_{\mathcal{L}}]$ such that $L(M_{\mathrm{union}}) = L(M_1) \cup L(M_2)$, and for any AFL operation σ other than union there is an algorithm to construct $M \in \mathrm{TM}[P_{\mathcal{L}}]$ such that $L(M) = \sigma(L(M_1))$.*

Well known examples of constructive full AFL's are the regular and the context-free languages. An example of a more general constructive full AFL is the family of languages recognized by (one-way, single head) k-iterated pushdown automata, $k \geq 1$, [6]. It is easy to verify that any (k-iterated) pushdown automaton can be simulated by a Turing machine where the transition relation satisfies a suitably defined syntactic property that forces the work tape to simulate a (k-iterated) pushdown store. It seems that any full AFL consisting only of recursive languages that is defined by a "reasonable" machine model could be characterized in the above way. The family of recursively enumerable languages is a full AFL that is not a constructive full AFL.

Corollary 3.2. *Let $\mathcal{L}$ be a constructive full AFL. Given a TGR system $\varrho = (T, \Sigma, n_1, n_2)$ where T is regular and $L \in \mathcal{L}$, the membership problem for $\varrho^{(*)}(L)$ is decidable.*

The set of useful templates of a TGR system $\varrho = (T, \Sigma, n_1, n_2)$ with an initial language L is the set $T \cap \mathrm{subw}(\varrho^{(*)}(L)) \cap \Sigma^{\geq 2n_1 + n_2}$ [4]. Thus by Theorem 3.1:

Corollary 3.3. *Given a TGR system $\varrho = (T, \Sigma, n_1, n_2)$ where T is regular and a context-free initial language L, we can effectively decide whether or not a given template is useful on (L, ϱ).*

Corollary 3.4. *Let $\mathcal{L}$ be a constructive full AFL. Given a TGR system $\varrho = (T, \Sigma, n_1, n_2)$ where T is regular and $L \in \mathcal{L}$, we can effectively decide whether or not a given template is useful on (L, ϱ).*

To conclude this section we make a couple of remarks on limitations in attempting to extend the previous results. The 2-iterated pushdown automata recognize the indexed languages [1] and, thus, from Corollary 3.2 we get a decidability result for the membership problem when the initial language is an indexed language. However, there is no known polynomial time parsing algorithm for general indexed languages and Corollary 3.1 cannot be extended for the case where the initial language is indexed.

4   Effective Closure Properties

We would now like to attack the question of, given $\varrho = (T, \Sigma, n_1, n_2)$, with T regular, and L recognized by a pushdown automaton (respectively, a finite automaton), can we effectively construct a pushdown automaton (respectively, a finite automaton) which recognizes $\varrho^{(*)}(L)$? Note that in the former case it is known that $\varrho^{(*)}(L)$ is context-free (and in the latter case regular) [4] but the results are non-constructive.

We first need to provide some details from [4]. The main non-constructive proof from this paper shows that, for an arbitrary TGR system $\varrho = (T, \Sigma, n_1, n_2)$ with T regular, and an arbitrary full AFL $\mathcal{L}$ the following holds: If $L \in \mathcal{L}$, then $\varrho^{(*)}(L) \in \mathcal{L}$. The proof of this result relies on two auxiliary results, the first one of which is the following:

Proposition 4.1. *(Theorem 4.2 of [4]) Let $\varrho = (T, \Sigma, n_1, n_2)$ be a TGR system and let $L \subseteq \Sigma^*$. Let T_u be the useful subset of ϱ on L. If T is a regular language, then T_u is also regular.*

The proof of the above result [4] is not constructive, even in the case where we have some effective representation for L. However, the proof does give some information as to the structure of the DFA which accepts T_u. If Q is the state set of a DFA which accepts T, then the proof creates a finite set of automata $\mathcal{X}_{T,L}$, each automaton with a state set of size $q_{T,L} = (|Q| + 1)^n \cdot (|\Sigma| + 1)^{n-1}$ where $n = 2n_1 + n_2 - 1$. Moreover, the proof establishes that one of these automata accepts T_u, but does not tell us which one is the correct automaton.

Indeed, let $\varrho = (T, \Sigma, n_1, n_2)$ be a TGR system where T is regular and let $\mathcal{L}$ be a constructive full AFL, and let $L \in \mathcal{L}$. Then, by Corollary 3.4, we can decide whether or not a given template is useful on (L, ϱ). Consider $T_u \cap \Sigma^{\leq 2 \cdot q_{T,L}}$, the finite set of all words which are useful on (L, ϱ) and which are of length less than or equal to $2 \cdot q_{T,L}$. Using Corollary 3.4 we can now effectively determine this set. In addition, for each automaton $M = (Q, \Sigma, q_0, F, \delta) \in \mathcal{X}_{T,L}$, we can check whether or not $T_u \cap \Sigma^{\leq 2 \cdot q_{T,L}} = L(M) \cap \Sigma^{\leq 2 \cdot q_{T,L}}$.

Claim. $T_u \cap \Sigma^{\leq 2 \cdot q_{T,L}} = L(M) \cap \Sigma^{\leq 2 \cdot q_{T,L}}$ if and only if $T_u = L(M)$.

Proof of the claim. It is sufficient to show the implication from left to right. According to Proposition 6.3 of [5], the following is true: Let M_1, M_2 be two DFAs with state sets Q_1, Q_2 respectively. Then $L(M_1) = L(M_2)$ whenever for all $s \in \Sigma^*$ such that $|s| < |Q_1| + |Q_2|$ we have $s \in L(M_1)$ if and only if $s \in L(M_2)$.

Assume by contradiction that $T_u \neq L(M)$. But there exists $M' \in \mathcal{X}_{T,L}$ (also with a state set of size $q_{T,L}$) such that $L(M') = T_u$, and hence

$$L(M') \cap \Sigma^{\leq 2 q_{T,L}} = T_u \cap \Sigma^{\leq 2 q_{T,L}} = L(M) \cap \Sigma^{\leq 2 q_{T,L}}.$$

However, according to the proposition from [5], this implies $L(M') = L(M)$, a contradiction. This concludes the proof of the claim.

By the above claim, we can find from $\mathcal{X}_{T,L}$ the correct automaton which accepts T_u. Hence, we can effectively construct a deterministic finite automaton which accepts T_u. Thus we have shown that the following holds:

Lemma 4.1. *Let $\mathcal{L}$ be a constructive full AFL. Given $\varrho = (T, \Sigma, n_1, n_2)$ with T regular and $L \in \mathcal{L}$, we can construct a DFA for the useful subset of ϱ on L.*

Corollary 4.1. *Let $\mathcal{L}$ be a constructive full AFL. Given $\varrho = (T, \Sigma, n_1, n_2)$ with T regular and $L \in \mathcal{L}$, we can effectively find a regular set of templates T_1 such that if $\varrho_1 = (T_1, \Sigma, n_1, n_2)$ then $\varrho_1^{(*)}(L) = \varrho^{(*)}(L)$ and ϱ_1 is useful on L.*

The second result from [4] that turns out to be useful is the following:

Proposition 4.2. *(Theorem 4.1 of [4]) If $\mathcal{L}$ is a full AFL, $\varrho = (T, \Sigma, n_1, n_2)$ is a TGR system and $L, T \in \mathcal{L}$, $L \subseteq \Sigma^*$, are such that ϱ is useful on L, then $\varrho^{(*)}(L) \in \mathcal{L}$.*

The proof of Proposition 4.2 in [4] establishes that $\varrho^{(*)}(L)$ is in $\mathcal{L}$ by showing that $\varrho^{(*)}(L)$ is obtained from L using a finite number of operations that can be expressed as compositions of AFL operations. This gives the following:

Corollary 4.2. *Let $\mathcal{L}$ be a constructive full AFL. Given a TGR system $\varrho = (T, \Sigma, n_1, n_2)$ where $T \in \mathcal{L}$, an intial language $L \in \mathcal{L}$, $L \subseteq \Sigma^*$, such that ϱ is useful on L, we can effectively construct (a Turing machine in $\mathrm{TM}[P_\mathcal{L}]$ for) $\varrho^{(*)}(L) \in \mathcal{L}$.*

Now we are ready to prove the main result of this section.

Theorem 4.1. *Let $\mathcal{L}$ be a constructive full AFL. Given $L \in \mathcal{L}$ and a TGR system $\varrho = (T, \Sigma, n_1, n_2)$ where T is regular, we can effectively construct (a Turing machine in $\mathrm{TM}[P_\mathcal{L}]$ for) the language $\varrho^{(*)}(L)$ (which is always in $\mathcal{L}$).*

Proof. By Corollary 4.1 we can effectively find a regular set of templates T_1 $(\subseteq T)$ such that if $\varrho_1 = (T_1, \Sigma, n_1, n_2)$ then ϱ_1 is useful on L and $\varrho_1^{(*)}(L) = \varrho^{(*)}(L)$.

Since any full AFL contains all regular languages, we have $T_1 \in \mathcal{L}$. Now, by Corollary 4.2, given L and T_1 we can effectively construct a Turing machine for $\varrho_1^{(*)}(L)$ and we are done. ∎

Since the regular and the context-free languages are examples of constructive full AFL's, as particular cases Theorem 4.1 implies that if ϱ is a TGR system with a regular set of templates, given a finite automaton (respectively, a pushdown automaton) for a language L, we can effectively construct a finite automaton (respectively, a pushdown automaton) for the language $\varrho^{(*)}(L)$.

Finally, it can be noted that Theorem 4.1 relies on Corollary 4.1 and Corollary 3.4 (that in turn relies on Corollary 3.2), and these results use brute-force constructions that basically enumerate all words up to a given length. It would be interesting to know whether for a regular initial language L and a regular set of templates there is some reasonably efficient algorithm to construct a (not necessarily deterministic) finite automaton for $\varrho^{(*)}(L)$.

References

1. Aho, A.V.: Indexed grammars – an extenion of context-free grammars. Journal of the ACM **15** (1968) 647–671
2. Daley, M., Ibarra, O., Kari, L., McQuillan, I., Nakano, K.: Closure and decision properties of some language classes under ld and dlad bio-operations. J. Automata, Languages and Combinatorics **8** (2003) 477–498
3. Daley, M., McQuillan, I.: Template-guided DNA recombination. Theoret. Comput. Sci. **330** (2005) 237–250
4. Daley, M., McQuillan, I.: Useful templates and iterated template-guided DNA recombination in ciliates. Theory of Computing Systems, in press. E-print available doi:10.1007/s00224-005-1206-6
5. Eilenberg, S.: Automata, Languages, and Machines, volume A. Academic Press, Inc., New York, NY, (1974)
6. Engelfriet, J.: Iterated stack automata and complexity classes. Information and Computation **95** (1991) 21–75
7. Ginsburg, S.: Algebraic and Automata-Theoretic Properties of Formal Languages. North-Holland, Amsterdam (1975)
8. Head, T., Pixton, D.: Splicing and regularity, to appear. Available at www.math.binghamton.edu/dennis/Papers
9. McQuillan, I., Salomaa, K., Daley, M.: Iterated TGR languages: Membership problem and effective closure properties. Queen's School of Computing Technical Report No. 2006-513. Available at www.cs.queensu.ca/TechReports
10. Prescott, D.M.: Genome Gymnastics: Unique modes of DNA evolution and processing in ciliates. Nature Reviews Genetics **1** (2000) 191–198
11. Prescott, D.M., Ehrenfeucht, A., Rozenberg, G.: Template guided recombination for IES elimination and unscrambling of genes in stichotrichous ciliates. J. Theoretical Biology **222** (2003) 323-330
12. Rozenberg, G., Salomaa, A. (eds.): Handbook of Formal Languages, Vols. 1–3. Springer Verlag, (1997)

On the Negation-Limited Circuit Complexity of Sorting and Inverting k-tonic Sequences

Takayuki Sato[1], Kazuyuki Amano[2], and Akira Maruoka[3]

[1] Dept. of Information Engineering, Sendai National College of Technology
Chuo 4-16-1, Ayashi, Aoba, Sendai 989-3128, Japan
taka@info.sendai-ct.ac.jp
[2] Dept. of Computer Science, Gunma University
Tenjin 1-5-1, Kiryu, Gunma 376-8515, Japan
amano@cs.gunma-u.ac.jp
[3] Graduate School of Information Sciences, Tohoku University
Aoba 6-6-05, Aramaki, Sendai 980-8579, Japan
maruoka@ecei.tohoku.ac.jp

Abstract. A binary sequence $x_1, \ldots, x_n$ is called k-tonic if it contains at most k changes between 0 and 1, i.e., there are at most k indices such that $x_i \neq x_{i+1}$. A sequence $\neg x_1, \ldots, \neg x_n$ is called an *inversion* of $x_1, \ldots, x_n$. In this paper, we investigate the size of a negation-limited circuit, which is a Boolean circuit with a limited number of NOT gates, that sorts or inverts k-tonic input sequences. We show that if $k = O(1)$ and $t = O(\log \log n)$, a k-tonic sequence of length n can be sorted by a circuit with t NOT gates whose size is $O((n \log n)/2^{ct})$ where $c > 0$ is some constant. This generalizes a similar upper bound for merging by Amano, Maruoka and Tarui [4], which corresponds to the case $k = 2$. We also show that a k-tonic sequence of length n can be inverted by a circuit with $O(k \log n)$ NOT gates whose size is $O(kn)$ and depth is $O(k \log^2 n)$. This reduces the size of the negation-limited inverter of size $O(n \log n)$ by Beals, Nishino and Tanaka [6] when $k = o(\log n)$. If $k = O(1)$, our inverter has size $O(n)$ and depth $O(\log^2 n)$ and contains $O(\log n)$ NOT gates. For this case, the size and the number of NOT gates are optimal up to a constant factor.

1   Introduction

To derive a strong lower bound on the size of a Boolean circuit for a function in NP is one of the most challenging open problems in theoretical computer science. But so far, the best known lower bound is only a linear in the number of input variables. This is quite contrast to the case of monotone circuit, which consists only of AND and OR gates, no NOT gates. Exponential lower bounds on the size of monotone circuits for explicit functions have been derived (e.g., [2,5,8,13]).

This motivates us to study the complexity of circuits with a limited number of NOT gates, which are usually called the negation-limited circuits. About a half

D.Z. Chen and D.T. Lee (Eds.): COCOON 2006, LNCS 4112, pp. 104–115, 2006.

century ago, Markov [11] proved that $r = \lceil \log(n+1) \rceil$ NOT gates are enough to compute any function on n variables, and that there is a function that requires r NOT gates to compute[1]. Beals, Nishino and Tanaka [6] constructed a circuit with r NOT gates that computes the *inverter* $\mathsf{Inv}_n(x_1, \ldots, x_n) = (\neg x_1, \ldots, \neg x_n)$ whose size is $O(n \log n)$. Thus, for every function f, the size of a smallest circuit with at most r NOT gates that computes f is at most $2\mathrm{Size}(f) + O(n \log n)$, where $\mathrm{Size}(f)$ is the size of a smallest circuit for f. This shows that restricting the number of NOT gates in a circuit to $O(\log n)$ entails only a small blowup in circuit size. Recently, several lower bounds on the size of a negation-limited circuit for an explicit function were obtained [3,4], and the relationship between the number of NOT gates and circuit size was also studied [9,14]. However, it is still unclear the effect on circuit complexity of restricting the number of NOT gates available.

In the first half of the paper (Section 3), we focus on the negation-limited circuit complexity of the *sorting* function, which is a function that sorts n binary inputs. This is motivated by the result of Amano, Maruoka and Tarui [4] showing that for every $t = 0, \ldots, \log \log n$, the size complexity of the *merging* function with t NOT gates is $\Theta((n \log n)/2^t)$. Roughly speaking, the size of a smallest circuit for merging is halved when the number of available NOT gates increases by one. The merging function is a function that takes two presorted binary sequences each of length n as inputs and merges into a sorted sequence of length $2n$. The merging function can be viewed as the special case of the sorting function in which an input is restricted to the form of the concatenation of two sorted sequences. Interestingly, it is known that both merging and sorting have monotone circuit complexity of $\Theta(n \log n)$ and (non-monotone) circuit complexity of $\Theta(n)$. So it is natural to consider the negation-limited circuit complexity of sorting, or an intermediate function between merging and sorting.

In this paper, we parameterize a binary sequence with the number of changes of the values when it is read from left to right. Formally, a binary sequence $x_1, \ldots, x_n$ is called k-*tonic* if there are at most k indices i such that $x_i \neq x_{i+1}$. The k-tonic sorting function is a function that outputs a sorted sequence of $x_1, \ldots, x_n$ if an input is k-tonic, and arbitrarily otherwise. The merging function can be regarded as the 2-tonic sorting function since input sequences $x_1 \geq \cdots \geq x_n$ and $y_1 \geq \cdots \geq y_n$ are 2-tonic if we reorder them to $x_1, \ldots, x_n, y_n, \ldots, y_1$. We show that if k is a constant, the k-tonic sorting function can be computed by a circuit with $t(\leq \log n)$ NOT gates whose size is $O((n \log n)/2^{ct})$ for some constant $1 > c > 0$. This can be viewed as a generalization of a similar upper bound for the merging function in [4], which corresponds to the case $k = 2$.

In the second half of the paper (Section 4), we investigate the negation-limited complexity of the *inverter* Inv_n. As described before, Beals, Nishino and Tanaka [6] constructed an inverter of size $O(n \log n)$ and depth $O(\log n)$ that contains $\lceil \log(n+1) \rceil$ negation gates. In the same paper [6], they stated the following question as an open problem (which is credited to Turán in [6]) : is the size of any $c \log n$ depth inverter using $c \log n$ NOT gates superlinear?

---

[1] All logarithms in this paper are base 2.

We give the construction of an inverter for k-tonic sequences whose size is $O(kn)$ and depth is $O(k \log^2 n)$ that contains $O(k \log n)$ NOT gates. If $k = O(1)$, our inverter has size $O(n)$ and depth $O(\log^2 n)$ and contains $O(\log n)$ NOT gates. This shows that the answer of Turán's problem is "no" if we relax the depth requirement from $O(\log n)$ to $O(\log^2 n)$ and restrict the inputs to k-tonic sequence with k being a constant. Both of our results suggest that limiting the number of changes in an input sequence may boost the power of NOT gates in a computation of Boolean functions.

2   Preliminaries and Results

A *circuit* is a combinational circuit that consists of AND gates of fan-in two, OR gates of fan-in two and NOT gates. In particular, a circuit without NOT gates is called *monotone* circuit. The *size* of a circuit C is the number of gates in C.

Let F be a collection of m Boolean functions $f_1, f_2, \ldots, f_m$. The *circuit complexity* of F, denoted by $\text{Size}(F)$, is the size of a smallest circuit that computes F. The *monotone circuit complexity* of F, denoted by $\text{Size}_{mon}(F)$, is the size of a smallest circuit that computes F. Following Beals et al. [6], we call a circuit including at most t NOT gates a t-*circuit*. The t-*negation limited circuit complexity* of F, denoted by $\text{Size}_t(F)$, is the size of a smallest t-circuit that computes F. If F cannot be computed by a t-circuit, then $\text{Size}_t(F)$ is undefined.

For a binary sequence x, the length of x is denoted by $|x|$. For $x = (x_1, \ldots, x_t) \in \{0,1\}^t$, $(x)_2$ denotes the integer whose binary representation is x where x_1 is the most significant bit, i.e., $(x)_2 = \sum_{i=1}^{t} x_i 2^{t-i}$. The number of 1's in a binary sequence x is denoted by $\sharp_1(x)$. For two integers $a < b$, $[a, b]$ denotes the set $\{a, a+1, \ldots, b\}$. The set $[1, n]$ is simply denoted by $[n]$.

Definition 1. *The sorting function on n inputs, denoted by Sort_n, is a collection of Boolean functions that sorts an n-bit binary sequence $x_1, \ldots, x_n$, i.e.,*

$$\text{Sort}_n(x_1, \ldots, x_n) = (z_1, \ldots, z_n),$$

such that $z_1 \geq \cdots \geq z_n$ and $\sum_i x_i = \sum_i z_i$. The merging function Merge_n is a collection of Boolean functions that merges two presorted binary sequences $x_1 \geq \cdots \geq x_n$ and $y_1 \geq \cdots \geq y_n$ into a sequence $z_1 \geq \cdots \geq z_{2n}$, i.e., $z_i = 1$ if and only if the total number of 1's in the input sequences is at least i. $\square$

The following results are known for the complexities of sorting and merging.

Theorem 1. *[12,1] All of the following are true:*

- *$Size(\text{Sort}_n) = \Theta(n)$ and $Size_{mon}(\text{Sort}_n) = \Theta(n \log n)$.*
- *$Size(\text{Merge}_n) = \Theta(n)$ and $Size_{mon}(\text{Merge}_n) = \Theta(n \log n)$.*

For the merging function, there is a clear tradeoff between the size of a circuit and the number of NOT gates.

Theorem 2. *[4] For every $0 \leq t \leq \log \log n$, $Size_t(\mathsf{Merge}_n) = \Theta((n \log n)/2^t)$.*

So it is interesting to consider whether such a tradeoff exists for more general functions.

Definition 2. *A turning point of a binary sequence $x_1, \ldots, x_n$ is an index i such that $x_i \neq x_{i+1}$. A binary sequence is called k-tonic if it has at most k turning points.*

Note that every n-bit binary sequence is $(n-1)$-tonic, and that input sequences to the merging function $x_1 \geq \cdots \geq x_n$ and $y_1 \geq \cdots \geq y_n$ can be regarded as 2-tonic if we reorder the sequences to $x_1, \ldots, x_n, y_n, \ldots, y_1$. Thus, we can define an "intermediate" function between merging and sorting based on the notion of k-tonic.

Definition 3. *A function $\{0,1\}^n$ to $\{0,1\}^n$ whose output is equal to the output of Sort_n for every k-tonic input is called a k-sorting function and is denoted by Sort_n^k. Note that the output of Sort_n^k is arbitrary if an input is not k-tonic.*

We show in Section 3 that a small number of NOT gates can reduce the size of a circuit for Sort_n^k, which extends the results on the upper bounds in Theorem 2. Precisely, we will show:

Theorem 3. *Suppose that $k = O(\log n)$ and $t \leq \log n$. Then there exists a constant c such that $Size_{ctk^2}(\mathsf{Sort}_n^k) = O(kn + (n \log n)/2^t)$. In particular, if $k = O(1)$ and $t = O(\log \log n)$, then $Size_t(\mathsf{Sort}_n^k) = O((n \log n)/2^{c't})$ for some constant $c' > 0$.*

In Section 4, we give the construction of an inverter for k-tonic sequences.

Definition 4. *An inverter with n binary inputs, denoted by Inv_n, is defined by*

$$\mathsf{Inv}_n(x_1, x_2, \ldots, x_n) = (\neg x_1, \neg x_2, \ldots, \neg x_n).$$

A function $\{0,1\}^n$ to $\{0,1\}^n$ whose output is equal to the output of Inv_n for every k-tonic input is called a k-tonic inverter and is denoted by Inv_n^k. Note that the output of Inv_n^k is arbitrary if an input is not k-tonic.

We will show:

Theorem 4. *The function Inv_n^k can be computed by a circuit of size $O(kn)$ and of depth $O(k \log^2 n)$ that contains $O(k \log n)$ NOT gates. In particular, if $k = O(1)$, then Inv_n^k can be computed by a linear size circuit of depth $O(\log^2 n)$ that contains $O(\log n)$ NOT gates.*

We remark that we need $\Omega(\log n)$ NOT gates to compute Inv_n^k even if $k = 1$. This can be easily proved by the result of Markov [11] (see also [6]). Thus, for the case $k = O(1)$, the size and the number of NOT gates are optimal up to a constant factor.

3   Negation-Limited Sorter for k-tonic Sequences

In this section, we describe the construction of a negation-limited circuit for the k-sorting function to prove Theorem 3.

As for the construction of a linear size sorter by Muller and Preparata [12], the construction is in two stages: the first computes the binary representation of the number of 1's in inputs, and the second generates appropriate outputs from this representation. Throughout this section, we assume that the length n of an input sequence is $n = 2^l$ for some natural number l.

Definition 5. *A counter* Count_n *is a function from* $\{0,1\}^n$ *to* $\{0,1\}^{\log n+1}$ *that outputs the binary representation of the number of 1's in an input sequence. A decoder* Decode_n *is a function from* $\{0,1\}^{\log n+1}$ *to* $\{0,1\}^n$ *such that* $\mathsf{Decode}_n(u) = (x_1, \ldots, x_n)$ *with* $x_1 = \cdots = x_{(u)_2} = 1$ *and* $x_{(u)_2+1} = \cdots = x_n = 0$.

It is obvious that $\mathsf{Sort}_n(x) = \mathsf{Decode}_n(\mathsf{Count}_n(x))$, and thus the size of a circuit for Sort_n is given by the sum of the sizes of circuits for Decode_n and Count_n. The linear sized sorter by Muller and Preparata [12] follows from:

Theorem 5. *[12]* $Size_n(\mathsf{Count}_n) = \Theta(n)$ *and* $Size_{mon}(\mathsf{Decode}_n) = \Theta(n)$.

Since there is a *monotone* circuit for Decode_n whose size is linear, one may think that it is sufficient to focus on the construction of a negation-limited circuit for Count_n. However, the last bit of the output of Count_n is the parity function, and so we need $\log n$ NOT gates to compute it [11]. In order to avoid to use such a large number of NOT gates, we only compute a limited number of significant bits of the number of 1's in inputs at the first stage of the construction.

Definition 6. *A t-counter, denoted by* $\mathsf{Count}_{n,t}$, *is a function from* $\{0,1\}^n$ *to* $\{0,1\}^{t+n/2^t}$ *defined as*

$$\mathsf{Count}_{n,t}(x) = (z, u),$$

where $z \in \{0,1\}^t$ *is the t most significant bits of the binary representation of the number of 1's in x and* $u = (u_1, \ldots, u_{n/2^t}) \in \{0,1\}^{n/2^t}$ *is a sorted sequence* $u_1 \geq u_2 \geq \cdots \geq u_{n/2^t}$ *such that* $\sharp_1(x) = (z)_2 \cdot n/2^t + \sharp_1(u)$. *A function whose output coincides with* $\mathsf{Count}_{n,t}(x)$ *for every k-tonic sequence x is denoted by* $\mathsf{Count}_{n,t}^k$.

A t-decoder, denoted by $\mathsf{Decode}_{n,t}$, *is a function from* $\{0,1\}^{t+n/2^t}$ *to* $\{0,1\}^n$ *defined as: For any binary sequence z of length t and any sorted sequence u of length $n/2^t$,*

$$\mathsf{Decode}_{n,t}(z, u) = (y_1, \ldots, y_n)$$

such that $y_1 = \cdots = y_w = 1$ *and* $y_{w+1} = \cdots = y_n = 0$ *where* $w = (z)_2 \cdot n/2^t + \sharp_1(u)$.

Note that $\mathsf{Sort}_n^k(x) = \mathsf{Decode}_{n,t}(\mathsf{Count}_{n,t}^k(x))$. We first construct a negation-limited circuit for $\mathsf{Count}_{n,t}^k$.

Theorem 6. *Suppose that $k = O(\log n)$ and $t \leq \log n$. Then there exists a constant c such that $\mathrm{Size}_{ctk^2}(\mathsf{Count}_{n,t}^k) = O(kn + (n \log n)/2^t)$.*

Proof. We first show an algorithm for computing $\mathsf{Count}_{n,t}^k$, and then we will describe a construction of a circuit which follows the algorithm.

A binary sequence is called *clean* if it consists of 0's only or 1's only, otherwise it is called *dirty*. Let x be an input sequence for $\mathsf{Count}_{n,t}^k$. The key observation to the algorithm is the fact that if we divide a k-tonic sequence x into $2k$ blocks, then at least half of them are clean. For simplicity, we suppose that $|x| = n = 2^l$ and $k = 2^a$ for some natural numbers l and a with $l > a$.

Algorithm C. This algorithm takes a binary sequence x of length n as an input and outputs (z, u) satisfying $\mathsf{Count}_{n,t}^k(x) = (z, u)$.

C1. For each $i = 1, 2, \ldots, t$, do the following:
   1. Divide x into $2k$ blocks of equal length: $B_1, B_2, \ldots, B_{2k}$.
   2. Let $p_1, p_2, \ldots, p_k \in [2k]$ be the first k indices of clean blocks.
   3. For each $j \in [k]$, let $c_{i,j} = 1$ if B_{p_j} is all 1's and $c_{i,j} = 0$ if B_{p_j} is all 0's.
   4. Let $\tilde{x}$ be a sequence of length $|x|/2$ obtained from x by removing $B_{p_1}, B_{p_2}, \ldots, B_{p_k}$ (i.e., the first k clean blocks).
   5. Substitute x by $\tilde{x}$.
C2. Let z_H and z_L be two binary sequences of length t and of length a such that

$$(z_H)_2 = \sum_{j=1}^{k} (c_{1,j} c_{2,j} \cdots c_{t,j})_2 \text{ div } 2^a, \qquad (z_L)_2 = \sum_{j=1}^{k} (c_{1,j} c_{2,j} \cdots c_{t,j})_2 \text{ mod } 2^a,$$

   where "div" and "mod" denote the quotient and remainder of two integers.
C3. Let u^1 be a sorted sequence of length $n/2^t$ that contains $(z_L)_2 \cdot 2^{l-(a+t)}$ 1's, and u^2 be a sorted sequence of length $n/2^t$ obtained by sorting x.
C4. If $\sharp_1(u^1) + \sharp_1(u^2) \geq n/2^t$, then let z be a sequence of length t such that $(z)_2 = (z_H)_2 + 1$ and let u be a sorted sequence of length $n/2^t$ that contains $\{\sharp_1(u^1) + \sharp_1(u^2) - n/2^t\}$ 1's. Otherwise, let $z = z_H$ and let u be a sorted sequence of length $n/2^t$ that contains $\{\sharp_1(u^1) + \sharp_1(u^2)\}$ 1's.
C5. Output (z, u).

In the following we show the correctness of the above algorithm. Consider the i-th iteration of the for loop at step C1 of the algorithm. Let x and $\tilde{x}$ be binary sequences before and after the i-th iteration. Suppose that a sequence x is k-tonic. This means that there are at most k dirty blocks, or equivalently, at least k clean blocks in $B_1, \ldots, B_{2k}$. So we can always choose k indices $p_1, \ldots, p_k$. It is easy to check that the sequence $\tilde{x}$ is also k-tonic. Since each block has length $n/(k \cdot 2^i) = 2^{l-(a+i)}$, it is obvious that

$$\sharp_1(x) = \sum_{j=1}^{k} c_{i,j} \cdot 2^{l-(a+i)} + \sharp_1(\tilde{x}).$$

By summing the above equation over $i = 1, \ldots, t$, the number of 1's in an initial input sequence is given by

$$\sum_{j=1}^{k}(c_{1,j}c_{2,j}\cdots c_{t,j})_2 \cdot 2^{l-(a+t)} + \sharp_1(u^2) = (z_H)_2 \cdot 2^{l-t} + (z_L)_2 \cdot 2^{l-(a+t)} + \sharp_1(u^2)$$

$$= (z_H)_2 \cdot 2^{l-t} + \sharp_1(u^1) + \sharp_1(u^2)$$

$$= (z)_2 \cdot 2^{l-t} + \sharp_1(u).$$

This completes the proof of the correctness of the algorithm.

Now we describe the construction of a circuit along the algorithm C starting from step C1, which is the most complex part of the construction. As for the above discussion, we first concentrate on the i-th iteration of the for loop, and so describe the construction of a circuit that takes sequences $B_1, \ldots, B_{2k}$ each of length $n/(k \cdot 2^i)$ as an input and outputs $c_{i,j}$ for $j \in [k]$ and $B_{s_1}, \ldots, B_{s_k}$ where $(s_1, \ldots, s_k) = [2k] \backslash (p_1, \ldots, p_k)$.

Given $(B_1, \ldots, B_{2k})$, we put $B^{(0)} = (B_1^{(0)}, \ldots, B_{2k}^{(0)}) = (B_1, \ldots, B_{2k})$. For each $p \in [k]$, define $B^{(p)} = (B_1^{(p)}, \ldots, B_{2k-p}^{(p)})$ as

$$B_q^{(p)} = \begin{cases} B_q^{(p-1)} & \text{if all of } B_1^{(p-1)}, \ldots, B_q^{(p-1)} \text{ are dirty,} \\ B_{q+1}^{(p-1)} & \text{there exists a clean block in } B_1^{(p-1)}, \ldots, B_q^{(p-1)}. \end{cases} \quad (1)$$

In other words, $B^{(p)}$ is a sequence obtained from $B^{(p-1)}$ by removing the first clean block in it. Then $B^{(p)}$ is equal to a sequence obtained from $B^{(0)}$ by removing the first p clean blocks. Hence $B^{(k)} = (B_1^{(k)}, \ldots, B_k^{(k)})$ is a desired sequence.

Now we introduce two types of auxiliary Boolean functions. For $p \in [0, k]$ and for $q \in [2k - p]$, let $\mathsf{Is\_Clean}_q^{(p)}$ be a function that outputs 1 if and only if $B_q^{(p)}$ is clean and let $\mathsf{Exist\_Clean}_q^{(p)}$ be a function that outputs 1 if and only if there exists a clean block in $B_1^{(p)}, \ldots, B_{q-1}^{(p)}$. These functions can be easily computed in a following way:

$$\mathsf{Is\_Clean}_q^{(0)} = \left(\bigwedge_{v \in B_q^{(0)}} v \right) \vee \left(\bigvee_{v \in B_q^{(0)}} v \right), \quad (\text{for } q \in [2k]),$$

$$\mathsf{Exist\_Clean}_0^{(p)} = 0, \quad (\text{for } p \in [0, k]),$$

$$\mathsf{Exist\_Clean}_q^{(p)} = \mathsf{Exist\_Clean}_{q-1}^{(p)} \vee \mathsf{Is\_Clean}_q^{(p)}, \quad (\text{for } p \in [0, k], q \in [2k - p]),$$

$$\mathsf{Is\_Clean}_q^{(p)} = (\mathsf{Is\_Clean}_{q+1}^{(p-1)} \wedge \mathsf{Exist\_Clean}_q^{(p-1)})$$

$$\vee (\mathsf{Is\_Clean}_q^{(p-1)} \wedge \overline{\mathsf{Exist\_Clean}_q^{(p-1)}}), (\text{for } p \in [k], q \in [2k - p]).$$

Thus by Eq. (1), for each $l = 1, 2, \ldots$, the l-th bit of $B_q^{(p)}$ is given by

$$(B_{q+1}^{(p-1)}[l] \wedge \mathsf{Exist\_Clean}_q^{(p-1)}) \vee (B_q^{(p-1)}[l] \wedge \overline{\mathsf{Exist\_Clean}_q^{(p-1)}}),$$

where $B[l]$ denotes the l-th bit of the block B. We also have

$$c_{i,j} = \bigvee_{q=1}^{k+1} \left(\mathsf{Is\_Clean}_q^{(j-1)} \wedge \overline{\mathsf{Exist\_Clean}_{q-1}^{(j-1)}} \wedge B_q^{(j-1)}[1] \right),$$

where $B_q^{(j-1)}[1]$ denotes the first bit of the block $B_q^{(j-1)}$.

Now we estimate the number of gates needed to compute these functions. Let $n_i = n/(2^{i-1})$, which is the length of an input sequence at the beginning of the i-th iteration of step C1. We use NOT gates at the computation of $\mathsf{Is\_Clean}_q^{(0)}$ for each $q \in [2k]$ and $\overline{\mathsf{Exist\_Clean}_q^{(p)}}$ for each $p \in [0, k]$ and $q \in [2k - p]$. So the number of NOT gates we need is at most $2k + 2k^2 \leq 3k^2$. The total number of gates is easily shown to be $O(k^2 + kn_i)$. Summing these over $i = 1, \ldots, t$, we need at most $3tk^2$ NOT gates and $\sum_{i=1}^{t} O(k^2 + kn_i) = O(tk^2 + kn)$ gates in total to simulate step C1 of the algorithm.

In step C2, all we have to do is to compute the addition of k integers of t bits. Since it is well known that the addition of two t-bit integers can be computed by a circuit of linear size (see e.g., [16, Chapter 3]), the number of gates needed to compute the addition of k integers of t bits is $O(k(t + \log k))$. The term $t + \log k$ here comes from the fact that the summand has at most $t + \log k$ digits. Here we use kt NOT gates, which is equal to the number of total input variables.

In step C3, we obtain u^1 as $\mathsf{Decode}_{n/2^t}(0z_L0^{l-(a+t)})$, which can be computed by a monotone circuit of size $O(n)$ by Theorem 5, and obtain u^2 as $\mathsf{Sort}_{n/2^t}(x)$, which can be computed by a monotone circuit of size $O((n/2^t) \log(n/2^t)) = O((n \log n)/2^t)$ by using the AKS-sorting network [1].

We can now proceed to step C4. Let $\tilde{u}$ be a sorted sequence of the concatenation of u^1 and u^2, which can be computed by a monotone circuit of size $O((2n/2^t) \log(2n/2^t)) = O((n \log n)/2^t)$ by using the AKS-sorting network [1]. Let $w \in \{0, 1\}$ be the $n/2^t$-th bit of $\tilde{u}$. Then $w = 1$ if and only if $\sharp_1(u^1) + \sharp_1(u^2) \geq n/2^t$. Thus, the desired sequence u is obtained by taking the first half of $\tilde{u}$ if $w = 0$ and the last half of $\tilde{u}$ if $w = 1$, which can be computed as $u_i = \overline{w}\tilde{u}_i \vee w\tilde{u}_{i+n/2^t}$ where u_i and $\tilde{u}_i$ denote the i-th bit of u and $\tilde{u}$, respectively. Clearly, z is given by the binary representation of $(z_H)_2 + w$ which can be computed by a t-bit adder. All these can be computed by a circuit of size $O((n \log n)/2^t)$ with $O(t)$ NOT gates.

The following table summarizes the number of gates used in each step.

Step	NOT gates	Total Size
C1	$O(tk^2)$	$O(tk^2 + kn)$
C2	kt	$O(k(t + \log k))$
C3	0	$O((n \log n)/2^t)$
C4	$O(t)$	$O((n \log n)/2^t)$

By summing these numbers, we conclude that the number of NOT gates in our circuit is $O(tk^2)$, and the total size is

$$O(tk^2 + kn + k(t + \log k) + (n \log n)/2^t) = O(kn + (n \log n)/2^t).$$

Here we use the assumption that $k = O(\log n)$ and $t \le \log n$. This completes the proof of the theorem. □

We can now proceed to the construction of a circuit for $\mathsf{Decode}_{n,t}$.

Theorem 7. *Suppose that* $t \le \log n$. *Then* $Size_{mon}(\mathsf{Decode}_{n,t}) = O(n)$.

Proof. Let $z \in \{0,1\}^t$ and $u \in \{0,1\}^{n/2^t}$ be inputs to $\mathsf{Decode}_{n,t}$. For such inputs, the output of $\mathsf{Decode}_{n,t}$ should be

$$\overbrace{11\cdots 11}^{(z)_2 \cdot n/2^t} u_1 \cdots u_{n/2^t}00\cdots 00.$$

For a binary sequence S, $S[i]$ denotes the i-th bit of S. Let A be an n-bit binary sequence given by 2^t copies of u. Put $B = \mathsf{Decode}_n(0z0^{l-t})$ and $C = \mathsf{Decode}_n(0z1^{l-t})$ Recall that $n = 2^l$. Let D be an n-bit binary sequence given by $D[i] = (A[i] \vee B[i]) \wedge C[i]$ for $i \in [n]$. Then the sequence D is

$$\overbrace{11\cdots 11}^{(z)_2 \cdot n/2^t} u_1 \cdots u_{n/2^t-1}00\cdots 00,$$

which is very close to the desired sequence, i.e., it misses the last bit of u. This discrepancy is fixed by putting $D[in/2^t] = (D[in/2^t - 1] \wedge u_{n/2^t}) \vee D[in/2^t]$ for each $i \in [2^t]$. Since Decode_n has a linear size monotone circuit (Theorem 5), the sequence D can also be computed by a monotone circuit of linear size. □

Theorem 3 follows immediately from Theorems 6 and 7.

4   Negation-Limited Inverter for k-tonic Sequences

In this section, we describe the construction of a negation-limited circuit for the k-tonic inverter to prove Theorem 4. Throughout this section, we suppose that the length n of an input is $2^a - 1$ for some natural number a. We first introduce several auxiliary functions.

Definition 7. *Let* $b \in \{0,1\}$. *Let* $\mathsf{Left}_n^b : \{0,1\}^n \to \{0,1\}^a$ *be the collection of Boolean functions defined as* $\mathsf{Left}_n^b(x) = p$ *if* $x_1 = \cdots = x_{(p)_2-1} = 1 - b$ *and* $x_{(p)_2} = b$, *i.e.,* p *is the binary representation of the smallest index* i *with* $x_i = b$. *If there are no* b's *in* x, *then the output of* Left_n^b *is unspecified. Let* $\mathsf{Decode}_n^b : \{0,1\}^a \to \{0,1\}^n$ *be the collection of Boolean functions defined as* $\mathsf{Decode}_n^b(p) = b^{(p)_2-1}(1-b)^{n-(p)_2+1}$. *If* p *is all* 0's *then the output of* Decode_n^b *is unspecified. Let* Or_n *and* And_n *denote the functions that output bitwise OR and AND of two input sequences, respectively.*

Lemma 8. *For each* $b \in \{0,1\}$, Left_n^b *can be computed by a circuit of size* $O(n)$ *and of depth* $O(\log^2 n)$ *that contains* $O(\log n)$ *NOT gates.*

Proof. We first give an algorithm to compute Left_n^1.

Algorithm L. This algorithm takes an n-bit binary sequence $x = (x_1, \ldots, x_n)$ as an input and outputs $p = (p_1, \ldots, p_a)$ which satisfies $x_1 = \cdots = x_{(p)_2 - 1} = 0$ and $x_{(p)_2} = 1$.

L1. Let $x^0 = x$ and $flag_0 = 1$.
L2. For $i = 1, \ldots, a - 1$ do the following:

   1. $p_i = \neg(\displaystyle\bigvee_{j=1}^{2^{a-i}-1} x_j^{i-1}) \wedge flag_{i-1}$,

   2. $x_j^i = \neg p_i x_j^{i-1} \vee p_i x_{j+2^{a-i}}^{i-1}$ (for $j = 1, \ldots, 2^{a-i} - 1$),

   3. $flag_i = \neg(p_i \wedge x_{2^{a-i}}^{i-1}) \wedge flag_{i-1}$,

L3. $p_a = x_1^{a-1} \wedge flag_{a-1}$.
L4. Outputs $(p_1, \ldots, p_a)$.

We now consider the correctness of algorithm L. We focus on the i-th iteration of step L2. In the case $p_i = 1$, since the number of 0-bits before the leftmost 1 in x^{i-1} is at least $2^{a-i} - 1$, we can show $(p_1 \cdots p_{i-1} 1 0^{a-i})_2 \leq (p)_2 \leq (p_1 \cdots p_{i-1} 1 1^{a-i})_2$. In particular, if $p_i = 1$ and $x_{2^{a-i}}^{i-1} = 1$ (which implies $flag_i = 0$), then the number of 0-bits before the leftmost 1 is equal to $2^{a-i} - 1$. Hence $(p)_2 = (p_1 \cdots p_{i-1} 1 0^{a-i})_2$, i.e., $p_{i+1}, \ldots, p_a$ should be all 0's. This will be satisfied since $flag_i = 0$. In the case $p_i = 0$, the number of 0-bits before the leftmost 1 in x^{i-1} is at most $2^{a-i} - 2$. Thus $(p_1 \cdots p_{i-1} 0 0^{a-i})_2 \leq (p)_2 \leq (p_1 \cdots p_{i-1} 0 1^{a-i})_2$. Therefore algorithm L outputs p_i correctly.

We now estimate the size of a circuit. For the i-th iteration of the for loop at step L2, p_i can be computed by a circuit of size 2^{a-i} and depth $a - i + 2$ with one NOT gate. A sequence x_j^i can be obtained by a circuit of size $3 \cdot (2^{a-i} - 1) = 3 \cdot 2^{a-i} - 3$ with one NOT gate, and $flag_i$ can be computed by using three gates including one NOT gate. For each i, step L2 can be done by a circuit of size $4 \cdot 2^{a-i} + 1$ with three NOT gates and depth $a - i + 4$. We only need one AND gate at step L3. Therefore, algorithm L can be simulated by a circuit of size $\sum_{i=1}^{a-1}(4 \cdot 2^{a-i} + 1) + 1 = 4(2^{a-1} - 1) + (a - 1) + 1 = O(n)$ with $3(a - 1) = O(\log n)$ NOT gates and of depth $\sum_{i=1}^{a-1}(a - i + 4) + 1 = O(a^2) = O(\log^2 n)$. The construction of a circuit for Left_n^0 is similar to that for Left_n^1 and is omitted. $\square$

Lemma 9. *For each $b \in \{0, 1\}$, Decode_n^b can be computed by a circuit of size $O(n)$ and of depth $O(\log n)$ that contains $O(\log n)$ NOT gates.*

Proof. It is obvious that $\mathsf{Decode}_n^1(p) = \mathsf{Decode}_n(q)$ with $(q)_2 = (p)_2 - 1$, and $\mathsf{Decode}_n^0(p)$ is equal to the reverse of $\mathsf{Decode}_n(q')$ with $(q')_2 = n - (p)_2 + 1$. We can easily see that each of q and q' can be computed by a circuit of size $O(\log n)$ and of depth $O(\log n)$ with $O(\log n)$ NOT gates. Since it is well known that Decode_n has a linear size $O(\log n)$ depth monotone circuit [12], we can obtain a desired circuit for Decode_n^b. $\square$

Proof. (of Theorem 4) As for the proof of Theorem 6, we first show an algorithm to compute Inv_n^k. Suppose that x is a k-tonic sequence starting with "0".

Algorithm I. This algorithm takes an n-bit binary sequence $x = (x_1, \ldots, x_n)$ as an input and outputs $z = (z_1, \ldots, z_n)$ where $z_i = \neg x_i$ if x is k-tonic.

I1. Let $x^0 = x$ and $z^0 = 1^n$.
I2. For $i = 1, \ldots, k$ do the following:
  If i is odd then
   1. $s^i = \mathsf{Decode}_n^1(\mathsf{Left}_n^1(x^{i-1}))$,
   2. $x^i = \mathsf{Or}_n(x^{i-1}, s^i)$,
   3. $z^i = \mathsf{And}_n(z^{i-1}, s^i)$,
  else
   1. $s^i = \mathsf{Decode}_n^0(\mathsf{Left}_n^0(x^{i-1}))$,
   2. $x^i = \mathsf{And}_n(x^{i-1}, s^i)$,
   3. $z^i = \mathsf{Or}_n(z^{i-1}, s^i)$,
I3. Outputs $z = z^k$.

The correctness of the algorithm I can be verified as follows: For some non-negative integers $p_0, \ldots, p_k \geq 0$, we can write x^0 as

$$x^0 = 0^{p_0} 1^{p_1} 0^{p_2} 1^{p_3} 0^{p_4} \cdots 0^{p_k}.$$

Then we have

$$s^1 = 1^{p_0} 0^{p_1} 0^{p_2} 0^{p_3} 0^{p_4} \cdots 0^{p_k},$$
$$x^1 = 1^{p_0} 1^{p_1} 0^{p_2} 1^{p_3} 0^{p_4} \cdots 0^{p_k},$$
$$z^1 = 1^{p_0} 0^{p_1} 0^{p_2} 0^{p_3} 0^{p_4} \cdots 0^{p_k},$$
$$s^2 = 0^{p_0} 0^{p_1} 1^{p_2} 1^{p_3} 1^{p_4} \cdots 1^{p_k},$$
$$x^2 = 0^{p_0} 0^{p_1} 0^{p_2} 1^{p_3} 0^{p_4} \cdots 0^{p_k},$$
$$z^2 = 1^{p_0} 0^{p_1} 1^{p_2} 1^{p_3} 1^{p_4} \cdots 1^{p_k}.$$

Note that x^1 is a $k - 1$ tonic sequence starting with $1^{p_0 + p_1}$ and x^2 is a $k - 2$ tonic sequence starting with $0^{p_0 + p_1 + p_2}$. Similarly, we can show that x^i is a $k - i$ tonic sequence and that

$$z^i = 1^{p_0} 0^{p_1} \cdots b^{p_i} b^{p_{i+1}} \cdots b^{p_k}.$$

Hence

$$z^k = 1^{p_0} 0^{p_1} 1^{p_2} 0^{p_3} 1^{p_4} \cdots 0^{p_k},$$

which is a desired output.

We now estimate the size of a circuit. For each iteration of the for loop at step I2, a sequence s^i can be computed by a circuit of size $O(n)$ and depth $O(\log^2 n)$ with $O(\log n)$ NOT gates, and sequences x^i and z^i can be computed by n gates and depth 1 without NOT gates. Hence the size and depth of an entire circuit are $O(kn)$ and $O(k \log^2 n)$, respectively. The total number of NOT gates is clearly $O(k \log n)$. $\square$

We finally remark that if we can improve the depth of our circuit for Left_n^b to $O(\log n)$, then we will have a negation-limited k-tonic inverter of depth $O(k \log n)$ which gives a negative answer to Turán's problem for the case $k = O(1)$.

Acknowledgment

The authors would like to thank Eiji Takimoto for helpful discussions and encouragement. This work was supported in part by Grant-in-Aid for Scientific Research on Priority Areas "New Horizons in Computing" from MEXT of Japan.

References

1. M. AJTAI, J. KOMÓS AND E. SZEMERÉDI, An $O(n \log n)$ Sorting Network, *Proc. 15th STOC*, pp. 1–9, 1983.
2. N. ALON AND R.B. BOPPANA, The Monotone Circuit Complexity of Boolean Functions, *Combinatorica*, 7(1), pp. 1–22, 1987.
3. K. AMANO AND A. MARUOKA, A Superpolynomial Lower Bound for a Circuit Computing the Clique Function with At Most $(1/6) \log \log n$ Negation Gates, *SIAM J. Comput.*, 35(1), pp. 201–216, 2005.
4. K. AMANO, A. MARUOKA AND J. TARUI, On the Negation-Limited Circuit Complexity of Merging, *Discrete Applied Mathematics*, 126(1), pp. 3–8, 2003.
5. A.E. ANDREEV, On a Method for Obtaining Lower Bounds for the Complexity of Individual Monotone Functions, *Sov. Math. Dokl.*, 31(3), pp. 530–534, 1985.
6. R. BEALS, T. NISHINO AND K. TANAKA, More on the Complexity of Negation-Limited Circuits, *Proc. 27th STOC*, pp. 585–595, 1995.
7. M.J. FISCHER, The Complexity of Negation-Limited Network–A Brief Survey, *LNCS*, 33, pp. 71–82, 1974.
8. D. HARNIK, R. RAZ, Higher Lower Bounds on Monotone Size, *Proc. 32nd STOC*, pp. 378–387, 2000.
9. S. JUKNA, On the Minimum Number of Negations Leading to Super-Polynomial Savings, *Inf. Process. Lett.*, 89(2), pp. 71–74, 2004.
10. E. A. LAMAGNA, The Complexity of Monotone Networks for Certain Bilinear Forms, Routing Problems, Sorting and Merging, *IEEE Trans. of Comput.*, 28(10), pp. 773–782, 1979.
11. A.A. MARKOV, On the Inversion Complexity of a System of Functions, *J. ACM*, 5, pp. 331–334, 1958.
12. D. E. MULLER AND F. P. PREPARATA, Bounds to Complexities of Networks for Sorting and Switching, *J. ACM*, 22, pp. 195–201, 1975.
13. A.A. RAZBOROV, Lower Bounds on the Monotone Complexity of Some Boolean Functions, *Soviet Math. Dokl.*, 281, pp. 798–801, 1985.
14. S. C. SUNG AND K. TANAKA, An Exponential Gap with the Removal of One Negation Gates, *Inf. Process. Let.*, 82(3), pp. 155–157, 2002.
15. K. TANAKA AND T. NISHINO, On the Complexity of Negation-Limited Boolean Networks, *SIAM J. Comput.*, 27(5), pp. 1334–1347, 1998.
16. I. WEGENER, *The Complexity of Boolean Functions*, Wiley-Teubner, 1987.

Robust Quantum Algorithms with ε-Biased Oracles

Tomoya Suzuki*, Shigeru Yamashita,
Masaki Nakanishi, and Katsumasa Watanabe

Graduate School of Information Science, Nara Institute of Science and Technology
{tomoya-s, ger, m-naka, watanabe}@is.naist.jp

Abstract. This paper considers the quantum query complexity of ε-biased oracles that return the correct value with probability only $1/2 + \varepsilon$. In particular, we show a quantum algorithm to compute N-bit OR functions with $O(\sqrt{N}/\varepsilon)$ queries to ε-biased oracles. This improves the known upper bound of $O(\sqrt{N}/\varepsilon^2)$ and matches the known lower bound; we answer the conjecture raised by the paper [1] affirmatively. We also show a quantum algorithm to cope with the situation in which we have no knowledge about the value of ε. This contrasts with the corresponding classical situation, where it is almost hopeless to achieve more than a constant success probability without knowing the value of ε.

1 Introduction

Quantum computation has attracted much attention since Shor's celebrated quantum algorithm for factoring large integers [2] and Grover's quantum search algorithm [3]. One of the central issues in this research field has been the *quantum query complexity*, where we are interested in both upper and lower bounds of a necessary number of oracle calls to solve certain problems [4,5,6]. In these studies, oracles are assumed to be *perfect*, i.e., they return the correct value with certainty.

In the classical case, there have been many studies (e.g., [7]) that discuss the case of when oracles are *imperfect* (or often called *noisy*), i.e., they may return incorrect answers. In the quantum setting, Høyer et al. [8] proposed an excellent quantum algorithm, which we call the *robust quantum search algorithm* hereafter, to compute the OR function of N values, each of which can be accessed through a quantum "imperfect" oracle. Their quantum "imperfect" oracle can be described as follows: When the content of the query register is x ($1 \leq x \leq N$), the oracle returns a quantum pure state from which we can measure the correct value of $f(x)$ with a constant probability. This noise model naturally fits into quantum subroutines with errors. (Note that most existing quantum algorithms have some errors.) More precisely, their algorithm robustly computes N-bit OR

---

* Currently with Center for Semiconductor Research & Development, Toshiba Cooporation.

D.Z. Chen and D.T. Lee (Eds.): COCOON 2006, LNCS 4112, pp. 116–125, 2006.
© Springer-Verlag Berlin Heidelberg 2006

functions with $O(\sqrt{N})$ queries to an imperfect oracle, which is only a constant factor worse than the perfect oracle case. Thus, they claim that their algorithm does not need a serious overhead to cope with the imperfectness of the oracles. Their method has been extended to a robust quantum algorithm to output all the N bits by using $O(N)$ queries [9] by Buhrman et al. This obviously implies that $O(N)$ queries are enough to compute the parity of the N bits, which contrasts with the classical $\Omega(N \log N)$ lower bound given in [7].

It should be noted that, in the classical setting, we do not need an overhead to compute OR functions with imperfect oracles either, i.e., $O(N)$ queries are enough to compute N-bit OR functions even if an oracle is imperfect [7]. Nevertheless, the robust quantum search algorithm by Høyer et al. [8] implies that we can still enjoy the quadratic speed-up of the quantum search when computing OR functions, even in the imperfect oracle case, i.e., $O(\sqrt{N})$ vs. $O(N)$. However, this is not true when we consider the probability of getting the correct value from the imperfect oracles *explicitly* by using the following model: When the query register is x, the oracle returns a quantum pure state from which we can measure the correct value of $f(x)$ with probability $1/2 + \varepsilon_x$, where we assume $\varepsilon \leq \varepsilon_x$ for any x and we know the value of ε. In this paper, we call this imperfect quantum oracle *an ε-biased oracle* (or a biased oracle for short) by following the paper [1]. Then, the precise query complexity of the above robust quantum search algorithm to compute OR functions with an ε-biased oracle can be rewritten as $O(\sqrt{N}/\varepsilon^2)$, which can also be found in [9]. For the same problem, we need $O(N/\varepsilon^2)$ queries in the classical setting since $O(1/\varepsilon^2)$ instances of majority voting of the output of an ε-biased oracle is enough to boost the success probability to some constant value. This means that the above robust quantum search algorithm does not achieve the quadratic speed-up anymore if we consider the error probability explicitly.

Adcock et al. [10] first considered the error probability explicitly in the quantum oracles, then Iwama et al. [1] continued to study ε-biased oracles: they show the lower bound of computing OR is $\Omega(\sqrt{N}/\varepsilon)$ and the matching upper bound when ε_x are the same for all x. Unfortunately, this restriction to oracles obviously cannot be applied in general. Therefore, for the general biased oracles, there have been a gap between the lower and upper bounds although the paper [1] conjectures that they should match at $\Theta(\sqrt{N}/\varepsilon)$.

Our Contribution. In this paper, we show that the robust quantum search can be done with $O(\sqrt{N}/\varepsilon)$ queries. Thus, we answer the conjecture raised by the paper [1] affirmatively, meaning that we can still enjoy the quantum quadratic speed-up to compute OR functions even when we consider the error probability explicitly. The overhead factor of $1/\varepsilon^2$ in the complexity of the original robust quantum search (i.e., $O(\sqrt{N}/\varepsilon^2)$) essentially comes from the classical majority voting in their recursive algorithm. Thus, our basic strategy is to utilize *quantum amplitude amplification and estimation* [11] instead of majority voting to boost the success probability to some constant value. This overall strategy is an extension of the idea in the paper [1], but we carefully perform the quantum amplitude amplification and estimation in quantum parallelism with appropriate accuracy to avoid the above-mentioned restriction to oracles assumed in [1].

In most existing (classical and quantum) algorithms with imperfect oracles, it is implicitly assumed that we know the value of ε. Otherwise, it seems impossible to know when we can stop the trial of majority voting with a guarantee of a more than constant success probability of the whole algorithm. However, we show that, in the quantum setting, we can construct a robust algorithm even when ε is unknown. More precisely, we can estimate unknown ε with appropriate accuracy, which then can be used to construct robust quantum algorithms. Our estimation algorithm also utilizes quantum amplitude estimation, thus it can be considered as an interesting application of quantum amplitude amplification, which seems to be impossible in the classical setting.

2 Preliminaries

In this section we introduce some definitions, and basic algorithms used in this paper.

The following unitary transformations are used in this paper.

Definition 1. *For any integer $M \geq 1$, a quantum Fourier transform $\mathbf{F}_M$ is defined by* $\mathbf{F}_M : |x\rangle \longmapsto \dfrac{1}{\sqrt{M}} \displaystyle\sum_{y=0}^{M-1} e^{2\pi i x y / M} |y\rangle \ \ (0 \leq x < M).$

Definition 2. *For any integer $M \geq 1$ and any unitary operator $\mathbf{U}$, the operator $\Lambda_M(\mathbf{U})$ is defined by*

$$|j\rangle|y\rangle \longmapsto \begin{cases} |j\rangle \mathbf{U}^j |y\rangle & (0 \leq j < M) \\ |j\rangle \mathbf{U}^M |y\rangle & (j \geq M). \end{cases}$$

Λ_M *is controlled by the first register $|j\rangle$ in this case. $\Lambda_M(\mathbf{U})$ uses $\mathbf{U}$ for M times.*

In this paper, we deal with the following biased oracles.

Definition 3. *A quantum oracle of a Boolean function f with bias ε is a unitary transformation O_f^{ε} or its inverse $O_f^{\varepsilon \dagger}$ such that*

$$O_f^{\varepsilon}|x\rangle|0^{m-1}\rangle|0\rangle = |x\rangle(\alpha_x |w_x\rangle|f(x)\rangle + \beta_x |w_x'\rangle|\overline{f(x)}\rangle),$$

where $|\alpha_x|^2 = 1/2 + \varepsilon_x \geq 1/2 + \varepsilon$ for any $x \in [N]$. Let also $\varepsilon_{\min} = \min_x \varepsilon_x$.

Note that $0 < \varepsilon \leq \varepsilon_{\min} \leq \varepsilon_x \leq 1/2$ for any x. In practice, ε is usually given in some way and $\varepsilon_{\min}$ or ε_x may be unknown. Unless otherwise stated, we discuss the query complexity with a given biased oracle O_f^{ε} in the rest of the paper.

We can also consider *phase flip oracles* instead of the above-defined *bit flip oracles*. A (perfect) phase flip oracle is defined as a map: $|x\rangle|0^{m-1}\rangle \longmapsto (-1)^{f(x)}|x\rangle|0^{m-1}\rangle$, which is equivalent to the corresponding bit flip oracle in the perfect case, since either oracle can be easily simulated by the other oracle with a pair of Hadmard gates. In a biased case, however, the two oracles cannot always be converted to each other. We need to take care of interference of the work registers, i.e., $|w_x\rangle$ and $|w'_x\rangle$, which are dealt with carefully in our algorithm.

Now we briefly introduce a few known quantum algorithms often used in following sections. In [11], Brassard et al. presented amplitude amplification as follows.

Theorem 4. *Let $\mathcal{A}$ be any quantum algorithm that uses no measurements and $\chi : \mathbb{Z} \rightarrow \{0,1\}$ be any Boolean function that distinguishes between success or fail (good or bad). There exists a quantum algorithm that given the initial success probability $p > 0$ of $\mathcal{A}$, finds a good solution with certainty using a number of applications of $\mathcal{A}$ and $\mathcal{A}^{-1}$, which is in $O(\frac{1}{\sqrt{p}})$ in the worst case.*

Brassard et al. also presented amplitude estimation in [11]. We rewrite it in terms of phase estimation as follows.

Theorem 5. *Let $\mathcal{A}, \chi$ and p be as in Theorem 4 and $\theta_p = \sin^{-1}(\sqrt{p})$ such that $0 \le \theta_p \le \pi/2$. There exists a quantum algorithm $Est\_Phase(\mathcal{A}, \chi, M)$ that outputs $\tilde{\theta}_p$ such that $|\theta_p - \tilde{\theta}_p| \le \frac{\pi}{M}$, with probability at least $8/\pi^2$. It uses exactly M invocations of $\mathcal{A}$ and χ, respectively. If $\theta_p = 0$ then $\tilde{\theta}_p = 0$ with certainty, and if $\theta_p = \pi/2$ and M is even, then $\tilde{\theta}_p = \pi/2$ with certainty.*

Our algorithm is based on the idea in [1], which makes use of the amplitude amplification. We refer interested users to [11] and [1].

3 Computing OR with ε-Biased Oracles

In this section, we assume that we have information about bias rate of the given biased oracle: a value of ε such that $0 < \varepsilon \le \varepsilon_{\min}$. Under this assumption, in Theorem 9 we show that N-bit OR functions can be computed by using $O(\sqrt{N}/\varepsilon)$ queries to the given oracle O_f^ε. Moreover, when we know $\varepsilon_{\min}$, we can present an optimal algorithm to compute OR with O_f^ε. Before describing the main theorem, we present the following key lemma.

Lemma 6. *There exists a quantum algorithm that simulates a single query to an oracle $O_f^{1/6}$ by using $O(1/\varepsilon)$ queries to O_f^ε if we know ε.*

To prove the lemma, we replace the given oracle O_f^ε with a new oracle $\tilde{O}_f^\varepsilon$ for our convenience. The next lemma describes the oracle $\tilde{O}_f^\varepsilon$ and how to construct it from O_f^ε.

Lemma 7. *There exists a quantum oracle $\tilde{O}_f^\varepsilon$ that consists of one O_f^ε and one $O_f^{\varepsilon\,\dagger}$ such that for any $x \in [N]$ $\tilde{O}_f^\varepsilon|x, 0^m, 0\rangle = (-1)^{f(x)}2\varepsilon_x|x, 0^m, 0\rangle + |x, \psi_x\rangle$, where $|x, \psi_x\rangle$ is orthogonal to $|x, 0^m, 0\rangle$ and its norm is $\sqrt{1 - 4\varepsilon_x{}^2}$.*

Proof. We can show the construction of $\tilde{O}_f^\varepsilon$ in a similar way in Lemma 1 in [1].
□

Now, we describe our approach to Lemma 6. The oracle $O_f^{1/6}$ is simulated by the given oracle O_f^ε based on the following idea. According to [1], if the query register $|x\rangle$ is not in a superposition, phase flip oracles can be simulated with sufficiently large probability: by using amplitude estimation through $\tilde{O}_f^\varepsilon$, we can estimate the value of ε_x, then by using the estimated value and applying amplitude amplification to the state in (7), we can obtain the state $(-1)^{f(x)}|x, 0^m, 0\rangle$ with high probability. In Lemma 6, we essentially simulate the phase flip oracle by using the above algorithm in a superposition of $|x\rangle$. Note that we convert the phase flip oracle into the bit flip version in the lemma.

We will present the proof of Lemma 6 after the following lemma, which shows that amplitude estimation can work in quantum parallelism. *Est_Phase* in Theorem 5 is straightforwardly extended to *Par_Est_Phase* in Lemma 8, whose proof can be found in [12].

Lemma 8. *Let $\chi : \mathbb{Z} \to \{0,1\}$ be any Boolean function, and let $\mathcal{O}$ be any quantum oracle that uses no measurements such that $\mathcal{O}|x\rangle|\mathbf{0}\rangle = |x\rangle\mathcal{O}_x|\mathbf{0}\rangle = |x\rangle|\Psi_x\rangle = |x\rangle(|\Psi_x^1\rangle + |\Psi_x^0\rangle)$, where a state $|\Psi_x\rangle$ is divided into a good state $|\Psi_x^1\rangle$ and a bad state $|\Psi_x^0\rangle$ by χ. Let $\sin^2(\theta_x) = \langle\Psi_x^1|\Psi_x^1\rangle$ be the success probability of $\mathcal{O}_x|\mathbf{0}\rangle$ where $0 \leq \theta_x \leq \pi/2$. There exists a quantum algorithm $Par\_Est\_Phase(\mathcal{O}, \chi, M)$ that changes states as follows: $|x\rangle|\mathbf{0}\rangle|\mathbf{0}\rangle \longmapsto |x\rangle \otimes \sum_{j=0}^{M-1} \delta_{x,j}|v_{x,j}\rangle|\tilde{\theta}_{x,j}\rangle$, where*

$$\sum_{j:|\theta_x - \tilde{\theta}_{x,j}| \leq \frac{\pi}{M}} |\delta_{x,j}|^2 \geq \frac{8}{\pi^2} \quad \text{for any } x, \text{ and } |v_{x,i}\rangle \text{ and } |v_{x,j}\rangle \text{ are mutually orthonormal vectors for any } i, j. \text{ It uses } \mathcal{O} \text{ and its inverse for } O(M) \text{ times.}$$

Proof. (of Lemma 6)

We will show a quantum algorithm that changes states as follows: $|x\rangle|\mathbf{0}\rangle|\mathbf{0}\rangle \longmapsto |x\rangle(\alpha_x|w_x\rangle|f(x)\rangle + \beta_x|w_x'\rangle|\overline{f(x)}\rangle)$, where $|\alpha_x|^2 \geq 2/3$ for any x, using $O(1/\varepsilon)$ queries to O_f^ε. The algorithm performs amplitude amplification following amplitude estimation in a superposition of $|x\rangle$.

At first, we use amplitude estimation in parallel to estimate ε_x or to know how many times the following amplitude amplification procedures should be repeated. Let $\sin\theta = 2\varepsilon$ and $\sin\theta_x = 2\varepsilon_x$ such that $0 < \theta, \theta_x \leq \pi/2$. Note that $\Theta(\theta) = \Theta(\varepsilon)$ since $\sin\theta \leq \theta \leq \frac{\pi}{2}\sin\theta$ when $0 \leq \theta \leq \pi/2$. Let also $M_1 = \left\lceil \frac{3\pi(\pi+1)}{\theta} \right\rceil$ and χ be a Boolean function that divides a state in (7) into a good state $(-1)^{f(x)}2\varepsilon_x|0^{m+1}\rangle$ and a bad state $|\psi_x\rangle$. The function χ checks only whether the state is $|0^{m+1}\rangle$ or not; therefore, it is implemented easily. By Lemma 8,

$$Par\_Est\_Phase(\tilde{O}_f^\varepsilon, \chi, M_1) \text{ maps } |x\rangle|\mathbf{0}\rangle|\mathbf{0}\rangle|\mathbf{0}\rangle \longmapsto |x\rangle \otimes \sum_{j=0}^{M-1} \delta_{x,j}|v_{x,j}\rangle|\tilde{\theta}_{x,j}\rangle|\mathbf{0}\rangle,$$

where $\sum_{j:|\theta_x - \tilde{\theta}_{x,j}| \leq \frac{\theta}{3(\pi+1)}} |\delta_{x,j}|^2 \geq \frac{8}{\pi^2}$ for any x, and $|v_{x,i}\rangle$ and $|v_{x,j}\rangle$ are mutually orthonormal vectors for any i, j. This state has the good estimations of θ_x in the third register with high probability. The fourth register $|0\rangle$ remains large enough to perform the following steps.

The remaining steps basically perform amplitude amplification by using the estimated values $\tilde{\theta}_{x,j}$, which can realize a phase flip oracle. Note that in the following steps a pair of Hadmard transformations are used to convert the phase flip oracle into our targeted oracle.

Based on the de-randomization idea as in [1], we calculate $m^*_{x,j} = \left[\frac{1}{2}\left(\frac{\pi}{2\tilde{\theta}_{x,j}} - 1\right)\right]$, $\theta^*_{x,j} = \frac{\pi}{4m^*_{x,j}+2}$, $p^*_{x,j} = \sin^2(\theta^*_{x,j})$ and $\tilde{p}_{x,j} = \sin^2(\tilde{\theta}_{x,j})$ in the superposition, and apply an Hadmard transformation to the last qubit. Thus we have

$$|x\rangle\left(\sum_{j=0}^{M-1} \delta_{x,j}|v_{x,j}\rangle|\tilde{\theta}_{x,j}\rangle|m^*_{x,j}\rangle|\theta^*_{x,j}\rangle|p^*_{x,j}\rangle|\tilde{p}_{x,j}\rangle \otimes |0^{m+1}\rangle|0\rangle \otimes \frac{1}{\sqrt{2}}(|0\rangle + |1\rangle)\right).$$

Next, let $\mathbf{R} : |p^*_{x,j}\rangle|\tilde{p}_{x,j}\rangle|0\rangle \rightarrow |p^*_{x,j}\rangle|\tilde{p}_{x,j}\rangle\left(\sqrt{\frac{p^*_{x,j}}{\tilde{p}_{x,j}}}|0\rangle + \sqrt{1 - \frac{p^*_{x,j}}{\tilde{p}_{x,j}}}|1\rangle\right)$ be a rotation and let $\mathbf{O} = \tilde{O}^\varepsilon_f \otimes \mathbf{R}$ be a new oracle. We apply $\mathbf{O}$ followed by $\Lambda_{M_2}(\mathbf{Q})$, where $M_2 = \left[\frac{1}{2}\left(\frac{3\pi(\pi+1)}{2(3\pi+2)\theta} + 1\right)\right]$ and $\mathbf{Q} = -\mathbf{O}(\mathbf{I} \otimes \mathbf{S}_0)\mathbf{O}^{-1}(\mathbf{I} \otimes \mathbf{S}_\chi)$; $\mathbf{S}_0$ and $\mathbf{S}_\chi$ are defined appropriately. Λ_{M_2} is controlled by the register $|m^*_{x,j}\rangle$, and $\mathbf{Q}$ is applied to the registers $|x\rangle$ and $|0^{m+1}\rangle|0\rangle$ if the last qubit is $|1\rangle$. Let $\mathbf{O}_x$ denote the unitary operator such that $\mathbf{O}|x\rangle|0^{m+1}\rangle|0\rangle = |x\rangle\mathbf{O}_x|0^{m+1}\rangle|0\rangle$. Then we have the state (From here, we write only the last three registers.)

$$\sum_{j=0}^{M-1} \frac{\delta_{x,j}}{\sqrt{2}}\left(|0^{m+1}\rangle|0\rangle|0\rangle + \mathbf{Q}^{m_{x,j}}_x\mathbf{O}_x\left(|0^{m+1}\rangle|0\rangle\right)|1\rangle\right), \tag{1}$$

where $\mathbf{Q}_x = -\mathbf{O}_x\mathbf{S}_0\mathbf{O}^{-1}_x\mathbf{S}_\chi$ and $m_{x,j} = \min(m^*_{x,j}, M_2)$ for any x, j. We will show that the phase flip oracle is simulated if the third register $|\tilde{\theta}_{x,j}\rangle$ has the good estimation of θ_x and the last register has $|1\rangle$. Equation (1) can be rewritten as

$$\sum_{j=0}^{M-1} \frac{\delta_{x,j}}{\sqrt{2}}\left(|0^{m+1}, 0\rangle|0\rangle + \left((-1)^{f(x)}\gamma_{x,j}|0^{m+1}, 0\rangle + |\varphi_{x,j}\rangle\right)|1\rangle\right),$$

where $|\varphi_{x,j}\rangle$ is orthogonal to $|0^{m+1}, 0\rangle$ and its norm is $\sqrt{1 - \gamma^2_{x,j}}$. Suppose that the third register has $|\tilde{\theta}_{x,j}\rangle$ such that $|\theta_x - \tilde{\theta}_{x,j}| \leq \frac{\theta_x}{3(\pi+1)}$. It can be seen that $m_{x,j} \leq M_2$ if $|\theta_x - \tilde{\theta}_{x,j}| \leq \frac{\theta_x}{3(\pi+1)}$. Therefore, $\mathbf{Q}_x$ is applied for $m^*_{x,j}$ times, i.e., the number specified by the fourth register. Like the analysis of Lemma 2 in [1], it is shown that $\gamma_{x,j} \geq \sqrt{1 - \frac{1}{9}}$.

Finally, applying an Hadmard transformation to the last qubit again, we have the state

$$\sum_{j=0}^{M-1} \frac{\delta_{x,j}}{2}\left((1+(-1)^{f(x)}\gamma_{x,j})|0^{m+2}\rangle|0\rangle + (1-(-1)^{f(x)}\gamma_{x,j})|0^{m+2}\rangle|1\rangle + |\varphi_{x,j}\rangle(|0\rangle - |1\rangle)\right).$$

If we measure the last qubit, we have $|f(x)\rangle$ with probability

$$\sum_{j=0}^{M-1} \left(\left| \frac{\delta_{x,j}(1+\gamma_{x,j})}{2} \right|^2 + \left| \frac{\delta_{x,j}\sqrt{1-\gamma_{x,j}^2}}{2} \right|^2 \right) \geq \frac{1}{2} \sum_{j:|\theta_x - \tilde{\theta}_{x,j}| \leq \frac{\theta}{3(\pi+1)}} |\delta_{x,j}|^2 (1+\gamma_{x,j}) \geq \frac{2}{3}.$$

Thus, the final quantum state can be rewritten as $|x\rangle(\alpha_x|w_x\rangle|f(x)\rangle + \beta_x|w_x'\rangle|\overline{f(x)}\rangle$, where $|\alpha_x|^2 \geq 2/3$ for any x.

The query complexity of this algorithm is the cost of amplitude estimation M_1 and amplitude amplification M_2, thus a total number of queries is $O(\frac{1}{\theta}) = O(\frac{1}{\varepsilon})$. Therefore, we can simulate a single query to $O_f^{1/6}$ using $O(\frac{1}{\varepsilon})$ queries to O_f^ε. $\square$

Now, we describe the main theorem to compute OR functions with quantum biased oracles.

Theorem 9. *There exists a quantum algorithm to compute N-bit OR with probability at least 2/3 using $O(\sqrt{N}/\varepsilon)$ queries to a given oracle O_f^ε if we know ε. Moreover, if we know $\varepsilon_{\min}$, the algorithm uses $\Theta(\sqrt{N}/\varepsilon_{\min})$ queries.*

The upper bound is derived from Lemma 6 and [8] straightforwardly. Also, Theorem 6 in [1] can prove the lower bound $\Omega(\sqrt{N}/\varepsilon_{\min})$.

4   Estimating Unknown ε

In Sect.3, we described algorithms by using a given oracle O_f^ε when we know ε. In this section, we assume that there is no prior knowledge of ε.

Our overall approach is to estimate ε (in precise $\varepsilon_{\min}$) with appropriate accuracy in advance, which then can be used in the simulating algorithm in Lemma 6. We present the estimating algorithm in Theorem 12 after some lemmas, which are used in the main theorem.

Lemma 10. *Let $\mathcal{O}$ be any quantum algorithm that uses no measurements such that $\mathcal{O}|x\rangle|0\rangle = |x\rangle|\Psi_x\rangle = |x\rangle(|\Psi_x^1\rangle + |\Psi_x^0\rangle)$. Let $\chi : \mathbb{Z} \to \{0,1\}$ be a Boolean function that divides a state $|\Psi_x\rangle$ into a good state $|\Psi_x^1\rangle$ and a bad state $|\Psi_x^0\rangle$ such that $\sin^2(\theta_x) = \langle \Psi_x^1|\Psi_x^1\rangle$ for any x ($0 < \theta_x \leq \pi/2$). There exists a quantum algorithm $Par\_Est\_Zero(\mathcal{O}, \chi, M)$ that changes states as follows:*

$$|x\rangle|0\rangle|0\rangle \to |x\rangle \otimes (\alpha_x|u_x\rangle|1\rangle + \beta_x|u_x'\rangle|0\rangle),$$

where $|\alpha_x|^2 = \dfrac{\sin^2(M\theta_x)}{M^2 \sin^2(\theta_x)}$ for any x. It uses $\mathcal{O}$ and its inverse for $O(M)$ times.

Par_Est_Zero can be based on Par_Est_Phase. We omit the proof. See [12] for more details.

Lemma 11. *Let $\mathcal{O}$ be any quantum oracle such that $\mathcal{O}|x\rangle|\mathbf{0}\rangle|0\rangle = |x\rangle(\alpha_x|w_x\rangle|1\rangle + \beta_x|u_x\rangle|0\rangle)$. There exists a quantum algorithm $Chk\_Amp\_Dn(\mathcal{O})$ that outputs $b \in \{0,1\}$ such that $b = 1$ if $\exists x; |\alpha_x|^2 \geq \frac{9}{10}$, $b = 0$ if $\forall x; |\alpha_x|^2 \leq \frac{1}{10}$, and $b =$ don't care otherwise, with probability at least $8/\pi^2$ using $O(\sqrt{N}\log N)$ queries to $\mathcal{O}$.*

Proof. Using $O(\log N)$ applications of $\mathcal{O}$ and majority voting, we have a new oracle $\mathcal{O}'$ such that $\mathcal{O}'|x\rangle|\mathbf{0}\rangle|0\rangle = |x\rangle(\alpha'_x|w'_x\rangle|1\rangle + \beta'_x|u'_x\rangle|0\rangle)$, where $|\alpha'_x|^2 \geq 1 - \frac{1}{16N}$ if $|\alpha_x|^2 \geq \frac{9}{10}$, and $|\alpha'_x|^2 \leq \frac{1}{16N}$ if $|\alpha_x|^2 \leq \frac{1}{10}$. Note that work bits $|w'_x\rangle$ and $|u'_x\rangle$ are likely larger than $|w_x\rangle$ and $|u_x\rangle$.

Now, let $\mathcal{A}$ be a quantum algorithm that makes the uniform superposition $\frac{1}{\sqrt{N}}\sum_x |x\rangle|\mathbf{0}\rangle|0\rangle$ by the Fourier transform $\mathbf{F}_N$ and applies the oracle $\mathcal{O}'$. We consider (success) probability p that the last qubit in the final state $\mathcal{A}|\mathbf{0}\rangle$ has $|1\rangle$. If the given oracle $\mathcal{O}$ satisfies $\exists x; |\alpha_x|^2 \geq \frac{9}{10}$ (we call Case 1), the probability p is at least $\frac{1}{N} \times (1 - \frac{1}{16N}) \geq \frac{15}{16N}$. On the other hand, if $\mathcal{O}$ satisfies $\forall x; |\alpha_x|^2 \leq \frac{1}{10}$ (we call Case 2), then the probability $p \leq N \times \frac{1}{N} \times \frac{1}{16N} = \frac{1}{16N}$. We can distinguish the two cases by amplitude estimation as follows.

Let $\tilde{\theta}_p$ denote the output of the amplitude estimation $Est\_Phase(\mathcal{A}, \chi, \lceil 11\sqrt{N}\rceil)$. The whole algorithm $Chk\_Amp\_Dn(\mathcal{O})$ performs $Est\_Phase(\mathcal{A}, \chi, \lceil 11\sqrt{N}\rceil)$ and outputs whether $\tilde{\theta}_p$ is greater than $0.68/\sqrt{N}$ or not. We will show that it is possible to distinguish the above two cases by the value of $\tilde{\theta}_p$. Let $\theta_p = \sin^{-1}(\sqrt{p})$ such that $0 \leq \theta_p \leq \pi/2$. Note that $x \leq \sin^{-1}(x) \leq \pi x/2$ if $0 \leq x \leq 1$. Theorem 5 says that in Case 1, the Est_Phase outputs $\tilde{\theta}_p$ such that

$$\tilde{\theta}_p \geq \theta_p - \frac{\pi}{11\sqrt{N}} \geq \sqrt{\frac{15}{16N}} - \frac{\pi}{11\sqrt{N}} > \frac{0.68}{\sqrt{N}},$$

with probability at least $8/\pi^2$. Similarly in Case 2, the inequality $\tilde{\theta}_p < \frac{0.68}{\sqrt{N}}$ is obtained.

$Chk\_Amp\_Dn(\mathcal{O})$ uses $\mathcal{O}$ for $O(\sqrt{N}\log N)$ times since $Chk\_Amp\_Dn(\mathcal{O})$ calls the algorithm $\mathcal{A}$ for $\lceil 11\sqrt{N}\rceil$ times and $\mathcal{A}$ uses $O(\log N)$ queries to the given oracle $\mathcal{O}$. $\qquad\square$

Theorem 12. *Given a quantum biased oracle O_f^ε, there exists a quantum algorithm $Est\_Eps\_Min(O_f^\varepsilon)$ that outputs $\tilde{\varepsilon}_{\min}$ such that $\varepsilon_{\min}/5\pi^2 \leq \tilde{\varepsilon}_{\min} \leq \varepsilon_{\min}$ with probability at least $2/3$. The query complexity of the algorithm is expected to be $O\left(\frac{\sqrt{N}\log N}{\varepsilon_{\min}}\log\log\frac{1}{\varepsilon_{\min}}\right)$.*

Proof. Let $\sin(\theta_x) = 2\varepsilon_x$ and $\sin(\theta_{\min}) = 2\varepsilon_{\min}$ such that $0 < \theta_x, \theta_{\min} \leq \frac{\pi}{2}$. Let χ also be a Boolean function that divides the state in (7) into a good state $(-1)^{f(x)}2\varepsilon_x|0^{m+1}\rangle$ and a bad state $|\psi_x\rangle$. Thus $Par\_Est\_Zero(\tilde{O}_f^\varepsilon, \chi, M)$ in Lemma 10 makes the state $|x\rangle \otimes (\alpha_x|u_x\rangle|1\rangle + \beta_x|u'_x\rangle|0\rangle)$ such that $|\alpha_x|^2 = \frac{\sin^2(M\theta_x)}{M^2\sin^2(\theta_x)}$. As stated below, if $M \in o(1/\theta_x)$, then $|\alpha_x|^2 \geq 9/10$. We can use Chk_Amp_Dn to check whether there exists x such that $|\alpha_x|^2 \geq 9/10$. Based on these facts, we present the whole algorithm $Est\_Eps\_Min(O_f^\varepsilon)$.

Algorithm(*Est_Eps_Min(O_f^ε)*)

1. Start with $\ell = 0$.
2. Increase ℓ by 1.
3. Run $Chk\_Amp\_Dn(Par\_Est\_Zero(\tilde{O}_f^\varepsilon, \chi, 2^\ell))$ for $O(\log \ell)$ times and use majority voting. If "1" is output as the result of the majority voting, then return to Step 2.
4. Output $\tilde{\varepsilon}_{\min} = \frac{1}{2}\sin\left(\frac{1}{5\cdot 2^\ell}\right)$.

Now, we will show that the algorithm almost keeps running until $\ell > \left\lfloor \log_2 \frac{1}{5\theta_{\min}} \right\rfloor$. We assume $\ell \leq \left\lfloor \log_2 \frac{1}{5\theta_{\min}} \right\rfloor$. Under this assumption, a proposition $\exists x; |\alpha_x|^2 \geq \frac{9}{10}$ holds since the equation $\varepsilon_{\min} = \min_x \varepsilon_x$ guarantees that there exists some x such that $\theta_{\min} = \theta_x$ and $|\alpha_x|^2 = \frac{\sin^2(2^\ell \theta_x)}{2^{2\ell}\sin^2(\theta_x)} \geq \cos^2(\frac{1}{5}) > \frac{9}{10}$ when $2^\ell \leq \frac{1}{5\theta_x}$. Therefore, a single Chk_Amp_Dn run returns "1" with probability at least $8/\pi^2$. By $O(\log \ell)$ repetitions and majority voting, the probability that we obtain "1" increases to at least $1 - \frac{1}{5\ell^2}$. Consequently, the overall probability that we return from Step 3 to Step 2 for any ℓ such that $\ell \leq \left\lfloor \log_2 \frac{1}{5\theta_{\min}} \right\rfloor$ is at least $\prod_{\ell=1}^{\left\lfloor \log_2 \frac{1}{5\theta_{\min}} \right\rfloor} \left(1 - \frac{1}{5\ell^2}\right) > \frac{2}{3}$. This inequality can be obtained by considering an infinite product expansion of $\sin(x)$, i.e., $\sin(x) = x\prod_{n=1}^{\infty}\left(1 - \frac{x^2}{n^2\pi^2}\right)$ at $x = \pi/\sqrt{5}$. Thus the algorithm keeps running until $\ell > \left\lfloor \log_2 \frac{1}{5\theta_{\min}} \right\rfloor$, i.e., outputs $\tilde{\varepsilon}_{\min}$ such that $\tilde{\varepsilon}_{\min} = \frac{1}{2}\sin\left(\frac{1}{5\cdot 2^\ell}\right) \leq \frac{1}{2}\sin(\theta_{\min}) = \varepsilon_{\min}$, with probability at least $2/3$.

We can also show that the algorithm almost stops in $\ell < \left\lceil \log_2 \frac{2\pi}{\theta_{\min}} \right\rceil$. Since $\frac{\sin^2(M\theta)}{M^2\sin^2(\theta)} \leq \frac{\pi^2}{(2M\theta)^2}$ when $0 \leq \theta \leq \frac{\pi}{2}$, $|\alpha_x|^2 = \frac{\sin^2(2^\ell \theta_x)}{2^{2\ell}\sin^2(\theta_x)} \leq \frac{1}{16}$ for any x if $2^\ell \geq \frac{2\pi}{\theta_{\min}}$. Therefore, in Step 3, "0" is returned with probability at least $8/\pi^2$ when $\ell \geq \left\lceil \log_2 \frac{2\pi}{\theta_{\min}} \right\rceil$. The algorithm, thus, outputs $\tilde{\varepsilon}_{\min} = \frac{1}{2}\sin\left(\frac{1}{5\cdot 2^\ell}\right) \geq \frac{1}{2}\sin(\frac{\theta_{\min}}{10\pi}) \geq \frac{\varepsilon_{\min}}{5\pi^2}$ with probability at least $8/\pi^2$.

Let $\tilde{\ell}$ satisfy $\left\lfloor \log_2 \frac{1}{5\theta_{\min}} \right\rfloor < \tilde{\ell} < \left\lceil \log_2 \frac{2\pi}{\theta_{\min}} \right\rceil$. If the algorithm runs until $\ell = \tilde{\ell}$, its query complexity is

$$\sum_{\ell=1}^{\tilde{\ell}} O(2^\ell \sqrt{N}\log N \log \ell) = O(2^{\tilde{\ell}}\sqrt{N}\log N \log \tilde{\ell}) = O\left(\frac{\sqrt{N}\log N}{\varepsilon_{\min}}\log\log\frac{1}{\varepsilon_{\min}}\right),$$

since $2^{\tilde{\ell}} \in \Theta\left(\frac{1}{\theta_{\min}}\right) = \Theta\left(\frac{1}{\varepsilon_{\min}}\right)$. $\square$

5   Conclusion

In this paper, we have shown that $O(\sqrt{N}/\varepsilon)$ queries are enough to compute N-bit OR with an ε-biased oracle. This matches the known lower bound while affirmatively answering the conjecture raised by the paper [1]. The result in this paper implies other matching bounds such as computing parity with $\Theta(N/\varepsilon)$ queries. We also show a quantum algorithm that estimates unknown value of ε with an ε-biased oracle. Then, by using the estimated value, we can construct a robust algorithm even when ε is unknown. This contrasts with the corresponding classical case where no good estimation method seems to exist.

Until now, unfortunately, we have had essentially only one quantum algorithm, i.e., the robust quantum search algorithm [8], to cope with imperfect oracles. (Note that other algorithms, including our own algorithm in Theorem 9, are all based on the robust quantum search algorithm [8].) Thus, it should be interesting to seek another *essentially different* quantum algorithm with imperfect oracles. If we find a new quantum algorithm that uses $O(T)$ queries to imperfect oracles with constant probability, then we can have a quantum algorithm that uses $O(T/\varepsilon)$ queries to imperfect oracles with an ε-biased oracle based on our method. This is different from the classical case where we need an overhead factor of $O(1/\varepsilon^2)$ by majority voting.

References

1. Iwama, K., Raymond, R., Yamashita, S.: General bounds for quantum biased oracles. IPSJ Journal **46**(10) (2005) 1234–1243
2. Shor, P.W.: An algorithm for quantum computation: discrete log and factoring. In: Proc. 35th Annual IEEE Symposium on Foudations of Computer Science. (1994) 124–134
3. Grover, L.K.: A fast quantum mechanical algorithm for database search. In: STOC. (1996) 212–219
4. Ambainis, A.: Quantum lower bounds by quantum arguments. J. Comput. Syst. Sci. **64**(4) (2002) 750–767
5. Beals, R., Buhrman, H., Cleve, R., Mosca, M., de Wolf, R.: Quantum lower bounds by polynomials. In: Proc. 39th Annual IEEE Symposium on Foudations of Computer Science. (1998) 352–361
6. Boyer, M., Brassard, G., Høyer, P., Tapp, A.: Tight bounds on quantum searching. Proc. of the Workshop on Physics of Computation: PhysComp'96 (1996) LANL preprint, http://xxx.lanl.gov/archive/quant-ph/9605034.
7. Feige, U., Raghavan, P., Peleg, D., Upfal, E.: Computing with Noisy Information. SIAM J. Comput. **23**(5) (1994) 1001–1018
8. Høyer, P., Mosca, M., de Wolf, R.: Quantum search on bounded-error inputs. In: ICALP. (2003) 291–299
9. Buhrman, H., Newman, I., Röhrig, H., de Wolf, R.: Robust polynomials and quantum algorithms. In: STACS. (2005) 593–604
10. Adcock, M., Cleve, R.: A quantum Goldreich-Levin Theorem with cryptographic applications. In: STACS. (2002) 323–334
11. Brassard, G., Høyer, P., Mosca, M., Tapp, A.: Quantum amplitude amplification and estimation. In: Quantum Computation & Information. Volume 305 of AMS Contemporary Mathematics Series Millenium Volume. (2002) 53–74
12. Suzuki, T., Yamashita, S., Nakanishi, M., Watanabe, K.: Robust quantum algorithms with ε-biased oracles. Technical Report LANL preprint, http://xxx.lanl.gov/archive/quant-ph/0605077. (2006)

The Complexity of Black-Box Ring Problems

V. Arvind, Bireswar Das, and Partha Mukhopadhyay

Institute of Mathematical Sciences
C.I.T Campus, Chennai 600 113, India
{arvind, bireswar, partham}@imsc.res.in

Abstract. We study the complexity of some computational problems on finite black-box rings whose elements are encoded as strings of a given length and the ring operations are performed by a black-box oracle. We give a polynomial-time quantum algorithm to compute a basis representation for a given black-box ring. Using this result we obtain polynomial-time quantum algorithms for several natural computational problems over black-box rings.

1 Introduction

Finite rings often play an important role in the design of algebraic algorithms. Berlekamp's randomized algorithm for factoring univariate polynomials over finite fields is a classic example [VZG03]. More recently, as explained in [AS05], the celebrated AKS primality test [AKS04] can be cast in a ring-theoretic framework. Lenstra's survey [Le92] gives other algorithmic examples. Recently, [AS05, KS05] have shown that Graph Isomorphism and Integer Factoring are polynomial-time reducible to Ring Isomorphism, where the rings are input in the *basis representation* (defined in Section 2).

As pointed out in [AS05], the representation of the finite ring is crucial to complexity of Ring Isomorphism. In this paper, we explore the complexity of ring-theoretic problems where the finite rings are given by a *black-box* (definitions are in Section 2). In a sense, a black-box ring is representation free. This model is motivated by finite black-box groups introduced by Babai and Szemerédi [BS84, Ba92] and intensively studied in algorithmic group theory. It turns out, surprisingly, that there is a polynomial-time *quantum* algorithm to obtain a basis representation for a given black-box ring. Thus, upto quantum polynomial time, the two representations are equivalent. A key procedure we use is an almost-uniform random sampling algorithm for finite black-box rings. Our algorithm is quite simple as compared to Babai's sampling algorithm for black-box groups [Ba91]. Additionally, if the characteristic of the ring is small (polynomially bounded in the input size), then we actually have an NC sampling algorithm. In contrast, for black-box groups it is still open if there is an NC sampler [Ba91].

It is an open question whether there is a randomized polynomial-time algorithm to recover a basis representation from the black-box oracle. The main obstacle is *additive independence testing* in a black-box ring R: given $r_1, r_2, \cdots, r_k \in R$

D.Z. Chen and D.T. Lee (Eds.): COCOON 2006, LNCS 4112, pp. 126–135, 2006.
© Springer-Verlag Berlin Heidelberg 2006

is there a nontrivial solution to $\sum_{i=0}^{l} x_i r_i = 0$. There is no known classical polynomial-time algorithm for this problem. However, it fits nicely in the hidden subgroup framework and we can solve it in quantum polynomial time as the additive group of R is abelian. As application we obtain quantum algorithms for some black-box rings problems in Sections 5 and 6.

2   Preliminaries

A *finite ring* is a triple $(R, +, *)$, where R is a finite nonempty set such that $(R, +)$ is a commutative group and $(R, *)$ is a semigroup, such that $*$ distributes over addition. A *subring* R' is a subset of R that is a ring under the same operations. Let $S \subset R$. The subring *generated* by S is the smallest subring $\langle S \rangle$ of R containing S. Thus, if $R = \langle S \rangle$ then every element of R can be computed by an arithmetic circuit that takes as input the generators from S and has the ring operations $+$ and $*$ as the gate operations. It is easy to see that every finite ring R has a generator set of size at most $\log |R|$.

A *ring oracle* R takes queries of the form $(q, x, y, +)$, $(q, x, y, *)$, (q, x, addinv), and (q, addid) where q, x, y are strings of *equal length* over Σ. The response to each of these queries is either a string of length $|q|$ or a symbol indicating invalid query. Let $R(q)$ be the set of $x \in \Sigma^{|q|}$ for which (q, x, addinv) is a valid query. Then R is a ring oracle if $R(q)$ is either empty or a ring with ring operations described by the responses to the above queries (where the response to (q, addid) is the string encoding additive identity). The oracle R defines the rings $R(q)$. The subrings of $R(q)$, given by generator sets will be called *black-box rings*.

A *basis representation* of a finite ring R [Le92, KS05] is defined as follows: the additive group $(R, +)$ is described by a direct sum $(R, +) = \mathbb{Z}_{m_1} e_1 \oplus \mathbb{Z}_{m_2} e_2 \oplus \cdots \oplus \mathbb{Z}_{m_n} e_n$, where m_i are the additive orders of e_i. Multiplication in R is specified by the products $e_i e_j = \sum_{k=1}^{n} \gamma_{ij}^k e_k$, for $1 \leq i, j \leq n$, where $\gamma_{ijk} \in \mathbb{Z}_{m_k}$.

Details about the classical and quantum complexity classes discussed in this paper can be found in [BDG88a, BDG88b, BV97].

3   Random Sampling from a Black-Box Ring

In this section we present a simple polynomial-time sampling algorithm that samples almost uniformly from finite black box rings. Let R be a black-box ring generated by S.

We will describe a randomized algorithm that takes S as input and with high probability computes an *additive generating set* T for $(R, +)$. I.e. every element of R is expressible as a sum of elements of T.

Using this additive generator set T it turns out that we can easily sample from $(R, +)$. We first prove this fact in the following lemma.

Lemma 1. *Let R be a finite black-box ring given by an additive generator set $\{r_1, r_2, \cdots, r_n\}$. Then there is a polynomial-time almost uniform sampling algorithm for R using $O(n \log(|R|/\epsilon))$ ring additions and $O(n \log(|R|/\epsilon))$ random bits.*

Proof. Let $k_1, k_2, \cdots, k_n$ be the additive orders of $\{r_1, r_2, \cdots, r_n\}$ in R. Define the onto homomorphism $\xi : \mathbb{Z}_{k_1} \times \mathbb{Z}_{k_2} \times \cdots \mathbb{Z}_{k_n} \longrightarrow R$ as $\xi(x_1, x_2, \cdots, x_n) = \sum_{i=1}^{n} x_i r_i$. Suppose we can almost uniformly sample from $\mathbb{Z}_{k_1} \times \mathbb{Z}_{k_2} \times \cdots \mathbb{Z}_{k_n}$. Let $(x_1, x_2, \cdots, x_n)$ be a sample point from $\mathbb{Z}_{k_1} \times \mathbb{Z}_{k_2} \times \cdots \mathbb{Z}_{k_n}$. Since ξ is an onto homomorphism, $\xi^{-1}(r)$ has the same cardinality for each $r \in R$. Hence, $\xi(x_1, x_2, \cdots, x_n)$ is an almost uniformly distributed random element from R.

Thus, it suffices to show that we can almost uniformly sample from $\mathbb{Z}_{k_1} \times \mathbb{Z}_{k_2} \times \cdots \mathbb{Z}_{k_n}$. Notice that we do not know the k_i's. But we know an upper bound, namely 2^m, for each of $k_1, k_2, \ldots, k_n$. Take a suitably large $M > 2^m$ to be fixed later in the analysis. The sampling is as follows: pick $(x_1, x_2, \cdots, x_n)$ uniformly at random from $[M]^n$ and output $\sum x_i r_i$. Let $(a_1, a_2, \cdots, a_n) \in \mathbb{Z}_{k_1} \times \mathbb{Z}_{k_2} \times \cdots \mathbb{Z}_{k_n}$ and let $p = \text{Prob}[x_i \equiv a_i \bmod k_i, 1 \leq i \leq n]$.

The x_i for which $x_i \equiv a_i \bmod k_i$ are precisely $a_i, a_i + k_i, \cdots, a_i + k_i \lfloor (M - a_i)/k_i \rfloor$. Let $M_i' = \lfloor (M - a_i)/k_i \rfloor$. Then $p = (\prod_i M_i')/M^n$. Clearly, $p \leq \prod_i (1/k_i)$. Furthermore, it is also easy to check that $p \geq (1 - 2^{m+1}/M)^n \cdot \prod_i (1/k_i)$. Choose $M > (n2^{m+1})/\epsilon$. Then $p \geq (1 - \epsilon) \prod_i (1/k_i)$, implying that $\sum x_i r_i$ is ϵ-uniformly distributed in R. The number of ring additions required is $O(n \log((n2^{m+1})/\epsilon))$ which is $O(n \log(|R|/\epsilon))$. The number of random bits used is also $O(n \log(|R|/\epsilon))$.

Let $R = \langle S \rangle$ be a black-box ring. Denote by $\hat{R}$ the additive subgroup of $(R, +)$ generated by S. I.e. $\hat{R}$ is the smallest additive subgroup of $(R, +)$ containing S. Notice that $\hat{R}$ could be a proper subset of R, and $\hat{R}$ need not be a subring of R in general.

Lemma 2. *Let $R = \langle S \rangle$ be a black-box ring, and $\hat{R}$ be the additive subgroup of $(R, +)$ generated by S. Then $\hat{R} = R$ if and only if $\hat{R}$ is closed under the ring multiplication: i.e. $\hat{R}r \subseteq \hat{R}$ for each $r \in S$.*

Proof. If $\hat{R} = R$ then the condition is obviously true. Conversely, notice that $\hat{R}r \subseteq \hat{R}$ for each $r \in S$ implies that $\hat{R}$ is closed under multiplication and hence $\hat{R} = R$.

Theorem 1. *There is a randomized algorithm that takes as input a black-box ring $R = \langle S \rangle$ and with high probability computes an additive generating set for $(R, +)$ and runs in time polynomial in the input size.*

Proof. The algorithm starts with S and proceeds in stages by including new randomly picked elements into the set at every stage. Thus, it computes a sequence of subsets $S = S_1 \subseteq S_2 \subseteq \ldots \subseteq S_\ell$, where ℓ will be appropriately fixed in the analysis. Let H_i denote the additive subgroup generated additively by S_i for each i. Notice that $H_1 = \hat{R}$. We now describe stage i of the procedure where, given S_i, the algorithm will compute S_{i+1}. First, notice that for each $r \in S$, $H_i r$ is a subgroup of $(R, +)$ that is additively generated by $\{xr \mid x \in S_i\}$. Thus, we can use Lemma 1 to ϵ-uniformly sample in polynomial time an element x_{ir} from $H_i r$, for each $r \in S$ (for a suitable ϵ to be chosen in the analysis). We now define the set $S_{i+1} = S_i \cup \{x_{ir} \mid r \in S\}$. Clearly, if ℓ is polynomially bounded then the above sampling procedure outputs S_ℓ in polynomial time. It thus remains to analyze the probability that S_ℓ additively generates $(R, +)$.

Claim. For $\ell = 4m+1$ and $\epsilon = 1/2^m$ the probability that S_ℓ additively generates $(R, +)$ is at least $1/6$.

Proof of Claim. The proof is a simple application of Markov's inequality. We define indicator random variables $Y_i, 1 \leq i \leq 4m$ as follows: $Y_i = 1$ if $H_i = H_{i+1}$ and $H_i \neq R$, and $Y_i = 0$ otherwise. Let $Y = \sum_{i=1}^{4m} Y_i$. First, we bound the expected value of each Y_i. If $H_i = R$ then clearly $E[Y_i] = 0$. Suppose $H_i \neq R$. By Lemma 2 there is an $r \in S$ such that $H_i r \not\subseteq H_i$. As $H_i r$ is an additive group it follows that $H_i r \cap H_i$ is a proper subgroup of $H_i r$ and hence $|H_i r \cap H_i| \leq |H_i r|/2$. Therefore, for a random $x \in H_i r$ the probability that it lies in H_i is at most $1/2$. Since x_{ir} is ϵ-uniformly distributed we have $\text{Prob}[Y_i = 1] \leq \text{Prob}[x_{ir} \in H_i] \leq 1/2(1+1/2^m)$. Putting it together, we get $\mu = E[Y] \leq 2m(1+1/2^m) \leq 2.5m$ for $m > 1$. Now, by Markov's inequality $\text{Prob}[Y > 3m] \leq \text{Prob}[Y > 3\mu/2.5] \leq 5/6$.

Combining Theorem 1 with Lemma 1 we immediately obtain the main theorem of this section.

Theorem 2. *There is a polynomial-time almost uniform sampling algorithm from black-box rings that takes as input $R = \langle S \rangle$ and $\epsilon > 0$, runs in time polynomial in input size and $\log(1/\epsilon)$ and outputs an ϵ-uniform random element from the ring R.*

Remark. We note that if the characteristic of the ring R is unary (in input size) then it is possible to modify the above polynomial-time sampling algorithm into an NC sampling algorithm.

Let $R = \langle r_1, r_2, \ldots, r_n \rangle$ be a black-box ring with elements encoded as strings in Σ^m. Examining the proof of Theorem 1 it is easy to see that every element $r \in R$ can be computed by an arithmetic circuit C_r (a straight-line program) that takes as input the generators $r_1, r_2, \ldots, r_n$ and has gates labeled $+$ and $*$ corresponding to the ring operations, such that C_r evaluates to r, and the size of the circuit C_r is $O(m^3 n^3)$. This is analogous to the reachability lemma for finite black-box groups [BS84].

Lemma 3 (ring reachability lemma). *Let $R = \langle r_1, r_2, \ldots, r_n \rangle$ be a black-box ring with elements encoded as strings in Σ^m. For every $r \in R$ there is an arithmetic circuit C_r of size $O(m^3 n^3)$ that has gates labeled by ring operations $+$ and $*$, takes as input $r_1, r_2, \ldots, r_n$ and evaluates to r.*

4   Quantum Algorithm for Finding a Basis Representation

In this section we describe a quantum polynomial-time algorithm that takes a black-box ring and computes a basis representation for it. The algorithm is Monte Carlo with small error probability.

Theorem 3. *There is a quantum polynomial-time algorithm that takes a black-box ring as input and computes a basis representation for the ring with small error probability.*

Proof. Let $R = \langle S \rangle$ be the input black-box ring. By the algorithm in Theorem 1 we first compute an additive generating set $\{r_1, r_2, \cdots, r_n\}$ for R. We first claim that there is a quantum polynomial-time algorithm for computing the additive orders d_i for each r_i. I.e. d_i is the least positive integer such that $d_i r_i = 0$, $1 \leq i \leq n$. To see this notice that 2^m is an upper bound on $|R|$, where m is the length of encodings of elements in R. Thus, the problem of computing d_i is precisely the period finding problem that can be solved in quantum polynomial-time by applying Shor's algorithm [Shor97].

The next step is to extract an *additively independent* set T of generators from $\{r_1, r_2, \cdots, r_n\}$ which will serve as the basis for R in its basis representation. Computing such a subset can be easily done using ideas from Cheung and Mosca in [CM01]. The idea is to first decompose $(R, +)$ as the direct sum of it's Sylow subgroups. This decomposition uses Shor's algorithm. Then each of the Sylow subgroups can further be decomposed into direct sum of cyclic groups by solving instances of hidden subgroup problem.

Finally, it remains to express the products rr', for $r, r' \in T$, as integer linear combinations of elements of T. We can again use Shor's period-finding quantum algorithm to compute the additive order d of rr'. Then we define a homomorphism $\varphi : \mathbb{Z}_d \times \mathbb{Z}_{d_{i_1}} \times \cdots \times \mathbb{Z}_{d_{i_\ell}} \to (R, +)$ as $\varphi(a, a_1, a_{i_1}, \cdots, a_{i_\ell}) = -arr' + \sum_{j=1}^{\ell} a_{i_j} r_{i_j}$. By applying [CM01] we can find an additive generating set for $\text{Ker}(\varphi)$. We can express the generating set for $\text{Ker}(\varphi)$ in terms of the basis for the ring R. Let M be the integer matrix whose columns are the generators of $\text{Ker}(\varphi)$. Let M_h be the corresponding *Hermite Normal Form* for M that can be computed in deterministic polynomial time. For expressing rr' as an integer linear combination of the basis elements, we need to seek a vector of the form $(1, x_1, x_2, \cdots, x_\ell)$ in the column space of M_h. Thus, $(1, 1)^{\text{th}}$ entry of M_h has to be an invertible element in the ring $\mathbb{Z}_d$. Let its inverse be λ. If C_1 is the first column of M_h, it is easy to see that λC_1 is a solution of the form $(1, x_1, x_2, \cdots, x_m)$ using which we can express rr' as an integer linear combination of the basis.

5   Testing if a Black-Box Ring Is a Field

In this section we describe a simple quantum polynomial time algorithm that takes a black-box ring as input and tests if it is a field. This result can be seen as a sort of generalization of primality testing: the ring $\mathbb{Z}_n$ is a field if and only if n is a prime. However, the black-box setting for the problem presents obstacles, like finding the additive order of elements, that seem hard for classical (randomized) polynomial time computation.

Theorem 4. *There is a quantum polynomial-time algorithm with small error probability for testing if a given black-box ring is a field.*

Proof. Let R be the input black-box ring. Applying the algorithm in Theorem 3 we obtain with high probability a basis representation for R: $(R, +) = \mathbb{Z}_{m_1} e_1 \oplus \mathbb{Z}_{m_2} e_2 \oplus \cdots \oplus \mathbb{Z}_{m_n} e_n$, and $e_i e_j = \sum_{k=1}^{n} \gamma_{ijk} e_k, \gamma_{ijk} \in \mathbb{Z}_{m_k}$.

Clearly, R is a field only if all m_i's are equal to a prime p. Using the AKS primality testing [AKS04] (or one of the polynomial time randomized tests) we check if p is prime. If not then the input is rejected. Thus, the basis representation can be written as $(R, +) = \mathbb{F}_p e_1 \oplus \mathbb{F}_p e_2 \oplus \cdots \oplus \mathbb{F}_p e_n$.

We next compute the minimal polynomial of e_1 over $\mathbb{F}_p$. This can be easily done in deterministic polynomial time. Suppose the minimal polynomial is $m_1(x)$ with degree d_1. Then in deterministic polynomial time we can test if $m_1(x)$ is irreducible over $\mathbb{F}_p$ [VZG03]. If it is not then the input R is rejected. Otherwise, $\mathbb{F}_p(e_1) = \{a_0 + a_1 e_1 + a_2 e_1^2 + \cdots + a_{d_1-1} e_1^{d_1-1} |$ for $1 \leq i \leq d_1 - 1$, $a_i \in \mathbb{F}_p\}$ is a finite field isomorphic to $\mathbb{F}_{p^{d_1}}$.

With the above step as the base case, inductively we assume that at the i-th step of the algorithm we have computed the finite field $\mathbb{F}_p(e_1, e_2, \ldots, e_i)$ contained in R with a basis $\{v_1, v_2, \cdots, v_k\}$ where each v_i is expressed as an $\mathbb{F}_p$-linear combination of $\{e_1, e_2, \cdots, e_n\}$. Let $d = \prod_{t=1}^{i} d_t$, where d_t is the degree of the minimal polynomial of e_t over $\mathbb{F}(e_1, \ldots, e_{t-1})$ for each t. By induction hypothesis $\mathbb{F}_p(e_1, e_2, \cdots, e_i) \cong \mathbb{F}_{p^d}$. Proceeding inductively, at the $i+1$-th step we again compute the minimum polynomial $m_{i+1}(x)$ of e_{i+1} over $\mathbb{F}_p(e_1, e_2, \cdots, e_i)$. Using the product relations defining the basis representation for R, it is easy to see that this computation will also boil down to solving a system of linear equations over $\mathbb{F}_p$. Also, we will similarly be able to check in polynomial time whether the obtained minimal polynomial is irreducible over $\mathbb{F}_{p^d}$ [VZG03].

We continue this procedure for n steps and if in none of the steps the above algorithm rejects the input, we conclude that R is a field. Clearly, if the basis representation for R is correct (which it is with high probability), the algorithm will correctly decide.

In the above theorem the power of quantum computation is used only to recover a basis representation for R. If R is already in basis representation then field testing is in P. We now give a classical complexity upper bound for the field testing.

Theorem 5. *Testing if a black-box ring is a field is in* $AM \cap coNP$.

Proof. A finite ring $R = \langle r_1, \ldots, r_n \rangle$ is not a field if and only if it has zero divisors: nonzero elements $a, b \in R$ whose product $ab = 0$. An NP test for this would be to guess small circuits C_a and C_b (using Lemma 3) for zero divisors a and b verifying their product $ab = 0$ using the black-box oracle. Thus the problem is in coNP. We now show that the problem is in AM. Merlin will send the basis representation for R to Arthur as follows: Merlin sends a basis $\{u_1, u_2, \cdots, u_l\}$ for $(R, +)$ along with their pairwise products in terms of generators. Also Merlin sends each generator r_i as a linear combination of the basis elements u_j. Arthur can now easily verify that $\{u_1, u_2, \cdots, u_l\}$ is a generating set for $(R, +)$ and that the product relations are correct. It remains to verify that $\{u_1, u_2, \cdots, u_l\}$ is *additively independent*. Merlin sends the additive orders d_i of u_i for each i, with the prime factorizations of d_i using which Arthur can verify that d_i are the additive orders. Now, to verify that $\{u_1, u_2, \cdots, u_\ell\}$ is additively independent it suffices to check that the $|R| = \prod_{i=1}^{\ell} d_i$. By a result of Babai [Ba92], order verification of black-box groups is in AM. This protocol can clearly be applied to $(R, +)$.

5.1   An Application of the Chebotarëv Density Theorem

We now briefly explore a somewhat different problem related to testing if a given ring is a field: suppose we are given a basis $e_1, e_2, \ldots, e_n$ along with product relations $e_i e_j = \sum_{k=1}^{n} \gamma_{ijk} e_k$ for integers γ_{ijk}. The question we ask here is whether there is *some* prime p such that *modulo p* the above is a basis representation for the finite field $\mathbb{F}_{p^n}$. We need some algebraic number theory to develop a polynomial-time randomized algorithm to test if there is such a prime.

First, notice that by using the product relations, we can as before compute the minimal polynomials m_i of the e_i *over the rationals* $\mathbb{Q}$. We can check in polynomial time that the m_i are all indeed irreducible over $\mathbb{Q}$ (using the LLL algorithm). Because if m_i are not irreducible over $\mathbb{Q}$ then they are not irreducible modulo any prime and we can reject the input in that case. Now, we use the product relations to compute the tower of fields $\mathbb{Q} \subseteq \mathbb{Q}(e_1) \subseteq \cdots \subseteq \cdots \subseteq \mathbb{Q}(e_1, \ldots, e_n)$. In fact, by the *primitive element theorem*, starting with $f_1 = e_1$ we can compute in polynomial time a primitive element f_i for the field $\mathbb{Q}(e_1, \ldots, e_i)$ as an integer linear combination of $e_1, \ldots, e_i$. Finally, we will obtain f_n such that $\mathbb{Q}(f_n) = \mathbb{Q}(e_1, e_2, \cdots, e_n)$. Using the product relation the problem of finding the minimal polynomial of f_n over $\mathbb{Q}$ reduces to solving a system of linear equations over $\mathbb{Q}$ which can be done in polynomial time. Let $f(x)$ be this minimal polynomial, which has to be irreducible of degree n. Thus, we have $\mathbb{Q}(f_n) \cong \mathbb{Q}[x]/(f(x))$. Let ℓ denote the lcm of the denominators of coefficients of $f(x)$. Then taking $e = l f_n$, we observe that e has a monic minimal polynomial $g(y)$ with integer coefficients and $\mathbb{Q}(f_n) = \mathbb{Q}(e) = \mathbb{Q}[y]/(g(y))$. Now our goal is to test if there is a prime p such that $g(y)$ is irreducible modulo p, so that $\mathbb{Z}_p[y]/g(y)$ is $\mathbb{F}_{p^n}$ and hence the given basis representation modulo p is $\mathbb{F}_{p^n}$.

Let L be the splitting field of $g(y)$. Consider its Galois group $\mathrm{Gal}(L/\mathbb{Q}) = G$ which is fully described by its action on the roots of $g(y)$. Thus G can be seen as a subgroup of S_n. Clearly, any $\sigma \in G$ is a product of disjoint cycles. If the length of these cycles is $n_1, n_2 \ldots, n_k$ such that $n_1 \leq n_2 \leq \ldots \leq n_k$ we say that σ has *cycle pattern* $(n_1, \ldots, n_k)$. Let p be a prime which does not divide the discriminant of g. Then, modulo p, the polynomial $g(y)$ has no multiple roots. If we factorize g modulo p (using Berlekamp's algorithm) we get $g(y) = g_1(y) g_2(y) \cdots g_k(y)$, where g_i are distinct, irreducible and degrees of the g_i is a partition of n. Writing the degrees d_i of g_i in increasing order we obtain the *decomposition pattern* $(d_1, d_2, \ldots, d_k)$ of g modulo p.

Now, the Frobenius density theorem (which is a weaker form of the Chebotarëv density theorem) tells us that the number of primes with a given decomposition is close to number of permutations in G having that particular cycle structure. Assuming GRH, the bounds are tight enough to be algorithmically applied. We describe this theorem in a form tailored to our question.

Let $C = (n_1, n_2, \ldots, n_k)$ be any cycle pattern and let $C(G)$ be the subset of G consisting of permutations with cycle pattern C. Notice that $C(G)$ is closed under conjugation. Let $\pi_C(x)$ be the number of primes $p \leq x$ such that p is unramified in L and the decomposition pattern of g modulo p is C. Then we have the following theorem which is a restatement of the Frobenius density theorem using [SS97, Lemma 3].

Theorem 6. *Let L be the splitting field of an irreducible polynomial $g(y) \in \mathbb{Z}[y]$. Let d be the discriminant of L. Assuming GRH, there are absolute constants α, β such that if $x \geq \alpha(\log|d|)^\alpha$, then $\pi_C(x) \geq \beta \frac{\#C(G)}{\#G} \frac{x}{\ln x}$.*

Since we are seeking a prime p such that $g(y)$ is irreducible modulo p, let C_0 be the cycle pattern (n). First we can see by an easy counting argument that if $C(G)$ is nonempty then $\#C(G) \geq \frac{\#G}{n}$. Let size$(g)$ denote the size of the polynomial g. Then it follows that $\log|d| \leq (n+1)!^2 \cdot \text{size}(g)$. We choose $x = \lceil \alpha(\log|d|)^\alpha \rceil$ which is a polynomial-sized integer. By the above theorem it follows that $g(y)$ is irreducible modulo p for a random prime $p \leq x$ with probability at least β/n if there is such a prime at all. This gives us a simple randomized polynomial time procedure that finds such a prime p if it exists with nonnegligible success probability (assuming GRH).

6   Complexity of Nilradical

Let R be a commutative ring. An element $x \in R$ is *nilpotent* if $x^n = 0$ for some $n > 0$. In a commutative ring R, the set of all nilpotent elements of R form an ideal $N(R)$ called the *nilradical* of R. The nilradical is crucial to the structure of rings and it plays an important role in decomposing finite rings. In this section, we show that the nilradical of *commutative* black-box rings can be computed in quantum polynomial time.

Let $R = \langle S \rangle$ be a commutative black-box ring. By Lemma 1 we first compute an additive generating set for R with high probability. Let T denote the computed additive generating set.

Applying Cheung & Mosca's ideas, as explained in Theorem 3, in quantum polynomial time we can compute from T an additive *independent* generating set for $(R, +)$. Call this generating set $T' = \{r_1, r_2, \ldots, r_\ell\}$. Now, using Shor's algorithm we can find their additive orders d_i with high probability, implying that $|R| = \prod_{i=1}^n d_i$. Again by Shor's integer factoring algorithm we compute the prime factorization $|R| = p_1^{\alpha_1} \cdots p_k^{\alpha_k}$. Let $n_i = \frac{|R|}{p_i^{\alpha_i}}$ for $1 \leq i \leq k$.

By elementary group theory, we know that the additive p_i-Sylow subgroups $(R_i, +)$ of $(R, +)$ is additively generated by $T_i = \{n_i r | r \in T'\}$. It is easy to see that R_i's are actually subrings of R (the p_i-Sylow subrings), in fact even ideals and furthermore $R = R_1 \oplus \cdots \oplus R_k$ is a direct sum ring decomposition with $R_i R_j = 0, \forall i \neq j$. Thus, the nilradical N of R is given by $N = N_1 \oplus N_2 \oplus \cdots \oplus N_k$, where each N_i is the nilradical of R_i. Thus, it suffice to explain how to compute an additive generating set for the nilradical of a ring R s.t $|R| = p^\alpha$ for some prime p.

As explained in Theorem 3 we compute a basis representation for R in quantum polynomial time. Let $e_1, e_2, \ldots, e_t$ be the basis elements with additive orders p^{α_i} $(1 \leq i \leq t)$ respectively. Then $R \cong \mathbb{Z}_{p_1^{\alpha_1}} e_1 \oplus \cdots \oplus \mathbb{Z}_{p_t^{\alpha_t}} e_t$ with $e_i e_j = \sum_{k=1}^{t} \gamma_{ijk} e_k$ is the basis representation.

Since $p^\alpha r = 0 \; \forall r \in R$, it easily follows that $pR = \{pr | r \in R\}$ is a subring of R contained in the nilradical. Indeed pR is an ideal of R. Thus R/pR is also a finite ring. Moreover, we can easily write down its basis representation as follows: Let $f_i = e_i + pR$, $1 \leq i \leq t$. Then $R/pR = \mathbb{F}_p f_1 \oplus \cdots \oplus \mathbb{F}_p f_t$ where the products $f_i f_j$ are $e_i e_j + pR$ and can be expressed as an $\mathbb{F}_p$-linear combination of the f_i's. The following lemma is easy to see.

Lemma 4. *N is the nilradical of R if and only if N/pR is the nilradical of R/pR.*

Thus, if we can find a basis for the nilradical N/pR of R/pR as linear combinations of the f_i's we can easily pull back into N by replacing the basis elements f_i's by e_i's. Therefore, we have reduced the problem to finding the nilradical of an $\mathbb{F}_p$-algebra R given in basis representation. The proof of the next lemma will be given in the full version.

Lemma 5. *Given an $\mathbb{F}_p$-algebra R in basis representation, its nilradical can be computed in deterministic polynomial time.*

Continuing with the original problem, let S'' be the pullback of S' w.r.t the homomorphism $\phi : R \to R/pR$ (namely replace f_i by e_i). Then $S'' \cup \{pe_i | 1 \leq i \leq t\}$ is an additive generating set for the nilradical of R. Putting it together, we have proved the following theorem.

Theorem 7. *The nilradical of a black-box ring can be computed in quantum polynomial time.*

Also similar kind of techniques as of Theorem 5 easily suggests the following result about the classical complexity of nilradical testing.

Theorem 8. *Let R be a black-box ring and I be an ideal of R given by a generator set. Testing if I is the nilradical of R is in $\mathrm{AM} \cap \mathrm{coAM}$.*

The *square-free part* of a positive integer n is the product of all distinct prime factors of n. We now observe that computing the nilradical of a black-box ring is harder than computing the square-free part of an integer.

Lemma 6. *Computing the square free part of an integer n is polynomial time Turing reducible to computing the nilradical of $\mathbb{Z}_n$.*

Proof. Let $n = p_1^{\alpha_1} p_2^{\alpha_2} \cdots p_k^{\alpha_k}$. Then the square-free part of n is $s = p_1 p_2 \ldots p_k$. An x in $\mathbb{Z}_n$ is in the nilradical N if and only if $x^m = 0 (\mathrm{mod}\, n)$, which is possible if and only if s divides x. Now, suppose we have an algorithm that computes a generator set T for N, where T generates N as a ring. Let $n_1 \in T$ be any element. Then $n_1 < n$ and s divides n_1. We again apply the algorithm to find a generator set for the nilradical N_1 of $\mathbb{Z}_{n_1}$. Continuing thus, we obtain a sequence of integers $n > n_1 > n_2 \ldots > n_t$ where n_i divides n_{i-1} for each i and each n_i is a multiple of s. Thus, this sequence is of length at most $\log n$ and must terminate at some $n_t = s$, which we can detect since the nilradical of $\mathbb{Z}_s$ is $\{0\}$.

References

[AKS04]   MANINDRA AGRAWAL, NEERAJ KAYAL, AND NITIN SAXENA. PRIMES is in P. *Annals of Mathematics*, 160(2):781-793, 2004.

[AS05]   MANINDRA AGRAWAL AND NITIN SAXENA. Automorphisms of Finite Rings and Applications to Complexity of Problems.*STACS'05, Springer LNCS 3404*. 1-17, 2005.

[Ba91]   L. BABAI.Local Expansion of Vertex-Transitive Graphs and Random Generation in Finite Groups.*STOC* 1991: 164-174.

[Ba92]   L. BABAI. Bounded Round Interactive Proofs in Finite Groups. *SIAM J. Discrete Math.*, 5(1): 88-111 1992.

[BDG88a] J. L. BALCÁZAR, J. DÍAZ, AND D. GABARRÓ. Structural Complexity I. *ETACS Monographs on Theoretical Computer Science, Springer Verlag.* 1988 (I).

[BDG88b] J. L. BALCÁZAR, J. DÍAZ, AND D. GABARRÓ. Structural Complexity II. *ETACS Monographs on Theoretical Computer Science, Springer Verlag.* 1990 (II).

[BM88]   L. BABAI, S. MORAN, Arthur-Merlin Games: A Randomized Proof System, and a Hierarchy of Complexity Classes. *Journal Comput. Syst. Sciences*, 36(2): 254-276 (1988).

[BMc74]   BERNARD R. MCDONALD, Finite Rings with Identity. *Marcel Dekker, Inc.*, 1974.

[BS84]   L. BABAI AND E. SZEMERÉDI, On the complexity of matrix group problems I, *In Proc. 25th IEEE Sympos. on the Foundation of Computer Science*, pp.229-240 (1984).

[BV97]   U. VAZIRANI AND E. BERNSTEIN, Quantum Complexity Theory. *Special issue on Quantum Computation of the Siam Journal of Computing*, Oct.(1997).

[CM01]   KEVIN K.H. CHEUNG, MICHELE MOSCA Decomposing Finite Abelian Groups. *Los Alamos Preprint Archive*, quant-ph/0101004, 2001.

[Ebr89]   WAYNE EBERLY, Computations for algebras and group representations, PhD thesis. *University of Toronto*, (1989).

[FR85]   K. FRIEDL AND L. RÓNYAI, Polynomial time solutions for some problems in computational algebra. *in Proc. 17th Ann. Symp. Theory of Computing*, 153-162,(1985).

[KS05]   NEERAJ KAYAL, NITIN SAXENA, On the Ring Isomorphism and Automorphism Problems. *IEEE Conference on Computational Complexity*, 2-12, 2005.

[Le92]   H. W. LENSTRA JR., Algorithms in algebraic number theory. *Bulletin of the AMS*, 26(2): 211-244, 1992.

[Shor97] PETER SHOR, Polynomial time algorithms for prime factorization and discrete logarithms on a quantum computer. *SIAM Journal on Computing*, 26(5):1484-1509, 1997.

[SS97]   T. SANDERS AND M. A. SHOKROLLAHI, Deciding properties of polynomials without factoring. *Proc. 38th IEEE Foundations of Computer Science*, 46-55, 1997.

[VZG03]   JOACHIM V.Z GATHEN AND JÜRGEN GERHARD, Modern Computer Algebra. *Cambridge University Press, 2nd Ed.*, 2003 .

Lower Bounds and Parameterized Approach for Longest Common Subsequence

Xiuzhen Huang

Department of Computer Science,
Arkansas State University,
P.O. Box 9, State University,
Arkansas 72467, USA
xzhuang@csm.astate.edu

Abstract. In this paper, different parameterized versions of the LONGEST COMMON SUBSEQUENCE (LCS) problem are extensively investigated and computational lower bound results are derived based on current research progress in parameterized computation. For example, with the number of sequences as the parameter k, the problem is unlikely to be solvable in time $f(k)n^{o(k)}$, where n is the length of each sequence and f is any recursive function. The lower bound result is asymptotically tight in consideration of the dynamic programming approach of time $O(n^k)$. Computational lower bounds for polynomial-time approximation schemes (PTAS) for the LCS problem are also derived. It is shown that the LCS problem has no PTAS of time $f(1/\epsilon)n^{o(1/\epsilon)}$ for any recursive function f, unless all SNP problems are solvable in subexponential time. Compared with former results on this problem, this result has its significance. Finally a parameterized approach for the LCS problem is discussed, which is more efficient than the dynamic programming approach, especially when applied to large scale sequences.

1 Introduction

A string s is a subsequence of a string s' if s can be obtained from s' by deleting some characters in s'. For example, "ac" is a subsequence of "atcgt". Given a set of strings over an alphabet Σ, the LONGEST COMMON SUBSEQUENCE problem is to find a common subsequence that has the maximum length. The alphabet Σ may be of fixed size or of unbounded size. The LONGEST COMMON SUBSEQUENCE (LCS) problem is a well-known optimization problem because of its applications, especially in bioinformatics. The fixed alphabet version of the problem is of particular interest considering the importance of sequence comparison (e.g. multiple sequence alignment) in the fixed size alphabet world of DNA and protein sequences. (Note that in computational biology, DNA sequences are in a four-letter alphabet, and protein sequences are in a twenty-letter alphabet).

D.Z. Chen and D.T. Lee (Eds.): COCOON 2006, LNCS 4112, pp. 136–145, 2006.

We study the LONGEST COMMON SUBSEQUENCE problem in parameterized computation in this paper. We first give a brief review on parameterized complexity theory and some recent progress on parameterized intractability. A *parameterized problem* Q is a decision problem consisting of instances of the form (x, k), where the integer $k \geq 0$ is called the *parameter*. The parameterized problem Q is *fixed-parameter tractable* [15] if it can be solved in time $f(k)|x|^{O(1)}$, where f is a recursive function[1]. Certain NP-hard parameterized problems, such as VERTEX COVER, are fixed-parameter tractable, and hence can be solved practically for small parameter values [10]. On the other hand, the inherent computational difficulty for solving many other NP-hard parameterized problems with even small parameter values has motivated the theory of *fixed-parameter intractability* [15]. The W-hierarchy $\bigcup_{t \geq 1} W[t]$ has been introduced to characterize the inherent level of intractability for parameterized problems. Examples of $W[1]$-hard problems include problems such as CLIQUE and DOMINATING SET. It has become commonly accepted that no $W[1]$-hard problem can be solved in time $f(k)n^{O(1)}$ for any function f, i.e., $W[1] \neq FPT$. $W[1]$-hardness has served as the hypothesis for fixed-parameter intractability.

Based on the W[1]-hardness of the CLIQUE algorithm, computational intractability of problems in computational biology has been derived [2,3,16,17,23,25]. For example, in [25], the author point out that "unless an unlikely collapse in the parameterized hierarchy occurs, this (This refers to the results proved in [25] that the problems LONGEST COMMON SUBSEQUENCE and SHORTEST COMMON SUPERSEQUENCE are $W[1]$-hard) rules out the existence of exact algorithms with running time $f(k)n^{O(1)}$ (i.e., exponential only in k) for those problems. This does not mean that there are no algorithms with much better asymptotic time-complexity than the known $O(n^k)$ algorithms based on dynamic programming, e.g., algorithms with running time $n^{\sqrt{k}}$ are not deemed impossible by our results."

Recent investigation in [7,8] has derived stronger computational lower bounds for well-known NP-hard parameterized problems. For example, for the CLIQUE problem, which asks if a given graph of n vertices has a clique of size k, it is proved that unless an unlikely collapse occurs in parameterized complexity theory, the problem is not solvable in time $f(k)n^{o(k)}$ for *any* function f. Note that this lower bound is asymptotically tight in the sense that the trivial algorithm that enumerates all subsets of k vertices in a given graph to test the existence of a clique of size k runs in time $O(n^k)$. Based on the hardness of the CLIQUE problem, lower bound results for a number of computational biology problems have been derived [18,9].

In this paper, we extensively investigate different parameterized versions of the LONGEST COMMON SUBSEQUENCE problem. Our results for the problem strengthen the results in the literature (such as [25]) significantly and advance our understanding on the complexity of the problems.

---

[1] In this paper, we always assume that complexity functions are "nice" with both domain and range being non-negative integers and the values of the functions and their inverses can be easily computed.

2   Terminologies in Approximation

We provide some basic terminologies for studying approximation algorithms and its relationship with parameterized complexity. For a reference of the theory of approximation, the readers are referred to [1].

An *NP optimization problem* Q is a 4-tuple (I_Q, S_Q, f_Q, opt_Q), where

1. I_Q is the set of input instances. It is recognizable in polynomial time;

2. For each instance $x \in I_Q$, $S_Q(x)$ is the set of feasible solutions for x, which is defined by a polynomial p and a polynomial time computable predicate π (p and π only depend on Q) as $S_Q(x) = \{y : |y| \leq p(|x|)$ and $\pi(x, y)\}$;

3. $f_Q(x, y)$ is the objective function mapping a pair $x \in I_Q$ and $y \in S_Q(x)$ to a non-negative integer. The function f_Q is computable in polynomial time;

4. $opt_Q \in \{\max, \min\}$. Q is called a *maximization problem* if $opt_Q = \max$, and a *minimization problem* if $opt_Q = \min$.

An *optimal solution* y_0 for an instance $x \in I_Q$ is a feasible solution in $S_Q(x)$ such that $f_Q(x, y_0) = opt_Q\{f_Q(x, z) \mid z \in S_Q(x)\}$. We will denote by $opt_Q(x)$ the value $opt_Q\{f_Q(x, z) \mid z \in S_Q(x)\}$.

An algorithm A is an *approximation algorithm* for an NP optimization problem $Q = (I_Q, S_Q, f_Q, opt_Q)$ if, for each input instance x in I_Q, A returns a feasible solution $y_A(x)$ in $S_Q(x)$. The solution $y_A(x)$ has an *approximation ratio* $r(n)$ if it satisfies the following condition:

$$opt_Q(x)/f_Q(x, y_A(x)) \leq r(|x|) \text{ if } Q \text{ is a maximization problem}$$
$$f_Q(x, y_A(x))/opt_Q(x) \leq r(|x|) \text{ if } Q \text{ is a minimization problem}$$

The approximation algorithm A has an *approximation ratio* $r(n)$ if for any instance x in I_Q, the solution $y_A(x)$ constructed by the algorithm A has an approximation ratio bounded by $r(|x|)$.

Definition 1. *An NP optimization problem Q has a polynomial-time approximation scheme (PTAS) if there is an algorithm A_Q that takes a pair (x, ϵ) as input, where x is an instance of Q and $\epsilon > 0$ is a real number, and returns a feasible solution y for x such that the approximation ratio of the solution y is bounded by $1 + \epsilon$, and for each fixed $\epsilon > 0$, the running time of the algorithm A_Q is bounded by a polynomial of $|x|$.*

An NP optimization problem Q can be parameterized in a natural way as follows. The following definition offers the possibility to study the relationship between the approximability and the parameterized complexity of NP optimization problems.

Definition 2. *Let $Q = (I_Q, S_Q, f_Q, opt_Q)$ be an NP optimization problem. The parameterized version of Q is defined as follows:*

(1) If Q is a maximization problem, then the parameterized version of Q is defined as $Q_\geq = \{(x, k) \mid x \in I_Q \wedge opt_Q(x) \geq k\}$;

(2) If Q is a minimization problem, then the parameterized version of Q is defined as $Q_\leq = \{(x, k) \mid x \in I_Q \wedge opt_Q(x) \leq k\}$.

3 Lower Bound Results for LCS

In the following we derive the lower bounds for the exact algorithms for the parameterized versions of the LONGEST COMMON SUBSEQUENCE (LCS) problem. We also extend the techniques and derive the lower bounds for the approximation algorithms for the optimization versions of the problem.

3.1 Formal Problem Definitions

Several parameterized versions of the LCS problem are discussed in [2,3,17,25]. We present the four parameterized versions of the problem.

The **LCS-k** problem:
Instance: a set $S = \{s_1, s_2, ..., s_k\}$ of strings over an alphabet Σ, and an integer $\lambda > 0$, where the alphabet Σ is of unbounded size.
Parameter: k.
Question: is there a string $s \in \Sigma^*$ of length λ, which is a subsequence of each string in S?

The **FLCS-k** problem:
Instance: a set $S = \{s_1, s_2, ..., s_k\}$ of strings over an alphabet Σ, and an integer $\lambda > 0$, where the alphabet Σ is of fixed size.
Parameter: k.
Question: is there a string $s \in \Sigma^*$ of length λ, which is a subsequence of each string in S?

The **LCS-λ** problem:
Instance: a set $S = \{s_1, s_2, ..., s_k\}$ of strings over an alphabet Σ, and an integer $\lambda > 0$, where the alphabet Σ is of unbounded size.
Parameter: λ.
Question: is there a string $s \in \Sigma^*$ of length λ, which is a subsequence of each string in S?

The **FLCS-λ** problem:
Instance: a set $S = \{s_1, s_2, ..., s_k\}$ of strings over an alphabet Σ, and an integer $\lambda > 0$, where the alphabet Σ is of fixed size.
Parameter: λ.
Question: is there a string $s \in \Sigma^*$ of length λ, which is a subsequence of each string in S?

The following results on the parameterized complexity of these parameterized problems are known:

- The LCS-k problem is W[t]-hard for $t \geq 1$ [3].
- The FLCS-k problem is W[1]-hard [25].
- The LCS-λ problem is W[2]-hard [3].
- The FLCS-λ problem is in FPT [25].

In particular, we are interested in the FLCS-k problem and the LCS-λ problem, which we discuss in the following sections.

3.2   FLCS-k

In [25], the FLCS-k problem is proved to be $W[1]$-hard. Unless $W[1] = \text{FPT}$, for the FLCS-k problem, the $W[1]$-hardness result rules out the existence of algorithms of time $f(k)n^{O(1)}$ for any function f, where k is the number of strings. In the conclusion of [25], the author pointed out that the $W[1]$-hardness of FLCS-k does not exclude the possibility of having an algorithm of time, say $O(n^{\sqrt{k}})$, which is much more efficient than the $O(n^k)$ time dynamic programming algorithm for the FLCS-k problem.

However, it is proved that

Theorem 1 ([9]). *The FLCS-k problem has no algorithm of time $f(k)n^{o(k)}$ for any function f, unless all SNP problems are solvable in subexponential time.*

Interested readers are referred to [9,18] for a detailed proof of this result.

The class SNP introduced by Papadimitriou and Yannakakis [22] contains many well-known NP-hard problems including, for any fixed integer $q \geq 3$, CNF q-SAT, q-COLORABILITY, q-SET COVER, and VERTEX COVER, CLIQUE, and INDE-PENDENT SET [19]. It is commonly believed that it is unlikely that all problems in SNP are solvable in subexponential time[2].

We define an optimization problem FLCS-k_{opt} and its corresponding parameterized problem FLCS'-k.

> The **FLCS-k_{opt}** problem:
> given a set $S = \{s_1, s_2, ..., s_l\}$ of strings over a fixed alphabet Σ, and an integer $\lambda > 0$, try to find a string $s \in \Sigma^*$ of length λ maximizing the size of a subset S' of S, such that s is a common subsequence of all the strings in S'.

By our definition, the parameterized version of the optimization problem FLCS-k_{opt} is

> The **FLCS'-k** problem:
> Instance: given a set $S = \{s_1, s_2, ..., s_l\}$ of strings over a fixed alphabet Σ, and an integer $\lambda > 0$.
> Parameter: an integer k, $0 < k \leq l$.
> Question: is there a string $s \in \Sigma^*$ of length λ such that s is a common subsequence of at least k strings in the set S?

From the definitions of the two parameterized problems FLCS-k and FLCS'-k, we can see that FLCS-k is a special case of FLCS'-k. There is a trivial linear fpt-reduction from FLCS-k to FLCS'-k: given an instance I_1 of FLCS-k, $I_1 = (S_1 = \{s_1, s_2, ..., s_k\}, \lambda$ and the parameter k), we build an instance I_2 of FLCS'-k, $I_2 = (S_2 = \{s_1, s_2, ..., s_k\}, \lambda$ and the parameter k), which asks if there is a string $s \in \Sigma^*$ of length λ that is a common subsequence of at least k strings

---

[2] A recent result showed the equivalence between the statement that all SNP problems are solvable in subexponential time, and the collapse of a parameterized class called *Mini*[1] to *FPT* [14].

(i.e., all strings) in the set S_2. Obviously, the instance I_2 is a yes-instance for the problem FLCS'-k if and only if the instance I_1 is a yes-instance for the problem FLCS-k, .

Theorem 2 ([8,9]). *Suppose that a problem Q_1 has no algorithm of time $f(k)n^{o(k)}$ for any function f, and that Q_1 is linear fpt-reducible to Q_2. Then the problem Q_2 has no algorithm of time $f'(k)n^{o(k)}$ for any function f'.*

By the above linear fpt-reduction, Theorem 1 and Theorem 2, we have

Lemma 1. *The FLCS'-k problem has no algorithm of time $f(k)n^{o(k)}$ for any function f, unless all SNP problems are solvable in subexponential time.*

Theorem 3 ([8,9]). *Let Q be an NP optimization problem. If the parameterized version of Q has no algorithm of time $f(k)n^{o(k)}$, then Q has no PTAS of running time $f(1/\epsilon)n^{o(1/\epsilon)}$ for any function f, unless all problems in SNP are solvable in subexponential time.*

Therefore, by Lemma 1 and Theorem 3, we have

Theorem 4. *The FLCS-k_{opt} problem has no PTAS of time $f(1/\epsilon)n^{o(1/\epsilon)}$ for any function f, unless all SNP problems are solvable in subexponential time.*

3.3 LCS-λ

The LCS-λ problem is proved to be $W[2]$-hard in [2,3]. Therefore, unless $W[2]$ = FPT, for the LCS-λ problem, there is no algorithm of time $f(\lambda)n^{O(1)}$ for any function f. We prove

Theorem 5. *The LCS-λ problem has no algorithm of time $f(\lambda)n^{o(\lambda)}$ for any function f, unless all SNP problems are solvable in subexponential time.*

Proof. We first give an linear fpt-reduction from DOMINATING SET to the LCS-λ problem. Based on the linear fpt-reduction, the lower bound result for DOMINATING SET [8] and Theorem 2, the theorem is proved.

The fpt-reduction from DOMINATING SET to the LCS-λ problem in [3] for proving the LCS-λ problem is $W[2]$-hard is essentially an linear fpt-reduction.

Given a graph $G = (V, E)$, $|V| = n$, and a parameter λ, and suppose an ascending order of the vertices $\{u_1, u_2, ..., u_n\}$ of G, we will construct a set S of strings such that they have a common subsequence of length λ if and only if G has a dominating set of size λ. The alphabet is $\Sigma = \{a[i,j] : 1 \leq i \leq \lambda, 1 \leq j \leq n\}$. We use the notations: $\Sigma_i = \{a[i,j] : 1 \leq j \leq n\}$, $\Sigma[t,u] = \{a[i,j] : (i \neq t)$ or $(i = t$ and $j \in N[u])\}$.

If $\Gamma \subseteq \Sigma$, let ($\uparrow \Gamma$) be the string of length $|\Gamma|$ which consists of one occurrence of each symbol in Γ in ascending order, and let ($\downarrow \Gamma$) be the string of length $|\Gamma|$ which consists of one occurrence of each symbol in Γ in descending order.

The set S consists of the following strings.

Control strings:

$X_1 = \Pi_{i=1}^{\lambda}(\uparrow \Sigma_i),$
$X_2 = \Pi_{i=1}^{\lambda}(\downarrow \Sigma_i).$
Check strings: For $u = 1, ..., n$:
$X_u = \Pi_{i=1}^{\lambda}(\uparrow \Sigma[i, u]),$
We observe that any sequence C of length λ that is a common subsequence of both control strings must consist of exactly one symbol from each Σ_i in ascending order. For such a sequence C we may associate the set V_c of vertices represented by C: if $C = a[1, u_1]...a[\lambda, u_\lambda]$, then $V_c = \{u_i : 1 \leq i \leq \lambda\} = \{x : \exists i \; a[i, x] \in C\}$.

We will prove that if C is also a subsequence of the check strings $\{X_u\}$, then V_c is a dominating set in G. Let $u \in V(G)$ and fix a substring C_u of X_u, with $C_u = C$. We have the fact [3]:

> **Fact.** For some index j, $1 \leq j \leq \lambda$, the symbol $a[j, u_j]$ occurs in the $(\uparrow \Sigma[j, u])$ portion of X_u, thus $u_j \in N[u]$ by the definition of $\Sigma[j, u]$.

By the above fact, if C is a subsequence of the control and check strings, then every vertex of G has a neighbor in V_c, that is, V_c is a dominating set in G.

On the other hand, if $D = \{u_1, .., u_\lambda\}$ is a dominating set in G with $u_1 < ... < u_\lambda$, then the sequence $C = a[1, u_1]...a[\lambda, u_\lambda]$ is easily seen to be a common subsequence of the strings in S.

The reduction from DOMINATING SET to LCS-λ is an linear fpt-reduction. □

Formally, we give the definition of the optimization problem LCS-λ_{opt}.

> The **LCS-λ_{opt}** problem:
> given a set $S = \{s_1, s_2, ..., s_k\}$ of strings over an alphabet Σ of unbounded size, try to find a string $s \in \Sigma^*$ of maximum length such that s is a common subsequence of all the strings in S.

By our definition, the parameterized version of the optimization problem LCS-λ_{opt} is

> The **LCS'-λ** problem:
> Instance: given a set $S = \{s_1, s_2, ..., s_k\}$ of strings over an alphabet Σ of unbounded size.
> Parameter: an integer $\lambda > 0$.
> Question: is there a string $s \in \Sigma^*$ of length at least λ such that s is a common subsequence of all strings in the set S?

Since that there is a string s of length at least λ such that s is a common subsequence of all strings in S is equivalent to that there is a string s of length exactly λ such that s is a common subsequence of all strings in S, the two problems LCS-λ and LCS'-λ are equivalent. By Theorem 5, the problem LCS'-λ has no algorithm of time $f(\lambda)n^{o(\lambda)}$ for any function f, unless all SNP problems are solvable in subexponential time. This result plus Theorem 3 gives us the following theorem:

Theorem 6. *The LCS-λ_{opt} problem has no PTAS of time $f(1/\epsilon)n^{o(1/\epsilon)}$ for any function f, unless all SNP problems are solvable in subexponential time.*

In [20], the authors showed that the LCS-λ_{opt} problem is inherently hard to approximate in the worst case. In particular, they proved that there exists a constant $\delta > 0$ such that, the LCS-λ_{opt} has no polynomial time approximation algorithm with performance ratio n^δ, unless P = NP. It is obvious to see that this lower bound holds only when the objective function value λ is larger than n^d for a constant $d > 0$. In particular, the lower bound result in [20] does not apply to the case when the value of λ is small. For example, in case $\lambda = n^\delta$, a trivial common subsequence of length one is a ratio-n^δ approximation solution. This implies that for the LCS problem, when the length λ of the common subsequence is a small function of n, no strong lower bound result as that of [20] has been derived.

On the other hand, our lower bound result in Theorem 6 for the LCS problem can be applied when the length of the common subsequence λ is any small function of the length n of each string.

4   Parameterized Approach for LCS

Given k sequences with each sequence of length n, we discuss in this section a parameterized approach, which choose a proper parameter, the diagonal band width b. The time complexity of the approach is $O(b * n^{(k-1)})$.

The parameterized approach for finding the longest common subsequence of two given sequences is of time $O(bn)$, where n is the length of the given sequence, b is the parameter, the value of the diagonal band width. This is a great improvement over the well known dynamic programming approach of time $O(n^2)$. Especially when the length of the given sequence n is very large and the two given sequences are very similar, the parameterized approach with a small value of the diagonal band width b can find the optimal solution more efficiently.

The banded alignment idea has been investigated in [6], but the parameterized approach here incorporates the idea of how to guarantee to find the optimal solution, which is discussed in [26]. To illustrate the basic idea of the parameterized approach, consider the case of two given sequences s_1 and s_2 with the same length n. The well known dynamic programming approach for solving the LCS problem is to build a two dimensional table where each entry represents the length of the longest common subsequence between the corresponding prefix of s_1 and the corresponding prefix of s_2 [11]. There are n^2 entries of the two dimensional table. Consider a diagonal band with width b of entries starting from the middle diagonal. The basic idea of the parameterized approach is to ignore entries outside the diagonal band. If an alignment goes outside of the diagonal band with width b, it is easy to see that the corresponding longest common subsequence cannot have a length of more than $n - b$. This is because the search loses one pair of match each time it moves one entry away from the diagonal. Therefore, if the search stays within the diagonal band with width b and finally gets a common subsequence of length at least $n - b$, it is guaranteed that this solution is optimal. That is, it finds the longest common subsequence of the two given sequences s_1 and s_2. Since this parameterized approach needs to fill up a band with width b of the two dimensional table, it takes linear time $O(bn)$, with b as the parameter.

Our experiment results show the efficiency of the parameterized approach. Especially when the two given sequences are very similar, one could pick a relatively small value for the band b in order to achieve the optimal solution, i.e., the longest common subsequences of the given two sequences.

5   Summary

In this paper computational lower bounds on the running time of the algorithms for different parameterized versions of the LONGEST COMMON SUBSEQUENCE (LCS) problem are extensively investigated. It is proved that the problem FLCS-k is unlikely to have an algorithm of time $f(k)n^{o(k)}$, where n is the length of the sequence, k is the total number of sequences and f is any recursive function. In consideration of the known upper bound of $O(n^k)$, we point out that the lower bound result is asymptotically tight. Computational lower bounds for polynomial-time approximation schemes (PTAS) for the optimization versions of the LCS problem are also derived. We then discuss a parameterized approach for the problem. Compared with the well known dynamic programming approach, the parameterized approach is much more efficient, especially when it is applied to find the longest common subsequence of very large scale sequences, which is common in sequence comparisons in bioinformatics.

References

1. G. Ausiello, P. Crescenzi, G. Gambosi, V. Kann, A. Marchetti-Spaccamela, and M. Protasi, *Complexity and Approximation, Combinatorial Optimization Problems and Their Approximability Properties*, New York: Springer-Verlag, (1999).
2. H. L. Bodlaender, R. G. Downey, M. R. Fellows, M. T. Hallett, and H. T. Wareham, Parameterized complexity analysis in computational biology, *Computer Applications in the Biosciences*, vol. 11, pp. 49-57, (1995).
3. H. L. Bodlaender, R. G. Downey, M. R. Fellows, and H. T. Wareham, The parameterized complexity of sequence alignment and consensus, *Theoretical Computer Science*, vol. 147, pp. 31-54, (1995).
4. L. Cai and J. Chen, On fixed-parameter tractability and approximability of NP optimization problems, *Journal Of Computer and System Sciences*, vol. 54, pp. 465-474, (1997).
5. M. Cesati and L. Trevisan, On the efficiency of polynomial time approximation schemes, *Information Processing Letters*, vol. 64, pp. 165-171, (1997).
6. K. M. Chao, W. R. Pearson, and W. Miller, Aligning two sequences within a specific diagonal band, *Computer Applications in the Biosciences*, vol. 8, pp. 481-487, (1992).
7. J. Chen, B. Chor, M. Fellows, X. Huang, D. Juedes, I. Kanj, and G. Xia, Tight lower bounds for parameterized NP-hard problems, *Information and Computation*, vol. 201, pp. 216-231, (2005).

8. J. Chen, X. Huang, I. Kanj, and G. Xia, Linear FPT reductions and computational lower bounds, in *Proc. of the 36th ACM Symposium on Theory of Computing*, pp. 212-221, (2004).

9. J. Chen, X. Huang, I Kanj and G. Xia, W-hardness linear FPT-reductions: structural properties and further applications, in *proceedings of the Eleventh International Computing and Combinatorics Conference (COCOON 2005), Lecture Notes in Computer Science*, vol. 3595, pp. 975-984, (2005).

10. J. Chen, I. Kanj, and W. Jia, Vertex Cover: Further observations and further improvements, *Journal of Algorithms*, vol. 41, pp. 280-301, (2001).

11. T. H. Cormen, C. E. Leiserson, R. L. Rivest, and C. Stein, *Introduction to Algorithms*, Second Edition, MIT Press, (2001).

12. X. Deng, G. Li, Z. Li, B. Ma, and L. Wang, A PTAS for distinguishing (sub)string selection, *Lecture Notes in Computer Science*, vol. 2380, pp. 740-751, (2002).

13. X. Deng, G. Li, Z. Li, B. Ma, and L. Wang, Genetic design of drugs without side-effects, *SIAM Journal on Computing*, vol. 32, pp. 1073-1090, (2003).

14. R. G. Downey, V. Estivill-Castro, M. R. Fellows, E. Prieto, and F. A. Rosamond, Cutting Up is Hard to Do: the Parameterized Complexity of k-Cut and Related Problems, Electr. Notes Theor. Comput. Sci. 78: (2003).

15. R. Downey and M. Fellows, *Parameterized Complexity*, Springer, New York, (1999).

16. M. Fellows, J. Gramm, and R. Niedermeier, Parameterized intractability of motif search problems, *Lecture Notes in Computer Science*, vol. 2285, pp. 262-273, (2002).

17. M. Hallett, *An Integrated Complexity Analysi of Problems for Computational Biology*, Ph.D. Thesis, University of Victoria, (1996).

18. X. Huang, *Parameterized Complexity and Polynomial-time Approximation Schemes*, Ph.D. Dissertation, Texas A&M University, (2004).

19. R. Impagliazzo, R. Paturi, and F. Zane, Which problems have strongly exponential complexity? *Journal Of Computer and System Sciences*, vol. 63, pp. 512-530, (2001).

20. T. Jiang and M. Li, On the approximation of shortest common supersequence and longest common subsequences, *SIAM Journal on Computing*, vol. 24, pp. 1112-1139, (1995).

21. M. Li, B. Ma, and L. Wang, On the closest string and substring problems, *Jounal of the ACM*, vol. 49, pp. 157-171, (2002).

22. C. Papadimitriou and M. Yannakakis, Optimization, approximation, and complexity classes, *Journal Of Computer and System Sciences*, vol. 43, pp. 425-440, (1991).

23. C. Papadimitriou and M. Yannakakis, On limited nondeterminism and the complexity of VC dimension, *Journal Of Computer and System Sciences*, vol. 53, 161-170, (1996).

24. C. Papadimitriou and M. Yannakakis, On the complexity of database queries, *Journal Of Computer and System Sciences*, vol. 58, pp. 407-427, (1999).

25. K. Pietrzak, On the parameterized complexity of the fixed alphabet shortest common supersequence and longest common subsequence problems, *Journal Of Computer and System Sciences*, vol. 67, pp. 757-771, (2003).

26. S.-H. Sze, *Lectures notes of Special Topics in Computational Biology*, Texas A&M University, (2002).

Finding Patterns with Variable Length Gaps
or Don't Cares

M. Sohel Rahman[1,*,**], Costas S. Iliopoulos[1,***], Inbok Lee[2],
Manal Mohamed[1], and William F. Smyth[3,†]

[1] Algorithm Design Group
Department of Computer Science, King's College London
Strand, London WC2R 2LS, England
{sohel, csi, manal}@dcs.kcl.ac.uk
http://www.dcs.kcl.ac.uk/adg
[2] School of Computer Science and Engineering
Seoul National University, Seoul, Korea
iblee@theory.snu.ac.kr
[3] Algorithms Research Group,
Department of Computing and Software,
McMaster University, Canada
smyth@mcmster.ca

Abstract. In this paper we have presented new algorithms to handle the pattern matching problem where the pattern can contain variable length gaps. Given a pattern P with variable length gaps and a text T our algorithm works in $O(n + m + \alpha \ \log(max_{1<=i<=l}(b_i - a_i)))$ time where n is the length of the text, m is the summation of the lengths of the component subpatterns, α is the total number of occurrences of the component subpatterns in the text and a_i and b_i are, respectively, the minimum and maximum number of don't cares allowed between the ith and $(i+1)$st component of the pattern. We also present another algorithm which, given a suffix array of the text, can report whether P occurs in T in $O(m + \alpha \ \log \log n)$ time. Both the algorithms record information to report all the occurrences of P in T. Furthermore, the techniques used in our algorithms are shown to be useful in many other contexts.

1 Introduction

The classical pattern matching problem is to find all the occurrences of a given pattern P of length m in a text T of length n, both being sequences of characters drawn from a finite character set Σ. This problem is interesting as a fundamental computer science problem and is a basic need of many applications, such as text

---

* Supported by the Commonwealth Scholarship Commission in the UK under the Commonwealth Scholarship and Fellowship Plan (CSFP).
** On Leave from Department of CSE, BUET, Dhaka-1000, Bangladesh.
*** Supported by EPSRC and Royal Society grants.
† Supported in part by an NSERC grant.

D.Z. Chen and D.T. Lee (Eds.): COCOON 2006, LNCS 4112, pp. 146–155, 2006.

retrieval, music retrieval, computational biology, data mining, network security, among many others. Several of these applications require, however, more sophisticated forms of searching. Pattern matching has been generalized to searching with error bounds, e.g. Hamming distance [3,7,14], edit distance [5,14,17]. These variations of the original problems are known as approximate pattern matching and in many, if not most, practical cases it is the approximate version of the pattern matching problem that turns out to be the most applicable one. Several other pattern matching problems have been considered within the approximate paradigm. Fischer and Paterson [8] generalized pattern matching to include don't cares: given a pattern P and a text T, either of which may contain don't cares, denoted $*$, the goal is to output all occurrences of P in T. The new dimension in the problem was the introduction of don't cares, also known as gaps in the literature, which matches any character in the alphabet i.e. $*$ matches every $a \in \Sigma$. Fischer and Paterson presented an algorithm that runs in time $O(n \log n \log \Sigma)$ which was subsequently improved to $O(n \log n)$ by Cole and Hariharan [6]. The don't care paradigm has been extended to several other problems as well. In this paper we are interested in a more general version of this problem where the pattern contains variable length gaps whose length lies within a certain range. We first define the problem more formally.

Problem 1. We are given a pattern P which consists of subpatterns P_i, $1 \le i \le l$. Each subpattern P_i is a string over the alphabet Σ. We are also given for each $1 \le i \le l-1$ two parameters a_i and b_i which imposes constraints on the variable length gaps consisting of don't care characters in P. In particular, a_i and b_i indicate, respectively, the least and highest number of don't cares allowed in the gaps between the two subpatterns P_i and P_{i+1}. The problem is to find the pattern P in a given text T.

As per the definition we can define a pattern P as follows:

$$P = P_1 *^{a_1,b_1} P_2 *^{a_2,b_2} ...P_{l-1} *^{a_{l-1},b_{l-1}} P_l.$$

Note that we define m to be the summation of the lengths of the subpatterns i.e. $m = \sum_{1<=i<=l} |P_i|$. We can think of the gaps as a distance constraint between the neighboring occurrences of the component subpatterns. So if we have a subpattern P_i in position x_i of the text, the occurrence of a P_{i+1} should start within the range $[(x_i + |P_i| + a_i), (x_i + |P_i| + b_i)]$. In other words, any occurrence of P_{i+1} in the text (at position x_{i+1}) is valid if and only if there is an occurrence of P_i in the text at position x_i such that $x_{i+1} \in \{(x_i + |P_i| + a_i), ..., (x_i + |P_i| + b_i)\}$. Note, however that a particular occurrence of a P_{i+1} may turn out to be valid for more than one occurrences of P_i.

Example 1. Suppose we are given a text T and a pattern P as follows.

$T = ACCGAGTGCGTGGACAAACTACGATTGTGGAATGAACT$

$P = AC *^{3,7} GTG *^{4,5} AACT$

As we can see from Figure 1, if we allow (arbitrary) gaps between the subpatterns of P then we can find a number of matches. However only 3 of those obey the gap constraints defined by the pattern P. Let us discuss the matter further for better comprehension. In this case the subpatterns and the gap constraints are defined as follows:

$P_1 = AC$, $P_2 = GTG$ and $P_3 = AACT$.
$(a_1, b_1) = (3, 7)$ and $(a_2, b_2) = (4, 5)$.

Now, in the first two instances although the first gap constraint is met the second one is violated. The third instance is, however, a valid match because it obeys both the gap constraints and so are the 5th and the 6th instances in the figure. The 4th instance on the other hand violates the first gap constraint, again, and hence is not a valid match. To illustrate another point lets take a closer look at the 5th and 6th instances in the figure. Note that the same occurrence of P_2 here is valid for 2 different occurrences of P_1 resulting, ultimately, in two different occurrences of P in T.

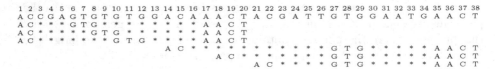

Fig. 1. Valid and invalid matches of Example 1

Applications for this problem include information retrieval, data mining and computational molecular biology. For a detailed discussion on the motivation of this problem we refer the readers to [15,16].

This paper presents new algorithms to solve Problem 1 defined above. We also show some generalizations of our approach in devising the algorithms that seem to be useful in related problems. In particular we first present an $O(n + m + \alpha \ \log(max_{1<=i<=l}(b_i - a_i)))$ algorithm which finds out whether a given P occurs in T where n is the length of the text, m is the summation of the lengths of the component sub patterns, α is the total number of occurrences of the component subpatterns in the text and a_i and b_i are, respectively, the minimum and maximum number of don't cares allowed between the ith and the $(i+1)$st component. It also records the information required to report all the occurrences in $O(l\beta)$ time where β denotes the number of occurrences of the (whole) pattern P in T. In our algorithm we use the famous Aho-Corasick pattern matching algorithm [1] to find out the occurrences of the subpatterns. However, as will be evident later, the technique used in this algorithm can be used with any pattern matching algorithms as long as the occurrences of the component patterns are reported in sorted order. We then present another algorithm using a suffix array to solve the problem. Given a suffix array this algorithm can report whether P is in T in $O(m + \alpha \ \log \log n)$ time. It also records the information to report all the occurrences of the pattern. Furthermore, the technique used in this algorithm can be used with any pattern matching algorithm that doesn't give the occurrences of a pattern in sorted order.

The rest of the paper is organized as follows. In Section 2 we give a very brief literature review and then we discuss our algorithms in Section 3 and Section 4. Finally we conclude in Section 5.

2   Previous Work

The simplest approach is to represent the pattern with variable length gaps as a regular expression and then apply regular expression matching (see [15,9] for details). However, as is pointed out in [15] this approach is too general.

Applying bit-vector simulation of a Nondeterministic Finite Automaton (NFA), based on the Shift-Or algorithm [4] for pattern matching, Navarro and Raffinot [16] presented an $O(m'n/w)$ time algorithm to solve the problem where $m' = \sum_{1 \le i \le l} |P_i| + \sum_{1 \le i \le l-1} |b_i|$ and w is the word length of the target machine. They also presented an $O(m'nk/w)$ time algorithm allowing approximate occurrences of the component subpatterns with error bound k whereas the algorithm of Akutsu [2] for the same problem required $O(mn \log n)$ time, independent of the error bound k. In [2] a combination of a balanced search tree and traditional dynamic programming approach for approximate pattern matching is used.

Lee et. al [15] presented an algorithm that reports whether a given P is in T in $O(nl + m + \alpha)$ time[1]. They also considered approximate occurrences of the sub-patterns in the text allowing k_i errors for each subpattern P_i presenting an $O(mn)$ algorithm. Inspired by [16] they also explored the bit vector approach and presented a $O(m'nk_{max}/w)$ algorithm where $k_{max} = max_{1 \le i \le l}(k_i)$.

3   Pattern Matching Algorithm

In this section we present our algorithm for finding a pattern with variable length gaps. We first outline the steps of the algorithm in Algorithm 1 and then a detailed description with time complexity analysis follows.

3.1   Analysis of the Algorithm

In this subsection we analyze the algorithm outlined above. The algorithm basically builds over the Aho-Corasick pattern matching machine which works in $O(n + m + \alpha)$ time. On top of that, as is indicated in Step 3, we need to perform search operations to validate the occurrences of the subpatterns and we need to do that as efficiently as possible. This is done as follows. Once an occurrence of a P_{i+1} is reported we want to find the P_i (preferably all of them) for which this P_{i+1} is valid. Since PMM outputs the occurrences of the patterns in the dictionary in sorted order, we employ the idea of binary search to find out the range(s) in which P_{i+1} lies in. Observe that to check the validity of a particular P_{i+1} we just need to consider the P_i's reported so far by the PMM. Preferably

---

[1] In [15] the running time is reported to be $O(nl + m)$. But a closer look will reveal that in order to calculate the "candidate ranges" $O(\alpha)$ time is required. However, since $\alpha = O(nl)$ it can be omitted from the running time.

Algorithm 1

1: Build an Aho-Corasick automaton for the dictionary $D = \{P_i | 1 \leq i \leq l\}$
2: Using the above automaton start pattern matching on text T employing the Aho-Corasick Pattern Matching Machine (PMM). We slightly modify the PMM algorithm so that it reports the start positions of the patterns in the dictionary.
3: As soon as the PMM reports a match of a P_{i+1} it searches the valid ranges calculated from the so far reported occurrences of P_i's and if this P_{i+1} is not within any of the valid ranges then it is discarded. Otherwise the valid range (for a possible occurrence of P_{i+2}) is calculated from P_{i+1} and stored. Also with the help of pointers we keep track of the P_i for which this P_{i+1} is valid. Note that there may be more than one such P_i and depending on what is required as a solution we may need to keep track of all of them. Note also that, initially, each occurrence of P_1 is stored along with the calculated valid ranges from them. We use $s_i{}^j$ and $e_i{}^j$ to denote, respectively, the start and end of the valid range calculated from $P_i{}^j$, the jth occurrence of P_i.
4: After the scanning of the text by PMM is complete we just need to explore the pointers to find the occurrences. Note that as soon as we get a valid P_l we can report that P occurs in T.

---

we want to keep track of all the P_i's for which a P_{i+1} is valid. To do that efficiently we also observe that for a particular P_{i+1} we need only to keep track of the first and last occurrences of P_i for which the P_{i+1} under consideration is valid. Now, how do we find out the first and last occurrences of P_i's for which a P_{i+1} is valid? We are given a set of ranges (one for each $P_i{}^j$) defined by $s_i{}^j$ and $e_i{}^j$, $1 \leq j \leq k_i$ where k_i is the number of occurrences of P_i so far. Let x_{i+1} denote the position of P_{i+1} in the text. To achieve our goal we need to find out the highest $s_i{}^j$ that is less than or equal to x_{i+1} and the lowest $e_i{}^j$ that is greater than or equal to x_{i+1}. Let $s_i{}^{j_p} = max\{s_i{}^j | s_i{}^j \leq x_{i+1} \text{ and } 1 \leq j \leq k_i\}$ and $e_i{}^{j_q} = min\{e_i{}^j | e_i{}^j \geq x_{i+1} \text{ and } 1 \leq j \leq k_i\}$. Then it is easy to observe that for all $P_i{}^j$ such that $q \leq j \leq p$ the P_{i+1} under consideration is valid.

$$x_{i+1}$$

$$
\begin{array}{cccccccccccccccccccc}
1 & 2 & 3 & 4 & 5 & 6 & 7 & 8 & 9 & 10 & 11 & 12 & 13 & 14 & 15 & 16 & 17 & 18 & 19 & 20 & \dots \\
P_i & P_i & P_i & P_i & P_i & P_i & P_i & P_i & P_i & P_i & P_i & P_{i+1} & & & & & & & & & \\
\times & \times & \times & \checkmark & \checkmark & \checkmark & \checkmark & \checkmark & \times & \times & \times & & & & & & & & & &
\end{array}
$$

Fig. 2. The positions of P_i for which a P_{i+1} can be valid for the pattern: $\dots P_i *^{3,7} P_{i+1} \dots$. Here we have $x_{i+1} - b_i - 1 = 4$ and $x_{i+1} - a_i - 1 = 8$

As regards the computational effort needed for a particular search it is easy to see that in the worst case we may need $O(\log \alpha)$ effort. However we can do a lot better. Recall that between a P_i and a valid P_{i+1} there must be at least a_i and at most b_i don't care characters. So by careful observation we can see that we just need to check $b_i - a_i + 1$ positions, namely from the position $x_{i+1} - b_i - 1$ to the position

$x_{i+1} - a_i - 1$ (Figure 2). So with a bit of careful programming we can reduce the computational effort down to $O(\log(max_{1<=i<=l}(b_i - a_i)))$ per search. So in the worst case, for the search, we may need to spend $O(\alpha \log(max_{1<=i<=l}(b_i - a_i)))$. However, we believe that in the average case it will need much less computational effort because a large number of the occurrences would be invalid and discarded.

3.2   Reporting the Occurrences

So far what we have done is to find whether the given pattern P exists in the given text T. It is clear that as soon as the algorithm finds one valid P_l we can straightaway report that P is in T and terminate the algorithm. However, in the worst case this would require the same time complexity as $O(n + m + \alpha \log(max_{1<=i<=l}(b_i - a_i)))$. This will happen, for example, if there is only one valid occurrence of P_l and if that is the last occurrence among all the occurrences of all the other component patterns. We, on the other hand, would also like to report the occurrences of P in T if indeed P exists in T. Note, however, that the definition of an occurrence of a pattern P with variable length gaps is not as obvious as that of a "normal" pattern. Normally an occurrence of a normal pattern is reported by either the start position or the end position of that occurrence of the pattern in the text. For example if we want to report the occurrences of the ("normal") pattern ACCTA in the text CACCTAGGTACC-TACCTAGG, we can either report $\{2, 10, 14\}$ or $\{6, 14, 18\}$ i.e., respectively, the start positions or the end positions of each of the occurrences of the pattern. However, this convention, if followed for the patterns with variable length gaps, does not convey the complete information about the occurrence because of the so called "elasticity" in the gaps.

Let us consider Figure 1 once again to get a clear idea. First assume that the pattern we are looking for is $P = AC *^{3,11} GTG *^{4,8} AACT$. This means each of the instances reported in the figure represents a valid match. Now any successful matching algorithm would report $\{20, 38\}$ or $\{1, 14, 18, 21\}$ assuming that it has to report, respectively, the start or end positions of the occurrences. However none of the above conveys the full information. Even reporting both the start and the end positions of the occurrence of the pattern does not give us the full picture as is evident from the first three instances in Figure 1 all of which have the same start and end positions! So it turns out that to completely define an occurrence we need to give the start (or end) position of each of the component subpatterns. Fortunately, the pointer information that we keep determines an implicit graph which can be traversed to enumerate all occurrences of the pattern P in $O(l\beta)$ time where l is, as defined before, the number of component patterns and β is the number of occurrences of the complete pattern P in text T. How this is done is as follows. Each $P_{i+1}{}^k$ keeps two backward pointers each pointing, respectively, to, $P_i{}^q$ and $P_i{}^p$ indicating that for all $P_i{}^j$ such that $q \le j \le p$ the $P_{i+1}{}^k$ is valid. Thus, implicitly, each $P_{i+1}{}^k$ has $p - q + 1$ edges each pointing to a $P_i{}^j, q \le j \le p$. So in effect, we get an implicit graph, which is partitioned into l sets $(V_i, 1 \leq i \leq l)$ of vertices and there can exist edges only between the adjacent partitions; more specifically there can exist directed edges from a vertex in V_{i+1} to a vertex in V_i $(1 \leq i \leq l-1)$ only. It is clear that to report all the occurrences of P we have to enumerate all possible paths of length l of the implicit directed graph. Traversing the graph in depth first manner and with a bit of clever programming we can do this in $O(l\beta)$ time as desired.

3.3 Generalization

It seems that the technique we have used in our algorithm can be generalized to some extent. We can divide our pattern matching algorithm into two separate sections. In the first section we report all the occurrences of the component patterns. In the second section we do the searching to construct the implicit graph. This would work well as long as we use an algorithm for the first part that gives the occurrences in sorted order. Suffix array (or tree), however, doesn't usually give the occurrences in sorted order and hence can not be used here directly. But they have the advantage of having the capability to handle a number of patterns in online fashion by preprocessing the text. It seems that we may need to do some more work to use suffix array (or tree) in our algorithm. We discuss this in a later section. Also, by using appropriate pattern matching algorithm we can even allow approximate matching of the component patterns.

We can think of another generalization from another point of view. It turns out that we can even report any path that is of length less than l using appropriate algorithm. This implies that we can report, if required, the occurrences where some (or perhaps specific) components of the patterns are missing leading to partial matching. To achieve this however, we can't discard any P_i as is done in Step 3 of our algorithm.

4 Using Suffix Array

As is indicated in Section 3.3, the algorithm discussed above cannot be applied directly if the output of the occurrences are generated by a suffix array. This is because suffix array doesn't report the occurrences in sorted order which would prevent us from doing the binary search. Note that if the maximum "elasticity" in gaps i.e. $max_{1<=i<=l}(b_i - a_i)$ is a constant, we can perform a linear search instead of the binary search with slightly asymptotically worse running time. However, it should be noted here that the sorted order also helps us in keeping all the pointer information in a compact form (namely by keeping only the two pointers) and it helps us later in reporting the occurrences as well. Note that in our algorithm we only require that $x_i{}^j < x_i{}^{j+1}$ for $1 \leq i \leq l$ and $1 \leq j \leq k_i$ where k_i is the number of occurrences of P_i. So for each P_i, $1 \leq i \leq l$, as soon as we compute the occurrences of P_i from the suffix array we can sort them and then apply the algorithm as before. So for sorting we have to spend an additional $O(\Sigma_{1 \leq i \leq l} k_i \, \log k_i) = O(\alpha \, \log \alpha)$ time. However we propose another method. In this method we use an elegant data structure invented by Emde Boas [18] that allows us to maintain a sorted list of integers in the range $[1..n]$ in $O(\log \log n)$ time per insertion and deletion. In addition to that it can return next(i) (successor element of i in the list) and prev(i) (predecessor element of i in the list) in constant time. We first present the algorithm for patterns with fixed gaps i.e. when for all $1 \leq i \leq l - 1$, $a_i = b_i$. We will see that we can generalize from that easily.

Algorithm 2

1: Build a suffix array for the text T.
2: For each P_i, $1 \leq i \leq l$ compute the occurrences of P_i in T. Let Occ_i be the collection of the occurrences of P_i and let $|Occ_i| = k_i$.
3: This step is repeated for each pair of P_i and P_{i+1}, $1 \leq i \leq l - 1$. In each round we first create an Emde Boas data structure, EB_i and insert $x_i{}^j + |P_i| - 1 + a_i$ (note that $a_i = b_i$) for all $1 \leq j \leq k_i$. It is clear that for any $P_{i+1}{}^r$ $(1 \leq r \leq k_{i+1})$ to be valid, we must have $x_{i+1}^r - 1 = x_i{}^j + |P_i| - 1 + a_i$ for some $1 \leq j \leq k_i$ and hence if any valid x_{i+1}^r is inserted in EB_i, we must have $x_{i+1}^r - prev(x_{i+1}^r) = 1$. So in this way for $1 \leq r \leq k_{i+1}$, we first insert x_{i+1}^r in EB_i and check whether it is valid or not; if it is valid we keep a pointer to appropriate $|P_i{}^j|$ and then delete it from EB_i.

4.1   Analysis of the Algorithm

In this section we are going to analyze the running time of algorithm 2. The suffix array construction can be done in $O(n)$ [10,12,11], and we can get all the occurrences of the component patterns in $O(m + \alpha)$. Now we need to concentrate on Step 3. In each round in Step 3 we first insert k_i elements requiring $O(k_i \log \log n)$ time. Then for each of the k_{i+1} elements we perform insert, delete and prev operations requiring $O(2k_{i+1} \log \log n)$ time. Since there are in total $l - 1$ rounds, the total computational effort spent is $O(\Sigma_{1 \leq i \leq l-1} \log \log n(k_i + 2k_{i+1})) = O(\alpha \log \log n)$.

4.2   Handling the Variable Length Gaps

The algorithm presented above works for patterns with fixed gaps i.e. when $a_i = b_i$ for all $1 \leq i \leq l - 1$. In order to handle the elasticity in the gaps we need to modify Step 3 of the algorithm slightly and as we shall see this modification will not affect the asymptotic running time of the algorithm. Recall that for a given P_{i+1}^r our goal is to find all $P_i{}^j$ for which P_{i+1}^r is valid. In what follows we discuss the modification for a particular round in Step 3. Instead of one, we create 2 Emde Boas data structure $EB_i{}^{start}$ and $EB_i{}^{end}$. For each $1 \leq j \leq k_i$ we set $s_i{}^j = x_i{}^j + |P_i| + (a_i - 1)$ and $e_i{}^j = x_i{}^j + |P_i| + (b_i + 1)$. This essentially means if P_{i+1}^r is valid then $s_i{}^j < x_{i+1}^r < e_i{}^j$ for some $1 \leq j \leq k_i$. Now for all $1 \leq j \leq k_i$ we insert $s_i{}^j$ in $EB_i{}^{start}$ and $e_i{}^j$ in $EB_i{}^{end}$. This time, unlike what we did in Algorithm 2, for each $s_i{}^j$ and $e_i{}^j$ inserted we also save the corresponding value

j for a reason that will be clear as we proceed. Let us suppose that we have a function F which, given $s_i{}^j$ or $e_i{}^j$ as parameter, returns the corresponding j. Note that we can implement this very easily and in the implementation we do not need a function at all. Now for each $x_{i+1}{}^r$, $1 \le r \le k_{i+1}$ we do the following: We insert $x_{i+1}{}^r$ in both $EB_i{}^{start}$ and $EB_i{}^{end}$. In $EB_i{}^{start}$ we look for $F(prev(x_{i+1}{}^r))$ $(= p$, say$)$ and in $EB_i{}^{end}$ we look for $F(next(x_{i+1}{}^r))$ $(= q$, say$)$. Observe that $P_{i+1}{}^r$ is valid for all $P_i{}^j$ such that $min(p,q) \le j \le max(p,q)$. So here again we just need to keep track of two pointers. Finally it is easy to observe that the asymptotic running time is not increased. So, given a suffix array we can solve whether P is in T in $O(m + \alpha \log\log n)$ time. The disadvantage of this running time is its dependency on n. We, however, would like to point out that $\log\log n$ is a very mild term. For example for a text size of 1024GB $\log\log n \approx 5.32$.

5 Conclusion

In this paper we have presented new algorithms to handle pattern matching problems where the pattern can contain variable length gaps. We have presented an $O(n + m + \alpha \log(max_{1 <= i <= l}(b_i - a_i)))$ algorithm (Algorithm 1) that given a pattern P with variable length gaps and a text T reports whether P is in T. The algorithm by Lee et al. needed a running time of $O(nl + m + \alpha)$. The algorithm also records the information needed to report all the occurrences of P in the form of an implicit graph. We can report all the occurrences by traversing the graph in $O(l\beta)$ time where reporting the occurrences mean reporting the start (or end) position of each of the component patterns. Although we have used the Aho-Corasic pattern matching machine to find out the occurrences of all the component subpatterns, the technique used in our algorithm is equally applicable with any pattern matching algorithm as long as it reports the occurrences in sorted order. As a result we can use the same technique to solve a number of variants of the problem just by using the appropriate pattern matching algorithm instead of the Aho-Chorasic algorithm used here. The running time of these new algorithms would be $O(\chi + \alpha \log(max_{1 <= i <= l}(b_i - a_i)))$ where $O(\chi)$ is the running time of the pattern matching algorithm used. We also present another technique that would work for pattern matching algorithms that doesn't report occurrences of patterns in sorted order, e.g. suffix array. Given a suffix array our algorithm (Algorithm 2) can report whether P is in T in $O(m + \alpha \log\log n)$ time. Again, like Algorithm 1, Algorithm 2 records information needed to report all the occurrences of P in T in the form of an implicit graph.

Acknowledgement

The authors would like to express their gratitude to the reviewers for their useful comments.

References

1. Aho, A., Corasick, M.: Efficient string matching: an aid to bibliographic search. Communications of the ACM, 18 333–340, 1975.
2. Akutsu, T.: Approximate string matching with variable length don't care characters. IEICE Trans. Information and Systems, E79-D:1353–1354, 1996.
3. Amir, A., Lewenstein, M., Porat E.: Faster algorithms for string matching with k mismatches. In Proceedings of the Symposium on Discrete Algorithms (SODA 2000), pages 794–803, 2000.
4. Baeza-Yates, R., Gonnet, G.: A new approach to text searching. Communications of the ACM, 35:74–82, 1992.
5. Cole, R., Hariharan, R.: Approximate string matching: a faster simpler algorithm. In Proceedings of the Symposium on Discrete Algorithms (SODA 1998), pages 463–472, 1998.
6. Cole, R., Hariharan, R.: Verifying candidate matches in sparse and wildcard matching. In Proceedings of the Symposium on Theory of Computing (STOC 2002), pages 592–601, 2002.
7. Galil, Z., Giancarlo, R.: Improved string matching with k mismatches. SIGACT News, 17(4):52–54, 1986.
8. Fischer, M. J., Paterson, M. S.: String matching and other products. Technical report, Massachusetts Institute of Technology, Cambridge, MA, 1974.
9. Gusfield, D.: Algorithms on strings, trees, and sequences. Cambridge University Press, 1997.
10. Kärkkäinen, J., Sanders, P.: Simple linear work Suffix Array Construction. In Proceedings 30th Internat. Colloq. Automata, Languages and Programming (2003), Lecture Notes in Computer Science, 2719, pages 943–955, 2003.
11. Ko, P., Aluru, S.: Space Efficient Linear Time Construction of Suffix Arrays. In Proceedings of the Combinatorial Pattern Matching, 14th Annual Symposium (CPM 2003), Lecture Notes in Computer Science, 2676, pages 200–210, 2003.
12. Kim, D. K., Sim, J. S., Park, H., Park, K.: Linear-Time Construction of Suffix Arrays. In Proceedings of the Combinatorial Pattern Matching, 14th Annual Symposium (CPM 2003), Lecture Notes in Computer Science, 2676, pages 186–199, 2003.
13. Landau, G. M., Vishkin, U.: Efficient string matching with k mismatches. Theoretical Computer Science, 43:239–249, 1986.
14. Landau, G. M., Vishkin, U.: Fast parallel and serial approximate string matching. Journal of Algorithms, 10(2):157–169, 1989.
15. Lee, I., Apostolico, A., Iliopoulos, C. S., Park, K. : Finding approximate occurrence of a pattern that contains gaps. In Proceedings of the 14th Australasian Workshop on Combinatorial Algorithms (AWOCA 2003), pages 89–100, 2003.
16. Navarro, G., Raffinot, M.: Fast and simple character classes and bounded gaps pattern matching. Journal of Computational Biology, 10(6):903–923, 2003.
17. Sahinalp, S. C., Vishkin, U.: Efficient approximate and dynamic matching of patterns using a labeling paradigm. In Proceedings of the Symposium on Foundations of Computer Science, pages 320–328, 1996.
18. van Emde Boas, P.: Preserving order in a forest in less than logarithmic time and linear space. Information Processing Letters, 6:80–82, 1977.

The Matrix Orthogonal Decomposition Problem in Intensity-Modulated Radiation Therapy[*]

Xin Dou[1], Xiaodong Wu[1,2], John E. Bayouth[2], and John M. Buatti[2]

[1] Dept. of Electrical and Computer Engineering, University of Iowa, Iowa City,
Iowa 52242, USA
{xdou, xiaodong-wu}@engineering.uiowa.edu
[2] Dept. of Radiation Oncology, University of Iowa, Iowa City, Iowa 52242, USA
{john-bayouth, john-buatti}@uiowa.edu.

Abstract. In this paper, we study an interesting matrix decomposition problem that seeks to decompose a "complicated" matrix into two "simpler" matrices while minimizing the sum of the horizontal complexity of the first sub-matrix and the vertical complexity of the second sub-matrix. The matrix decomposition problem is crucial for improving the "step-and-shoot" delivery efficiency in Intensity-Modulated Radiation Therapy, which aims to deliver a highly conformal radiation dose to a target tumor while sparing the surrounding normal tissues. Our algorithm is based on a non-trivial graph construction scheme, which enables us to formulate the decomposition problem as computing a minimum s-t cut in a 3-D geometric multi-pillar graph. Experiments on randomly generated intensity map matrices and on clinical data demonstrated the efficiency of our algorithm.

1 Introduction

In this paper, we study an interesting *matrix orthogonal decomposition* problem arising in *intensity-modulated radiation therapy* (IMRT) [15]. IMRT is a modern cancer therapy technique that aims to deliver a highly conformal radiation dose to a target tumor while sparing the surrounding normal tissues. The prescribed dose distribution of radiation is commonly described by an *intensity map* (IM), which is specified by a set of nonnegative integers on a 2-D grid (see Figure 1(a)). The number in a grid cell indicates the amount (in unit) of radiation to be delivered. The delivery is done by a set of cylindrical radiation beams orthogonal to the IM grid.

An advanced tool today for IM delivery is the multileaf collimator (MLC) [15]. An MLC consists of many pairs of tungsten alloy leaves of the same rectangular

---

[*] This research was supported in part by a faculty start-up fund from the University of Iowa, and in part by a fund from the American Cancer Society through an Institutional Research Grant to the Holden Comprehensive Cancer Center, the University of Iowa, Iowa City, IA, USA.

D.Z. Chen and D.T. Lee (Eds.): COCOON 2006, LNCS 4112, pp. 156–165, 2006.

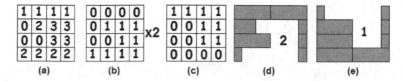

Fig. 1. (a) An example of intensity map. (b) and (c) MLC apertures used to deliver the IM in (a). (d) and (e) The corresponding collimator configurations in (b) and (c).

shape and size. The leaves can move left and right to form a rectilinear region, called an *MLC-aperture*. Each MLC-aperture is associated with an integer representing the radiation units delivered by its radiation beam.

One of the most popular IMRT delivery technique [14] is called the *static leaf sequencing* (SLS) or *step-and-shoot* approach [5, 15, 18] Mathematically, the "step-and-shoot" delivery planning can be viewed as the following matrix decomposition problem: Given an intensity map M (i.e., a matrix), decompose M into the form of $M = \sum_{i=1}^{\kappa} \alpha_i S_i$, where S_i is a special 0-1 matrix specifying an MLC-aperture, α_i is the amount of radiation delivered through S_i, and κ is the number of MLC-apertures used to deliver M (see Figure 1). (The reader is referred to [18, 1, 7, 4] for more details on the step-and-shoot IMRT technique.) There are two obvious measures for the quality of the step-and-shoot delivery: (1) the *beam-on time* which is given by $\sum_{i=1}^{\kappa} \alpha_i$, and (2) the number of MLC-apertures used. The beam-on time is the actual time that the patient is exposed under the radiation beams. Minimizing beam-on time is crucial to reduce the patient's risk under irradiation and to reduce the delivery error caused by the tumor motion [1]. On the other hand, minimizing the number of MLC-apertures used for each IM (hence, minimizing the treatment time of each IM) is also important because it not only lowers the treatment cost for each patient but also enables hospitals to treat more patients [4].

To deliver the IMs, in current SLS method MLC leaves move along one direction (say, horizontally or vertically) during the entire delivery process. This uni-direction delivery may not fully utilize the capacity of the advanced MLC, which is rotatable. In fact, in order to improve the efficiency of the IMRT delivery, it was proposed recently to rotate the MLC between the delivery of the MLC-apertures for an IM [9, 2, 8].

In this paper, we propose to use two orthogonal directions to deliver an IM (i.e., horizontal and vertical) and formulate the following **matrix orthogonal decomposition (MOD)** problem: Given an $m \times n$ non-negative integer matrix $A = (a_{i,j}) \in \mathbb{Z}^{+m \times n}$ (i.e., an IM) and an integer $\lambda \geq 1$, find two matrices (i.e., sub-IMs) $Q = (q_{i,j}), R = (r_{i,j}) \in \mathbb{Z}^{+m \times n}$ such that:

(1) $A = \lambda Q + R$,
(2)   the sum of the *horizontal complexity* $C_H(Q)$ of Q and the *vertical complexity* $C_V(R)$ of R is minimized, where

$$C_H(Q) = \sum_{i=1}^{m} \left(q_{i,1} + \sum_{j=2}^{n} \max(0, q_{i,j} - q_{i,j-1}) \right)$$

$$C_V(R) = \sum_{j=1}^{n} \left(r_{1,j} + \sum_{i=2}^{m} \max(0, r_{i,j} - r_{i-1,j}) \right) \qquad (1)$$

Then, the sub-IM Q and R are delivered in two orthogonal directions. The rational behind this decomposition is based on the following observations. The beam-on time $T_{bot}(B[i])$ for delivering each row $B[i]$ of B equals to $\left(b_{i,1} + \sum_{j=2}^{n} \max(0, b_{i,j} - b_{i,j-1}) \right)$ [7]. The horizontal complexity $C_H(Q)$ measures the total beam-on time of all rows of the IM Q when it is delivered horizontally, while the vertical complexity $C_V(R)$ is the total beam-on time of all columns of R when it is delivered vertically. Hence, the complexity of an IM that we use is closely related to the beam-on time of the IM. It is helpful to note that two IMs A and B with $A = \lambda \cdot B$ for some integer $\lambda > 1$, can be delivered by the same set of MLC-apertures. By adding the factor λ, it is very likely to reduce the total number of MLC-apertures since this can reduce the elements in R and thus the number of MLC-apertures used to deliver R. Most of current approaches for the SLS problem are based on a method for reducing the intensity level of IM matrices, then compute a set of MLC-apertures for the IM matrices with a smaller maximum intensity level [18, 4, 11, 12, 13, 3]. Our decomposition results in two "simpler" sub-IMs with smaller maximum intensity level, which, in turn, yields a more efficient delivery plan using fewer MLC-apertures and/or less total beam-on time.

We model the MOD problem as a minimum s-t cut problem. As an approach of partitioning, the minimum s-t cut has been extensively used. For example, several medical image segmentation techniques based on minimum s-t cuts were developed by, to name a few, Boykov and Jolly [19], Kim and Zabih [20], and Wu and Chen [16].

To our best knowledge, no previous work specifically for solving the matrix orthogonal decomposition problem discussed in this paper was known before. The closely related work is Chen $et~al.$'s optimal linear time algorithm [3] for partitioning an IM matrix A into two sub-IMs of the form $\lambda \cdot Q + R$, without introducing new delivery error while minimizing the maximum intensity level of the sub-IM R.

In this paper, we develop an $T(mn\lfloor \frac{H}{\lambda} \rfloor, mn\lfloor \frac{H}{\lambda} \rfloor)$ time algorithm for the IM matrix orthogonal decomposition problem, where $T(n', m')$ is the time for computing a minimum s-t cut in an edge-weighted directed graph with $O(n')$ vertices and $O(m')$ edges. Our algorithm is based on a non-trivial graph construction scheme, which enables us to formulate the decomposition problem as computing a minimum s-t cut in a 3-D geometric multi-pillar graph (defined in Section 2.1). Experiments on randomly generated IM matrices and on clinical data are performed.

2   Our Algorithm for the IM Orthogonal Decomposition Problem

This section presents our efficient IM matrix orthogonal decomposition (MOD) algorithm. We model the MOD problem as a minimum s-t cut problem on a 3-D geometric multi-pillar graph by a complicated graph transformation scheme.

2.1   Modeling the MOD Problem

We define a 3-D geometric *multi-pillar graph* $G = (V, E)$ on a 2-D $m \times n$ grid Γ from the given IM matrix $A = (a_{i,j})_{m \times n}$ and the integer $\lambda > 0$, as follows.

Let $g(i,j)$ ($0 < i \leq m$ and $0 < j \leq n$) denote a grid point in Γ. For each grid point $g(i,j) \in \Gamma$, there is a set $Col(i,j)$ of $\lfloor \frac{a_{i,j}}{\lambda} \rfloor + 2$ (defined as *height* $h_{i,j}$ of the pillar) vertices in G corresponding to $a_{i,j}$ of the IM matrix A; $Col(i,j) = \{g(i,j,k) \mid k = 1, 2, \ldots, h_{i,j}\}$, called the (i,j)-*pillar* of G (see Figure 2(a) and (b) for an example). In addition, we add two dumbing vertices, a source s and a sink t, in G since we want to formulate our MOD problem as computing a minimum s-t cut in G.

For the ease to introducing edges in G, we here give some notation. We say that two pillars $Col(i,j)$ and $Col(i',j')$ are *adjacent* to each other if $|i-i'|+|j-j'| = 1$. For each pillar $Col(i,j)$, $g(i,j,1)$ (resp., $g(i,j,h_{i,j})$) is called the *base* (resp., *top*) vertex of the pillar. For every vertex $g(i,j,k)$ in G with $i < m$ and $0 < k < h_{i,j}$, we define its *lower neighbor* and its *upper neighbor* on the pillar $Col(i+1,j)$: (1) if $1 \leq (\lfloor \frac{a_{i+1,j}-a_{i,j}}{\lambda} \rfloor + k) \leq h_{i+1,j} - 1$, the lower neighbor of $g(i,j,k)$ is $g(i+1, j, \lfloor \frac{a_{i+1,j}-a_{i,j}}{\lambda} \rfloor + k)$; (2) if $g(i+1,j,k')$ is the lower neighbor of $g(i,j,k)$ and $k' < h_{i+1,j}$, the upper neighbor of $g(i,j,k)$ is $g(i+1,j,k'+1)$. Intuitively, the upper neighbor of $g(i,j,k)$ is the vertex on $Col(i+1,j)$ immediately "above" the lower neighbor of $g(i,j,k)$.

We are now ready to put directed edges in G. We introduce four subsets, E_{vt}, E_{hz}, E_q, and E_r, of directed edges into G, which are used to realize different parts of the complexity equation.

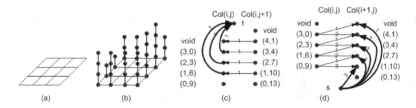

Fig. 2. (a) A 2-D grid. (b) Multi-pillar vertices of the IM in Figure 1. (c) Illustrating E_{hz} (thin edges) and E_q (thick edges) of the case $a_{i,j} = 9$, $a_{i,j+1} = 13$, and $\lambda = 3$. (d) Illustrating E_{vt} (thin) and E_r (thick) of the case $a_{i,j} = 9$, $a_{i+1,j} = 13$, and $\lambda = 3$.

- *The edges in E_{vt}*: Consider pillar $Col(i,j)$ and $Col(i+1,j)$, for $0 < i < m$ and $0 < j \leq n$. For each non-base vertex $g(i,j,k)$, two directed edges are put in E_{vt}: (1) a *lower edge* to its lower neighbor, and (2) an *upper edge* to its upper neighbor (see Figure 2(d)). The weight of the lower edge is $(\lambda - [a_{i+1,j} - a_{i,j}]\%\lambda)$ (note that "%" denotes a modulate operation), and the weight of the upper edge is $([a_{i+1,j} - a_{i,j}]\%\lambda)$. Meanwhile, for the base vertex $g(i,j,1)$, we put an *upper-base edge* with a weight of $([a_{i+1,j} - a_{i,j}]\%\lambda)$ to its upper neighbor. If the lower neighbor $g(i+1,j,lw)$ of $g(i,j,k)$ is not a base vertex, a set of directed edges (called the *lower-base edges*) from $g(i,j,k)$ to $g(i+1,j,k')$ for every $2 \leq k' \leq lw$ is introduced into E_{vt}; the weight of each of these edge is λ. Note that all the above edges are added only when the corresponding neighbor exists and the neighbor is not the base vertex.
- *The edges in E_{hz}*: Consider pillar $Col(i,j)$ and $Col(i,j+1)$, for $0 < i \leq m$ and $0 < j < n$. For each non-base vertex $g(i,j,k)$ on $Col(i,j)$, if $k < \min\{h_{i,j}, h_{i,j+1}\}$, we put an edge from $g(i,j,k)$ to $g(i,j+1,k)$ with a weight of 1. (see Figure 2(c)). If the height $h_{i,j+1}$ of $Col(i,j+1)$ is larger than the height $h_{i,j}$ of $Col(i,j)$, a directed edge of weight 1 is also introduced from each vertex $g(i,j+1,k)$ on the pillar $Col(i,j+1)$ to the top vertex $g(i,j,h_{i,j})$ of $Col(i,j)$, for $k = h_{i,j}, \ldots, h_{i,j+1} - 1$.
- *The edges in E_q*: For each non-base vertex $g(i,1,k)$ of every pillar $Col(i,1)$, $i = 1, 2, \ldots, m$, we put a directed edge of weight 1 from $g(i,1,k)$ to the sink t (see Figure 2(c) when $j = 1$).
- *The edges in E_r*: The top vertex of each pillar $Col(1,j)$, for $j = 1, 2, \ldots, n$, has a directed edge with a weight of $[a_{1,j}\%\lambda]$ from the source s. Additionally, For each non-base, non-top vertex $g(1,j,k)$ of every pillar $Col(1,j)$ we add a directed edge of weight λ from the source s. Figure 2(d) shows an example for this construction when $i = 1$.

In addition, we introduce two more sets of edges, E_{mo} and E_{ad}, into G. The set of edges in E_{mo} is used to guarantee the monotonicity property of the result. While the edges in E_{ad} is employed to avoid the degeneracy of the solution.

- *The edges in E_{mo}*: On each pillar $Col(i,j)$, an edge of weight $+\infty$ is added from every vertex $g(i,j,k)$ to vertex $g(i,j,k-1)$ for $k = 2, 3, \ldots, h_{i,j}$.
- *The edges in E_{ad}*: An edge of weight $+\infty$ is put in E_{ad} from the source s to the base vertex of each pillar. Meanwhile, an edge of weight $+\infty$ is added from the top vertex of each pillar to the sink t.

Hence, the edge set E of G is $E_{vt} \cup E_{hz} \cup E_q \cup E_r \cup E_{mo} \cup E_{ad}$. We thus complete the construction of the multi-pillar graph G.

2.2   Computing an Optimal Matrix Orthogonal Decomposition

The graph G thus constructed allows us to find the optimal matrix orthogonal decomposition for the given IM matrix A, by computing a minimum-weight s-t cut in G. In order to do that, below we prove that following facts: (1) Any valid s-t cut $\mathcal{C}$ (i.e., the total edge weight $w(\mathcal{C})$ of $\mathcal{C}$ is finite) defines a feasible

decomposition of A (i.e., $A = \lambda \cdot Q + R$), such that $C_H(Q) + C_V(R) = w(\mathcal{C})$;
(2) any feasible decomposition of $A = \lambda \cdot Q + R$ specifies a valid s-t cut $\mathcal{C}$ in G,
such that $w(\mathcal{C}) = C_H(Q) + C_V(R)$. Consequently, a valid s-t cut in G with the
minimum total edge weight can be used to specify an optimal matrix orthogonal
decomposition of A.

We first argue that any valid s-t cut in G corresponds to a feasible decomposition
of A and any feasible decomposition of A corresponds to a valid s-t cut in G.

We have these two obvious observations:

Observation 1. *For a valid s-t cut $\mathcal{C} = (S, \bar{S})$ in G, the base vertices of all
pillars are included in the source set S and the all top vertices of pillars are
included in the sink set $\bar{S}$.*

Observation 2. *For a valid s-t cut $\mathcal{C} = (S, \bar{S})$ in G, if a vertex $g(i, j, k) \in
Col(i, j)$ is in the source set S, each vertex $g(i, j, k')$ with $k' < k$ is also in S;
if a vertex $g(i, j, k) \in Col(i, j)$ is in the sink set $\bar{S}$, every vertex $g(i, j, k')$ with
$k' > k$ is also in the sink set $\bar{S}$.*

Thus, we can define a matrix $D = (d_{i,j})_{m \times n}$, $d_{i,j} \in \mathbb{Z}^+, 1 \le d_{i,j} \le h_{i,j} - 1$
to describe a valid s-t cut $\mathcal{C} = (S, \bar{S})$ in G, such that for each pillar $Col(i, j)$,
$S \cap Col(i, j) = \{g(i, j, k) \mid k = 1, 2, \ldots d_{i,j}\}$ and $\bar{S} \cap Col(i, j) = \{g(i, j, k) \mid k =
d_{i,j}+1, d_{i,j}+2, \ldots h_{i,j}\}$. Then, a feasible decomposition of A, with $A = \lambda \cdot Q + R$,
can be defined, as follows. For every pair (i, j) ($1 \le i \le m$ and $1 \le j \le n$),
$q_{i,j} = d_{i,j} - 1$ (Note that $r_{i,j}$ is uniquely defined by $q_{i,j}$).

On the other hand, given a feasible decomposition $A = \lambda \cdot Q + R$, a valid s-t
cut in G can be specified by letting $d_{i,j} = q_{i,j} + 1$ for every pair $(i, j) \in \Gamma$. Hence,
the following lemma holds.

Lemma 1. *Any valid s-t cut in G has a one-to-one correspondence to a feasible
decomposition of the IM matrix A.*

Next, we show that the total edge weight $w(\mathcal{C})$ of $\mathcal{C}$ equals to the complexity of
the decomposition.

From Observations 1 and 2, edges in E_{mo} or in E_{ad} cannot be in $\mathcal{C}$. We thus
only need to consider edges in E_{vt}, E_{hz}, E_q, and E_r. Actually, we are able to
show that the total edge weight of the intersection of $\mathcal{C}$ with E_{vt}, E_{hz}, E_q, and
E_r, equals to $\sum_{j=1}^{n} \sum_{i=2}^{m} \max(0, r_{i,j} - r_{i-1,j})$, $\sum_{i=1}^{m} \sum_{j=2}^{n} \max(0, q_{i,j} - q_{i,j-1})$,
$\sum_{i=1}^{m} q_{i,1}$, and $\sum_{j=1}^{n} r_{1,j}$, respectively.

Lemma 2. *For a valid s-t cut $\mathcal{C} = (S, \bar{S})$ in G, the total edge weight of $\mathcal{C} \cap E_{vt}$
equals to $\sum_{j=1}^{n} \sum_{i=2}^{m} \max(0, r_{i,j} - r_{i-1,j})$.*

Proof. In the construction of the edge set E_{vt}, all edges are added between two
adjacent pillars on the same column of Γ, we thus can first consider the edges
that are between pillars $Col(i, j)$ and $Col(i + 1, j)$, and sum on the whole grid.

Recall our construction scheme and the constraint of range of k (the starting
vertex must be in the source set and the ending vertex must be in the sink set),
the number of lower edges in the cut $\mathcal{C}$ is,

$$\max\left\{0, \min\left\{h_{i+1,j} - 1 - \lfloor\frac{a_{i+1,j} - a_{i,j}}{\lambda}\rfloor, q_{i,j} + 1\right\}\right.$$
$$\left. - \max\left\{2, q_{i+1,j} + 2 - \lfloor\frac{a_{i+1,j} - a_{i,j}}{\lambda}\rfloor\right\} + 1\right\}. \tag{2}$$

For the upper edges between $Col(i,j)$ and $Col(i+1,j)$, in a similar way, we can calculate that the number of such edges in the cut $\mathcal{C}$ is

$$\max\left\{0, \min\left\{h_{i+1,j} - 1 - \lfloor\frac{a_{i+1,j} - a_{i,j}}{\lambda}\rfloor, q_{i,j} + 1\right\}\right.$$
$$\left. - \max\left\{1, q_{i+1,j} + 1 - \lfloor\frac{a_{i+1,j} - a_{i,j}}{\lambda}\rfloor\right\} + 1\right\}. \tag{3}$$

The number of the upper-base and lower-base edges between $Col(i,j)$ and $Col(i+1,j)$ that are in the cut $\mathcal{C}$ is

$$\max(\lfloor\frac{a_{i+1,j} - a_{i,j}}{\lambda}\rfloor - q_{i+1,j}, 0). \tag{4}$$

When $0 \leq a_{i+1,j} - a_{i,j} < \lambda$ or $\lfloor\frac{a_{i+1,j} - a_{i,j}}{\lambda}\rfloor = 0$, number of edges in equation (2), (3), and (4) can be reduced to $\max(0, q_{i,j} - q_{i+1,j})$, $\max(0, q_{i,j} - q_{i+1,j} + 1)$, and 0. Thus the total weight of these edges can be calculated as $\max(r_{i+1,j} - r_{i,j}, 0)$.

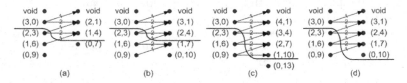

Fig. 3. Examples illustrating the proof of Lemma 2. (a) An example with $\lfloor\frac{a_{i+1,j} - a_{i,j}}{\lambda}\rfloor < 0$, wherein $a_{i,j} = 9$, $a_{i+1,j} = 7$, $q_{i,j} = 2$, $q_{i+1,j} = 0$, and $\lambda = 3$. (b) Increasing $a_{i+1,j}$ to 10 and $q_{i+1,j}$ to 1, $r_{i,j}$ will not be changed, neither are the edges across the cut. (c) An example with $\lfloor\frac{a_{i+1,j} - a_{i,j}}{\lambda}\rfloor > 0$, wherein $a_{i,j} = 9$, $a_{i+1,j} = 13$, $q_{i,j} = 2$, $q_{i+1,j} = 0$, and $\lambda = 3$. (d) Decreasing $a_{i+1,j}$ to 10 and keeping $q_{i+1,j}$ unchanged, $r_{i+1,j}$ is decreased by 3, but an edge of weight 3 can counteract this change.

When $\lfloor\frac{a_{i+1,j} - a_{i,j}}{\lambda}\rfloor > 0$, we can decrease $a_{i+1,j}$ by $\lambda\lfloor\frac{a_{i+1,j} - a_{i,j}}{\lambda}\rfloor$ and $q_{i+1,j}$ by $\min\left\{q_{i+1,j}, \lfloor\frac{a_{i+1,j} - a_{i,j}}{\lambda}\rfloor\right\}$ (to make sure that $q'_{i+1,j} \geq 0$) to $a'_{i+1,j}$ and $q'_{i+1,j}$, respectively. Observe that the case for $q_{i+1,j} \geq \lfloor\frac{a_{i+1,j} - a_{i,j}}{\lambda}\rfloor$ is the same as the case for $\lfloor\frac{a_{i+1,j} - a_{i,j}}{\lambda}\rfloor < 0$. However, if $q_{i+1,j} < \lfloor\frac{a_{i+1,j} - a_{i,j}}{\lambda}\rfloor$, the new $r'_{i+1,j} = a'_{i+1,j} - \lambda q'_{i+1,j}$ will be $(\lfloor\frac{a_{i+1,j} - a_{i,j}}{\lambda}\rfloor - q_{i+1,j}) \times \lambda$ less than the actual $r_{i+1,j}$. The term $\max\left\{\lfloor\frac{a_{i+1,j} - a_{i,j}}{\lambda}\rfloor - q_{i+1,j}, 0\right\} \times \lambda$ can then counteract the change. Hence, in this case, we again have the total weight of the edges in the intersection of the

s-t cut $\mathcal{C}$ and the edges between $Col(i,j)$ and $Col(i+1,j)$, is $\max(r_{i+1,j}-r_{i,j}, 0)$. Figure 3 (c) and (d) illustrate the essential idea using an example.

When $\lfloor \frac{a_{i+1,j}-a_{i,j}}{\lambda} \rfloor < 0$, the situation is similar and Figure 3 (a) and (b) show an example to illustrate the idea.

Taking all the above possibilities into account, we conclude that the total weight of the edges in the intersection of the s-t cut $\mathcal{C}$ and the edges between $Col(i,j)$ and $Col(i+1,j)$, is $\max(r_{i+1,j} - r_{i,j}, 0)$.

By considering all pairs of adjacent pillars on the same columns of Γ, we have $w(\mathcal{C} \cap E_{vt}) = \sum_{j=1}^{n} \sum_{i=2}^{m} \max(0, r_{i,j} - r_{i-1,j})$. Thus, Lemma 2 follows. □

Using a similar argument as for Lemma 2, we have the following lemmas.

Lemma 3. *For a valid s-t cut $\mathcal{C} = (S, \bar{S})$ in G, the total edge weight of $\mathcal{C} \cap E_{hz}$ equals to $\sum_{i=1}^{m} \sum_{j=2}^{n} \max(0, q_{i,j} - q_{i,j-1})$.*

Lemma 4. *For a valid s-t cut $\mathcal{C} = (S, \bar{S})$ in G, the total edge weight of $\mathcal{C} \cap E_q$ equals to $\sum_{i=1}^{m} q_{i,1}$.*

Lemma 5. *For a valid s-t cut $\mathcal{C} = (S, \bar{S})$ in G, the total edge weight of $\mathcal{C} \cap E_r$ equals to $\sum_{j=1}^{n} r_{1,j}$.*

Putting Lemmas 2 - 5 all together, we have the following fact.

Lemma 6. *For any valid s-t cut $\mathcal{C}$ in G and its specified decomposition of A, with $A = \lambda \cdot Q + R$, we have $w(\mathcal{C}) = C_H(Q) + C_V(R)$.*

From Lemmas 1 and 6, an minimum-weight s-t $\mathcal{C}^*$ in G can be used to define an optimal matrix orthogonal decomposition of A, with $A = \lambda \cdot Q^* + R^*$, such that $C_H(Q^*) + C_V(R^*)$ is minimized. Note that $|V| = O(mn\lfloor \frac{H}{\lambda} \rfloor)$ and $|E| = O(mn\lfloor \frac{H}{\lambda} \rfloor)$, where H is the largest intensity level in the IM matrix A. Denote by $T(n', m')$ the time for finding a minimum s-t cut in an edge-weighted directed graph with $O(n')$ vertices and $O(m')$ edge. We have our main result.

Theorem 3. *The MOD problem can be solved in $T(mn\lfloor \frac{H}{\lambda} \rfloor, mn\lfloor \frac{H}{\lambda} \rfloor)$ time.*

3   Experiment Results

To evaluate our algorithm, we performed some statistical studies using 1000 randomly generated 15×15 IM matrices each with intensity levels range from 4 to 64 in powers of 2. The number of MLC-apertures are computed using Xia and Verhey's algorithm [18] without considering interleaf motion constraint.

Table 1 shows percentage of IMs getting improved and the average results (both beam-on time and number of MLC-apertures) before and after performing our decomposition method (the average is calculated based only on those IMs getting improved). We observed that our MOD algorithm generated as much as 38.1% less MLC-apertures and 33.3% less beam-on time than single direction delivery.

Table 1. The average beam-on time and the number of MLC-apertures

	# of MLC-apertures				beam-on time	
	%improved	avg before	avg after	%improved	avg before	avg after
4	9%	5.98±0.60	5.78±0.44	25%	7.75±0.99	7.32±0.80
8	24%	9.07±0.64	8.46±0.66	66%	17.09±1.96	15.47±1.56
16	25%	12.04±0.71	11.44±0.58	73%	35.33±4.27	32.05±3.95
32	45%	14.91±0.85	14.20±0.69	81%	69.71±8.93	63.41±6.46
64	54%	18.16±0.94	17.11±0.72	94%	144.15±18.23	129.04±13.08

We have also experimented with some real medical data sets available to us. 77% IMs that we tested on got improved number of MLC-apertures. Our MOD algorithm produced as much as 27.3% less MLC-apertures with an average of 13.1% comparing with the SLS method using a single direction for delivery.

The experiments are performed on a Pentium-D 2.8GHz computer with 3.5GB of memory. We used a program provided by Matlab to compute the minimum s-t cut in a graph, and expected to have a much faster execution time by implementing the minimum cut algorithm using C. The average execution time of decomposition is shown in Table 2. Our experiments on randomly generated IMs and on the clinical data demonstrated the efficiency of our MOD algorithm. Although the worst cast running time of our MOD algorithm is pseudo-polynomial with respect to the maximum intensity level H of the IM matrix, its practical execution time on real medical data is expected to be quite short, since on the medical data sets used in current clinical treatments, the maximum intensity level of an IM matrix is rarely larger than 100 and is mostly about tens.

Table 2. Execution Times (in seconds)

Maximum intensity	Size 10×10		Size 15×15	
level (H)	$\lambda = 1$	$\lambda = \lfloor \sqrt{H} \rfloor$	$\lambda = 1$	$\lambda = \lfloor \sqrt{H} \rfloor$
4	0.3125	0.1955	0.9925	0.6330
8	0.6090	0.3360	2.2420	1.2815
16	1.6570	0.4135	5.7425	1.3360
32	5.1560	0.7265	19.1880	2.6635

References

[1] N. Boland, H.W. Hamacher, and F. Lenzen, Minimizing Beam-on Time in Cancer Radiation Treatment Using Multileaf Collimators, *Networks*, 43(4):226-240, 2004.
[2] Y. Chen, Q. Hou, and J.M. Galvin, A graph-searching method for MLC leaf sequencing under constraints, *Med. Phys.*, 31(2004), pp. 1504-11.
[3] D.Z. Chen, X.S. Hu, S. Luan, and C. Wang, Mountain Reduction, Block Matching, and Medical Applications. *Proc. of the 21st Annual Symposium on Computational Geometry (SoCG)*, Pisa, Italy, pp. 35-44, 2005.

[4] D.Z. Chen, X.S. Hu, S. Luan, C. Wang, and X. Wu. Geometric Algorithms for Static Leaf Sequencing Problems in Radiation Therapy. *International Journal of Computational Geometry and Applications*, 14(5):311-339, 2004.

[5] J. Dai and Y. Zhu. Minimizing the Number of Segments in a Delivery Sequence for Intensity-Modulated Ratiation Therapy with Multileaf Collimator. *Med. Phys.*, 28(10):2113-2120, 2001.

[6] N. Dogan, L.B. Leybovich, A. Sethi, and B. Emami, Automatic Feathering of Split Fields for Step-and-Shoot Intensity Modulated Radiation Therapy, *Phys. Med. Biol.*, 48:1133-1140, 2003.

[7] K. Engel, A New Algorithm for Optimal Multileaf Collimator Field Segmentation, *http://www.trinity.edu/aholder/HealthApp/oncology/papers/paperlist.html*, March 2003.

[8] B. Hardemark, H. Rehbinder, and J. Lof, Rotating the MLC Between Segments Improves Performance in step-andshoot IMRT delivery, Presentation at the AAPM 45th Meeting (1014 August 2003), *Med. Phys.*, 30(2003).

[9] T. Kalinowski, Reducing the number of monitor units in multileaf collimator field segmentation. *Physics in Medicine and Boilogy* 50:1147–1161, 2005.

[10] M. Langer, V. Thai, and L. Papiez, Improved Leaf Sequencing Reduces Segments of Monitor Units Needed to Deliver IMRT Using Multileaf Collimators. *Med. Phys.* 28:2450-2458, 2001.

[11] L.D. Potter, S.X. Chang, T.J. Cullip, and A.C. Siochi, A Quality and Efficiency Analysis of the $IMFAST^{TM}$ Segmentation Algorithm in Head and Neck "Step & Shoot" IMRT Treatments, *Med. Phys.*, 29(3)(2002), pp. 275-283.

[12] W. Que, Comparison of Algorithms for Multileaf Collimator Field Segmentation, *Med. Phys.*, 26(1999), pp. 2390-2396.

[13] R. Svensson, P. Kallman, and A. Brahme, An Analytical Solution for the Dynamic Control of Multileaf Collimation, *Phys. in Med. and Biol.*, 39(1994), pp. 37-61.

[14] S.Webb. *The Physics of Conformal Radiotherapy Advances in Technology*. Bristol, Institute of Physics Publishing, 1997.

[15] S.Webb, *Intensity-Modulated Radiation Therapy*, Institute of Cancer Research and Royal Marsden NHS Trust, Jan. 2001.

[16] X. Wu and D.Z. Chen, Optimal Net Surface Problems with Applications, *lecture Notes in Computer Science*, Vol. 2380, Springer Verlag, *Proc. of the 29th International Colloquium on Automata, Languages and Programming (ICALP)*, Malaga, Spain, July 2002, pp. 1029-1042.

[17] Q. Wu, M. Arnfield, S. Tong, Y. Wu, and R. Mohan, Dynamic Splitting of Large Intensity-Modulated Fields, *Phys. Med. Biol.*, 45:1731-1740, 2000.

[18] P. Xia and L.J. Verhey., MLC Leaf Sequencing Algorithm for Intensity Modulated Beams with Multiple Static Segments. *Med. Phys.*, 25:1424-1434, 1998.

[19] Y. Boykov and M.-P. Jolly, Interactive Organ Segmentation Using Graph Cuts", *Proc. Medical Image Computing and Computer-Assisted Intervention (MICCAI)*, 276-286, 2000.

[20] J. Kim and R. Zabih, A Segmentation Algorithm for Contrast-Enhanced Images, *Proc. IEEE Int'l Conf. Computer Vision*, 502-509, 2003.

A Polynomial-Time Approximation Algorithm for a Geometric Dispersion Problem

Marc Benkert[1,*], Joachim Gudmundsson[2], Christian Knauer[3], Esther Moet[4], René van Oostrum[4], and Alexander Wolff[1,*]

[1] Faculty of Computer Science, Karlsruhe University, P.O. Box 6980,
D-76128 Karlsruhe, Germany
http://i11www.iti.uka.de/algo/group
[2] National ICT Australia Ltd[**], Sydney, Australia
joachim.gudmundsson@nicta.com.au
[3] Institute of Computer Science, Freie Universität Berlin, Germany
christian.knauer@inf.fu-berlin.de
[4] Department of Computing and Information Sciences, Universiteit Utrecht,
The Netherlands
{esther, rene}@cs.uu.nl

Abstract. We consider the problem of placing a maximal number of disks in a rectangular region containing obstacles such that no two disks intersect. Let α be a fixed real in $(0, 1]$. We are given a bounding rectangle P and a set $\mathcal{R}$ of possibly intersecting unit disks whose centers lie in P. The task is to pack a set $\mathcal{B}$ of m disjoint disks of radius α into P such that no disk in $\mathcal{B}$ intersects a disk in $\mathcal{R}$, where m is the maximum number of *unit* disks that can be packed. Baur and Fekete showed that the problem cannot be solved in polynomial time for $\alpha \geq 13/14$, unless $\mathcal{P} = \mathcal{NP}$. In this paper we present an algorithm for $\alpha = 2/3$.

1 Introduction

Obnoxious facility location problems consider the placement of facilities of which clients consider it undesirable to be in the proximity, for instance, nuclear power plants or garbage dumps. There are several models for and variations to the problem; see the survey by Cappanera [3]. We consider the following instance: we are given a bounding rectangle P, a set $\mathcal{R}$ of n points in P (the red points), and an integer k, and we should construct a set $\mathcal{B}$ of k (blue) points such that the minimum distance from a blue point to another point (either red or blue) is maximized over all points in $\mathcal{B}$. If the optimal distance is denoted by r_{opt}, then we can reformulate the problem as follows: we are given a set of n centers of possibly overlapping red disks with unknown radius r_{opt}, and we are to determine r_{opt}

* Supported by grant WO 758/4-2 of the German Research Foundation (DFG). Part of this work was done while M. Benkert visited NICTA on DAAD grant D/05/08276.
** Funded by the Australian Government's Backing Australia's Ability initiative, in part through the Australian Research Council.

D.Z. Chen and D.T. Lee (Eds.): COCOON 2006, LNCS 4112, pp. 166–175, 2006.

and to find a set of k blue disks with radius r_{opt} such that no blue disk overlaps any other disk, whether red or blue.

The problem of packing objects into a bounded region is one of the classic problems in mathematics and theoretical computer science, see for example the monographs [11,13] which are solely devoted to this problem, and the survey by Tóth [12]. In this paper we consider problems related to packing disks into a polygonal region. As pointed out by Baur and Fekete in [2], even when the structure of the region and the objects are simple, only very little is known, see for example [6,9].

Consider the following decision problem corresponding to our optimization problem: we are given $\mathcal{R}$ and k and a radius r, and we must decide whether $r > r_{opt}$ or $r \leq r_{opt}$. In the latter case, we must also give a set $\mathcal{B}$ of k blue disk with radius r such that no blue disk overlaps any other disk. If we had an algorithm at our disposal that solves the decision problem in polynomial time, then we could solve the original optimization problem in polynomial time by applying Megiddo's parametric search [10]. Unfortunately, the decision problem is known to be $\mathcal{NP}$-complete [5]. Therefore we are looking for an algorithm that approximates the decision problem in the following sense: if m disks of radius r can be placed, then our algorithm places m disks of radius αr, for some fixed $\alpha \in (0, 1]$. If $m < k$, then we know that $r > r_{opt}$, and if $m \geq k$, then either $r \leq r_{opt}$, or $r > r_{opt}$ and $\alpha r \leq r_{opt}$. In other words, placement of at least k disks of radius less than αr_{opt} is guaranteed, and we may or may not be fortunate for radii in between αr_{opt} and r_{opt}.

Obviously, we would like to maximize α while keeping a polynomial running time. Given such an algorithm, we can use it to compute an α-approximation to the original optimization problem, again by using parametric search, albeit in a somewhat non-standard way.

By rescaling r to 1, we can regard the decision problem as that of packing $m \geq k$ unit disks into a rectangle that is already partially covered by n unit disks. In this paper, we consider the following problem:

Problem 1 (APPROXSIZE). *Let $\alpha \in (0, 1]$ be a fixed real. Given a bounding rectangle P and a set $\mathcal{R}$ of possibly intersecting unit disks whose centers lie in P, pack $\geq m$ non-intersecting disks of radius α into P, where m is the maximal number of unit disks that can be packed in P.*

Note that we do not know the value of m a priori. For $\alpha = 1/2$, the problem can be solved by placing disks with radius $1/2$ greedily, i.e., as long as there is space to place a disk, we place one at an arbitrary feasible position. The following simple charging argument shows that we will place at least m disk of radius $1/2$ in this way. Consider an arbitrary placement of m unit disks, and charge a disk C with radius $1/2$ to a unit disk D if the center c of C lies inside D. After the greedy algorithm has finished, all of the m unit disks have a charge of at least one. Otherwise, we can place a disk with radius $1/2$ in an uncharged unit disk such that their centers coincide, and this contradicts the termination condition of the greedy algorithm.

In their pioneering work [7] Hochbaum and Maas gave a polynomial-time approximation scheme (PTAS) for the problem of packing a maximal number of unit disks into a region. The problem is known to be $\mathcal{NP}$-complete [5]. Even though the corresponding geometric dispersion problem looks very similar, inapproximability results have been shown. Baur and Fekete [2] proved hardness results for a variety of geometric dispersion problems, and their results can be modified to our setting with a bit of effort. Specifically, they showed that APPROXSIZE cannot be solved in polynomial time for any radius that exceed $13/14$, unless $\mathcal{P} = \mathcal{NP}$. Furthermore, for the case when the objects are squares, Baur and Fekete gave an $O(\log k \cdot n^{40})$-time $2/3$-approximation algorithm, where k is the number of squares. However, since a square is a simpler shape and easier to pack than a disk their approach cannot be generalized to disks. The main contribution of this paper is a polynomial-time $2/3$-approximation algorithm. Actually, we conjecture that $2/3$ is indeed the largest value for which the problem can be solved in polynomial time.

APPROXSIZE has applications in non-photorealistic rendering system, where 3D models are to be rendered in an oil painting style, as well as in random examinations of, e.g., soil or water.

2   Algorithm Outline

We now give a rough outline of our algorithm DISKPACKING. We use the term r-disk as shorthand for a disk of radius r. For $r > 0$ and a set $R \subseteq \mathbb{R}^2$ let the r-freespace $\mathcal{F}_r(R)$ of R be the set of the centers of all r-disks that are completely contained in R. By $\mathcal{F}_r^{\otimes}(R)$ we denote the Minkowski sum of $\mathcal{F}_r(R)$ and an r-disk.

We first compute the sets $\mathcal{F}_1 = \mathcal{F}_1(P \setminus \bigcup \mathcal{R})$ and $\mathcal{F}_1^{\otimes} = \mathcal{F}_1^{\otimes}(P \setminus \bigcup \mathcal{R})$, see Fig. 1. Then, we apply the PTAS of Hochbaum and Maas [7] to $\mathcal{F}_1^{\otimes}$. For any positive integer t, the PTAS packs in $O(n^{t^2})$ time at least $(1 - 1/t)^2 \cdot m$ unit disks into $\mathcal{F}_1^{\otimes}$, where m is the maximum number of unit disks that can be packed into $\mathcal{F}_1^{\otimes}$ and n is the minimum number of unit squares whose union covers $\mathcal{F}_1^{\otimes}$. Setting $t = 25$ we obtain in $O(n^{625})$ time a set $\mathcal{B}$ of $m' \geq 12/13 \cdot m$ unit disks.

Note that the approximation scheme by Hochbaum and Maas can be modified such that the algorithm is strongly polynomial with respect to the size of our input. If the number of disks that can be packed is not polynomial in the size of P and $\mathcal{R}$ then there must exist a huge empty square region within P. This can be "cut out" and packed almost optimally by using a naïve approach. The added error obtained is bounded by $O(1/\tilde{n}^2)$ where $\tilde{n}$ is the optimal number of disks that can be packed in the square. This step can be repeated until there are no more huge empty squares.

Given $\mathcal{B}$ we compute a set $\mathcal{B}_{2/3}$ of disks of radius $2/3$ that has cardinality at least $13/12 \cdot m' \geq m$ and is contained in $P \setminus \bigcup \mathcal{R}$. We obtain $\mathcal{B}_{2/3}$ in two steps. First, we compute a sufficiently large matching in the nearest-neighbor graph $G = (\mathcal{B}, E)$ of $\mathcal{B}$ with respect to a metric $\text{dist}(\cdot, \cdot)$ that we will specify later. Second, we define a region for each pair of matching unit disks such that we can place three $2/3$-disks in each region (see Fig. 2) and all regions are pairwise

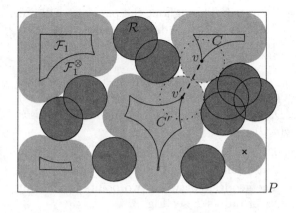

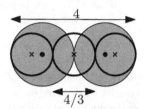

Fig. 1. The 1-freespace $\mathcal{F}_1^{\otimes}$ (light shaded) and a shortcut vv' (dashed) between the connected components C and C' of $\mathcal{F}_1$

Fig. 2. Packing three 2/3-disks in the region spanned by a pair of unit disks

disjoint. For each unmatched unit disk D we place a 2/3-disk D' such that the centers d and d' of D and D', respectively, coincide.

In the next sections we describe each step of Algorithm DISKPACKING in more detail.

3   The Freespace and a Metric on Unit Disks

We briefly recall the setting. We are given a set $\mathcal{R}$ of unit disks whose centers lie in a rectangle P. The disks in $\mathcal{R}$ are allowed to intersect. We first compute the freespace $\mathcal{F}_1$ of $P \setminus \bigcup \mathcal{R}$. According to Kedem et al. [8] the union of s disks can be computed in $O(s \log^2 s)$ time and its complexity is linear in s. Applying their algorithm to the disks in $\mathcal{R}$ scaled by a factor of 2 and intersecting the resulting union with P, we can compute $\mathcal{F}_1$ in $O(|\mathcal{R}| \log^2 |\mathcal{R}|)$ time, where $|\mathcal{R}|$ is the cardinality of $\mathcal{R}$.

Next, we want to introduce a metric $\mathrm{dist}(\cdot, \cdot)$ on unit disks in $\mathcal{F}_1^{\otimes}$. With the current definition of $\mathcal{F}_1$ we have the problem that two unit disks centered on points in different connected components of $\mathcal{F}_1$ can intersect. We solve this problem by considering a superset $\mathcal{F}_1^+$ of $\mathcal{F}_1$ that connects close components of $\mathcal{F}_1$. By $|p, q|$ we denote the Euclidean distance between two points p and q in the plane.

Definition 1. *Let $\mathcal{C}_1$ be the set of connected components of $\mathcal{F}_1$, and let $C, C' \in \mathcal{C}_1$. Let v and v' be vertices on the boundaries of C and C', respectively. We say that the line segment vv' is a shortcut if $|v - v'| \le \sqrt{11} \cdot 2/3 \approx 2.21$. Let $\mathcal{S}(C, C')$ be the set of all shortcuts induced by C and C'. We set $\mathcal{F}_1^+ = \mathcal{F}_1 \cup \bigcup_{C, C' \in \mathcal{C}_1; \, s \in \mathcal{S}(C, C')} s$.*

Figure 1 depicts $\mathcal{F}_1$, $\mathcal{F}_1^\otimes$, and a shortcut vv'. Throughout the paper we will use upper-case letters to denote disks and the corresponding lower-case letters to denote their centers. Now, we are ready to define our metric for a connected component of $\mathcal{F}_1^+$, see Fig. 3.

Definition 2. *Let D and D' be unit disks in $\mathcal{F}_1^\otimes$, with centers d and d', respectively. The distance $\mathrm{dist}(D, D')$ of D and D' is the length of the geodesic $g(d, d')$ of d and d' in $\mathcal{F}_1^+$. The tunnel $T(D, D')$ of D and D' is the union of all r-disks in $P \setminus \bigcup \mathcal{R}$ centered at points of $g(d, d')$ with $r \leq 1$.*

It is easy to see that any 2/3-disk $D_{2/3}$ centered at a point of $g(d, d')$ does not intersect any disk in $\mathcal{R}$. (This will also follow from Lemma 2.) Thus $D_{2/3}$ is contained in the tunnel $T(D, D')$. The geodesic between two points in $\mathcal{F}_1^+$ can only consist of line segments and arcs of radius 2, see Fig. 3.

Recall that our algorithm computes a matching in the nearest-neighbor graph $G = (\mathcal{B}, E)$ induced by the metric $\mathrm{dist}(\cdot, \cdot)$ on the set $\mathcal{B}$ of unit disks that we get from the PTAS by Hochbaum and Maas. For each pair $\{D, D'\}$ in the matching we define a region $T_{2/3}(D, D')$ into which we will then place three 2/3-disks as in Fig. 2. An obvious way to define $T_{2/3}(D, D')$ would be to take the union of all 2/3-disks centered at points of the geodesic between d and d' in $\mathcal{F}_{2/3}$. Our definition is not as straight-forward, but will simplify the proof that $T_{2/3}(D, D')$ and $T_{2/3}(F, F')$ are disjoint if D, D', F, and F' are pairwise disjoint. This is needed to ensure that the 2/3-disks that we will place in the tunnels $T_{2/3}$ are disjoint.

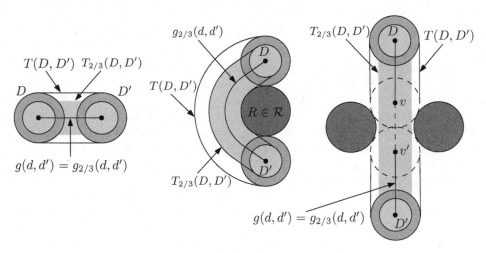

Fig. 3. The geodesic $g(d, d')$. Left: unrestricted. Center: obstacle R. Right: shortcut.

Definition 3. *Let D and D' be unit disks in $\mathcal{F}_1^\otimes$. Let $g_{2/3}(d, d')$ be a geodesic from d to d' in $\mathcal{F}_{2/3}(T(D, D'))$. Then, the 2/3-tunnel $T_{2/3}(D, D')$ of D and D' is the union of all 2/3-disks centered at points of $g_{2/3}(d, d')$.*

According to Chang et al. [4] the geodesics $g(d, d')$ and $g_{2/3}(d, d')$ from d to d' can be computed in $O(|\mathcal{R}|^2 \log |\mathcal{R}|)$ time.

4   The Nearest-Neighbor Graph

Recall that m is the maximum number of disjoint unit disks that fit into $\mathcal{F}_1^{\otimes}$. For $t = 25$ the $(1 - 1/t)^2$-approximation of Hochbaum and Maas [7] yields a set $\mathcal{B}$ of $m' \geq 12m/13$ unit disks in $\mathcal{F}_1^{\otimes}$. Our plan is to compute the nearest-neighbor graph $G = (\mathcal{B}, E)$ induced by the metric $\mathrm{dist}(\cdot, \cdot)$, find a matching of sufficient size in G, and finally place three $2/3$-disks in the $2/3$-tunnel $T_{2/3}(C, D)$ for each pair $\{C, D\}$ in the matching. We show that if we place another $2/3$-disk for each unmatched disk in $\mathcal{B}$, then we place at least $13m'/12 \geq m$ disks of radius $2/3$ by this approach.

By construction, two unit disks D_1 and D_2 whose centers lie in different components of $\mathcal{F}_1^{+}$ have an empty intersection, so we can consider each connected component of $\mathcal{F}_1^{+}$ separately.

After running the algorithm of Hochbaum and Maas we greedily add to $\mathcal{B}$ disjoint unit disks in $\mathcal{F}_1^{\otimes} \setminus \bigcup \mathcal{B}$ until no more disks can be added. This is needed to ensure the following lemma:

Lemma 1. *The nearest-neighbor graph $G = (\mathcal{B}, E)$ (w.r.t. dist) is planar and has maximum degree 6.*

Proof. Let $C \in \mathcal{B}$ be an arbitrary unit disk, let $C' \in \mathcal{B}$ be the nearest neighbor of C in $\mathcal{B}$, and let $\mathcal{D} = \{D_1, \ldots, D_k\} \subseteq \mathcal{B}$ be the neighbors of C in $\mathcal{B}$ for which C is their nearest neighbor. If $k \leq 5$ then the degree bound obviously holds, thus we only have to consider the case when $k \geq 6$. For each disk D_i, $1 \leq i \leq k$, place a unit disk D_i' with center on $g(c, d_i)$ such that $|c, d_i'| = 2$, i.e., D_i' touches C. From the definition of the nearest-neighbor graph it follows that every point on $g(c, d_i)$ is closer to C or D_i than to any other unit disk in $\mathcal{B} \setminus \{D_i\}$. As a result the set $D_1', \ldots, D_k'$ and C' has to be disjoint. Using a simple packing argument, $k = 6$ and $C' \in \mathcal{D}$ follows and thus the degree bound stated in the lemma holds.

Finally, G is planar since no two edges in a nearest-neighbor graph can intersect. $\qquad\square$

From now on we will call $\{C, D\} \subseteq \mathcal{B}$ a *nearest pair* if either (C, D) or (D, C) is an edge in G, i.e., either D is the nearest disk in $\mathcal{B}$ to C or C is the nearest disk in $\mathcal{B}$ to D. For every nearest pair $\{C, D\}$ we define $\mathcal{A}(C, D)$ to be $C \cup D \cup T_{2/3}(C, D)$. As the nearest pair $\{C, D\}$ is a potential candidate to become a matching pair, we want to ensure that we can use $\mathcal{A}(C, D)$ to pack three $2/3$-disks in it such that all the packed $2/3$-disks are pairwise disjoint. Thus, we have to prove:

(i) three $2/3$-disks fit into $\mathcal{A}(C, D)$ and
(ii) for any nearest pair $\{E, F\}$ where C, D, E and F are pairwise disjoint $\mathcal{A}(C, D) \cap \mathcal{A}(E, F) = \emptyset$.

Note that we do not have to care whether, e.g., $\mathcal{A}(C, D)$ intersects $\mathcal{A}(C, E)$ because the matching will choose at most one pair out of $\{C, D\}$ and $\{C, E\}$. Three $2/3$-disks obviously fit into $\mathcal{A}(C, D)$ since C and D do not intersect, thus, (i) is fulfilled. The remaining part of the paper will focus on proving (ii).

We split the proof into two parts. The first part shows that $T_{2/3}(C, D)$ does not intersect any disk other than C and D. The second part shows that no two 2/3-tunnels $T_{2/3}(C, D)$ and $T_{2/3}(E, F)$ intersect. We start with two technical lemmas that we need to prove the first part.

Lemma 2. *Let C and D be two unit disks in $\mathcal{F}_1^\otimes$. If $|c, d| \leq \frac{2}{3}\sqrt{11}$ then $g_{2/3}(c, d)$ is a line segment.*

Proof. Let $T'_{2/3}(C, D)$ be the Minkowski sum of a 2/3-disk and the line segment cd, see Fig. 4a. If $g_{2/3}(c, d)$ is not a line segment, then a disk E in $\mathcal{B} \cup \mathcal{R}$ intersects $T'_{2/3}(C, D)$. We establish a lower bound on $|c, d|$ for this to happen. Note that C, D and E are pairwise disjoint as C and D are disks in $\mathcal{B}$.

Clearly, the minimum distance between c and d is attained if E and $T'_{2/3}(C, D)$ only intersect in a single point and furthermore, both E and C as well as E and D intersect in a single point. This means that $|c, e| = |d, e| = 2$. Moreover, the Euclidean distance between e and the straight-line segment cd is $1 + \frac{2}{3} = \frac{5}{3}$. By Pythagoras' theorem we calculate $|c, d|$ to be at least $\frac{2}{3}\sqrt{11}$. This means that $T'_{2/3}(C, D)$ is contained in $P \backslash \mathcal{R}$. Even if C and D belong to different components of $\mathcal{F}_1$, by Definition 1 they are connected via a shortcut. Thus, $g_{2/3}(c, d)$ is a line segment. $\square$

Lemma 3. *Let D and E be two unit disks in $\mathcal{F}_1^\otimes$ that are infinitesimally close to each other. Then $\text{dist}(D, E) \leq \frac{2}{3}\pi$.*

Proof. For simplification we assume that D and E touch, as illustrated in Fig. 4a. The curve $g(D, E)$ attains its longest length if there is an obstacle disk R that touches D and E and no shortcut could be taken. In this case $g(D, E)$ describes an arc of radius 2 and 60°, thus its length is $\frac{1}{6} \cdot 2 \cdot 2\pi = \frac{2}{3}\pi$. $\square$

Now, we are ready to prove the first part:

Lemma 4. *Let $\{C, D\} \subseteq \mathcal{B}$ be a nearest pair. No disk of $\mathcal{B} \cup \mathcal{R} \backslash \{C, D\}$ intersects $T_{2/3}(C, D)$.*

Proof. From the definition of freespace and Definitions 2 and 3 it immediately follows that neither $T(C, D)$ nor $T_{2/3}(C, D)$ are intersected by a disk in $\mathcal{R}$. Thus, it remains to prove that apart from C and D no disk in $\mathcal{B}$ intersects $T_{2/3}(C, D)$.

W.l.o.g. let C be the nearest disk in $\mathcal{R}$ to D. The proof is done by contradiction: we assume that there is a disk $E \in \mathcal{B}$ that intersects $T_{2/3}(C, D)$.

First, we move a unit disk on $g(C, D)$ from the position of D to the first position in which it hits E, denote the disk in this position by $\overline{D}$, see Fig. 4a where $D = \overline{D}$ holds. Note that $\overline{D}$ does not necessarily lie entirely within $\mathcal{F}_1^\otimes$. However, according to Lemma 2 (C, $\overline{D}$ and E are disjoint), the Euclidean distance between c and $\overline{d}$ is at least $\frac{2}{3}\sqrt{11}$. We prove that the geodesic $g(\overline{d}, e)$ within $\mathcal{F}_1^+$ is of length less than $g(c, \overline{d})$. This contradicts C being the nearest neighbor of D.

In the remainder of the proof we show that $g(c, \overline{d})$ is larger than $g(d, \overline{e})$, regardless of whether $\overline{d}$ is in $\mathcal{F}_1$ or not. The details, especially of the latter case, are rather technical and somewhat unpleasant, and due to space limitations we omit them here. They can be found in the full version of this paper [1]. $\square$

Lemma 4 settles that no other disks apart from C and D intersect $T_{2/3}(C, D)$. We still have to show that no two $\frac{2}{3}$-tunnels $T_{2/3}(C, D)$ and $T_{2/3}(E, F)$ intersect.

Theorem 1. *Let $\{C, D\}, \{E, F\} \subseteq \mathcal{B}$ be two nearest pairs such that C, D, E and F are pairwise disjoint, it holds that $T_{2/3}(C, D) \cap T_{2/3}(E, F) = \emptyset$.*

Proof. The proof is by contradiction again. Assume that $T_{2/3}(C, D)$ and $T_{2/3}(E, F)$ intersect. First, we exclude the scenario in which even the geodesics $g_{2/3}(c, d)$ and $g_{2/3}(e, f)$ intersect. If $g_{2/3}(c, d)$ and $g_{2/3}(e, f)$ intersect it immediately follows that $g(c, d)$ and $g(e, f)$ also intersect. Note that this comprises the case in which $g(c, d)$ and $g(e, f)$ take the same shortcut. Let i be one of the intersection points of $g(c, d)$ and $g(e, f)$. By appropiately exchanging parts of $g(c, d)$ and $g(e, f)$ having endpoints c, d, e, f and i, it is easy to construct a curve that at least contradicts one of $g(c, d)$ and $g(e, f)$ to be a geodesic and thus one of the pairs $\{C, D\}$ or $\{E, F\}$ to be nearest. Thus, we can assume that only $T_{2/3}(C, D)$ and $T_{2/3}(E, F)$ intersect. Obviously, it is enough to prove the theorem for the case in which the tunnels intersect in a single point p, see Fig. 4b. Thus, p lies in $\mathcal{F}_1$ and neither $g(c, d)$ nor $g(e, f)$ takes a shortcut containing p and we can w.l.o.g. assume that no shortcut is taken at all (comparing with one of the disks of $\mathcal{D}(S)$ on $g(c, d)$ or $g(e, f)$ instead of C, D, E or F for the used shortcut S). Again, we show that $\{C, D\}$ and $\{E, F\}$ can't be nearest pairs at the same time.

We observe that at least one of the disks $\{C, D, E, F\}$ intersects the unit disk P with center p; otherwise there would be another disk in $\mathcal{B}$ located in the space between C, D, E and F which would immediately contradict $\{C, D\}$ as well as $\{E, F\}$ being nearest pairs. W.l.o.g. let C be a disk that intersects P.

Let p_{CD} be the point on $g_{2/3}(C, D)$ such that $|p, p_{CD}| = 2/3$, see Fig. 4b. Define p_{EF} correspondingly. We assume that there is a vicinity of p_{CD} and p_{EF} in which $g_{2/3}(C, D)$ and $g_{2/3}(E, F)$ are arcs. The case when one vicinity of p_{CD} and p_{EF} is a straight-line is easier and can be treated with similar arguments.

The curvature of $g_{2/3}(C, D)$ and $g_{2/3}(E, F)$ in a vicinity of p_{CD} and p_{EF} induces the existence of two disks $R, S \in \mathcal{R}$ as illustrated in Fig. 4b. Since R and S forces the curvature of $g_{2/3}(C, D)$ and $g_{2/3}(E, F)$ we may introduce the following coordinate system. The origin is p and the coordinates of r and s are $(0, \frac{7}{3})$ and $(0, -\frac{7}{3})$, respectively.

As a consequence of Lemma 2 we get that each geodesic $g_{2/3}$ starts with a straight-line segment of length at least $\frac{1}{3}\sqrt{11} \approx 1.11$. Thus, the curvature of $g_{2/3}(C, D)$ in p_{CD} infers that $|c, p_{CD}| \geq 1.11$ holds, which means that C either lies completely to the left of the y-axis or to the right. This holds analogously for the other disks. W.l.o.g. we assume that C and E lie to the left of the y-axis and D and F lie to the right, see Fig. 4b.

Note that we have to take care which relationship inferred that $\{C, D\}$ and $\{E, F\}$ are nearest pairs, e.g. C could be the nearest neighbor of D or D could be the nearest neighbor of C. We will prove the following:

(i) $\text{dist}(C, E) < \text{dist}(E, F)$
(ii) $\text{dist}(C, E) < \text{dist}(C, D)$
(iii) $\text{dist}(D, F) < \text{dist}(C, D)$

Item (i) says that C is closer to E than F is. Thus, in order for $\{E, F\}$ to be a nearest pair, E must be the nearest neighbor of F. We use this fact to show that (ii) and (iii) hold. Together, (ii) and (iii) comprise the contradiction: (ii) says that D is not the nearest neighbor of C, while (iii) says that C is not the nearest neighbor of D. Hence, $\{C, D\}$ cannot be a nearest pair.

The proofs of (i)—(iii) can be found in the full version of this paper [1].   □

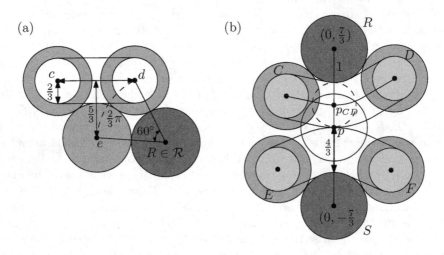

Fig. 4. Illustrations for (a) the proof of Lemma 2 and 3. (b) the proof of Theorem 1.

5   The Set $\mathcal{B}_{2/3}$

After computing $\mathcal{B}$ and the nearest neighbor graph $G = (\mathcal{B}, E)$, we compute a matching in G. Let $m' = |\mathcal{B}|$ be the number of unit disks in $\mathcal{B}$. Recall that G is planar and has degree at most 6. We show that we can find a matching in which the number of matched disks is at least $\frac{1}{6} \cdot m'$. Observe that G can consist of more than one connected component. We look at each connected component separately. Let C be a connected component and let c be the number of disks that it contains. Clearly, C contains a spanning tree of degree at most 6. It is easy to see that there is a matching in C that matches at least $\frac{1}{6} \cdot c$ disks. Doing this for each connected component yields a matching in G that contains at least $\frac{1}{6} \cdot m'$ matched disks.

According to Theorem 1 and Lemma 4 we can pack three 2/3-disks in $\mathcal{A}(C, D)$ for every matched pair $\{C, D\}$ such that these 2/3-disks are pairwise disjoint. For each of the remaining unmatched disks we pack one 2/3-disk in each disk. Let set of all disks packed as above be $\mathcal{B}_{2/3}$. By construction, there are no interferences between these sets belonging to different connected components of $\mathcal{F}_1^+$. The cardinality of $\mathcal{B}_{2/3}$ is at least $\frac{1}{6} \cdot \frac{3}{2} \cdot m' + \frac{5}{6} \cdot m' = \frac{13}{12} \cdot m'$. Since the cardinality of $\mathcal{B}$ is at least $\frac{12}{13} \cdot m$, the set $\mathcal{B}_{2/3}$ contains at least m 2/3-disks and we can conclude with the following theorem:

Theorem 2. *Algorithm* DISKPACKING *is a polynomial-time 2/3-approximation for the problem* APPROXSIZE.

6   Conclusion

Naturally, our result is purely of theoretic interest. The bottleneck for the running time is the application of Hochbaum and Maas' PTAS with approximation factor $(1 - 1/13)$. To obtain an algorithm with better running time, it seems unavoidable to use a completely different approach. For future work it would also be desirable to narrow the gap between the approximation factor of 2/3 of our algorithm and the inapproximability result of 13/14 of Baur and Fekete [2]. We conjecture that, unless $\mathcal{P} = \mathcal{NP}$, the lower bound of 2/3 is indeed tight.

References

1. M. BENKERT, J. GUDMUNDSSON, C. KNAUER, E. MOET, R. VAN OOSTRUM AND A. WOLFF. A polynomial-time approximation algorithm for a geometric dispersion problem. Technical Report 2005-8, Universität Karlsruhe, May 2006. Available at http://www.ubka.uni-karlsruhe.de/indexer-vvv/ira/2006/8
2. C. BAUR AND S. P. FEKETE. Approximation of Geometric Dispersion Problems. *Algorithmica*, 30:450–470, 2001.
3. P. CAPPANERA. A survey on obnoxious facility location problems. Technical Report TR-99-11, University of Pisa, 1999.
4. E.-C. CHANG, S. W. CHOI, D. KWON, H. PARK AND C.-K. YAP. Shortest path amidst disc obstacles is computable. In proceedings of the 21st *ACM Symposium on Computational Geometry*, 2005.
5. R. J. FOWLER, M. S. PATERSON AND S. L. TANIMOTO. Optimal packing and covering in the plane are NP-compete. *Information Processing Letters*, 12:133–137, 1981.
6. Z. FÜREDI. The densest packing of equal circles into a parallel strip. *Discrete & Computational Geometry*, 6:95–106, 1991.
7. D. HOCHBAUM AND W. MAAS. Approximation Schemes for Covering and Packing Problems in Image Processing and VLSI. *Journal of the ACM*, 32:130–136, 1985.
8. K. KEDEM, R. LIVNE, J. PACH AND M. SHARIR. On the union of Jordan regions and collision-free translational motion amidst polygonal obstacles. *Discrete & Computational Geometry*, 1:59–71, 1986.
9. C. MARANAS, C. FLOUDAS AND P. PARDALOS. New results in the packing of equal circles in a square. *Discrete Mathematics*, 128:187–293, 1995.
10. N. MEGIDDO. Applying parallel computation algorithms in the design of serial algorithms. *Journal of the ACM*, 30(4):852–865, 1983.
11. C. A. ROGERS. Packing and Covering. *Cambridge University Press*, 1964.
12. G. F. TÓTH. Packing and Covering. In Handbook of Discrete and Computational Geometry, 2nd Edition J. E. Goodman and J. O'Rourke, editors CRC Press LLC, 2004.
13. C. ZONG AND J. TALBOT. Sphere Packings. Springer-Verlag, 1999.

A PTAS for Cutting Out Polygons with Lines

Sergey Bereg[1], Ovidiu Daescu[1,*], and Minghui Jiang[2,**]

[1] Department of Computer Science, University of Texas at Dallas,
Richardson, Texas 75083-0688, USA
{besp, daescu}@utdallas.edu
[2] Department of Computer Science, Utah State University,
Logan, Utah 84322-4205, USA
mjiang@cc.usu.edu

Abstract. We present a simple $O(m+n^6/\epsilon^{12})$ time $(1+\epsilon)$-approximation algorithm for the problem of cutting a convex n-gon out of a convex m-gon with line cuts of minimum total cutting length. This problem was introduced by Overmars and Welzl in the First Annual ACM Symposium on Computational Geometry in 1985. We also present a constant approximation algorithm for the generalized problem of cutting two disjoint convex polygons out of a convex polygon.

1 Introduction

We consider a problem introduced by Overmars and Welzl [6] in 1985:

Given a polygonal piece of paper Q with a polygon P drawn on it, cut P out of Q in the cheapest possible way.

Fig. 1. (a) A cutting sequence. (b) The cost.

A *cut* is a line that runs through the piece of paper Q but not the polygon P. We refer to Figure 1. Each cut divides the piece of paper into several pieces (two pieces when Q is convex); we continue to cut the piece that contains P until, after a cutting sequence, it finally "becomes" P, that is, P is cut out of Q. The *cost* of a cut is the length of the segment marking the intersection between the cutline and the "current" piece of paper that contains P. Our optimization goal is to find a cutting sequence with the minimum total cost.

---

* Supported by NSF grant CCF-0430366.
** Corresponding author. Supported by USU research fund A13501.

D.Z. Chen and D.T. Lee (Eds.): COCOON 2006, LNCS 4112, pp. 176–185, 2006.

With line cuts, this problem has a solution only if the polygon P is convex, which we will assume. To cut out non-convex polygons, other types of cuts such as rays and segments [3,2,8] must be used instead.

We also assume that the piece of paper Q is convex. When Q is non-convex, Overmars and Welzl [6] have shown pathological cases in which no optimal cutting sequence exists if Q is considered to be topologically closed. Even when Q is convex, Bhadury and Chandrasekaran [1] have shown that the optimal solution may reside in the algebraic extension of the input data field, which confirms the difficulty of solving the problem exactly.

Due to the difficulty of obtaining exact solutions [6,1], recent research efforts on this problem have been focused mostly on finding good approximations. Bhadury and Chandrasekaran [1] first gave an approximation scheme with pseudo-polynomial running time, that is, the running time of their algorithm is polynomial only if the input data is encoded in unary. Dumitrescu [4,5] presented the first polynomial-time approximation algorithm with an $O(\log n)$ approximation factor. Subsequently, Daescu and Luo [2] designed a constant-factor approximation algorithm, and Tan [8] improved the approximation factor further to $5 + 2\sqrt{2}$.

In this paper, we present a simple polynomial-time approximation scheme for the problem of cutting out polygons with lines. Our main result is an $O(m + n^6/\epsilon^{12})$ time $(1 + \epsilon)$-approximation algorithm for cutting out a convex n-gon P out of a convex m-gon Q. Note that this problem is "inherently algebraic in nature" and "the best that can be done is to provide approximately optimal solutions" [1].

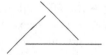

Fig. 2. Cutting out three polygons with lines is not always possible

The problem of cutting out polygons with lines can be generalized. Given a convex polygon Q and a set of k disjoint convex polygons $\{P_i\}_{i=1..k}$ in Q, how to cut out every P_i with lines such that the total cutting length is minimized? We denote by $(P_1, \cdots, P_k, Q)$ this generalized problem, and by (P, Q) the original problem. The related cuttability problem, that is, whether $\{P_i\}$ can be cut out of Q by lines, was studied by Pach and Tardos [7]. As we can see from Figure 2, a cutting sequence does not always exist even when $k = 3$. However, when $k = 2$, there is always a solution.

This paper is organized as follows. In Section 2, we introduce the preliminaries. In Section 3, we present the algorithm. In Section 4, we give the correctness proof and the analysis. In Section 5, we study the generalized problem of cutting out two convex polygons out of a convex polygon.

2   Preliminaries

We first introduce some definitions. Given a polygon P, we denote by $|P|$ the perimeter of P, and by ∂P the boundary of P. A cut is a *tangent cut* if it is tangent to (touches but does not cut through) the polygon P. A tangent cut is a *vertex cut* if it touches a single vertex of P; otherwise, if a tangent cut runs along an edge of P, it is an *edge cut*. The (at most two) vertices of P that a tangent cut touches are the *anchor* vertices of the tangent cut. A vertex cut has one anchor vertex; an edge cut has two anchor vertices.

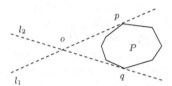

Fig. 3. The intersection angle

We refer to Figure 3. Let l_1 and l_2 be two tangent cuts with different anchor vertices such that l_1 precedes l_2 in the cutting sequence and that l_2 intersects l_1. Denote by (l_1, l_2) the intersection o of the two cuts. Let p (respectively, q) be an anchor vertex of l_1 (respectively, l_2). The *angle* of the intersection (l_1, l_2) is $\angle poq$ ($\angle poq \leq \pi$).

Our algorithm is based on three simple techniques. The first technique is dynamic programming. For a restricted case of the problem, in which only the n edge cuts along the n edges of P are allowed in a cutting sequence, Overmars and Welzl [6] presented an $O(m + n^3)$ time exact algorithm based on dynamic programming. This algorithm is immediate from the observation that, after the first edge cut, the concatenation of the remaining $n - 1$ uncut edges of P forms a convex polygonal chain with both ends on ∂Q; every subsequent edge cut reduces the problem to two independent sub-problems.

The second technique is sampling-and-rounding. Overmars and Welzl [6] have shown that, for any problem instance, there exists an optimal cutting sequence of $O(n)$ tangent cuts. Instead of restricting the cutting sequence to edge cuts only, our algorithm uses an extended set of candidate cuts including both the n edge cuts and some carefully chosen vertex cuts. If every vertex cut in the optimal sequence is close to a candidate cut, then the optimal sequence of candidate cuts gives a good approximation of the optimal cutting sequence. (The optimal sequence of candidate cuts can be found by the same dynamic programming algorithm since a vertex can be viewed as a zero-length edge and a vertex cut is a degenerated edge cut.) The sampling-and-rounding technique was first used by Bhadury and Chandrasekaran [1] in their pseudo-polynomial-time approximation algorithm. We choose our candidate cuts carefully to avoid the pseudo-polynomial running time.

The third technique is that, since polynomial-time approximation algorithms for this problem already exist [4,2,8], we can run any of these algorithms as a preprocessing step to get an estimate of the optimal solution, then use the estimate to guide our selection of candidate cuts.

3 The Algorithm

Our algorithm consists of three steps: the estimation step, the sampling step, and the dynamic programming step.

In the estimation step, we run Tan's algorithm [8] to get an estimate EST of the cost OPT of the optimal cutting sequence such that $OPT \leq EST \leq c \cdot OPT$, where c is the constant approximation factor. Our choice of Tan's algorithm in this estimation step is arbitrary; other approximation algorithms [2,4] can be used instead, though they may incur a slight increase in the overall running time.

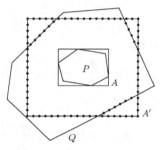

Fig. 4. Portal points

In the sampling step, we first construct a set V of portal points. We refer to Figure 4. Let A be the smallest axis-parallel rectangle that contains P. Let A' be an axis-parallel rectangle with length $a + 2EST$ and width $b + 2EST$, where a and b are the length and the width of A. The two rectangles A and A' are centered at the same point; the distances between their corresponding edges are exactly EST. The portal points are placed on the boundary of $Q \cap A'$ and on the boundary of A' to discretize them into polygonal chains with maximum segment length $\delta \cdot EST$, where $\delta = \Theta(\epsilon^2/n)$.

We next construct a set E of candidate cuts. First put the n edge cuts into E, then, for each vertex p of P, and for each portal point $v \in V$, if the line through both p and v specifies a valid tangent cut anchored at p (that is, the line does not cut through P), put the tangent cut into E. We call the candidate cuts constructed so far the *type-1 cuts*.

We refer to Figure 5. For every two type-1 cuts l_1 and l_2 intersecting at a point o and anchored at two vertices q and r respectively (q, r, and o may coincide), we compute, for every vertex p of P outside the triangle $\triangle oqr$, all the vertex cuts anchored at p such that the length of the cutline's segment between l_1 and

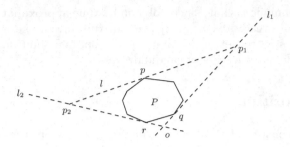

Fig. 5. Type-2 cuts

l_2 is (locally) minimal. We put these cuts into the candidate set E and called them the *type-2 cuts*.

We show how to find these type-2 cuts. Assume without loss of generality that the line equations for l_1, l_2, and l are

$$y = 0, \qquad y = ax, \qquad \frac{y - y_0}{x - x_0} = k.$$

The intersection p_1 of l and l_1 is at

$$x_1 = x_0 - \frac{y_0}{k}, \qquad y_1 = 0.$$

The intersection p_2 of l and l_2 is at

$$x_2 = \frac{y_0 - ax_0}{a - k} + x_0, \qquad y_2 = a\left(\frac{y_0 - ax_0}{a - k} + x_0\right).$$

Therefore, we have

$$|p_1 p_2|^2 = (x_2 - x_1)^2 + (y_2 - y_1)^2 = \left(\frac{y_0 - ax_0}{a - k} + \frac{y_0}{k}\right)^2 + a^2\left(\frac{y_0 - ax_0}{a - k} + x_0\right)^2.$$

It follows that

$$\frac{\mathrm{d}}{\mathrm{d}k}|p_1 p_2|^2 = 2\left(\frac{y_0 - ax_0}{a - k} + \frac{y_0}{k}\right)\left(\frac{y_0 - ax_0}{(a - k)^2} - \frac{y_0}{k^2}\right) + 2a^2\left(\frac{y_0 - ax_0}{a - k} + x_0\right)\frac{y_0 - ax_0}{(a - k)^2}.$$

The equation $\frac{\mathrm{d}}{\mathrm{d}k}|p_1 p_2|^2 = 0$ can be simplified to a polynomial equation of k with degree four, whose roots can be computed in constant time on a RAM machine.

Finally, in the dynamic programming step, we use dynamic programming [6] to find a sequence of candidate cuts from E with the minimum cost. This cutting sequence is our approximate solution.

4 Proof and Analysis

We prove by construction that our algorithm indeed gives a $(1+\epsilon)$-approximation. Given an optimal cutting sequence S of $O(n)$ tangent cuts, the existence of which is proved by Overmars and Welzl [6], we construct a sequence S' of $O(n)$ candidate cuts to approximate S.

We initialize S' with the same sequence of cuts in S. Each cut in S' forms exactly two intersections, one on each side of its anchor vertex (or, in the case of an edge cut, vertices), with either ∂Q or the preceding cuts in S'. We say that an intersection is *sharp* if its angle is less than $\theta = \Theta(\epsilon)$. Let l be the first cut in S' that forms a sharp intersection with a preceding cut. We consider the only two cases:

1. The cut l forms only one sharp intersection with a preceding cut l_1.
2. The cut l forms two sharp intersections with two preceding cuts l_1 and l_2.

The cut l_1 must intersect ∂Q on the side that l intersects it. Suppose on the contrary that l_1 intersects a preceding cut l_0 instead of ∂Q, then the intersection (l_0, l_1) both precedes and is sharper than the intersection (l_1, l), which contradicts our choice of l. By the same argument, the cut l_2 (in the second case) must intersect ∂Q on the side that l intersects it.

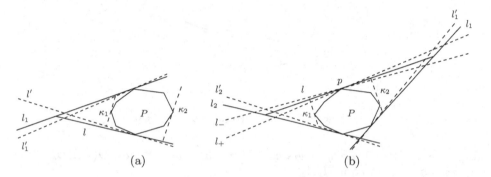

Fig. 6. (a) The first case. (b) The second case.

We refer to Figure 6(a) for the first case. We rotate the two cuts l and l_1 toward each other (the angle of their intersection increases) to their nearest type-1 cuts l' and l_1'. We then find two additional type-1 cuts κ_1 and κ_2 that are near-perpendicular to l' (that is, at most angle δ away from the perpendicular position), one on each side of P, and insert them immediately after l' in S'. We call these two cuts the *shortcuts*. Because of the shortcuts, no subsequent cuts in S' can form sharp intersections.

We refer to Figure 6(b) for the second case. We rotate the two cuts l_1 and l_2 toward l (the angles of both (l, l_1) and (l, l_2) increase) to their nearest type-1 cuts l_1' and l_2'. If l is a vertex cut anchored at the vertex p, let l_- and l_+ be the two nearest candidate cuts anchored at the same vertex p such that l is between l_- and l_+. We round l to $l' \in \{l_-, l_+\}$ such that the segment of l' between the two lines l_1' and l_2' is shorter. (Note that this is the only place in our construction that a cut in S may be rounded to a type-2 cut; as a result, S' includes at most one type-2 cut.) We then find the two type-1 cuts (shortcuts) κ_1 and κ_2 that are near-perpendicular to l', and insert them immediately after l' in S'. Again, because of the shortcuts, no subsequent cuts in S' can form sharp intersections. The maximum number of sharp intersections in the sequence S' is either one (in the first case) or two (in the second case).

Finally, for each remaining (not yet rounded) vertex cut in S', we round it to its nearest type-1 cut in E. The cutting sequence S' thus obtained includes the same n edge cuts that are originally in the optimal sequence S, so it is a valid cutting sequence that indeed cuts P out of Q. Because of the rounding and the additional shortcuts, S' may include redundant cuts and may have a topology different from that of S; however, this should not be of any concern since we are only interested in the cost of the approximation.

We next show that the cost of S' is close to the cost of S. In the following analysis, we charge the cost of each tangent cut to its two intersections with either ∂Q or the preceding cuts in the sequence. For a vertex cut, the cost of an intersection is the distance from the intersection to the anchor vertex; for an edge cut, the cost is the distance from the intersection to the midpoint of the two anchor vertices.

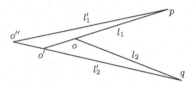

Fig. 7. Non-sharp intersection

Lemma 1. *The rounding incurs an additional cost of at most $O(\epsilon/n)OPT$ for a non-sharp intersection (l_1, l_2).*

Proof. Two type-1 cuts corresponding to a common anchor vertex v and two consecutive portal points on the boundary of A' are at most angle δ away from each other, since the distance from v to either portal point is at least EST and the distance between the two portal points is at most $\delta \cdot EST$. The type-1 cuts in E can be ordered into a cycle according to the slopes of their cutlines; two consecutive cuts in the cycle are at most angle δ away from each other. (Note that the edge cuts and the portal points on the boundary of $Q \cap A'$ make the sampling even denser.) Every tangent cut in S can be rotated in either direction for an angle of at most δ to a type-1 cut.

We refer to Figure 7 for the worst case that l_1 and l_2 rotate away from each other to l_1' and l_2'. The two cuts l_1 and l_2 are anchored at vertices p and q, respectively. Let $o = (l_1, l_2)$, $o' = (l_1, l_2')$, and $o'' = (l_1', l_2')$. We have $\angle poq \geq \theta$, $\angle opo'' \leq \delta$, and $\angle oqo' \leq \delta$. Since $\theta = \Theta(\epsilon)$ and $\delta = \Theta(\epsilon^2/n)$, we have

$$\frac{|oo'|}{|oq|} = \frac{\sin \angle oqo'}{\sin(\angle poq - \angle oqo')} = O\left(\frac{\angle oqo'}{\angle poq}\right) = O(\delta/\theta) = O(\epsilon/n).$$

It follows that $|oo'| = O(\epsilon/n)|oq|$. By the same argument, we can show that $|o'o''| = O(\epsilon/n)|o'p| = O(\epsilon/n)|op| + O(\epsilon^2/n^2)|oq|$. Therefore, we have

$$|o''q| \leq |oq| + |oo'| + |o'o''| = |oq| + O(\epsilon/n)(|op| + |oq|) \leq |oq| + O(\epsilon/n)OPT. \quad \square$$

Lemma 2. *The rounding incurs an additional cost of at most $O(\epsilon^2/n)OPT$ for an intersection between a cut l and ∂Q.*

Proof. The cost of the intersection between l and ∂Q is at most OPT, which is bounded by EST; therefore, the intersection is inside the rectangle A'. Because of our choice of the portal points on the boundary of the polygon $Q \cap A'$, the rounding of l to l' moves the intersection for a distance of at most $\delta \cdot EST = O(\epsilon^2/n)OPT$. If l' intersects a preceding candidate cut in S' instead of ∂Q, the rounding incurs an even less additional cost. $\quad \square$

The previous two lemmas considered $O(n)$ intersections, the total additional cost of which is $O(n)O(\epsilon/n)OPT = O(\epsilon)OPT$.

Lemma 3. *The rounding incurs no additional cost for the sharp intersections.*

Proof. We refer to Figure 6(a) for the first case. Let v be the anchor vertex of both l and l'. The distance from the intersection (l_1, l) to v is at least the distance from the intersection (l_1', l') to v, because we rotate l and l_1 toward each other to l' and l_1'. The cost of the intersection (l_1', l') is therefore at most the cost of the intersection (l_1, l). (The cost of the cut l_1 is charged to the intersection between l_1 and ∂Q; it is not considered here.)

We refer to Figure 6(b) for the second case. The same argument shows that the total cost of the two intersections (l_1', l) and (l_2', l) is at most the total cost of the two intersections (l_1, l) and (l_2, l). Rotating l to l' in either direction would increase the cost of one of its two intersections. However, our choice of the type-2 cuts in E ensures that, when l' is between two candidate cuts l_- and l_+, the total cost of l' reaches the minimum at either l_- or l_+. Therefore, the total cost of the two intersections (l_1', l) and (l_2', l) is at least the total cost of the two intersections (l_1', l') and (l_2', l'). $\quad \square$

Lemma 4. *The cost of the two shortcuts is at most $O(\epsilon)OPT$.*

Proof. We refer to Figure 6(a) for the first case. Let $u = (l_1, l)$ and $u' = (l_1', l')$, and let v be the anchor vertex of l and l'. The distance d from u' to a shortcut (either κ_1 or κ_2) is at most $|u'v| + |P|$. Since $|u'v| \leq |uv| \leq OPT$ and $|P| = O(OPT)$, we have $d = O(OPT)$. The angle of the intersection (l_1', l') is at most $\theta + 2\delta = \Theta(\epsilon) + \Theta(\epsilon^2/n) = \Theta(\epsilon)$. Therefore, the cost of the two shortcuts is at most $O(d)\Theta(\epsilon) = O(\epsilon)OPT$. The analysis for the second case is similar. $\quad \square$

The previous four lemmas together imply that the cost of our constructed sequence S' is at most $1 + O(\epsilon)$ times the cost of the optimal sequence S. Adjusting the parameters for $\delta = \Theta(\epsilon^2/n)$ and $\theta = \Theta(\epsilon)$, we have a $(1 + \epsilon)$-approximation.

We now analyze the running time of our algorithm. For the two rectangles A and A', we have $|A| = O(|P|) = O(OPT)$ and $|A'| = |A| + 8EST$. Since

$EST = \Theta(OPT)$, we have $|A'| = O(OPT)$. It follows that the perimeter of the polygon $Q \cap A'$ is $O(OPT)$. The number of portal points is therefore $\frac{O(OPT)}{\delta \cdot EST} = O(1/\delta) = O(n/\epsilon^2)$. The number of type-1 cuts is $O(n/\epsilon^2)$. The number of type-2 cuts is $O(n^2/\epsilon^4)$. The dynamic programming algorithm [6] takes $O(m + (n^2/\epsilon^4)^3) = O(m + n^6/\epsilon^{12})$ time. We have the following theorem.

Theorem 1. *For the problem of finding an optimal line cutting sequence to cut out a convex n-gon P out of a convex m-gon Q, we have an $O(m + n^6/\epsilon^{12})$ time $(1 + \epsilon)$-approximation algorithm.*

5   Cutting Out Two Polygons

Given an instance (P_1, P_2, Q), let H be the convex hull of P_1 and P_2. A cutting sequence for (P_1, P_2, Q) consists of four parts (parts 1, 3, and 4 may be empty):

1. A sequence that cuts a polygon Q' out of Q such that $H \subseteq Q'$.
2. A single *dividing cut* that cuts Q' into two polygons Q_1 and Q_2 such that $P_1 \subseteq Q_1$ and $P_2 \subseteq Q_2$.
3. A sequence that cuts P_1 out of Q_1.
4. A sequence that cuts P_2 out of Q_2.

Denote by $OPT_{(P_1,P_2,Q)}$ the optimal cost for (P_1, P_2, Q); denote by $OPT_{(P,Q)}$ the optimal cost for (P, Q). We have the following inequality:

$$OPT_{(P_1,P_2,Q)} \geq OPT_{(P_1,P_2,Q')} \geq \max\{OPT_{(P_1,Q')}, OPT_{(P_2,Q')}\}$$
$$\geq \max\{OPT_{(P_1,Q_1)}, OPT_{(P_2,Q_2)}\} \geq \max\{|P_1|, |P_2|\}.$$

Denote by C_p the infimum of the length of a chord of Q through a point $p \in Q$, and define $C_P = \max_{v \in V(P)} C_v$ for a convex polygon $P \subseteq Q$. Dumitrescu [4] proved that C_P is a lower bound for $OPT_{(P,Q)}$. Using essentially the same argument, we obtain the following inequality:

$$OPT_{(P_1,P_2,Q)} \geq \max\{C_{P_1}, C_{P_2}\} \geq C_H.$$

Theorem 2. *There is a polynomial-time constant approximation for (P_1, P_2, Q).*

Proof. Dumitrescu [4] presented a "separation algorithm" for (P, Q) that cuts a polygon Q' out of Q using a sequence of tangent cuts with a cost of at most $c_1 C_P + c_2 |P|$ such that $P \subseteq Q'$ and $|Q'| \leq c_3 |P|$, where c_1, c_2, and c_3 are three constants. Let $c' = \max\{2(c_1 + c_2 + 1), c_3\}$. We consider two cases.

Case one: $|H| > c' OPT_{(P_1,P_2,Q)}$. For two points $p_1 \in \partial P_1$ and $p_2 \in \partial P_2$, the distance $|p_1 p_2|$ is at most $|H|/2$. On the other hand, the distance between two points on ∂P_1 (respectively, ∂P_2) is at most $|P_1|/2$ (respectively, $|P_2|/2$). Therefore, the distance $|p_1 p_2|$ is at least $(|H| - |P_1| - |P_2|)/2 > (c' OPT_{(P_1,P_2,Q)} - OPT_{(P_1,P_2,Q)} - OPT_{(P_1,P_2,Q)})/2 \geq (c_1 + c_2) OPT_{(P_1,P_2,Q)}$.

We can use Dumitrescu's separation algorithm to cut a polygon Q_1 out of Q such that $P_1 \subseteq Q_1$ and $|Q_1| \leq c_3 |P_1|$. Since this cutting sequence includes only cuts tangent to P_1 and has a cost of at most $c_1 C_{P_1} + c_2 |P_1| \leq (c_1 + c_2) OPT_{(P_1, P_2, Q)}$, these cuts can never cut through (or even reach) P_2. Since $P_1 \subseteq Q_1$ and $|Q_1| \leq c_3 |P_1| \leq c' OPT_{(P_1, P_2, Q)} < |H|$, there must be a dividing cut in this sequence that separates P_1 from P_2. Let Q_2 be the polygon that contains P_2 after that dividing cut. We then cut P_1 out of Q_1 and cut P_2 out of Q_2 independently using a constant approximation algorithm [2,8], the cost of which is at most $cOPT_{(P_1, Q_1)} + cOPT_{(P_2, Q_2)} \leq 2cOPT_{(P_1, P_2, Q)}$. The total cost is at most $(c_1 + c_2 + 2c) OPT_{(P_1, P_2, Q)}$.

Case two: $|H| \leq c' OPT_{(P_1, P_2, Q)}$. We first use Dumitrescu's separation algorithm to cut a polygon Q' out of Q such that $H \subseteq Q'$ and $|Q'| \leq c_3 |H|$, then cut Q' into Q_1 and Q_2 with an arbitrary dividing cut, and finally cut P_1 out of Q_1 and cut P_2 out of Q_2. The separation algorithm incurs a cost of at most $c_1 C_H + c_2 |H| \leq c_1 OPT_{(P_1, P_2, Q)} + c_2 c' OPT_{(P_1, P_2, Q)} = (c_1 + c_2 c') OPT_{(P_1, P_2, Q)}$. The cost of the dividing cut is at most $|Q'| \leq c_3 |H| \leq c_3 c' OPT_{(P_1, P_2, Q)}$. The cost for the two subproblems (P_1, Q_1) and (P_2, Q_2) is at most $2cOPT_{(P_1, P_2, Q)}$. The total cost is at most $(c_1 + c_2 c' + c_3 c' + 2c) OPT_{(P_1, P_2, Q)}$. □

Acknowledgment

Minghui Jiang thanks Binhai Zhu for suggesting this problem to work on and for initial discussion.

References

1. J. Bhadury and R. Chandrasekaran. Stock cutting to minimize cutting length. *European Journal of Operations Research*, 88:69–87, 1996.
2. Ovidiu Daescu and Jun Luo. Cutting out polygons with lines and rays. *International Journal of Computational Geometry & Applications*, 16(2-3):227–248, 2006. (Preliminary version in *Proceedings of the 15th Annual International Symposium on Algorithms and Computation (ISAAC'04)*, LNCS 3341, pages 669–680, 2004.)
3. Erik D. Demaine, Martin L. Demaine, and Craig S. Kaplan. Polygons cuttable by a circular saw. *Computational Geometry: Theory and Applications*, 20:69–84, 2001.
4. Adrian Dumitrescu. An approximation algorithm for cutting out convex polygons. *Computational Geometry: Theory and Applications*, 29:223–231, 2004. (Preliminary version in *Proceedings of the 14th Annual ACM-SIAM Symposium on Discrete Algorithms (SODA'03)*, pages 823–827, 2003.)
5. Adrian Dumitrescu. The cost of cutting out convex n-gons. *Discrete Applied Mathematics*, 143(1-3):353–358, 2004.
6. Mark H. Overmars and Emo Welzl. The complexity of cutting paper. In *Proceedings of the 1st Annual ACM Symposium on Computational Geometry (SoCG'85)*, pages 316–321, 1985.
7. János Pach and Gábor Tardos. Cutting glass. *Discrete Computational Geometry*, 24:481–495, 2000.
8. Xuehou Tan. Approximation algorithms for cutting out polygons with lines and rays. In *Proceedings of the 11th International Computing and Combinatorics Conference (COCOON'05)*, LNCS 3595, pages 534–543, 2005.

On Unfolding Lattice Polygons/Trees and Diameter-4 Trees*

Sheung-Hung Poon

Department of Mathematics and Computer Science, TU Eindhoven,
5600 MB, Eindhoven, The Netherlands
spoon@win.tue.nl

Abstract. We consider the problems of straightening polygonal trees and convexifying polygons by continuous motions such that rigid edges can rotate around vertex joints and no edge crossings are allowed. A tree can be *straightened* if all its edges can be aligned along a common straight line such that each edge points "away" from a designated leaf node. A polygon can be *convexified* if it can be reconfigured to a convex polygon. A *lattice tree* (resp. *polygon*) is a tree (resp. polygon) containing only edges from a square or cubic lattice. We first show that a 2D lattice chain or a 3D lattice tree can be straightened efficiently in $O(n)$ moves and time, where n is the number of tree edges. We then show that a 2D lattice tree can be straightened efficiently in $O(n^2)$ moves and time. Furthermore, we prove that a 2D lattice polygon or a 3D lattice polygon with simple shadow can be convexified efficiently in $O(n^2)$ moves and time. Finally, we show that two special classes of diameter-4 trees in two dimensions can always be straightened.

Keywords: Comput. geom., unfolding, straightening, convexifying.

1 Introduction

Graph reconfiguration problems have wide applications in contexts including robotics, molecular conformation, animation, wire bending, rigidity and knot theory. The motivation for reconfiguration problems of lattice graphs arises in applications in molecular biology and robotics. For instance, the bonding-lengths in molecules are often similar [6,11,12], as are the segments of robot arms.

A *unit tree* (resp. *unit polygon*) is a tree (resp. polygon) containing only edges of unit length. An *orthogonal tree* (resp. *orthogonal polygon*) is a tree (resp. polygon) containing only edges parallel to coordinate-axes. A *lattice tree* (resp. *lattice polygon*) is a tree (resp. polygon) containing only edges from a square or cubic lattice. Note that a lattice tree or polygon is basically a unit orthogonal tree or polygon. A graph is *simple* if non-adjacent edges do not intersect. We consider the problem about the reconfiguration of a simple chain, polygon, or

---

* This research was supported by the Netherlands' Organisation for Scientific Research (NWO) under project no. 612.065.307.

D.Z. Chen and D.T. Lee (Eds.): COCOON 2006, LNCS 4112, pp. 186–195, 2006.

tree through a series of continuous motions such that the lengths of all graph edges are preserved and no edge crossings are allowed. A tree can be *straightened* or *flattened* if all its edges can be aligned along a common straight line such that each edge points "away" from a designated leaf node. In particular, a chain can be straightened if it can be stretched out to lie on a straight line. A polygon can be *convexfied* if it can be reconfigured to a convex polygon. We say a chain or tree is *locked* if it cannot be straightened. We say a polygon is *locked* if it cannot be convexified. We consider one *move* in the reconfiguation as a continuous monotonic change for the joint angle at some vertex.

In four dimensions or higher, a polygonal tree can always be straightened, and a polygon can always be convexified [7]. In two dimensions, a polygonal chain can always be straightened and a polygon can always be convexified [9,14,5]. However, there are some trees in two dimensions that can lock [3,8,13]. In three dimensions, even a 5-chain can lock [4]. Alt et al. [2] showed that deciding the reconfigurability for trees in two dimensions and for chains in three dimensions is PSPACE-complete. However the problem of deciding straightenability for trees in two dimensions and for chains in three dimensions remains open. Due to the complexity of the problems in two and three dimensions, some special classes of trees and polygons have been considered. Poon [13] showed that a unit tree of diameter 4 in two dimensions can always be straightened. In their paper, they posed a challenging open question whether a unit tree in either two or three dimensions can always be straightened.

The rest of this paper is organized as follows. We define some technical terms used in our paper in Section 2. We present efficient algorithms to straighten lattice chains and trees and to convexify lattice polygons in both two and three dimensions, respectively, in Sections from 3 to 6. In Section 7, we show that two special classes of diameter-4 trees in two dimensions can always be straightened. Finally, we conclude with some conjectures in Section 8.

2   Definitions

Let P be unit tree or polygon in two or three dimensions. Define a small value $\epsilon = \frac{1}{100n}$, where n is the number of edges in P. We call point q is *convergent* to p if q is within distance ϵ from p. A unit edge is called *convergent* to a lattice edge if any point on the edge is within distance ϵ from a particular lattice edge. Such a unit edge is called a *near-lattice edge*, and the particular lattice edge is called its *core edge*. A em core vertex is a vertex of some core edge. A *near-lattice tree* (resp. *near-lattice polygon*) is a tree (resp. polygon) that contains only near-lattice edges. Suppose P is a near-lattice tree or polygon. The *core* of P, denoted by $K(P)$, is the union of core edges for all edges in P. A *spring* in P is the set of edges in P converging to a common lattice edge. A spring with only one edge is called a *singleton*. A *leaf spring* is a spring with its core edge possessing a leaf vertex in the core of P.

A near-lattice tree is called *folded* if its core contains a single lattice edge. A near-lattice polygon is called *nearly folded* if its core is a lattice rectangle of unit width. Remark that the definition of a *nearly-folded* polygon is due to our unfolding algorithm, in which we convert a given lattice polygon to such a polygon, which can then be convexified straightforwardly.

3   2D Lattice Chains and 3D Lattice Trees

2D lattice chains. Given a simple 2D lattice chain $P = p_0p_1 \ldots p_n$. Starting from an end edge, say p_0p_1, we fold up the whole chain, edge by edge. At step i, a spring S_i containing a zig-zag path from p_0 to p_{i+1} is formed such that p_ip_{i+1} is a lattice edge staying at its original position, and $\angle p_j = \epsilon^2$ for $1 \le j \le i$. Vertices $p_{i-1}, p_{i-3}, \ldots$ all lie in a lattice cell, say σ. Step $i + 1$ of the algorithm tries to combine the spring S_i and the edge $e = p_{i+1}p_{i+2}$ to form a new spring S_{i+1}. We need to consider several cases depending on the position of e. If p_{i+1} is non-straight and e does not lie on σ, we rotate S_i around vertex p_{i+1}, away from σ, until p_ip_{i+1} makes an angle ϵ^2 with e, and we are done. See Figure 1(a) for illustration. Otherwise, we first rotate e around p_{i+2} to the side containing σ until

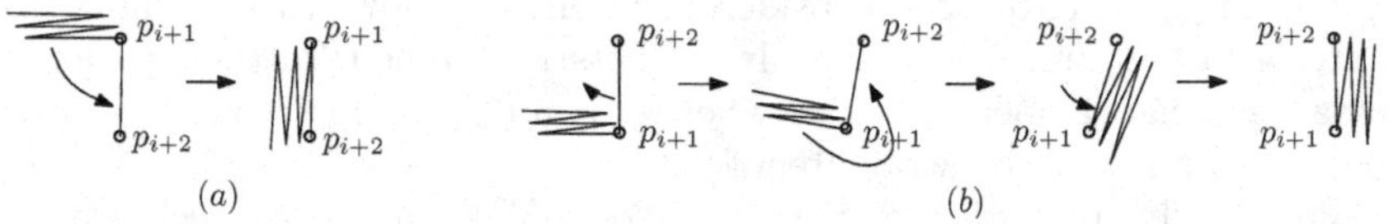

(a)   (b)

Fig. 1. Folding 2D chains

e makes an angle of $\pi/12$ with its original position, and it is safe now to rotate S_i around p_{i+1} by sweeping through the side not containing S_i until $\angle p_ip_{i+1}p_{i+2} = \epsilon^2$. Then we move e back to its original position. See Figure 1(b) for illustration. It is clear that it takes only constant number of moves to construct S_{i+1} from S_i. Thus the whole chain can be folded up, in $O(n)$ moves and time, into a zig-zag path, which can then be straightened straightforwardly.

Theorem 1. *A 2D lattice chain can be straightened in $O(n)$ moves and time.*

3D lattice trees. As in the previous section, we can unfold 3D lattice chains in the same manner. In fact, we can even unfold 3D lattice trees in a similar fashion. we do this in a bottom-up fashion according to the given tree structure. The folding process starts from the leaves of the given tree. In each step, we fold up each set of all leaf springs incident to a common internal core vertex v to the internal core edge incident to v. Each time when we fold up a spring towards an edge, we keep the "tail" of the spring away from its moving direction. It is clear that folding each leaf spring takes constant number of moves. Thus we obtain the following theorem.

Theorem 2. *A 3D lattice tree can be straightened in $O(n)$ moves and time.*

4   2D Lattice Trees

Given a 2D lattice tree P. We consider a leftmost vertex r of P as its root. We consider the parent of the root r as the lattice point to the left of r. Our algorithm proceeds by pulling P to the left successively until the whole tree is straightened. Each pulling step moves each vertex along its edge connecting to its parent until it is within distance ϵ^2 to its parent in the previous step. This step is repeated n times so that, finally, P is straightened. Figure 2 shows the execution of the algorithm on a small tree. Step i generates a new polygon P_i. We assume $P = P_0$. First, we can show the following lemma.

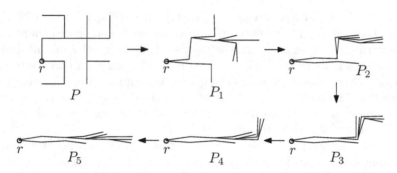

Fig. 2. Straightening a lattice tree by pulling it to the left successively

Lemma 1. *During step i of the algorithm, suppose v is moving on the edge $v_{i-1}u_{i-1}$, where v_{i-1} and u_{i-1} are the positions of v and its parent just after step $i-1$. Then*

(i) Each vertex of P_i is within distance $i\epsilon^2$ to a lattice vertex.
(ii) Each vertex v is within distance $i\epsilon^2$ to the core lattice edge of e.
(iii) No edge crossings can occur.

Proof. Suppose, just after step i, v stops at position v_i. Note that $d(v_i, u_{i-1}) \leq \epsilon^2$ due to the algorithm.

(i) It is clear that any vertex of P_0 is a lattice point. Assume any vertex of P_{i-1} is within distance $(i-1)\epsilon^2$ to a lattice vertex U. Consider any vertex v_i of P_i. We know that $d(v_i, u_{i-1}) \leq \epsilon^2$. As $d(u_{i-1}, U) \leq (i-1)\epsilon^2$, we have $d(v_i, U) \leq i\epsilon^2$.

(ii) Suppose the core edge of $u_{i-1}v_{i-1}$ is UV, where u_{i-1} and v_{i-1} converge to U and V, respectively. Since $d(v_i, U) \leq i\epsilon^2$ and $d(v_{i-1}, V) \leq (i-1)\epsilon^2$, we have $d(v, UV) \leq i\epsilon^2$ as v moves along segment $v_{i-1}v_i$.

(iii) (*Sketch*) Consider a moving edge $e = uv$, where u is the parent of v. We show that e cannot cross other moving edges. Let the parent of u_{i-1} in P_{i-1} be t_{i-1}, which converges to lattice point T. By part (ii), we have $d(u, TU) \leq i\epsilon^2$. We consider two cases depending on whether T, U and V are collinear. If they are, then any point p on e is within distance $i\epsilon^2$ from TUV. Otherwise, T, U and V are not collinear. Then any point p on e either lies in the lattice cell with

T, U and V as its vertices or is within distance $i\epsilon^2$ from TUV. Also $d(e, TUV) \leq 1/\sqrt{2} + 2i\epsilon^2 \leq 2/3 < 1 - 2\epsilon$. Then it is not hard to see that e cannot cross other moving edges. The details are omitted in this abstract. ⊟

As each pulling step takes $O(n)$ moves and time, the whole algorithm takes $O(n^2)$ moves and time.

Theorem 3. *A 2D lattice tree can be straightened in $O(n^2)$ moves and time.*

5   2D Lattice Polygons

Given a simple 2D lattice polygon P. Start with any lattice edge e of P. We label the edges of P in counter-clockwise order with consecutive numbers by starting with labeling edge e with the number 1. An edge of P is called an *odd edge* if its label number is odd; otherwise, it is called an *even edge*. Note that a 2D lattice polygon has even number of edges. We suppose that, throughout the entire motion, the parity of each edge remains fixed. A vertex is called *straight* if it is collinear with both its preceding and following vertices on the polygon. Otherwise, it is called *non-straight*.

A *block* of a lattice polygon is a rectangle of width one such that its left, top and bottom sides coincides with the edges of the given polygon. A *collapsible block* of a lattice polygon is a block such that its right side complementing the given polygon is a single segment. Such a segment is called the *opening segment*, or simply the *opening*, of the corresponding collapsible block. And the two end-points of an opening segment are called *opening vertices*. The path between its two opening vertices on a collapsible block is called *a collapsible path*. A block or path of a near-lattice polygon is *collapsible* if the corresponding block or path of its core lattice polygon is collapsible.

The parity of a spring is defined as the parity of its two end edges. Suppose we walk along the edges of a near-lattice polygon in anti-clockwise order. We call a non-singleton spring *left-twisted* if its edges run to the left; otherwise it is called *right-twisted*. See Figure 3(a) for examples. A near-lattice polygon is called *consistently twisted* if odd and even springs have opposite directions of twisting. See Figure 3(b) for an example of a consistently-twisted near-lattice polygon. We first need the following two lemmas.

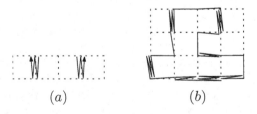

(a)   (b)

Fig. 3. (a) A left-twisted spring and a right-twisted spring, respectively. (b) A consistently twisted near-lattice polygon.

Lemma 2. *In a non-nearly-folded near-lattice polygon, there is a collapsible block; more precisely, the block with the smallest height is a collapsible block.*

Proof. Suppose to the contrary that the block B with the smallest height is not collapsible. Then its opening contains at least two segments. Thus there is some block B' with its left side between these two segments. Obviously the height of B' is shorter than that of B. This contradicts our assumption.   ⊡

Lemma 3. *In a consistently-twisted near-lattice polygon, consider a collapsible block B such that all its non-singleton springs lie on its left side. Then*

(i) All its non-singleton springs can be transformed into one non-singleton spring on any edge on the left side of the block with consistent twisting.

(ii) Its corresponding collapsible path can be transformed to a consistently-twisted near-lattice path convergent to the opening segment of B.

Proof. (Sketch)
   (*i*) Note that the non-singleton springs lie only on the left side of B. A non-singleton spring can be transformed into a singleton by moving its remaining edges to the adjacent spring. Repeating this process results in only one non-singleton spring on the left side of B. See Figure 4(a) for illustration.
   (*ii*) We need to consider several cases depending on different directions of the incident edges of its two opening vertices. Note that according to part (*i*), we can assume the left side of B contains only one non-singleton spring, which reduces a lot of cases we need to consider. Figure 4(b) shows the collapse of a specific collapsible block. Other cases can be handled in a similar way, whose details are omitted in this abstract.   ⊡

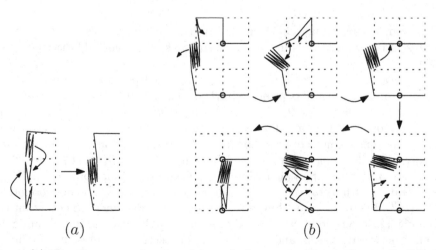

(a)   (b)

Fig. 4. (*a*) Transform several springs into a single non-singleton spring on the left side of B. (*b*) Collapse a collapsible block.

Lemma 2 says that a near-lattice polygon which is not nearly folded always contains a collapsible block, which can then be collapsed using Lemma 3 in $O(n)$ moves and time. Hence, each step of our algorithm is to select the block with smallest height in the polygon to collapse. Note that all the blocks of the polygon can be maintained in a priority queue with their heights as keys. After $O(n)$ collapsing steps, we end up with a nearly folded polygon, which clearly can be convexified in $O(n)$ moves and time.

Theorem 4. *A 2D lattice polygon can be straightened in* $O(n^2)$ *moves and time.*

6   3D Orthogonal Chains and Polygons

A 3D 5-chain can lock [4]. We can simulate this chain by an orthogonal 9-chain as shown in Figure 5(a), where each of its two end edges is longer than the union of its internal edges. Furthermore, by doubling this 9-chain, we obtain a 3D locked orthogonal unknotted 20-gon as shown in Figure 5(b).

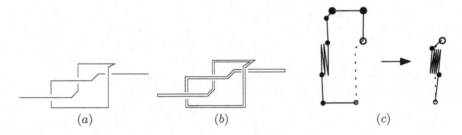

Fig. 5. (a) A 3D orthogonal locked 9-chain. (b) A 3D orthogonal locked 20-gon. (c) Collapse a 3D collapsible block.

Theorem 2 implies that a 3D lattice chain can be straightened in $O(n)$ moves. We present below an algorithm to straighten a class of unknotted lattice polygons in 3D.

A *3D polygon with simple shadow* is a simple 3D polygon whose shadow is a simple polygon when it is projected orthogonally onto some plane, which we assume to be xy-plane. A *3D polygon with simple projection* is a 3D polygon with simple shadow such that any line parallel to z-axis intersects the polygon with at most one single connected component. Alberto-Calvo et al. [1] showed that a 3D polygon with simple projection can be convexified in $O(n + T)$ time, where T is the running time of an algorithm to convexify the planar projection. There are several algorithms to convexify a planar polygon [5,9,14]. The best bound for T is $O(n^{79})$ due to the algorithm by Cantarella et al. [5], where the constant is a polynomial in the ratio between the maximum edge length and initial minimum distance between a vertex and an edge. In this section, we present an efficient algorithm to convexify a 3D lattice polygon with simple shadow in $O(n^2)$ moves and time. In this abstract, we only sketch the main idea.

Suppose P is the given 3D lattice polygon with simple shadow. Consider any vertical lattice plane π parallel to z-axis. Let P_π be the intersection of π and P. Our algorithm starts by collapsing the blocks in each P_π. This step is similar to what we do for the 2D case. The difference is that we need to consider more types of blocks. We define a *vertical block* as a block of width one with its opening on its left or right side, and a *horizontal block* as a block of height one with its opening on its top or bottom side. Note that a block in 2D case is basically a vertical back with its opening on its right side. A *collapsible block* is defined similarly as the 2D case. This step of our algorithm searches for any collapsible vertical or horizontal block to collapse for each P_π. It takes $O(n^2)$ moves and time, and ends up with a near-lattice polygon P' whose core is a polygon with simple projection.

In order to collapse the edges of P', we need to define the three dimensional version of blocks. A *3D block* of P' is a 3D box B with width one along x-axis or y-axis such that the intersection of B and the core of P' projects orthogonally to a two dimensional block on xy-plane. A *collapsible 3D block* is a 3D block whose orthogonal projection is a collapsible 2D block. Figure $5(c)$ shows the result of collapsing a 3D collapsible block, which can be done in $O(n)$ time. Let lattice polygon Q be projection of the core of P' onto xy-plane. Each 2D block in Q corresponds to a 3D block in P'. In this step, our algorithm finds each 3D collapsible block by identifying its corresponding 2D collapsible block in its orthogonal projection. As there are at most $O(n)$ 3D collapsible blocks to consider, the running time for this step is again $O(n^2)$.

Theorem 5. *A 3D lattice polygon with simple shadow can be straightened in $O(n^2)$ moves and time.*

7   Diameter-4 Trees in 2D

Let T be a polygonal tree of diameter 4 in the plane, with o as its central node. We call the edges incident to o *back edges*, and the rest *front edges*. A *UB-tree* is a tree of diameter 4 with all its back edges of unit length. An *SB-tree* is a tree of diameter 4 with all its back edges not longer than their corresponding front edges. In this section, we show that a UB-tree or SB-tree in two dimensions can always be straightened.

We first to define some technical terms. A *branch* is a path from o to a leaf of T. We define $E(B)$ to be the extension ray of the back edge of B. The *direct straightening* of branch $B = ouv$ means to rotate the front edge uv around the back vertex u until it aligns along $E(B)$ by sweeping through the smaller angle. We denote $S(B)$ to be the swept region for directly straightening B. The *direct collapse* of branch $B = ouv$ means to rotate the front edge uv around the back vertex u by sweeping through the smaller angle until it makes an arbitrarily small angle with ou. We say B' *follows* B if B' intersects $E(B)$, and B' and B are branches of the same turn. We say B' *directly covers* B if B' intersects $E(B) \cup S(B)$, and B and B' are branches of opposite turns. We then define B' *covers* B if there exists some branch B'' following B or simply being B such that

B' directly covers B''. We define a branch B' *mutually covers* another branch B if B' covers B, and vice versa. Branch B' *non-mutually covers* branch B if B' covers B, but B does not cover B'.

We categorize the branches into three groups:

- (Group I) branches falling on a maximal following cycle,
- (Group II) those falling on non-mutual covering sequences, and
- (Group III) those including mutually covered branches and branches covered by some straightened branches.

See Figure 6 for an example of different groups of branches in a UB-tree. Group I branches are those with all of their vertices dotted, Group III branches are bold, and the remained branches are Group II branches.

Our algorithm consists of three phases, in which we straighten the three groups of branches in the above order respectively. In the following cycle (if there is), the front edge of some of its branches B can be swept through $S(B)$ by passing through an angle of $\Omega(\frac{1}{n})$. Thus Phase I needs $O(n^2)$ moves to straighten such a following cycle. In Phase II, we observe that the last branch (including its following branches) of a maximal non-mutual covering sequence can be straightened directly. We repeat this peeling process to straighten all Group II branches. Finally, we consider the final phase of our algorithm. Group III branches have an impor-

Fig. 6. A UB-tree

tant property that their front edges are shorter than their corresponding back edges. This implies that all branches with a common back edge can all be collapsed directly altogether. Then all the collapsed branches can be rotated to pack inside a small wedge, say a quadrant. Now the collapsed branches can be drawn out of the quadrant one by one to be straightened directly. This completes our algorithm. The algorithm can be implemented to take $O(n^2)$ moves and $O(n^3 \log n)$ time. We summarize in the following theorem.

Theorem 6. *A UB-tree or SB-tree of diameter 4 in two dimensions can be straightened in $O(n^2)$ moves and in $O(n^3 \log n)$ time.*

8   Conclusion

We present efficient algorithms to straighten lattice chains and trees and to convexify lattice polygons in both two and three dimensions, respectively. In particular, we only manage to show that a special class of lattice polygons in three dimensions can be convexified. We believe that any unknotted lattice polygon in three dimensions can always be convexified. We are currently investigate in this direction. It is also open whether an unknotted unit polygon in three dimensions can always be straightened. Furthermore, it is open whether a unit tree in two or three dimensions can always be straightened [13]. In Section 6, we show that an orthogonal chain in three dimensions can lock. However, it is unknown whether an orthogonal tree in two dimension can lock. In fact, we conjecture that it can always be straightened.

References

1. J. Alberto-Calvo, D. Krizanc, P. Morin, M. Soss, G. Toussaint. Convexifying polygons with simple projections. *Information Processing Letters*, 80(2):81-86, 2001.
2. H. Alt, C. Knauer, G. Rote, and S. Whitesides. The Complexity of (Un)folding. In *Proc. 19th ACM Symposium on Computational Geometry (SOCG)*, 164–170, 2003.
3. T. Biedl, E. Demaine, M. Demaine, S. Lazard, A. Lubiw, J. O'Rourke, S. Robbins, I. Streinu, G. Toussaint, and S. Whitesides. A Note on Reconfiguring Tree Linkages: Trees can Lock. *Discrete Applied Mathematics*, volume 117, number 1-3, pages 293-297, 2002.
4. T. Biedl, E. Demaine, M. Demaine, S. Lazard, A. Lubiw, J. O'Rourke, M. Overmars, S. Robbins, I. Streinu, G. Toussaint, and S. Whitesides. Locked and Unlocked Polygonal Chains in Three Dimensions. *Discrete & Computational Geometry*, volume 26, number 3, pages 269-281, October 2001.
5. J. Cantarella, E.D. Demaine, H. Iben, and J. O'Brien. An Energy-Driven Approach to Linkage Unfolding. In *Proceedings of the 20th Annual ACM Symposium on Computational Geometry (SoCG 2004)*, 134–143, 2004.
6. H.S. Chan and K.A. Dill. The protein folding problem. *Physics Today*, pages 24–32, February 1993.
7. R. Cocan and J. O'Rourke. Polygonal Chains Cannot Lock in 4D. *Computational Geometry: Theory & Applications*, 20, 105–129, 2001.
8. R. Connelly, E. Demaine, and G. Rote. Infinitesimally Locked Self-Touching Linkages with Applications to Locked Trees. In *Physical Knots: Knotting, Linking, and Folding of Geometric Objects in R^3*, edited by J. Calvo, K. Millett, and E. Rawdon (editors), 2002, pages 287–311, American Mathematical Society.
9. R. Connelly, E.D. Demaine, and G. Rote. Straightening Polygonal Arcs and Convexifying Polygonal Cycles. *Discrete & Computational Geometry*, volume 30, number 2, 205–239, 2003.
10. E. D. Demaine, S. Langerman, J. O'Rourke, and J. Snoeyink. Interlocked Open Linkages with Few Joints. In *Proc. 18th Annual ACM Symposium on Computational Geometry (SoCG)*, Barcelona, Spain, June 5-7, pages 189–198, 2002.
11. K.A. Dill. Dominant forces in protein folding. *Biochemistry*, 29(31), 7133–7155, August 1990.
12. B. Hayes. Protototeins. *American Scientist*, 86, 216–221, 1998.
13. S.-H. Poon. On Straightening Low-Diameter Unit Trees. In *Proc. 13th International Symposium on Graph Drawing*, 519–521, 2005.
14. I. Streinu. A combinatorial approach for planar non-colliding robot arm motion planning. In *Proc. 41st ACM Annual Symposium on Foundations of Computer Science (FOCS)*, 443–453, 2000.

Restricted Mesh Simplification Using Edge Contractions

Mattias Andersson[1], Joachim Gudmundsson[2], and Christos Levcopoulos[1]

[1] Department of Computer Science, Lund University, Box 118, 221 00 Lund, Sweden
`mattias@cs.lth.se, christos@cs.lth.se`
[2] National ICT Australia Ltd., IMAGEN Program, Locked Bay 9013, Alexandria
NSW 1435, Australia
`Joachim.Gudmundsson@nicta.com.au`

Abstract. We consider the problem of simplifying a triangle mesh using edge contractions, under the restriction that the resulting vertices must be a subset of the input set. That is, contraction of an edge must be made onto one of its adjacent vertices. In order to maintain a high number of contractible edges under this restriction, a small modification of the mesh around the edge to be contracted is allowed. Such a contraction is denoted a 2-step contraction. Given m "important" points or edges it is shown that a simplification hierarchy of size $O(n)$ and depth $O(\log(n/m))$ may be constructed in $O(n)$ time. Further, for many edges not even 2-step contractions may be enough, and thus, the concept is generalized to k-step contractions.

1 Introduction

In computer graphics objects are commonly represented using triangle meshes. One important problem regarding these meshes is how to efficiently simplify them, while maintaining a good approximation of the original mesh. As an example, scanners often produce information redundant meshes containing millions of points and triangles. Further, often the simplification should be performed in several rounds, such that a level-of-detail hierarchy is constructed. One application of such a hierarchy is that an appropriate level may be chosen depending on viewing distance, as finer details tend to be unnecessary as the distance increases. Other applications include progressive transmission and efficient storing. It is common to represent the level-of-detail hierarchy as a directed, acyclic and hierarchical graph, where each level in the graph corresponds to a level in the level-of-detail hierarchy, and where each node in the graph corresponds to a triangle, in the natural way. The first, top-most, level in the graph corresponds to the input mesh. When a contraction is made two triangles disappear, and one or more triangles are affected in such a way that their appearance change. In the graph this is represented with edges between disappearing triangles at some level i, and the affected triangles at level $i + 1$. The efficiency of a simplification algorithm is directly related to the size [3,11] and depth of the hierarchy

D.Z. Chen and D.T. Lee (Eds.): COCOON 2006, LNCS 4112, pp. 196–204, 2006.
© Springer-Verlag Berlin Heidelberg 2006

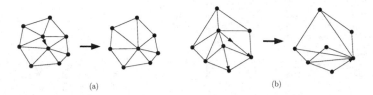

Fig. 1. (a) A valid edge contraction (merge-operation). (b) An invalid edge contraction.

graph that it produces. Simplification algorithms constructing hierarchies of size $O(n)$ and depth $O(\log n)$ have been presented for several problem variants [1,2,4,10].

Mesh simplification is generally regarded as a mature field (see [5,8] for surveys), consisting of several suggested methods and problem variants. In this paper the method of iteratively contracting edges [1,6,7,9] is considered, where contractions are made such that no crossing edges result from the contraction. This method is examined under the restriction that the set of output points is required to be a subset of the input points, i.e., contraction of an edge must be done onto one of its adjacent vertices. In order to maintain a high number of contractible edges a small modification of the mesh around an edge to be contracted is allowed. This method is denoted a 2-step contraction, and it is shown that a hierarchical graph may be produced, under the restriction that m "important" points or edges may not be contracted, of size $O(n)$ and depth $O(\log(n/m))$. Further, in order to enable contraction of edges that are not even 2-step contractible the concept is generalized to k-step contractions. We show that k is upper bounded by either $deg(v) - 4$ of a vertex v to be contracted, or by the number of concave corners on the link of v (see Section 2 for definition). Each step in this method is geomorphic, i.e., it is visually smooth. Further, as will be seen the relations of adjacency between triangles change temporarily during a k-step contraction, but are restored once it is complete.

Note that many of the results in previous papers were achieved using only 1-step contractions. However, in those cases there were no restrictions on the set of output points. Note also that in this paper we only consider vertices in the plane, and as such the results mainly apply to terrains, and not to the general setting of arbitrary objects in 3-dimensional space. However, the ambition is to extend the results to such a setting in the future.

2   Contracting in k Steps

As input we are given a planar triangulation T. We can assume that the outer hull of T is a rectangle, as illustrated in Fig. 2a.

The aim is to simplify T by iteratively performing edge contractions, as shown in Fig. 1a, where an edge (u, v) can be contracted such that u is moved to v, or v is moved to u. A problem that often occurs during edge contractions of

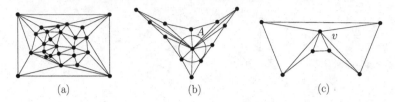

Fig. 2. (a) A planar triangulation with a rectangular outer hull is given as input. (b) No edge in region A can be contracted, unless using k-step contractions. (c) Example of a vertex v that is not 1-step contractible.

triangulations is that the resulting graph might not be a planar triangulation. An edge contraction is said to be *valid* if the resulting graph is still a planar triangulation (see Figure 1a), and invalid (see Figure 1b) otherwise. In this paper we consider the problems of defining valid contractions and computing valid contractions. Below some basic operations and notations are defined:

Definition 1. *Given a vertex v, the link of v, or $lk(v)$ is the cycle through the neighbors of v, where the edges of the cycle lie on triangles incident on v.*

Definition 2. *Two basic operations:*

- *A merge operation contracts one vertex v onto another vertex u', both connected by an edge in T, into one vertex u', as illustrated in Fig. 3c-d.*
- *Given vertices s, t and u, let $C(s,t,u)$ be the vertices (s and t included) of the chain of $lk(v)$ which connects s and t, and includes u. A split-and-merge on v and u, using s and t (illustrated in Fig. 3a-c) denotes a split-and-merge where v is split into two vertices, v and v_1, where v_1 is connected to $C(s,t,u) \cup v$, and v is connected to $lk(v) \setminus C(s,t,u) \cup s \cup t$. After this split v_1 is merged on u. The split-and-merge operation is said to be* valid *if the triangulation is planar at each step of the operation.*

Note that an edge contraction is obtained by a single merge operation. Below we define 1-step contractible using the merge concept, and we then generalize this concept into k-step contractible.

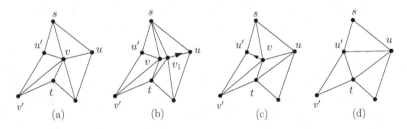

Fig. 3. Illustrating a 2-step contraction of a degree 6 node v

Definition 3. *If two vertices u and v, connected by an edge in the triangulation T can be merged into a new vertex w placed at u such that the contracted triangulation still is planar, then v is said to be* 1-step contractible *(at u).*

Definition 4. *A vertex v is said to be* k-step contractible *if and only if one can perform at most k − 1 valid split-and-merge operations followed by a 1-step contraction of v.*

With regards to guaranteeing a hierarchical simplification graph of small size and depth, mainly 2-step contractions will be considered. Fig. 3 shows a vertex v that is 2-step contractible since a valid split-and-merge operation is followed by a 1-step contraction of v. However, depending on complexity, some areas of an object may require more triangles for good approximation than others. (For example, the nose of a face, versus the more flat cheek.) Thus, it would be desirable to be able to choose local areas in which to contract. Problem is, 2-step contractions may not always be sufficient. For example, in the structure shown in Fig. 2b, if one wants to perform k-step contractions which change or contract only edges inside the area A, then k has to be at least proportional to the number of edges inside A (divided by some constant). Thus, the generalized concept of a k-step contraction is needed, and below (Theorem 1 and 2) upper bounds on k are shown.

Theorem 1. *Any vertex v, not on the hull of T, with degree at most k is q-step contractible, where* $q = \max\{1, k - 4\}$.

Proof. The theorem is proven by induction on the degree of v.

Base case: Vertices of degree at most four can easily be 1-step contracted. We thus assume that v has degree five, which immediately implies that $lk(v)$ contains five points. Consider the interior of $lk(v)$. If there exists a corner v' of $lk(v)$ which can see all other corners of $lk(v)$ then v is 1-step contractible at v', and thus, the theorem holds. Next, since $lk(v)$ has at least three convex corners, $lk(v)$ has at most two concave corners. If $lk(v)$ has only one concave corner u then this corner must see all the vertices of $lk(v)$ and, hence, v is 1-step contractible to u. If $lk(v)$ has two concave corners we have two cases, as shown in Fig. 4. Note

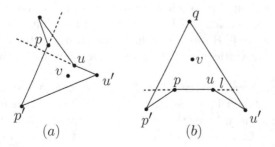

Fig. 4. The two cases of Theorem 1

that edges between v and $lk(v)$ are not included in order to avoid cluttering of the figure. In the first case, Fig. 4a, the concave corner points p and u are not incident, while in the second, Fig. 4b, they are.

First case. First note that p and u must lie inside the triangle defined by the three points of the convex corners in $lk(v)$, and also that p and u always see each other. There exist two incident convex corner points p' and u', such that p' is incident on p and u' is incident on u. It is straightforward to see that p sees all points of $lk(v)$ if the edge (p, p') does not cross the line-extension of the line (u, u'), as p then can see u'. The same holds for u, the edge (u, u') and the line extension of (p, p'). However, both cases can not occur simultaneously as this implies that the edges (u, u') and (p, p') must cross. Thus, either p or u can see all of $lk(v)$.

Second case. Consider the one convex corner point q not incident on either p or u. Since q is connected to both p' and u', q will see all corners of $lk(v)$ if and only if q sees both p and u. Next, consider the line-extension l of the edge (p, u). Since p and u are concave corners p' and u' must lie on the same side of l. Further, q must connect to p' and u' such that p' and u' form convex angles and p and u form concave angles. This means that q must lie on the opposite side of p and u, with regards to l, which immediately implies that q can see both p and u.

Induction hypothesis: Assume that the theorem holds for all vertices of degree at most $m - 1$. *Induction step:* Assume that v has degree m. There exists a vertex on the link of v that can see at least four consecutive vertices of the link including itself. Denote these vertices $u_1, \ldots, u_4$. Now a valid split-and-merge on v and u_3, using u_1 and u_4 can be performed. Note that the result of this split-and-merge will be one 'flipped' edge. The degree of v is now $m - 1$, thus applying the induction hypothesis on v proves the theorem.   □

Note that if v has degree more than six then it might not be 1-step contractible as shown in Fig. 2c. Next, the following corollary can be shown using the fact that any planar graph has total degree at most $6n - 12$, and the fact that there can be at most $n - 3$ points of degree six (full proof omitted).

Corollary 1. *At least two edges in T are 1-step contractible.*

Note that if only few edges are 1-step contractible then almost all vertices in T must have degree 6. The bound stated in Corollary 1 is probably very conservative. If the number of 1-step contractible edges is small that implies that almost all vertices of T have degree 6. However, we have not been able to construct any examples where almost all vertices have degree 6 while simultaneously being not 1-step contractible.

Next, we present an alternative bound on k. As only concave corners restrict visibility, intuitively it should be easier to contract a vertex with few concave corners on its link. Assume w.l.o.g. that v has c concave vertices on $lk(v)$, as shown in Fig. 5. Let s_1 be any concave vertex farthest from v and order the concave vertices $s_1, \ldots s_c$ as they appear clockwise around v (in this context, let $i + 1 = 1$ if $i = c$, and let $i - 1 = c$ if $i = 1$). Next, let β_i denote the angle

$\angle s_{i-1}s_is_{i+1}$ and let α_i denote the angle $\angle s_ivs_{i+1}$. Further, let $C(s_i)$ denote the subchain of $lk(v)$ clockwise from s_i to s_{i+1} and let $C_P(s_i)$ denote the convex polygon bounded by $C(s_i)$ and the edge (s_i, s_{i+1}). The following observation will be needed (proof omitted):

Observation 1. *If $\alpha_i \leq 180°$, then the two consecutive concave vertices s_i and s_{i+1} must see each other.*

Further, the theorem below is shown using a valid split-and-merge which reduces the number of concave corners by at least one in the resulting $lk(v)$. Such a split-and-merge will be denoted a *reducing* split-and-merge, for which the following observation can be made (proof omitted).

Observation 2. *If $\beta_i \leq 180°$, $v \notin C_P(s_{i-1})$ and $v \notin C_P(s_i)$, then a split-and-merge on v and s_i, using s_{i-1} and s_{i+1}, is reducing.*

The following theorem can now be shown.

Theorem 2. *Every vertex v, not on the hull of T, with at most c concave vertices on its link is c-step contractible.*

Proof. The theorem is proven by induction on c.
Base cases: If $c = 1$ then let s be the concave vertex of v on $lk(v)$. Obviously, v can be 1-step contracted at s, thus the theorem holds for $c = 1$.

If $c = 2$ then let s_1 and s_2 be the two concave vertices on the link of v. It is easily seen that a split-and-merge to s_1 followed by merging v to s_2 can be performed, thus v is 2-step contractible.

Induction hypothesis: Assume that the theorem holds for all vertices of degree at most $c - 1$.

Induction step: We show that there always exists reducing split-and-merge, and thus, applying the induction hypothesis on v proves the theorem. Consider the following cases:

Case 1: $\alpha_c \leq 180°$ and $\alpha_1 \leq 180°$: See Fig. 5a. In this case we immediately have that $v \notin C_P(s_c)$ and $v \notin C_P(s_1)$ since both α_c and α_1 have degree at most $180°$. Further, since s_1 is a point at furthest distance from v, it holds that $\beta_1 \leq 180°$, and thus, a split-and merge on v and s_1, using s_c and s_2, is reducing (Observation 2).

Case 2: $\alpha_c > 180°$ or $\alpha_1 > 180°$: See Fig. 5b. Assume w.l.o.g. that $\alpha_c > 180°$ which immediately implies that $v \in C_P(s_c)$. Since s_1 and s_c both are visible from v there can be no other point s_i, $i \neq c$, such that $v \in C_P(s_i)$. Let l_i and l'_i be the lines through s_i and v, and through s_i and s_{i+1}, respectively. Let H_1 be the area defined by lines l_1 and l'_1, as shown in Fig. 5(b). H_c is defined correspondingly using lines l_c and l'_c. Assume that H_1 contains a vertex $u \in C(s_c)$, and consider a split-and-merge on v and s_1, using u and s_2. Note that all vertices between u and s_1 are visible to s_1, since $H_1 \cup H_c$ can contain vertices from $C(s_c)$ only. The same holds between s_1 and s_2 since $v \notin C_P(s_1)$ (see above). Further, the definition of l_1 guarantees that v will remain in the resulting $lk(v)$ after the

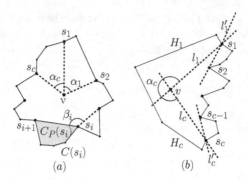

Fig. 5. An illustration of the two cases of Theorem 2. Figure (a) illustrates Case 1, and Figure (b) illustrates Case 2.

split-and-merge, and line l'_1 guarantees that the corner defined by edges (u, s_1) and (s_1, s_2) is convex. Thus, if H_1 contains a vertex $u \in C(s_c)$, the above split-and-merge must be reducing. Correspondingly, a split-and-merge on v and s_c, using u and s_{c-1} will be reducing if H_c contains a vertex $u \in C(s_c)$. Finally, note that the area $H_1 \cup H_c$ must actually contain a vertex from $C(s_c)$ since $\alpha_c > 180$ and $C(s_c)$ connects s_c with s_1. □

Allowing k-step contractions increases the flexibility of simplification since it allows a greater fraction of the edges to be contracted, as shown by lower bound below, which follows from Theorem 1 and the fact that the total degree is bounded (proof omitted).

Observation 3. *At least* $(\frac{k-1}{k+2})n$ *vertices are k-step contractible, for any* $k \geq 2$.

Note that a similar bound can be obtained using Theorem 2 instead. However, we have not been able to improve the bound in Observation 3 using this theorem.

3   The Hierarchical Graph

In this section we show that using 2-step contractions we can achieve a hierarchical graph, as defined in the introduction, of size $O(n)$ and depth $O(\log(n/m))$, given m *important* points or edges that may not be contracted. In order to do this several edges must be simultaneously contracted in each *round*, that is, at each level of the graph. Next, note that a previously valid 1-step contractible edge might become invalid after other edges have been contracted, as shown in Fig. 6. In order to avoid this problem, for the purpose of finding simultaneously contractable edges, we consider independent edges. Let S'_2 be the set of 2-step contractible vertices of degree at most six. Combining Theorem 1 and Observation 3 it is straightforward to see that $|S'_2| \geq \frac{n}{4}$. Since a vertex in S'_2 has at most six neighbors we can choose at least $\frac{n}{4 \cdot 7} = \frac{n}{28}$ vertices from S'_2 such that none of these chosen vertices has a neighbor from S'_2. Thus, there exists a constant fraction γ of independent 2-step contractible vertices, and the following theorem can be shown.

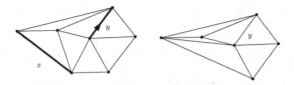

Fig. 6. Initially x and y are contractible, but after x has been contracted y is no longer contractible

Theorem 3. *Given m important points $S'' \subset S$ in a triangulation T one can perform $O(\log(n/m))$ rounds of 2-step contractions to obtain a triangulation T' of a point set S' with complexity $O(m)$ such that $S'' \subseteq S' \subset S$.*

Proof. Let n_i denote the number of vertices before round i and consider an arbitrary constant $\delta < \gamma$. Perform rounds until $m \geq \delta n_i$, that is until the resulting point set S' have complexity $O(m)$. This is possible, since as long as $m \leq \delta n_i$, there are at least $\gamma n_i - \delta n_i = (\gamma - \delta)n_i$ 2-step contractible vertices remaining, containing no important point. Thus, T' can be obtained using at most $O(\log_{\frac{1}{\gamma-\delta}} n - \log_{\frac{1}{\gamma-\delta}} m) = O(\log(n/m))$ rounds of contractions.   □

Lemma 1. *Using rounds of 2-step contractions a hierarchical graph of size $O(n)$ and depth $O(\log(n/m))$, given m important points, may be produced in $O(n)$ time.*

Proof. Note that the above theorem immediately enables the construction of hierarchical graph of depth $O(\log(n/m))$. Next, consider the size. Note that the number of nodes in the hierarchical graph is $O(n)$ and only 2-step contractible vertices of degree at most six are used during the rounds of contractions. This means that at most four triangles are affected by a contraction, which implies that each node in the hierarchical graph has at most four incident edges. Thus, the hierarchical graph has size $O(n)$.

Next, consider the time complexity of creating the hierarchical graph. Note that Theorem 3 was shown using only 2-step contractible edges of constant degree (at most six). Thus, in each round i the set of γn_i independent 2-step contractible edges can be found in $O(n_i)$ time. This means, since $n_i \leq n\gamma^{i-1}$, that the total running time is $O(n + n\gamma + n\gamma^2 + \ldots + n\gamma^{O(\log(n/m))}) = O(n)$.   □

Finally, note that the above results also hold for m important edges (or m edges and vertices, in total), since each important edge restricts possible contraction for only a constant (two) number of vertices.

References

1. S. Cheng, T. Dey and S. Poon. Hierarchy of Surface Models and Irreducible Triangulations. *Computational Geometry Theory and Applications (CGTA)*, 27:135-150, 2004.
2. M. de Berg, K. T. G. Magillo. On Levels of Detail in Terrains. In Proc. *ACM Symposium on Computational Geometry*, pp.26-27, 1995.

3. L. De Floriani, P. Magillo and E. Puppo. Building and Traversing a Surface at Variable Resolution. In Proc. *IEEE Visualization'97*, pp.103-110, 1997.
4. C. A. Duncan, M. T. Goodrich and S. G. Kobourov. Planarity-Preserving Clustering and Embedding for Large Planar Graphs. In Proc. *Graph Drawing '99*, pp.186-196, 1999.
5. M. Garland. Multiresolution Modelling: Survey and Future Opportunities. *Eurographics '99*, State of the Art Report (STAR).
6. S. Gumhold, P. Borodin and R. Klein. Intersection Free Simplification In Proc. *4th Israel-Korea Bi-National Conference on Geometric Modeling and Computer Graphics*, pp 11-16, 2003
7. P. Heckbert and M. Garland. Surface Simplification Using Quadric Error Metrics. In Proc. *SIGGRAPH'97*, pp 209-216, 1997
8. P. Heckbert and M. Garland. Survey of Polygonal Surface Simplification Algorithms. Multiresolution Surface Modelling Course, *SIGGRAPH'97*
9. H. Hoppe. Progressive Meshes. In Proc. *SIGGRAPH'96*, pp. 99-108, 1996
10. D. G. Kirkpatrick. Optimal Search in Planar Subdivisions. *SIAM Journal of Computing*, 12:28-35, 1983.
11. J. C. Xia, J. El-Sana and A. Varshney. Dynamic View-Dependent Simplification for Polygonal Models. In Proc. *IEEE Visualization'96*, pp.327-334, 1996.

Enumerating Non-crossing Minimally Rigid Frameworks

David Avis[1], Naoki Katoh[2,*], Makoto Ohsaki[2,*],
Ileana Streinu[3,**], and Shin-ichi Tanigawa[2]

[1] School of Computer Science, McGill University, Canada
avis@cs.macgill.ca
[2] Department of Architecture and Architectural Engineering, Kyoto University,
Kyoto 615-8450 Japan
{ohsaki, naoki, is.tanigawa}@archi.kyoto-u.ac.jp
[3] Dept. of Comp. Science, Smith College, Northampton, MA 01063, USA
streinu@cs.smith.edu

Abstract. In this paper we present an algorithm for enumerating without repetitions all the non-crossing generically minimally rigid bar-and-joint frameworks (simply called non-crossing Laman frameworks) on a given generic set of n points. Our algorithm is based on the reverse search paradigm of Avis and Fukuda. It generates each output graph in $O(n^4)$ time and $O(n)$ space, or, with a slightly different implementation, in $O(n^3)$ time and $O(n^2)$ space. In particular, we obtain that the set of all non-crossing Laman frameworks on a given point set is connected by flips which remove an edge and then restore the Laman property with the addition of a non-crossing edge.

1 Introduction

Let $G = (V, E)$ be a graph with vertices $\{1, \ldots, n\}$ and m edges. G is a *minimally rigid graph* (also called *Laman graph*) if $m = 2n - 3$ and every subset of $n' \leq n$ vertices spans at most $2n' - 3$ edges. An embedding $G(\boldsymbol{p})$ of the graph G on a set of points $\boldsymbol{p} = \{p_1, \cdots, p_n\} \subset R^2$ is a mapping of the vertices V to points in the Euclidian plane $i \mapsto p_i \in \boldsymbol{p}$. The edges $ij \in E$ are mapped to straight line segments $p_i p_j$. An embedding $G(\boldsymbol{p})$ is *non-crossing* if no pair of segments $p_i p_j$ and $p_k p_l$ corresponding to non-adjacent edges $ij, kl \in E, i, j \notin \{k, l\}$ have a point in common.

Laman graphs embedded on *generic* point sets are called *Laman frameworks* and has the special property of being *minimally rigid* [14,10], when viewed as bar-and-joint frameworks with fixed edge-lengths, which motivates the tremendous interest in their properties.

In this paper we give an algorithm for enumerating all the *non-crossing Laman frameworks* embedded on a *given generic point set*.

* Supported by JSPS Grant-in-Aid for Scientific Research on priority areas of New Horizons in Computing.
** Supported by NSF grant CCF-0430990 and NSF-DARPA CARGO CCR-0310661.

D.Z. Chen and D.T. Lee (Eds.): COCOON 2006, LNCS 4112, pp. 205–215, 2006.

To the best of our knowledge, this is the first algorithm proposed for enumerating (without repetitions, in polynomial time and without using additional space) all the non-crossing generically minimally rigid frameworks. We achieve $O(n^4)$ time per graph in $O(n)$ space (or, with a slightly different implementation, in $O(n^3)$ time and $O(n^2)$ space) by using reverse search.

The reverse search enumeration technique of Avis and Fukuda [2,3] has been successfully applied to a variety of combinatorial and geometric enumeration problems. The necessary ingredients to use the method are an implicitly described connected graph on the objects to be generated, and an implicitly defined spanning tree in this graph. In this paper we supply these ingredients for the problem of generating Laman frameworks.

Relevant to the historical context of our work are the results of Bereg [6,7] using reverse search combined with data-specific lexicographic orderings to enumerate triangulations and pointed pseudo-triangulations of a given point set. We notice in passing that there exist several other algorithms for enumerating (pseudo-)triangulations [9,8,1], but they are based on different techniques.

Also relevant is the pebble game algorithm of Jacobs and Hendrickson [11] for 2-dimensional rigidity, see also [5]. Our complexity analysis relies on the recent results, due to Lee et al. [15,16], regarding the detailed data-structure complexity of finding and maintaining rigid components during the pebble game algorithm. Indeed, the time-space trade-off of our algorithm is inherited from [16].

We briefly discuss now the difference between generating non-crossing Laman frameworks as opposed to pointed pseudo-triangulations. A *pointed pseudo-triangulation* is a special case of a non-crossing Laman framework on a given point set [19], where every vertex in the embedding is incident to an angle larger than π. Pseudo-triangulations are connected via simple *flips*, in which the removal of any non-convex-hull edge leads to the choice of a *unique* other edge that can replace it, in order to restore the pointed pseudo-triangulation property. The flip graph of all pointed pseudo-triangulations is a connected subgraph of the graph of all the Laman frameworks. In fact, it is the one-skeleton of a polytope [18], and the reverse search technique can be directly applied to it. Bereg's efficient algorithm makes use of specific properties of pointed pseudo-triangulations which do not extend to arbitrary non-crossing Laman frameworks. In particular, remove-add flips are *not unique*, relative to the removed edge, in the case of non-crossing Laman frameworks. Moreover, it is not even known whether the set of all the non-crossing Laman frameworks on a points set is connected via these flips.

We describe now briefly how this problem came to our attention via the work of the third author. Graph theoretical approaches are widely used in *structural mechanics* [12], where the edges and vertices in the graph represent the bars and rotation-free joints of a structure called a *truss*. It is well-known [4] that the stiffest truss under static loads is statically determinate, a concept directly related to the previously defined Laman-graph property. Ohsaki et al. [17] presented a method for generating multi-stable flexible bar-joint system, and found that the optimal structure is statically determinate. Since the optimal topology is found by removing unnecessary nodes and members from the highly connected initial structure, the computational cost can be reduced if the candidate set of Laman frameworks are first enumerated. In a related direction, Kawamoto et al. [13] presented a method based on the enumeration of *planar* graphs to find an optimal mechanism. None of these papers give a general approach for the systematic enumeration of rigidity-constrained structures - which is the topic of our paper.

2   Preliminaries

Besides the definition given in the introduction, Laman graphs can be characterized in various ways. In particular, Laman graph on n vertices has an inductive construction, called *Henneberg construction* [20]. Start from an edge for $n = 2$. At each step, add a new vertex in one of the following two ways:

Henneberg I: add a new vertex and connect it to two old vertices via two new edges.
Henneberg II: remove an old edge, add new vertex, and connect it to two endpoints of the removed edge and to some other vertex.

A *mechanism* is a flexible framework obtained by removing one or more edges from a generic Laman framework. Its *number of degrees of freedom* or *dof*s, is the number of removed edges. We will encounter mostly *1-degree-of-freedom (1dof) mechanisms*, which arise from a Laman framework by the removal of one edge. Similarly, considering a Laman graph, the graph obtained by removing one edge from it can be called the *graph of 1dof*. In particular, a mechanism with k dofs has exactly $2n - 3 - k$ edges, and each subset of n' vertices spans at most $2n' - 3$ edges. A subset of some n' vertices spanning *exactly* $2n' - 3$ edges is called a *rigid block*. A *maximal block* is called a *rigid component* or a *body*. The edge set of a mechanism is partitioned into rigid components, with some components possibly containing just one edge each. Two rigid components are either disjoint (vertex-wise), or may share one vertex. A *joint* is a vertex shared by at least two components. These properties, as well as efficient algorithms for computing rigid components in Laman mechanisms, can be found in [11, 5, 15, 16].

The Laman frameworks on a generic point set form the set of *bases* of the *generic rigidity matroid* on K_n, see [10]. The bases have all the same size $2n - 3$. Bases may be related via the *base exchange* operation, which we will call a *flip* between two Laman frameworks. Two Laman frameworks L_1 and L_2 are connected by a flip if their edge sets agree on $2n - 4$ positions. The flip is given by the pair of edges (e_1, e_2) not common to the two bases, $e_1 \in L_1 \setminus L_2$, $e_2 \in L_2 \setminus L_1$. Using flips, we can define a graph whose nodes are *all* the Laman frameworks on n vertices, and whose edges correspond to flips. It is well-known that the graph whose nodes are the bases of a matroid connected via flips, is connected. But a priori, the subset of *non-crossing Laman framworks* may not necessarily be. We will prove this later in Section 4.

Revese search is a memory efficient method for visiting all the nodes of a connected graph that can be defined implicitly by an adjacency oracle. It can be used whenever a spanning tree of the graph is defined implicitly by a *parent* function which is defined for each vertex of the graph except a prespecified *root*. Iterating the parent function leads to a path to the root from any other vertex in the graph. The set of such paths defines a spanning tree, known as the *search tree*.

3   The Search Tree

Given a set of points p on the plane. Let $\mathcal{G}$ be a set of all the non-crossing Laman frameworks on p. In this section we define the main structure required by reverse search, a *search tree* on $\mathcal{G}$. We choose a certain Laman framework G_{root} to be the root. Then we define a *parent* for every $G \in \mathcal{G} \setminus \{G_{root}\}$. To show that the parent function defines a search tree we associate an *index* to every Laman framework, such that the parent function always returns a Laman framework with smaller index. This gives a forest structure. To prove that it is connected and thus a search tree, we will show in Section 4 that the parent of every non-root node indeed exists.

Root. We choose the root of the search tree to be a *greedy pseudo-triangulation* corresponding to a fixed direction. The simplest way to define it is relative to the horizontal direction (x-axis).

We first sort the points by x-coordinate and label them as $\{1, 2, \cdots, n\}$ in this order. Then we construct a *Henneberg I* pointed pseudo-triangulation as follows. Start with the edge 12 and continue for $n-2$ steps. At each step, the next vertex (in x-sorted order) is added (vertex $i + 2$ at step i, $i = 1, \cdots, n - 2$), together with the two *tangents* from point p_{i+2} to the (convex hull of) the framework constructed so far. Fig. 1(a) illustrates the root, and Fig. 1(b) gives an example of a non-root framework.

Index. Given $G \in \mathcal{G} \setminus \{G_{root}\}$, we define its *index* as a pair $index(G) = (c, d)$, where $c \in \{2, \cdots, n\}$ and $d = \in \{1, \cdots, n\}$ are, respectively, the label of the *critical vertex* of G and the *critical degree* of the critical vertex, defined below.

The *critical vertex* (with respect to the root) is the largest label of a vertex whose incident edges differ from the corresponding set of incident edges in the

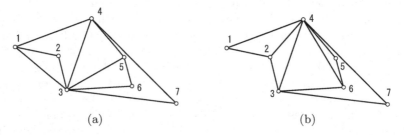

(a)   (b)

Fig. 1. (a)The root non-crossing Laman framework on a set of 7 points. (b)A non-root Laman framework.

root framework. Since all the vertices with labels at least $c + 1$ have the same edges as their counterparts in the root framework, it follows easily that the subset of G spanned by the vertices $\{1, \cdots, c\}$ is Laman. An edge in G is called *non-root* if it doesn't exist in G_{root}. The *critical degree* is the number of non-root edges incident to the critical vertex. For example, the non-root framework in Fig. 1(b) has index $(6, 1)$. We use the index as a measure of how far a node (Laman framework) is from the root, whose index is defined to be $(1, 0)$.

Parent rule. The *parent* of a node (Laman framework) is defined in terms of its critical vertex via a certain *Remove-add flip*. The removed edge which does not exist in the *root* will be incident to the critical vertex, and the added edge will be chosen so that it will decrease the index. In general there will be several choices of such flips. To uniquely define the parent, we will use a lexicographic ordering of these flips. The efficiency of the *parent* function depends on the lexicographic ordering. We will discuss it in Section 5. The correctness of the *parent* definition follows from our Main Theorems:

Theorem 1. *Every non-root non-crossing Laman framework has a parent whose index is smaller than that of the current node.*

The proof of the above theorem relies on properties of non-crossing Laman frameworks described in the next section. Based on Theorem 1, we will propose a reverse search algorithm by the standard techniques developed by [2, 3]. The detailed description of the algorithm and the complexity analysis will be given in Section 5.

Theorem 2. *The set $\mathcal{G}$ can be reported in $O(n^3)$ time per non-crossing Laman framework using $O(n^2)$ space and in $O(n^4)$ time using $O(n)$ space.*

4   Remove-Add Flips in Non-crossing Laman Frameworks

This section contains the proof of Theorem 1 which follows from a sequence of Lemmas (whose proofs are omitted).

In a graph, a subgraph induced by a vertex set $V' \subset V$ is a *cut* if its removal disconnects the graph.

Lemma 1. *A rigid component cannot be a cut in a graph of 1dof.*

The following statements are better understood if we forget for a moment the geometry of the embedding of the non-crossing Laman framework *in the Euclidean plane*, and think of G as a spherical (topological) embedding.

Proposition 1. *In a topologically embedded planar Laman graph, each face cycle is simple and subdivides the sphere into two disk-like regions.*

Lemma 1 shows that a rigid component cannot disconnect the graph, therefore for each cycle in a rigid component, only one (but not both) of the two (spherical) disk-like regions induced by it may contain vertices not in the rigid component.

Since a face is empty, if its boundary cycle is part of a rigid component, we will say that the face *belongs* to the rigid component. Notice that a rigid component may thus contain the outer face. We obtain:

Lemma 2. *In a topologically embedded planar Laman graph, the union of all the faces belonging to the same rigid component form a topological disk.*

If the removed edge does not belong to the outer face of a planar embedding of G, then the complement of G' may fall inside an interior face and the union of the (embedded) interior faces of G' may look like an annulus. See Fig. 2.

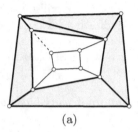 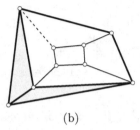

(a)   (b)

Fig. 2. Removing the dotted edge results in a 1dof mechanism. (a) The removed edge is interior. The shaded rigid component is not a topological disk (the other rigid components are just bars). In (b), the outer rigid component is made out of all the bold bars, and each of the other bars is a separate component.

Lemma 2 implies that the faces of the non-crossing framework can be divided into two categories: *rigid*, if they belong to a rigid component and *flexible*, otherwise. A rigid component *bounds* a flexible face, if they share at least one edge. Since Laman frameworks are 2-connected, all the rigid components are made out of simple faces, and therefore the boundary of a rigid component is also a simple cycle. For flexible faces, this may not be true. The next lemmas give some useful properties of flexible faces:

Lemma 3. *All the flexible faces of a 2-connected 1dof non-crossing mechanism have at least four bounding rigid components.*

Lemma 4. *If the 1dof non-crossing mechanism is not 2-connected, then all but one face are as in Lemma 3. The unique exception is a special flexible face incident to exactly two components, and whose bounding cycle is not simple.*

The situations described in Lemma 4 are illustrated in Fig. 3. From these properties of non-crossing Laman frameworks and 1dof mechanisms, we now sketch the proof for our main theorem.

Proof (of Theorem 1). Let $G \in \mathcal{G} \setminus G_{root}$ and let p_c be the critical vertex of G. Then the subgraph induced by $\{1, 2, \ldots, c\}$ is still a Laman graph (whose framework embedded on $\{p_1, \ldots, p_c\}$ is denoted by G'). From the definition of the critical vertex, p_c is always on the outer face in G' and there is at least one

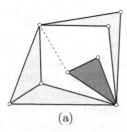
(a)

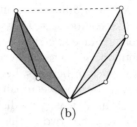
(b)

Fig. 3. The special type of flexible face which is incident to two rigid components only. (a) Unbounded and (b) Bounded face. Notice that the original graph is 2- but not 3-connected, and the removed dotted edge is incident to the cut pair.

non-root edge incident to p_c. Then we delete the non-root edge $p_c p_i$ from G' in the *parent* function. It suffices to prove that there always exists an edge in $e \notin G'$ satisfying: (i) after inserting e to $G' \setminus \{p_c p_i\}$, we obtain a non-crossing Laman framework, and (ii) e is disjoint from p_c or a root edge incident with p_c by which resulting framework has a smaller index. The proof has two cases depending on whether $p_c p_i$ is on the outer face or not.

Case 1: $p_c p_i$ is not on the outer boundary. In this case 1dof mechanism $G' \setminus \{p_c p_i\}$ satisfies the properties proven in the previous lemmas.

(**1-a**) When $G' \setminus \{p_c p_i\}$ is 2-connected. From Lemma 3 all flexible faces are incident to at least four rigid components (one of which may be annulus-like component) and four distinct joints. We call a pair of joints *incident* if they belong to the same component. Consider the geodesic paths between any pair of non-incident joints, and call them *geodesic diagonals* of the faces. There are at least two such geodesic diagonals in each face, since each face has at least four joints. A geodesic diagonal may lie entirely on the boundaries of the rigid component incident to the face, in which case it must also go through the joints between those rigid component, or may have at least one segment lying inside the face. Moreover, such a segment may go between vertices belonging to the same rigid component, or may go between two different components. In this last case, the segment connects two distinct rigid components. We say such segment *free*. We claim that there exists at least one free segment, on at least one geodesic diagonal path in all the flexible faces of the mechanism.

Suppose that there is more than one flexible face in $G' \setminus \{p_c p_i\}$. Since each of them has at least one free segment, there always exists at least two free segements one of which is not incident to p_c. Then, we consider the case where there is only one flexible face in $G' \setminus \{p_c p_i\}$. Note that all the joints are on the outer face and the flexible face. If all four joints are the convex hull vertices, then the interior face contains two distinct *free* segments, and one of which is not incident to p_c. Otherwise, suppose the interior face contains only one free segment $p_c p_i$. Then, there exists one free segment in the outer face which is certainly not incident to p_c.

(**1-b**) When $G' \setminus \{p_c p_i\}$ is 1-connected. From Lemma 4, $G' \setminus \{p_c p_i\}$ has an annulus-like component (see Fig.3(a)). Then its inner boundary is a polygon P, and P must consist of at least three *vertices* including p_c. Let p_a and p_b be two of these vertices different from p_c. Since the vertex p_i does not belong to this annulus-like component, there is another disk-like body lying inside P to which p_i belongs. Since two rigid components cannot have more than one vertex (joint) in common, at least one vertex of p_a and p_b does not belong to the rigid component to which p_i belongs. Let p_a be such a vertex. Consider the two paths inside the flexible face between pairs of vertices (p_c, p_i) and (p_a, p_i) respectively: they must contain at least two distinct line segments lying entirely inside the face (diagonals). These two paths connect vertices lying on two distinct rigid components (one is the annulus-like component), so each of them can be added as a bar to create a non-crossing Laman framework. Furthermore, the diagonal between p_i and p_a is not incident to p_c.

Case2: $p_c p_i$ is on the outer boundary. At least one of the upper and lower hull edges incident to p_c is missing in G'. Without loss of generality, we assume the upper hull edge is missing, and let p_c^{up} be the endpoint of upper hull edge other than p_c. Suppose that p_c and p_c^{up} belong to distinct rigid components. Adding $p_c p_c^{up}$ to $G' \setminus \{p_c p_i\}$ creates another non-crossing Laman framework which satisfies the desired properties. Then, we consider the case where p_c and p_c^{up} belong to the same rigid component. Consider adding an auxiliary vertex $\bar{p}$ incident to p_c and p_c^{up} such that $G' \cup \{\bar{p}p_c, \bar{p}p_c^{up}\}$ is a non-crossing Laman framework (it can always be constructed by a *Henneberg I*), we can show the existence of at least one *free* segment not incident to p_c and $\bar{p}$ by the same way as in Case 1. Deleting $\bar{p}$ and adding this free edge produces a non-crossing Laman framework.

This completes the proof of Theorem 1: the search tree for reverse search is well defined. $\square$

Remark. Notice that the *Flip* operation is not always valid when the removed edge is on the convex hull (see Fig. 3(b)).

5   Algorithm and Its Analysis

In this section we give a more detailed version of the algorithm for enumerating non-crossing Laman frameworks and analyze its running time. We start by introducing some notations. For each vertex p_i, an *upper-hull edge* (*lower-hull edge*) of p_i is defined as the upper (lower) convex hull edge of $\{p_1, \ldots, p_i\}$ incident to p_i, and let p_i^{up} (p_i^{low}) denote the other endpoint of the upper-hull (lower-hull) edge of p_i. Note that $p_i p_i^{up}$ and $p_i p_i^{low}$ are both root edges.

When we refer to an edge $p_i p_j$, the label of p_i is assumed to be larger than that of p_j. We define an edge ordering on the set of all possible edges in such a way that: (i) an edge $p_i p_j$ precedes an edge $p_k p_l$ whenever $i < k$, and (ii) when $i = k$ holds, $p_i p_i^{low}$ precedes all the other edges incident to p_i, and $p_i p_i^{up}$ is ordered next to $p_i p_i^{low}$. If neither $p_i p_j$ nor $p_i p_l$ is a root edge, $p_i p_j$ precedes $p_i p_l$ when $j < l$. We use the notations $p_i p_j \prec p_k p_l$ when $p_i p_j$ precedes $p_k p_l$, and $p_i p_j = p_k p_l$ when they coincide. According to the edge ordering, all possible edges are labelled appropriately. For an edge set A, we use the notations $\max\{e \mid e \in A\}$ and $\min\{e \mid e \in A\}$ to denote the largest and smallest labelled edges in A, respectively.

Parent operation. We define the following *paren function* $f_{parent} : \mathcal{G} \setminus \{G_{root}\}$ $\rightarrow \mathcal{G}$ based on the proof of Theorem 1:

Definition 1. *(Parent function) For $G' \in \mathcal{G}$ with $G' \neq G_{root}$, let $p_{c'}$ be the critical vertex in G'. $G = G' \setminus \{e_1'\} \cup \{e_2'\}$ is the parent of G', where*
- $e_1' = \max\{e \mid e \in G' \setminus G_{root}\}$, *and*
- $e_2' = \max\{e \in K_n \mid e \preceq p_{c'}p_{c'}^{up}$, *and* $G' \setminus \{e_1'\} \cup \{e\} \in \mathcal{G}\}$.

Note that the edge e_1' to be deleted is incident to $p_{c'}$ from the definition of the critical vertex in Section 3. We can prove the following lemma:

Lemma 5. *Time complexity of the parent function is $O(n^2)$ time in $O(n^2)$ space or alternatively $O(n^3)$ time using $O(n)$ space.*

Finding a next child. Given $G \in \mathcal{G}$. Let L_G and L_{K_n} be the list of edges of G and K_n ordered lexicographically, let $\delta(G)$ and $\delta(K_n)$ be the number of elements of L_G and L_{K_n} and let $L_G(i)$ and $L_{K_n}(i)$ be the i-th elements of L_G and L_{K_n}, respectively. Then, we define the *adjacency* function such that, for $e_1 \in G$ and $e_2 \in K_n$,

$$Adj(G, i, j) := \begin{cases} G \setminus \{e_1\} \cup \{e_2\} & \text{if } G \setminus \{e_1\} \cup \{e_2\} \in \mathcal{G}, \\ null & \text{otherwise}, \end{cases}$$

where $e_1 = L_G(i)$ and $e_2 = L_{K_n}(j)$.

Based on the algorithm in [2,3], we describe our algorithm in Algorithm 1. The while-loop from line 4 to 14 has $\delta(L_G) \cdot \delta(L_{K_G})$ iterations which requires $O(n^5)$

Algorithm 1. Reverse Search

1: $G_{root} :=$ the *root* of the search tree;
2: $G := G_{root}$; $i, j := 0$; $Output(G)$;
3: **repeat**
4:　**while** $i \leq \delta(L_G)$ **do**
5:　　$i := i + 1$;
6:　　**while** $j \leq \delta(L_{K_G})$ **do**
7:　　　$j := j + 1$;
8:　　　**if** $Adj(G, i, j) \neq null$ and $f_{parent}(Adj(G, i, j)) = G$ **then**
9:　　　　$G := Adj(G, i, j)$; $i, j := 0$;
10:　　　　$Output(G)$;
11:　　　　**go to** line 4;
12:　　　**end if**
13:　　**end while**
14:　**end while**
15:　**if** $G \neq G_{root}$ **then**
16:　　$G' := G$; $G := f_{parent}(G)$;
17:　　determine integers pair (i, j) such that $Adj(G, i, j) = G'$
18:　　$i := i - 1$
19:　**end if**
20: **until** $G = G_{root}$, $i = \delta(L_G)$ and $j = \delta(L_{K_G})$

---

time if simply checking the line 8. In order to improve $O(n^5)$ time to $O(n^3)$ time we claim the followings: For every integer i that represents a removing edge, preparing a data structure of [15, 16] in $O(n^2)$ time and in $O(n^2)$ space,

(i) it can be checked in $O(1)$ time whether $Adj(G, i, j)$ returns *null* or not for each integer j, and

(ii) given an integer j_0 with $0 \leq j_0 \leq \delta(L_{K_G})$, the minimum integer $j > j_0$ satisfying $f_{parent}(Adj(G, i, j)) = G$ can be found in $O(n^2)$ time with $O(n^2)$ space if such an edge exists.

Thus, we achieve $O(n^3)$ time algorithm since we take $O(n^2)$ time for each integer i. (Also, we can show $O(n^4)$ time algorithm with $O(n)$ space by similar way.) By this we proved Theorem 2.

6   Conclusions

We presented an algorithm for enumerating all the non-crossing minimally rigid graphs embedded on a generic point set. While our main focus in this paper is to show that this can be done in polynomial time, we expect that the complexity of our algorithm may be improved via a combination of more sophisticated data structures and further insights into the specific properties of *non-crossing* Laman frameworks (as opposed to just Laman frameworks).

References

1. O. Aichholzer, G. Rote, B. Speckmann, and I. Streinu. The zig-zag path of a pseudo-triangulation. In *Proc. 8th International Workshop on Algorithms and Data Structures (WADS)*, LNCS 2748, pages 377–388, 2003. Springer Verlag.
2. D. Avis and K. Fukuda. A pivoting algorithm for convex hulls and vertex enumeration of arrangements and polyhedra. *Discrete Comput. Geom.*, 8:295–313, 1992.
3. D. Avis and K. Fukuda. Reverse search for enumeration. *Discrete Applied Mathematics*, 65(1-3):21–46, March 1996.
4. M. P. Bendsøe and O. Sigmund. *Topology Optimization: Theory, Methods and Applications*. Springer, 2003.
5. A. Berg and T. Jordán. Algorithms for graph rigidity and scene analysis. *Proc. of 11th Annual European Symposium on Algorithms (ESA)*, LNCS 2832, pages 78–89. Springer, 2003.
6. S. Bespamyatnikh. An efficient algorithm for enumeration of triangulations. *Comput. Geom. Theory Appl.*, 23(3):271–279, 2002.
7. S. Bereg. Enumerating pseudo-triangulations in the plane. *Comput. Geom. Theory Appl.*, 30(3):207–222, 2005.
8. H. Brönnimann, L. Kettner, M. Pocchiola, and J. Snoeyink. Enumerating and counting pseudo-triangulations with the greedy flip algorithm. In *Proc. ALENEX*, Vancouver, Canada, 2005.
9. A. Dumitrescu, B. Gärtner, S. Pedroni, and E. Welzl. Enumerating triangulation paths. *Comput. Geom. Theory Appli.*, 20(1-2):3–12, 2001.
10. J. Graver, B. Servatius, and H. Servatius. *Combinatorial Rigidity*. Graduate Studies in Mathematics vol. 2. American Mathematical Society, 1993.

11. D. J. Jacobs and B. Hendrickson. An algorithm for two-dimensional rigidity percolation: the pebble game. *J. Comput. Physics*, 137:346–365, November 1997.
12. A. Kaveh. *Structural Mechanics: Graph and Matrix Methods*. Research Studies Press, Somerset, UK, 3rd edition, 2004.
13. A. Kawamoto, M. Bendsøe, and O. Sigmund. Planar articulated mechanism design by graph theoretical enumeration. *Struct Multidisc Optim*, 27:295–299, 2004.
14. G. Laman. On graphs and rigidity of plane skeletal structures. *Journal of Engineering Mathematics*, 4:331–340, 1970.
15. A. Lee and I. Streinu. Pebble game algorihms and sparse graphs. In *Proc. EURO-COMB*, Berlin, September 2005.
16. A. Lee, I. Streinu, and L. Theran. Finding and maintaining rigid components. In *Proc. Canad. Conf. Comp. Geom.*, Windsor, Canada, August 2005.
17. M. Ohsaki and S. Nishiwaki. Shape design of pin-jointed multi-stable compliant mechanisms using snapthrough behavior. *Struct. Multidisc. Optim.*, 30:327–334, 2005.
18. G. Rote, F. Santos, and I. Streinu. Expansive motions and the polytope of pointed pseudo-triangulations. In J. P. Boris Aronov, Saugata Basu and M. Sharir, editors, *Discrete and Computational Geometry - The Goodman-Pollack Festschrift*, Algorithms and Combinatorics, pages 699–736. Springer Verlag, Berlin, 2003.
19. I. Streinu. Pseudo-triangulations, rigidity and motion planning. *Discrete Comput. Geom.*, 34:587–635, December 2005.
20. T. S. Tay and W. Whiteley. Generating isostatic frameworks. *Structural Topology*, 11:21–69 1985.

Sequences Characterizing k-Trees

Zvi Lotker[1], Debapriyo Majumdar[2],
N.S. Narayanaswamy[3], and Ingmar Weber[2]

[1] Centrum voor Wiskunde en Informatica, Amsterdam
Z.Lotker@cwi.nl
[2] Max-Planck-Institut für Informatik, Saarbrücken
{deb, iweber}@mpi-inf.mpg.de
[3] Indian Institute of Technology Madras, Chennai
swamy@shiva.iitm.ernet.in

Abstract. A non-decreasing sequence of n integers is the degree sequence of a 1-tree (i.e., an ordinary tree) on n vertices if and only if there are least two 1's in the sequence, and the sum of the elements is $2(n-1)$. We generalize this result in the following ways. First, a natural generalization of this statement is a necessary condition for k-trees, and we show that it is not sufficient for any $k > 1$. Second, we identify non-trivial sufficient conditions for the degree sequences of 2-trees. We also show that these sufficient conditions are *almost* necessary using bounds on the partition function $p(n)$ and probabilistic methods. Third, we generalize the characterization of degrees of 1-trees in an elegant and counter-intuitive way to yield integer sequences that characterize k-trees, for all k.

1 Introduction

1.1 Degree Sequence and Characterization

Definition 1. *The* degree sequence *of an undirected graph $G = (V, E)$ is the list of degrees of its nodes, with duplication, sorted in non-decreasing order. A* graphic sequence *is a sequence of integers which is the degree sequence of a simple undirected graph. That is, a graph that does not contain loops or parallel edges. Graph G realizes a degree sequence Δ if Δ is the degree sequence of G.*

The basic sequence recognition problem is to determine whether a sequence of integers is a graphic sequence at all. This problem was solved half a century ago by Havel [9], Hakimi [7] and Erdös and Gallai [3]. Their solutions are constructive. That is, if the sequence is graphic, they show how to construct a simple graph that realizes it.

For a specific graph class $\mathcal{C}$, there can be two types of classification results. The first type is a global classification, where we are given a sequence Δ and need to determine whether *every* simple graph that realizes Δ belongs to $\mathcal{C}$. The second type is an existential classification, where we need to determine whether there *exists* a graph in $\mathcal{C}$ that realizes Δ and, if so, to construct one.

D.Z. Chen and D.T. Lee (Eds.): COCOON 2006, LNCS 4112, pp. 216–225, 2006.

Hammer and Simone [8] studied split graphs, which are graphs that have the property that their node set can be partitioned into a clique and an independent set. Their results imply that if G is a split graph, then any graph with the same degree sequence as G is also a split graph. Furthermore, the degree sequences that are realized by split graphs can be identified in linear time. Another example of a sequence recognition result was conjectured by Erdös et al. [15] and proved by Li et al. [12]. The problem is to find the minimal value $\sigma(k, n)$ such that every graphic sequence of length n without zero terms that sums to $\sigma(k, n)$ can be realized by a graph that contains a clique of size $k + 1$. This value was shown to be $\sigma(k, n) = (k - 1)(2n - k) + 2$. A related result is the Turán number [16] $ex(k, n)$ which is the smallest integer such that a graph with n nodes and $ex(n, k)$ edge is guaranteed to contain a clique of size $k + 1$ [4].

In this paper we consider the characterization problem of k-trees.

1.2 k-Trees and Previous Work

Definition 2. *A* k-tree *is recursively defined as follows.*

1. *A complete graph with $k + 1$ nodes is a k-tree.*
2. *If G is a k-tree and the nodes $v_1, \ldots, v_k$ form a k-clique in G, then the graph obtained by adding a node to G and connecting it by an edge to each of $v_1, \ldots, v_k$ is a k-tree.*

A 1-tree is a tree, hence this definition generalizes the notion of a tree. The minimum degree of a node in a k-tree is k, and in the context of a k-tree, by "leaf" we mean a node of degree k. Given an input graph, it can be determined in time $O(kn)$ whether this graph is a k-tree [5,13]. Every k-tree has treewidth k, and in fact k-trees are instrumental in one of the definitions of treewidth [14]. *Degree sets* of k trees have been studied extensively by Duke and Winkler [1,2,18]. Note that, while degree sequences are ordered in non-decreasing order, the degree set has no sequence information, nor the number of times a certain number may be used as the degree of a vertex. In this sense, characterizing degree sequences is harder than characterizing degree sets. In particular they show that degree sets of 2-trees are indeed characterized by the degree sets of 2-caterpillars, see Definition 6, which are a subclass of 2-trees. In [1], Duke and Winkler show that if D is any finite set of positive integers, which includes 1, then D is the set of vertex degrees (for a slightly different but equivalent definition of "degree") of some k-tree for k=2,3, and 4, and that there is precisely one such set, $D = \{1, 4, 6\}$, which is not the set of degrees of any 5-tree. They also show for each $k \geq 2$ that such a set D is the set of degrees of some k-tree, provided only that D contains some element d, which satisfies $d \geq k(k - 1) - 2 \lfloor \frac{k}{2} \rfloor + 3$.

However, prior to our work, degree sequences only of trees were characterized:

Theorem 1 (Folklore). *A degree sequence $\Delta = <d_1, d_2, \ldots, d_n>$ can be realized by a tree iff:*

1. *$1 \leq d_i \leq n - 1$ for all $1 \leq i \leq n$.*
2. *$\sum_{i=1}^{n} d_i = 2n - 2$.*

1.3 Our Work and Results

This work follows from an effort to characterize degree sequences of 2-trees. Theorem 1 shows that the necessary conditions on the degree sequence of a tree are indeed sufficient. A natural generalization of this theorem would be that, for all $k \geq 0$, the necessary conditions for the degree sequence of a k-tree, see Definition 3, are sufficient. However, in this paper (see Section 2, we show that the conjecture is false for all $k \geq 2$. Following this, in Section 3, we identify the *right* generalization of degree sequences in a way that helps to characterize such sequences that correspond to k-trees. The generalization lies in viewing the degree sequence of a graph in a slightly different way; *the entries of a degree sequence count the number of 2-cliques(edges) that contain a 1-clique(a vertex)*.

While we show that plausible k-sequences (see Definition 3) do not characterize k-trees, we present some fundamental results on them for $k = 2$. In Section 4 we show that if a plausible 2-sequence contains a 3, then it is the degree sequence of 2-tree. In this proof, we identify a structure of a 2-tree that makes it possible to output such a tree in linear time. Having shown that a plausible 2-sequence which contains a 3 is the degree sequence of a 2-tree, we show in Section 5 that almost every plausible 2-sequence contains a 3 and hence almost every plausible 2-sequence is realizable. This proof is based on the idea that for a certain number n, each plausible 2-sequence corresponds to a partition of $2n - 7$. We then use bounds on the partition function $p(n)$ [10,11,17], the integer function that counts the number of partition of n, to prove the claim.

Throughout the paper, the symbol n usually denotes the size of a k-tree or a degree sequence. We sometimes identify nodes by their degree. For example, by "adding a 3", we mean "adding a node of degree 3".

2 Non-realizable k-Sequences

For 1-trees it turns out that the necessary conditions on the degree sequence are indeed sufficient. The natural conjecture would be that the same holds for k-trees too, for $k \geq 2$. In Lemma 1 we show that this conjecture is false for k-trees by exhibiting one class of sequences that satisfy the necessary conditions but are not realizable by k-trees. To show this we define *plausible k-sequences* as those that satisfy the necessary conditions.

Definition 3. *A sequence of integers* $\Delta = < d_1, d_2, \ldots, d_n >$ *is a* plausible k-sequence *if the following conditions hold:*

1. *$d_i \leq d_{i+1}$ for all $1 \leq i < n$.*
2. *$d_n \leq n - 1$.*
3. *$d_1 = d_2 = k$.*
4. *$\sum_{i=1}^{n} d_i = k(2n - k - 1)$.*

Lemma 1. *For every $k > 1$, for every integer n such that $b = \frac{k(n+1)}{k+2}$ is a positive integer, the plausible k-sequence $d_1 = d_2 = \ldots = d_{n-k-2} = k, d_{n-k-1} = \ldots = d_n = b$ is not the degree sequence of any k-tree.*

Proof. Consider a k-tree T corresponding to the said plausible k-sequence. Let $L \subseteq T$ be the set of all nodes of degree k. Now $T - L$ induces a k-tree on $k + 2$ nodes, which has two non-adjacent nodes, say a and b, of degree k. Now, no matter in what order we add the vertices of L to obtain the k-tree T from the k-tree $T - L$, we will never be able to *equalize* the degrees of $T - L$. The proof is by an averaging argument, and exploits the fact that a and b are not adjacent. Let us consider the following two vertex sets $A = \{a, b\}$, and $B = T - L - A$. In each step of a construction of T from $T - L$, we show that the average degree of vertices in B is more than the average degree of vertices in A. Clearly, in $T - L$, the average degree in A is k, and in B it is $k + 1$. Whenever a new vertex is added, it must be adjacent to at least $k - 1$ vertices in B and at most one vertex in A. Therefore, after adding m vertices, the average degree of A will be at most $k + \frac{m}{2}$, and the average degree of vertices in B will be at least $k + 1 + \frac{m(k-1)}{k}$. So the degrees of vertices in $T - L$ can never become all equal. Therefore, $d_1 = \ldots = d_{n-k-2} = k, d_{n-k-1} = \ldots = d_n = b$ is not the degree sequence of a k-tree.   □

3   Integer Sequences That Characterize k-Trees

Definition 4. *The $(k, k + 1)$-degree of a k-clique C in a graph G is defined as the number of $(k + 1)$-cliques in G which contain C. The $(k, k + 1)$-degree sequence of a graph G is the list of $(k, k + 1)$-degrees of the k-cliques in G, with duplicates, sorted in non-decreasing order.*

The $(1, 2)$-degree sequence of a graph is its degree sequence, and its $(2, 3)$-degree sequence can be thought of as the *edge-triangle* degree sequence.

Definition 5. *For $n \geq k + 1$, a sequence of integers $\Delta = < d_1, d_2, \ldots, d_r >$ is a $(k, k + 1)$-sequence if the following conditions hold:*

1. $r = k + 1 + (n - k - 1)k$.
2. $d_i \leq d_{i+1}$ *for all* $1 \leq i < r$.
3. *If* $n = k + 1$, $d_i = 1$ *for* $1 \leq i \leq k + 1$. *If* $n > k + 1$, *then* $d_i = 1$ *for* $1 \leq i \leq 2k$.
4. $\sum_{i=1}^{r} d_i = (k + 1)(n - k)$.

The following two lemma follow from the definition of a $(k, k + 1)$-sequence, and are used in the proof of Theorem 3, which is our main theorem.

Lemma 2. *For $n = k + 1$, the $(k, k + 1)$-sequence is unique and every element is a 1. For $n = k + 2$, the $(k, k + 1)$-sequence is unique; $d_1 = d_2 = \ldots = d_{2k} = 1, d_{2k+1} = 2$.*

Lemma 3. *Let $r > k + 1$. Let $< d_1, \ldots, d_r >$ be a $(k, k + 1)$-sequence and let l be the smallest integer such that $d_l > 1$. If $< d_{k+1}, \ldots, d_l - 1, \ldots, d_r >$ is the $(k, k + 1)$-degree sequence of a k-tree, then $< d_1, \ldots, d_r >$ is the $(k, k + 1)$-degree sequence of a k-tree.*

Theorem 2. *Let $\Delta = < d_1, d_2, \ldots, d_r >$ be a sequence of integers. Then Δ is the $(k, k + 1)$-degree sequence of a k-tree iff Δ is a $(k, k + 1)$-sequence.*

*Proof.*First we prove the necessary condition. In a k-tree on n vertices, the number of k-cliques, denoted by r, is $k(n-k)+1$. Further, the sum of the entries in the $(k, k+1)$-degree sequence $< d_1, d_2, \ldots, d_r >$ is $\sum_{i=1}^{r} d_i = (k+1)(n-k)$. The proofs of these claims are by induction on n. The base case is for $n = k+1$; in this case there are k k-cliques, a unique $k+1$-clique, and the sum of the degrees is $k+1$. To complete the induction, if we assume that these formulas hold for n, proving that they hold for $n+1$ follows by simple arithmetic. We now prove the property on the entries of the degree sequence. If $n = k+1$, as observed before, there are k k-cliques, and a unique $k+1$-clique. So the degree sequence is $d_1 = d_2 = \ldots = d_{k+1} = 1$. For the case of $n > k+1$, we observe a simple invariant maintained in every k-tree: there are two vertices of degree k, this property is easily seen in the inductive construction of k-trees. Further, in a k-tree a vertex of degree k is present in exactly k k-cliques. Each of these k-cliques is contained in the unique $k+1$-clique induced by the vertex and all its neighbors. The entries corresponding to these k k-cliques are 1 in the $(k, k+1)$-degree sequence. Since there are two vertices of degree k in any k-tree, it follows that there are $2k$ 1's in the $(k, k+1)$-degree sequence. Therefore, $d_1 = d_2 = \ldots = d_{2k} = 1$.

We prove the sufficient condition by induction on the length of the $(k, k+1)$-sequence. Let us consider a $(k, k+1)$-sequence $d_1, d_2, \ldots, d_r$. If $r = k+1$, then the corresponding k-tree is the clique of $k+1$ vertices. If $r = 2k+1$, then the corresponding k-tree has $k+2$ vertices in which there is a k-clique, and two non-adjacent vertices are both adjacent to each vertex in the k-clique. There are no other vertices and edges in the graph. Therefore, $(k, k+1)$-sequences of length $k+1$ and $2k+1$ can be realized by k-trees, which is the base case for our induction. Let us consider the case when $r > 2k+1$. Let l be the smallest integer such that $d_l > 1$. Clearly, $l > 2k$. We show that $d_{k+1}, \ldots, d_l - 1, \ldots, d_r$ is a $(k, k+1)$-sequence. The sum of the degrees is clearly $(k+1)(n-1-k)$. We only need to show that $d_{k+1} = d_{k+2} = \ldots = d_{3k} = 1$. If we assume not that is we assume that $d_{k+1} = \ldots d_b = 1, b < 3k$. Then it follows that we have $k(n-1-k)+1-b+k$ entries in the sequence which are more than 1. Further, we also know that the sum of these entries is $(k+1)(n-1-k)-b+k$. It now follows that $(k+1)(n-1-k)-b+k \geq 2k(n-1-k)+2-2b+2k$, that is $b \geq (n-1-k)(k-1)+k+2$. If $b < 3k$, then it follows that $2k-2 > (n-1-k)(k-1)$, which in turn implies that $n < 2+(k+1)$, that is $n = k+1$ or $n = k+2$. This means $r \leq 2k+1$, a contradiction to the fact that we are considering $r > 2k+1$. Therefore our assumption that $d_{k+1}, \ldots, d_r$ is not a $(k, k+1)$-sequence is wrong. Inductively, $d_{k+1}, \ldots, d_l - 1, \ldots, d_r$ is the $(k, k+1)$-degree sequence of a k-tree. By Lemma 3 it now follows that $d_1, \ldots, d_r$ is also the $(k, k+1)$-degree sequence of a k-tree. Hence the characterization is complete.   □

4   Sufficient Conditions for 2-Trees

In this section we present our main results on the sufficient conditions on the degree sequence of 2-trees.

We call a 2-tree, which contains exactly two leaves, a 2-*chain*. In a 2-tree T, a *pruning sequence* is a minimal sequence of degree 2 nodes of T such that after removing these nodes according to the sequence, we get a 2-chain. The process of applying a pruning sequence to a 2-tree is called pruning.

Definition 6. *A 2-caterpillar is either a 3-clique, or a 2-tree with a pruning sequence.*

Definition 7. *For each $l \geq 1$, a $[d_1, d_2, \ldots, d_l]$-path is a path $v_1, \ldots, v_l$ such that for $1 \leq i \leq l$, the degree of v_i is d_i. For $l = 2$, we refer to a $[d_1, d_2]$-path as a $[d_1, d_2]$-edge.*

Theorem 3. *If a plausible 2-sequence contains at least one 3, then it is the degree sequence of a 2-tree. Furthermore, if $n > 4$ then there is 2-tree realizing this degree sequence in which there is a $[2, 3, min_1]$-path, where $min_1 = d_l$ and $d_l \geq 4$ but $d_{l-1} < 4$. If $l < n$, then there is even a $[2, 3, min_1, min_2]$-path. Here $min_2 = d_{l+1}$, i.e., min_2 is the next degree in the sequence.*

The proof of this theorem will be by induction on n, i.e., the number of vertices. In the induction step, certain boundary cases can occur. These special cases are dealt with by the following lemmas.

Lemma 4. *If in a plausible 2-sequence $d_{n-1} < 4$, then the sequence contains exactly two 2's and $d_n = n - 1$. Further, such a sequence is the degree sequence of a 2-tree. In this special case, Theorem 3 holds.*

Proof. Since $\sum_{i=1}^{n} d_i = 4n - 6$, and the fact that $d_1 = d_2 = 2$, it follows that $\sum_{i=3}^{n} d_i = 4n - 10$. Hence, $\sum_{i=3}^{n-1} d_i \geq 3n - 9 = 3(n-3)$ as $d_n \leq n-1$. Therefore, the average value of $\{d_3, \ldots, d_{n-1}\}$ is at least 3. Since $d_{n-1} < 4$ it follows that $\sum_{i=1}^{n-1} d_i = 2t + 3(n - t - 1) = 3n - t - 3$, where t is the number of 2's in $d_1, \ldots, d_{n-1}$. Therefore, $d_n = n + t - 3$. Since $d_n \leq n - 1$, it follows that $t \leq 2$. Therefore, $t = 2$, and consequently, $d_3 = \ldots = d_{n-1} = 3, d_n = n - 1$. This sequence is trivially realized by a "fan": a central node of degree $n - 1$, which is surrounded by nodes of degree 3 with a node of degree 2 at either end of this ring. □

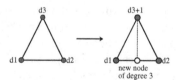   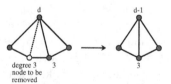

Fig. 1. Inserting a node of degree 3 to a $[d_1, d_2, d_3]$-triangle and changing the degree of only one node from d_3 to $d_3 + 1$

Fig. 2. Deleting a node of degree 3 from a $[3, 3, d]$-triangle and changing the degree of only one node from d to $d - 1$

Lemma 5. *A plausible 2-sequence with exactly two 2's in the sequence will also contain at least one 3. Such a sequence is the degree sequence of a 2-tree. In this special case, Theorem 3 holds. In fact, the nodes of high degrees $h_1, h_2, \ldots, h_r$, where the h_j are the subset of the degrees d_i with $d_i \geq 4$, can be arranged in any arbitrary order such that there is a $[2, 3, h_{j_1}, \ldots, h_{j_r}]$ path.*

Proof. In the explicit construction we will, first, reduce all nodes of degree > 4 to degree 4 and then remove the "appropriate" number of 3's from the sequence, namely, a node of degree $4 + x$ corresponds to x nodes of degree 3, as the sum is fixed at $4n - 6$ and there are only two 2's. This, in the end, leaves a sequence of the form $2, 2, 3, 3, 4, 4, \ldots, 4$ with exactly two 2's, two 3's and the same number of 4's as nodes of high degree in the original sequence. This sequence is then realizable by a "straight chain". See Figure 3 for an illustration. Once we have this basic backbone, we fix a $[2, 3, 4, 4, 4, \ldots, 4]$-path and identify the 4's with the desired degrees $h_{j_1}, \ldots, h_{j_r}$. For each such node of intended degree h_{j_s} we then insert $(h_{j_s} - 4)$ 3's into the 2-tree, as illustrated in Figure 1. Figure 3 illustrates the whole process.   □

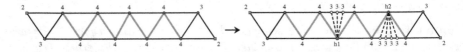

Fig. 3. Inserting $(h_i - 4)$ degree 3 nodes on a $[4, \ldots, 4]$-path (shown in bold), changing the degree of degree 4 nodes to h_i, if $h_i > 4$. The nodes are labelled by their degrees.

Lemma 6. *If $d_n = n - 1$ in a plausible 2-sequence, then the sequence is the degree sequence of a 2-tree. In fact, the nodes of high degrees $h_1, h_2, \ldots, h_r$, where the h_j are the subset of the degrees d_i where $d_i \geq 4$, can be arranged in any arbitrary order such that, if the 2-sequence contains a 3, there is a $[2, 3, \ldots, 3, h_{j_1}, \ldots, h_{j_r}]$ path passing through all nodes of degree 3 consecutively or, if the 2-sequence does not contain a 3, there is $[2, h_{j_1}, \ldots, h_{j_r}]$ path.*

Proof. Note that the combination of the conditions of Lemmas 5 and 6 leads to the very strict conditions of Lemma 4. So we can assume that there are at least three 2's in the sequence. The proof of the lemma is by induction. The induction starts at $n = 5$ with the only plausible sequences $< 2, 2, 2, 4, 4 >$ and $< 2, 2, 3, 3, 4 >$, both of which are realizable as desired by inspection. Now suppose the lemma holds for up to n. Note that we can always assume that h_{j_1} is not the maximum degree $n - 1$ (for n nodes) as, if the sequence is realizable, the node of maximum degree will be connected to *all* other nodes and can thus be inserted anywhere along an existing path. So, we can first move it to "the end" by assuming $h_{j_r} = n - 1$. If there is only one node of high degree ≥ 4, then we are also in the case of Lemma 4. Now, given a 2-sequence with $n + 1$ degrees and $d_{n+1} = n$ and $4 \leq h_{j_1} < n$, simply remove a 2, as there are at least three 2's by the comment before, and reduce both the maximum degree d_{n+1} and the degree h_{j_1} by one and apply induction. As the node of maximum degree, which now has degree $n-1$, is still connected to all remaining nodes it is, in particular, connected to the node of degree $h_{j_1} - 1$. Hence, we can put a leaf back on top of the $[n-1, h_{j_1} - 1]$-edge to get back the original degree sequence.   □

The following observation, illustrated in Figure 2, will allow us to reduce the sequences of 3's along the path to a single 3.

Observation 1. *Given a* $[3, 3]$*-edge as part of a* $[3, 3, d]$*-triangle in a 2-tree, we can remove one of the two 3's while also reducing* d *to* $d - 1$ *and we obtain another 2-tree.*

Corollary 1. *If* $d_n = n - 1$ *in a plausible 2-sequence, then the sequence is the degree sequence of a 2-tree. In fact, if the 2-sequence contains at least one 3, then the nodes of high degrees* $h_1, h_2, \ldots, h_r$*, where the* h_j *are the subset of the degrees* d_i *with* $d_i \geq 4$*, can be arranged in any arbitrary order such that there is a* $[2, 3, h_{j_1}, \ldots, h_{j_r}]$ *path. Thus, in particular, for* $d_n = n - 1$ *Theorem 3 holds.*

Proof. Note that if, in the case where $d_n = n - 1$, we have a $[3, 3]$-edge, then both corresponding nodes must also be connected to the central node of degree $n - 1$. Thus, using the observation above, we can remove one of the nodes of degree 3 along with its edge connected to the node of degree $n - 1$ (now becoming $n - 2$) and bridge its other two edges thereby leaving all other degrees unchanged. If we now put a leaf on top of the other leaf, which is not involved in the desired path, then this 2 becomes a 3, we insert another 2 (the newly added leaf) and the central node of degree $n - 2$ goes back to degree $n - 1$. Using this trick repeatedly, we can remove any sequence of 3's along the path to a single 3.   □

With these lemmas, we can now prove the Theorem 3.

Proof. The statement is proved constructively by induction on n. The statement holds for $n = 4$. At each step, if we ever get left with (a) only one high degree greater than 3, (b) only two 2's or (c) $d_n = n - 1$, then we refer to the lemmas above, namely Lemma 4, Lemma 5 and Corollary 1

Case 1. Assume that we have no 4's in the sequence, so $\min_1 > 4$ and $\min_2 > 4$. Then reduce $\min_1$ and $\min_2$ by 1 and remove a 2 from the sequence. This gives another plausible 2-sequence and there will also remain at least one 3. So, by the induction hypothesis, construct a 2-tree with a $[2, 3, \min_1 - 1, \min_2 - 1]$-path. Observe that by reducing the two minima among the vertices of degree more than 3, they will still remain the minima, as $\min_1, \min_2 > 4$. Add a vertex to this 2-tree and connect it to the two last nodes on this path. This gives a 2-tree realizing our original sequence of length n.

Case 2. Now assume we have at least one 4 in the sequence, so $\min_1 = 4$. Then reduce a 3 to a 2, reduce a 4 to a 3 and remove a 2. Again, this will give a plausible 2-sequence of shorter length with at least one 3. Observe that $\min_2$ has now become the smallest high degree. By induction, we then get a $[2, 3, \min_2, x]$-path for some x. Add a vertex and connect it to the first two nodes on this path. This then gives a $[2, 3, 4, \min_2]$-path.   □

5   Almost Every Plausible 2-Sequence Is Realizable

Let the partition function $p(n)$ give the number of ways for writing a positive integer n as a sum of positive integers, where the order of addends is not considered. From [10,11,17], we know the following asymptotic formula.

Theorem 4. *As $n \to \infty$, $p(n) \to \dfrac{\exp(\pi\sqrt{2n/3})}{4\sqrt{3}n}$.*

Lemma 7. *The number of plausible 2-sequences of size n is at most $p(2n - 6)$.*

Proof. The lemma follows from the fact that every plausible 2-sequence Δ of size n defines a unique partition of the number $2n - 6$. It is because, by subtracting 2 from each number of Δ, we get a monotonic sequence of n non-negative numbers, whose sum is $(4n - 6) - 2n = 2n - 6$. $\square$

Lemma 8. *The number of plausible 2-sequences of size n containing at least one 3 is greater than $p(2n - 7) - 2n \cdot p(n)$.*

Proof. Let $\Delta = < d_i >_{i=1}^n$ be a plausible 2-sequence of size n containing at least one 3. Since the sum of all d_i's is $4n - 6$ and since Δ contains at least two 2's and one 3, the sum of the remaining $n - 3$ elements in Δ is $4n - 13$. Now, since all the elements are bigger than or equal to 2, Δ defines a partition of the number $(4n - 13) - 2(n - 3) = 2n - 7$ into $n - 3$ blocks. However, not all partitions $(b_1, \ldots, b_l)$ of $2n - 7$ correspond to plausible 2-sequences. There are two types of partitions which do not correspond to plausible sequences. First, the partition may contain more than $n - 3$ blocks and thus cannot correspond to a 2-sequence of size n; we call such a partition a "long partition". Below we show that the total number of such partitions is bounded from above by $np(n)$. Since the order of the partition is not considered, we can assume that a partition is sorted non-increasing order. Therefore, if the partition is "long", then $b_{n-2} = 1$, because $b_i \geq 1$ for all $i \leq n - 3$ and $\sum_{i=1}^{n-2} b_i \leq 2n - 7$. Therefore, $b_{n-2+i} \leq 1$ for all $i \geq 1$ and there is a unique j, determined by the sum of $b_1, \ldots, b_{n-3}$ such that $b_l = 0$ for $l \geq j$. Hence, a "long" partition is determined by the sum S of the first $n - 3$ elements of the partition. Since $n - 5 = 2n - 7 - (n - 2)$, it follows that $n - 3 \leq S \leq 2n - 8$ and therefore the number of long partitions is exactly $\sum_{i=1}^{n-5} p(i)$. Finally, since $p(n) > p(n - 5 - i)$ for all $i = 1, 2, \ldots, n - 5$, it follows that the number of long partitions is less than $np(n)$.

The other type of partitions of $2n - 7$ that do not correspond to plausible k-sequences are those for which the biggest number is greater than $n - 3$, leading to a 2-sequence which violates the maximum condition. The sum of the rest of the numbers in the partition is between 1 and $n - 5$. Therefore the number of such partitions is at most $\sum_{i=1}^{n-5} p(i)$, which is again less than $np(n)$. Thus the lemma follows. $\square$

Theorem 5. *Almost every plausible 2-sequence is realizable by a 2-tree.*

Proof. By Theorem 3, it is enough to show that almost every plausible 2-sequence contains a 3. Consider a random experiment that picks a sequence randomly from the set of all plausible 2-sequences. Denote by A an event that the picked sequence contains a 3. By Lemma 7 and 8, we have $\Pr[A] \geq \frac{p(2n-7)-2n\cdot p(n)}{p(2n-6)}$. From Theorem 4 we know that the right hand side approaches 1 as n approaches ∞, so we have $\lim_{n\to\infty} \Pr[A] = 1$, hence the result. $\qquad\square$

References

1. R.A. Duke and P.M. Winkler. *Degree Sets of k-Trees: Small k. Israel Journal of Mathematics*, Vol. 40, Nos 3-4, 1981.
2. R.A. Duke and P.M. Winkler. *Realizability of almost all degree sets by k-trees. Congressus Numerantium*, Vol. 35 (1982), pp. 261-273.
3. P. Erdös and T. Gallai. Graphs with prescribed degree of vertices (Hungarian). *Mat. Lapok*, 11:264–274, 1960.
4. P. Erdös and M. Simonovits. Compactness results in extremal graph theory. *Combinatorica*, 2:275–288, 1982.
5. E.C. Freuder. Complexity of k-tree structured constraint satisfaction problems. In *Proc. of the 8th National Conference on Artificial Intelligence*, 1990.
6. M.C. Golumbic. Algorithmic Graph Theory and Perfect Graphs. Academic Press, 1980
7. S.L. Hakimi. On the realizability of a set of integers as degrees of the vertices of a graph. *J. SIAM Appl. Math.*, 10:496–506, 1962.
8. P.L. Hammer and B. Simeone. The splittance of a graph. *Combinatorica*, 1:275–284, 1981.
9. V. Havel. A remark on the existence of finite graphs (Czech). *Casopis Pest. Mat.*, 80:477–480, 1955.
10. G. H. Hardy and S. Ramanujan. *Une formule asymptotique pour le nombres des partitions de n. Comptes Rendus Acad. Sci. Paris*, Ser. A, 2 Jan. 1917.
11. G. H. Hardy and S. Ramanujan. *Asymptotic formulae in combinatory analysis. Proc. London Math. Soc.*, 17:75115, 1918.
12. Z.X. Song J.S. Li and R. Luo. The Erdös-Jacobson-Lehel conjecture on potentially p_k-graphic sequences is true. *Science in China, Ser. A*, 41:510–520, 1998.
13. Claudia Marcela Justel and Lilian Markenzon. Incremental evaluation of computational circuits. In *Proc. of the Second International Colloquium Journes d'Informatique Messine: JIM'2000*, 2000.
14. Ton Kloks. *Treewidth*. Universiteit Utrecht, 1993.
15. M. S. Jacobson P. Erdös and J. Lehel. Graphs realizing the degree sequences and their respective clique numbers. *Y. Alavi et al., ed. Graph Theory, Combinatorics and Applications*, 1:439–449, 1991.
16. P. Turán. *On an extremal problem in graph theory. Mat. Fiz. Lapok*, 48:436–452, 1941.
17. Ya. V. Uspensky. *Asymptotic expressions of numerical functions occurring in problems concerning the partition of numbers into summands. Bull. Acad. Sci. de Russie*, 14(6):199218, 1920.
18. P.M. Winkler. *Graphic Characterization of k-Trees. Congressus Numeratium*, Vol. 33 (1981), pp. 349-357.

On the Threshold of Having a Linear Treewidth in Random Graphs

Yong Gao

The Irving K. Barber School of Arts and Sciences
University of British Columbia Okanagan, Kelowna, Canada V1V 1V7
yong.gao@ubc.ca

Abstract. The concept of tree-width and tree-decomposition of a graph plays an important role in algorithms and graph theory. Many NP-hard problems have been shown to be polynomially sovable when restricted to the class of instances with a bounded tree-width. In this paper, we establish an improved lower bound on the threshold for a random graph to have a linear treewidth, which improves the previous result by Kloks [1].

1 Introduction

The concept of tree-width and tree-decomposition of a graph plays an important role in the study of algorithm and graph theory [2,1]. Many NP-hard problems have been shown to be polynomially sovable when restricted to the class of instances with a bounded tree-width. Dynamic programming algorithms based on the tree-decomposition of graphs have found many applications in research field such as artificial intelligence[3,4].

Over the past ten years, there has been much interest in the study of the phase transitions and typical-case behavior of problem instances randomly-generated from some probability distribution [5,6,7,8]. Randomly-generated problem instances have been widely used as benchmarks to test the efficiency of algorithms and to gain insight on the problem hardness.

The theory of random graphs pioneered by the work of Erdös [9] deals with the phase transitions and threshold phenomena of various graph properties such as the connectivity, the colorability, and the size of (connected) components. In [1], Kloks studied the threshold phenomenon for a random graph to have a treewidth linear to the number of vertices, and proved that **whp** a random graph $G(n, m)$ with $\frac{m}{n} > 1.18$ has a treewidth linear in n. Kloks commented that it was not known whether his lower bound 1.18 can be further improved and that the treewidth of a random graph $G(n, m)$ with $\frac{1}{2} < \frac{m}{n} < 1$ is unknown [1]. In [10], it is shown that the threshold phenmena of having a linear treewidth can be used to explain the results of many empirical studies in AI. To the best knowledge of the author, no further result has been obtained regarding the lower bound on the treewidth of random graphs since Kloks' work.

D.Z. Chen and D.T. Lee (Eds.): COCOON 2006, LNCS 4112, pp. 226–234, 2006.

The purpose of this paper is to establish a lower bound on the threshold for a random graph to have a linear treewidth, which improves the previous result by Kloks.

2   Preliminary and the Main Result

The *treewidth* can be defined in several equivalent ways. The one that is the easiest to state is via the *k-tree* defined recursively as follows ([1]):

1. A clique with k+1 vertices is a k-tree;
2. Given a k-tree T_n with n vertices, a k-tree with $n+1$ vertices is constructed by adding to T_n a new vertex and connecting it to a k-clique of T_n.

A graph is called a partial k-tree if it is a subgraph of a k-tree. The treewidth $tw(G)$ of a graph G is the minimum value k for which G is a partial k-tree.

Let $\mathcal{P} = (\Omega, \mathcal{A}, \mathrm{Pr})$ be a probability space where Ω is a sample space, $\mathcal{A}$ is a σ-field, and Pr is a probability measure. Throughout this paper, we will use the following notations:

$$\mathcal{E}_{\mathcal{P}}[X] : \text{ the expectation of a random variable } X;$$
$$\sigma_{\mathcal{P}}^2[X] : \text{ the variance of a random variable } X;$$
$$I_A : \text{ the indicator function of an event } A \in \mathcal{A}.$$

When the probability space is clear from the context, we will suppress the subscripts and simply write $\mathcal{E}[X], \sigma^2[X]$, and I_A.

Let $\{\mathcal{P}_n = (\Omega_n, \mathcal{A}_n, \mathrm{Pr}_n), n \geq 1\}$ be a sequence of probability spaces and let $\{A_n \in \mathcal{A}_n, n \geq 1\}$ be a sequence of events. We say that $\{A_n \in \mathcal{A}_n, n \geq 1\}$ occur *with high probability* (**whp**) if $\lim_n \mathrm{Pr}_n\{A_n\} = 1$.

A random graph $G(n, m)$ over n vertices is defined to be a graph with m edges selected uniformly and randomly without replacement from all the $\binom{n}{2}$ possible edges [11]. Kloks proved that **whp** , a random graph $G(n, m)$ with $\frac{m}{n} > 1.18$ has a treewidth linear in n. Kloks commented that it was not known whether his lower bound 1.18 can be further improved and that the treewidth of a random graph $G(n, m)$ with $\frac{1}{2} < \frac{m}{n} < 1$ is unknown [1]. To my best knowledge, no further result has been obtained regarding the treewidth of $G(n, m)$ since Kloks' work.

Our main result is an improved lower bound on the threshold of having a linear treewidth. The improvement comes from two factors: (1) the use of a new combinatorial construct to make better use of the first moment method; and (2) the use of a random graph equivalent to $G(n, m)$ that makes it possible to have a more accurate estimation of some quantity.

Theorem 1. *For any $\frac{m}{n} = c > 1.081$, there is a constant $\delta > 0$ such that*

$$\lim_n \mathrm{Pr}\left\{ tw(G(n, m)) > \delta n \right\} = 1. \tag{1}$$

We will be working on a random graph model $\overline{G}(n, m)$ that is slightly different from $G(n, m)$ in that the m edges are selected independently and uniformly with replacement. It turns out that as far as the property of having a linear treewidth is concerned, the two random graph models are equivalent. This is due to the following observations:

1. There are only $o(n)$ duplicated edges in $\overline{G}(n, m)$. In fact, let I_e be the indicator function of the event that the potential edge $e \in V^2$ is duplicated and write $I = \sum\limits_{e \in V^2} I_e$. We have

$$\mathcal{E}[I_e] = \sum_{r \geq 2} \binom{m}{r} \frac{1}{N^r} (1 - \frac{1}{N})^{m-r} = O(\frac{1}{n^2}), \quad \text{where } N = \binom{n}{2}.$$

And thus, $\mathcal{E}[I] = O(1)$. On the other hand, we have for any pair of potential edges e_1 and e_2,

$$\mathcal{E}[I_{e_1} I_{e_2}] \leq \mathcal{E}[I_{e_1}] \mathcal{E}[I_{e_2}]$$

since I_{e_1} and I_{e_2} are negatively correlated. It follows that the variance of I is also $O(1)$, and therefore **whp** $I = o(n)$.

2. Due to the symmetry of the sampling space, a graph consisting of the first $m - o(n)$ non-duplicated edges of $\overline{G}(n, m)$ has the same distribution as $G(n, m - o(n))$.

3. For any graph G and its super-graph G' such that G' has $o(n)$ more edges than G, we have

$$tw(G') = tw(G) + o(n).$$

This is because adding one edge to a graph increases the treewidth of the graph at most by one.

Based on these observations, we will continue to use the notation $G(n, m)$ instead of $\overline{G}(n, m)$ in the rest of the paper, but with the understanding that the m edges are selected independently and uniformly with replacement.

3   Proof of the Result

As the first step to prove theorem 1, we introduce the following concept which will be used to provide a necessary condition for a graph to have a treewidth of certain size:

Definition 1. *Let $G(V, E)$ be a graph with $|V| = n$. A partition $\mathbf{W} = (S, A, B)$ of V is said to be a rigid and balanced l-partition if the following conditions are satisfied:*

1. $|S| = l + 1$;
2. $\frac{1}{3}(n - l - 1) \leq |A|, |B| \leq \frac{2}{3}(n - l - 1)$; and

3. S separates A and B, i.e., there are no edges between vertices of A and vertices of B; and

4. If $|B| > |A|$, then any vertex v in B is not isolated in B, i.e., there exists at least another vertex in B that is adjacent to v.

A partition that satisfies the first three conditions in the above definition is called a *balanced partition* and was used by Kloks in his proof of the 1.18 lower bound. The rigid and balanced partition generalizes Kloks's balanced partition by requiring that any vertex in the larger subset of a partition cannot be moved to the other subset of the partition, and hence the word "rigid".

Lemma 1. *Any graph with a treewidth $l > 4$ must have a rigid and balanced l−partition.*

Proof. From [1], any graph with a treewidth $l > 4$ must have a partition, say $\mathbf{W} = (S, A, B)$, that satisfies the first three conditions in Definition 1. If this partition does not satisfy the fourth condition, then we can move the vertices that are isolated in B one by one to A until either $|B| = |A|$ or there is no more isolated vertex in B.

The following lemma gives an upper bound on the conditional probability for a partition $\mathbf{W} = (S, A, B)$ to be rigid given that the partition is balanced.

Lemma 2. *Let $G(n, m), c = \frac{m}{n}$, be a random graph and let $\mathbf{W} = (S, A, B)$ be a partition such that $|S| = l + 1, |A| = a$, and $|B| = b$. Assume that $b = tn$ and $b > a$. Then for n sufficiently large,*

$$\Pr \{ \mathbf{W} \text{ is rigid} \mid \mathbf{W} \text{ is balanced} \} \leq \left(\frac{1}{e} \right)^{r(t)n} \tag{2}$$

where

$$r(t) = \frac{t^2}{2c} \left(\frac{1}{e} \right)^{\frac{4ct}{1 - 2t(1 - t)}}.$$

Proof. Conditional on that $\mathbf{W}$ is a balanced partition of $G(n, m)$, each of the m edges can only be selected from the set of edges

$$E_W = V^2 \setminus \{(u, v) : u \in A, v \in B\}.$$

Notice that

$$s \equiv |E_W| = \frac{n(n - 1)}{2} - ba = \frac{n(n - 1)}{2} - tn(n - tn - (l + 1)).$$

Let I_v be the indicator function of the event that the vertex $v \in B$ is isolated in B and write $I = \sum_{v \in B} I_v$. Then, the random variable I is a function of the m outcomes when selecting the m edges of the random graph $G(n, m)$. For any two sets of outcome $(w_1, \cdots, w_m)$ and $(\overline{w}_1, \cdots, \overline{w}_m)$ that only differ at the i-th coordinate, i.e., the edges of two corresponding graphs are the same except for the i-th edge, we have

$$|I(w_1, \cdots, w_m) - I(\overline{w}_1, \cdots, \overline{w}_m)| \leq 2.$$

This is because changing one edge either increases or decreases the number of isolated vertices at most by two. Thus, applying McDiarmid's inequality [12] gives us

$$\Pr\{\mathbf{W} \text{ is rigid} \mid \mathbf{W} \text{ is balanced}\} = \Pr\{I = 0 \mid \mathbf{W} \text{ is balanced}\}$$
$$\leq \Pr\{I - \mathcal{E}[I] \leq -\mathcal{E}[I]\}$$
$$\leq \left(\frac{1}{e}\right)^{\frac{2\varepsilon^2[I]}{4cn}}.$$

By the definition of the random variable I, the term $\mathcal{E}[I]$ is

$$b\left(1 - \frac{b-1}{s}\right)^{cn} = tn\left(1 - \frac{tn-1}{n(n-1)/2 - tn(n-tn-l-1)}\right)^{cn}.$$

Formula (2) follows.   □

We need two more lemmas on the behavior of some functions that will be used in the proof of Theorem 1.

Lemma 3. *For any $c > 1$, the function $r(t)$ in Lemma 2 is monotone-decreasing on $[\frac{1}{2}, \frac{2}{3}]$.*

Proof. Taking the derivative of the function

$$\log(r(t)) = 2\log(t) - \frac{4ct}{1 - 2t + 2t^2},$$

we have

$$\frac{1}{r(t)}r'(t) = \frac{2(1 - 2t + 2t^2)^2 - 4c(t - 2t^2 + 2t^3) - 4c(-2t^2 + 4t^3)}{t(1 - 2t + 2t^2)^2}.$$

Now consider the numerator of the right-hand-side of the above, i.e., the function

$$h(t) = 2(1 - 2t + 2t^2)^2 - 4c(t - 2t^2 + 2t^3) - 4c(-2t^2 + 4t^3).$$

The monotonicity of the function $r(t)$ can be established if we can show that $h(t) \leq 0, \forall t \in [\frac{1}{2}, \frac{2}{3}]$. Since we have $h(\frac{1}{2}) = \frac{1}{2} - c < 0$ and $h(\frac{2}{3}) = \frac{50}{81} - \frac{144}{81}c < 0$, it is enough to show that $h(t)$ itself is monotone. The first and second derivatives of the function $h(t)$ are respectively

$$h'(t) = 4(-2 + 8t - 12t^2 + 8t^3) - 4c(1 - 8t + 18t^2)$$

and

$$h''(t) = 4[(8 - 24t + 24t^2) - c(-8 + 36t)].$$

Notice that as a quadratic polynomial, $h''(t) = 4(24t^2 - (24 + 36c)t + 8(1+c))$ can be shown to be always less than 0 for any $t \in [\frac{1}{2}, \frac{2}{3}]$. Since $h'(\frac{1}{2}) = -4c(1+\frac{1}{2}) < 0$, it follows that $h'(t) < 0, \forall t \in [\frac{1}{2}, \frac{2}{3}]$. Therefore $h(t)$ is monotone as required.   □

Lemma 4. *Let $g(t)$ be a function defined as*

$$g(t) = \frac{(1 - 2t + 2t^2 + 2\delta t)^c}{t^t(1 - t)^{1-t}} \tag{3}$$

where $c > 1$ and $\delta > 0$ are constants. Then, for small enough δ, $g(t)$ is monotone-increasing on $[\frac{1}{2}, \frac{2}{3}]$.

Proof. Consider the function $h(t) = \log g(t)$

$$h(t) = c \log(1 - 2t + 2t^2 + 2\delta t) - t \log t - (1 - t) \log(1 - t).$$

We have

$$h'(t) = c\frac{-2 + 4t + 2\delta}{1 - 2t + 2t^2 + 2\delta t} - \log t + \log(1 - t)$$

and $h'(\frac{1}{2}) \geq 0$. The second-order derivative of $h(t)$ is

$$h''(t) = \frac{c}{(1 - 2t + 2t^2 + 2\delta t)^2 t(1 - t)} \times z(t, \delta)$$

where

$$z(t, \delta) = 4(1 - 2t + 2t^2 + 2\delta t)(1 - t)t - (4t - 2 + 2\delta)^2(1 - t)t - (1 - 2t + 2t^2 + 2\delta t)^2.$$

First, assume that $\delta = 0$. On the interval $[\frac{1}{2}, \frac{2}{3}]$, we have

$$(4t - 2 + 2\delta)^2 \leq (4 \times \frac{2}{3} - 2)^2 = \frac{4}{9},$$

$$\frac{2}{9} \leq t(1 - t) \leq \frac{1}{2}(1 - \frac{1}{2}) = \frac{1}{4}$$

and

$$\frac{1}{2} \leq (1 - 2t + 2t^2 + 2\delta t)^2 \leq (1 - 2 \times \frac{2}{3} + 2 \times (\frac{2}{3})^2)^2 = \frac{5}{9}.$$

It follows that

$$z(t, \delta = 0) \geq 4 \times \frac{1}{2}\frac{2}{9} - \frac{1}{9} - (\frac{5}{9})^2 = \frac{2}{81} > 0.$$

Since the family of functions $z(t, \delta), \delta > 0$ are uniformly continuous on $[\frac{1}{2}, \frac{2}{3}]$, we have that for small enough δ, $z(t, \delta) > 0$. Therefore, the second-order derivative $h''(t)$ is always larger than zero. And so is $h'(t)$ (recall that $h'(\frac{1}{2}) > 0$). It follows that $h(t)$ is monotone-increasing, and so is $g(t)$. $\square$

Proof of Theorem 1

Proof. Let $\mathbf{W} = (S, A, B)$ be a partition of the vertices of $G(n, m)$ such that $|S| = l + 1 = \beta n, |B| \geq |A|, |B| = b = tn$, with $\frac{1}{2} \leq t \leq \frac{2}{3}$. Let $I_{\mathbf{W}}$ be the indicator function of the event that $\mathbf{W}$ is a rigid and balanced l-partition of $G(n, m)$. We have

$$\begin{aligned}
\mathcal{E}[\, I_{\mathbf{W}} \,] &= \Pr\{\, \mathbf{W} \text{ is rigid and balanced} \,\} \\
&= \Pr\{\, \mathbf{W} \text{ is balanced} \,\} \Pr\{\, \mathbf{W} \text{ is rigid} \mid \mathbf{W} \text{ is balanced} \,\}
\end{aligned} \tag{4}$$

From Lemma 2, we know that

$$\Pr\{\, \mathbf{W} \text{ is rigid} \mid \mathbf{W} \text{ is balanced} \,\} \leq \left(\frac{1}{e}\right)^{r(t)n}$$

By the definition of a balanced partition,

$$\begin{aligned}
\Pr\{\, \mathbf{W} \text{ is balanced} \,\} &= \left(1 - \frac{tn(n - tn - \beta n)}{n(n-1)/2}\right)^{cn} \\
&= \left[1 - 2t + 2t^2 + 2t\beta + O(1/n)\right]^{cn}.
\end{aligned} \tag{5}$$

This is because in order for $\mathbf{W}$ to be a balanced partition of $G(n, m)$, each of the m independent trials can only select an edge from the set of vertex pairs $V^2 \setminus \{(u, v) : u \in A, v \in B\}$. Write

$$\phi_1(t) = \left[1 - 2t + 2t^2 + 2t\beta + O(1/n)\right]^c,$$

$$\phi_2(t) = \left[\left(\frac{1}{e}\right)^{\frac{1}{c}r(t)}\right]^c$$

and

$$\phi(t) = \phi_1(t)\phi_2(t)$$

so that we have

$$\mathcal{E}[\, I_{\mathbf{W}} \,] = [\phi(t)]^n.$$

Let $I = \sum_{\mathbf{W}} I_{\mathbf{W}}$ be the number of rigid and balanced l-partitions of the random graph $G(n, m)$ where the sum is taken over all such possible partitions. For a partition (S, A, B), there are $\binom{n}{\beta n}$ ways to choose the vertex set S with $|S| = \beta n$. For a fixed vertex set S, there are $\binom{n-\beta n}{b}$ ways ($\frac{1}{2}n \leq b \leq \frac{2}{3}n$) to choose the pair (A, B) such that one of them has the size b. Therefore,

$$\begin{aligned}
\mathcal{E}[\, I \,] &= \sum_{\mathbf{W}} \mathcal{E}[\, I_{\mathbf{W}} \,] \\
&\leq \binom{n}{\beta n} \sum_{\frac{1}{2}n \leq b \leq \frac{2}{3}n} \binom{n - \beta n}{b} [\phi(\tfrac{b}{n})]^n \\
&\leq \binom{n}{\beta n} \sum_{\frac{1}{2}n \leq b \leq \frac{2}{3}n} \binom{n}{b} [\phi(\tfrac{b}{n})]^n.
\end{aligned}$$

By Stirling's formula, we have for n large enough

$$\mathcal{E}[I] \leq \left(\frac{1}{\beta^\beta(1-\beta)^{1-\beta}}\right)^n \sum_{\frac{1}{2}n \leq b \leq \frac{2}{3}n} \left(\frac{\phi_1(\frac{b}{n})\phi_2(\frac{b}{n})}{\frac{b}{n}^{\frac{b}{n}}(1-\frac{b}{n})^{1-\frac{b}{n}}}\right)^n$$

By Lemma 3,

$$\phi_2\left(\frac{b}{n}\right) \leq \phi_2\left(\frac{2}{3}\right) = \left[\left(\frac{1}{e}\right)^{\frac{2}{9c^2}}\left(\frac{1}{e}\right)^{4.8c}\right]^c$$

By Lemma 4,

$$\frac{\phi_1(\frac{b}{n})}{\frac{b}{n}^{\frac{b}{n}}(1-\frac{b}{n})^{1-\frac{b}{n}}} \leq \frac{\phi_1(\frac{2}{3})}{(\frac{2}{3})^{\frac{2}{3}}(\frac{1}{3})^{\frac{1}{3}}} = \frac{(\frac{5}{9}+\frac{4}{3}\beta)^c}{(\frac{2}{3})^{\frac{2}{3}}(\frac{1}{3})^{\frac{1}{3}}}.$$

Therefore,

$$\mathcal{E}[I] \leq O(n)\left(\frac{1}{\beta^\beta(1-\beta)^{1-\beta}}\right)^n \left(\frac{[(\frac{5}{9}+\frac{4}{3}\beta)(\frac{1}{e})^{\frac{2}{9c^2}e^{4.8c}}]^c}{(\frac{2}{3})^{\frac{2}{3}}(\frac{1}{3})^{\frac{1}{3}}}\right)^n.$$

From the above, it can be shown that for sufficiently small β and $c > 1.081$,

$$\mathcal{E}[I] \leq O(n)\gamma^n$$

with $0 < \gamma < 1$. The theorem then follows from Markov's inequality and Lemma 1:

$$\lim_n \Pr\{tw(G(n,m)) < \beta n\} \leq \lim_n \Pr\{I > 0\} \leq \lim_n \mathcal{E}[I] = 0. \qquad \square$$

4   Further Discussion

We conjecture that the threshold of having a linear treewidth is less than one (actually close to $1/2$) based on the size of the "giant" component in a random graph. Recall that Lemma 1 says that in order for a graph to have a treewidth $\leq l-1$, the graph must have a balanced partition $\mathbf{W} = (S, A, B)$ such that $|S| = l$ and $\frac{1}{3}(n-l) \leq |A|, |B| \leq \frac{2}{3}(n-l)$.

Consider the random graph $G(n,m)$ with $1/2 < \frac{m}{n} < 1$ on the set V of vertices. Let $S \subset V$ be a subset of vertices and assume that $|S| = \beta n$ with β small enough. Then, the induced subgraph $G_{V \setminus S}(n,m)$ is a random graph with the edges-vertices ratio c slightly less than m/n. Let

$$t(c) = \frac{1}{2c} \sum_{k=1}^{\infty} \frac{k^{k-1}}{k!}(2ce^{-2c})^k$$

and $1 - p_S(n)$ be the probability that the size of the largest component of $G_{V \setminus S}(n,m)$ is in the order of $(1 - t(c))n$. The famous result on the size of the giant component in a random graph, see e.g. [11], indicates that $p_S(n)$ tends to zero. It is also true that $(1 - t(c))$ is larger than $2/3$ even for c well below 1. Notice that the probability for $G(n, m)$ to have a balanced partition of the form (S, A, B) is less than $p_S(n)$; the existence of such a balanced partition implies that the components of the induced subgraph $G_{V \setminus S}(n, m)$ are all of size less than $\frac{2}{3}n$. Since there are $\binom{n}{\beta n}$ such S, we could have shown that the threshold of having a linear treewidth is less than one if the probability $p_S(n)$ is exponentially small. Unfortunately, we currently do not know yet if such an exponential upper bound for $p_S(n)$ exists.

Acknowledgment

The author thanks Professor J. Culberson for many invaluable discussions. Supported in part by NSERC Discovery Grant RGPIN 327587-06 and a startup grant from the University of British Columbia Okanagan.

References

1. Kloks, T.: Treewidth: Computations and Approximations. Springer-Verlag (1994)
2. Bodlaender, H.L.: A tourist guide through treewidth. Technical report, Technical Report RUU-CS-92-12, Department of Computer Science, Utrecht University (1992)
3. Dalmau, V., Kolaitis, P., Vardi, M.Y.: Constraint satisfaction, bounded treewidth, and finite-variable logics. In: Proceedings of Principles and Practices of Constraint Programming (CP-2002), Springer (2002) 310–326
4. Dechter, R., Fattah, Y.: Topological parameters for time-space tradeoff. Artificial Intelligence **125** (2001) 93–118
5. Cheeseman, P., Kanefsky, B., Taylor, W.M.: Where the *really* hard problems are. In: Proceedings of the 12th International Joint Conference on Artificial Intelligence, Morgan Kaufmann (1991) 331–337
6. Cook, S., Mitchell, D.G.: Finding hard instances of the satisfiability problem: A survey. In Du, Gu, Pardalos, eds.: Satisfiability Problem: Theory and Applications. Volume 35 of DIMACS Series in Discrete Mathematics and Theoretical Computer Science. American Mathematical Society (1997)
7. Achlioptas, D., Beame, P., Molloy, M.: A sharp threshold in proof complexity. In: ACM Symposium on Theory of Computing. (2001) 337–346
8. Kirousis, L., Kranakis, E.: Special issue on typical case complexity and phase transitions. Discrete Applied Mathematics **153** (2005)
9. Erdös, P., Renyi, A.: On the evolution of random graphs. Publ. Math. Inst. Hungar. Acad. Sci. **5** (1960) 17–61
10. Gao, Y.: Phase Transitions and Typical-case Complexity: easy (hard) aspects of hard (easy) problem. PhD thesis, Department of Computing Science, University of Alberta, Edmonton, Canada (2005)
11. Bollobas, B.: Random Graphs. Cambridge University Press (2001)
12. McDiarmid, C.: On the method of bounded differences. In: Surveys in Combinatorics. London Mathematical Society Lecture Note Series, vol. 141. Cambridge Univ. Press (1989) 148–188

Reconciling Gene Trees with Apparent Polytomies*

Wen-Chieh Chang and Oliver Eulenstein

Department of Computer Science, Iowa State University, Ames, IA 50011, USA
{wcchang, oeulenst}@cs.iastate.edu

Abstract. We consider the problem of reconciling gene trees with a species tree based on the widely accepted Gene Duplication model from Goodman *et al.* Current algorithms that solve this problem handle only binary gene trees or interpret polytomies in the gene tree as true. While in practice polytomies occur frequently, they are typically not true. Most polytomies represent unresolved evolutionary relationships. In this case a polytomy is called *apparent*. In this work we modify the problem of reconciling gene and species trees by interpreting polytomies to be apparent, based on a natural extension of the Gene Duplication model. We further provide polynomial time algorithms to solve this modified problem.

1 Introduction

In order to predict the function of genes it is critical to distinguish between speciation and duplication events in the genes' common evolutionary history [1,2]. Duplication events, which are pervasive in many gene families, typically result in incongruence between evolutionary histories of genes and the histories of the species from which the genes were sampled from. Evolutionary histories of either genes or species are represented through rooted phylogenetic trees (where every internal node has at least two children) and we refer to them as either *gene* or *species trees* respectively. An example for incongruence between a gene and its species tree that is caused by gene duplication is depicted in Fig. 1.

The gene duplication (GD) model from Goodman *et al.* [3] infers gene duplication events and losses from the incongruence of a given gene and species tree. While the basic GD model has been widely accepted and utilized through efficient algorithms [4,5,6,7,8], the model is constrained by its interpretation of internal nodes that have more then two children called polytomies. Polytomies can be either 'true' or 'apparent' [9,10]. A polytomy is true if all of its children diverged from it at the same time. A polytomy is apparent when it replaces some phylogenetic subtree that could not be fully resolved in the evolutionary history. In practice, most gene trees contain numerous weakly supported or completely unresolved evolutionary relationships that may be represented most accurately by apparent polytomies. The original GD model is confined to true polytomies.

---

* This research is supported by NSF Grant No. 0334832.

D.Z. Chen and D.T. Lee (Eds.): COCOON 2006, LNCS 4112, pp. 235–244, 2006.

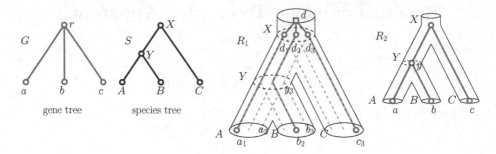

Fig. 1. The reconciled trees R_1 and R_2 explain the inconsistencies between the gene tree G and the species tree S assuming that r is a true and an apparent multifurcation respectively. The reconciled tree R_1 "truly" duplicates node d into the copies d_1, d_2, and d_3 in species X. Each of the copies evolves along the species tree and follows speciation events. The embedding of G into R_1 is highlighted by the solid edges. The reconciled tree R_2 explains the inconsistency assuming that r is an apparent multifurcation which is replaced by the "unknown" topology of of the species tree S without any duplication event.

Since true polytomies are rare evolutionary events, available algorithms for the GD model were mostly designed for only fully binary input trees.

In this work we introduce the first algorithm that infers gene duplications and losses from gene trees by interpreting polytomies as apparent. We (i) show a natural modification of the basic GD model for apparent polytomies, (ii) formulate the Reconciliation problem that infers duplications and losses from the extended GD model, (iii) present an overview of structural properties of the extended GD model, and (vi) derive from these properties a polynomial time algorithm that solves the reconciliation problem.

1.1   Previous Work and Interpreting the GD Model

Goodman *et al.* [3] introduced the GD model to infer gene duplication and losses for a given rooted gene and species tree. This work was formalized and later refined by Page [4] and others [11,12,6,13,14,15,16].

Given a gene tree G and a species tree S, the GD model assumes that a surjective (onto) mapping from the leaves of the gene tree to the leaves of the species tree is provided. This *leaf association function* maps each leaf gene node to the species which it was sampled from. An LCA mapping function LCA : $V(G) \rightarrow V(S)$ can be computed in linear time on a PRAM [13] (see also [17]) from the leaf association function where LCA(g) is the most recent species in S that theoretically can contain gene g. The node LCA(g) is the *species of g*.

Theoretically, the idea of the basic GD model is to infer all possible gene trees from the species tree S by allowing genes to either duplicate in two or more copies or to speciate. Duplication can only take place in one species at a time, thus the duplicated gene and its copies have the same species. For example in Fig. 1 gene d is duplicated into three copies d_1, d_2 and d_3 and all genes belong to the same species. Speciation occurs when a gene in a species evolves into one gene for each child of the species. For example in Fig. 1 gene y_3 in species Y speciates into gene a_3 and b_3 in the species A and B respectively. A gene tree that is derived from the species tree by either duplicating or speciating genes in nodes of a species tree is called a duplication tree. For example the tree R_1 in Fig. 1 is a duplication tree.

Formally D is a *duplication tree* of S, if there exists a function $\mathsf{Dup} : V(D) \rightarrow V(S)$ such that: (i) a leaf in D maps to a leaf in S, and (ii) every internal node u in D is either a duplication or a speciation node. Let $\mathsf{Ch}_T(x)$ denote the children of a node x in a rooted tree T. The node u is a *duplication node* (or d-node) if $\mathsf{Dup}(\mathsf{Ch}_D(u)) = \{\mathsf{Dup}(u)\}$, and it is a *speciation node* (or s-node) if $\mathsf{Dup}(\mathsf{Ch}_D(u)) = \mathsf{Ch}_S(\mathsf{Dup}(u))$. We also call $\mathsf{Dup}(u)$ the *species of u*.

Some of the duplication trees are evolutionary compatible with the gene tree. In this case, the gene tree G can be embedded into the duplication tree D through an *embedding function* $\mathsf{Emb} : V(G) \rightarrow V(D)$ that preserves the pairwise least common ancestor relations in G. Fig. 1 depicts an example where gene tree G can be embedded (solid lines) into the duplication tree R_1. Duplication trees that allow such an embedding of the gene tree are *explanation trees*. Explanation trees are evolutionary compatible with the gene tree and thus explain the incompatibility between the gene tree and the species tree through gene duplication and losses. A *loss* is a maximum subtree in E that has no embedding from the gene tree G. E.g. in Fig. 1 the subtree rooted at node z_3 of the reconciled tree R_1 is a loss.

Of particular biological interest are two special types of explanation trees for a given gene and a species tree.

Node reconciled trees are explanation trees with the minimum number of nodes [4]. It was shown by [6] and later by [16] that a reconciled tree is uniquely determined by the given gene and species trees.

Dup-loss reconciled trees are explanation trees with the minimum reconciliation cost. The reconciliation cost of an explanation tree E is the duplication cost of E plus the number of losses in E. The *duplication cost* is the overall number of copies minus one for each d-node in E.

Node and dup-loss reconciled trees and their reconciliation cost can be computed in polynomial time for complete binary gene and species trees [4,7,5]. Node reconciled trees and their reconciliation cost can be computed for general gene and species trees under true polytomies [6].

In the case of complete binary gene and species trees the definitions of node and dup-loss reconciled trees are equivalent [16,18].

This is not true for gene and species trees with apparent polytomies as is it is shown in Fig. 1. Thus we consider node and dup-loss reconciled trees for our extended GD model.

1.2   Presented Work

In this paper, we modify the basic GD model to interpret polytomies in gene trees as apparent. Apparent polytomies represent unknown phylogenetic subtrees.

Thus, the idea of the modified GD model is to replace the polytomies by complete binary subtrees. Therefore we consider the set $\mathcal{G}$ of all gene trees that we construct from the given gene tree G by replacing star trees, which are the polytomies and their children, with more refined trees. Let $\mathsf{Exp}_{G,S}$ be the set that contains the explanation trees for each combination of a gene tree in $\mathcal{G}$ and the species tree S in the basic GD model. Equivalently $\mathsf{Exp}_{G,S}$ can be described as the duplication trees into which the gene tree G can be embedded using a relaxed embedding function. The *relaxed embedding* is defined similar to the embedding for the basic GD model, but preserves only the tree order, rather then the pairwise least common ancestor relations. Fig. 1 depicts an example. The solid lines in the reconciled tree R_1 represent an embedding under preserving the pairwise least common ancestor relations. In contrast the solid lines in the reconciled tree R_2 represent an embedding that preserves the tree order, but not the pairwise least common ancestor relations.

As we show, node and dup-loss reconciled trees are not necessarily unique thus their definitions are not equivalent. Given a gene and its species tree, the *node reconciliation problem* is to find a node reconciled tree and the *dup-loss reconciliation problem* is to find a dup-loss reconciled tree. In this paper we show that both problems are solvable in polynomial time (an asymptotic upper bound is provided in Section 3.4).

1.3 Outline

We solve the node reconciled tree problem through a divide-and-conquer approach that divides the node reconciled tree problem into independent subproblems that we solve directly through dynamic programming. Section 2 presents an overview of the divide-and-conquer approach and Sections 3 introduces a dynamic programming solution. We briefly describe a similar solution for the dup-loss problem in Section 4. Due to space requirements we refer the reader for most of our proofs to the technical report of Chang and Eulenstein [18].

Let G be a gene tree, S its species tree, and R a node reconciled tree from G to S. Further let (i)LCA be the LCA mapping from G to S, (ii) Dup specify R, and (iii) Emb be the relaxed embedding from G into R. The subtree of a tree T rooted at node $v \in V(T)$ is denoted as T_v.

2   Divide-and-Conquer

The gene duplication problem (GDP) is formally defined as:

Instance: A gene tree G, a species tree S and a leaf association A from G to S.

Find: A reconciled tree R from G to S w.r.t. A.

GDP can be divided into independent subproblems based on the following theorem.

Theorem 1 (LCA-theorem). *For every gene g in the gene tree, it holds that* $\mathsf{LCA}(g) = \mathsf{Dup}(\mathsf{Emb}(g))$ *if R is a reconciled tree.*

The theorem shows that the species of a gene in G and the species of its embedding in the reconciled tree R are identical. This allows to partition the edges of the gene and species tree into independent subproblems of the node reconciled tree problem, and the edges of the reconciled tree into solutions to these subproblems.

Consider a partition of the edges in the gene tree G into star trees (parent-child edges). Each star tree rooted at node g defines the edges of a subtree in the species tree, called the *environment of g in the species tree S*. This subtree is defined to be rooted at the node $LCA(g)$ where the subtrees rooted at $LCA(c)$ for every child c of g are removed. Similarly, we define the *environment of g in the reconciled tree R*.

As a result we partition the original problem instance, the gene and species tree, into subproblems consisting of a start tree in G and its environment in S. The subproblem, called *the core problem*, is defined as follows.

Instance: A star-tree G where $C = Le(G)$, a tree S and a mapping function
   $M : V(G) \rightarrow V(S)$ where $M(Root(G)) = LCA_S(M(C))$.
Find: A reconciled tree from G to S w.r.t. M.

Our claim is that the solution to the core problem is the environment of g in R. A cut-and-paste argument verifies this claim. Suppose the environment of g in R is a not a solution, then there exists an explanation tree that solves the given subproblem with fewer nodes. Replacing the environment of g in R with this explanation tree results in an explanation tree for the original problem that has fewer nodes then R.

Following the same argument, the solutions to the core problems can be combined to obtain a solution to the original GDP.

3   Solving the Core Problem

Here we outline a dynamic programming approach for solving the core problem. To show that solutions to the core problem exhibit optimal substructure we prove that every solution contains node reconciled trees of a particular form, called *normal form*. The normal form allows us to describe the size of a node reconciled tree recursively that can be then computed by dynamic programming.

3.1   A Reconciled Tree in Normal Form

A node reconciled tree R is in *normal form* if for any d-node d in $V(R)$ with copies $d_1, \ldots, d_k$ the following two properties are satisfied:

1. **The property of normalized duplication:** There exists no d-node in $V(R_{d_i})$ for $2 \leq i \leq k$.
2. **The property of normalized embedding:** For any node u in $V(R_{d_i}) \cap Emb(V(G))$ $(1 \leq i < k)$, there exists a distinct embedded node v in $V(R_{d_j}) \cap Emb(V(G))$ $(i < j \leq k)$ where $Dup(v)$ is an ancestor of $Dup(u)$ in S.

We describe the effect on R if both properties are satisfied. Consider a duplication node d in R and its copies $d_1, \ldots, d_k$ ordered from left to right, and let C be the set of all children in the star-tree G that are embedded into the subtree rooted at d. Each subtree rooted at $d_1, \ldots, d_k$ exists because it contains an embedding from nodes in C. Thus the subtrees partition the set C into k non-empty sets, $C_1, \ldots, C_k$, based on the nodes that are embedded into each subtree. Recall that R satisfies the first property. Thus the subtrees rooted at $d_2, \ldots, d_k$, called *non-duplication subtrees*, do not contain any duplication. It follows that the species of the genes in a non-duplication subtree form an *anti-chain* in S (no two elements are on a same path). Now the second property requires that the elements in the anti-chains in S for each non-duplication subtree are ordered by $\leq$. We define $S_i \leq S_j$ if for any $i \in S_i$ and $j \in S_j$ that are on a same path, and i has smaller or equal depth then j in S. We refer to the gene nodes in C that form the ordered anti-chains in each subtree rooted at $d_2, \ldots, d_k$ as *layers*. An example o layers is depicted in Fig. 2.

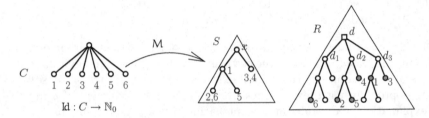

Fig. 2. The problem instance is simplified to a star gene tree where C is the leaf set. Without loss of generality, a total ordering (Id) in C is assumed. The mapping function M is essentially the LCA mapping function. Layers in C are $\{1,3\}$, $\{2,5,4\}$ and $\{6\}$. Please note that in this example, R is not an reconciled tree since it is not optimal.

Theorem 2 (reconciled tree in normal form). *There exists a reconciled tree in normal form.*

The above theorem warrants the existence of a node reconciled tree in normal form in all problem instances. It is shown by an algorithm that transforms any arbitrary reconciled tree into a reconciled tree that satisfies the two properties. A similar approach shows that for the core problem, a reconciled tree in normal form has at most one d-node mapped to a single species node through Dup. This implication helps us to reconstruct an optimal solution in a dynamic programming approach. Furthermore, Theorem 2 identifies an optimal substructure in a node reconciled tree to form a recursive formula which will be introduced later. Because of Theorem 2, we will simply refer to a reconciled tree in normal from as a reconciled tree.

3.2   Optimal Substructure

In order to describe the substructure recursively, we introduce a notation to represent a partition of (a subset of) C. If v is a node in $V(R)$, then C_v is the subset of C mapped to R_v through Emb, and the definition can be extended recursively to a subset of C. The same notation applies to a node in S. As shown in Fig. 2, we can use C_{d_1}, C_{d_2} and C_{d_3} to denote the layers $\{6\}$, $\{2,5,4\}$ and $\{1,3\}$ respectively. And the set $\{2,5,4\}$ can be further partitioned into layers $\{4\}$ and $\{2,5\}$ if necessary. Note that we can generalize the definition of layers with the imposed total ordering in C and the ordered anti-chains to describe them uniquely. That is, we call C_{d_3} the first layer of C_x, denoted by $\text{Layer}(C_x, 1)$, where $x = \text{Dup}(d)$, C_{d_2} the second layer of C_x, denoted by $\text{Layer}(C_x, 2)$, and C_{d_1} the third layer of C_x, denoted by $\text{Layer}(C_x, 3)$. We also denote $C_{d_1} \cup C_{d_2}$ as $\text{Remain}(C_x, 1)$, which represents $C_x - \text{Layer}(C_x, 1)$ and are called *remains*, to emphasize that each non-duplication subtree is embedded by a layer. Note that the numbers of non-empty layers and remains are always finite and no greater than $|C|$.

The linear ordering introduced in C simply gives a unique structure among layers and remains, and allows us to find an equivalent reconciled tree. With the generalization in mind, the following theorem concludes the optimal substructure to describe a reconciled tree recursively.

Theorem 3 (layer, remain embedding). *Let v be a vertex in R. It holds that C_v is either a layer or a remain of $C_{\text{Dup}(v)}$, and R_v is a reconciled tree from $G_{|C_v}$ to $S_{\text{Dup}(v)}$.*

The theorem can be shown by using the recursive structure introduced in the normal form, combining the definitions of layers and remains properly. In the next step, we show how to apply the theorem to formulate the substructure recursively.

3.3   Recursive Solution

In the case of node reconciled trees, the objective is to minimize the size of a reconciled tree, which means that we need to count the nodes first. In a top-down fashion, assuming we know whether a node is an s-node or a d-node (and the number of its copies), the formula can be illustrated by Fig. 3 as there are exactly two cases.

For the first case, suppose we know that the genes in C are embedded into the subtree R_d of a node reconciled tree in normal form rooted at a duplication node d with copies $d_1, \ldots, d_k$. The first $(k-1)$ layers of C are embedded into the subtrees $R_{d_2}, \ldots, R_{d_k}$. Then $\text{Size}(R_d, C)$ is defined as the number of nodes in R_d if we embed C into it. We can then write $\text{Size}(R_d, C)$ as follows

$$\text{Size}(R_d, C) = \text{Size}(R_{d_1}, C_1) + (k-1) \cdot \text{Size}(S_{d'}) + 1 \tag{1}$$

where C_1 is essentially $\text{Remain}(C, k-1)$ and d' is the species $\text{Dup}(d)$ in S. $\text{Size}(S_{d'})$ is the size of each non-duplication subtree, since they do not contain any duplication, and it is given by S. We have $(k-1)$ such subtrees. An additional node is counted for the root of R_d, and $\text{Size}(R_{d_1}, C_1)$ accounts for the size of the remaining subtree rooted at R_{d_1}. Furthermore, as implied by the normal

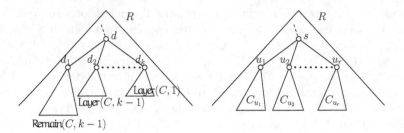

Fig. 3. Each complete subtree of a reconciled tree R has a layer or remain embedded. In the case that s is an s-node in R; $C_{u_1}, \ldots, C_{u_r}$ form a partition of C_s. In the case that d is a d-node in R; $R_{d_2} \ldots R_{d_k}$ are embedded with the firs $k-1$ layers of C_d (denoted as C above).

form properties, d_1 in R has to be an s-node, which automatically advances the recursion into the second scenario.

In the second scenario, also illustrated in Fig. 3, if a node s in R is an s-node, we know exactly its children, as in the species tree. Therefore the objective function $\mathsf{Size}(R_s, C)$ can be expressed similarly to Equation (1)

$$\mathsf{Size}(R_s, C) = \sum_{u \in \mathsf{Ch}(s)} \mathsf{Size}(R_u, C_u) \qquad (2)$$

The above two equations describe the size of a reconciled tree precisely in a top-down recursive fashion. By reversing the direction, we can find an optimal solution (as they are not necessarily unique) from bottom up. In the next step, we present the general idea behind the optimization.

3.4   Optimal Node Reconciled Tree

The goal of optimization is essentially finding the optimal number of copies of each node in a reconciled tree. Since a d-node has at least two copies, we can generalize the notion by saying an s-node has exactly one copy. As demonstrated in the previous discussion, the number of allowed copies for each node is bounded by the number of non-empty layers in the remain, which may be reduced by a duplication event of some ancestor node. Hence, the solution can be found by trying all possible subproblems in a given subtree then finding the optimal number of copies.

The process can be memoized using an $O(|S| \cdot |C|)$ table for all feasible combinations of vertices and their non-empty remains. One also needs to know $|S_v|$ for each $v \in V(S)$, in which a DFS traversal is sufficient. All leaf nodes can be initialized immediately since they are all base cases. Each cell stores two values: $\mathsf{Size}(v, C'_v)$ and $k(v, C'_v)$, where C'_v is a non-empty remain of v, as the solution to the subproblem under the given condition. Once the table is filled up in a bottom-up fashion, $\mathsf{Size}(\mathsf{Root}(S), C)$ represents $|R|$. The whole optimization process takes $O(|S| \cdot |C|^2)$ time as each cell has at most $|C|$ possible values for k (and those values are not all necessarily feasible).

3.5   Constructing a Node Reconciled Tree

It suffices to build a node reconciled tree R, if the optimal k is found for each $v \in V(S)$. Starting at $r = \text{Root}(S)$, $k(r, C)$ determines whether the root of the resulting node reconciled tree is a d-node or an s-node. For $v \in \text{In}(S)$, if $k(v, C'_v) = k'$ is optimal, we know there are k' copies of the node x in R such that $\text{Dup}(x) = v$ and the embedding nodes in children of x are also determined accordingly for the next step. Essentially by backtracking $k(v, C'_v)$, there is a duplication function, and we also know the unique layer structures embedded into each non-duplication subtree, which gives an embedding function.

4   Dup-Loss Reconciled Tree

In order to minimize the duplication and loss cost, it is necessary to know how to calculate losses, since the duplication cost in a reconciled tree R is clear. As losses are closely related to the layers, a lookup table associates the number of losses in each subtree of S and a given layer can be calculated recursively as a part of pre-processing. The loss table is of size $O(|S| \cdot |C|)$, and we claim it can be filled up properly in $O(|S| \cdot |C|)$ time. The recursion is similar to Equation (1) and (2), but it computes the duplication and loss cost instead of the size of a subtree. Hence, to determine a dup-loss reconciled tree takes $O(|S| \cdot |C|^2)$ time as well.

5   Outlook

Of biological interest is the extension of the GD model that considers in addition to apparent polytomies in the gene tree also apparent topologies in the species tree. This extension requires a biologically meaningful definition of a reconcile tree. Using the relaxed embedding of our extended GD model for the reconciliation, a solution similar to the presented one might be sufficient.

Further *dup reconciled trees* that are explanation trees with the minimum duplication cost under the extended GD model might be valuable to biologists. A modification of Theorem 1 should allow to compute dup reconciled trees using our presented approach.

References

1. Page, R.D.M., Holmes, E.C.: Molecular Evolution: A Phylogenetic Approach. Blackwell Science Ltd, Osney Mead, Oxford (1998)
2. Felsenstein, J.: Inferring phylogenies. Sinauer Associates, Inc., Sunderland, Massachusetts (2004)
3. Goodman, M., Czelusniak, J., Moore, G.W., Romero-Herrera, A.E., Matsuda, G.: Fitting the gene lineage into its species lineage, a parsimony strategy illustrated by cladograms constructed from globin sequences. Systematic Zoology **28** (1979) 132–163
4. Page, R.D.M.: Maps between trees and cladistic analysis of historical associations among genes, organisms, and areas. Systematic Biology **43**(1) (1994) 58–77

5. Eulenstein, O.:   A linear time algorithm for tree mapping (1997) `http://taxonomy.zoology.gla.ac.uk/rod/genetree/maths/maths.html`, (access date: February 19, 2006), Arbeitspapire der GMD, 1046.
6. Eulenstein, O.:   Vorhersage von Genduplikationen und deren Entwicklung in der Evolution.   Ph.d. dissertation, Bonn University (1998) `http://www.bi.fraunhofer.de/publications/research/1998/020/Text.pdf`, GMD Research Series, 20.
7. Ma, B., Li, M., Zhang, L.: On reconstructing species trees from gene trees in term of duplications and losses. In: RECOMB. (1998) 182–191
8. Zmasek, C.M., Eddy, S.R.: A simple algorithm to infer gene duplication and speciation events on a gene tree. Bioinformatics $17(9)$ (2001) 821–828
9. Maddison, W.P.: Reconstructing character evolution on polytomous cladgrams. Cladistics 5 (1989) 365–377
10. Slowinski, J.B.: Molecular polytomies. Molecular Phylogenetics and Evolution $19(1)$ (2001) 114–120
11. Guigó, R., Muchnik, I., Smith, T.F.: Reconstruction of ancient molecular phylogeny. Molecular Phylogenetics and Evolution $6(2)$ (1996) 198–213
12. Mirkin, B., Muchnik, I., Smith, T.F.: A biologically meaningful model for comparing molecular phylogenies. Journal of Computational Biology 2 (1995) 493–507
13. Zhang, L.: On a mirkin-muchnik-smith conjecture for comparing molecular phylogenies. Journal of Computational Biology $4(2)$ (1997) 177–187
14. Chen, K., Durand, D., Farach-Colton, M.: Notung: Dating gene duplications using gene family trees. In: RECOMB. (2000) 96–106
15. Bonizzoni, P., Vedova, G.D., Dondi, R.: Reconciling gene trees to a species tree. In: CIAC2003 - Italian Conference on Algorithms and Complexity, Rome, Italy (2003)
16. Górecki, P., Tiuryn, J.: On the structure of reconciliations. In: Recomb Comparative Genomics Workshop 2004. Volume 3388. (2004)
17. Bender, M.A., Farach-Colton, M.: The lca problem revisited. Latin American Theoretical Informatics (2000) 88–94 Apr.
18. Chang, W.C., Eulenstein, O.: Reconciling gene trees with apparent polytimies. Technical report, Department of Computer Science, Iowa State University (2006)

Lower Bounds on the Approximation of the Exemplar Conserved Interval Distance Problem of Genomes*

Zhixiang Chen[1], Richard H. Fowler[1], Bin Fu[2], and Binhai Zhu[3]

[1] Department of Computer Science, University of Texas-American, Edinburg,
TX 78541-2999, USA
chen@panam.edu, fowler@cs.panam.edu
[2] Department of Computer Science, University of New Orleans, New Orleans,
LA 70148 and Research Institute for Children, 200 Henry Clay Avenue,
New Orleans, LA 70118, USA
fu@cs.uno.edu
[3] Department of Computer Science, Montana State University, Bozeman,
MT 59717-3880, USA
bhz@cs.montana.edu

Abstract. In this paper we present several lower bounds on the approximation of the exemplar conserved interval distance problem of genomes. We first prove that the exemplar conserved interval distance problem cannot be approximated within a factor of $c \log n$ for some constant $c > 0$ in polynomial time, unless P=NP. We then prove that it is NP-complete to decide whether the exemplar conserved interval distance between any two sets of genomes is zero or not. This result implies that the exemplar conserved interval distance problem does not admit any approximation in polynomial time, unless P=NP. In fact, this result holds even when a gene appears in each of the given genomes at most three times. Finally, we strengthen the second result under a weaker definition of approximation (which we call weak approximation). We show that the exemplar conserved interval distance problem does not admit a weak approximation within a factor of m, where m is the maximum length of the given genomes.

1 Introduction

A central problem in the genome comparison and rearrangement area is to compute the number (i.e., genetic distances) and the actual sequence of genetic operations needed to convert a source genome to a target genome. This problem originates from evolutionary molecular biology. In the past, typical genetic distances studied include edit [10], signed reversal [13,9,1] and breakpoint [17],

---

* This research is supported in part by FIPSE Congressional Award P116Z020159, NSF CNS-0521585, Louisiana Board of Regents under contract number LEQSF(2004-07)-RD-A-35 and MSU-Bozeman's Short-term Professional Development Leave Program.

D.Z. Chen and D.T. Lee (Eds.): COCOON 2006, LNCS 4112, pp. 245–254, 2006.

conserved interval [3,4], etc. (It was Sturtevant and Dobzhansky who came up with the idea of signed reversal and breakpoint distance, though implicitly, in 1936 [16].) Recently, conserved interval distance was also proposed to measure the similarity of multiple sequences of genes [3]. For an overview of the research performed in this area, readers are referred to [8,7] for a comprehensive survey.

Until a few years ago, in genome rearrangement research, people always assumed that each gene appears in a genome exactly once. Under this assumption, the genome rearrangement problem is essentially the problem of comparing and sorting signed/unsigned permutations [8,7]. However, this assumption is very restrictive and is only justified in several small virus genomes. For example, this assumption does not hold on eukaryotic genomes where paralogous genes exist [12,15]. Certainly, it is important in to compute genomic distances efficiently, e.g., by Hannenhalli and Pevzner's method [8], when no gene duplications arise; on the other hand, one might have to handle this gene duplication problem as well. A few years ago, Sankoff proposed a way to select, from the duplicated copies of genes, the common ancestor gene such that the distance between the reduced genomes (*exemplar genomes*) is minimized [15]. He also proposed a general branch-and-bound algorithm for the problem [15]. Recently, Nguyen, Tay and Zhang used a divide-and-conquer method to compute the exemplar breakpoint distance empirically [12]. As these problem seemed to be hard, theoretical research was followed almost immediately. It was shown that computing the signed reversals and breakpoint distances between exemplar genomes are both NP-complete [5]. Recently, Blin and Rizzi further proved that computing the conserved interval distance between exemplar genomes is NP-complete [4]; moreover, it is NP-complete to compute the minimum conserved interval matching (i.e., without deleting the duplicated copies of genes). There has been no formal theoretical results, before Nguyen [11] and our recent work [6], on the approximability of the exemplar genomic distance problems except the NP-completeness proofs [5,4]. Nguyen [11] proved that exemplar breakpoint distance cannot be approximated within constant ratio in polynomial time unless $P = NP$. Actually, the result was proved through a reduction from the set cover problem. This work was announced in [12].

In [6], we present the first set of inapproximability and approximation results for the Exemplar Breakpoint Distance problem, given two genomes each containing only one sequence of genes drawn from n identical gene families. (Some of the results hold subsequently for the Exemplar Reversal Distance problem.) For the One-sided Exemplar Breakpoint Distance Problem, which is also known to be NP-complete, we obtain a factor-$2(1 + \log n)$, polynomial-time approximation. The approximation algorithm follows the greedy strategy for Set-Cover, but constructing the family of sets is non-trivial and is related to a new problem of *longest constrained common subsequences* which is related to but different from the recently studied *constrained longest common subsequences* [2].

2   Preliminaries

In the genome comparison and rearrangement problem, we are given a set of genomes, each of which is a signed sequence of genes. The order of the genes corresponds to their positions on the linear chromosome and the signs correspond to which of the two DNA strands the genes are located. While most of the past research are under the assumption that each gene occurs in a genome once, this assumption is problematic in reality for eukaryotic genomes or the likes where duplications of genes exist [15]. Sankoff proposed a method to select an *exemplar genome*, by deleting redundant copies of a gene, such that in an exemplar genome any gene appears exactly once; moreover, the resulting exemplar genomes should have a property that certain genetic distance between them is minimized [15].

The following definitions are very much following those in [3,4]. Given n *gene families* (alphabet) $\mathcal{F}$, a genome G is a sequence of elements of $\mathcal{F}$ such that each element is with a sign ($+$ or $-$). In general, we allow the repetition of a gene family in any genome. Each occurrence of a gene family is called a *gene*, though we will not try to distinguish a gene and a gene family if the context is clear. Given a genome $G = g_1 g_2 ... g_m$ with no repetition of any gene, we say that gene g_i *immediately precedes* g_j if $j = i + 1$. Given genomes G and H, if gene a immediately precedes b in G but neither a immediately precedes b nor $-b$ immediately precedes $-a$ in H, then they constitute a *breakpoint* in G. The *breakpoint distance* is the number of breakpoints in G (symmetrically, it is the number of breakpoints in H).

The number of a gene g appearing in a genome G is called the cardinality of g in G, written as $card(g, G)$. A gene in G is called *trivial* if g has cardinality exactly 1; otherwise, it is called *non-trivial*. In this paper, we assume that all the genomes we discuss could contain both trivial and non-trivial genes. A genome G is called *r-repetitive*, if all the genes from the same gene family appear at most r times in G. A genome G is called a *k-span* genome, if all the genes from the same gene family are within distance at most k in G. For example, $G = -adc - bdaeb$ is 2-repetitive and it is a 5-span genome.

Given a genome $G = g_1 g_2 \cdots g_m$, an interval $[g_i, g_j]$ is simply the substring $g_i g_{i+1} \cdots g_j$ (which will also be denoted as $G[i, j]$). For example, given $G' = bdc - ag - e - fh, G'' = bdce - gafh$, between the two intervals $I_1 = dc - ag - e - f$ and $I_2 = dce - gaf$, there are 2 breakpoints $c - a$ and $-e - f$. A *signed reversal* on a genome G simply reverses the order and signs of all the elements in an interval of G. In the previous example, if a signed reversal operation is conducted in I_1 on G', then we obtain a new genome $G^* = bfe - ga - c - dh$. (All the reversals concerned in this paper are signed reversals. Henceforth, we simply use *reversal* to make the presentation simpler.) The *reversal distance* between genomes G and H is the minimum number of reversals to transfer G into H.

Given a genome G over $\mathcal{F}$, an *exemplar genome* of G is a genome G' obtained from G by deleting duplicating genes such that each gene family in G appears exactly once in G'. For example, let $G = bcaadagef$ there are two exemplar genomes: $bcadgef$ and $bcdagef$.

Given a set of genomes $\mathcal{G}$ and two gene families $a, b \in \mathcal{F}$, an interval $[a, b]$ is a *conserved interval* of $\mathcal{G}$ if (1) a precedes b or $-b$ precedes $-a$ in any genome in $\mathcal{G}$; and (2) the set of unsigned genes (i.e., ignoring signs) between a and b are the same for all genomes in $\mathcal{G}$. Let $\mathcal{G} = \{G_1, G_2\}$, where $G_1 = bc - ag - e - fdh, G_2 = b - ce - gaf - dh$, there are three conserved intervals between G_1 and G_2: $[e, a], [b, h]$ and $[-a, g]$.

Given two sets of genomes $\mathcal{G}$ and $\mathcal{H}$, the *conserved interval distance* between $\mathcal{G}$ and $\mathcal{H}$ is defined as

$$d(\mathcal{G}, \mathcal{H}) = N_{\mathcal{G}} + N_{\mathcal{H}} - 2N_{\mathcal{G} \cup \mathcal{H}},$$

where $N_{\mathcal{G}}$ (resp. $N_{\mathcal{H}}$ and $N_{\mathcal{G} \cup \mathcal{H}}$) is the number of conserved intervals in $\mathcal{G}$ (resp. $\mathcal{H}$ and $\mathcal{G} \cup \mathcal{H}$). Continuing the example in the previous paragraph, let $\mathcal{H} = \{H_1, H_2\}$, where $H_1 = b - cg - af - edh, H_2 = bagcdefh$, then there are two conserved intervals between H_1 and H_2: $[b, h]$ and $[a, c]$. There is only one conserved interval in $\mathcal{G} \cup \mathcal{H}$: $[b, h]$. Therefore, $d(\mathcal{G}, \mathcal{H}) = 3 + 2 - 2 \times 1 = 3$.

If $\mathcal{G}$ and $\mathcal{H}$ are both a singleton, i.e., $\mathcal{G}$ contains only a genome G, and $\mathcal{H}$ contains only a genome H, then we simply use the notation $d(G, H) = N_G + N_H - 2N_{G \cup H}$ to stand for $d(\mathcal{G}, \mathcal{H})$. Note that when only one genome G is considered, every interval in G is a conserved interval. This implies that when G (resp. H) has n trivial genes, then $d(G, H) = 2\binom{n}{2} - 2N_{G \cup H}$.

The Exemplar Conserved Interval Distance Problem, denoted as the *ECID problem*, is defined as follows:

Instance: Two sets of genomes $\mathcal{G}$ and $\mathcal{H}$, each genome is of length $O(m)$ and covers n identical gene families (i.e., it contains at least one gene from each of the n gene families); an integer K.

Question: Are there respective exemplar genomes $\mathcal{G}^*$ of $\mathcal{G}$ and $\mathcal{H}^*$ of $\mathcal{H}$, such that the conserved interval distance distance $d(\mathcal{G}^*, \mathcal{H}^*)$ is at most K?

In the next three sections, we present lower bounds on the approximation of the optimization version of the ECID problem, namely, to compute or approximate the minimum value K in the above formulation. Given a minimization problem Π, let the optimal solution of Π be OPT. We say that an approximation algorithm $\mathcal{A}$ provides a *performance guarantee* of α for Π if for every instance I of Π, the solution value returned by $\mathcal{A}$ is at most $\alpha \times OPT$. (Usually we say that $\mathcal{A}$ is a factor-α approximation for Π.) Typically we are interested in polynomial time approximation algorithms.

In many biological problems, the optimal solution value OPT could be zero. (For example, in some minimum recombination haplotype reconstruction problems the optimal solution could be zero.) In that case, if computing such a zero optimal solution value is NP-complete then the problem does not admit *any* approximation unless P=NP. However, in reality one would be happy to obtain a solution with value one or two. Due to this reason, we relax the above (traditional) definition of approximation to a *weak approximation*. Given a minimization problem Π, let the optimal solution of Π be OPT. We say that a weak approximation algorithm B provides a *performance guarantee* of α for

Π if for every instance I of Π, the solution value returned by B is at most $\alpha \times (OPT + 1)$.

3   A $c \log n$ Lower Bound on Approximating ECID

Theorem 1. *It is NP-complete to approximate the Exemplar Conserved Interval Distance problem within a factor of $c \log n$ for some constant $c > 0$.*

Proof. We use a reduction from the Dominating Set problem to the ECID problem for two sets of genomes $\mathcal{G} = \{G_1, G_2\}$ and $\mathcal{H} = \{H_1, H_2\}$ that will be constructed from the given graph.

Let $T = (V, E)$ be any given graph with $V = \{v_1, v_2, \cdots, v_n\}$ and $E = \{e_1, e_2, \cdots, e_m\}$. We assume that vertices and edges in T are sorted by their corresponding indices. We construct two sets of genomes $\mathcal{G} = \{G_1, G_2\}$ and $\mathcal{H} = \{H_1, H_2\}$ as follows. For each $v_i \in V$, we have four corresponding genes v_i^1, v_i^2, v_i^3, and v_i^4. We enforce a rule that v_i^k is incident to v_j^l if and only if $k = l$ and v_i is incident to v_j in T. Let $\mathcal{B}_i^j$ be the sorted sequence of vertices incident to v_i^j and $\bar{\mathcal{B}}_i^j$ be the unsigned reversal of $\mathcal{B}_i^j$. ("|" is not a gene and is used for readability purpose.)

$$G_1 = v_1^1 \mathcal{B}_1^1 v_1^3 | v_1^2 \mathcal{B}_1^2 v_1^4 | \cdots | v_{n-1}^1 \mathcal{B}_{n-1}^1 v_{n-1}^3 | v_{n-1}^2 \mathcal{B}_{n-1}^2 v_{n-1}^4 | v_n^1 \mathcal{B}_n^1 v_n^3 | v_n^2 \mathcal{B}_n^2 v_n^4$$

$$G_2 = -v_1^3 \bar{\mathcal{B}}_1^1 - v_1^1 | - v_1^4 \bar{\mathcal{B}}_1^2 - v_1^2 | \cdots | - v_{n-1}^3 \bar{\mathcal{B}}_{n-1}^1 - v_{n-1}^1 | - v_{n-1}^4 \bar{\mathcal{B}}_{n-1}^2 - v_{n-1}^2 |$$
$$-v_n^3 \bar{\mathcal{B}}_n^1 - v_n^1 | - v_n^4 \bar{\mathcal{B}}_n^2 - v_n^2$$

$$H_1 = v_1^1 \mathcal{B}_1^1 v_1^3 | \cdots | v_{n-1}^1 \mathcal{B}_{n-1}^1 v_{n-1}^3 | v_n^1 \mathcal{B}_n^1 v_n^3 | v_1^2 - v_1^4 | \cdots | v_{n-1}^2 - v_{n-1}^4 | v_n^2 - v_n^4$$

$$H_2 = -v_1^3 \bar{\mathcal{B}}_1^1 - v_1^1 | \cdots | - v_{n-1}^3 \bar{\mathcal{B}}_{n-1}^1 - v_{n-1}^1 | - v_n^3 \bar{\mathcal{B}}_n^1 - v_n^1 |$$
$$-v_n^4 v_n^2 | - v_{n-1}^4 v_{n-1}^2 | \cdots | - v_1^4 v_1^2$$

Fig. 1 shows a simple graph with six vertices $v_1, v_2, \ldots, v_6$. The corresponding genomes for this graph are given as follows.

$$G_1 = v_1^1 v_3^1 v_1^3 | v_1^2 v_3^2 v_1^4 | v_2^1 v_3^1 v_2^3 | v_2^2 v_3^2 v_2^4 | v_3^1 v_1^1 v_2^1 v_5^1 v_3^3 | v_3^2 v_1^2 v_2^2 v_5^2 v_3^4 |$$
$$v_4^1 v_5^1 v_6^1 v_4^3 | v_4^2 v_5^2 v_6^2 v_4^4 | v_5^1 v_3^1 v_4^1 v_6^1 v_5^3 | v_5^2 v_3^2 v_4^2 v_6^2 v_5^4 | v_6^1 v_4^1 v_5^1 v_6^3 | v_6^2 v_4^2 v_5^2 v_6^4 |$$

$$G_2 = -v_1^3 v_3^1 - v_1^1 | - v_1^4 v_3^2 - v_1^2 | - v_3^2 v_3^1 - v_2^1 | - v_2^4 v_3^2 - v_2^2 |$$
$$-v_3^3 v_5^1 v_2^1 v_1^1 - v_3^1 | - v_3^4 v_5^2 v_2^2 v_1^2 - v_3^2 | - v_4^3 v_6^1 v_5^1 - v_4^1 | - v_4^4 v_6^2 v_5^2 - v_4^2 |$$
$$-v_5^3 v_6^1 v_4^1 v_3^1 - v_5^1 | - v_5^4 v_6^2 v_4^2 v_3^2 - v_5^2 | - v_6^3 v_5^1 v_4^1 - v_6^1 | - v_6^4 v_5^2 v_4^2 - v_6^2 |$$

$$H_1 = v_1^1 v_3^1 v_1^3 | v_2^1 v_3^1 v_2^3 | v_3^1 v_1^1 v_2^1 v_5^1 v_3^3 | v_4^1 v_5^1 v_6^1 v_4^3 | v_5^1 v_3^1 v_4^1 v_6^1 v_5^3 | v_6^1 v_4^1 v_5^1 v_6^3 |$$
$$v_1^2 - v_1^4 | v_2^2 - v_2^4 | v_3^2 - v_3^4 | v_4^2 - v_4^4 | v_5^2 - v_5^4 | v_6^2 - v_6^4$$

$$H_2 = -v_1^3 v_3^1 - v_1^1 | - v_2^3 v_3^1 - v_2^1 | - v_3^3 v_5^1 v_2^1 v_1^1 - v_3^1 |$$
$$-v_4^3 v_6^1 v_5^1 - v_4^1 | - v_5^3 v_6^1 v_4^1 v_3^1 - v_5^1 | - v_6^3 v_5^1 v_4^1 - v_6^1 |$$
$$-v_6^4 v_6^2 | - v_5^4 v_5^2 | - v_4^4 v_4^2 | - v_3^4 v_3^2 | - v_2^4 v_2^2 | - v_1^4 v_1^2$$

Claim A. The given graph T has a dominating set of size K if and only if there are exemplar genomes g_i for G_i and h_i for H_i for $i = 1, 2$, such that we have, letting $\mathcal{G}^* = \{g_1, g_2\}$ and $\mathcal{H}^* = \{h_1, h_2\}$,

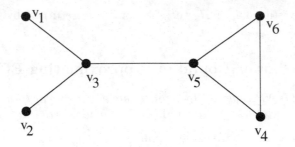

Fig. 1. Illustration of a simple graph for the reduction

$$(1) \quad N_{\mathcal{G}^*} = 2K$$
$$(2) \quad N_{\mathcal{H}^*} = K$$
$$(3) \quad N_{\mathcal{G}^* \cup \mathcal{H}^*} = K$$

Note that (1), (2) and (3) together imply $d(\mathcal{G}^*, \mathcal{H}^*) = K$.

We now prove the above claim. The "only if part" is easy. We only show the proof for (2) as the proofs for (1) and (3) would be similar. If T has a dominating set of size K, then for all those v_j which is not in the dominating set we delete all B_j^1 in H_1 and $\mathcal{B}_j^1$ in H_2. For those remaining B_i^1 in H_1 and $\mathcal{B}_i^1$ in H_2, we delete the duplications (say v_i^1) consistently in H_1 and H_2. It is easy to see that the conserved intervals we have are $[v_i^1, v_i^3]$ in H_1 and $[-v_i^3, -v_i^1]$ in H_2, which correspond to the vertices in the dominating set.

The "if part" is slightly more tricky. Assume that (1), (2) and (3) are all true. In this case, we only need to focus on (2), i.e., $N_{\mathcal{H}^*} = K$. First, notice that the second half of H_1 and H_2 (i.e., those involved with $v_j^2 - v_j^4$ or their unsigned reversals) will not contribute anything to the number of conserved intervals in $\mathcal{H}^*$. Notice also that only these $[v_i^1, v_i^3]$ from the first half of H_1 can possibly form conserved intervals with the corresponding $[-v_i^3, -v_i^1]$ from the first half of H_2. If the number of conserved intervals in $\mathcal{H}^*$ is K, then those K conserved intervals must come from $[v_i^1, v_i^3]$ in h_1 and $[-v_i^3, v_i^1]$ in h_2. Moreover, if there is any deletion in B_i^1 in H_1 and in $\mathcal{B}_i^1$ in H_2, then the deletion has to be consistent. If v_j^1 appears in B_i^1 and $\mathcal{B}_i^1$, then unless it appears in B_l^1 and $\mathcal{B}_l^1$ we must keep them to avoid extra conserved intervals in the form of $[v_j^1, v_j^3]$ in H_1 and $[-v_j^3, -v_j^1]$ in H_2. Therefore, from the K conserved intervals in $\mathcal{H}^*$ we can construct the K vertices which form the dominating set for T.

Let $opt(T)$ denote the size of the minimum dominating set of the graph T, and let $opt(\mathcal{G}, \mathcal{H})$ denote the minimum exemplar conserved interval distance between $\mathcal{G}$ and $\mathcal{H}$. It follows from Claim A that $opt(T) = opt(\mathcal{G}, \mathcal{H})$. The size of T is $|V| + |E| = n + m$. It is easy to see that the size of $\mathcal{G}$ and $\mathcal{H}$ is at most $8(n + m)$. Raz and Safra [14] proved that the Dominating Set Problem cannot be approximated within a factor of $c_1 \log(n + m)$ from some constant $c_1 > 0$. Let $c = c_1/4$. If there is an algorithm that can approximate the exemplar conserved interval distance problem within a factor of $c_1 \log(|\mathcal{G}| + |\mathcal{H}|)$, where

$|\mathcal{G}|$ (resp. $|\mathcal{H}|$) denotes size of $\mathcal{G}$ (resp. $\mathcal{H}$), i.e., the number of genes in it. Then, this algorithm can be used to solve the Dominating Set Problem: the returned exemplar conserved interval distance for $opt(\mathcal{G}, \mathcal{H})$ is also for $opt(T)$. Let $app(T)$, which is $app(\mathcal{G}, \mathcal{H})$, denote the result returned by the algorithm. Then, we have

$$app(T) = app(\mathcal{G}, \mathcal{H}) \le c\log(8(n+m))opt(\mathcal{G}, \mathcal{H}) = c\log(8(n+m))opt(T)$$
$$\le 4c\log|T|opt(T) = c_1\log|T|opt(T)$$

Hence, $opt(T)$ can be approximated within a factor of $c_1 \log|T|$, a contradiction to the result obtained by Raz and Safra [14]. Therefore, the exemplar conserved interval distance problem cannot be approximated with a factor of $c\log(|\mathcal{G}|+|\mathcal{H}|)$ for a constant $c > 0$. □

4   The Zero Exemplar Conserved Interval Distance Problem

Recently, Chen, Fu and Zhu proved in [6] that the zero exemplar breakpoint distance problem is NP-complete. Following the spirit of [6], in this section we shall consider the *zero exemplar conserved interval distance problem*, i.e., the problem of deciding whether the exemplar conserved interval distance between any two given sets of genomes $\mathcal{G}$ and $\mathcal{H}$ is zero or not. We shall show that this problem, like the zero exemplar breakpoint distance problem, is also NP-complete.

Lemma 1. *Let G and H be two genomes such that each has n trivial genes and the set of genes in G is the same as the set of genes in H. (In other word, G is a signed permutation of H.) Then, the conserved interval distance between G and H is zero, i.e., $d(G, H) = 0$, if and only if either $G = H$ or G is the signed reversal of H.*

Proof. It follows from the given condition that $d(G, H) = 2\binom{n}{2} - 2N_{G \cup H}$. If $G = H$ or G is a signed reversal of H, then every two genes in G form a conserved interval in G and H. Thus, $N_{G \cup H} = \binom{n}{2}$. This implies $d(G, H) = 0$.

Now, suppose $d(G, H) = 0$. Then, we have $N_{G \cup H} = \binom{n}{2}$, i.e., every two genes in G form a conserved interval in G and H. We can prove by induction on n that either $G = H$ or G is the singed reversal of H. The details are omitted due to space limit. □

Theorem 2. *Given any two genomes G and H which are both 3-repetitive, it is NP-complete to decide whether the exemplar conserved interval distance between G and H is zero or not.*

Proof. It is easy to see that this ZECID problem is in NP. To prove its NP-hardness, we will construct a reduction from the 3SAT problem to the ZECID problem, following the reduction for proving the NP-hardness for the zero breakpoint distance problem in [6].

Let $F = f_1 \bigwedge f_2 \bigwedge \cdots \bigwedge f_q$ be a conjunctive normal form, where each f_i is a 3-disjunctive clause like $(x_1 \bigvee x_4 \bigvee \neg x_7)$. We construct two genomes G and H such that F is satisfiable iff G and H have zero exemplar conserved interval distance.

We consider $f_i, 1 \le i \le q$, as names of genes. Assume that F has n boolean variables $x_i, 1 \le i \le n$. Let $G = S_1 g_1 S_2 g_2 \cdots g_{n-1} S_n$ and $H = S_1^* g_1 S_2^* g_2 \cdots g_{n-1} S_n^*$, where $g_1, \cdots, g_{n-1}$ are peg genes that occur only once in G or H. For $1 \le i \le n$, $S_i = f_{i_1} \cdots f_{i_u} f_{j_1} \cdots f_{j_v}$ and $S_i^* = f_{j_1} \cdots f_{j_v} f_{i_1} \cdots f_{i_u}$, where $f_{i_1}, \cdots, f_{i_u}$ are the clauses containing x_i, and $f_{j_1}, \cdots, f_{j_v}$ are the clauses containing $\neg x_i$. Since each clause has at most 3 literals, S and H are 3-repetitive.

Following the approach in [6], if F is satisfiable, then we have an exemplar genomes G' and H' such that $G' = H'$. Hence, by Lemma 1 we have $d(G', H') = 0$. If there are two exemplar genomes G'' and H'' such that $d(G'', H'') = 0$, then by Lemma 1 we have $G'' = H''$, because G'' and H'' contain all unsigned genes in the set $\{f_1, \cdots, f_q, g_1, \cdots, g_{n-1}\}$ and no genes are repetitive. If S_i becomes empty in G'' then we can assign a value to x_i arbitrarily. Otherwise, we assign $x_i = 1$ if it becomes a subsequence of $f_{i_1} \cdots f_{i_u}$ in G'', or we assign $x_i = 0$ if it becomes a subsequence of $f_{j_1} \cdots f_{j_v}$. It is easy to verify that such a truth assignment will make F true.   □

Example. $F = (x_1 \vee \neg x_2 \vee x_4) \bigwedge (\neg x_1 \vee x_3 \vee x_4) \bigwedge (x_2 \vee x_3 \vee \neg x_4) \bigwedge (\neg x_1 \vee \neg x_2 \vee \neg x_3)$, where $F_1 = (x_1 \vee \neg x_2 \vee x_4)$, $F_2 = (\neg x_1 \vee x_3 \vee x_4)$, $F_3 = (x_2 \vee x_3 \vee \neg x_4)$, and $F_4 = (\neg x_1 \vee \neg x_2 \vee \neg x_3)$.
$G = F_1 F_2 F_4 g_1 F_3 F_1 F_4 g_2 F_2 F_3 F_4 g_3 F_1 F_2 F_3$ and
$H = F_2 F_4 F_1 g_1 F_1 F_4 F_3 g_2 F_4 F_2 F_3 g_3 F_3 F_1 F_2$.
$d(G'', H'') = 0$, with $G'' = H'' = F_4 g_1 F_3 g_2 g_3 F_1 F_2$, corresponds to the truth assignment that $x_1 = \text{False}(0)$, $x_3 = \text{False}(0)$ or $\text{True}(1)$, and $x_2 = x_4 = \text{True}(1)$.

Corollary 1. *Given any two sets of genomes $\mathcal{G}$ and $\mathcal{H}$, it is NP-complete to decide whether the exemplar conserved interval distance between $\mathcal{G}$ and $\mathcal{H}$ is zero or not.*

Theorem 2 and the above corollary imply that the ECID problem does not admit any approximation unless P=NP—if such a polynomial-time approximation existed then it would be able to decide whether G and H have zero exemplar conserved interval distance in polynomial time hence contradicting Theorem 2.

5   Weak Inapproximability Bound

Let $opt(\mathcal{G}, \mathcal{H})$ be the optimal exemplar conserved interval distance between $\mathcal{G}$ and $\mathcal{H}$. We also use $d(X, Y)$ to denote the minimum conserved interval distance between two genomes X and Y, where X and Y do not have to be exemplar. We also adopt a similar approach as in [6] but with some more involved analysis. We obtain the following inapproximability bounds under a much weaker model of approximation. Notice that the m factor in the bounds here are stronger than the $m^{1-\epsilon}$ factor in the bounds in [6] for exemplar break point distance problem.

Theorem 3. *Let $g(x) : N \to N$ be a function computable in polynomial time. If there is a polynomial time algorithm such that given two genomes $\mathcal{G}$ and $\mathcal{H}$ of length at most m it can return exemplar genomes G and H satisfying $d(G, H) \le g(m)opt(\mathcal{G}, \mathcal{H}) + m$, then P=NP.*

Proof. Let f be a given CNF formula. Let $G(f), H(f)$ be the genomes as constructed in Theorem 2 such that f is satisfiable if and only if $d(G(f), H(f)) = 0$. Let $|G(f)| = |H(f)| = u$, i.e., the number of all the genes occurred in $G(f)$ (or $H(f)$). Let $\Sigma(S)$ be the alphabet of a sequence S. If Σ_i is a different set of letters with $|\Sigma_i| = |\Sigma(S)|$, we define $S(\Sigma_i)$ to be a new sequence obtained by replacing all letters in S, in one to one fashion, by those in Σ_i.

For $M \geq 1$, Let $\Sigma_1, \Sigma_2, \cdots, \Sigma_M$ be M disjoint sets of letters of size $|\Sigma(G(f))|$. Let $G_1 = G(f)(\Sigma_1), G_2 = G(f)(\Sigma_2), \cdots, G_M = G(f)(\Sigma_M)$ be the sequences derived from $G(f)$. Let $H_1 = H(f)(\Sigma_1), H_2 = H(f)(\Sigma_2), \cdots, H_M = H(f)(\Sigma_M)$ be the sequences derived from $H(f)$.

Define $\mathcal{G} = G_1 s_1 G_2 s_2 \cdots G_M s_M$ and $\mathcal{H} = H_1 s_1 H_2 s_2 \cdots H_M s_M$, where s_i is a peg gene appearing only once in $\mathcal{G}$ and $\mathcal{H}$, respectively. Let $m = |\mathcal{G}| = |\mathcal{H}|$. In fact, m is the number of all the genes in $\mathcal{G}$ (or $\mathcal{H}$).

Assume that some polynomial time algorithm $\mathcal{A}$ outputs respectively two exemplar genomes G and H of $\mathcal{G}$ and $\mathcal{H}$, and $d(G, H) \leq g(m)d(\mathcal{G}, \mathcal{H}) + m$, we can then decide if f is satisfiable by checking whether $d(G, H) \leq m$. If f is satisfiable, as in the proof of Theorem 2, two identical exemplar genomes can be obtained from $\mathcal{G}$ and $\mathcal{H}$. Hence, we have $d(\mathcal{G}, \mathcal{H}) = 0$ by Lemma 1. This implies that $d(G, H) \leq m$. If f is not satisfiable, then from Theorem 2, $d(G_i, H_i) \geq 1$; namely, there is at least one conserved interval in G_i but not in H_i. This implies one of the following is true: (1) $a \cdots b$ in G_i but $b \cdots a$ in H_i; (2) $c \in [a, b]$ in G_i but $c \notin [a, b]$ in H_i; and (3) $c \notin [a, b]$ in G_i but $c \in [a, b]$ in H_i. For case (1), for any d in $G_j s_j$, $j \neq i$, either $[a, d]$ or $[d, a]$ is a conserved interval in G_j but not in H_j. Similarly, for any e in $H_j s_j$, $j \neq i$, either $[a, e]$ or $[e, a]$ is a conserved interval in H_j but not in G_j. Thus, in this case, we have at least $(u + 1)(M - 1)$ conserved interval in either $\mathcal{G}$ or $\mathcal{H}$ but not in both. Hence, we have $d(\mathcal{G}, \mathcal{H}) \geq 2(u + 1)(M - 1)$. It follows from some similar analysis that $d(\mathcal{G}, \mathcal{H}) \geq 2(u + 1)(M - 1)$ is true for the other two cases. Therefore, in either of the three cases, when $M \geq 2$, we have $d(\mathcal{G}, \mathcal{H}) \geq 2(u+1)(M-1) > (u+1)M = m$. Since G, H are exemplar genomes of $\mathcal{G}$ and $\mathcal{H}$, we have $d(G, H) > m$.　　□

Corollary 2. *If there is a polynomial time algorithm such that given $\mathcal{G}$ and $\mathcal{H}$ of length at most m it can return exemplar genomes G and H satisfying $d(G, H) \leq m[opt(\mathcal{G}, \mathcal{H}) + 1]$, then P=NP.*

This negative result shows that even under a much weaker model, unless P=NP, it is not possible to obtain a good approximation to the optimal exemplar conserved interval distance problem.

6   Concluding Remarks

We prove several lower bounds on the approximation of the Exemplar Conserved Interval Distance problem. Although it seems that the general problem does not admit any approximation, good approximation may exist for special cases of genomes, and good heuristics may perform well empirically or on average. It would be interesting to study some meaningful special cases. For example, in real-world datasets repetitions of genes are typically pegged and not very far away [12]. Are these cases easier to solve/approximate?

References

1. V. Bafna and P. Pevzner, Sorting by reversals: Genome rearrangements in plant organelles and evolutionary history of X chromosome, *Mol. Bio. Evol.*, 12:239-246, 1995.
2. S. Bereg and B. Zhu. RNA multiple structural alignment with longest common subsequences. *Proc. 11th Intl. Ann. Comput. and Combinatorics (COCOON'05)*, LNCS 3595, pp. 32-41, 2005.
3. A. Bergeron and J. Stoye. On the similarity of sets of permutations and its applications to genome comparison. *Proc. 9th Intl. Ann. Comput. and Combinatorics (COCOON'03)*, LNCS 2697, pp. 68-79, 2003.
4. G. Blin and R. Rizzi. Conserved interval distance computation between non-trivial genomes. *Proc. 11th Intl. Ann. Comput. and Combinatorics (COCOON'05)*, LNCS 3595, pp. 22-31, 2005.
5. D. Bryant. The complexity of calculating exemplar distances. *In D. Sankoff and J. Nadeau, editors, Comparative Genomics: Empirical and Analytical Approaches to Gene Order Dynamics, Map Alignment, and the Evolution of Gene Families*, pp. 207-212. Kluwer Acad. Pub., 2000.
6. Z. Chen, B. Fu and B. Zhu, The approximability of the exemplar breakpoint distance problem, Proceedings of the Second Intl. Conf. Algorithmic Aspects in Information and Management (AAIM'06), LNCS 4041, pp. 291-302. Springer, 2006.
7. O. Gascuel, editor. *Mathematics of Evolution and Phylogeny.*Oxford University Press, 2004.
8. S. Hannenhalli and P. Pevzner. Transforming cabbage into turnip: polynomial algorithm for sorting signed permutations by reversals. *J. ACM,* 46(1):1-27, 1999.
9. C. Makaroff and J. Palmer. Mitochondrial DNA rearrangements and transcriptional alternatives in the male sterile cytoplasm of Ogura radish. *Mol. Cell. Biol.,* 8:1474-1480, 1988.
10. M. Marron, K. Swenson and B. Moret. Genomic distances under deletions and insertions. *Theoretical Computer Science,* **325**(3):347-360, 2004.
11. C.T. Nguyen, Algorithms for calculating exemplar distances, Honors Thesis, School of Computing, National University of Singapore, 2005.
12. C.T. Nguyen, Y.C. Tay and L. Zhang. Divide-and-conquer approach for the exemplar breakpoint distance. *Bioinformatics,* **21**(10):2171-2176, 2005.
13. J. Palmer and L. Herbon. Plant mitochondrial DNA evolves rapidly in structure, but slowly in sequence. *J. Mol. Evolut.,* **27**:87-97, 1988.
14. R. Raz and S. Safra. A sub-constant error-probability low-degree test, and sub-constant error-probability PCP characterization of NP. In *Proc. 29th ACM Symp. on Theory Comput. (STOC'97)*, pages 475-484, 1997.
15. D. Sankoff. Genome rearrangement with gene families. *Bioinformatics,* **16**(11):909-917, 1999.
16. A. Sturtevant and T. Dobzhansky. Inversions in the third chromosome of wild races of *drosophila pseudoobscura*, and their use in the study of the history of the species. *Proc. Nat. Acad. Sci. USA,* 22:448-450, 1936.
17. G. Watterson, W. Ewens, T. Hall and A. Morgan. The chromosome inversion problem. *J. Theoretical Biology,* **99**:1-7, 1982.

Computing Maximum-Scoring Segments in Almost Linear Time

Fredrik Bengtsson and Jingsen Chen

Department of Computer Science and Electrical Engineering
Luleå University of Technology
S-971 87 Luleå
Sweden

Abstract. Given a sequence, the problem studied in this paper is to find
a set of k disjoint continuous subsequences such that the total sum of all
elements in the set is maximized. This problem arises naturally in the
analysis of DNA sequences. The previous best known algorithm requires
$\Theta(n \log n)$ time in the worst case. For a given sequence of length n, we
present an almost linear-time algorithm for this problem. Our algorithm
uses a disjoint-set data structure and requires $O(n\alpha(n, n))$ time in the
worst case, where $\alpha(n, n)$ is the inverse Ackermann function.

1 Introduction

In the analysis of biomolecular sequences, one is often interested in finding biologically meaningful segments, e.g. GC-rich regions, non-coding RNA genes,
transmembrane segments, and so on [1,2,3]. Fundamental algorithmic problems
arising from such sequence analysis are to locate consecutive subsequences with
high score (given a suitable scoring function on the subsequences). In this paper,
we present an almost linear time algorithm that takes a sequence of length n
together with an integer k and computes a set of k non-overlapping segments
of the sequence that maximizes the total score. If a general segment scoring
function (other than the sum of the segments as in this paper) is used, then the
problem of finding a k-cover with maximum score can be solved in $O(n^2 k)$ time
[4,5]. When the score of a segment is the sum of the elements in the segment,
the previous best known algorithm runs in $\Theta(n \log n)$ time [1]. Our new algorithm requires $O(n\alpha(n, n))$ time in the worst case, where $\alpha(n, n)$ is the inverse
Ackermann function.

The problem studied can be viewed as a generalization of the classical *maximum sum subsequence problem* introduced by Bentley [6]. The latter is to find
the continuous segment with largest sum of a given sequence and can be solved in
linear time [7,8]. Several other generalizations of this classical problem have been
investigated as well. For example when one is interested in not only the largest,
but also k continuous largest subsequences (for some parameter k) [9,10,11,12].
Other generalizations arising from bioinformatics is to look for an interesting
segment (or segments) with constrained length [13,14,15].

D.Z. Chen and D.T. Lee (Eds.): COCOON 2006, LNCS 4112, pp. 255–264, 2006.
© Springer-Verlag Berlin Heidelberg 2006

The paper is organized as follows. In Section 2 the problem is defined. Section 3 gives an overview of our algorithm while the details of the algorithm is presented in Section 4. The analysis of our algorithm appears in Section 5 and the paper is concluded with some open problems in Section 6.

2   Problem and Notations

Given a sequence $X = \langle x_1, x_2, \ldots, x_n \rangle$ of real numbers, let $X_{i,j}$ denote the subsequence of consecutive elements of X starting at index i and ending at index j; i.e. the *segment* $X_{i,j} = \langle x_i, x_{i+1}, \ldots, x_j \rangle$. A segment $X_{i,j}$ is *positive* (*negative*) if its *score* (*sum* or *value*) $\sum_{\ell=i}^{j} x_\ell > 0 \ (< 0)$. Call $X_{i,j}$ a *non-negative run* (*negative run*) of X if

- $x_\ell \geq 0$ (or < 0) for all $i \leq \ell \leq j$;
- $x_{i-1} < 0$ (or ≥ 0) if $i > 1$; and
- $x_{j+1} < 0$ (or ≥ 0) if $j < n$.

Given an integer $1 \leq k \leq n$, a *k-cover* for the sequence X is a set of k disjoint non-empty segments of X. The *score* (*sum* or *value*) of a k-cover is determined by adding up the sums of each of its segments.

Definition 1. *An* optimal *k*-cover *for a given sequence* X *is a k-cover of* X *whose score is the maximum over all possible k-covers of* X.

3   The Algorithm

By generalizing the recursive relation between maximum-scoring segments [1], we are able to design an algorithm that computes an optimal k-cover in almost linear time; the complexity of our algorithm is independent of the parameter k. The main idea is to transform the input sequence into a series of sequences of decreasing lengths under cover-preserving. We employ the union-find algorithmic technique to achieve the efficiency.

Our algorithm consists of three phases: Preprocessing, Partitioning, and Concatenating. The preprocessing phase deals with some trivial cases of the problem and simplifies the input for the rest of the algorithm. The latter two phases choose subsequences of which candidate segments to the optimal k-cover are examined and will be executed in an iterative fashion.

A special class of sequences plays an important role in our algorithm design, namely alternating sequences.

Definition 2. *A sequence* $Y = \langle y_1, y_2, \ldots, y_m \rangle$ *of real numbers is an* alternating sequence *if* m *is odd,* $y_1, y_3, \ldots,$ *and* y_m *are positive, and* $y_2, y_4, \ldots,$ *and* y_{m-1} *are negative. An alternating sequence is an* a-sequence *if its elements are mutually distinct.*

The following fact implies that finding an optimal k-cover in some alternating sequence needs only constant time.

Observation 1. *All the positive elements of an alternating sequence of length $2k - 1$ represent an optimal k-cover of the sequence.*

3.1   Preprocessing

This phase processes special cases of the problem and, if needed, finds the segmentation of the input into runs. The segmented input will be of alternating type and work as input for the two phases of our algorithm that follows. We treat the preprocessing step here separately from the other two steps of the algorithm, which simplifies the presentation of the latter. However, it does not make the problem to be solved simple.

The problem becomes trivial when $k \geq m$ (the number of positive runs of the input sequence). For the case when $k < m$, the following segmentation of the input is performed:

1. Find all the non-negative runs and negative runs of X.
2. Construct a new sequence Y from X by replacing every run with its score.
3. Negative elements at both ends of Y (if any) are removed from Y.

By storing the indices of all the runs of X, one can easily refer each element of Y to its corresponding segment of X. Clearly, the whole preprocessing phase takes at most $O(n)$ time in the worst case. Furthermore, any optimal k-cover of Y corresponds to an optimal k-cover of X, if $k < m$. In fact,

Proposition 1. *Given a sequence X of real numbers, if the number of non-negative runs of X is at least k, then there is an optimal k-cover of X such that any member of the optimal cover is either an entire non-negative run of X or a concatenation of neighboring runs of X.*

The sequence resulting from the segmentation is an *alternating sequence* $Y = \langle y_1, y_2, \ldots, y_m \rangle$, where $m \leq n$. Moreover, we can define a new order ($\prec$) on the elements of Y such that $y_i \prec y_j \Leftrightarrow \{y_i < y_j \vee y_i = y_j \wedge i < j\}$. However, throughout the paper we will use the standard notation to simplify presentation. Therefore, after $O(n)$ preprocessing time, we consider only alternating sequences of mutually distinct real numbers (that is, *a-sequences*) in the rest of our algorithm. The following property of such sequences characterizes optimal covers of the sequences.

Observation 2. *Let Y be an alternating sequence of length m and $k \leq \lceil m/2 \rceil$. Then, each member of an optimal k-cover of Y will always be a segment of Y of odd length.*

3.2   Partitioning and Concatenating

In this subsection, we give an overview of these two phases of our algorithm, whereafter a detailed presentation follows. Our approach in computing an optimal k-cover for a given a-sequence Y is to construct a series of a-sequences from Y while the lengths of the sequences are decreasing and the optimal k-cover remains the same.

For a given a-sequence Y of length m and an integer k, $1 \le k \le m$, we consider only the case when $k < \lceil m/2 \rceil$. Otherwise, the solution is straightforward. Let Y_t $(t = 1, 2, \cdots)$ be an a-sequence of length m_t associated with a working sequence S_t of length n_t, where $Y_1 = Y$ and $S_1 = Y$. The working sequence contains elements of Y that are currently interesting for the algorithm. In the following, each element of Y_t and S_t refers to the block of concatenated elements that the element currently corresponds to. The tth iteration of our algorithm (in particular, the partitioning and concatenating procedure) is as follows. Let ξ_0 be the largest absolute value of the elements in Y and let $r_1 = \lceil \|S_1\|/2 \rceil$.

Input: Y_t, S_t, a threshold r_t, and a pivot ξ_{t-1}, where $k < \lceil m_t/2 \rceil$
Output: Y_{t+1}, S_{t+1}, r_{t+1}, and ξ_t, where $m_{t+1} \le m_t$ and $n_{t+1} < n_t$.

1. Partition
   (a) Compute the $(r_t)^{th}$ largest absolute value ξ_t of all the elements in S_t.
   (b) Let D_t be the sequence containing all the elements of S_t whose absolute value is less than or equal to ξ_t. Preserve the ordering among the elements from Y_t; the indices of the elements are in increasing order.
2. Concatenation
   (a) Let Y_t' be the sequence resulting from, for each element y in D_t, repeatedly replacing some blocks in Y_t of odd lengths around y with their score until every element has an absolute value not less than ξ_t. Let k' be the number of positive elements in Y_t'.
   (b) If $k < k'$ (that is, we merged too few blocks in the previous step), then
       − $S_{t+1} \leftarrow \langle$ All the elements now in S_t whose absolute value lies between ξ_t and ξ_{t-1}; if some elements now belong to the same block, then just insert one of them into $S_{t+1} \rangle$
       − $r_{t+1} \leftarrow \lceil \|S_{t+1}\| /2 \rceil$
       − $Y_{t+1} \leftarrow Y_t'$.
   (c) If $k > k'$ (that is, we merged too many blocks in the previous step), then $S_{t+1} \leftarrow D_t$, $r_{t+1} \leftarrow \lceil \|S_{t+1}\|/2 \rceil$, and $Y_{t+1} \leftarrow Y_t$.
   (d) If $k = k'$, then we are done.

The goal is to eventually construct an a-sequence of length $2k - 1$ and hence the optimal k-cover is found due to Observation 1. With a careful implementation, the lengths of the a-sequences constructed will gradually decrease. In accomplishing our task within the desired time bound, we cannot afford to actually construct all the sequences Y_t $(t = 1, 2, \cdots)$. In such case, we may end up with an algorithm that takes $\Omega(m \log m)$ time in the worst case.

In the following, we show how to implement the algorithm efficiently. Actually, we never construct Y_t, but operate directly on Y all the time.

4   Algorithmic Details

Recall that the input now is an a-sequence $Y = \langle y_1, y_2, \ldots, y_m \rangle$ and an integer k, where $k < \lceil m/2 \rceil$. To implement the above algorithm efficiently, we will employ a disjoint-set data structure [16] augmented with extra information such as indices and scores of blocks. In addition to the standard disjoint-set data structure, we store the following extra fields at the leader node of each set:

- The index, in Y, of the leader
- The range, in Y, of the largest block created to which the leader belongs
- The score of the block

4.1   Union-Find

Initially, for $i = 1, 2, \ldots, m$, we perform $\texttt{MakeSet}(y_i)$ with $\{i, (i, i), y_i\}$ as extra information. Also, let $\texttt{FindSet}(y_i)$ return this extra information. For any two elements x and y in Y, the operation $\texttt{Union}(x, y)$ is performed as follows.
$\texttt{Union(x,y)}$

1. Let $(i, (i_1, i_2), s_x) = \texttt{FindSet}(x)$.
2. Let $(j, (j_1, j_2), s_y) = \texttt{FindSet}(y)$.
3. If $i = j$, then return; else let ℓ be the new leader index decided by the disjoint-set data structure (i.e., ℓ is either i or j).
4. The new extra information will be $(\ell, (\min\{i_1, j_1\}, \max\{i_2, j_2\}), s_x + s_y)$.

In the above procedure, if $j_1 - i_2 = 1$ (that is, the blocks joined are adjacent), the extra information maintained for each block will represent the intended extra information in the previous subsection. This is the case that is needed by our algorithm later on.

For the simplicity, for any y in Y, denote by $b(y)$ and $v(y)$ the index, in Y, of its leader and the score of its block, respectively; i.e., $(b(y), (j_1, j_2), v(y)) = \texttt{FindSet}(y)$. It is important to emphasize that any block created with $\texttt{Union}$s can be represented by an arbitrary element of Y from within the block. Any intermediate sequence used during the process contains only the original element from Y. Consider now the tth iteration of the phases for $t = 1, 2, \cdots$.

4.2   Partition

In this step, we partition the current input S_t according to the given threshold r_t. Notice that each element in S_t corresponds to some block of Y. Therefore, it is necessary to map from the elements stored in S_t to the blocks created by the previous concatenations. Thus, one $\texttt{FindSet}$ is done on each element of S_t.

Now, we can do the desired selection on all the absolute values obtained using a worst-case linear-time selection algorithm [16]. After that, a partition is performed around the pivot, ξ_t, and all the elements of S_t whose absolute value is less than or equal to ξ_t are included in a sequence D_t. Observe that for each comparison done on a pair of elements, one must do $\texttt{FindSet}$ operations first. The output of the partition step is (D_t, ξ_t). Thereafter, a series of replacements is applied to D_t in order to create a new a-sequence of smaller length.

4.3　Concatenation

In this subsection, we focus on the problem of constructing shorter a-sequences from a given a-sequence. One approach is to repeatedly concatenate blocks and replace them with their score. However, one cannot just choose an arbitrary block and then do the replacement, because this could potentially yield incorrect results. Among all possible replacements of blocks, one special kind of the replacements, the *merge*, will do the job.

The Merge Operation. A *merge* operation only applies to segments of length three. In particular, a *merge* can apply to a segment $\langle y_{i-1}, y_i, y_{i+1} \rangle$ only if $|y_i| < \min\{|y_{i-1}|, |y_{i+1}|\}$. The result of a merge on the segment $\langle y_{i-1}, y_i, y_{i+1} \rangle$, denoted by $merge(y_{i-1}, y_i, y_{i+1})$, is a new block with a value equal to $y_{i-1} + y_i + y_{i+1}$; realizable with $\texttt{Union}(\texttt{Union}(y_{i-1}, y_i), y_{i+1})$. Call such an operation a *merge around* y_i. Specially, a *merge around* y_1 implies $\texttt{Union}(y_1, y_2)$ if $|y_1| < |y_2|$ and a *merge around* y_m implies $\texttt{Union}(y_{m-1}, y_m)$ if $|y_{m-1}| < |y_m|$. Hence, the elements y_1 and y_m can be treated in the same way as other elements. In general, for any y in Y, let $(b(y), (i, j), s) = \texttt{FindSet}(y)$. A merge around y is then the merge operation on the segment $\langle y_{i-1}, y_{b(y)}, y_{j+1} \rangle$ (i.e., $\texttt{Union}(\texttt{Union}(y_{i-1}, y_{b(y)}), y_{j+1})$) if applicable. In this case, such a merge is also called a *merge around* y' for any y' in the block $Y_{i,j}$. Hence, we use the term element y to mean both the original element in Y and interchangeably the longest block created containing y. Clearly,

Proposition 2. *Let* y *be an element in the block resulting from* $merge(y_{i_1}, y_{i_2}, y_{i_3})$. *Then,* $|v(y)| \geq \max\{|v(y_{i_1})|, |v(y_{i_2})|, |v(y_{i_3})|\}$.

Proposition 3. *Let* M *be the set of merge operations performed on a given alternating sequence* $Y = \langle y_1, y_2, \ldots, y_m \rangle$ *and* Y_M *the sequence (called the* compact *version of* Y *under* M*) constructed from* Y *by replacing each merged block with a singleton element. If no merge is done around neither* y_1 *nor* y_m, *then* Y_M *is also an alternating sequence.*

Moreover, if Y is an a-sequence, so is Y_M. Hence, by repeatedly merging blocks one can obtain some compact version of Y, particularly a version with no smaller elements (that is, all its elements have absolute values greater than ξ_t).

Repeated Merges on D_t. Obviously, in order to ensure that there is no element y in Y with $|v(y)| \leq \xi_t$, at least all the elements in D_t must be involved in some merges. For each element in D_t and all the newly formed blocks (regarded as new elements of Y_t for some t), we need to decide whether a merge operation *will* be performed around it. For y in Y, define $Test(y) = true$ if a merge *can* be done around y according to the definition of the merge operation and $|v(y)| \leq \xi_t$; otherwise, $Test(y) = false$. Let $D_t = \langle d_1, d_2, \ldots, d_{n_t} \rangle$. Basically, we traverse D_t from d_1 to d_{n_t} and, for each element, determines if it should be merged. The current element is merged repeatedly until it is larger than the pivot. When it is, then its left neighbour is checked again to see if it should be merged again.

Notice that each merge operation creates a compact version of Y with the length decreased by 2. Therefore, the number of positive elements in the current version, Y_t of Y, (after all the merges done) can easily be counted when doing merges. This means that we do not need to actually construct the compact versions Y_{t+1} of Y at the moment. Only when we finally find an a-sequence of length $2k-1$, that compact version of Y is then computed; which costs in the worst case $O(m)$.

Furthermore, if a merge around y_1 (or y_m) was performed during the process, then a new block with negative score (which is either the prefix or suffix of Y) appears. The reason for the block being negative is that $y_1 < y_2$, otherwise the merge would not have occurred. Such a block can immediately be removed, because it is always unnecessary to include it in the solution to the current compact version Y_{t+1} of Y. Thus, if the block $\langle y_1 y_2 \rangle$ is selected for merge, we can effectively remove this block without doing any merge. The merge around y_m is analogous. Hence, Y_{t+1} is an a-sequence as well.

t^{th} **Iteration.** The goal of the tth iteration is to construct implicitly a new a-sequence Y_{t+1} of length m_{t+1} from an a-sequence Y_t of length m_t. From the construction, we know that $m_{t+1} \leq m_t$. The equality holds when there are too few positive elements in the compact version of Y after the merges. In this case, we cancel all the merges performed in this iteration. In order to be able to cancel the merge operations performed earlier, we record and store all changes made to the disjoint-set data structure. We need only store all the changes made in the current iteration. This is because that if there are not too few positive elements in the resulting sequence, we will never need to cancel the previous merge operations.

Observe that the iteration works on the working sequence S_t associated with Y_t, decides whether a merge should be performed around every element, and produces a new working sequence S_{t+1} associated with Y_{t+1}. Fortunately, the working sequences get shorter after every iteration. In fact, the lengths of such sequences decrease very fast, which implies that the number of iterations performed in our algorithm is not too many.

From Propositions 2 and 3, the lengths of a-sequences (i.e., the compact versions of Y) will eventually decrease to $2k-1$; say the last one is $Y_{t'}$. We will show in the next section that the optimal k-cover of Y_{t+1} corresponds to an optimal k-cover of Y_t, and thus is represented by all the positive elements in $Y_{t'}$.

5   Analysis

Roughly speaking, the following task, called $Concatenate(Y, \xi)$, is performed during the concatenation step of our algorithm when applying to a given alternating sequence Y and a positive real ξ:

Assume that we have already the partition set $D = \{y \in Y : |v(y)| \leq \xi\}$. The task is to check for each element $y \in D$ to determine whether a merge can be done around it. If a merge is done, then the element y (actually, its corresponding block) will have a new value $v(y)$ due to the merge. An element is deleted from D only if $|v(y)| > \xi$ after merge(s). This process continues until D becomes empty. Let Z be the sequence constructed from Y by replacing every merged block with a singleton element having the score of the block as its value; denoted by $Z = Concatenate(Y, \xi)$.

5.1   Correctness

Before establishing the correctness of our algorithm, we investigate the recursive behavior of the optimal k-cover for alternating sequences. An element (or a block) y in the input sequence Y is *included* in an optimal k-cover $\mathcal{C}$ for the input if y is either a member of $\mathcal{C}$ or a segment of some member of $\mathcal{C}$. The following fact can be obtained directly from a nice relationship between optimal covers presented by Csűrös [1].

Lemma 1. *If $k < \lceil m/2 \rceil$ and y_i is an element with the smallest absolute value in Y, then either the entire segment $\langle y_{i-1}, y_i, y_{i+1} \rangle$ is included in an optimal k-cover of Y or none of y_{i-1}, y_i, or y_{i+1} is.*

Since y_i has the minimum absolute value, one can always perform a merge operation around it; call such a merge a *min-merge* operation. The $\Theta(m \log m)$-time algorithm for finding an optimal k-cover proposed by Csűrös [1] performs only min-merge operations. Our algorithm goes further by using general merge operations, which leads to an almost-linear-time solution. Actually, the elements included in the optimal k -cover computed by our algorithm are exactly the same as those selected by Csűrös' algorithm for being included in some optimal k-cover.

For a given alternating sequence Y of length m and an integer k, $k < \lceil m/2 \rceil$, the procedure to construct an optimal k-cover of Y by repeated min-merge operations [1] actually implies the following recursive constructions. For any integer q, $k \leq q < \lceil m/2 \rceil$, an optimal q-cover of Y can be computed as follows: While the number of positive elements in the sequence is greater than q repeatedly join segments each of length 3 by performing min-merges. Call the sequence obtained Z_q and let M_q be the set of min-merge operations performed (ordered by the time at which the operation is executed). Then, an optimal k-cover of Y can be found with further min-merge operations starting from Z_q. Moreover, the optimal q-cover computed by such a procedure is exactly the same as the one resulted from our algorithm. More precisely,

Lemma 2. *Let ℓ be the number of positive elements of $Z = Concatenate(Y, \xi)$, where $\xi > 0$ is a given pivot. If $\ell < \lceil m/2 \rceil$ and an optimal ℓ-cover of Y is constructed from Y by repeated min-merge operations, resulting in the sequence Z_ℓ, then Z_ℓ is the same as the sequence $Z = Concatenate(Y, \xi)$.*

The proof of the above lemma will be included in the full version of this paper. Notice that $Z = Concatenate(Y, \xi)$ is an alternating sequence as well. Also, an alternating sequence with ℓ positive elements forms an optimal ℓ-cover of the sequence. Then, we have that

Corollary 1. *Let ℓ be the number of positive elements of $Z = Concatenate(Y, \xi)$, where $\xi > 0$ is a given real number. If $\ell < \lceil m/2 \rceil$, then the optimal ℓ-cover of Z corresponds to an optimal ℓ-cover of Y.*

Observe that our algorithm try to compute the number of positive elements, ℓ, of the sequence resulting from the concatenation step for a given pivot ξ. We want to ensure that $k \leq \ell < \lceil m/2 \rceil$ and try to decrease the value of ℓ by a recursive computation of the pivot value. The correctness of our algorithm thus follows.

Theorem 1. *Given a positive real number ξ. Let $1 \leq k < \lceil m/2 \rceil$ and $Z = Concatenate(Y, \xi)$. If $k \leq \ell$ (the number of positive elements of Z) and $\ell < \lceil m/2 \rceil$, then there is an optimal k-cover of Z that corresponds to an optimal k-cover of Y.*

5.2 Complexity

Given a sequence $X = \langle x_1, x_2, \ldots, x_n \rangle$ of real numbers, the preprocessing step of our algorithm takes $O(n)$ time. After that, the iterations have been run on the working sequences S_t for $t = 1, 2, \cdots$, where $S_1 = Y$ (the segmented version of X). The time (except for that consumed by the disjoint-set data structure) needed for the tth iteration of our algorithm (in particular, the partitioning and concatenating procedure) is $O(\|S_t\|)$. Notice from the design of our algorithm that $\|S_{t+1}\| \leq \frac{2}{3}\|S_t\|$ and $\|S_1\| = \|Y\| = m \leq n$ (the proof will appear in the full version of the paper). Hence, the time complexity of our algorithm (excluding cost for union-finds) satisfies the recurrence $T(n) = T\left(\frac{2}{3}n\right) + O(n)$ and thus equals $O(n)$.

Moreover, the number of union-find operations performed during the tth iteration is also $O(\|S_t\|)$. This implies that the total number of disjoint-set operations executed by our algorithm is $\sum_{t \geq 1} O(\|S_t\|) = O(\sum_{t \geq 1} \|S_t\|) = O(n)$. All these operations cost thus $O(n \cdot \alpha(n, n))$ in the worst case, where $\alpha(n, n)$ is the inverse Ackermann function. To sum up,

Theorem 2. *Given a sequence X of n real numbers and an integer $1 \leq k \leq n$, the problem of computing an optimal k-cover of the sequence can be done in $O(n \cdot \alpha(n, n))$ time in the worst case, where $\alpha(n, n)$ is the inverse Ackermann function.*

6 Conclusions

The problem of computing the maximum-scoring segments of a given sequence has been studied. We show how to solve the problem in $O(n\alpha(n, n))$ time in the worst case. Of course, a linear-time algorithm for this problem is desirable. Many algorithmic problems arising in the analysis of DNA sequences are of this flavor. Namely, one is interested in finding segments with constraints on the length of the segments and/or with different scoring functions. Both theoretical and practical efficient algorithms for these problems are interesting.

References

1. Csűrös, M.: Maximum-scoring segment sets. IEEE/ACM Transactions on Computational Biology and Bioinformatics **1**(4) (2004) 139–150
2. Fariselli, P., Finelli, M., Marchignoli, D., Martelli, P., Rossi, I., Casadio, R.: Maxsubseq: An algorithm for segment-length optimization. The case study of the transmembrane spanning segments. Bioinformatics **19** (2003) 500–505
3. Huang, X.: An algorithm for identifying regions of a DNA sequence that satisfy a content requirement. Computer Applications in the Biosciences **10** (1994) 219–225
4. Auger, I.E., Lawrence, C.E.: Algorithms for the optimal identification of segment neighbourhoods. Bulletin of Mathematical Biology **51**(1) (1989) 39–54
5. Bement, T.R., Waterman, M.S.: Locating maximum variance segments in sequential data. Mathematical Geology **9**(1) (1977) 55–61
6. Bentley, J.L.: Programming pearls: Algorithm design techniques. Communications of the ACM **27** (1984) 865–871
7. Bentley, J.L.: Programming pearls: Perspective on performance. Communications of the ACM **27** (1984) 1087–1092
8. Smith, D.: Applications of a strategy for designing divide-and-conquer algorithms. Science of Computer Programming **8** (1987) 213–229
9. Bae, S.E., Takaoka, T.: Algorithms for the problem of k maximum sums and a VLSI algorithm for the k maximum subarrays problem. In: Proceedings of the 7th International Symposium on Parallel Architectures, Algorithms and Networks. (2004) 247–253
10. Bae, S.E., Takaoka, T.: Improved algorithms for the k-maximum subarray problem for small k. In: Proceedings of the 11th Annual International Conference on Computing and Combinatorics. Volume 3595 of LNCS. (2005) 621–631
11. Bengtsson, F., Chen, J.: Efficient algorithms for k maximum sums. In: Proceedings of the 15th Annual International Symposium Algorithms and Computation. Volume 3341 of LNCS. (2004) 137–148 Revised version to appear in Algorithmica.
12. Lin, T.C., Lee, D.T.: Randomized algorithm for the sum selection problem. In: In Proceedings of the 16th Annual Internatinal Symposium on Algorithms and Computation. Volume 3827 of LNCS. (2005) 515–523
13. Bergkvist, A., Damaschke, P.: Fast algorithms for finding disjoint subsequences with extremal densities. In: Proceedings of the 16th Annual International Symposium on Algorithms and Computation. Volume 3827 of LNCS. (2005) 714–723
14. Chung, K.M., Lu, H.I.: An optimal algorithm for the maximum-density segment problem. In: Proceedings of 11th Annual European Symposium on Algorithms. Volume 2832 of LNCS. (2003) 136–147
15. Ruzzo, W.L., Tompa, M.: A linear time algorithm for finding all maximal scoring subsequences. In: Proceedings of the 7th Annual International Conference on Intelligent Systems for Molecular Biology. (1999) 234–241
16. Cormen, T.H., Leiserson, C.E., Rivest, R.L.: Introduction to Algorithms. The MIT Press (1990)

Enumerate and Expand: New Runtime Bounds for Vertex Cover Variants[*]

Daniel Mölle, Stefan Richter, and Peter Rossmanith

Dept. of Computer Science, RWTH Aachen University, Germany
{moelle, richter, rossmani}@cs.rwth-aachen.de

Abstract. The enumerate-and-expand paradigm for solving NP-hard problems has been introduced and applied to some VERTEX COVER variants in a recently published preliminary paper. In this paper we improve on the runtime for CONNECTED VERTEX COVER, obtaining a bound of $O^*(2.7606^k)$,[1] and use the technique in order to gain the fastest known method for counting the number of vertex covers in a graph, which takes $O^*(1.3803^k)$ time.

1 Introduction

In the recently published preliminary paper "Enumerate and Expand: Improved Algorithms for Connected Vertex Cover and Tree Cover" [12], a new paradigm for the exact solution of NP-hard problems has been introduced and applied to some VERTEX COVER variants. More specifically, significantly improved upper bounds on the parameterized complexity of CONNECTED VERTEX COVER and TREE COVER have been found. In the meantime, our first bound for CONNECTED VERTEX COVER, namely $O^*(3.2361^k)$, was improved on by two different algorithms: Fernau and Manlove designed an $O^*(2.9316^k)$ algorithm [8], and we refined the original algorithm — again using the enumerate-and-expand technique — obtaining a running time of $O^*(2.7606^k)$.

In this paper, we first describe the design and analysis of our new algorithm for CONNECTED VERTEX COVER. As another application of the enumerate-and-expand paradigm, we develop a new algorithm that counts the vertex covers of size k in a graph in $O^*(1.3803^k)$ time, beating the previously best bound of $O^*(1.4656^k)$. This old bound has been around for a couple of years, but remained unpublished until recently (see, e.g, [7]).

VERTEX COVER arguably constitutes the most intensely studied problem in parameterized complexity, which is reflected by a long history of improved runtime bounds [1,15,3,16,2] culminating in the algorithm by Chen, Kanj, and Xia with running time $O(1.2738^k + kn)$ [4]. This naturally led to the investigation of

---

[*] Supported by the DFG under grant RO 927/6-1 (TAPI).
[1] The O^*-notation is equivalent to the well-known Landau notation, except that polynomial factors may be suppressed. For instance, $2^k k^2 n^3 = O^*(2^k)$.

D.Z. Chen and D.T. Lee (Eds.): COCOON 2006, LNCS 4112, pp. 265–273, 2006.

generalizations such as CONNECTED VERTEX COVER, CAPACITATED VERTEX COVER, and MAXIMUM PARTIAL VERTEX COVER.

Many approaches to the solution of these problems rely on the enumeration of vertex covers. However, it is easy to establish a lower bound of 2^k on the number of different vertex covers of size at most k in a graph [6,12]. In order to accelerate such algorithms, we thus take a different approach: Instead of going through all the minimal vertex covers, we just enumerate subsets of vertex covers that can easily be expanded to complete vertex covers. We call this method *Enumerate and Expand*.

The rest of this section basically amounts to reviewing important parts of the enumerate-and-expand method and its early applications [12]. Recall that, given a graph $G = (V, E)$ and a natural number k, the problem VERTEX COVER asks for a set $C \subseteq V$, $|C| \leq k$, such that every edge is incident to at least one node from C. CONNECTED VERTEX COVER introduces the additional constraint that the subgraph $G[C]$ induced by C be connected.

Definition 1. *Let C be a graph class. A C-cover for a graph $G = (V, E)$ is a subset $C \subseteq V$ such that $G[V \setminus C] \in C$.*

That is, a C-cover does not have to cover all edges, but the uncovered part of the graph must be in C. For instance, C is a vertex cover if and only if C is an $\mathcal{I}$-cover, where $\mathcal{I}$ denotes the class of all graphs without edges.

Definition 2. *Let $G = (V, E)$ a graph, and $k \in \mathbf{N}$. A family of subsets of V is k-representative if it contains a subset of every vertex cover C of G with $|C| \leq k$.*

The concept of k-representative families is substantial for the enumerate-and-expand method: Assume we want to find all node sets $S \subseteq V$, $|S| \leq k$, such that S fulfills a certain property P in the input graph $G = (V, E)$. If each such S constitutes a vertex cover for the input graph, then it suffices to enumerate the members of a k-representative family and expand them accordingly (i.e., with respect to P). For suitable families and properties, it turns out that the expansion can be done rather efficiently because the enumerated node sets allow us to simplify the graphs to be processed.

Theorem 1. *Let $G = (V, E)$ be a graph, $k \in \mathbf{N}$, and C a graph class that contains all graphs with degree at most d and has a linear-time membership test. A k-representative family of C-covers for G can be enumerated in $O(\zeta_d^k k^2 + kn)$ time, where ζ_d is the unique positive root of the polynomial $z^{d+1} - z^d - 1$.*

This theorem can be proven using well-known kernelization techniques and a simple algorithm that branches on nodes of maximum degree [12].

Let $\mathcal{D}_d$ be the class of all graphs with maximum degree d and $\mathcal{M} := \mathcal{D}_1$. Clearly, a graph is in $\mathcal{M}$ if and only if its edges constitute a matching. In the aforementioned paper [12], CONNECTED VERTEX COVER was solved by first enumerating a k-representative family of $\mathcal{M}$-covers, and then expanding these into connected vertex covers. This was done with the help of the following lemma

Table 1. Approximate values of some ζ_d's

d	0	1	2	3	4	5	6
ζ_d	2	1.6181	1.4656	1.3803	1.3248	1.2852	1.2555

using a reduction to instances of a restricted Steiner tree problem. Here, $S(n, k)$ refers to the time needed to obtain a minimum Steiner tree for a k-terminal set that also constitutes a vertex cover in a network of n nodes.

Lemma 1. [12, Lem. 1] *The following problem can be solved in $S(n, k)$ time:*

> **Input:**   A graph $G = (V, E)$, an $\mathcal{M}$-cover C, a number $k \in \mathbf{N}$
> **Parameter:** k
> **Question:** Is there a connected vertex cover $\hat{C} \supseteq C$
>   of size at most k?

An upper bound on $S(n, k)$ is $O^*(2^k)$ where only small additional polynomial factors are involved [12], making the overall algorithm quite practical. The application of the specialized Steiner tree algorithm is crucial in achieving a low running time, because the commonly used generic Steiner tree algorithm by Dreyfus and Wagner [5] takes $O^*(3^k)$ steps. This running time has recently been improved to $O^*((2 + \varepsilon)^k)$ [11,13], but the hidden polynomials are prohibitively large for small ε.

Abstracting from the details of the reduction, the following runtime bound has been obtained for CONNECTED VERTEX COVER, where $\phi = (1 + \sqrt{5})/2$ denotes the golden ratio:

Corollary 1. [12, Cor. 3] *The decision problem* CONNECTED VERTEX COVER *can be solved in $O^*(\phi^k S(n, k)) = O^*((2\phi)^k)$ steps.*

This bound amounts to approximately 3.2361^k. In the upcoming section we will improve this running time to 2.7606^k.

2   Connected Vertex Cover Revisited

There are two key ideas involved in speeding up the computation of connected vertex covers. The first one is to enumerate $\mathcal{D}_3$-covers instead of $\mathcal{M}$-covers. While enumerating all k-representative $\mathcal{M}$-covers takes about 1.6181^k steps, only 1.3803^k steps are required in the case of $\mathcal{D}_3$-covers. This implies a much more complicated expansion phase, a price we have to pay. Whereas removing an $\mathcal{M}$-cover from a graph leaves isolated edges and nodes, an arbitrary cubic graph remains on deletion of a $\mathcal{D}_3$-cover.

Of course, we might be lucky and still get a very simple cubic graph. For these $\mathcal{D}_3$-covers, the complexity lies in the enumeration. Otherwise, for complicated cubic graphs, we continue to refine these covers into $\mathcal{M}$-covers. If this needs to be done, however, we can prove that these $\mathcal{M}$-covers induce subgraphs with only few components. It fortunately turns out that we may replace these components by single nodes, thus reducing the size of the connected vertex cover we are looking for. Amazingly enough, this particular effect outweighs the extra effort put into the enumeration of $\mathcal{M}$-covers.[2]

Table 2. Our new algorithm for CONNECTED VERTEX COVER

---

forall D_3-covers C_3 in G of size at most k **do**
    forall $\mathcal{M}$-covers C_1 in $G[V - C_3]$ of size at most $k - |C_3|$ **do**
        1. Let I and M be the set of isolated nodes and the matching (that is, the set of edges) in the remaining graph $G[V - C_3 - C_1]$, respectively. Let $\mathcal{A}$ be the set of components in $G[C_1]$.
        2. Construct a graph G' from G by copying all of $G[V - C_1 - C_3]$ and mapping every $A \in \mathcal{A}$ onto a single node v_A, where v_A and any node $v \in V - C_1 - C_3$ are connected by an edge in G' iff there is an edge between some node in A and v in G.
        3. Let $d = |V[G]| - |V[G']| = \sum_A (|V[A]| - 1)$.
        4. Employ the reduction to the Steiner tree problem from Lemma 1: In G', subdivide every edge from M, and mark all the v_A as well as all the subdivision nodes as terminals.
        5. If $|\mathcal{A}| \leq 3k - 2|C_3| - 3|C_1| - |M|$, then compute an optimum Steiner tree, and **return** *true* if it has at most $k - d$ nodes.
    od
od
return *false*

In the remainder of the section, we present the algorithm (see Table 2) in detail and prove its correctness as well as a runtime bound. Justification for the replacement of entire components by single placeholder nodes is brought forth by the following lemma.

Lemma 2. *Let C be an $\mathcal{M}$-cover for a graph $G = (V, E)$. There is an optimum Steiner tree T for any terminal set $Y \supseteq C$ on the unit-cost network over G such that each component of $G[C]$ remains connected in $T[C]$.*

Proof. Consider some optimum Steiner tree T' that does not meet the requirements. Then there is a component of $G[C]$ that constitutes $p \geq 2$ components $V_1, \ldots, V_p$ in $T'[C]$. Moreover, it is easy to see that two such components, say V_1 and V_2, are connected in $G[C]$ by some edge e. On the other hand, they are connected in T' via a path containing an edge f that is not in $G[C]$. Exchanging f for e changes neither cost nor connectivity. Obviously, this process can be iterated until the modified Steiner tree T meets the requirements. $\qquad\square$

---

[2] The technique can easily be modified so as to use $\mathcal{D}_2$- or $\mathcal{D}_4$-covers, but $\mathcal{D}_3$-covers yield the best runtime bounds.

Theorem 2. *The algorithm from Table 2 solves* CONNECTED VERTEX COVER.

Proof. Recall that the old $O^*(3.2361^k)$ algorithm enumerates all $\mathcal{M}$-covers C with $|C| \leq k$ and tries to expand them to connected vertex covers using the reduction to the Steiner tree problem from Lemma 1. There are three differences between the old and the new algorithm:

In the revised approach, $\mathcal{M}$-covers are enumerated by going through all feasible $\mathcal{D}_3$-covers and then through all feasible $\mathcal{M}$-covers on the remaining graph. It is easy to see that this yields exactly the same covers (and, on a side note, does not even affect the running time consumed for enumeration).

Secondly, the components $A \in \mathcal{A}$ are mapped onto single nodes, and thus a different instance of the Steiner tree problem is constructed. However, Lemma 2 implies that, if S denotes the set of nodes used to subdivide the edges in M, there is a Steiner tree with k nodes for the terminal set $C \cup S$ in G iff there is a Steiner tree with $k - d$ nodes for the terminal set $X := \{ v_A \mid A \in \mathcal{A} \} \cup S$.

Finally, the Steiner tree algorithm is only employed if $|\mathcal{A}| \leq 3k - 2|C_3| - 3|C_1| - |M|$. This makes sense because otherwise there cannot be a connected vertex cover of size k in G that contains $C_1 \cup C_3$. In order to establish this claim we show that there cannot be a connected vertex cover of size $k - d$ in G' that contains X.

Let $X_1 := \{ v_A \in X \mid V[A] \subseteq C_1 \}$ the set of nodes that stem from components in $G[C_1 \cup C_3]$ only containing nodes from C_1. A connected vertex cover C for G' that contains X must connect all the nodes from X_1, of which there are at least $|X| - |C_3|$. This can only be done by adding nodes from I and M. Observe that every node from I can be adjacent to at most three nodes from X_1, and every node from $V[M]$ can be adjacent to at most two nodes from X_1, in G'. Assume that C contains k_I nodes from I and k_M nodes from $V[M]$. Then, $|X| - |C_3| \leq |X_1| \leq 3k_I + 2k_M$. Moreover, for the corresponding connected vertex cover of size k in G we have that $k = k_I + k_M + |C_1| + |C_3|$, as well as $|M| \leq k_M$. From these three inequalities and $|\mathcal{A}| = |X|$, the claim $|\mathcal{A}| \leq 3k - 2|C_3| - 3|C_1| - |M|$ follows. $\square$

Theorem 3. *The running time of the algorithm from Table 2 is $O^*((2\zeta_3)^k)$.*

Proof. Let C_3 be a $\mathcal{D}_3$-cover and C_1 an $\mathcal{M}$-cover of $G[V - C_3]$. Then $C_3 \cup C_1$ is an $\mathcal{M}$-cover of G with corresponding matching M. Assume moreover that C is a connected vertex cover with $C_3 \cup C_1 \subseteq C$. If our algorithm tries to find C by using $C_3 \cup C_1$ and M, it first replaces $C_3 \cup C_1$ by contracting all components in $G[C_1 \cup C_3]$ into single nodes. Let us call the resulting set X and the modified graph G'. It then computes a Steiner tree for the terminal set consisting of X and $|M|$ additional nodes from the subdivision of M. The running time to compute this Steiner tree is $O^*(2^{|X|+|M|})$.

We proceed by bounding the size of X. Since C is a *connected* vertex cover, $G'[C]$ is connected. Again, assume that C contains k_I nodes from I and k_M nodes from $V[M]$. Remember that every node in $G'[V - C_3]$ has maximum degree three. If such a node is in C, it can connect to up to three nodes from X, reducing the number of components by up to three. If it is incident to an edge in M, then the

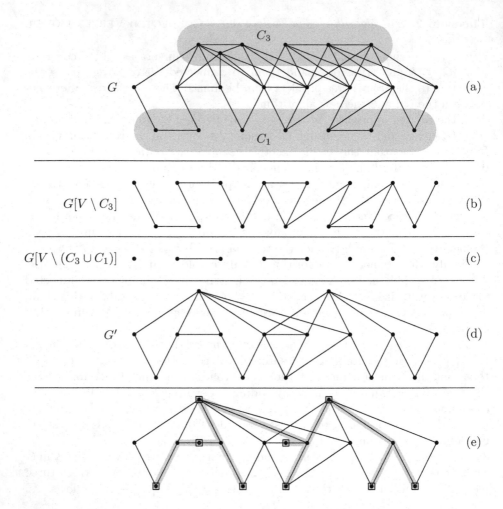

Fig. 1. (a) An example graph G with $\mathcal{D}_3$-cover C_3 and subsequent $\mathcal{M}$-cover $C_3 \cup C_1$. (b) Deleting C_3 leaves a cubic graph. (c) Deleting C_1 now leaves a graph from $\mathcal{M}$. (d) We are looking for a minimum connected vertex cover in G containing $C_3 \cup C_1$. The seven components of $G[C_3 \cup C_1]$ in G are mapped onto seven nodes in G' that constitute the set X. (e) A connected vertex cover that is a superset of X needs to connect the nodes in X and to cover the edges in M. This can be done optimally by a minimum Steiner tree for X and the subdivision nodes for M, as depicted by the emphasized edges. The resulting connected vertex covers have size eleven in G' and 17 in G.

number of components is reduced by up to two. Hence, $|C_1| \le 3k_I + 2k_M$ and $|X| \le |C_3| + 3k_I + 2k_M$.

The time to enumerate all $\mathcal{D}_3$-covers of size $|C_3|$ and then all $\mathcal{M}$-covers of size $|C_1|$ is $O^*(\varsigma_3^{|C_3|} \phi^{|C_1|})$. The running time of our algorithm is therefore bounded by

$$\sum_{|C_1|,|C_3|,|X|,k_M} O^*\big(\zeta_3^{|C_3|}\phi^{|C_1|}2^{|X|+k_M}\big).$$

In order to obtain a worst-case bound of the form $O^*(\alpha^k)$ for some real number α, we proceed by expressing the problem of maximizing the above term as a linear program, normalized by setting $k = 1$:

$$\text{maximize} \quad |C_3|\ln\zeta_3 + |C_1|\ln\phi + (|X| + k_M)\ln 2$$

$$\text{subject to} \quad |X| \le |C_3| + 3k_I + 2k_M$$
$$|X| \le |C_3| + |C_1|$$
$$k = k_I + k_M + |C_1| + |C_3|$$
$$k \ge |C_3| + |C_1| + |M|$$
$$k = 1$$
$$0 \le k_I, k_M, |C_1|, |C_3|, |M|$$

Leaving out the constraint $|M| \le k_M \le 2|M|$ does not make a difference. Solving the linear program reveals that the maximum is assumed for $|C_3| = k$, $|C_1| = 0$, $|X| = k$, $k_M = 0$. Consequently, the total running time of our algorithm is bounded by $O^*((2\zeta_3)^k) = O^*(2.7606^k)$. $\qquad\square$

3 Counting Vertex Covers

Another rather obvious application of the enumerate-and-expand method lies in counting the number of vertex covers. That is, enumerate a k-representative set of partial covers and find out into how many appropriate vertex covers they can each be expanded. We just need to make sure that the expansion sets do not overlap. However, partial covers are generated by recursively branching on a node v: So in one branch, only covers that contain v will be considered, while in the other branch no cover can contain v.

Depending on the kind of partial covers enumerated, different running times emerge: Simply enumerating all vertex covers will cost $O^*(2^k)$ time. In the case of $\mathcal{M}$-covers, the running time is $O^*(\phi^k)$, and $O^*(\zeta_2^k)$ with $\mathcal{D}_2$-covers, because counting the number of vertex covers on partial two-trees can be done in polynomial time [7,12,14]. Obviously, enumerating $\mathcal{D}_3$-covers is even faster. However, the remaining graph might be an arbitrary cubic graph. Unfortunately, VERTEX COVER is known to be NP-complete even on cubic graphs. Surprisingly, calculating the number of vertex covers in a cubic graph is, though exponential in k, still fast enough to beat the above approaches.

Theorem 4. *The number of vertex covers of size k in a graph can be calculated in $O^*(\zeta_3^k) = O^*(1.3803^k)$ steps.*

Proof. Let us first show that a cubic graph containing a vertex cover of size at most k has treewidth at most $(\frac{1}{3} + \varepsilon)k$ for any $\varepsilon > 0$ and that a corresponding tree decomposition can be computed in polynomial time.

Assume that G is a cubic graph. Without loss of generality we can assume furthermore that all nodes have degree three because contracting a degree-two node does not change any treewidth larger than two. Now if there is a vertex cover of size k, then there can be at most $3k$ edges and consequently no more than $2k$ nodes. Fomin and Høie [10] have shown that a tree decomposition of width at most $(\frac{1}{6}+\varepsilon)n$ can be found for any cubic graph with n nodes and arbitrary $\varepsilon > 0$ in polynomial time. Given a graph together with a tree decomposition of width w, standard dynamic programming techniques yield an $O^*(2^w)$ algorithm for deciding VERTEX COVER [17]. It is easy to adapt this algorithm for counting all vertex covers of a given size.

The overall algorithm proceeds as follows: Go through all the members C of a family of k-representative $\mathcal{D}_3$-covers, compute the number of vertex covers of size $k - |C|$ in $G[V \setminus C]$ in time $O^*(2^{(1/3+\varepsilon)(k-|C|)})$ for each C, and finally sum up all the results. The overall runtime can be bounded by

$$\sum_{|C|=0}^{k} O^*\big(\zeta_3^{|C|} 2^{(1/3+\varepsilon)(k-|C|)}\big) = O^*(\zeta_3^k)$$

if $\varepsilon \leq 0.13$, because then $2^{1/3+\varepsilon} < \zeta_3$. □

4   Concluding Remarks

The exact solution of NP-hard problems has lately been receiving increased attention, as outlined in the survey by Fomin, Grandoni, and Kratsch [9]. Against this backdrop, the enumerate-and-expand method is a new tool for the development of exact and parameterized algorithms. We have demonstrated its capabilities in two non-trivial applications, deriving the best known runtime bounds for solving CONNECTED VERTEX COVER and counting vertex covers of a given size.

References

1. R. Balasubramanian, M. R. Fellows, and V. Raman. An improved fixed parameter algorithm for vertex cover. *Information Processing Letters*, 65(3):163–168, 1998.
2. L. S. Chandran and F. Grandoni. Refined memorization for vertex cover. *Information Processing Letters*, 93:125–131, 2005.
3. J. Chen, I. A. Kanj, and W. Jia. Vertex cover: Further observations and further improvements. *Journal of Algorithms*, 41:280–301, 2001.
4. J. Chen, I. A. Kanj, and G. Xia. Simplicity is beauty: Improved upper bounds for vertex cover. Technical Report TR05-008, School of CTI, DePaul University, 2005.
5. S. E. Dreyfus and R. A. Wagner. The Steiner problem in graphs. *Networks*, 1:195–207, 1972.
6. H. Fernau. On parameterized enumeration. In *Proc. of 8th COCOON*, number 2387 in LNCS, pages 564–573. Springer, 2002.
7. H. Fernau. *Parameterized Algorithmics: A Graph-Theoretic Approach*. Habilitation thesis, Universität Tübingen, 2005.

8. H. Fernau and D. F. Manlove. Vertex and edge covers with clustering properties: Complexity and algorithms. Technical Report TR-2006-210, Dept of Computing Science, University of Glasgow, Apr. 2006.

9. F. V. Fomin, F. Grandoni, and D. Kratsch. Some new techniques in design and analysis of exact (exponential) algorithms. *EATCS Bulletin*, 87:47–77, 2005.

10. F. V. Fomin and K. Høie. Pathwidth of cubic graphs and exact algorithms. Technical Report 298, Department of Informatics, University of Bergen, May 2005.

11. B. Fuchs, W. Kern, D. Mölle, S. Richter, P. Rossmanith, and X. Wang. Dynamic programming for minimum steiner trees. *Theory of Computing Systems*, 2006. To appear.

12. D. Mölle, S. Richter, and P. Rossmanith. Enumerate and expand: Improved algorithms for connected vertex cover and tree cover. 2006. To appear in Proc. of 1st CSR.

13. D. Mölle, S. Richter, and P. Rossmanith. A faster algorithm for the Steiner tree problem. In *Proc. of 23rd STACS*, number 3884 in LNCS, pages 561–570. Springer, 2006.

14. R. Niedermeier. *Invitation to fixed-parameter algorithms*. Habilitation thesis, Universität Tübingen, 2002.

15. R. Niedermeier and P. Rossmanith. Upper bounds for Vertex Cover further improved. In *Proc. of 16th STACS*, number 1563 in LNCS, pages 561–570. Springer, 1999.

16. R. Niedermeier and P. Rossmanith. On efficient fixed parameter algorithms for Weighted Vertex Cover. *Journal of Algorithms*, 47:63–77, 2003.

17. J. A. Telle and A. Proskurowski. Algorithms for vertex partitioning problems on partial k-trees. *SIAM Journal on Discrete Mathematics*, 10(4):529–550, 1997.

A Detachment Algorithm
for Inferring a Graph from Path Frequency

Hiroshi Nagamochi

Department of Applied Mathematics and Physics,
Kyoto University,
Yoshida Honmachi, Sakyo, Kyoto 606-8501, Japan
nag@amp.i.kyoto-u.ac.jp

Abstract. Inferring a graph from path frequency has been studied as
an important problem which has a potential application to drug design.
Given a multiple set g of strings of labels with length at most K, the
problem asks to find a vertex-labeled graph G that attains a one-to-one
correspondence between g and the set of sequences of labels along all
paths of length at most K in G. In this paper, we prove that the problem
with $K = 1$ can be formulated as a problem of finding a loopless and
connected detachment, based on which an efficient algorithm for solving
the problem is derived. Our algorithm also solves the problem with an
additional constraint such that every vertex is required to have a specified
degree.

1 Introduction

Kernel methods have been popular tools for designing classifiers such as support vector machines. In kernel methods, a set of objects (or data) in the target
problem are mapped to a space, called a *feature space*, where an object is transformed into a vector with real coordinates, and a kernel function is defined as an
inner product of two feature vectors. Recently, a feature space has been used in a
new approach in order to design or choose a desired (possibly unknown) object
[3]. As in kernel methods, given objects mapped to points in a feature space,
this approach searches a point y in the feature space using a suitable objective
function, and then maps this point back to an object in the input space, where
the object mapped back is called a *pre-image* of the point. Given a mapping ϕ
from an input space to a feature space and a point y in the feature space, the
pre-image problem asks to find an object x with $y = \phi(x)$ in the input space. The
pre-image problem for graphs is very important because it has a potential application to drug design and elucidation of chemical structures from mass/NMR
spectra data, and has been studied by several researchers [3,9].

Akutsu and Fukagawa [1] started the theoretical aspect of the problem of
inferring graphs from *path frequency*. In this case, a feature vector g is a multiple
set of strings of labels with length at most K which represents path frequency
(i.e., the numbers of occurrences of vertex-labeled paths of length at most K).
Given a feature vector g, they considered the problem of finding a vertex-labeled

D.Z. Chen and D.T. Lee (Eds.): COCOON 2006, LNCS 4112, pp. 274–283, 2006.

graph G that attains a one-to-one correspondence between g and the set of sequences of labels along all paths of length at most K in G. For the problem of inferring a tree, they gave dynamic programming algorithms that runs in polynomial time in n when K and the number of labels are bounded by constants, where n denotes the size of an output graph. They also proved that the problem is strongly NP-hard even for planar graphs and $K = 4$ [2]. Afterwards, they extended their dynamic programming algorithms to the problem of inferring a graph in a restricted class of outerplanar graphs. However, the time complexity of these dynamic programming algorithms is a polynomial of n whose exponent is exponential in K and the number of labels. Furthermore, in an inferred graph, vertices with the same label may have different degrees.

In this paper, we consider the problem of inferring a multigraph from a feature vector g of path frequency with $K = 1$. We show that the problem can be formulated as a problem of finding loopless and connected *detachments* of graphs, and give an efficient algorithm based on matroid intersection in discrete optimization. Our algorithm tests whether there exists a solution to a given vector g or not in $O(\min\{|g|2^{|g|}, n^{3.5} + m\})$ time, and delivers a solution (if any) in $O(n^{3.5} + m)$ time, where $|g|$ is the number of nonzero entries in an input vector g and n and m are the numbers of vertices and edges of a multigraph to be constructed. In particular, testing the feasibility of a vector g can be executed in constant time if the number of labels is bounded by a constant. We next introduce a graph inference problem with an additional constraint such that every vertex is required to have a specified degree. We prove that this graph inference problem can also be solved in $O(\min\{n + |g|2^{|g|}, n^{3.5} + m\})$ time.

2   Graph Inference Problem

This section defines problems of inferring graphs from path frequency.

A graph is called a *multigraph* if it is allowed to have multiple edges and self-loops; otherwise it is called *simple*. A multigraph having no self-loops is called *loopless*. A multigraph G with a vertex set V and an edge set E is denoted by (V, E). The vertex set and edge set of a given multigraph G may be denoted by $V(G)$ and $E(G)$, respectively.

Let $\mathbf{Z}_+$ denote the set of nonnegative integers. Let Σ be a set of labels, Σ^k be the set of all sequences of k labels in Σ, and $\Sigma^{\leq k} = \cup_{1 \leq j \leq k}\Sigma^j$. Let $\mathcal{F}_k(\Sigma)$ denote the set of all vectors g whose coordinate indexed by $t \in \Sigma^{\leq k+1}$ is an nonnegative integer (i.e., g is a mapping from $\Sigma^{\leq k+1}$ to $\mathbf{Z}_+$). A vector $g \in \mathcal{F}_k(\Sigma)$ may be called a *feature vector*. Let $g(t)$ denote the entry of $g \in \mathcal{F}_k(\Sigma)$ indexed by $t \in \Sigma^{\leq k+1}$.

A multigraph H is called Σ-*labeled* if each vertex $v \in V(H)$ is labeled by a label $\ell(v) \in \Sigma$. Let H be a loopless Σ-labeled multigraph. For a walk $\pi = (v_0, e_1, v_1, e_2, v_2, \ldots, e_h, v_h)$ in H, let $\ell(\pi)$ denote the sequence of the vertex labels in π, i.e., $\ell(\pi) = \ell(v_0)\ell(v_1)\ldots\ell(v_h)$. For a label sequence t over Σ, let $occ(t, H)$ denotes the number of walks π such that $\ell(\pi) = t$. The *feature vector* $f_K(H)$ *of level* K in H is a vector $g \in \mathcal{F}_K(\Sigma)$ such that $g(t) = occ(t, H)$ for all

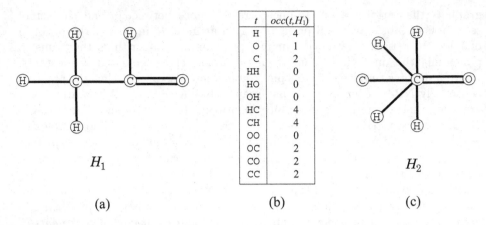

t	$occ(t,H_1)$
H	4
O	1
C	2
HH	0
HO	0
OH	0
HC	4
CH	4
OO	0
OC	2
CO	2
CC	2

H_1 H_2

(a)   (b)   (c)

Fig. 1. (a) A (Σ, ρ)-labeled multigraph H_1, where $\Sigma = \{\mathtt{H}, \mathtt{O}, \mathtt{C}\}$, $\rho(x) = 1$ if $\ell(x) = \mathtt{H}$, $\rho(x) = 2$ if $\ell(x) = \mathtt{O}$, and $\rho(x) = 4$ if $\ell(x) = \mathtt{C}$, respectively; (b) $occ(t, H_1) = occ(t, H_2)$ for all sequences $t \in \Sigma^{\leq 2}$; (c) A Σ-labeled multigraph H_2

$t \in \Sigma^{\leq K+1}$, i.e.,

$$f_K(H) = (occ(t, H))_{t \in \Sigma^{\leq K+1}}.$$

For example, Figure 1(a) shows a loopless Σ-labeled multigraph H_1, where $\Sigma = \{\mathtt{H}, \mathtt{O}, \mathtt{C}\}$, Figure 1(b) gives $occ(t, H_1)$ for all $t \in \Sigma^{\leq 2}$, and we have $f_1(H_1) = (4, 1, 2, 0, 0, 0, 4, 4, 0, 2, 2, 2)$. Figure 1(c) shows a different loopless Σ-labeled multigraph H_2 such that $f_1(H_2) = f_1(H_1)$.

For a given feature vector $g \in \mathcal{F}_K(\Sigma)$, there may be no Σ-labeled multigraph H with $f_K(H) = g$. Different Σ-labeled graphs H and H' may have the same feature vector $f_K(H) = f_K(H') = g$, as observed in Fig. 1.

Akutsu and Fukagawa [1] formulated the following important problem.

Graph Inference from Path Frequency (GIPF). Given a feature vector $g \in \mathcal{F}_K(\Sigma)$, output a loopless and connected Σ-labeled multigraph H with $f_K(H) = g$. If there does not exist such H, then output "no solution."

An inferred graph needs to meet a degree constraint in some applications. In this paper, we define a degree-constrained graph inference problem as follows. A *valence-sequence* ρ is a function $\rho : V(H) \rightarrow \mathbf{Z}_+$. A Σ-labeled multigraph H is called (Σ, ρ)-*labeled* if $deg(x; H) = \rho(x)$ for each $x \in V(H)$.

Figure 1(a) shows a (Σ, ρ)-labeled multigraph H_1 for valence-sequence ρ such that $\rho(x) = 1$ if $\ell(x) = \mathtt{H}$, $\rho(x) = 2$ if $\ell(x) = \mathtt{O}$, and $\rho(x) = 4$ if $\ell(x) = \mathtt{C}$, respectively. Note that multigraph H_2 in Fig. 1(c) is not (Σ, ρ)-labeled.

Graph Inference from Path Frequency and Label Valence (GIFV). Given a feature vector $g \in \mathcal{F}_K(\Sigma)$ and a valence-sequence ρ, output a loopless and connected (Σ, ρ)-labeled multigraph H with $f_K(H) = g$. If there does not exist such H, then output "no solution."

3   Detachments in Multigraphs

3.1   Multigraphs and Matroids

A singleton set $\{x\}$ may be simply written as x. Let $G = (V, E)$ be a multigraph which may have self-loops. For two subsets $X, Y \subset V$ (not necessarily disjoint), $E(X, Y; G)$ denotes the set of edges e joining a vertex in X and a vertex in Y (i.e., $e = \{u, v\}$ satisfies $u \in X$ and $v \in Y$), and $d(X, Y; G)$ denotes $|E(X, Y; G)|$. Note that $E(X, Y; G)$ includes all self-loops $\{u, u\}$ with $u \in X \cap Y$ if any. We may write $E(X, V - X; G)$ and $d(X, V - X; G)$ as $E(X; G)$ and $d(X; G)$, respectively. Note that $d(u, v; G)$ is the number of multiple edges with end vertices u and v in G. The *degree* of a vertex v is defined to be $deg(v; G) = d(v; G) + 2d(v, v; G)$. For a multigraph $G = (V, E)$ and a subset $X \subseteq E$ (resp., $X \subseteq V$), let $G - X$ denotes the multigraph obtained by removing the edges in X (resp., the vertices in X together with incident edges) from G.

Let $c(G)$ denote the number of components in a multigraph G. Removing k edges from G increases the number of components at most by k. Hence we have:

Lemma 1. *For a multigraph* $G = (V, E)$ *and a subset* $E' \subseteq E$, $c(G - E') \leq c(G) + |E'|$. $\qquad\square$

We here review the definition and some important property of matroids (see [10] for more on matroid theory). For a finite set S, let $\mathcal{I}$ be a family of subsets of S. System $(S, \mathcal{I})$ is called a *matroid* if it satisfies three conditions (i) $\emptyset \in \mathcal{I}$, (ii) If $I \in \mathcal{I}$, then any subset I' of I also belongs to $\mathcal{I}$, and (iii) For any $I_1, I_2 \in \mathcal{I}$ with $|I_1| < |I_2|$, there is an element $e \in I_2 - I_1$ such that $I_1 \cup \{e\} \in \mathcal{I}$. For a set $I \in \mathcal{I}$ of a matroid $\mathcal{M} = (S, \mathcal{I})$ and an element $e \in S - I$ with $I \cup \{e\} \notin \mathcal{I}$, the set of elements $e' \in I \cup \{e\}$ such that $I \cup \{e\} - e' \in \mathcal{I}$ is called a *circuit* and is denoted by $C(I, e)$. The *rank function* r of a matroid $\mathcal{M} = (S, \mathcal{I})$ is defined as a function $r : 2^S \to \mathbf{Z}_+$ such that $r(S')$ is the maximum cardinality $|I|$ of a member $I \in \mathcal{I}$ with $I \subseteq S'$. Given two matroids $\mathcal{M}_1 = (S, \mathcal{I}_1)$ and $\mathcal{M}_1 = (S, \mathcal{I}_2)$ on the same set S, finding a maximum common member $I^* \in \mathcal{I}_1 \cap \mathcal{I}_2$ is known as the matroid intersection problem. It is not difficult to observe that $|I| \leq r_1(S') + r_2(S - S')$ holds for every $I \in \mathcal{I}_1 \cap \mathcal{I}_2$ and $S' \subseteq S$, where r_i is the rank function of $\mathcal{M}_i$, $i = 1, 2$.. Edmonds has proven the following min-max theory.

Theorem 1. [6] $\max\{|I| \mid I \in \mathcal{I}_1 \cap \mathcal{I}_2\} = \min\{r_1(S') + r_2(S - S') \mid S' \subseteq S\}$. $\quad\square$

3.2   Detachments

Let G be a multigraph which may have self-loops. A *detachment* H of G is a multigraph with $E(H) = E(G)$ such that $V(H)$ can be partitioned into $|V(G)|$ subsets W_v, $v \in V(G)$ in such a way that G is obtained from H by contracting each subset W_v into a single vertex v.

Given a function $r : V(G) \to \mathbf{Z}_+$, a detachment $H = (\cup_{v \in V(G)} W_v, E(G))$ of G is called an *r-detachment* of G if $|W_v| = r(v)$, $v \in V(G)$, where we denote

$W_v = \{v^1, v^2, \ldots, v^{r(v)}\}$. In other words, H is obtained from G by splitting each vertex $v \in V(G)$ into $r(v)$ copies of v, where each edge $\{u, v\} \in E(G)$ joins some vertices $u^i \in W_u$ and $v^j \in W_v$. A self-loop $\{u, u\}$ in G may be mapped to a self-loop $\{u^i, u^i\}$ or a non-loop edge $\{u^i, u^j\}$ in a detachment H of G. Note that $d(W_u, W_v; H) = d(u, v; G)$ holds for all $u, v \in V(G)$.

For example, an r-detachment of graph G_g in Fig. 2(a) is shown in Fig. 2(c), where $r(\texttt{H}) = 4$, $r(\texttt{O}) = 1$ and $r(\texttt{C}) = 2$.

For a function $r : V(G) \to \mathbf{Z}_+$, an r-degree specification is a set ρ of vectors $\rho(v) = (\rho_1^v, \rho_2^v, \ldots, \rho_{r(v)}^v)$, $v \in V(G)$ such that $\sum_{1 \le i \le r(v)} \rho_i^v = deg(v; G)$. An r-detachment H of G is called a ρ-detachment if each $v \in V$ satisfies $deg(v^i; H) = \rho_i^v$ for all $v^i \in W_v = \{v^1, v^2, \ldots, v^{r(v)}\}$. For a subset $X \subseteq V(G)$, $r(X)$ denotes $\sum_{v \in X} r(v)$. Nash-Williams [11] obtained the following characterization of connected r-detachments of G which are allowed to have self-loops.

Theorem 2. [11] *Let $G = (V, E)$ be a multigraph and $r : V \to \mathbf{Z}_+$. Then there exists a connected r-detachment H of G if and only if*

$$r(X) + c(G - X) - d(X, V; G) \le 1 \text{ for every nonempty subset } X \subseteq V. \quad (1)$$

Furthermore, if G has a connected r-detachment then there exists a connected ρ-detachment H_ρ of G for every r-degree specification ρ.   □

The theorem does not characterize the necessary and sufficient condition for a given multigraph G to have a *loopless* connected r-detachment or ρ-detachment H.

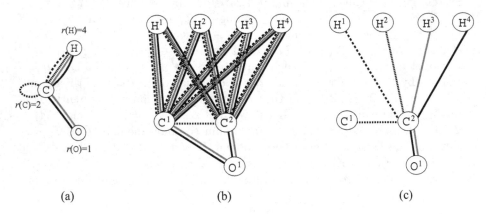

Fig. 2. (a) A multigraph G_g obtained from the vector $g \in \mathcal{F}_1(\{\texttt{H}, \texttt{O}, \texttt{C}\})$ in Fig. 1(c); (b) An r-expansion $\hat{H}(G_g)$ of G_g, where $r(\texttt{H}) = 4$, $r(\texttt{O}) = 1$ and $r(\texttt{C}) = 2$; (c) An r-detachment of G_g

3.3   Loopless Detachments

In this subsection, we give an efficient algorithm for computing a loopless and connected ρ-detachment of a given multigraph. For this, we derive the necessary and sufficient conditions for a given multigraph to have *loopless* connected r- and ρ-detachments as follows.

Theorem 3. *Let $G = (V, E)$ be a multigraph and $r : V \to \mathbf{Z}_+$. Then:*

(i) *There exists a loopless and connected r-detachment H of G if and only if (1) holds and $r(v) \geq 2$ for each self-loop $\{v, v\} \in E$.*

(ii) *Whether (1) holds or not can be tested in $O(\min\{r(V)^{3.5} + |E|, \ r(V)^{1.5} |E| r_{max}^2\})$ time, and a multigraph H in (i) if any can be obtained in $O(\min \{r(V)^{3.5} + |E|, r(V)^{1.5} |E| r_{max}^2\})$ time, where $r_{max} = \max_{v \in V} r(v)$.*   □

Theorem 4. *Let $G = (V, E)$ be a multigraph, $r : V \to \mathbf{Z}_+$, and ρ be an r-degree specification. Then:*

(i) *G has a loopless and connected ρ-detachment H_ρ if and only if it hold (1) and*

$$1 \leq \rho_i^v \leq d(v; G) + d(v, v; G) \text{ for all } v^i \in W_v \text{ and } v \in V. \qquad (2)$$

(ii) *Given a loopless and connected r-detachment H of G, a loopless and connected ρ-detachment H_ρ can be constructed in $O(|E| \min\{r(V)^2, |E| r_{max}^2\})$ time.*   □

Proof of Theorem 3. We first prove Theorem 3. First consider the necessity of Theorem 3(i). If $r(v) = 1$ for some self-loop $\{v, v\} \in E$, then clearly G cannot have a loopless r-detachment. Assume that there is a connected r-detachment $H = (\cup_{v \in V} W_v, E(G))$ of $G = (V, E)$. Let X be an arbitrary nonempty subset of V. For $X_H = \cup_{v \in X} W_v$ and $E' = E(X_H, V(H); H)$, each vertex in X_H has no incident edge in graph $H - E'$, and $c(H - E') = |X_H| + c(H - X_H) \geq r(X) + c(G - X)$ holds. Since $1 = c(H) \geq c(H - E') - |E'|$ holds by Lemma 1 and $|E'| = d(X_H, V(H); H) = d(X, V; G)$ holds, we have $1 \geq r(X) + c(G - X) - d(X, V; G)$, which implies the necessity of Theorem 3(i).

We now show the sufficiency of Theorem 3(i). Given a multigraph $G = (V, E)$ and a function r in Theorem 3, we define an *r-expansion* as a multigraph $\hat{H}(G) = (W = \cup_{v \in V} W_v, F)$ such that its vertex set W is the union of $|V|$ disjoint vertex subsets $W_v = \{v^1, v^2, \ldots, v^{r(v)}\}$, $v \in V$ and its edge set F is the union of $|E|$ disjoint edge subsets F_e, $e \in E$ defined by

$$F_e = \{\{u^i, v^j\} \mid u^i \in W_u, \ v^j \in W_v\} \text{ if } e = \{u, v\} \in E \ (u \neq v),$$
$$F_e = \{\{u^i, u^j\} \mid u^i, u^j \in W_u, \ i \neq j\} \text{ if } e = \{u, u\} \in E.$$

Note that $|W| = r(V)$ and $|F| = \sum_{\{u,v\} \in E : u \neq v} r(u) r(v) + \sum_{\{u,u\} \in E} r(u)(r(u) - 1)/2 = O(|E| r_{max}^2)$ hold, and that the resulting multigraph (W, F) is loopless since $|W_u| = r(u) \geq 2$ holds for any self-loop $e = \{u, u\} \in E$ by the assumption on r. Any subset $F' \subseteq F$ such that $|F' \cap F_e| = 1$, $e \in F$ can be viewed as a loopless r-detachment (W, F') of G.

We here introduce a partition matroid $\mathcal{M}_1 = (F, \mathcal{I}_1)$ with $\mathcal{I}_1 = \{I \subseteq F \mid |I \cap F_e| \leq 1 \ \forall e \in E\}$ and the graphic matroid $\mathcal{M}_2 = (F, \mathcal{I}_2)$ of $\hat{H}(G)$, i.e., $\mathcal{I}_2 = \{I \subseteq F \mid I$ contains no cycle in $\hat{H}(G)\}$. Observe that, for any loopless r-detachment (W, F') of G, its maximal forest $F'' \subseteq F'$ (i.e., a maximal subset of F' having no cycle) satisfies

$$c((W, F'')) = |W| - |F''| \text{ and } F'' \in \mathcal{I}_1 \cap \mathcal{I}_2.$$

In particular, $c((W, F'')) = |W| - |F''| = 1$ if (W, F') is connected. Therefore, it suffices to show that $\mathcal{I}_1 \cap \mathcal{I}_2$ contains a subset I^* with $|I^*| = |W| - 1$ if (1) holds, since this implies that $c((W, I^*)) = |W| - |I^*| = 1$ and that a loopless and connected r-detachment (W, F') is obtained from I^* by adding $|E| - |I^*|$ more edges choosing an arbitrary edge $e' \in F_e$ for each $e \in E$ with $I^* \cap F_e = \emptyset$ so that $|F' \cap F_e| = 1$ holds for all $e \in E$.

By Theorem 1, the maximum cardinality $|I|$ of a member $I \in \mathcal{I}_1 \cap \mathcal{I}_2$ is equal to $\min\{r_1(F') + r_2(F - F') \mid F' \subseteq F\}$, where r_i is the rank function of $\mathcal{M}_i$, $i = 1, 2$. We can prove the next property, as shown in [11].

Lemma 2. If (1) holds, then

$$r_1(F') + r_2(F - F') \geq r(V) - 1 \text{ for every subset } F' \subseteq F. \tag{3}$$

Proof. Omitted due to space limitation. □

Given a multigraph G and a function r, we compute a member $I^* \in \mathcal{I}_1 \cap \mathcal{I}_2$ with the maximum cardinality $|I^*|$. If (1) holds, then $|I^*| = \min\{r_1(F') + r_2(F - F') \mid F' \subseteq F\} \geq r(V) - 1$ must hold by this lemma and Theorem 1, and G admits a loopless and connected r-detachment. This shows the sufficiency of (1), proving Theorem 3(i).

To test whether (1) holds or not, we compute a maximum common member $I^* \in \mathcal{I}_1 \cap \mathcal{I}_2$. This can be done in polynomial time by using the matroid intersection algorithm in [5]. The time complexity is reduced to $O(\min\{r(V)^{3.5}, r(V)^{1.5}|E|r_{max}^2\})$ by utilizing the structure of the problem (the detail is omitted due to space limitation). This proves Theorem 3(ii).

Proof of Theorem 4. Next we prove Theorem 4. We first consider the necessity of Theorem 4(i). Assume that G has a loopless and connected ρ-detachment H_ρ. It is easy to see that $1 \leq deg(v^i) = \rho_i^v$ holds for all $v^i \in W_v$ and $v \in V$. If $\rho_i^v = deg(v^i) > d(v; G) + d(v, v; G)$ holds, then at least one self-loop in $E(v, v; G)$ must be incident to v^i. Hence $\rho_i^v \leq d(v; G) + d(v, v; G)$ necessarily holds.

To show the sufficiency of Theorem 4(i), we again consider an r-detachment H of G as a spanning subgraph $H = (W, F')$ of the r-expansion $\hat{H}(G) = (W, F)$ such that $|F' \cap F_e| = 1$ for every $e \in E$. Given an r-degree specification ρ in Theorem 4, we show that F' can be modified into a ρ-detachment of G. Let $D(H)$ denote the difference $\sum_{v \in V} \sum_{1 \leq i \leq r(v)} |deg(v^i; H) - \rho_i^v|$.

Lemma 3. Let $H = (W, F')$ be a connected spanning subgraph of the r-expansion $\hat{H}(G)$ such that $|F' \cap F_e| = 1$ for every $e \in E$. If $D(H) > 0$, then one of the following (i) and (ii) holds:

(i) There are edges $e_a \in F' \cap F_e$ and $e_b \in F_e - F'$ for some edge $e \in E$ such that $H' = (W, (F' - e_a) \cup \{e_b\})$ remains connected and $D(H') = D(H) - 2$ holds.

(ii) There are edges $e_a \in F' \cap F_e$, $e_b \in F_e - F'$ $e'_a \in F' \cap F_{e'}$ and $e'_b \in F_{e'} - F'$ for some edges $e, e' \in E$ such that $H' = (W, (F' - \{e_a, e'_a\}) \cup \{e_b, e'_b\})$ remains connected and $D(H') = D(H) - 2$ holds.

Proof. Omitted due to space limitation.   □

After modifying F' into $(F' - e_a) \cup \{e_b\}$ by edges e_a and e_b in (i) of this lemma, the resulting $H = (W, F')$ remains connected and satisfies $|F' \cap F_e| = 1$ for every $e \in E$, and the difference $D(H)$ reduces by 2. Analogously for the modification by (ii) of the lemma. Therefore by repeating these procedures until $D(H)$ becomes zero, we obtain a loopless and connected ρ-detachment $H = (W, F')$ of G. This proves Theorem 4(i). Since $D(H) \leq 2|E|$, and the modification is applied $O(|E|)$ times. We represent a multigraph $H = (W, F')$ as an edge-weighted simple graph. Then the connectivity of two vertices in H can be tested in $O(\min\{r(V)^2, |E|r_{max}^2\})$ time, and we can obtain a loopless and connected r-detachment of G in $O(|E| \min\{r(V)^2, |E|r_{max}^2\})$ time, proving Theorem 4(ii).

4   Inferring Multigraphs

We are ready to prove our results on graph inference. Given a feature vector $g \in \mathcal{F}_K(\Sigma)$, let g_k denote the vector which consists of entries $g(t)$, $t \in \Sigma^k$, $|g_k|$ denote the number of nonzero entries in g_k, and let $V_k = \{t \in \Sigma^k \mid g(t) \geq 1\}$, where $|g_k| = |V_k|$ holds. We assume that a given feature vector g is represented only by its positive entries, since otherwise it would require unnecessarily large space complexity to store many zero entries. Let $|g|$ denote the number of nonzero entries in g, and let $n = \sum_{t \in \Sigma^1} g(t)$, $p = \max_{t \in \Sigma^1} g(t)$ and $m = \sum_{t \in \Sigma^2} g(t)$. Thus, $g \in \mathcal{F}_1(\Sigma)$ is given by $O(|g| \log n)$ space.

A feature vector $g \in \mathcal{F}_1(\Sigma)$ is called *valid* with respect to Σ if it satisfies that $V_2 \subseteq V_1 \times V_1$, $g(uv) = g(vu)$ for all $uv \in V_2$, and $g(uu)$ is an even integer and $g(u) \geq 2$ for all $uu \in V_2$.

Theorem 5. *Given an instance $I = g \in \mathcal{F}_1(\Sigma)$ of GIPF, the feasibility of I can be tested in $O(\min\{|g|2^{|g_1|}, n^{3.5} + m, n^{1.5}mp^2\})$ time, and a solution of I (if any) can be constructed in $O(\min\{n^{3.5} + m, n^{1.5}mp^2\})$ time.*

Proof. Given a feature vector $g \in \mathcal{F}_1(\Sigma)$, we can check whether or not g is valid with respect to Σ in $O(|g_1| + |g_2|) = O(|g|)$ time. If g is not valid, then we see that there is no loopless Σ-labeled multigraph H with $f_1(H) = g$. Consider the case where g is valid. By regarding $V_1 = \{t \in \Sigma^1 \mid g(t) \geq 1\}$ and $V_2 = \{t \in \Sigma^2 \mid g(t) \geq 1\}$ as a vertex set and an edge set, we construct a multigraph $G_g = (V = V_1, E = V_2)$ such that $d(u, v; G) = g(uv)(= g(vu))$ for all $u, v \in V$ with $u \neq v$ and $d(u, u; G) = g(uu)$ for all $u \in V$, where a set of edges $E(u, v; G_g)$ is stored as a single edge weighted by integer $d(u, v; G)$. Let $r(v) := g(v)$, $v \in V$. Since g is valid, such a multigraph G_g exists and $r(v) \geq 2$ holds for each self-loop $\{v, v\} \in E$. We see that any loopless and connected Σ-labeled multigraph H with $f_1(H) = g$ is a loopless and connected r-detachment of G_g. We test whether there exists an r-detachment H of G_g or not and find such a solution H to I if any. This can be done in $O(\min\{r(V)^{3.5} + |E|, r(V)^{1.5}|E|r_{max}^2\}) = O(\min\{n^{3.5} + m, n^{1.5}mp^2\})$ time by Theorem 3(ii). Note that the feasiblity of I can also be tested by checking (1) for all possible subsets X of V. This takes $O(|g|2^{|g_1|})$ time since $c(G_g - X)$ can be computed in $O(|g|)$ time.   □

For example, given feature vector $g \in \mathcal{F}_1(\{H,O,C\})$ with $g(t) = occ(t, H_1)$ in Fig. 1(b), multigraph $G_g = (V, E)$ in this proof is given as in Fig. 2(a). An r-expansion $\hat{H}(G_g)$ is given in Fig 2(b), from which a loopless and connected r-detachment is obtained in Fig. 2(c), which is equivalent to graph H_2 in Fig. 1(c).

Corollary 1. *Given an instance $I = g \in \mathcal{F}_1(\Sigma)$ of GIPF for trees, the feasibility of I can be tested in $O(\min\{|g|2^{|g_1|}, n^{3.5}, n^{2.5}p^2\})$ time, and a solution of I (if any) can be constructed in $O(\min\{n^{3.5}, n^{2.5}p^2\})$ time.*

Proof. We can test if a given g satisfies $m = n - 1$ or not in $O(\min\{|g|, n\})$ time. If $m \neq n - 1$, then no Σ-labeled tree T with $f_2(T) = g$ exists. Otherwise (if $m = n - 1$) we apply Theorem 5 to obtain a connected Σ-labeled multigraph H with $f_1(H) = g$ if any, which must be a tree since $|V(H)| = n$ and $|E(H)| = m = n - 1$.   □

Analogously with Theorem 5 and Corollary 1, we have the next results.

Theorem 6. *Given an instance $I = (g \in \mathcal{F}_1(\Sigma), \rho)$ of GIFV, the feasibility of I can be tested in $O(\min\{n + |g|2^{|g_1|}, n^{3.5} + m, n^{1.5}mp^2\})$ time, and a solution of I (if any) can be constructed in $O(\min\{n^{3.5} + m, n^{1.5}mp^2\})$ time.*   □

Corollary 2. *Given an instance $I = (g \in \mathcal{F}_1(\Sigma), \rho)$ of GIFV for trees, the feasibility of I can be tested in $O(\min\{n + |g|2^{|g_1|}, n^{3.5}, n^{2.5}p^2\})$ time, and a solution of I (if any) can be constructed in $O(\min\{n^{3.5}, n^{2.5}p^2\})$ time.*   □

We close this section by making a remark that our algorithm can be used to reduce the search space for solving GIPF and GIFV for $K > 1$. Given a feature vector $g \in \mathcal{F}_K(\Sigma)$, suppose that there is a loopless and connected Σ-labeled multigraph H with $f_K(H) = g$. For such a graph H, we consider the k-*path graph* $H_k = (W_k, F_k)$ such that its vertex set W_k consists of all paths of length of k as vertices and its edge set F_k contains an edge $\{t, t'\}$ if and only if H has a path $(v_1, v_2, \ldots, v_{k+1})$ of length $k + 1$ with $t = (v_1, v_2, \ldots, v_k)$ and $t' = (v_2, v_3, \ldots, v_{k+1})$. Since H_k is a loopless and connected Σ^k-labeled multigraph, we can test whether g_k and g_{k+1} can have such H_k as in the case of g_1 and g_2 by suitably defining multigraph G_g (the detail is omitted due to space limitation). Hence if $g \in \mathcal{F}_K(\Sigma)$ has a pair of g_k and g_{k+1} ($1 \le k \le K$) that has no H_k, then we conclude that g has no solution H with $f_K(H) = g$.

5   Concluding Remarks

In this paper, we proved that the problem of inferring a multigraph from frequency of paths of length at most $K = 1$ can be solved efficiently by formulating it as the loopless and connected detachment problem. Our algorithm can handle the case where each vertex is required to have a specified degree. Our new approach will be applied to infer multigraphs/digraphs with a higher connectivity since the characterizations of k-edge-connected detachments of multigraphs/digraphs have already been obtained [4,7,8,11].

References

1. Akutsu, T., Fukagawa, D.: Inferring a graph from path frequency, Proc. 16th Symp. on Combinatorial Pattern Matching LNCS, vol. 3537 (2005) 371–382.
2. Akutsu, T., Fukagawa, D.: On inference of a chemical structure from path frequency, Proc. 2005 International Joint Conference of InCoB, AASBi and KSBI (BIOINFO2005), Busan, Korea (2005) 96–100.
3. Bakir, G. H., Zien, A., Tsuda, K.: Learning to find graph pre-images, In Proc. the 26th DAGM Symposium, LNCS, 3175 (2004) 253–261.
4. Berg, A. R., Jackson, B. Jordán, T: Highly edge-connected detachments of graphs and digraphs, J. Graph Theory **43** (2003) 67–77.
5. Cunningham, W. H.: Improved bounds for matroid partition and intersection algorithms, SIAM J. Computing **15** (1986) 948–957.
6. Edmonds, J.: Matroids, submodular functions, and certain polyhedra, in: Combinatorial Structures and Their Applications (R.K. Guy, H. Hanani, N. Sauer, and J. Schönheim, eds), Gordon and Breach, New York, 69–87, 1970.
7. Fukunaga, T, Nagamochi, H: Eulerian detachments with local-edge-connectivity, submitted to a journal.
8. Jackson, B., Jordán T.: Non-separable detachments of graphs, J. Combin. Theory (B) **87** (2003) 17–37.
9. Kashima, H., Tsuda, K., Inokuchi, A.: Marginalized kernels between labeled graphs, Proc. of the 20th Int. Conf. on Machine Learning (2003) 321–328.
10. Korte, B.,Vygen, J.: Combinatorial Optimization: Theory and Algorithms, Springer-Verlag, Berlin, Heidelberg, New York, 2000.
11. Nash-Williams, St. J. A.: Connected detachments of graphs and generalised Euler trails, J. Lond Math Soc **31** (1985) 17–29.

The d-Identifying Codes Problem for Vertex Identification in Graphs: Probabilistic Analysis and an Approximation Algorithm*

Ying Xiao[1], Christoforos Hadjicostis[2], and Krishnaiyan Thulasiraman[3]

[1] Packet Design Inc., Palo Alto, CA 94304, USA
yingxiao@ieee.org
[2] University of Illinois at Urbana-Champaign, IL 61801, USA
chadjic@uiuc.edu
[3] University of Oklahoma, Norman, OK 73019, USA
thulasi@ou.edu

Abstract. Given a graph $G(V, E)$, the identifying codes problem is to find the smallest set of vertices $D \subseteq V$ such that no two vertices in V are adjacent to the same set of vertices in D. The identifying codes problem has been applied to fault diagnosis and sensor based location detection in harsh environments. In this paper, we introduce and study a generalization of this problem, namely, the d-identifying codes problem. We propose a polynomial time approximation algorithm based on ideas from information theory and establish its approximation ratio that is very close to the best possible. Using analysis on random graphs, several fundamental properties of the optimal solution to this problem are also derived.

1 Introduction

Consider an undirected graph G with vertex set V and edge set E. A ball of radius $t \geq 1$ centered at a vertex v is defined as the set of all vertices that are at distance t or less from v. The vertex v is said to cover itself and all the vertices in the ball with v as the center. The identifying codes problem defined by Karpovsky et al. [9] is to find a minimum set D such that every vertex in G belongs to a unique set of balls of radius $t \geq 1$ centered at the vertices in D. The set D may be viewed as a code identifying the vertices and is called an identifying set. Two important applications have triggered considerable research on the identifying codes problem. One of these is the problem of diagnosing faulty processors in a multiprocessor system [9]. Another application is robust location detection in emergency sensor networks [13]. Next we briefly describe the application of identifying codes in fault diagnosis.

Consider a communication network modeled as an undirected graph G. Each vertex in the graph represents a processor and each edge represents the communication link connecting the processors represented by the end vertices. Some of

---

* This work was supported by the National Science Foundation under the ITR grant ECS-0426831.

D.Z. Chen and D.T. Lee (Eds.): COCOON 2006, LNCS 4112, pp. 284–298, 2006.

the processors could become faulty. To simplify the presentation let us assume
that at most one processor could become faulty at any given time. Assume that
a processor, when it becomes faulty, can trigger an alarm placed on an adjacent
processor. We would like to place alarms on certain processors that will facilitate
unique identification of the faulty processors. We would also like to place alarms
on as few processors as possible. If D is a minimum identifying set for the case
$t = 1$, then placing alarms on the processors represented by the vertices in the
set D will help us to uniquely identify the faulty processor. Notice that we only
need to consider $t = 1$ because if $t > 1$ is desired, we can proceed with G^t, the
tth power of G.

Karpovsky et al [9] have studied the identifying codes selection problem extensively and have established bounds on the cardinally of the identifying sets.
They have shown how to construct the identifying sets for specific topologies
such as binary cubes and trees. For arbitrary topology, [2] presents heuristic
approaches for a closely related problem that arises in selecting probes for fault
localization in communication networks. Several problems closely related to the
identifying codes problem have been studied in the literature. Some of these may
be found in [5], [6], [10], [11].

Karpovsky et al. [9] have shown that unique identification of vertices may not
always be possible for certain topologies. In other words, triggering of alarms
on a set of processors could mean that one of several candidate processors could
be faulty. Once such a set of possible faulty processors has been identified then
testing each processor in this set will identify the faulty processor. This motivates the generalization of the identifying codes problem to d-identifying codes
problem defined below. This generalization is similar to the introduction of t/s
diagnosable systems that generalize the t-diagnosable systems introduced by
Preparata, Metze and Chien [12]. An introduction to t-diagnosable systems and
their generalization may be found in [3], [4].

1.1 Definition of the d-Identifying Codes Problem

Consider an undirected graph $G(V, E)$ with each vertex $v \in V$ associated with
an integer cost $c(v) > 0$ and an integer weight $w(v) > 0$.

Let $N[v]$ be the set of vertices containing v and all its neighbors. For a subset
of vertices $S \subseteq V$, define the cost and weight of S as

$$c(S) = \sum_{v \in S} c(v) \text{ and } w(S) = \sum_{v \in S} w(v).$$

Two vertices $u, v \in V$ are distinguished by vertex w iff $|N[w] \cap \{u, v\}| = 1$. A
set of vertices $D \subseteq V$ is called an identifying set if (1) every unordered vertex
pair (u, v) is distinguished by some vertex in D and (2) D is a dominating set
of G, i.e., each vertex in G is adjacent to at least one vertex in D (we will relax
this requirement later).

Given $D \subseteq V$, define $I_D(v) = N[v] \cap D$ and an equivalence relation $u \equiv v$
iff $I_D(u) = I_D(v)$. The equivalence relation partitions V into equivalence classes
$V_D = \{S_1, S_2 \dots, S_l\}$ such that $u, v \in S_i \Longleftrightarrow I_D(u) = I_D(v)$.

For any $D \subseteq V$, let V_D be the equivalence classes induced by D. If D is a dominating set of G and $d \geq \max\{w(S_1), w(S_2) \ldots, w(S_l)\}$, then D is called a d-identifying set of G. The d-identifying codes problem is to find a d-identifying set $D \subseteq V$ with minimum cost.

Note that if $d = 1$ then the d-identifying codes problem reduces to the identifying codes problem if the vertex costs and weights are equal to unity. Also, whereas the cost of the d-identifying set is a measure of the cost of installing alarms, the value of d is a measure of the degree of uncertainty in the identification of faulty processors. Since the value of d is also a measure of the expenses involved in testing each processor in an equivalence class, d has to be set at a small value.

The identifying set must be a dominating set. However we can drop this requirement after a simple transformation of the graph, i.e., adding a new isolated vertex with weight d and a very big cost such that any cost aware algorithm will not include this vertex in the solution set. Thus it will be the only vertex not adjacent to the identifying set. So we will ignore the dominating set condition for the simplicity of presentation.

We denote $\ln x \equiv \log_e x$, $\lg x \equiv \log_2 x$.

1.2 Main Results

In this paper we introduce and study the d-identifying codes problem. We first propose an approximation algorithm inspired by a heuristic for the minimum probe selection problem [2] based on ideas from information theory. In Theorem 1, we establish the approximation ratio of our algorithm in terms of an entropy function $H(\cdot)$. As a byproduct of the analysis in Theorem 1, we derive in Corollary 1 a lower bound on the cost of the optimal solution. We then study the characteristics of the optimal entropy function that results in the approximation ratio of $1 + \ln d + \ln |V| + \ln(\lg |V|)$ for the d-identifying codes problem and of $1 + \ln |V| + \ln(\lg |V|)$ for the identifying codes problem in Theorem 1. We show that the approximation ratio of our algorithm is very close to the best possible for the d-identifying codes problem if $NP \notin DTIME(n^{\lg \lg n})$. We also derive several fundamental properties of the optimal solution using random graphs. Certain proofs are omitted to conserve space and a few are put in the appendix.

2 An Approximation Algorithm for the d-Identifying Codes Problem

2.1 A Greedy Algorithm

We first present the approximation algorithm solving the d-identifying codes problem as Algorithm 1 below based on ideas from information theory. Following information theoretical terminology, $H(V_S)$ is called the entropy defined on V_S which is the set of equivalence classes induced by S. Similarly, $I(V_S; v) = H(V_S) - H(V_{S+v})$ is called the information content of $v \in V - S$ w.r.t. S. We defer the definition of the entropy until Sect. 2.2. Actually, the framework of our greedy algorithm without specific entropy definition is applicable to a class of identifying codes problems whose detailed specifications can be hidden in the definition of the entropy. Based on the framework of the greedy algorithm, one only needs to focus on the design of entropy for other variations of the identifying codes problem, e.g., the strong identification codes problem [11]. However, finding the optimal entropy is not straightforward. On the contrary, it is usually the most tricky work.

Algorithm 1. Greedy Algorithm

1: Initialize $D = \emptyset$
2: **while** $H(D) > 0$ **do**
3:     Select vertex $v^* = \arg\max_{v \in V - D} I(V_D; v)/c(v)$
4:     $D \leftarrow D \cup \{v^*\}$.
5: **end while**

The time complexity of the above greedy algorithm is $O(n^2 T_H(n))$, where T_H is the time complexity function of the algorithm computing $H(\cdot)$. The following theorem is the main result on the approximation ratio of the greedy algorithm.

Theorem 1. *Denote V_D as the set of equivalence classes induced by $D \subseteq V$. Suppose an entropy function $H(\cdot)$ satisfies the following conditions:*

(a) $H(V_D) = 0$ for any d-identifying set D,
(b) If $H(V_S) \neq 0$, then $H(V_S) \geq 1$, and
(c) $I(V_S; v) \geq I(V_{S+u}; v)$ for all $u \neq v, S \subseteq V$,

then the greedy algorithm returns a d-identifying set D such that $c(D)/c(D^*) < \ln[H(V_\emptyset)] + 1$ (recall that by definition $V_\emptyset = V$), where $D^* = \{v_1^*, v_2^* \ldots, v_{|D^*|}^*\}$ is the minimum d-identifying set.

Proof. Suppose at the rth iteration, the greedy algorithm picks vertex v_r. Let D_r be the partial d-identifying set at the beginning of the rth iteration, $H_r = H(V_{D_r})$, and $D_r^* = D^* - D_r$. Note that $D_1 = \emptyset$ and $H_1 = H(V_\emptyset)$.

Since $D_r \cup D_r^*$ is a d-identifying set, $H(V_{D_r \cup D_r^*}) = 0$ by (a). Define $D_r^*(i) = \{v_1^*, v_2^* \ldots, v_i^*\}$, i.e., the first i values from D_r^*. Note that $D_r^*(0) = \emptyset$. We have

$$H(V_{D_r}) = H(V_{D_r}) - H(V_{D_r \cup D_r^*})$$

$$= \sum_{i=0}^{|D_r^*|-1} [H(V_{D_r \cup D_r^*(i)}) - H(V_{D_r \cup D_r^*(i+1)})] = \sum_{i=0}^{|D_r^*|-1} I(V_{D_r \cup D_r^*(i)}; v_{i+1}^*).$$

By (c), $I(V_{D_r \cup D_r^*(i)}; v_{i+1}^*) \leq I(V_{D_r \cup D_r^*(i-1)}; v_{i+1}^*) \cdots \leq I(V_{D_r}; v_{i+1}^*)$.
According to the greedy algorithm, $I(V_{D_r}; v_{i+1}^*)/c(v_{i+1}^*) \leq I(V_{D_r}; v_r)/c(v_r)$.

Hence

$$H_r = H(V_{D_r}) = \sum_{i=0}^{|D_r^*|-1} I(V_{D_r \cup D_r^*(i)}; v_{i+1}^*)$$

$$\leq \frac{c(D_r^*)}{c(v_r)} I(V_{D_r}; v_r) \leq \frac{c(D^*)}{c(v_r)} I(V_{D_r}; v_r).$$

Then we know that $H_{r+1} = H_r - I(V_{D_r}; v_r) \leq (1 - \frac{c(v_r)}{c(D^*)})H_r$.

Let the number of iterations of the greedy algorithm be $t = |D|$, where D is the solution returned by the greedy algorithm. We have

$$1 \leq H_t \leq \prod_{v \in D_t} (1 - \frac{c(v)}{c(D^*)})H_0 \leq \exp\{-\frac{c(D_t)}{c(D^*)}\}H_0.$$

The first inequality holds because of (b) (note that $D_t = D - v_t$) and the last inequality is true because $1 - x \leq e^{-x}$.

On the other hand, by (a), $H(V_D) = 0$,

$$I(V_{D_t}; v_t) = H(V_{D_t}) \leq \frac{c(D^*)}{c(v_t)} I(V_{D_t}; v_t) \Rightarrow c(v_t) \leq c(D^*).$$

So $\frac{c(D)}{c(D^*)} = \frac{c(D_t)+c(v_t)}{c(D^*)} = \frac{c(D_t)}{c(D^*)} + \frac{c(v_t)}{c(D^*)} \leq \ln[H(V_\emptyset)] + 1$. $\quad\square$

Using a similar argument, we can derive a lower bound on the cost of the minimum d-identifying set if the costs of all the vertices are equal.

Corollary 1. *Let $G(V, E)$ be a graph with n vertices with equal cost which are labeled such that $I(V_\emptyset; v_1) \geq I(V_\emptyset; v_2) \cdots \geq I(V_\emptyset; v_n)$. Then the optimal cost of the minimum d-identifying set, $OPT_d(G) \geq K$, where K is the smallest integer such that $\sum_{i=1}^{K} I(V_\emptyset; v_i) \geq H(V_\emptyset)$.*

2.2   Optimal Entropy Function

Let $f_d(\cdot)$ be some non-negative function (to be specified later) and $H_d(V_D) = \sum_{S \in V_D} f_d(w(S))$ and $H_d(\emptyset) = 0$, where $V_D = \{S_1, S_2, ...\}$ is the set of equivalence classes induced by $D \subseteq V$.

We first examine Condition (c) in Theorem 1, i.e., $I(V_S; v) \geq I(V_{S+u}; v)$ for any $u \neq v$, $S \subseteq V$. In Fig. 1, there are two cases. In Case 1, v is adjacent to all the vertices in T_{uv} and T_v. In Case 2, v is only adjacent to vertices in T_{uv}, where T is an equivalence class in V_S; T_{uv}, T_u, T_v, and T_0 is the set of vertices in T adjacent to both u and v, only u, only v, and none of u, v, respectively. In other words, $T_{uv} \cup T_u$, $T_v \cup T_0 \in V_{S+u}$ and $T_{uv}, T_u, T_v, T_0 \in V_{S+u+v}$. Let $i = w(T_{uv}), j = w(T_u), k = w(T_v)$, and $l = w(T_0)$.

It is easy to verify that the following conditions are necessary and sufficient for Theorem 1(a)-(c) to be true:

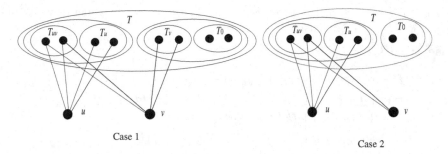

Case 1

Case 2

Fig. 1. Two cases. In Case 1, v is adjacent to all the vertices in T_{uv} and T_v. In Case 2, v is only adjacent to vertices in T_{uv}.

If $i, j, k, l \in \{0, 1, 2 \ldots\}$ and at most one of i, j, k, l is 0, then

$$f_d(i+j+k+l) - f_d(i+k) - f_d(j+l)$$
$$\geq [f_d(i+j) - f_d(i) - f_d(j)] + [f_d(k+l) - f_d(k) - f_d(l)], \qquad (1)$$
$$f_d(t) = 0, 0 \leq t \leq d, \text{ and} \qquad (2)$$
$$f_d(t) \geq 1, \forall t \geq d+1. \qquad (3)$$

Recall that the approximation ratio given in Theorem 1 is $\ln[H(V_\emptyset)] + 1 = \ln f_d(w(V)) + 1$. An entropy function is called optimal if it is the minimum function among all functions that satisfy (1)-(3). Because the approximation ratio is $\ln f_d(w(V)) + 1$, we are only interested in the order of the function and ignore the constant coefficients and constant terms in the function. Assume that $w(V)$ is large. We next construct optimal entropy functions. We first consider $d = 1$. For this special case, define

$$f_1(n) = n \lg n. \qquad (4)$$

Lemma 1. $f_1(n)$ satisfies (1)-(3).

Lemma 2. Given $d \geq 2$, the function defined below satisfies (1)-(3).

$$f_d(n) = \begin{cases} n \lg(n/d), n \geq d \\ 0, otherwise. \end{cases} \qquad (5)$$

Proof. Since $f_d(n)$ is a nondecreasing function and

$$f_d(d+1) = (d+1) \lg(1 + \frac{1}{d}) = \lg((1 + \frac{1}{d})^{d+1}) \geq \lg e > 1,$$

Condition (3) is true.

Condition (2) holds by definition of $f_d(n)$.

We next prove that Condition (1) holds.

If $i + j + k + l \leq d$, the proof is trivial. Without loss of generality, assume $i + j + k + l \geq d + 1$. Consider 5 cases:

Case 1: $i, j, k, l \geq d$.

$$f_d(i + j + k + l) + f_d(i) + f_d(j) + f_d(k) + f_d(l)$$
$$= (i + j + k + l) \lg((i + k + j + l)/d) + i \lg(i/d)$$
$$+ j \lg(j/d) + k \lg(k/d) + l \lg(l/d)$$
$$= (i + j + k + l) \lg(i + j + k + l) + i \lg i + j \lg j$$
$$+ k \lg k + l \lg l - 2(i + j + k + l) \lg d$$
$$\geq (i + k) \lg(i + k) + (j + l) \lg(j + l) + (i + j) \lg(i + j)$$
$$+ (k + l) \lg(k + l) - ((i + k) + (j + l) + (i + j) + (k + l)) \lg d$$
$$= f_d(i + k) + f_d(j + l) + f_d(i + j) + f_d(k + l)$$

Case 2: Precisely one of $i + k, j + l, i + j$, and $k + l$ is $\leq d$.

Due to the symmetry of i, j, k, l in the function, assume that $i + k \leq d$. We have $i \leq d$ and $k \leq d$. Let $g(n) = n \lg(n/d)$.

Therefore

$$f_d(i + j + k + l) + f_d(i) + f_d(j) + f_d(k) + f_d(l)$$
$$\geq g(i + j + k + l) + g(i) + g(j) + g(k) + g(l) - g(i) - g(k)$$
$$\geq g(j + l) + g(i + j) + g(k + l) + (g(i + k) - g(i) - g(k))$$
$$= f_d(j + l) + f_d(i + j) + f_d(k + l) + ((i + k) \lg \frac{i + k}{d} - i \lg \frac{i}{d} - k \lg \frac{k}{d})$$
$$\geq f_d(j + l) + f_d(i + j) + f_d(k + l)$$

Case 3: Precisely $i + k \leq d$ and $j + l \leq d$ or $i + j \leq d$ and $k + l \leq d$.

Assume $i + k \leq d$ and $j + l \leq d$.

In this case, $i, j, k, l \leq d$. It suffices to show that

$$f_d(i + j + k + l) \geq f_d(i + j) + f_d(k + l).$$

This is obviously true.

Case 4: Precisely $i + k \leq d$ and $i + j \leq d$ (ignore those equivalent cases).

We have $i, j, k \leq d$. Hence

$$f_d(i + j + k + l) + f_d(i) + f_d(j) + f_d(k) + f_d(l)$$
$$= f_d(i + j + k + l) + f_d(l)$$
$$\geq (i + j + k + l) \lg((i + j + k + l)/d) + l \lg(l/d)$$
$$= [(j + k + l) \lg(i + j + k + l) + l \lg l] +$$
$$[i \lg(i + j + k + l) - (i + j + k + l) \lg d - l \lg d]$$
$$\geq [(j + l) \lg(j + l) + (k + l) \lg(k + l)] - [((j + l) \lg d + (k + l) \lg d)]$$
$$= f_d(j + l) + f_d(k + l)$$

Case 5: All or 3 of the 4 terms, $i + k, j + l, i + j$, and $k + l$ are $\leq d$.
    The proof for this case is trivial.   □

Finally, we get the main results of this paper.

Theorem 2. *Using the entropy $H_d(V_D) = \sum_{S \in V_D} f_d(w(S))$ with $f_d(\cdot)$ as defined in (5), the greedy algorithm guarantees the approximation ratio of $1 + \ln d + \ln(|V| \lg |V|)$.*

Proof. Without loss of generality, assume $w(V) > d \geq max_{v \in V} w(v)$. We have
$$H_d(V_\emptyset) = f_d(w(V)) = w(V) \lg(w(V)/d) \leq d|V| \lg |V|.$$
The rest of the proof follows from Theorem 1.   □

Corollary 2. *For the identifying codes problem, our algorithm guarantees the approximation ratio of $1 + \ln |V| + \ln(\lg |V|)$.*

We next show that the function defined in (5) is optimal in asymptotic sense, i.e., the approximation ratio based on Theorem 1 cannot be improved by finding better entropy.

Lemma 3. $f(n) \geq \Theta(n \lg n)$.

Proof. Set $i = j = k = l = 2^i d$ in (1), we get

$$f_d(2^{i+2}d) - 2f_d(2^{i+1}d) \geq 2(f_d(2^{i+1}d) - 2f_d(2^i d)). \tag{6}$$

Solving the recurrence inequalities on i, we get

$$f_d(2^{i+2}d) \geq 2f_d(2^{i+1}d) + 2^i f(2d). \tag{7}$$

Hence, $f_d(2^i d) \geq i 2^{i-1} f(2d)$. Letting $n = 2^i d$ completes the proof.   □

2.3   Hardness of the d-Identifying Codes Problem

To study the hardness, i.e., the approximability of the d-identifying codes problem, we consider a subclass of d-identifying codes problem where the vertex costs and weights are 1. Since this subclass includes the identifying codes problem, d-identifying codes problem is at least as hard as identifying codes problem (here d is treated as a variable). On the other hand, an interesting question is whether the approximability of the d-identifying codes problem changes with some fixed d. For example, if the best approximation ratio for the identifying codes problem is ϕ, one may ask whether the 2-identifying codes problem is $\phi/2$ or 2ϕ approximable. The next lemma shows that the approximability will not change if d is a constant.

Lemma 4. *For any fixed $d \geq 2$, if there exists a polynomial time ϕ-approximation algorithm for the d-identifying codes problem, there also exists a polynomial time ϕ-approximation algorithm for the identifying codes problem.*

Lemma 4 means that for any fixed d, the d-identifying codes problem is at least as hard as the identifying codes problem in term of approximability. Thus, with an application of the results in [6], we have the following theorem.

Theorem 3. *For any given $d \geq 1$, the d-identifying codes problem with unit vertex costs and weights is not approximable within $(1 - \epsilon) \ln |V|$ unless $NP \in DTIME(n^{\lg \lg n})$.*

In view of Corollary 2, we can see that the approximation ratio of our algorithm is quite tight for the d-identifying codes problem where the vertex costs and weights are 1. Furthermore, we can expect that our approximation ratio is also very tight for general d-identifying codes problem as in the special case.

3   A Special Case with Unit Vertex Costs and Weights

In Sect. 2, we established the approximation ratio, i.e., the ratio of the cost of the approximation solution and the optimal cost of the d-identifying set. In this section we shall investigate the characteristics of the optimal solution itself. One of the goals of this study is to show that the approximation algorithm will return a small set (compared to the cardinality of the vertex set) even with the worst approximation ratio.

Since it is difficult to study the d-identifying codes problem with arbitrary vertex costs and weights, we shall only consider a special class of d-identifying codes problem in which the cost and weight of each vertex is 1. In this setting, the cost and weight of a set of vertices is just the cardinality of the set.

We shall next investigate the impact of d on the cardinality of the resultant solution. Let $OPT_1(G)$ and $OPT_d(G)$ be the cardinality of the minimum identifying set and the minimum d-identifying set, respectively. It can be shown that the value of $OPT_1(G)$ / $OPT_d(G)$ is unbounded.

Lemma 5. *Given $d \geq 2$ and $M > 0$, there exists a graph G such that*

$$OPT_1(G)/OPT_d(G) \geq M.$$

The graph used to prove Lemma 5 is rather artificial. So let's consider the size of d-identifying sets on average basis. To study the average characteristics, assumptions are needed on the distribution of graphs. Given the vertices of G, the cardinality of the d-identifying set is totally decided by the edges. So we assume that for any unordered pair of vertices there is an edge with probability p which is a constant. Notice that this is exactly the definition of a class of random graphs [1], [7].

For the sake of completeness, we first present Suen's inequality proved in [8], [14]. Let $A_1, A_2 \ldots, A_n$ be a set of events, and $X = \sum_{i=1}^{n} X_i$, where X_i is the indicator variable of event A_i ($X_i = 1$ if event A_i occurs and $X_i = 0$ otherwise). We use $i \sim j$ to indicate that events A_i and A_j are dependent. Denote $\mu = \sum_{i=1}^{n} E[X_i]$, $\Delta = \sum_{i,j:i \sim j} E[X_i X_j]$, and $\delta = \max_{1 \leq i \leq n} \sum_{j:j \sim i} E[X_j]$. Then $\Pr(X = 0) \leq \exp\{-\mu + \Delta e^{2\delta}\}$.

Let $P \equiv p^{d+1} + (1-p)^{d+1}$, $Q(i) \equiv p^{2d+2-i} + (1-p)^{2d+2-i}$, $R^+(p,d) \equiv \ln(1+ (\frac{1-p}{p})^{d+1})/\ln(\frac{1}{p})$, and $R^-(p,d) \equiv R^+(1-p,d)$.

The following lemma is easy to prove.

Lemma 6. $R^+(p,d)$ $(R^-(p,d))$ *is a strictly decreasing (increasing) function of p and a decreasing (decreasing) function of d for $p \in [1/2, 1)$ $(p \in (0, 1/2])$.*

Given $\epsilon \in (0,1)$, denote $p_\epsilon^- \in (0, 1/2]$, $p_\epsilon^+ \in [1/2, 1)$ as two values such that $R^-(d, p_\epsilon^-) = R^+(d, p_\epsilon^+) = \epsilon$. Since $R^-(1/2, d) = R^+(1/2, d) = 1$, $p_\epsilon^- < 1/2 < p_\epsilon^+$. It can be shown that $p_\epsilon^- + p_\epsilon^+ = 1$ and $p_\epsilon^- \to 0$ $(p_\epsilon^+ \to 1)$ if $\epsilon \to 0$.

Lemma 7. *If $1 \le i \le d$, $0 < \epsilon < 1$, and $p \in [p_\epsilon^-, p_\epsilon^+]$, then*

$$\left(\frac{d+1-\epsilon}{d+1-i}\right)\left(\frac{\ln Q(i)}{\ln P} - 1\right) > 1.$$

Theorem 4. *Given $0 < \epsilon < 1$, $\forall p \in [p_\epsilon^-, p_\epsilon^+]$, with high probability, there exists no d-identifying set of cardinality of $\frac{(d+1-\epsilon)\ln n}{\ln(1/P)}$ in $G(n,p)$ if n is sufficiently large.*

Proof. Let $c = \frac{(d+1-\epsilon)\ln n}{\ln(1/P)}$. it suffices to show that

$$\Pr(\text{There exists a } d\text{-identifying set of cardinality } c) = o(1) \to 0.$$

Consider a given set C of cardinality c. Let $S \subset V$ be a set of $d+1$ vertices, define event A_S: $\forall u, v \in S, I_C(u) = I_C(v)$. We can see that C is a d-identifying set iff no such event occurs for all S with $|S| = d+1$. Denote X_S to be the indicator variable for event A_S.

It can be seen that two events A_S and $A_{S'}$ are dependent iff $S \cap S' \ne \emptyset$.

Let $X = \sum_{S \subset V - C, |S| = d+1} X_S$.

Evidently, $\Pr(C \text{ is a } d\text{-identifying set}) \le \Pr(X = 0)$.

Assume $n - c - d - 1 \ge n/k$ for some small k (recall d is a constant). Then

$$\mu = \binom{n-c}{d+1}(p^{d+1} + (1-p)^{d+1})^c \ge (n-c-d-1)^{d+1}P^c$$

$$\ge (n/k)^{d+1}P^c = \exp\{(d+1)\ln n - (d+1)\ln k + c\ln P\}$$

$$= \exp\{(d+1)\ln n - (d+1)\ln k - (d+1-\epsilon)\ln n\}$$

$$= \exp\{\epsilon \ln n - (d+1)\ln k\} = \Theta(n^\epsilon),$$

$$\Delta = \sum_{i=1}^{d} \binom{n-c}{2d+2-i}\binom{2d+2-i}{d+1}(p^{2d+2-i} + (1-p)^{2d+2-i})^c,$$

and

$$\delta = \sum_{i=1}^{d} \binom{d+1}{i}\binom{n-c-d-1}{d+1-i}P^c$$

Denote $\theta = \min_{1 \le i \le d}\{(\frac{d+1-\epsilon}{d+1-i})(\frac{\ln Q(i)}{\ln P} - 1)\} - 1$. By Lemma 7, $\theta > 0$. We have

$$\frac{\Delta}{\mu} \le \sum_{i=1}^{d} \frac{\binom{n-c}{2d+2-i}\binom{2d+2-i}{d+1}}{\binom{n-c}{d+1}}(\frac{Q(i)}{P})^c = \sum_{i=1}^{d}\binom{n-c-d-1}{d+1-i}(\frac{Q(i)}{P})^c$$

$$\le \sum_{i=1}^{d}(\frac{ne}{d+1-i})^{d+1-i}(\frac{Q(i)}{P})^c$$

$$\le \sum_{i=1}^{d}\exp\{(d+1-i)(1-\frac{(d+1-\epsilon)}{(d+1-i)}(\frac{\ln Q(i)}{\ln P}-1))\ln n + d\}$$

$$\le \sum_{i=1}^{d}\exp\{-\theta(d+1-i)\ln n + d\} \le d\exp\{-\theta\ln n + d\} = \Theta(n^{-\theta}) = o(1).$$

Similarly, we can show that $e^{2\delta} = O(\exp\{n^{\epsilon-1}\}) \to 1$.
Hence $-\mu + \Delta e^{2\delta} = -\mu(1 - \frac{\Delta}{\mu}e^{2\delta}) \le \Theta(-n^{\theta})$ and
Pr(There exists a d-identifying code of cardinality c) $\le \binom{n}{c}\exp\{\Theta(-n^{\theta})\}$
Since $\binom{n}{c}\exp\{\Theta(-n^{\theta})\} = O(\exp\{\Theta(\ln^2 n - n^{\theta}))$,

Pr(There exists a d-identifying set of cardinality c) $= o(1) \to 0$.   □

Theorem 5. *For a set of vertices* $C \subseteq V$ *and* $|C| = \frac{(d+1+\epsilon)\ln n}{\ln(1/P)}$,

$$\lim_{n \to \infty} \Pr(C \text{ is a } d\text{-identifying set }) = 1.$$

Proof. Let $X = \sum_{S \subset V, |S|=d+1} X_S$, where X_S is defined as in the proof of Theorem 4. By Markov's inequality, we have,

$$\Pr(C \text{ is a } d\text{-identifying set}) = \Pr(X = 0) = 1 - \Pr(X \ge 1) \ge 1 - E(X).$$

$$E(X) = \sum_S E(X_S) = \sum_{i=0}^{d+1} \sum_{|S \cap C|=i} E(X_S)$$

$$= \sum_{i=0}^{d+1}\binom{|C|}{i}\binom{n-|C|}{d+1-i}P^{|C|-i}(p^{d+1-i})^i p^{i(i-1)/2}$$

$$\le n^{d+1}P^{|C|-d-1} \le \exp\{-\epsilon\ln n + (d+1)\ln(1/P)\} \to 0.$$ □

By Theorem 4 and Theorem 5, with high probability, the cardinality of minimum d-identifying set is approximately $(d+1)\ln n / \ln(1/P)$ when n is sufficiently large.

4   Summary

In this paper we introduced and studied the d-identifying codes problem that generalizes the identifying codes problem studied in [9]. This problem is of great theoretical and practical interest in several applications, in particular, fault diagnosis in multiprocessor systems and placement of alarms for robust identification of faulty components in sensor networks. The value of d associated with the identifying set is a measure of the degree of uncertainty in the identification of faulty processors. We presented an approximation algorithm and established its approximation ratio. This algorithm is a generalization of the heuristic presented in [2] but without analysis of the approximation ratio. Our analysis also provides a way to compute a lower bound on the cost of the optimum solution. We also established certain hardness results in terms of approximability of the d-identifying codes problem.

We performed a probabilistic analysis on random graphs assuming that vertex costs and weights are all equal. We established that a d-identifying set of certain cardinality exists with very high probability. We also showed that a d-identifying set of cardinality smaller than this number does not exist with a high probability.

References

1. Béla Bollobás. *Random graphs*. Academic Press, Inc, London, 1985.
2. M. Brodie, I. Rish, and S. Ma. Optimizing probe selection for fault localization. In *International Workshop on Distributed Systems: Operations and Management*, 2001.
3. A. Das and K. Thulasiraman. Diagnosis of t/s-diagnosable systems. In *International Workshop on Graph-Theoretic Concepts in Computer Science*, pages 193–205, 1990.
4. A. Das, K. Thulasiraman, and V. Agarwal. Diagnosis of t/(t+1)-diagnosable systems. *SIAM J. Comput.*, 23(5):895–905, 1994.
5. A. Frieze, R. Martin, J. Moncel, and K. Ruszinkóand C. Smyth. Codes identifying sets of vertices in random networks. submitted for publication, 2005.
6. Bjarni V. Halldórsson, Magnús M. Halldórsson, and R. Ravi. On the approximability of the minimum test collection problem. In *ESA*, pages 158–169, 2001.
7. S. Janson, T. Luczak, and A. Rucinski. *Random graphs*. Wiley, New York, 2000.
8. Svante Janson. New versions of Suen's correlation inequality. *Random Struct. Algorithms*, 13(3-4):467–483, 1998.
9. M. Karpovsky, K. Chakrabarty, and L. Levitin. On a new class of codes for identifying vertices in graphs. *IEEE Trans. on Information Theory*, 44(2):599–611, 1998.
10. M. Laifenfeld and A. Trachtenberg. Disjoint identifying-codes for arbitrary graphs. submitted to IEEE Symposium on Information Theory, 2005.
11. Tero Laihonen. Optimal codes for strong identification. *Eur. J. Comb.*, 23(3):307–313, 2002.
12. F. Preparata, G. Metze, and R. Chien. On the connection assignment problem of diagnosable systems". *IEEE Trans. on Electronic Computers*, 16:848–854, 1967.
13. S. Ray, R. Ungrangsi, F. Pellegrini, A. Trachtenberg, and D. Starobinski. Robust location detection in emergency sensor networks. In *INFOCOM*, 2003.
14. Stephen Suen. A correlation inequality and a Poisson limit theorem for nonoverlapping balanced subgraphs of a random graph. *Random Struct. Algorithms*, 1(2):231–242, 1990.

Appendix

Lemma 1. $f_1(n)$ satisfies (1)-(3).

Proof. Conditions (2)-(3) are trivial. We only show the proof to Condition (1). Let $i, j, k, l \geq 0$ and at most one of them be 0. Consider 2 cases.

Case 1: $i, j, k, l > 0$. It suffices to show that

$$(i + j + k + l) \lg(i + j + k + l) - (i + k) \lg(i + k) - (j + l) \lg(j + l)$$
$$\geq [(i + j) \lg(i + j) - i \lg i - j \lg j] + [(k + l) \lg(k + l) - k \lg k - l \lg l].$$

Equivalently, we will prove that

$$\lg\left(\frac{(i + j + k + l)^{(i+j+k+l)} i^i j^j k^k l^l}{(i + k)^{(i+k)} (j + l)^{(j+l)} (i + j)^{(i+j)} (k + l)^{(k+l)}}\right) \geq 0.$$

Define function

$$g(x) = \ln\left(\frac{(x + j + k + l)^{x+j+k+l} x^x j^j k^k l^l}{(x + k)^{x+k} (j + l)^{j+l} (x + j)^{x+j} (k + l)^{k+l}}\right).$$

It suffices to show that $\forall x > 0$, $g(x) \geq 0$. We have

$$g'(x_0) = \ln \frac{x_0(x_0 + j + k + l)}{(x_0 + k)(x_0 + j)} = 0 \Leftrightarrow x_0 = kj/l > 0.$$
$$g''(x_0) = l/[x_0(x_0 + j + k + l)] > 0.$$

Since

$$\frac{(x + j + k + l)^{x+j+k+l} x^x j^j k^k l^l}{(x + k)^{x+k} (j + l)^{j+l} (x + j)^{x+j} (k + l)^{k+l}}$$
$$= (1 + \frac{xl - jk}{(x + k)(x + j)})^x (1 - \frac{xl - jk}{(j + l)(x + j)})^j$$
$$\times (1 - \frac{xl - jk}{(x + k)(k + l)})^k (1 + \frac{xl - jk}{(j + l)(k + l)})^l,$$

$g(x_0) = \ln 1 = 0$ and hence $\forall x > 0, g(x) \geq g(x_0) = 0$.

Case 2: Precisely one of i, j, k, l is 0.

Without loss of generality, assume $l = 0$. It suffices to show that

$$\forall x \geq 0, h(x) = \ln\left(\frac{(i + j + x)^{i+j+x} i^i}{(i + x)^{i+x} (i + j)^{i+j}}\right) \geq 0.$$

Since $h'(x) = \ln(\frac{i+j+x}{i+x}) \geq 0$ and $h(0) = 0$, $\forall x \geq 0, h(x) \geq h(0) = 0$.   □

Lemma 4. For any fixed $d \geq 2$, if there exists a polynomial time ϕ-approximation algorithm for the d-identifying codes problem, there also exists a polynomial time ϕ-approximation algorithm for the identifying codes problem.

Proof. We first give a polynomial time ϕ-approximation algorithm for the identifying codes problem on $G(V, E)$ staring from the ϕ-approximation algorithm for the d-identifying codes problem:

1. Construct a graph $G'(V', E')$ defined as follows: Split each vertex $v \in V$ into d copies denoted as $v^d = \{v_1, v_2 \ldots, v_d\}$. For all $(u, v) \in E$, add edges to connect all vertices in u^d to all vertices in v^d and for all $v \in V$, add edges to join each pair of vertices in v^d (See Fig. 2). Formally,

$$V' = \bigcup_{v \in V} \{v_1, v_2 \ldots v_d\}, \text{ and}$$

$$E' = \{(u_i, v_j), i, j = 1, 2 \ldots, d | (u, v) \in E\}$$

$$\bigcup \{(v_i, v_j), i, j = 1, 2 \ldots, d) | v \in V\}.$$

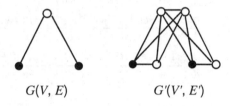

$$G(V, E) \qquad\qquad G'(V', E')$$

Fig. 2. Transformation from identifying codes problem to 2-identifying codes problem: The identifying set and d-identifying set consists of the solid vertices

2. Apply the ϕ-approximation algorithm to get a d-identifying set D_d on G'.
3. Return $D = \{v \in V | v^d \cap D_d \neq \emptyset\}$ as an identifying set on G.

The construction of G' takes $O(d^2 |E|)$ time with d as a constant.

We next show that the above procedure is a ϕ-approximation algorithm for the identifying codes problem.

Let D^* be an optimal solution to the identifying codes problem on G. It is easy to verify that $D'_d = \{v_1 | v \in D^*\}$ is a d-identifying set of G'. Denote D_d^* as an optimal d-identifying set of G'. We have

$$c(D_d^*) = |D_d^*| \leq |D'_d| = |D^*| = c(D^*).$$

Because in G', $\forall v \in V, v_1, v_2 \ldots, v_d \in V'$ has the same set of neighbors in G', there is no way to distinguish them in G'. Hence the set of equivalence classes of V' induced by D_d is simply $V'_{D_d} = \{v^d | v \in V\}$.

Observe that condensing all the vertices in v^d for all $v \in V$ into a single vertex v transforms G' back to G. So an identifying set of G can be formed by picking all the vertices whose corresponding set of vertices in G' contain at least one vertex in D_d. Hence the set D returned by the above procedure is an identifying set of G. Since the v^d's are pairwise disjoint,

$$c(D) = |D| \leq |D_d| = c(D_d) \leq \phi \cdot c(D_d^*) \leq \phi \cdot c(D^*). \qquad \square$$

Lemma 7. If $1 \leq i \leq d, 0 < \epsilon < 1$, and $p \in [p_\epsilon^-, p_\epsilon^+]$, then

$$(\frac{d+1-\epsilon}{d+1-i})(\frac{\ln Q(i)}{\ln P} - 1) > 1.$$

Proof. If $p = 1/2$, the proof is trivial. We now consider 2 cases.

Case 1: $1/2 < p \leq p_\epsilon$.

By Lemma 6, $\epsilon \leq R^+(p,d) \leq 1$ and $\forall 1 \leq i \leq d, R^+(p,d) > R^+(p, 2d+1-i)$. So

$$
\begin{aligned}
\frac{\ln Q(i)}{\ln P} - 1 &= \frac{\ln(p^{2d+2-i} + (1-p)^{2d+2-i})}{\ln(p^{d+1} + (1-p)^{d+1})} - 1 \\
&= \frac{(2d+2-i) - R^+(p, 2d+1-i)}{(d+1) - R^+(p,d)} - 1 \\
&= \frac{(d+1-i) + R^+(p,d) - R^+(p, 2d+1-i)}{(d+1) - R^+(p,d)} \\
&\geq \frac{d+1-i}{d+1-\epsilon} + \frac{R^+(p,d) - R^+(p, 2d+1-i)}{d+1-\epsilon}.
\end{aligned}
$$

Therefore,

$$(\frac{d+1-\epsilon}{d+1-i})(\frac{\ln Q(i)}{\ln P} - 1) \geq 1 + \frac{R^+(p,d) - R^+(p, 2d+1-i)}{d+1-i} > 1.$$

Case 2: $p_\epsilon \leq p < 1/2$.

$$\frac{\ln Q(i)}{\ln P} - 1 = \frac{(d+1-i) + R^-(p,d) - R^-(p, 2d+1-i)}{(d+1) - R^-(p,d)}.$$

The rest of the proof is the same as in Case 1.   □

Reconstructing Evolution of Natural Languages: Complexity and Parameterized Algorithms*

Iyad A. Kanj[1], Luay Nakhleh[2], and Ge Xia[3]

[1] School of Computer Science, Telecommunications and Information Systems,
DePaul University, 243 S. Wabash Avenue, Chicago, IL 60604-2301, USA
ikanj@cs.depaul.edu
[2] Department of Computer Science, Rice University, 6100 Main St., MS 132 Houston,
TX 77005-1892
nakhleh@cs.rice.edu
[3] Department of Computer Science, Lafayette College, Easton, PA 18042, USA
gexia@cs.lafayette.edu

Abstract. In a recent article, Nakhleh, Ringe and Warnow introduced *perfect phylogenetic networks*—a model of language evolution where languages do not evolve via clean speciation—and formulated a set of problems for their accurate reconstruction. Their new methodology assumes *networks*, rather than *trees*, as the correct model to capture the evolutionary history of natural languages. They proved the NP-hardness of the problem of testing whether a network is a perfect phylogenetic one for characters exhibiting at least three states, leaving open the case of binary characters, and gave a straightforward brute-force parameterized algorithm for the problem of running time $O(3^k n)$, where k is the number of bidirectional edges in the network and n is its size. In this paper, we first establish the NP-hardness of the binary case of the problem. Then we provide a more efficient parameterized algorithm for this case running in time $O(2^k n^2)$. The presented algorithm is very simple, and utilizes some structural results and elegant operations developed in this paper that can be useful on their own in the design of heuristic algorithms for the problem. The analysis phase of the algorithm is very elegant using amortized techniques to show that the upper bound on the running time of the algorithm is much tighter than the upper bound obtained under a conservative worst-case scenario assumption. Our results bear significant impact on reconstructing evolutionary histories of languages–particularly from phonological and morphological character data, most of which exhibit at most two states (i.e., are binary), as well as on the design and analysis of parameterized algorithms.

1 Introduction

Languages differentiate and divide into new languages via a process similar to how biological species divide into new species: communities separate (typically

---

* The first author was supported in part by DePaul University Competitive Research Grant.

D.Z. Chen and D.T. Lee (Eds.): COCOON 2006, LNCS 4112, pp. 299–308, 2006.

geographically), the language changes differently in each of the new communities, and in time people from separate communities can no longer understand each other. While this is not the only means by which languages change, it is this process which is referred to when we say, for example, "French and Italian are both descendants of Latin." The evolution of related languages is mathematically modeled as a rooted tree in which internal nodes represent the ancestral languages and the leaves represent the extant languages.

Reconstructing this process for various language families is a major endeavor within historical linguistics, but is also of interest to archaeologists, human geneticists, and physical anthropologists, for example, because an accurate reconstruction of how certain languages evolved can help answer questions about human migrations, the time that certain artifacts were developed, when ancient people began to use horses in agriculture, the identity of physically European mummies found in China, etc. (see in particular [7,13,18]). Various researchers [2,3,5,14] have noted that if communities are sufficiently separated after they diverge, then the inference of the phylogeny (i.e., evolutionary tree) for the languages can be inferred by comparing the characteristics of the languages (grammatical features, regular sound changes, and cognate classes for different basic meanings), and searching for "perfect phylogenies." However, the problem of determining if a perfect phylogeny exists, and then computing it, is NP-hard [1]. Consequently, efficient techniques for the inference of evolutionary trees for language families were not easily obtained. In the 1990's, various fixed-parameter approaches for the perfect phylogeny problem were developed (although inspired by the biological context rather than the linguistic one). Subsequently, Ringe and Warnow worked together to fully develop the methodology (character encoding and algorithmic techniques) needed to apply these algorithms to the Indo-European language family.

However, while the methodology seemed very clearly heading in the right direction, and even seemed to potentially answer many of the controversial problems in Indo-European evolution (see [10,12,14,15,16]), it became necessary to extend the model to address the problem of how characters evolve when the language communities remain in significant contact. To address this issue, Nakhleh *et al.* introduced the *perfect phylogenetic networks* (PPN) model in which languages do not evolve via a clean speciation process [8,9]. They proved the NP-hardness of the problem of testing whether a network is a perfect phylogenetic one for characters exhibiting at least three states, leaving open the case of binary characters, and gave a straightforward $O(3^k n)$ time parameterized algorithm for the problem [8], where k is the number of bidirectional edges in the network and n is its size.

In this paper we consider the binary case of the problem. This case is of prime interest on its own since it models the problem of reconstructing evolutionary histories of languages, particularly from phonological and morphological character data, most of which exhibit at most two states [6,11,12,14,16,17]. We first prove the NP-hardness of this problem. Then we present a branch-and-bound parameterized algorithm that solves the problem in $O(2^k n^2)$ time. The algorithm employs several interesting structural (network) operations that are very useful in the design of heuristic algorithms for the problem. When analyzed using the standard methods for analyzing parameterized branch-and-bound algorithms, and which usually work under a worst-case scenario assumption, the upper bound obtained on the size of the search tree of the algorithm is $O(3^k)$, matching the upper bound of the trivial brute-force algorithm. This worst-case analysis for a branch-and-search process is usually very conservative— the worst cases can appear very rarely in the entire process, while most other cases permit much better branching and reductions. Instead, we use amortized analysis to show that "expensive" operations can be balanced by efficient ones, and that the actual size of the search tree can be upper bounded by $O(2^k)$. The running time of the algorithm becomes $O(2^k n^2)$. The analysis phase of the algorithm is very elegant illustrating that parameterized algorithms perform much better than their claimed upper bounds, and suggesting that the standard approaches used in analyzing the size of the search tree for parameterized algorithms are very conservative. Most of the proofs in this paper are omitted for lack of space and are available in the technical report **05–007** at the following web address: http://www.cs.depaul.edu/research/technical.asp.

2 Inferring Evolutionary Trees

An evolutionary tree, or phylogeny, for a set L of taxa (i.e., species or languages) describes the evolution of the taxa in L from their most recent common ancestor. Each taxon in L corresponds to a leaf in the evolution tree. The Different types of data can be used as input to methods of tree reconstruction; "qualitative character" data, which reflect specific observable discrete characteristics of the taxa under study, are one such type of data. There are several ways of describing qualitative characters: as partitions of the set of taxa into equivalence classes, or as functions that map the taxa to the distinct states. Qualitative characters for languages are grammatical features, unusual sound changes, and cognate classes for different meanings. The assumption of the historical linguistic methodology is that these qualitative characters evolve in such a way that there is no back-mutation (when characters exhibit parallel evolution we can find most of it and exclude those characters). What this means is that when the state of the qualitative character changes in the evolutionary history of the set of languages, it changes to a state which does not exist anywhere else on earth at that time, nor has it appeared earlier. We now formalize this concept mathematically.

Suppose that T is a rooted tree describing the evolution of a set L of languages. Therefore the leaves in T are the languages in L. Suppose that a qualitative character α is defined for each of the languages in L as a function $\alpha : L \to Z$, where Z denotes the set of integers (i.e. each integer represents a possible state for α). That is, α is a labeling to the leaves in T. We say a qualitative character α is compatible (or "convex") on T if we can extend α to every internal node of the tree T, thus defining a qualitative character α', or a labeling to the internal nodes of T, so that for every state, the nodes in T having that specific state induce a connected subgraph of T. (In other words, $\forall z \in Z$, the set of nodes $\{v \in V(T) : \alpha'(v) = z\}$ induces a connected subgraph of T.)

A different way of casting the above problem which is more intuitive is the following. Given a rooted tree T whose leaves are labeled with integers, decide if the internal nodes in T can be labeled so that each set of nodes in T with the same label induces a connected subgraph of T.

Ringe and Warnow postulated that *all* properly encoded qualitative characters for the Indo-European data should be compatible on the true tree, if such a tree existed. Such a tree is called a *perfect phylogeny*. We have the following definition and theorem.

Definition 1. *Let C be a set of qualitative characters defined on a set L of languages. A tree T is a **perfect phylogeny** for C and L if every qualitative character in C is compatible on T.*

Theorem 1. *Let T be a phylogenetic tree on a set L of n languages, and assume that each language in L is assigned a state for α. Then we can test the compatibility of α on T in $O(n)$ time.*

The initial analysis of the Indo-European data done by Warnow and Ringe in [16] demonstrated that the IE linguistic data is, nevertheless, "almost perfect": they found a tree on which the proportion of compatible characters to incompatible characters was enormous. (Even this was quite surprising; the existence of a tree on which a large proportion of characters is compatible is extremely unlikely in biological data analysis.) This suggested that the basic approach was a good one but that the model had to be extended: A tree model is inappropriate and the evolutionary process is better represented as a "network" [8].

3 Phylogenetic Networks Compatibility: Preliminaries and Complexity

This model of how languages evolve on networks references an underlying rooted tree (modeling "genetic descent") to which bidirectional edges (modeling how linguistic characters can be transmitted through contact) are added. Therefore, the underlying tree is rooted, and the edges of that tree can be naturally oriented from parent to child, whereas the additional edges are by design bidirectional, since contact between language communities can result in the flow of linguistic characters in both directions. This model was formalized in [8] as follows.

Definition 2. *A* phylogenetic network *on a set L of languages is a rooted directed graph $N = (V, E)$ with the following properties:*

   (i) *$V = L \cup I$, where I denotes added nodes which represent ancestral languages, and L denotes the set of leaves of T.*
  (ii) *E can be partitioned between the edges of a tree $T = (V, E_T)$, and the set of "non-tree" edges or bidirectional edges $E' = E - E_T$. For more convenience in the notation, we will refer to a bidirectional edge by a b-edge. The edges in T are oriented from parent to child, and hence T is a directed rooted tree.*

(iii) N is "weakly acyclic", i.e., if N contains directed cycles, then those cycles contain only edges from E'.

(iv) Every internal node in N has at least two children in T.

Properties (iii) and (iv) above will be referred to as the phylogenetic networks properties.

For a phylogenetic network N, we denote by T_N the underlying tree of N. For a node $u \in N$, we denote by $label(u)$ the label of node u, and by $\pi(u)$ the parent of u in T_N. If e is a b-edge between two nodes u and v in the network N, then e has three possible statuses: (1) the edge e can be simply removed denoting that no transfer took place between the two ancestral languages representing u and v, (2) e can be directed from u towards v denoting that the transfer was from the ancestral language representing u to that representing v, or (3) e can be directed from v towards u denoting that the transfer was from the ancestral language representing v to that representing u. If e is directed from u towards v, then the network is transformed as follows. Remove the edge $(\pi(v), v)$ from N, and make u the new parent of v in the resulting network (that is, add the edge (u, v) as a tree edge to the resulting network). Similarly, if e is directed from v towards u, then the edge $(\pi(u), u)$ is removed from N, and the edge (v, u) is added. Note that if there are t b-edges in N, then the t b-edges induce $O(3^t)$ trees based on 3^t different statuses of the t edges. We denote by Γ the set of the trees induced by the t b-edges in N.

An assignment to the statuses of the edges in a network N whose leaves are labeled by a character is said to be *successful* if the character is compatible with the tree induced by this assignment. A *successful labeling* for a compatible tree is a labeling to the nodes of T in which all the nodes with the same label induce a connected subgraph of T.

Note that the order in which the b-edges that are incident on a certain node are assigned can potentially make a difference in the resulting tree.

Definition 3. *Let $N = (V, E)$ be a phylogenetic network on L and Γ be the set of trees induced by all the assignments to the b-edges in N. Let C be a set of characters defined on L, and let $c : L \to Z$ be a character in C. Then c is said to be* compatible on N *if c is compatible on at least one of the trees in Γ. N is called a* Perfect Phylogenetic Network *if all characters in C are compatible on N.*

The CHARACTER COMPATIBILITY ON PHYLOGENETIC NETWORKS problem, denoted henceforth by CCPN, was defined as follows [8].

CCPN
Given a phylogenetic network $N = (V, E)$ on a set L, and a set of characters C defined on L, decide if N is a perfect phylogenetic network.

This problem was shown to be NP-hard [8] for the case where each character has *at least* three states. We will consider the case of the CCPN problem in

which each character has exactly two states. This problem is called the BINARY
CHARACTER COMPATIBILITY ON PHYLOGENETIC NETWORKS, denoted henceforth by BCCPN. This problem is of prime interest on its own in the field of
linguistics (see [6,11,12,14,16,17]).

> **BCCPN**
> Given a phylogenetic network $N = (V, E)$ on a set L, and a set of
> characters C defined on L such that each character in C has two states
> (i.e., binary) decide if N is a perfect phylogenetic network.

Remark 1. Deciding if a network N is perfect phylogenetic on a set of characters
C reduces to deciding if every character $c \in C$ is compatible on N. Therefore,
without loss of generality, we will denote by BCCPN the problem of deciding
whether a given binary character c is compatible on N. The mentioning of c
becomes irrelevant in this case, and we will simply say N is compatible to denote
that the implicit (given) character c is compatible on N.

Theorem 2. BCCPN *is NP-complete.*

Theorem 2 implies that the CCPN problem is NP-complete as well by special-
ization, giving an alternative, yet different, proof to that in [8].

4   A Parameterized Algorithm for BCCPN

A parameterized problem is a set of pairs of the form (x, k) where x is the input instance and k is a positive integer called the *parameter*. A parameterized
problem is said to be *fixed-parameter tractable*, if the problem can be solved in
time $f(k)|x|^c$, where f is a computable function of the parameter k, $|x|$ is the
input size, and c is a constant independent of k [4]. The area of parameterized algorithms and complexity was introduced mainly in the work of Downey
and Fellows [4], and is based on the core observation that for many practical
occurrences of intractable problems some parameters remain small, even if the
problem instances are large.

Taking the advantage of the fact the the the number of b-edges in the phylogenetic network is small [9], the BCCPN problem can be naturally parameterized by the number of b-edges, k, in the phylogenetic network. We call this
problem the PARAMETERIZED BCCPN problem. It is easy to see that the PA-
RAMETERIZED BCCPN problem can be solved in $O(3^k n)$ time, where n is the
number of nodes in the phylogenetic network, by enumerating the status of every b-edge in the network, then checking whether the resulting induced tree is
compatible. We will significantly improve on this upper bound next. The algorithm we present is a decision algorithm deciding if the network is compatible
or not.

Assumption I. Let (N, k) be an instance of PARAMETERIZED BCCPN. If
there is at most one leaf in N of label 0 (resp. 1), then N is compatible. This

is true since if we label all the internal nodes in N with 1 (resp. 0), then every assignment to the b-edges in N is a successful assignment. Since these particular cases can be identified in $O(n)$ time, we will assume henceforth that at any stage of the algorithm, there are at least two leaves of label 0 and at least two leaves of label 1.

Definition 4. *Let N be a phylogenetic network. An internal node s in N is said to be a splitting node if there exists a successful assignment to the b-edges in N that results in a compatible tree T, such that there is a valid labeling for the nodes in T with all the nodes in the subtree rooted at s labeled with the same label, and all the other nodes in the tree labeled with the other (different) label.*

Definition 5. *Let N be a phylogenetic network and suppose that s is a splitting node in N. Let A be a successful assignment to the b-edges in N, and let T be the tree induced by A. The assignment A is said to respect the splitting node s, if there is a valid labeling for the nodes in T with all the nodes in the subtree rooted at s labeled with the same label, and all the other nodes in the tree labeled with the other (different) label.*

Remark 2. Observe that, if we assume the statements in **Assumption I**, then for any compatible phylogenetic network N there is at least one splitting node in N.

The main algorithm, **Phylogenetic_Compatibility**, which solves the PARAMETERIZED BCCPN problem is given in Figure 2. The algorithm **Phylogenetic_Compatibility** tries every node in N as the splitting node. For each node selected as the splitting node, it calls the subroutine **Is_Compatible** to check whether there exists a successful assignment to N that respects the selected splitting node. Thus, the subroutine **Is_Compatible** works under the assumption that the splitting node is given. The subroutine **Is_Compatible** utilizes the subroutines **Clean**, **Reduce**, and **Merge**, given in Figure 1. These subroutines apply some operations to reduce the network N, and also work under the assumption that the splitting node has been selected.

Proposition 1. *Let N be a phylogenetic network such that none of the operations **Reduce**, **Clean**, or **Merge** is applicable to N. Then there exist two nodes u and u' in N such that: (1) $label(u) = label(u')$, (2) (u, u') is a b-edge in N, and (3) all children of u and u' are leaves.*

We call a pair of nodes $\{u, u'\}$ satisfying the three conditions in Proposition 1 a *nice pair*. Proposition 1 establishes the existence of a nice pair in any phylogenetic network N to which none of the operations **Reduce**, **Clean**, or **Merge** is applicable. Now we are ready to present the main algorithm **Phylogenetic_Compatibility** which solves the PARAMETERIZED BCCPN problem. We will assume that

Clean$((u, u'))$

Precondition: $label(u) \neq label(u')$ and (u, u') is a b-edge

1. remove the b-edge (u, u') from N;

Reduce(u)

1. **if** u has two leaf-children with different labels **then reject**;
2. **if** all the children of u are leaves and there is no b-edge incident on u **then**
       if u is marked as the splitting node **then**
           if there is a leaf in N that is not a child of u
             and of the same label as the children of u **then reject**;
           else accept;
       else
           remove u and its children and replace them with a leaf l;
           label l with the same label as the children of u;
           add the tree edge $(\pi(u), l)$;
3. **if** u is unlabeled and has a labeled child w (w could be a leaf) with no
   b-edge incident on w **then**
       if w is marked as the splitting node **then** set $label(u) = 1 - label(w)$;
       else set $label(u) = label(w)$;
4. **if** u is labeled and has an unlabeled child w with no incident b-edge **then**
       if w is marked as the splitting node **then** set $label(w) = 1 - label(u)$;
       else set $label(w) = label(u)$;
5. **if** u is labeled and has at most one leaf-child **then**
       add two leaves as children to u of the same label as u;
6. **if** u has more than two leaves with the same label **then** remove all of them
   except two;

Merge$(\langle u, u' \rangle)$

Precondition: $label(\pi(u)) \neq label(u) = label(u')$ and (u, u') is a b-edge
1. cut off the tree edge $(\pi(u), u)$ from N;
2. remove the b-edge (u, u');
3. identify the two nodes u and u' (i.e., merge the two nodes into one
   new node);
4. let the new node be w; set $label(w) = label(u')$ and $\pi(w) = \pi(u')$ (add the
   tree edge $(\pi(u'), w)$);
5. make the children of both u and u' children of w;
6. shift all the b-edges that are incident on u and u' to make them incident on
   w without changing the other endpoints of the b-edges;
7. **if** u or u' is marked as the splitting node **then** mark the new node w as the
   splitting node;

Fig. 1. The subroutine **Merge**

Assumption I is valid before each operation performed by the algorithm and its
subroutines. The algorithm is given in Figure 2.

Theorem 3. *The algorithm* **Phylogenetic_Compatibility** *is correct.*

Is_Compatible (N, k)

1. if $k = 0$ and N is not compatible **then** reject;
2. **while Reduce** is applicable to a node in N apply it;
3. **if** any of **Clean** or **Merge** is applicable **then** apply it and **go to** step 1;
4. let $\{u, u'\}$ be a nice pair in N; {* assume without loss of generality that $label(u) = label(u') = 1$ *}
    Case 1. Both $\pi(u)$ and $\pi(u')$ are labeled
        remove the b-edge (u, u');
    Case 2. One of $\pi(u)$ and $\pi(u')$ is labeled, say $\pi(u)$. Branch as follows
        first side of the branch: set $label(\pi(u')) = 1$ and remove the
                                    b-edge (u, u');
        second side of the branch: set $label(\pi(u')) = 0$;
    Case 3. (Both $\pi(u)$ and $\pi(u')$ are unlabeled.) Branch as follows
        first side of the branch: set $label(\pi(u)) = 0$;
        second side of the branch: set $label(\pi(u')) = 0$;
        third side of the branch: set $label(\pi(u)) = label(\pi(u')) = 1$ and
                                    remove the b-edge (u, u');

Phylogenetic_Compatibility

Input: an instance (N, k) of PARAMETERIZED BCCPN where N is a
        phylogenetic network and k is a positive integer
Output: yes/no decision based on whether N is compatible or not
1. **for** every node s in N **do**
    1.1. $N' = N$;
    1.2. mark s as the splitting node in N';
    1.3. call **Is_Compatible** on (N', k);
    1.4. **if Is_Compatible** returns yes **then return** yes;
2. **return** (no);

Fig. 2. Is_Compatible and Phylogenetic_Compatibility

5 Analysis of the Algorithm Is_Compatible

To analyze the running time of the algorithm **Phylogenetic_Compatibility**, and since the algorithm **Phylogenetic_Compatibility** ends up calling the subroutine **Is_Compatible** $O(n)$ times, it suffices to analyze the running time of **Is_Compatible** and multiply it by $O(n)$. The subroutine **Is_Compatible** is a branch-and-bound process, and its execution can be depicted by a search tree. Therefore, the main step in the analysis is deriving an upper bound on the number of leaves in the search tree. The branches performed by the subroutine **Is_Compatible** can be classified into two branches: $(1, 1)$-branches and $(1, 1, 1)$-branches. The latter branch corresponds to an $O(3^k)$ upper bound on the size of the search tree, matching the bound of a trivial brute-force algorithm that enumerates each of the three statuses of every b-edge. Differing from the common analysis techniques based on the worst-case scenario, we use a novel way for analyzing the size of the search tree using amortized techniques, and obtain:

Lemma 1. *Let T be the search corresponding to the subroutine* **Is-Compatible** *on an instance* (N, k). *The number of leaves of* T *is* $O(2^k)$.

Theorem 4. *The* PARAMETERIZED BCCPN *problem can be solved in time* $O(2^k n^2)$ *where* n *is the number of nodes in the network.*

References

1. H. Bodlaender, M. Fellows, and T. Warnow. Two strikes against perfect phylogeny. In *Proceedings of ICALP'92*, LNCS, pages 273–283. Springer Verlag, 1992.
2. A.J. Dobson. Unrooted trees for numerical taxonomy. Unpublished manuscript.
3. A.J. Dobson. Lexicostatistical grouping. *Anthropological Linguistics*, 11:216–221, 1969.
4. R. Downey and M. Fellows. *Parameterized Complexity*. Springer, New York, 1999.
5. H.A. Gleason. Counting and calculating for historical reconstruction. *Anthropological Linguistics*, 1:22–32, 1959.
6. Russell D. Gray and Quentin D. Atkinson. Language-tree divergence times support the anatolian theory of indo-european origin. *Nature*, 426(6965):435–439, November 2003.
7. J.P. Mallory. *In Search of the Indo-Europeans*. Thames and Hudson, London, 1989.
8. L. Nakhleh. *Phylogenetic Networks*. PhD thesis, The University of Texas at Austin, 2004.
9. L. Nakhleh, D. Ringe, and T. Warnow. Perfect phylogenetic networks: A new methodology for reconstructing the evolutionary history of natural languages. *LANGUAGE*, 2005. In press.
10. D. Ringe. Some consequences of a new proposal for subgrouping the IE family. In B.K. Bergen, M.C. Plauche, and A. Bailey, editors, *24th Annual Meeting of the Berkeley Linguistics Society, Special Session on Indo-European Subgrouping and Internal Relations*, pages 32–46, 1998.
11. D. Ringe, T. Warnow, and A. Taylor. Indo-European and computational cladistics. *Transactions of the Philological Society*, 100(1):59–129, 2002.
12. D. Ringe, T. Warnow, A. Taylor, A. Michailov, and L. Levison. Computational cladistics and the position of Tocharian. In V. Mair, editor, *The Bronze Age and early Iron Age peoples of Eastern Central Asia*, pages 391–414. 1998.
13. R.G. Roberts, R. Jones, and M.A. Smith. Thermoluminescence dating of a 50,000-year-old human occupation site in Northern Australia. *Science*, 345:153–156, 1990.
14. A. Taylor, T. Warnow, and D. Ringe. Character-based reconstruction of a linguistic cladogram. In J.C. Smith and D. Bentley, editors, *Historical Linguistics 1995, Volume I: General issues and non-Germanic languages*, pages 393–408. Benjamins, Amsterdam, 2000.
15. T. Warnow. Mathematical approaches to comparative linguistics. *Proc. Natl. Acad. Sci.*, 94:6585–6590, 1997.
16. T. Warnow, D. Ringe, and A. Taylor. Reconstructing the evolutionary history of natural languages. Technical Report 95-16, Institute. for Research in Cognitive Science, Univ. of Pennsylvania, 1995.
17. T. Warnow, D. Ringe, and A. Taylor. Reconstructing the evolutionary history of natural languages. In *ACM-SIAM Symposium on Discrete Algorithms (SODA)*, pages 314–322, 1996.
18. J.P. White and J.F. O'Connell. *A Prehistory of Australia, New Guinea, and Sahul*. Academic Press, New York, 1982.

On Dynamic Bin Packing: An Improved Lower Bound and Resource Augmentation Analysis

Wun-Tat Chan[1,*], Prudence W.H. Wong[2,**], and Fencol C.C. Yung[1]

[1] Department of Computer Science, University of Hong Kong, Hong Kong
wtchan@cs.hku.hk, ccyung@graduate.hku.hk
[2] Department of Computer Science, University of Liverpool, UK
pwong@csc.liv.ac.uk

Abstract. We study the dynamic bin packing problem introduced by Coffman, Garey and Johnson [7]. This problem is a generalization of the bin packing problem in which items may arrive and depart from the packing dynamically. The main result in this paper is a lower bound of 2.5 on the achievable competitive ratio, improving the best known 2.428 lower bound [3], and revealing that packing items of restricted form like unit fractions (i.e., of size $1/k$ for some integer k), which can guarantee a competitive ratio 2.4985 [3], is indeed easier.

We also investigate the resource augmentation analysis on the problem where the on-line algorithm can use bins of size b (> 1) times that of the optimal off-line algorithm. An interesting result is that we prove $b = 2$ is both necessary and sufficient for the on-line algorithm to match the performance of the optimal off-line algorithm, i.e., achieve 1-competitiveness. Further analysis is made to give a trade-off between the bin size multiplier b and the achievable competitive ratio.

1 Introduction

Bin packing is a classical combinatorial optimization problem (see the surveys [11, 8, 5]). The objective is to pack a sequence of items into a minimum number of bins such that the total size of the items in a bin does not exceed the bin capacity. The on-line version of the problem assumes that items may arrive at arbitrary time and no advance information is known about the items not yet arrived. *Dynamic bin packing* (DBP) was introduced as a generalization of the on-line bin packing by Coffman, Garey and Johnson [7]. In this generalization, items may also depart at arbitrary time and both on-line and off-line algorithms are not allowed to move items from one bin to another. The goal is to minimize the maximum number of bins used over all time.

The performance of an on-line algorithm $\mathcal{A}$ is generally measured by its *competitive ratio* [2]. For our problem where a sequence of item arrivals and departures is given, the competitive ratio c is the worst case ratio between the

---

* This research is partly supported by Hong Kong RGC Grant HKU5172/03E.
** This research is partly supported by Nuffield Foundation Grant NAL/01004/G.

D.Z. Chen and D.T. Lee (Eds.): COCOON 2006, LNCS 4112, pp. 309–319, 2006.
© Springer-Verlag Berlin Heidelberg 2006

maximum number of bins used by $\mathcal{A}$ over all time and the maximum number of bins used by the optimal off-line algorithm (which knows the whole sequence in advance) over all time. Algorithm $\mathcal{A}$ is said to be *c-competitive*.

Coffman, Garey and Johnson [7] proved that the lower bound on the competitive ratio of any on-line algorithm on dynamic bin packing is 2.388 [1]. They also showed that a modified version of first-fit is 2.788-competitive [7]. Chan, Lam and Wong [3] improved the lower bound to 2.428 by considering only *unit fraction items*, where a unit fraction item is an item of size $1/k$ for some integer k. They also showed that for packing unit fraction items only, first-fit is 2.4985-competitive [3]. A natural question arises: Is packing items of restricted form, such as unit fraction items, as difficult as packing items of general form? In another aspect, *resource augmentation analysis* [15] has been studied in the context of on-line bin packing [12,13], in which an on-line algorithm can use bins of size b (> 1) times that of the optimal off-line algorithm. To our knowledge, there is no previous work on resource augmentation analysis for dynamic bin packing. We address the above questions in this paper.

Our contributions. This paper presents the following results on DBP.
1. We push up the lower bound on competitive ratio from 2.428 [3] to 2.5 [2], giving a negative answer to the question that packing unit fraction items is as difficult as packing general items because packing unit fraction items attains a competitive ratio 2.4985 [3] (< 2.5).
2. We investigate on resource augmentation analysis, showing an interesting result that doubling the bin size for the on-line algorithm is both necessary and sufficient to match the performance of the optimal off-line algorithm, i.e., to attain 1-competitiveness. Further analysis is made to give a trade-off between the bin size multiplier b (for $1 < b \leq 2$) and the achievable competitive ratio.

Related work. There is a long history of results for the classical bin packing problem and its variants [11,8,5]. The best upper bound and lower bound on the competitive ratio for on-line bin packing to date are 1.58889 [16] and 1.54014 [17], respectively. The upper bound reveals that dynamic bin packing is more difficult than on-line bin packing. For both dynamic and on-line bin packing, items of various restricted forms have been studied, which include unit fraction items [1,3], items of divisible sizes [6] (where each possible item size can be divided by the next smaller item size), and items of discrete sizes [4,10,9] (where possible item sizes are $\{1/k, 2/k, \cdots, j/k\}$ for some $1 \leq j \leq k$). Resource augmentation analysis for on-line bin packing has been studied [12,13]; matching upper and lower bounds (up to an additive constant) are given for bounded space bin packing [12] in which there is a limit on the number of opened bins that can be used at any time; better upper bound has been derived for (unbounded space)

---

[1] A variant of the problem is to assume a stronger off-line algorithm that can repack the current set of items into the minimum possible number of bins each time a new item arrives, in which case a stronger lower bound of 2.5 is achieved [7].

[2] There was a 2.5 lower bound [7] when the off-line algorithm can repack, our result achieves the same bound even when the off-line algorithm does not repack.

on-line bin packing [13]. Ivkovic and Lloyd studied the *fully dynamic bin packing problem* [14], which is a variant of dynamic bin packing that allows repacking of items for each item arrival or departure. They gave a 1.25-competitive on-line algorithm for the problem [14].

Notations. We now give a precise definition of the problem and the necessary notations for further discussion. In dynamic bin packing, items arrive and depart at arbitrary time. Each item comes with a size. We denote by s-*item* an item of size s. When an item arrives, it must be assigned to a bin immediately without exceeding the capacity of the assigned bin. At any time, the *load* of a bin is the total size of items currently assigned to that bin that have not yet departed. We denote by ℓ-*bin* a bin of load ℓ. Migration is not allowed, i.e., once an item is assigned to a bin, it cannot be moved to another bin. The objective is to minimize the maximum number of bins used over all time.

In the resource augmentation analysis (Sections 3 and 4), an on-line algorithm $\mathcal{A}$ is given size-b bins with $1 \leq b \leq 2$, while the optimal off-line algorithm uses size-1 bins. Consider any input sequence σ. Let $\mathcal{A}_b(\sigma, t)$ denote the number of size-b bins used at time t by $\mathcal{A}$, similarly, we have $\mathcal{O}_1(\sigma, t)$ for the optimal off-line algorithm. $\mathcal{A}$ is said to be c-competitive if there exists a constant k such that for any input sequence σ, $\max_t \mathcal{A}_b(\sigma, t) \leq c \cdot \max_t \mathcal{O}_1(\sigma, t) + k$.

Organization of the paper. In Section 2, we present the 2.5 lower bound. In Section 3, we show that doubling the bin size is both necessary and sufficient to achieve 1-competitiveness. In Section 4, we study the trade-off between bin size and competitive ratio. Finally, we give some concluding remarks in Section 5.

2   A 2.5 Lower Bound

In this section, we prove that no on-line algorithm can be better than 2.5-competitive. Consider any on-line algorithm $\mathcal{A}$. The adversary works in stages using items of various sizes including ϵ, $\frac{1}{6}$, $\frac{1}{3}$, $\frac{1}{2} - \frac{\epsilon}{4}$, $\frac{1}{2}$, $\frac{1}{2} + \frac{\epsilon}{4}$, $\frac{2}{3}$ and 1, where ϵ is a small constant to be defined later. Roughly speaking, the adversary first releases some items of a particular size. Depending on how $\mathcal{A}$ packs the items, the adversary lets some items depart and further releases some other items such that the total size of items present at any time is always the same (with some minor difference). The choices of items to be departed ensures that the space released from some departed items cannot be reused for newly arrived items, thus, forcing $\mathcal{A}$ to use more new bins. The adversary works as follows.

Let $\epsilon = \frac{1}{18k}$ for some large positive integer k. Recall that for any $s > 0, \ell > 0$, we denote by s-*item* an item of size s, and by ℓ-*bin* a bin of load ℓ. When we discuss how items are packed into bins, we denote by $x{:}\ell$ that there are x bins of load ℓ, and by $\{x_1{:}\ell_1, x_2{:}\ell_2, ...\}$ a packing configuration. The adversary releases items in stages such that $\mathcal{A}$ uses a maximum of $45k$ bins while the total size of items at any time is no more than $18k + 2$. The item sizes are chosen carefully to allow the optimal off-line algorithm to use at most $18k + 2$ bins.

Theorem 1. *No on-line algorithm is better than 2.5-competitive.*

Consider any on-line algorithm $\mathcal{A}$. Let n_i be the number of new bins used by $\mathcal{A}$ in Stage i. In Stage 1, $\frac{18k}{\epsilon}$ items of size ϵ are released, thus, $n_1 \geq 18k$. We distinguish between three cases: $n_1 \geq 24k$, $24k > n_1 \geq 21k$, and $21k > n_1 \geq 18k$. The first case is the easiest case and we skip the details. We focus on the second case and the third case can be handled in a similar way.

Case 2: $24k > n_1 \geq 21k$. We make the following observations.

Observation 1. *If $24k > n_1 \geq 21k$, then $\mathcal{A}$ uses at least $6k$ bins of load at least $\frac{2}{3} + \epsilon$, $12k$ bins at least $\frac{1}{2} + \epsilon$, and $15k$ bins at least $\frac{1}{3} + \epsilon$ at the end of Stage 1.*

Proof. Assume that there are less than $6k$ bins of load at least $\frac{2}{3} + \epsilon$. The remaining bins has a maximum load of $\frac{2}{3}$. Then the maximum load that has been packed is $< 6k + (24k - 6k)(\frac{2}{3}) = 18k$, contradicting that a total load of $18k$ has been released. The other cases are similar. □

In Stage 2, we let items depart until $\{6k:(\frac{2}{3} + \epsilon), 6k:(\frac{1}{2} + \epsilon), 3k:(\frac{1}{3} + \epsilon), 6k:\epsilon\}$. We then release $30k$ items of size $\frac{1}{3}$. If $\mathcal{A}$ packs a $\frac{1}{3}$-item into some bin B of load $\frac{1}{2} + \epsilon$, we depart a total size of $\frac{1}{6}$ of ϵ-item from B, making its load become $\frac{2}{3} + \epsilon$. For every two such bins, we further release one $\frac{1}{3}$-item. Repeat departing groups of ϵ-items of size $\frac{1}{6}$ and releasing items of size $\frac{1}{3}$ as long as $\mathcal{A}$ packs a $\frac{1}{3}$-item into a bin of load $\frac{1}{2} + \epsilon$. This process must terminate because once $\mathcal{A}$ packs a $\frac{1}{3}$-item into a bin, its load becomes $\frac{2}{3} + \epsilon$, meaning that it cannot accommodate another $\frac{1}{3}$-item. We assume that there are even number of $(\frac{1}{2} + \epsilon)$-bins that are packed with $\frac{1}{3}$-items; the other case is similar.

Let x and y be the number of $(\frac{1}{2} + \epsilon)$- and $(\frac{1}{3} + \epsilon)$-bins, respectively, that have been packed a $\frac{1}{3}$-item at the end of Stage 2. Let a_1 and a_2 be the number of new bins (used in Stage 2) that have been packed exactly one $\frac{1}{3}$-item and at least two $\frac{1}{3}$-items, respectively, i.e., $n_2 = a_1 + a_2$. The total load of all bins is $\leq (6k + x + y + 6k)(\frac{2}{3} + \epsilon) + (6k - x)(\frac{1}{2} + \epsilon) + (3k - y)(\frac{1}{3} + \epsilon) + a_2 + \frac{a_1}{3}$. This quantity must be $\geq 18k + 21k\epsilon$. Using the property that $x \leq 6k$ and $y \leq 3k$, we can derive that $a_1 + a_2 \geq 4k$.

We further consider two sub-cases: $a_1 + a_2 \geq 9k$ and $9k > a_1 + a_2 \geq 4k$.

Case 2.1: $a_1 + a_2 \geq 9k$. In Stage 3, we depart items until $\{21k:\epsilon, 9k:\frac{1}{3}\}$. Finally, release $15k$ items of size 1, thus, $n_3 = 15k$. The total number of bins used by $\mathcal{A}$ becomes $21k + 9k + 15k = 45k$.

Case 2.2: $9k > a_1 + a_2 \geq 4k$. Figure 1 shows the target configuration the adversary achieves. Using a similar idea as before, we can show that $a_2 > k$.

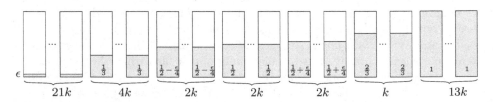

Fig. 1. The final configuration achieved by the adversary in Case 2.2

The remaining stages run as follows.

3. Depart items until
$$\{6k{:}(\tfrac{1}{2} + \epsilon),\ x{:}(\tfrac{1}{3} + \tfrac{1}{6} + \epsilon),\ (6k - x){:}(\tfrac{1}{2} + \epsilon),\ 3k{:}\epsilon,\ 6k{:}\epsilon,\ k{:}(\tfrac{1}{3} + \tfrac{1}{3}),\ 3k{:}\tfrac{1}{3}\ \}.$$
This is possible because $a_2 > k$ and $a_1 + a_2 \geq 4k$. We then release $20k$ items of size $\tfrac{1}{2} - \tfrac{\epsilon}{4}$. Since only bins of load ϵ and $\tfrac{1}{3}$ can accommodate one such item, $n_3 \geq (20k - 12k)/2 = 4k$.

4. Depart items until
$$\{\ 6k{:}(\tfrac{1}{2} + \epsilon),\ x{:}(\tfrac{1}{3} + \tfrac{1}{6} + \epsilon),\ (6k - x){:}(\tfrac{1}{2} + \epsilon),\ 3k{:}\epsilon,\ 6k{:}\epsilon,$$
$$k{:}(\tfrac{1}{3} + \tfrac{1}{3}),\ 3k{:}\tfrac{1}{3},\ 2k{:}(\tfrac{1}{2} - \tfrac{\epsilon}{4})\ \}.$$
Next we release $18k$ items of size $\tfrac{1}{2}$, making $n_4 \geq (18k - 14k)/2 = 2k$.

5. Depart items until
$$\{\ 6k{:}(\tfrac{1}{2} + \epsilon),\ x{:}(\tfrac{1}{3} + \tfrac{1}{6} + \epsilon),\ (6k - x){:}(\tfrac{1}{2} + \epsilon),\ 3k{:}\epsilon,\ 6k{:}\epsilon,$$
$$k{:}(\tfrac{1}{3} + \tfrac{1}{3}),\ 3k{:}\tfrac{1}{3},\ 2k{:}(\tfrac{1}{2} - \tfrac{\epsilon}{4}),\ 2k{:}\tfrac{1}{2}\ \}.$$
Releasing $16k$ items of size $\tfrac{1}{2} + \tfrac{\epsilon}{4}$ makes $n_5 \geq 16k - 14k = 2k$.

6. Depart items until
$$\{\ 6k{:}(\tfrac{1}{3} + \epsilon),\ x{:}(\tfrac{1}{3} + \epsilon),\ (6k - x){:}(\tfrac{1}{3} + \epsilon),\ 3k{:}\epsilon,\ 6k{:}\epsilon,$$
$$k{:}(\tfrac{1}{3} + \tfrac{1}{3}),\ 3k{:}\tfrac{1}{3},\ 2k{:}(\tfrac{1}{2} - \tfrac{\epsilon}{4}),\ 2k{:}\tfrac{1}{2},\ 2k{:}(\tfrac{1}{2} + \tfrac{\epsilon}{4})\ \}.$$
We then release $13k$ items of size $\tfrac{2}{3}$, making $n_6 \geq 13k - 12k = k$.

7. Depart items until
$$\{6k{:}\epsilon,\ x{:}\epsilon,\ (6k - x){:}\epsilon,\ 3k{:}\epsilon,\ 6k{:}\epsilon,\ k{:}\tfrac{1}{3},\ 3k{:}\tfrac{1}{3},\ 2k{:}(\tfrac{1}{2} - \tfrac{\epsilon}{4}),\ 2k{:}\tfrac{1}{2},\ 2k{:}(\tfrac{1}{2} + \tfrac{\epsilon}{4}),\ k{:}\tfrac{2}{3}\}.$$

A final of $13k$ items of size 1 are released, making $n_7 = 13k$. Totally, $\mathcal{A}$ uses $21k + 4k + 2k + 2k + 2k + k + 13k = 45k$ bins.

Case 3: $21k > n_1 \geq 18k$. Figure 2 shows the target configuration the adversary aims to achieve. We leave the details in the full paper.

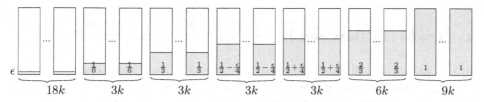

ϵ

| $18k$ | $3k$ | $3k$ | $3k$ | $3k$ | $6k$ | $9k$ |

with markings: $\tfrac{1}{6}$, $\tfrac{1}{6}$, $\tfrac{1}{3}$, $\tfrac{1}{3}$, $\tfrac{1}{2} - \tfrac{\epsilon}{4}$, $\tfrac{1}{2} - \tfrac{\epsilon}{4}$, $\tfrac{1}{2} + \tfrac{\epsilon}{4}$, $\tfrac{1}{2} + \tfrac{\epsilon}{4}$, $\tfrac{2}{3}$, $\tfrac{2}{3}$, 1, 1

Fig. 2. The final configuration achieved by the adversary in Case 3

3 1-Competitive if and Only if Size-2 Bins Are Used

In this section, we show that using size-2 bins is both necessary (Theorem 3) and sufficient (Theorem 2) to achieve 1-competitiveness. Any-fit (AF) is an algorithm that always packs a new item into a non-empty bin arbitrarily as long as the bin can accommodate the item.

Theorem 2. *Any fit algorithm with size-2 bins is 1-competitive.*

Proof. Suppose AF uses n size-2 bins for a sequence of items. When AF first uses n bins due to the arrival of a new item X, all the existing $n - 1$ bins must have a load greater than 1, otherwise, X can be packed into one of these bins and AF does not need to open a new bin. In other words, the total load of items is at least $n - 1 + s$ where s is the size of X. Any algorithm using unit-size bins needs at least n bins to pack all these items. Therefore, the maximum number of size-2 bins used by AF is at most that used by the optimal off-line algorithm. $\square$

Theorem 3. *No on-line algorithm can be 1-competitive by using size-x bins, for any $x < 2$.*

Proof. Suppose $x = 2 - \epsilon$, for some small $\epsilon > 0$. Let k be a positive integer such that $1/2^{k-1} \geq \epsilon > 1/2^k$. Notice the size satisfies the property $2 - 1/2^{k-1} \leq 2 - \epsilon = x < 2 - 1/2^k$. The adversary works in two phases.

In the first phase, release 2^{3k-1} items of size $1/2^k$. The total load of the items is 2^{2k-1} and all items can be packed into 2^{2k-1} unit-size bins. If the on-line algorithm uses more than 2^{2k-1} bins, we are done. So we only need to consider the case in which the on-line algorithm uses at most 2^{2k-1} bins. We are going to prove that the on-line algorithm uses at least 2^k bins with load at least $1 - 1/2^k$. Let Y be the number of such bins. Then the total load accommodated by the on-line algorithm is at most $Y(2 - 1/2^{k-1}) + (2^{2k-1} - Y)(1 - 1/2^{k-1}) = 2^{2k-1}(1 - 1/2^{k-1}) + Y$. This load cannot be smaller than the total load of items, i.e, $2^{2k-1}(1 - 1/2^{k-1}) + Y \geq 2^{2k-1}$. In other words, $Y \geq 2^{2k-1}/2^{k-1} = 2^k$.

In the second phase, we retain a load of $1 - 1/2^k$ in 2^k bins and let all other items depart. Then release $2^{2k-1} - (2^k - 1)$ items of size 1. Notice that none of these items can be packed into an existing bin because $1 + 1 - 1/2^k$ is greater than x, the size of the bin. Therefore, the total number of bins used by the on-line algorithm is $2^{2k-1} + 1$. It can be shown that the optimal off-line algorithm only needs 2^{2k-1} bins. We leave the details in the full paper. $\square$

4   Trade-Off Between Bin Size and Competitive Ratio

In this section, we discuss results where the on-line algorithm uses bins of size $1 < b < 2$. We first give a general lower bound for any on-line algorithm. Then we analyze the performance of first-fit (packs to the first bin that can fit), best-fit (heaviest loaded bin) and worst-fit (lightest loaded bin) giving their upper bounds.

4.1   General Lower Bound for $1 < b < 2$

In this section, we describe two adversaries, one gives better lower bound for $1 < b < 1.5$ and the other for $1.5 \leq b < 2$.

Lemma 1. *No on-line algorithm using size-b bins can be better than $\frac{2}{b}$-competitive.*

Proof. Consider any on-line algorithm $\mathcal{A}$. Let ϵ be a small constant and $k = \lfloor \frac{1}{\epsilon} \rfloor - 2$. The adversary runs in 3 stages.

1. Release ϵ-items of total size k. If $\mathcal{A}$ uses more than $\frac{2k}{b}$ bins, we are done. Otherwise, we claim that there must be at least $(\frac{2}{b} - 1)k$ bins with load $\geq b - 1 + \epsilon$, otherwise, the total possible load accommodated by $\mathcal{A}$ is less than $((\frac{2}{b} - 1)k - 1)b + (k + 1)(b - 1 + \epsilon) = k - 1 + (k + 1)\epsilon < k$, contradiction.
2. Depart items until $\{ k(\frac{2}{b} - 1):(b - 1 + \epsilon) \}$. Then release k items of size $\frac{b}{2} + \epsilon$. At most one such item can be packed into an existing bin or an empty bin. So, at least $k - (\frac{2}{b} - 1)k = 2k(1 - \frac{1}{b})$ new bins are opened.
3. Depart items until $\{ k(\frac{2}{b} - 1):(b - 1 + \epsilon), 2k(1 - \frac{1}{b}):(\frac{b}{2} + \epsilon) \}$. Finally, release $(\frac{2}{b} - 1)k$ items of size 1. None of the items can be packed into existing bins, thus, another $(\frac{2}{b} - 1)k$ new bins are opened. Number of bins used becomes $\frac{2k}{b}$.

We can show that the items can be packed into $k + O(1)$ unit-size bins.   □

Let m be the largest integer such that $b - \frac{1}{m} < 1$ and let $k = m!(m-1)!$. We define the functions $\alpha(i)$ and $\beta(i)$ for any positive integer $1 \leq i \leq m$ as follows. Let $\alpha(1) = k, \beta(i) = \sum_{j=1}^{i} \alpha(j)$,
$$\alpha(i) = \beta(i-1)\left(\frac{m+1-i}{m+2-i} - \frac{m-i}{m+1-i}\right).$$
In other words, $\alpha(i) = \frac{\beta(i-1)}{(m+2-i)(m+1-i)}$. E.g., $\alpha(2) = \frac{k}{m(m-1)}, \alpha(3) = \frac{k+\frac{k}{m(m-1)}}{(m-1)(m-2)}$.

Lemma 2. *No on-line algorithm using size-b bins can be better than $\frac{\beta(m)}{m!(m-1)!}$-competitive, where m is the largest integer such that $b - \frac{1}{m} < 1$.*

Proof. This adversary makes use of unit fraction items, i.e., in the form $1/w$, for some integer w. The following fact can be proved by simple arithmetic.

Fact 1. *Both $\alpha(i)$ and $\beta(i)$ are integer multiples of $(m-i+1)!(m-i)!$*

Consider any on-line algorithm $\mathcal{A}$. The adversary runs in m stages. In Stage 1, we release $\frac{1}{m}$-items up to a total size of k, i.e., km such items. For each stage $2 \leq i \leq m$, we depart some items released in previous stage and then release some $\frac{1}{m+1-i}$-items, such that in Stage i, the following invariants are maintained: (1) items of a total size $\beta(i-1)\frac{m+1-i}{m+2-i}$ depart and the same size of $\frac{1}{m+1-i}$-items are released, keeping the total size of items being k; and (2) $\mathcal{A}$ uses at least $\alpha(i)$ more new bins at the end of the stage. Stage i proceeds as follows.

a. Retain one $\frac{1}{m+2-i}$-item from each of the $\alpha(i-1)$ new bins used in Stage $i-1$ and let all other $\frac{1}{m+2-i}$-items depart, i.e., only retain $\alpha(i-1)$ such items.
b. Release items of size $\frac{1}{m+1-i}$ until the total size of all items is k.

By this adversary, we can prove by induction that the invariants hold. In other words, at the end of Stage m, $\mathcal{A}$ uses a total of at least $\beta(m)$ bins.

Consider the optimal off-line algorithm. Note that the total size of items at any time is kept at k. Furthermore, the total size of items of the same type of item is always an integer because of Fact 1, and since the item size is unit fraction, we can always pack the same type of item fully into the same unit-size bin. Therefore, the optimal off-line algorithm only needs k bins.   □

The lower bound below follows from Lemmas 1 and 2 (see Figure 3 for the trend).

Theorem 4. *No on-line algorithm using size-b bins is better than* $\max\{\frac{\beta(m)}{m!(m-1)!}, \frac{2}{b}\}$*-competitive, where m is the largest integer such that* $b-\frac{1}{m} < 1$.

4.2   Performance of First-Fit for $1 < b < 2$

In this section, we analyze the upper bound of the first-fit algorithm (FF) using size-b bins. To simplify the discussion, we refer to two properties pointed out by Coffman et al. [7]. (1) We can focus on the input sequences such that the maximum number of bins used by FF when the last item is packed and not before. (2) No non-empty bin ever becomes empty during the execution of FF on input sequences satisfying the first property. It can be shown [7] that the two properties are true because FF will work out the same packing for the modified input sequence, e.g., a modified sequence in which the items packed to that non-empty bin are removed. By the second property, we can label the non-empty bins by the order they became non-empty, i.e., bin i refers to the i-th bin used by FF, and the labels never change.

Theorem 5. *The competitive ratio of FF using size-b bins is at most* $\min\{\frac{2b+1}{2b-1}, \frac{5-b}{(2b-1)}, \frac{b^2+3}{b(2b-1)}\}$.

Proof. Let k denote the maximum number of bins used by the optimal off-line algorithm using unit-size bins. Suppose x is the last (maximum) bin that FF ever packs an item of size $\leq 1/2$. Let B be the last bin that FF opens with an item of size $\leq b/2$ and y be the number of bins (including B) with label $> x$ whose smallest item has a size in the range $(1/2, b/2]$ when FF opens B. Let $z = n - x - y$, where n is the maximum number of bins used by FF. We claim that the following inequalities hold.

$$(x-1)(b-1/2) \leq k \tag{1}$$
$$xb/2 + (y-1) \leq k \tag{2}$$
$$(x+y)(b-1) + zb/2 \leq k \tag{3}$$
$$x(b-1) + y/2 + zb/2 \leq k \tag{4}$$
$$y + z \leq k \tag{5}$$

Using the above inequalities, we can show that (i) $x + y + z \leq \frac{2b+1}{2b-1}k + 1$; (ii) $x + y + z \leq \frac{5-b}{(2b-1)}k + O(1)$; and (iii) $x + y + z \leq \frac{b^2+3}{b(2b-1)}k + O(1)$. $\square$

Figure 3 shows how the competitive ratio of FF varies with b. Notice that our formula in Theorem 5 reaches the value 1 when $b = 2$ matching Theorem 2; yet when $b = 1$, the value is 3, not matching the existing best upper bound of 2.788 [7]. We leave it as an open question to close the gap between the upper and lower bounds. We now state the performance of best-fit (BF) and worst-fit (WF) (see the upper three curves in Figure 3 and Table 1).

Theorem 6. *(1) BF using size-b bins is $\frac{1}{b-1}$-competitive, this bound is tight.*
(2) WF using size-b bins is $\frac{4}{b^2}$-competitive; on the other hand, its competitive ratio is no better than $\min\{\frac{2+b}{b}, \frac{b^2-8b+20}{4b}\}$.

Proof (Sketch). *Upper bound of BF.* Suppose BF uses a maximum of n bins. When BF first packs an item into bin n, the load of bin i for $i < n$ is at least $b-1$, otherwise, BF can pack the item into those bins instead of opening a new bin. Therefore, the optimal off-line algorithm needs $k \geq (n-1)(b-1)$ bins, and hence the competitive ratio of BF is at most $\frac{k}{b-1}$ since $n \leq \frac{k}{b-1} + 1$.

Upper bound of WF. Let k denote the maximum number of bins used by the optimal off-line algorithm using unit-size bins. Suppose WF uses a maximum of n bins. Let x be the largest integer such that there is no item of size $> \frac{b}{2}$ in bin x at the item instance where WF packs the first item into bin n. Let $y = n - x$. We claim the following inequalities hold.

$$yb/2 + x(b-1) \leq k \tag{6}$$
$$xb/2 \leq k \tag{7}$$

The inequalities can be proved in a similar way as in the analysis for FF. By the two inequalities and simple arithmetic, we can show that $x + y \leq 4/b^2 + 1$.

The lower bounds will be given in the full paper.   □

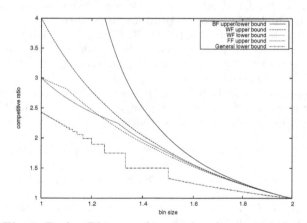

Fig. 3. Trade-off between bin size b and competitive ratio

Table 1. Summary of results for bin size $1 \leq b < 2$

Algorithm	upper bound	lower bound
BF	$\frac{1}{b-1}$	$\frac{1}{b-1}$
WF	$\frac{4}{b^2}$	$\min\{\frac{2+b}{b}, \frac{b^2-8b+20}{4b}\}$
FF	$\min\{\frac{2b+1}{2b-1}, \frac{5-b}{(2b-1)}, \frac{b^2+3}{b(2b-1)}\}$	$\max\{\frac{\beta(m)}{m!(m-1)!}, \frac{2}{b}\}$

5   Concluding Remarks

In this paper, we have shown a 2.5 lower bound for dynamic bin packing, revealing that dynamic bin packing of general items is more difficult than unit fraction items. An open question is to close the gap between this 2.5 lower bound and the 2.788 upper bound [7]. We believe it is possible to push down the upper bound by analyzing some modified version of FF. One can also analyze other algorithms, like the class of Harmonic algorithms [17], yet our preliminary study showed that some versions of Harmonic algorithm have a non-constant lower bound for DBP; further investigation on other variants of Harmonic algorithms is desirable. We also give the first resource augmentation analysis for dynamic bin packing, showing that doubling bin size is both necessary and sufficient to achieve 1-competitiveness. Trade-off between bin size and competitive ratio is also studied. Note that the formula derived for the upper bound of FF does not yet match the general lower bound. We are attempting to give tighter bounds to close the gap.

References

1. A. Bar-Noy, R. E. Ladner, and T. Tamir. Windows scheduling as a restricted version of bin packing. In J. I. Munro, editor, *SODA*, pages 224–233. SIAM, 2004.
2. A. Borodin and R. El-Yaniv. *Online Computation and Competitive Analysis*. Cambridge University Press, 1998.
3. W. T. Chan, T. W. Lam, and P. W. H. Wong. Dynamic bin packing of unit fractions items. In *ICALP*, pages 614–626. Springer, 2005.
4. E. G. Coffman, Jr., C. Courcoubetis, M. R. Garey, D. S. Johnson, P. W. Shor, R. R. Weber, and M. Yannakakis. Bin packing with discrete item sizes, Part I: Perfect packing theorems and the average case behavior of optimal packings. *SIAM J. Discrete Math.*, 13:38–402, 2000.
5. E. G. Coffman, Jr., G. Galambos, S. Martello, and D. Vigo. Bin packing approximation algorithms: Combinatorial analysis. In D.-Z. Du and P. M. Pardalos, editors, *Handbook of Combinatorial Optimization*. Kluwer Academic Publishers, 1998.
6. E. G. Coffman, Jr., M. Garey, and D. Johnson. Bin packing with divisible item sizes. *Journal of Complexity*, 3:405–428, 1987.
7. E. G. Coffman, Jr., M. R. Garey, and D. S. Johnson. Dynamic bin packing. *SIAM J. Comput.*, 12(2):227–258, 1983.
8. E. G. Coffman, Jr., M. R. Garey, and D. S. Johnson. Bin packing approximation algorithms: A survey. In D. S. Hochbaum, editor, *Approximation Algorithms for NP-Hard Problems*, pages 46–93. PWS, 1996.
9. E. G. Coffman, Jr., D. S. Johnson, L. A. McGeoch, P. W. Shor, and R. R. Weber. Bin packing with discrete item sizes, Part III: Average case behavior of FFD and BFD. in preparation.
10. E. G. Coffman, Jr., D. S. Johnson, P. W. Shor, and R. R. Weber. Bin packing with discrete item sizes, Part II: Tight bounds on first fit. *Random Structures and Algorithms*, 10:69–101, 1997.
11. J. Csirik and G. J. Woeginger. On-line packing and covering problems. In A. Fiat and G. J. Woeginger, editors, *On-line Algorithms—The State of the Art*, volume 1442 of *Lecture Notes in Computer Science*, pages 147–177. Springer, 1996.

12. J. Csirik and G. J. Woeginger. Resource augmentation for online bounded space bin packing. *J. Algorithms*, 44(2):308–320, 2002.
13. L. Epstein and R. van Stee. Online bin packing with resource augmentation. In G. Persiano and R. Solis-Oba, editors, *WAOA*, volume 3351 of *Lecture Notes in Computer Science*, pages 23–35. Springer, 2004.
14. Z. Ivkovic and E. L. Lloyd. Fully dynamic algorithms for bin packing: Being (mostly) myopic helps. *SIAM J. Comput.*, 28(2):574–611, 1998.
15. B. Kalyanasundaram and K. Pruhs. Speed is as powerful as clairvoyance. In *36th Annual Symposium on Foundations of Computer Science*, pages 214–221, 1995.
16. S. S. Seiden. On the online bin packing problem. *J. ACM*, 49(5):640–671, 2002.
17. A. van Vliet. An improved lower bound for on-line bin packing algorithms. *Inf. Process. Lett.*, 43(5):277–284, 1992.

Improved On-Line Broadcast Scheduling with Deadlines*

Feifeng Zheng[1], Stanley P.Y. Fung[2], Wun-Tat Chan[3], Francis Y.L. Chin[3],
Chung Keung Poon[4], and Prudence W.H. Wong[5]

[1] School of Management, Xi'an JiaoTong University, China
zhengff@mailst.xjtu.edu.cn
[2] Department of Computer Science, University of Leicester, United Kingdom
pyfung@mcs.le.ac.uk
[3] Department of Computer Science, The University of Hong Kong, Hong Kong, China
{wtchan, chin}@cs.hku.hk
[4] Department of Computer Science, City University of Hong Kong, China
ckpoon@cs.cityu.edu.hk
[5] Department of Computer Science, University of Liverpool, United Kingdom
pwong@csc.liv.ac.uk

Abstract. We study an on-line broadcast scheduling problem in which
requests have deadlines, and the objective is to maximize the weighted
throughput, i.e., the weighted total length of the satisfied requests. For
the case where all requested pages have the same length, we present
an online deterministic algorithm named BAR and prove that it is 4.56-
competitive. This improves the previous algorithm of Kim and Chwa [11]
which is shown to be 5-competitive by Chan et al. [4]. In the case that
pages may have different lengths, we prove a lower bound of $\Omega(\Delta/\log \Delta)$
on the competitive ratio where Δ is the ratio of maximum to minimum
page lengths. This improves upon the previous $\sqrt{\Delta}$ lower bound in [11,4]
and is much closer to the current upper bound of $(\Delta + 2\sqrt{\Delta} + 2)$ in [7].
Furthermore, for small values of Δ we give better lower bounds.

1 Introduction

Data broadcast scheduling is a core problem in many applications that involve
distribution of information from a server to a large group of receivers. In contrast to the traditional point-to-point mode of communication, broadcasting
technologies are employed so that different clients requesting the same data can
be satisfied simultaneously by only one broadcast. For example, in Hughes' DirecPC system [12], clients make requests over phone lines and the server satisfies
the requests through broadcasts via a satellite. Typical information that will be
broadcasted include movies (video on-demand), stock market quotation, traffic

---

* The work described in this paper was fully supported by grants from the Research
Grants Council of the Hong Kong SAR, China [CityU 1198/03E, HKU 7142/03E,
HKU 5172/03E], an NSF Grant of China [No. 10371094], and a Nuffield Foundation
Grant of UK [NAL/01004/G].

D.Z. Chen and D.T. Lee (Eds.): COCOON 2006, LNCS 4112, pp. 320–329, 2006.

and landmark information, etc. Very often, the information to be disseminated is time critical and thus it is important to meet the deadlines of the requests.

Motivated by these applications, we study the following *On-line Scheduling of Broadcasts* (Broadcasting): We are given a set of pages to be broadcasted to clients upon request. Each request r has four attributes, namely, $p(r)$: the requested page, $a(r)$: the arrival time, $d(r)$: the deadline by which the requested page has to be received in its entirety, and $w(r)$: the weight of the request. A request is not known until it arrives, i.e., at time $a(r)$. When it arrives, all $p(r)$, $d(r)$ and $w(r)$ become known. When the server broadcasts a page, all requests to the same page that have arrived will receive the content of the page simultaneously. Upon completion, each of these requests will be satisfied, provided that the completion time is before its respective deadline. The server is allowed to abort the current page it is broadcasting before its completion and start a new one. To satisfy an aborted request, the requested page has to be broadcasted again from the beginning. Thus it is an on-line scheduling problem with preemptions and restarts. Our goal is to maximize the total weighted throughput, i.e., the total weighted lengths of all satisfied requests.

Related Work. Most of the previous works on the problem of on-line broadcast scheduling concentrate on minimizing the flow time where the flow time of a request is the time elapsed between its arrival and its completion. For example, [10,5,6,8] studied the problem of minimizing the total flow time while Bartal and Muthukrishan [2] studied the minimization of the maximum flow time. Aksoy and Franklin [1] presented a practical parameterized algorithm and evaluated it with extensive experiments. While the flow time is important and related to how the clients perceive the responsiveness of the system, the objective of maximizing the throughput is crucial for applications in which requests are associated with deadlines. Jiang and Vaidya [9] considered the problem of maximizing the percentage of satisfied requests assuming knowledge of the requests distribution.

Kim and Chwa [11] were the first to design algorithms with provable worst case performance bounds for this problem. In particular, one of their results is a 5.828-competitive algorithm for our problem. Using a tighter analysis, Chan et al [4] showed that Kim and Chwa's algorithm actually has competitive ratio at most 5, through a reduction to a job scheduling problem with cancellations. It was shown [14] that we need new techniques to further improve the bound. For the case of different page lengths, let Δ be the ratio between the length of the longest and shortest page. There is a $(\Delta + 2\sqrt{\Delta} + 2)$-competitive algorithm [7] and a $\sqrt{\Delta}$ lower bound [11,4].

A related on-line interval scheduling problem is studied by Woeginger [13]. Translated into our terminologies, his problem is to schedule requests with tight deadlines (i.e., the length of time interval between its arrival time and deadline is exactly the length of the requested page) and pages have different lengths. He proved that when the page length and the weight of requests can be arbitrarily related, no deterministic algorithm can have constant competitive ratio. He went on to give a 4-competitive heuristic for several special cases in which the page length and the weight of requests satisfy certain relationship. In particular, the heuristic works for the case of unit page length and arbitrary weights. He then complemented his upper bound with several lower bounds, including a tight lower bound for this special case of unit page length. In some sense, our problem is a generalization of Woeginger's problem allowing non-tight intervals.

Our Results. In this paper we give three different results for Broadcasting. Our first contribution is to give an improved algorithm for the case of unit page length. We consider the deadlines of the requests, a parameter ignored by previous algorithms, in our scheduling decision. By considering the fact that some of the currently-serving requests might have distant deadlines and can be served later after the completion of new requests, we improve the competitive ratio from 5 [4] to 4.56. In Section 3 we describe this algorithm and its analysis.

Our second contribution is to give an improved lower bound for the case of different page lengths. We give a lower bound of $\Omega(\Delta/\log \Delta)$ on the competitive ratio, an improvement over the previous $\Omega(\sqrt{\Delta})$ bound and which almost matches the linear upper bound. This is discussed in Section 4.

All existing lower bound proofs for the case of different page lengths do not work very well when $\Delta > 1$ is small, and thus the lower bound for those cases is still 4, that being the lower bound for the unit page length case. In Section 5 we describe our third contribution of proving better lower bounds for these cases. The lower bound for the competitive ratio is improved to e.g., 4.245, 4.481, 4.707 and 5.873 for $\Delta = 2, 3, 4$ and 10 respectively. The result is obtained by extending the lower bound proof for the unit page length case [13] to the case of different page lengths.

Due to space constraints, most proofs are omitted from this version. They can be found in the full version of the paper.

2   Notations

We first state the problem formally. Assume there are n pages, $P_1, \ldots, P_n$, in the system. A request for some page may arrive in arbitrary time with arbitrary deadline. If a page is fully broadcasted, then a request for that page is *satisfied* if the requests arrive before the broadcast of the page, and the broadcast finishes before the deadline of the request. A broadcast can be aborted at any time, and can later be restarted from the beginning.

A schedule S is a sequence of broadcasts $J_1, J_2, \ldots$ where each broadcast J_i is a set of requests to the same page started being served at time $s(J_i)$. The broadcasts are indexed such that $s(J_i) < s(J_{i'})$ for $i < i'$. For convenience, we will write J_i to represent both the set of requests and the requested page. Let $l(J_i)$ be the length of the page broadcasted by J_i. If $s(J_i) + l(J_i) > s(J_{i+1})$, then the broadcast J_i is *aborted* by J_{i+1}; otherwise J_i is said to be *completed*. The *profit* of a completed request is its weight times the length of the page it requests. The profit of a broadcast J_i, denoted by $|J_i|$, is the sum of $w(r) \times l(J_i)$ over all r in J_i. We denote by $|S|$ the total profit of completed broadcasts in the schedule S, i.e., we only count those satisfied requests. The objective is to maximize the total profit of satisfied requests $|S|$ during a time period.

Given an input I (a set of requests) and an algorithm A, we denote by $S_A(I)$ and $S^*(I)$ the schedules produced by A and by an optimal offline algorithm on I respectively. When A and I are understood from the context, we will simply denote the schedules by S and S^* respectively. To gauge the quality of the schedules produced by an on-line algorithm, the competitive ratio analysis [3] is often used. The competitive ratio of algorithm A is defined as $r_A = \sup_I \frac{|S^*(I)|}{|S_A(I)|}$.

3   Unit Page Length: The BAR Algorithm

In this section, we consider the case where each page is of the same length. Thus we assume without loss of generality that broadcasting each page requires one unit of time. We present our BAR algorithm (for Bi-level Abortion Ratio) and prove that it is 4.56-competitive.

At any moment, there is a (possibly empty) pool of requests that have arrived and are waiting to be served. Requests in the pool will be discarded when their deadlines cannot be met even if the server starts to serve them immediately. The algorithm is triggered either when a broadcast is completed or when a new request arrives. In the former case, the server will pick the page with the largest total profit of requests in the pool to be broadcasted next. When a new request arrives while the server is broadcasting a page, the server will either continue the current broadcast J (and add the new requests to the pool), or abort J in favour of a new broadcast R with larger total profit of requests in the pool (including newly arrived requests and possibly part of J). The decision is made according to the relative profits of J and R. It will also consider the previous broadcast and the deadlines of requests currently in the system.

More precisely, let J_0 be the broadcast aborted by J. If J does not abort any broadcast, define $J_0 = \phi$ (the empty set). Let J' be the largest-profit set of requests that can be satisfied in a single broadcast after completing J, assuming no more requests arrive. Similarly R' denotes the largest-profit set of requests that can be satisfied in one broadcast after completing R, if we abort J and start R now. See Figure 1. Let α, β be some constants such that $1.5 \le \alpha < 2 \le \beta < 2.5$. The exact values will be determined later. If either one of the following conditions is satisfied, we abort J and start R:

C1: $\beta|J| \le |R|$ and $\beta^2|J_0| \le |R|$, or
C2: $\alpha|J| \le |R| < \beta|J|$, $\beta|J| + |J'| \le |R| + |R'|$, and $\beta|J_0| \le |J|$.

Otherwise, we continue with the broadcasting of J and add the new requests into the pool.

We give some intuitive rationale behind these conditions. In previous algorithms [11,4], the abortion is simply determined by considering whether $\beta|J| \le |R|$. This completely ignores the deadlines of the requests. The improvement of BAR comes from the introduction of [C2], which gives an alternative abortion condition with a lower threshold α and, as we show below, considers the deadlines of requests.

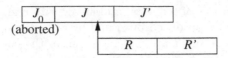

Fig. 1. BAR is broadcasting J and determining whether to abort J and start R

The first part of condition [C1] is the usual abortion condition by considering the profit of requests. The second part of [C1] enforces some technical properties that are required in bounding the profit of the requests, despite some abortions being caused by condition [C2].

The second part of condition [C2] utilizes deadlines of requests. Rather than directly comparing the deadlines of requests, we consider the total profit of requests that can later be satisfied (before their deadlines) in deciding whether to abort the page currently being broadcast. Suppose an abortion happens due to condition [C2]. Then some part of R' must come from J (which is aborted). Otherwise, the requests in R' come from the pool only, and since the broadcast J' will finish earlier than R', R' is a possible choice of J'. Hence $|R'| \leq |J'|$ and the condition $\beta|J| + |J'| \leq |R| + |R'|$ cannot be true. So $R' \cap J$ is not empty, and they must be requesting the same page. That means once condition [C2] happens, there must be some requests in J that have long deadlines, long enough to be completed after R is completed. Therefore, even though $|R|$ is only α times larger than $|J|$, we may still satisfy enough requests to achieve a good competitive ratio.

The third part of condition [C2] ensures there will not be two consecutive abortions caused by [C2], so that this weaker-threshold condition will not be used too often.

The remaining of this section is devoted to the proof of the following theorem.

Theorem 1. *BAR is 4.56-competitive for* Broadcasting *in the unit page length case.*

3.1 Basic Subschedules

We now elicit certain useful structures in a schedule produced by BAR. A sequence of broadcasts $(J_1, \ldots, J_k)$ is called a *chain* if J_i is aborted by J_{i+1} for all $i = 1, \ldots, k-1$ and J_1 is preceded by either an idle interval or a completed broadcast. A chain $C = (J_1, \ldots, J_k)$ in which J_k is completed is called a *basic subschedule*. Thus a basic subschedule consists of a sequence of zero or more aborted broadcasts followed by a completed broadcast; and the sequence cannot be extended at the front. Furthermore, the broadcast before an idle interval must be completed. Therefore, the whole schedule can be decomposed into a sequence of basic subschedules.

Consider an arbitrary basic subschedule, $B = (J_1, J_2, \ldots, J_k)$ where J_k is a completed broadcast and all the others are aborted ones. The total profit of requests satisfied by BAR in this basic subschedule is $|B| = |J_k|$. To analyze the competitive ratio, we will associate with B a carefully chosen set of requests satisfied by the offline optimal algorithm $\mathcal{O}$.

Consider a broadcast J by BAR. We can make use of condition [C1] and/or [C2] to argue that requests started by $\mathcal{O}$ while BAR is broadcasting J cannot be too large, if these requests are available in the pool maintained by BAR. Note, however, that $\mathcal{O}$ can also serve requests with arbitrarily large profits that have been satisfied by BAR before without violating [C1, C2]. Thus, we classify the requests satisfied by $\mathcal{O}$ into two types according to whether the request has been satisfied by BAR at the time $\mathcal{O}$ starts them. (Since $\mathcal{O}$ is an offline algorithm, we assume that it will never abort a broadcast. Thus, saying that a request is started by $\mathcal{O}$ is equivalent to saying that it will be satisfied by $\mathcal{O}$.) More precisely, we define J_i^*, for $i = 1, \ldots, k$, as the set of requests started by $\mathcal{O}$ within the interval $[s(J_i), s(J_{i+1}))$ but have not been satisfied by BAR before, where $s(J_i)$ is the start time of J_i and we take $s(J_{k+1}) = s(J_k) + 1$. Also, we define B^* as the set of requests in J_k that are started by $\mathcal{O}$ after the basic subschedule B. We will try to obtain an upper bound on $\sum_{i=1}^{k} |J_i^*| + |B^*|$.

Note that if a broadcast J_i is aborted by a broadcast J_{i+1} due to condition [C1], the ratio $|J_{i+1}|/|J_i|$ is at least β. However if the abortion is due to condition [C2], $|J_{i+1}|/|J_i|$ may be smaller than β. Nevertheless, we can still bound the profits of J_i's and J_i^*'s by geometric series in the following lemmas. Consider a chain $C = (J_1, \ldots, J_k)$. It is said to have *big endian* if $|J_k| \geq \beta|J_{k-1}|$; and *small endian* otherwise. If C has only one broadcast, we take $J_{k-1} = \phi$. Thus C will be considered to have big endian.

The following two lemmas bound the profits of J_i and J_i^*.

Lemma 1. *Consider a chain $C = (J_1, \ldots, J_k)$ with big endian. Then $|J_i| \leq |J_k|/\beta^{k-i}$ for all $i = 1, \ldots, k$.*

Lemma 2. *Consider a chain $C = (J_1, \ldots, J_k)$ with big endian. Then $|J_i^*| < |J_k|/\beta^{k-i-1}$ for all $i = 1, \ldots, k$. Hence $\sum_{i=1}^{k} |J_i^*| < \frac{\beta^2}{\beta-1}\left(1 - \frac{1}{\beta^k}\right)|J_k|$.*

The following lemma bounds the profit of requests served by $\mathcal{O}$ in a basic subschedule.

Lemma 3. *Consider a basic subschedule $B = (J_1, \ldots, J_k)$. If B has small endian,*

$$\sum_{j=1}^{k} |J_j^*| + |B^*| \leq \left(\beta + \frac{1}{\beta - 1}\right)\beta|J_{k-1}| + |J_k| + |J_k'| - \frac{\beta}{\beta - 1}|J_1|.$$

If B has big endian,

$$\sum_{j=1}^{k} |J_j^*| + |B^*| \leq \left(\alpha + \frac{\beta}{\beta - 1} + 1\right)|J_k| + |J_k'| - \frac{\beta}{\beta - 1}|J_1|.$$

Proof. Suppose B has small endian. We observe that the chain $(J_1, \ldots, J_{k-1})$ must have big endian by construction of BAR. (Note that $k \geq 2$ if B has small endian since $k = 1$ implies B has big endian.) By Lemma 2, we have $\sum_{j=1}^{k-1} |J_j^*| < \frac{\beta^2}{\beta-1} \left(1 - \frac{1}{\beta^{k-1}}\right) |J_{k-1}|$. Also, we have $|J_k^*| < \beta^2 |J_{k-1}|$ for otherwise, J_k^* would have aborted J_k due to condition [C1]. Thus $\sum_{j=1}^{k} |J_j^*| < \frac{\beta^3}{\beta-1} \left(1 - \frac{1}{\beta^k}\right) |J_{k-1}|$. As for B^*, we note that $\beta|J_{k-1}| + |J_{k-1}'| \leq |J_k| + |J_k'|$ since J_k aborts J_{k-1} by condition [C2]. Moreover, $|B^*| \leq |J_{k-1}'|$ because requests in B^* have deadlines no earlier than that of J_{k-1}' and they have not been satisfied by BAR at time $s(J_{k-1}')$. Hence $|B^*| \leq |J_k| - \beta|J_{k-1}| + |J_k'|$. Combining these two bounds, we have $\sum_{j=1}^{k} |J_j^*| + |B^*| \leq \frac{\beta^3}{\beta-1} \left(1 - \frac{1}{\beta^k}\right) |J_{k-1}| + (|J_k| - \beta|J_{k-1}| + |J_k'|) = \left(\beta + \frac{1}{\beta-1}\right) \beta|J_{k-1}| + |J_k| + |J_k'| - \frac{1}{(\beta-1)\beta^{k-3}} |J_{k-1}| \leq \left(\beta + \frac{1}{\beta-1}\right) \beta|J_{k-1}| + |J_k| + |J_k'| - \frac{\beta}{\beta-1} |J_1|$.

We omit the proof for the big endian case.   □

In Lemma 3, no matter B has big or small endian, we can bound $|J_k'|$ from above by the profit of the first broadcast in the basic subschedule after B. If B is followed by an idle interval, then we can actually argue that $|J_k'| = 0$. That is, we associate $|J_k'|$ with the basic subschedule following B. By the same token, B will also be associated with such value from the preceding basic subschedule. In the next subsection we will see how this association is used in the analysis.

3.2   Aggregated Subschedules

Let $B_i = (J_{i,1}, \ldots, J_{i,k_i})$ denote the i-th basic subschedule in a sequence of basic subschedules. For notational convenience define $J_{i,0} = \phi$, and $prev(B_i) = J_{i,k_i-1}$, i.e., the second last broadcast in a basic subschedule.

Lemma 4. *The last basic subschedule before an idle interval must have big endian.*

Based on the above lemma, we can partition the original schedule into a number of *aggregated subschedules*, each of which containing zero or more basic subschedules with small endians followed by one basic subschedule with big endian.

Consider an arbitrary aggregated subschedule, $A = (B_1, \ldots, B_m)$ where for $i = 1, \ldots, m$, $B_i = (J_{i,1}, \ldots, J_{i,k_i})$ is a basic subschedule with k_i broadcasts. Obviously, the total profit of requests satisfied by BAR in A is

$$|A| = \sum_{i=1}^{m} |J_{i,k_i}|. \tag{1}$$

Also, since $|J_{i,k_i}| \geq \alpha|prev(B_i)|$ for $i = 1, \ldots, m-1$, we have

$$|A| \geq \alpha \left(\sum_{i=1}^{m-1} |prev(B_i)| \right) + |J_{m,k_m}|. \tag{2}$$

Recall that B_i's have small endians for $i = 1, \ldots, m - 1$. We have $k_i \geq 2$ since basic subschedules with only one broadcast must have big endians by definition. By condition [C2], $\beta|prev(B_i)| + |J'| \leq |J_{i,k_i}| + |R'|$ where $|R'| \leq |J_{i+1,1}|$ because R' is a candidate set of requests to be served after J_{i,k_i} is completed. Hence we have $\beta|prev(B_i)| \leq |J_{i,k_i}| + |J_{i+1,1}|$, and together with (1),

$$|A| \geq \sum_{i=1}^{m-1} \beta|prev(B_i)| + |J_{m,k_m}| - \sum_{i=2}^{m} |J_{i,1}|. \tag{3}$$

On the other hand, the total profit of requests satisfied by $\mathcal{O}$ and associated with aggregated subschedule A is:

$$|A^*| = \left(\sum_{j=1}^{k_1} |J_{1,j}^*| + \cdots + \sum_{j=1}^{k_m} |J_{m,j}^*| \right) + (|B_1^*| + \cdots + |B_m^*|)$$

$$\leq \sum_{i=1}^{m-1} \left(\beta + \frac{1}{\beta - 1} \right) \beta|prev(B_i)| + \left(\alpha + \frac{\beta}{\beta - 1} \right) |J_{m,k_m}|$$

$$+ \sum_{i=1}^{m} |J_{i,k_i}| + \sum_{i=2}^{m} |J_{i,1}| + |J'_{m,k_m}| - \frac{\beta}{\beta - 1} \sum_{i=1}^{m} |J_{i,1}|$$

where the inequality follows from Lemma 3. Consider (1) + (2) $\times (\frac{\beta^2}{\alpha})$+ (3) $\times \frac{1}{\beta-1}$. After some algebraic manipulations (which we omit) we have

$$\left(1 + \frac{\beta^2}{\alpha} + \frac{1}{\beta - 1} \right) |A| \geq |A^*| + |J_{1,1}| - |J'_{m,k_m}| \tag{4}$$

as long as $\frac{\beta^2}{\alpha} + \frac{1}{\beta-1} \geq \frac{\beta}{\beta-1} + \alpha$. We can bound $|J'_{m,k_m}|$ from above by the profit of the first broadcast (i.e., $|J_{1,1}|$) in the next aggregated subschedule. Thus, if we have a sequence of aggregated subschedules $A_1, \ldots, A_l$, then from (4) we have

$$\left(\frac{\beta^2}{\alpha} + \frac{\beta}{\beta - 1} \right) (|A_1| + \cdots + |A_l|) \geq |A_1^*| + \cdots + |A_l^*| - |J'| \tag{5}$$

where J is the last broadcast in A_l. Since there is no more broadcast after A_l, $|J'| = 0$.

If the whole aggregated subschedule consists of only one basic subschedule with big endian, i.e., $A = (B_1)$, then $|A| = |J_{1,k_1}|$ and we can verify that inequality (5) still holds.

The condition $\frac{\beta^2}{\alpha} + \frac{1}{\beta-1} \geq \frac{\beta}{\beta-1} + \alpha$ can be satisfied by having $\alpha^2 + \alpha \leq \beta^2$, i.e., $\alpha \leq \sqrt{\beta^2 + \frac{1}{4}} - \frac{1}{2}$. Setting $\alpha = \sqrt{\beta^2 + \frac{1}{4}} - \frac{1}{2}$, the competitive ratio of BAR is

$$\frac{\sum_{i=1}^{l} |A_i^*|}{\sum_{i=1}^{l} |A_i|} \leq \frac{\beta}{\beta - 1} + \frac{\beta^2}{\alpha} = \frac{\beta}{\beta - 1} + \frac{\beta^2}{\sqrt{\beta^2 + \frac{1}{4}} - \frac{1}{2}} = \frac{3}{2} + \frac{1}{\beta - 1} + \sqrt{\beta^2 + \frac{1}{4}}.$$

This has a minimum value of approximately 4.561 attained when $\beta \approx 2.015$, and $\alpha \approx 1.576$.

4   Variable Page Length: An Improved Lower Bound

In this section we consider the case where the pages can have different lengths. We give a lower bound on the competitive ratio of any deterministic online algorithm for Broadcasting. Let Δ be the ratio between the length of the longest and shortest page.

Theorem 2. *The competitive ratio of any deterministic online algorithm for* Broadcasting *cannot be smaller than* $\Omega(\Delta/\log \Delta)$.

Proof. Assume that there are two pages, P and Q whose lengths are Δ and 1, respectively. Given any online algorithm $\mathcal{A}$, we construct a sequence of requests as follows. At time 0, a request for P arrives with deadline at time Δ, i.e., it has a tight deadline. The weight of the request is 1. There are at most $\lceil\Delta\rceil$ requests for Q, denoted by r_i for $0 \le i \le \lceil\Delta\rceil - 1$. r_i arrives consecutively, i.e., $a(r_i) = i$, and they all have tight deadlines, i.e., $d(r_i) = i + 1$. The weight of r_i, i.e., $w(r_i)$, is $\Delta^r(i + 1)^k$ where r and k are some constants which will be defined later. If $\mathcal{A}$ broadcasts Q at any time t, no more request of r_i arrives for $i > t$.

Now we analyze the performance of $\mathcal{A}$ against the optimal offline algorithm $\mathcal{O}$. There are two cases. (1) If $\mathcal{A}$ satisfies the request for P by broadcasting P at time 0, $\mathcal{O}$ will satisfy all requests r_i by broadcasting Q at time i for $0 \le i \le \lceil\Delta\rceil - 1$. Hence, we have $|\mathcal{A}| = \Delta$ and $|\mathcal{O}| = \sum_{i=0}^{\lceil\Delta\rceil-1} \Delta^r(i + 1)^k$. Since $\sum_{i=1}^{x} i^y = \Theta(x^{y+1}/(y + 1))$, $|\mathcal{O}|/|\mathcal{A}| = \Theta(\Delta^{r-1}\lceil\Delta\rceil^{k+1}/(k + 1)) \ge \Theta(\Delta^{r+k}/(k + 1))$.

(2) If $\mathcal{A}$ broadcasts Q at time t, only r_t can be satisfied. However, $\mathcal{O}$ can either satisfy the request for P by broadcasting P at time 0 or satisfy all r_i by broadcasting Q at time i for $0 \le i \le t$. We have $|\mathcal{A}| = \Delta^r(t + 1)^k$ and $|\mathcal{O}| = \max\{\Delta, \sum_{i=0}^{t} \Delta^r(i + 1)^k\}$. Hence, $|\mathcal{O}|/|\mathcal{A}| = \max\{\Delta, \Theta(\Delta^r(t+1)^{k+1}/(k+1))\}/\Delta^r(t + 1)^k = \max\{\Delta^{1-r}/(t+1)^k, \Theta((t+1)/(k+1))\}$. In order to minimize the ratio, $\mathcal{A}$ should choose $t = (\Delta^{1-r}(k + 1))^{1/(k+1)} - 1$. In that case, the ratio is $\Theta(\Delta^{(1-r)/(k+1)}/(k + 1)^{k/(k+1)}) \ge \Theta(\Delta^{(1-r)/(k+1)}/(k + 1))$.

In order to maximize the minimum ratio among the two cases, we let $r = (1 - k - k^2)/(k + 2)$. Hence, the competitive ratio is $\Theta(\Delta^{1-1/(k+2)}/(k + 1)) \ge \Theta(\Delta^{1-1/(k+2)}/(k + 2))$. We further let $k + 2 = \ln \Delta$ where the function $\Delta^{1-1/(k+2)}/(k + 2)$ achieves the maximum, i.e., $\Delta^{1-1/\ln\Delta}/\ln \Delta$. Since $\Delta^{1/\ln\Delta}$ is the constant e, i.e., the base of natural logarithm, we have proved that the competitive ratio is $\Omega(\Delta/\log \Delta)$.   □

5   Lower Bound for Small Δ

The current lower bound for the broadcasting problem is 4 when $\Delta = 1$ [13]. The $\Omega(\Delta/\log \Delta)$ lower bound we just proved as well as the previous $\sqrt{\Delta}$ one [4] gives very small lower bounds (much smaller than 4) for small values of Δ. In this section we give better lower bounds for this case.

Let α be the unique positive real root of the equation

$$2\alpha^2 - 4\alpha^{3/2} + 2\alpha + 1 = \sqrt{(\lfloor\Delta\rfloor - 1)^2 + 4\alpha^2} + \lfloor\Delta\rfloor. \tag{6}$$

The following table shows some values of α.

Δ	1	2	3	4	10	very large
α	4	4.245	4.481	4.707	5.873	$\sqrt{\Delta}$

Theorem 3. *For $\Delta \geq 2$, no deterministic algorithm for* Broadcasting *can be better than α-competitive, where α is the unique positive root of (6).*

The proof of this theorem uses a modified construction from the lower bound proof in [13], by adding requests with different lengths and carefully setting the arrival time of requests in a different way.

References

1. D. Aksoy and M. Franklin. Scheduling for large scale on-demand data broadcast. In *Proc. of IEEE INFOCOM*, pages 651–659, 1998.
2. Y. Bartal and S. Muthukrishnan. Minimizing maximum response time in scheduling broadcasts. In *Proc. 11th SODA*, pages 558–559, 2000.
3. A. Borodin and R. El-yaniv. *Online computation and competitive analysis*. Cambridge University Press, 1998.
4. Wun-Tat Chan, Tak-Wah Lam, Hing-Fung Ting, and Prudence W.H. Wong. New results on on-demand broadcasting with deadline via job scheduling with cancellation. In *10th COCOON*, LNCS 3106, pages 210–218, 2004.
5. J. Edmonds and K. Pruhs. Broadcast scheduling: when fairness is fine. In *Proc. 13th SODA*, pages 421–430, 2002.
6. T. Erlebach and A. Hall. NP-hardness of broadcast scheduling and inapproximability of single-source unsplittable min-cost flow. In *Proc. 13th ACM-SIAM SODA*, pages 194–202, 2002.
7. Stanley P. Y. Fung, Francis Y. L. Chin, and Chung Keung Poon. Laxity helps in broadcast scheduling. In *Proceedings of 9th Italian Conference on Theoretical Computer Science*, pages 251–264, 2005.
8. R. Gandhi, S. Khuller, Y.A. Kim, and Y.C. Wan. Algorithms for minimizing response time in broadcast scheduling. *Algorithmica*, 38(4):597–608, 2004.
9. S. Jiang and N. Vaidya. Scheduling data broadcasts to "impatient" users. In *Proc. ACM International Workshop on Data Engineering for Wireless and Mobile Access*, pages 52–59, 1999.
10. B. Kalyanasundaram, K. Pruhs, and M. Velauthapillai. Scheduling broadcasts in wireless networks. In *Proc. 8th ESA*, LNCS 1879, pages 290–301, 2000.
11. Jae-Hoon Kim and Kyung-Yong Chwa. Scheduling broadcasts with deadlines. *Theoretical Computer Science*, 325(3):479–488, 2004.
12. DirecPC Home Page. http://www.direcpc.com/.
13. Gerhard J. Woeginger. On-line scheduling of jobs with fixed start and end times. *Theoretical Computer Science*, 130:5–16, 1994.
14. Feifeng Zheng, Francis Y. L. Chin, Stanley P. Y. Fung, Chung Keung Poon, and Yinfeng Xu. A tight lower bound for job scheduling with cancellation. *Information Processing Letters*, 97(1):1–3, 2006.

A Tight Analysis of Most-Requested-First for On-Demand Data Broadcast

Regant Y.S. Hung and H.F. Ting[*]

Department of Computer Science,
The University of Hong Kong,
Pokfulam, Hong Kong
{yshung, hfting}@cs.hku.hk

Abstract. This paper gives a complete and tight mathematical analysis on the performance of the Most-Requested-First algorithm for on-demand data broadcast. The algorithm is natural and simple, yet its performance is surprisingly good in practice. We derive tight upper and lower bounds on MRF's competitiveness and thus reveal the exact competitive ratios of the algorithm under different system configurations. We prove that the competitive ratio of MRF is exactly $3 - \frac{\ell}{d}$ when the start-up delay d is a multiple of the page length ℓ; otherwise the ratio is 3.

1 Introduction

In an on-demand data broadcast system, clients make requests for data such as weather forecasting, stock prices, traffic information and sports results using various mobile devices such as notebooks, personal digital assistants (PDAs) and GPRS-enabled cellular phones. The server broadcasts the requested data at some time, and all pending requests on the data are satisfied with this single broadcast. The performance of such system depends critically on how the data pages are broadcast and hence finding a good broadcast scheduling algorithm is an important design issue. The First-Come-First-Serve (FCFS), Longest-Wait-First (LWF) and Most-Requested-First (MRF) are among the most popular scheduling algorithms and their empirical performances have been studied extensively [1, 4, 5, 15, 16, 18]. Recently, there are some theoretical analysis on these algorithms. Mao [14], and Hawkins and Mao [7] studied the competitiveness of FCFS. In particular, they showed that the competitive ratio of FCFS is no more than $\frac{m}{s}$ where m is the total number of pages that can be broadcast, and s is the number of servers in the system. On the other hand, Kalyanasundaram, Pruhs and Velauthapillai [11] showed that MRF and FCFS cannot be $O(1)$-competitive. For LWF, Edmonds and Pruhs [6] proved that it cannot be $O(1)$-competitive either. Furthermore, they studied resource augmentation for LWF; they showed that if LWF runs in a server that is six times faster than the one in which the optimal scheduler runs, then LWF is $O(1)$-competitive.

---

[*] This research was supported in part by Hong Kong RGC Grant HKU-7045/02E.

D.Z. Chen and D.T. Lee (Eds.): COCOON 2006, LNCS 4112, pp. 330–339, 2006.

	Upper bounds	Lower bounds
d is a multiple of page length ℓ	$3 - \frac{\ell}{d}$	$3 - \frac{\ell}{d}$
$d > \ell$ and is not a multiple of ℓ	3	3
$\frac{\ell}{2} < d < \ell$	4	2
$d \leq \frac{\ell}{2}$ and $\lceil \ell/d \rceil$ is a multiple s	$\lceil \frac{\ell}{d} \rceil$	$\lceil \frac{\ell}{d} \rceil$
$d \leq \frac{\ell}{2}$ and $\lceil \ell/d \rceil$ is not a multiple of s.	$2 \lceil \frac{\ell}{d} \rceil$	$\lceil \frac{\ell}{d} \rceil$

Fig. 1. The competitive ratio of MRF with startup delay d and number of servers s

We note that the above theoretical studies use the average response time as the performance measure and they assume that once a request is generated by a user, the request will be held until it is satisfied. This assumption is not without its critics. For example, Jiang and Vaidya [10] argued that this assumption is not always true; clients are impatient and they may leave with their requests unserved after waiting too long. To take this kind of behaviour into consideration, we studied a somewhat different model in which there is a *start-up delay guarantee*, or simply start-up delay. The model was introduced by Bar-Noy, Goshi and Ladner [2]. Roughly speaking, if the request for some page p arrives at time t, the system can accept this request by starting to broadcast the page p within the time interval $[t, t + d]$ and gain the profit from this request; otherwise the system gain nothing from it. Here, d is the start-up delay.

Note that for this model, the system throughput (i.e., the total profit gained by all those requests served before the start-up delay) is more important than the average response time. Therefore, we will use throughput as the performance measure. We studied the MRF algorithm for this model. In [8, 9], we consider the case when the start-up delay d is smaller than the page length ℓ. We prove that when $\frac{\ell}{2} < d < \ell$ then MRF is 4-competitive, and when $d \leq \frac{\ell}{2}$, MRF is about $\lceil \frac{\ell}{d} \rceil$-competitive[1]. We also derive nearly matching lower bounds for these two cases. In this paper, we study MRF for the case when $d \geq \ell$. It is surprising that we can derive a complete and tight mathematical analysis on the competitiveness of MRF for this case. We prove that the competitive ratio of MRF is exactly $(3 - \frac{\ell}{d})$ when d is a multiple of ℓ; otherwise, its competitive ratio is exactly 3. We summarize the results in Figure 1.

2   The Model and the MRF Algorithm

An on-demand data broadcast system is specified by the tuple $(d, S, \mathcal{P})$ where d is the start-up delay, S is a set of identical servers, $\mathcal{P}$ is a set of pages. As in [6, 9, 11], we assume that all the pages have the same length. Note that the assumption is

---

[1] Under the system throughput, not the average response time, performance measure.

not unrealistic and is applicable to systems using DNS servers. To simplify our analysis, we normalize the page length to one. Each server can broadcast a page at a time, and it takes one time unit to complete the broadcast. Once a server starts to broadcast a page at some time t, it cannot stop the broadcast of that page until it completes the broadcast at $t + 1$. In other words, preemptions are not allowed. (We note that there are also many stuides on systems that allow preemptions, e.g., [3, 12, 13, 17].) Every request r is associated with a *profit* $\rho(r)$. For any set of requests R, we let $\rho(R)$ denote the sum $\sum_{r \in R} \rho(r)$. If a request r arrives at time t, then the system will gain a profit of $\rho(r)$ from r if it starts to broadcast the requested page during the time interval $[t, t + d]$; after $t + d$, the request is expired and the system cannot gain any profit from it.

We study the problem of online scheduling for an on-demand data broadcast system. The input to the problem is a sequence σ of requests. Without loss of generality, we assume that the first request arrives at time 0. Given any schedule Ψ for serving σ, we say that a request r for some page P is *pending* at time i if it arrives at some time $t \in [i - d, i]$, and no server broadcasts P during $[t, i)$. We say that r is *accepted* by the server s at time i if r is pending at i and s broadcasts the page requested by r at i. In such case, we also say that r is accepted by the schedule Ψ, and is accepted by the algorithm constructing Ψ. We say that an online algorithm A for scheduling an on-demand data broadcast system has *competitive ratio* κ if for any input request sequence σ, we have $\rho(\text{Opt}, \sigma) \leq \kappa \rho(\text{A}, \sigma)$ where $\rho(\text{Opt}, \sigma)$ and $\rho(\text{A}, \sigma)$ are the total profits of the requests accepted by the optimal offline algorithm Opt and by the online algorithm A, respectively.

MRF is an online algorithm for scheduling on-demand data broadcast system $(d, S, \mathcal{P})$. It is very simple and works as follows: At times $0, 1, 2, \ldots$, MRF uses the $|S|$ servers to broadcast the $|S|$ pages that allow it to gain the most total profit. (Note that it takes one time unit for a server to broadcast one page, and if the pending requests ask for fewer than $|S|$ pages, some of the servers will be idle.)

3   The Case When $d > 1$ and Is Not an Integer

In this section, we prove that when the start-up delay d is greater than 1, MRF has competition ratio at most 3. The proof is rather straightforward. The more interesting result in this section is the construction of an input sequence that enables us to show that, with the additional requirement that d is not an integer, the competitive ratio of MRF is at least 3, which matches the upper bound.

Theorem 1. *When $d > 1$, the competitive ratio of MRF is at most 3.*

Proof. Consider any input request sequence σ. Suppose that Opt is an optimal offline algorithm. Let M and O be the sets of requests in σ that are accepted by MRF and Opt, respectively. Let $R = O - M$ be the set of requests accepted by Opt but not by MRF. Note that $\rho(O) \leq \rho(R) + \rho(M)$ (i.e., the total profit of the requests in O is no greater than the sum of those in R and M). Below, we show that $\rho(R) \leq 2\rho(M)$ and the theorem follows.

Let R_o be the set of requests in R that are accepted at time 0, and for any $i \geq 1$, let $R_i \subseteq R$ be the set of requests that are accepted during $(i - 1, i]$. We further divide R_i into two sets: R'_i is the set of requests in R_i that are pending at time i according to schedule of MRF, and $R''_i = R_i - R'_i$. For any integer $i \geq 0$, let M_i be the set of requests accepted by MRF at time i. By the design of MRF, we have $\rho(R_0) \leq \rho(M_0)$. Note that $\rho(R'_i) \leq \rho(M_i)$ because all the requests in R'_i are acceptable by MRF at i, and by design, M_i is the set of such requests with the largest profit. Finally, note that any request $r \in R''_i$ must be arrived before $i - 1$; otherwise, it is still pending at i and thus r should not be in R''_i. Hence, all requests in R''_i are acceptable by MRF at $i - 1$ and we can conclude that $\rho(R''_i) \leq \rho(M_{i-1})$. Therefore, $\rho(R) = \rho(R_o) + \sum_{i \geq 1} \rho(R'_i) + \sum_{i \geq 1} \rho(R''_i) \leq \rho(M_0) + \sum_{i \geq 1} \rho(M_i) + \sum_{i \geq 1} \rho(M_{i-1}) \leq 2\rho(M)$. □

Theorem 2. *When the start-up delay $d \geq 1$ is not an integer, the competitive ratio of MRF is at least 3.*

Proof. We establish the bound by constructing a difficult input request sequence for MRF. For ease of description, we assume that there is only one server in the system. We can show that this lower bound is also true for any number of servers c by creating $c - 1$ copies of the requests, each copy with different set of pages.

Suppose that $d = d_o + r$ where d_o is a positive integer and $0 < r < 1$. Consider the $2d_o + 1$ pages $P_0, P_1, \ldots, P_{2d_o}$. Let δ be a small value such that $0 < \delta < \min\{r, 1 - r\}$. For each integer $i \geq 0$, let R_i be the set containing the following three requests $r_{i,1}, r_{i,2}, r_{i,3}$ arriving respectively at time $i, i + \delta$, and $i + d_o + \delta$, all asking for the page P_{i_o} where $i_o = i \bmod (2d_o + 1)$. The requests $r_{i,2}, r_{i,3}$ have profit of 1, and the first request $r_{i,1}$ has profit of $1 + \epsilon$ where $\epsilon > 0$ is an extremely small value. (This ϵ is only for resolving the arbitrariness.) Our input is just the union of these R_i's, i.e., $\bigcup_{i \geq 0} R_i$. See Figure 2 for an example.

Consider the schedule that broadcasts, for each $i \geq 0$, the page P_{i_o} at time $i + d_o + \delta$ where $i_o = i \bmod (2d_o + 1)$. Note that this schedule is feasible because each page has length 1 and its broadcast takes one time unit. Furthermore, this schedule accepts all requests because for each i, the broadcast of P_{i_o} at time $i + d_o + \delta$ can accept all the three requests in R_i; the earliest request $r_{i,1}$ arrives at i, and is still pending at time $i + d = i + d_o + r > i + d_o + \delta$. Below, we will prove by induction that for any $i \geq 0$, MRF will broadcast P_{i_o} at time i and thus can only accept the first request $r_{i,1}$ in R_i. Then, the theorem follows.

It is true for the base case because $r_{0,1}$ arrives at time 0 and MRF accepts it immediately. Suppose that MRF accepts $r_{0,1}, r_{1,1}, \ldots, r_{i-1,1}$ at time $0, 1, 2, \ldots, i - 1$. For time i, we consider the requests in R_j for all $0 \leq j \leq i$ as follows.

1. When $0 \leq j < i - 2d_o$: Note that the three requests in R_j are expired at time i. In fact, the last request $r_{j,3}$, which arrives at time $j + d_o + \delta \leq i - 2d_o - 1 + d_o + \delta$, is expired at or before $(i - 2d_o - 1 + d_o + \delta) + d_o + r = i + \delta - (1 - r) < i$.
2. When $i - 2d_o \leq j < i$: By the induction hypothesis, the first request $r_{j,1}$ has already been accepted. If $j \leq i - d_o - 1$, then the request $r_{j,2}$, which arrives at time $j + \delta \leq i - d_o - 1 + \delta$, is expired at or before $i - d_o - 1 + \delta + d_o + r < i$. If $j \geq i - d_o$, then $r_{j,3}$ arrives after i because $j + d_o + \delta \geq i - d_o + d_o + \delta$. Hence, only one request in $R_j = \{r_{j,1}, r_{j,2}, r_{j,3}\}$ is pending at time i.
3. When $j = i$: R_j has exactly one request, namely $r_{i,1}$, arrives at time i.

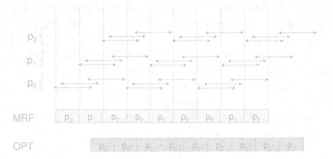

Fig. 2. An example with $d = 1.4$. The system has one server and three pages p_0, p_1 and p_2. The arrows show the durations of the requests.

Therefore, the set of requests pending at time i comprises exactly one request in each of $R_{i-2d_o}, R_{i-2d_o+1}, \ldots, R_i$. Note that for any $i - 2d_o \leq j \neq j' \leq i$, $j \bmod (2d_o + 1) \neq j' \bmod (2d_o + 1)$ and R_j and $R_{j'}$ are asking for different pages. Since for $1 \leq j < i$, $r_{j,1}$ has been accepted (induction hypothesis), MRF will broadcast P_{i_o} at time i to accept $r_{i,1}$ to get the maximum profit $1 + \epsilon$.   □

4   The Case When $d > 1$ and Is an Integer

In this section, we prove that when d is an integer, the competitive ratio of MRF is no more than $3 - \frac{1}{d}$. We also show that this bound is tight.

Consider any input request sequence σ. For any time interval I, let σ_I denote the sequence of requests in σ that arrive during I. For example, $\sigma_{[0,t)}$ is the sequence of requests in σ arriving before t, and $\sigma_{[t,t]}$ are those arriving at t. Let O be the set of requests that are accepted by Opt, the optimal offline algorithm, and for any integer $i \geq 0$, we let O_i be the set of requests that are accepted by Opt during $(i - 1, i]$. For any page $P \in \mathcal{P}$, define $O_{i,P} \subseteq O_i$ to be the subset of requests in O_i that ask for P. Define M, M_i and $M_{i,P}$ for MRF similarly. (Note that by design, MRF will only accept requests in M_i at time i.)

For any server $s \in S$, let $M(s)$ be the set of requests in M that are accepted by s. Note that $M_i \cap M(s)$ is the set of requests accepted by s at time i, and it is empty if MRF does not use s to accept any request at i. For any fixed $i \geq 0$, let $M_{i,\min}$ denote the set $M_i \cap M(s)$ with the minimum profit (over all servers $s \in S$). Note that for any page P that is broadcast by MRF at time i, $M_{i,P} = M_i \cap M(s)$ for some server s at i. The following fact is easy to verify.

Fact 1. *For any page P that is broadcast by MRF at time i, we have $\rho(M_{i,P}) \geq \rho(M_{i,\min})$ and $\rho(M_{i,\min}) = 0$ if MRF does not use all the $|S|$ servers at time i.*

For any integers $0 \leq i \leq j$, let $B_i^j(P)$ be the set of instances $t \in [i, j]$ such that MRF broadcasts page P at t. Since MRF only broadcasts a page only at some integral time, $B_i^j(P)$ includes only integers. Let $\bar{B}_i^j(P) = \{i, i+1, \ldots, j\} - B_i^j(P)$ be the set of instances $t \in [i, j]$ such that MRF does not broadcast page P at t. To derive an upper bound on $\rho(O)$ in terms of $\rho(M)$, we need the following lemma.

Lemma 1. *Suppose the start-up delay d is an integer. For any time i and page P, $\rho(O_{i,P}) \leq \sum_{j \in B_{i-d}^{i-1}(P)} \rho(O_{i,P} \cap M_{j,P}) + \sum_{j \in \bar{B}_{i-d}^{i-1}(P)} \frac{\rho(M_{j,\min})}{d} + \Delta(i,P)$, where $\Delta(i,P) = \rho(O_{i,P} \cap M_{i,P})$ if* MRF *broadcasts P at time i; otherwise, $\Delta(i,P) = \rho(M_{i,\min})$.*

Proof. We analyze how the requests in $O_{i,P}$ are served by MRF. Let $O_{i,P}^+$ be the set of requests in $O_{i,P}$ that are accepted by MRF at or before i, and let $O_{i,P}^- = O_{i,P} - O_{i,P}^+$ be those that are still not accepted at i. (These requests may already be expired, or may be accepted by MRF later after i.) Since all requests in $O_{i,P}$ are served by Opt during $(i-1,i]$, they must arrive after $i-d-1$. This implies the requests in $O_{i,P}^+$ can only be accepted by the broadcast of P by MRF during $(i-d-1,i]$. Together with the fact that MRF will only broadcast a page at integral time units, we conclude that $\rho(O_{i,P}^+)$ is equal to

$$\sum_{j \in B_{i-d}^{i}(P)} \rho(O_{i,P} \cap M_{j,P}) = \sum_{j \in B_{i-d}^{i-1}(P)} \rho(O_{i,P} \cap M_{j,P}) + \Delta_1(i,P) \qquad (1)$$

where $\Delta_1(i,P) = \rho(O_{i,P} \cap M_{i,P})$ if MRF broadcasts P at i (i.e., $i \in B_{i-d}^{i}(P)$); otherwise $\Delta_1(i,P) = 0$. To handle $O_{i,P}^-$, we note that $O_{i,P}^- = O_{i,P}^- \cap \sigma_{(i-d-1,i]}$ as all requests in $O_{i,P}$ arrive during $(i-d-1,i]$. We divide $O_{i,P}^-$ into two sets, $O_{i,P}^- \cap \sigma_{(i-d-1,i-d]}$ and $O_{i,P}^- \cap \sigma_{(i-d,i]}$, and we estimate their profits as follows.

For the first set, we claim that

$$\rho(O_{i,P}^- \cap \sigma_{(i-d-1,i-d]}) \leq \sum_{j \in \bar{B}_{i-d}^{i-1}(P)} \frac{\rho(M_{j,\min})}{d}. \qquad (2)$$

If $O_{i,P}^- \cap \sigma_{(i-d-1,i-d]} = \emptyset$, we have (2) immediately. Suppose that the set is not empty. Note that all requests in $O_{i,P}^- \cap \sigma_{(i-d-1,i-d]}$ are all pending at each time $j \in [i-d, i-1]$ because (i) its earliest request arrives after $i-d-1$ and its last request arrives at or before $i-d$, and (ii) none of its requests will be accepted before i. However, MRF did not accept them at time j for all $j \in \bar{B}_{i-d}^{i-1}(P)$; by the greedy nature of MRF, we conclude $\rho(O_{i,P}^- \cap \sigma_{(i-d-1,i-d]}) \leq \rho(M_{j,\min})$. Therefore, $\sum_{j \in \bar{B}_{i-d}^{i-1}(P)} \rho(O_{i,P}^- \cap \sigma_{(i-d-1,i-d]}) \leq \sum_{j \in \bar{B}_{i-d}^{i-1}(P)} \rho(M_{j,\min})$, or equivalently, $\rho(O_{i,P}^- \cap \sigma_{(i-d-1,i-d]}) \leq \sum_{j \in \bar{B}_{i-d}^{i-1}(P)} \rho(M_{j,\min})/|\bar{B}_{i-d}^{i-1}(P)|$. Finally, because of our assumption that $O_{i,P}^- \cap \sigma_{(i-d-1,i-d]}$ is not empty, we conclude $\bar{B}_{i-d}^{i-1}(P) = \{i-d, i-d+1, \ldots, i-1\}$ (any broadcast of P during $[i-d, i-1]$ will accept all requests in $O_{i,P}$ arriving during $(i-d-1,i-d]$). Hence, $|\bar{B}_{i-d}^{i-1}(P)| = d$ and (2) follows.

For the second set, let $\Delta_2(i,P) = 0$ if MRF broadcasts P at time i; otherwise let $\Delta_2(i,P) = \rho(M_{i,\min})$. We claim that

$$\rho(O_{i,P}^- \cap \sigma_{(i-d,i]}) \leq \Delta_2(i,P). \qquad (3)$$

If $O_{i,P}^- \cap \sigma_{(i-d,i]} = \emptyset$, (3) follows immediately. Suppose that the set is not empty. Similar to the above analysis, we observe that (i) all requests at the set are pending at i, and (ii) MRF does not broadcast P at i. By the greedy nature of

MRF, we conclude that $\rho(O_{i,P}^- \cap \sigma_{(i-d,i]}) \leq \rho(M_{i,\min})$, and since MRF does not broadcast P at i, we have $\Delta_2(i, P) = \rho(M_{i,\min})$; Inequality (3) follows.

From (1), (2), (3), and $\Delta(i, P) = \Delta_1(i, P) + \Delta_2(i, P)$, the lemma follows. $\square$

We are now ready to derive the upper bound on the competitive ratio.

Theorem 3. *When the start-up delay d is a positive integer, the competitive ratio of MRF is at most $3 - \frac{1}{d}$.*

Proof. For any $i \geq 0$, let $\mathtt{Opt}(i)$ and $\mathtt{MRF}(i)$ be the sets of pages broadcast by $\mathtt{Opt}$ and $\mathtt{MRF}$ during $(i-1, i]$, respectively. Note that by Lemma 1, we have $\rho(O) = \sum_{i \geq 0} \sum_{P \in \mathtt{Opt}(i)} \rho(O_{i,P})$, which is smaller than or equal to

$$\sum_{i \geq 0} \sum_{P \in \mathtt{Opt}(i)} \left(\sum_{j \in B_{i-d}^{i-1}(P)} \rho(O_{i,P} \cap M_{j,P}) + \sum_{j \in \bar{B}_{i-d}^{i-1}(P)} \frac{\rho(M_{j,\min})}{d} + \Delta(i, P) \right).$$

Hence, to prove the theorem, it suffices to prove $I_1 \leq (2 - \frac{1}{d})\rho(M)$ and $I_2 \leq \rho(M)$ where $I_1 = \sum_{i \geq 0} \sum_{P \in \mathtt{Opt}(i)} \left(\sum_{j \in B_{i-d}^{i-1}(P)} \rho(O_{i,P} \cap M_{j,P}) + \sum_{j \in \bar{B}_{i-d}^{i-1}(P)} \frac{\rho(M_{j,\min})}{d} \right)$ and $I_2 = \sum_{i \geq 0} \sum_{P \in \mathtt{Opt}(i)} \Delta(i, P)$.

We first prove $I_2 \leq \rho(M)$, which is easier. Consider any fixed $i \geq 0$. Let $C_i = \mathtt{Opt}(i) \cap \mathtt{MRF}(i)$. Note that for any $P \in C_i \subseteq \mathtt{MRF}(i)$, MRF broadcasts P at i and by definition, $\Delta(i, P) = \rho(O_{i,P} \cap M_{i,P})$. Hence,

$$\sum_{P \in C_i} \Delta(i, P) = \sum_{P \in C_i} \rho(O_{i,P} \cap M_{i,P}) \leq \sum_{P \in C_i} \rho(M_{i,P}). \tag{4}$$

For any $P \in \mathtt{Opt}(i) - C_i$, we claim that

$$\sum_{P \in \mathtt{Opt}(i) - C_i} \Delta(i, P) = \sum_{P \in \mathtt{Opt}(i) - C_i} \rho(M_{i,\min}) \leq \sum_{P \in \mathtt{MRF}(i) - C_i} \rho(M_{i,P}). \tag{5}$$

We have the first equality because $P \notin \mathtt{MRF}(i)$ and thus MRF does not broadcast P at i, and by definition, $\Delta(i, P) = \rho(M_{i,\min})$. For the second inequality, it obviously holds if $\rho(M_{i,\min}) = 0$. Suppose that $\rho(M_{i,\min}) > 0$. From Fact 1, we conclude that MRF uses all the $|S|$ servers at i and $|\mathtt{MRF}(i)| = |S|$. It follows that $|\mathtt{Opt}(i)| \leq |S| = |\mathtt{MRF}(i)|$. Since $C_i \subseteq \mathtt{Opt}(i)$ and $C_i \subseteq \mathtt{MRF}(i)$, we have $|\mathtt{Opt}(i) - C_i| = |\mathtt{Opt}(i)| - |C_i|$ and $|\mathtt{MRF}(i) - C_i| = |\mathtt{MRF}(i)| - |C_i|$. Therefore, $\sum_{P \in \mathtt{Opt}(i) - C_i} \rho(M_{i,\min}) = |\mathtt{Opt}(i) - C_i|\rho(M_{i,\min}) = (|\mathtt{Opt}(i)| - |C_i|)\rho(M_{i,\min}) \leq (|\mathtt{MRF}(i)| - |C_i|)\rho(M_{i,\min}) = (|\mathtt{MRF}(i) - C_i|)\rho(M_{i,\min}) = \sum_{P \in \mathtt{MRF}(i) - C_i} \rho(M_{i,\min}) \leq \sum_{P \in \mathtt{MRF}(i) - C_i} \rho(M_{i,P})$, and (5) follows.

From (4), (5), and $\sum_{P \in \mathtt{MRF}(i)} \rho(M_{i,P}) = \rho(M_i)$ and $\sum_{i \geq 0} \rho(M_i) = \rho(M)$, we conclude $I_2 = \sum_{i \geq 0} \sum_{P \in \mathtt{Opt}(i)} \Delta(i, P) \leq \rho(M)$

We now show $I_1 \leq (2 - \frac{1}{d})\rho(M)$. For any $i \geq 0$ and any page P, define the characteristic function $\delta_{\mathtt{Opt}}(i, P) = 1$ if $P \in \mathtt{Opt}(i)$, and $\delta_{\mathtt{Opt}}(i, P) = 0$ otherwise. Define $\delta_{\mathtt{MRF}}(i, P)$ for MRF similarly. We now rewrite I_1 using these characteristic functions such that we can estimate its value more easily. To make the manipulations more transparent, we let $G(j, P) = \delta_{\mathrm{MRF}}(j, P)\rho(O_{i,P} \cap M_{j,P})$ and $H(j, P) = (1 - \delta_{\mathrm{MRF}}(j, P))\frac{\rho(M_{j,\min})}{d}$. It can be verified that I_1 is equal to

$$\sum_{i \geq 0} \sum_{P \in \mathcal{P}} \delta_{\mathrm{Opt}}(i, P) \sum_{i-d \leq j \leq i-1} \left(\delta_{\mathrm{MRF}}(j, P)\rho(O_{i,P} \cap M_{j,P}) + (1 - \delta_{\mathrm{MRF}}(j, P))\frac{\rho(M_{j,\min})}{d} \right)$$

$$= \sum_{i \geq 0} \sum_{P \in \mathcal{P}} \sum_{i-d \leq j \leq i-1} (\delta_{\mathrm{Opt}}(i, P)G(j, P) + \delta_{\mathrm{Opt}}(i, P)H(j, P))$$

$$= \sum_{j \geq 0} \sum_{j+1 \leq i \leq j+d} \sum_{P \in \mathcal{P}} (\delta_{\mathrm{Opt}}(i, P)G(j, P) + \delta_{\mathrm{Opt}}(i, P)H(j, P))$$

$$= \sum_{j \geq 0} (Q_j + R_j + S_j)$$

where $Q_j = \sum_{j+1 \leq i \leq j+d} \sum_{P \in \mathcal{P}} \delta_{\mathrm{Opt}}(i, P)\frac{\rho(M_{j,\min})}{d}$,

$R_j = \sum_{j+1 \leq i \leq j+d} \sum_{P \in \mathcal{P}} \delta_{\mathrm{Opt}}(i, P)\delta_{\mathrm{MRF}}(j, P)\rho(O_{i,P} \cap M_{j,P})$, and

$S_j = \sum_{j+1 \leq i \leq j+d} \sum_{P \in \mathcal{P}} \delta_{\mathrm{Opt}}(i, P)\delta_{\mathrm{MRF}}(j, P)(-\frac{\rho(M_{j,\min})}{d})$.

Below, we show that for all $j \geq 0$, $Q_j + R_j + S_j \leq (2 - \frac{1}{d})\rho(M_j)$. Then, the inequality $I_1 \leq (2 - \frac{1}{d})\rho(M)$ follows.

For any time $j \geq 0$ and any page $P \in \mathcal{P}$, we say that P is *critical* for time j if (i) MRF broadcasts P at time j, and (ii) there is a time $i \in [j+1, j+d]$ such that Opt also broadcasts P at time i. Let Σ_j be the set of critical pages for time j. It can be verified that

$$\Sigma_j = \{P \in \mathcal{P} \mid \delta_{\mathrm{Opt}}(i, P)\delta_{\mathrm{MRF}}(j, P) = 1 \text{ for some } i \in [j+1, j+d]\},$$

and $P \subseteq \mathrm{MRF}(j)$. Furthermore, observe that for any page $P \in \mathcal{P}$, if $P \notin \Sigma_j$, then $\delta_{\mathrm{Opt}}(i, P)\delta_{\mathrm{MRF}}(j, P) = 0$ for all $i \in [j+1, j+d]$. Below, we bound Q_j, R_j and S_j.

Note that $Q_j = \sum_{j+1 \leq i \leq j+d} |\mathrm{Opt}(i)|\frac{\rho(M_{j,\min})}{d} \leq |S|\rho(M_{j,\min})$. For R_j, note that for any $P \notin \Sigma_j$, $\delta_{\mathrm{Opt}}(i, P)\delta_{\mathrm{MRF}}(j, P) = 0$ for all $i \in [j+1, j+d]$. Hence,

$$R_j \leq \sum_{j+1 \leq i \leq j+d} \sum_{P \in \Sigma_j} \rho(O_{i,P} \cap M_{j,P}) \leq \sum_{P \in \Sigma_j} \rho(M_{j,P}).$$

(We have the last inequality because $O_{i,P} \cap O_{i',P} = \emptyset$ for any $i \neq i'$.) For S_j, note that $\sum_{P \in \mathrm{MRF}(j) - \Sigma_j} \left(\rho(M_{j,P}) - \frac{\rho(M_{j,\min})}{d} \right)$ is non-negative and hence $S_j \leq \sum_{j+1 \leq i \leq j+d} \sum_{P \in \mathcal{P}} \delta_{\mathrm{Opt}}(i, P)\delta_{\mathrm{MRF}}(j, P)\left(-\frac{\rho(M_{j,\min})}{d} \right) + \sum_{P \in \mathrm{MRF}(j) - \Sigma_j} \left(\rho(M_{j,P}) - \frac{\rho(M_{j,\min})}{d} \right)$. Note that $\sum_{j+1 \leq i \leq j+d} \sum_{P \in \mathcal{P}} \delta_{\mathrm{Opt}}(i, P)\delta_{\mathrm{MRF}}(j, P)(-1) \leq \sum_{P \in \Sigma_j}(-1)$, and we conclude that $S_j \leq \sum_{P \in \Sigma_j} -\frac{\rho(M_{j,\min})}{d} + \sum_{P \in \mathrm{MRF}(j) - \Sigma_j} \left(\rho(M_{j,P}) - \frac{\rho(M_{j,\min})}{d} \right) = \sum_{P \in \mathrm{MRF}(j) - \Sigma_j} \rho(M_{j,P}) - |\mathrm{MRF}(j)|\frac{\rho(M_{j,\min})}{d}$.

Hence, $Q_j + R_j + S_j \leq \sum_{P \in \mathrm{MRF}(j)} \rho(M_{j,P}) + (|S| - \frac{|\mathrm{MRF}(j)|}{d})\rho(M_{j,\min})$. Note that if $\rho(M_{j,\min}) = 0$, then $Q_j + R_j + S_j = \sum_{P \in \mathrm{MRF}(i)} \rho(M_{j,P}) = \rho(M_j) < (2 - 1/d)\rho(M_j)$, as we have claimed. Otherwise, MRF has used all the $|S|$ servers and $|\mathrm{MRF}(i)| = |S|$, and our claim is still true because $Q_j + R_j + S_j \leq \rho(M_j) + (|S| - |S|/d)\rho(M_{j,\min}) = \rho(M_j) + (1 - \frac{1}{d})|S|\rho(M_{j,\min}) \leq (2 - \frac{1}{d})\rho(M_j)$. $\square$

Theorem 4. *When the start-up delay d is a positive integer, the competitive ratio of MRF is at least $3 - \frac{1}{d}$.*

Proof. As in the proof of Theorem 2, we assume without loss of generality that there is only one server in the system. To construct a difficult input instance for MRF, we consider $2d$ different pages $P_0, P_1, \ldots, P_{2d-1}$. For each integer $i \geq 0$, we create a set R_i of requests on page P_{i_o} where $i_o = i \bmod 2d$ as follows. If i is divisible by d, then R_i contains $d + 1$ requests arriving at time $i + 0.5, i + 1.5, \ldots, i + d + 0.5$; the first d requests have profit $1 + \epsilon$ and the last one has profit 1 where $\epsilon > 0$ is an extremely small value. If i is not divisible by d, then R_i contains exactly 2 requests arriving at $i + 0.5$ and $i + d + 0.5$, both have profit 1. Our input is the union of these R_i's, i.e., $\bigcup_{i \geq 0} R_i$.

Note that the schedule S that broadcasts, for each $i \geq 0$, the page P_{i_o} at time $i + d + 0.5$ where $i_o = i \bmod 2d$ accepts all requests in $\bigcup_{i \geq 0} R_i$. We claim for any integer $t \geq 1$, MRF accepts exactly one request at t (there is no request at time 0). Then, the theorem follows because for any time interval $[2id + 1, 2id + 2d]$ $(i \geq 0)$, MRF accepts $2d$ requests during this interval, and S accepts all the $6d - 2$ requests in $R_{2id+1}, R_{2id+2}, \ldots, R_{2id+2d}$.

We prove our claim by mathematical induction. We will prove something stronger. Consider any time $t = 2id + j$ where $i \geq 0$ and $1 \leq j \leq 2d$. We prove that if $1 \leq j \leq d$, MRF broadcasts the page P_0 at t, and if $d + 1 \leq j \leq 2d$, MRF broadcasts P_d at t. Furthermore, in both cases, MRF accepts one request. The base case is easy to verify. Suppose that it is true for times $1, 2, \ldots, t - 1$ and we consider the time $t = 2id + j$. Suppose that $d + 1 \leq j \leq 2d$ (the case when $1 \leq j \leq d$ can be analyzed similarly). To find out the requests pending at t, we can focus on the request sets R_m where $t - 2d \leq m \leq t - 1$ because at t (i) the requests in $R_0, R_1, \ldots, R_{t-2d-1}$ are all expired, and (ii) the requests in $R_t, R_{t+1}, \ldots$ have not arrived.

Note that in the interval $[t - 2d, t - 1] = [2id + j - 2d, 2id + j - 1]$, there are exactly two integers $m_o = 2id$ and $m_1 = 2id + d$ that are divisble by d. When m is neither m_o nor m_1, m is not divisible by d and by construction, R_m has only two requests. If $m \leq t - d - 1$, then the first request, which arrives at time $m + 0.5 \leq t - d - 0.5$, is expired at $m + 0.5 + d < t$, and if $m \geq t - d$, then the second request arrives at time $m + d + 0.5 \geq t + 0.5$, i.e., after t. Therefore, there is exactly one pending request from R_m at t, and it has profit of 1. For the case when $m = m_o$, note that the requests in $R_{m_o} = R_{2id}$ ask for P_0 and all but the last request arrive before $m_o + d = 2id + d$. Note that $m_o + d \leq t - 1$, by the induction hypothesis, MRF broadcasts P_0 at $m_o + d$ and accepts all except the last requests in R_{m_o}, and broadcast P_d at time $m_o + d + 1, \ldots, t - 1$. Therefore, R_{m_o} has one request (the last one) pending at t, and it has profit 1. For the case when $m = m_1 = 2id + d$, note that the requests in R_{2id+d} ask for page P_d. By the induction hypothesis, MRF broadcasts P_d at time $t - 1$ and the requests arrive at or before $t - 1$ are accepted. Therefore, there is exactly one request on P_d, the one in R_{m_1} arriving at $t - 1 + 0.5$, that is pending at t, and it has profit of $1 + \epsilon$. Finally, note that $R_{k-2d}, R_{k-2d+1}, \ldots, R_{k-1}$ are requesting for $2d$ different pages and from above discussion, each of them has exactly one request pending at time t. Furthermore, only the request on P_d has profit $1 + \epsilon$. Therefore, MRF will broadcast P_d at time t and accept one request. □

References

1. D. Aksoy and M. Franklin. Scheduling for large-scale on-demand data broadcasting. In *IEEE INFOCOM*, volume 2, pages 651–659, 1998.
2. A. Bar-Noy, J. Goshi, and R. Ladner. Off-line and on-line guaranteed start-up delay for media-on-demand with stream merging. In *Proceedings of the 15th Annual ACM Symposium on Parallel Algorithms and Architecture*, pages 164–173, 2003.
3. W.T. Chan, T.W. Lam, H.F. Ting, and W.H. Wong. New results on on-demand broadcasting with deadline via job scheduling with cancellation. In *Proceedings of the 10th Annual International Conference on Computing and Combinatorics*, pages 210–218, 2004.
4. A. Dan, D. Sitaram, and P. Shahabuddin. Scheduling policies for an on-demand video server with batching. In *Proceedings of the ACM Multimedia*, pages 15–23, 1994.
5. H.D. Dykeman, M.H. Ammar, and J.W. Wong. Scheduling algorithms for videotex systems under broadcast delivery. In *IEEE International Conference on Communications (ICC)*, pages 1847–1851, 1986.
6. J. Edmonds and K. Pruhs. A maiden analysis of Longest Wait First. In *Proceedings of the Fifteenth ACM-SIAM Symposium on Discrete Algorihtms*, pages 818–827, 2004.
7. A.T. Hawkins and W. Mao. On multi-channel data broadcast scheduling. In *The Second Workshop on Intelligent Multimedia Computing and Networking*, pages 915–918, 2002.
8. Regant Y.S. Hung. Scheduling online batching systems. Master's thesis, The University of Hong Kong, 2005.
9. Regant Y.S. Hung and H.F. Ting. Scheduling online batching systems: a competitiveness study on patience and sharing. In *Proceedings of the Latin American Theoretical Informatics Symposium*, pages 605–616, 2006.
10. S. Jiang and N.H. Vaidya. Scheduling data broadcast to "impatient" users. In *Proceedings of the 1st ACM International Workshop on Data Engineering for Wireless and Mobile Access (MobiDE)*, pages 52–59, 1999.
11. B. Kalyanasundaram, K. Pruhs, and M. Velauthapillai. Scheduling broadcasts in wireless networks. *Journal of Scheduling*, 4(6):339–354, 2000.
12. B. Kalyanasundaram and M. Velauthapillai. On-demand broadcasting under deadline. In *Proceedings of the 11th Annual European Symposium on Algorithms, volume 2832 of* Lecture Notes in Computer Science, pages 313–324, 2003.
13. J.H. Kim and K.Y. Chwa. Scheduling broadcasts with deadlines. *Theoretical Computer Science*, 325(3):479–448, 2004.
14. W. Mao. Competitive analysis of online algorithm for on-demand data broadcast scheduling. In *Proceedings of the IEEE International Symposium on Parallel Architectures, Algorithms, and Networks*, pages 292–296, 2000.
15. N.J. Sarhan and C.R. Das. A simulation-based analysis of scheduling policies for multimedia servers. In *Proceedings of the 36th Annual Simulations Symposium*, pages 183–190, 2003.
16. H. Shachnai and P. Yu. Exploring wait tolerance in effective batching for video-on-demand scheduling. *Multimedia Systems*, 6:382–394, 1998.
17. H.F. Ting. A near optimal scheduler for on-demand data broadcasts. In *Proceedings of the Sixth International Conference on Algorithms and Complexity*, 2006, to appear.
18. J.W. Wong. Broadcast delivery. *Proceedings of IEEE*, 76(12):1566–1577, 1988.

On Lazy Bin Covering and Packing Problems⋆

Mingen Lin, Yang Yang, and Jinhui Xu

Department of Computer Science and Engineering
University at Buffalo, the State University of New York
Buffalo, NY 14260, USA
{mlin6, yyang6, jinhui}@cse.buffalo.edu

Abstract. In this paper, we study two interesting variants of the classical bin packing problem, called *Lazy Bin Covering (LBC)* and *Cardinality Constrained Maximum Resource Bin Packing (CCMRBP)* problems. For the offline LBC problem, we first show its NP-hardness, then prove the approximation ratio of the First-Fit-Decreasing algorithm, and finally present an APTAS. For the online LBC problem, we give competitive analysis for the algorithms of Next-Fit, Worst-Fit, First-Fit, and a modified HARMONIC$_M$ algorithm. The CCMRBP problem is a generalization of the *Maximum Resource Bin Packing (MRBP)* problem [1]. For this problem, we prove that its offline version is no harder to approximate than the offline MRBP problem.

1   Introduction

Bin packing is a fundamental problem in combinatorial optimization and finds applications in many different areas. A long and rich history exists and many important results have been obtained [2,3,4,5,6,7,8]. In its most basic form, the bin packing problem seeks to pack items of size between zero and one into a minimum number of unit-sized bins. Depending on its applications, the bin packing problem could also have many other different forms. Recently, Boyar *et al.* studied an interesting variant of the classical bin packing problem, called *Maximum Resource Bin Packing (MRBP)* [1], which considers the bin packing problem from a reverse perspective and maximizes the total number of used bins. For instance, in its offline version, the MRBP problem requires an ordering of the packed bins such that no item in a later bin fits in an earlier bin. Motivated by this new problem, in this paper we first consider an interesting variant of the classical bin covering problem called *Lazy Bin Covering (LBC)*, and then study a generalization of the MRBP problem called *cardinality constrained MRBP (CCMRBP)*.

Lazy Bin Covering

The classical bin covering problem can be viewed as a "dual" problem of the bin packing problem. Its objective is to pack items of size between zero and one into

---

⋆ This research was partially supported by an NSF CARRER Award CCF-0546509.

D.Z. Chen and D.T. Lee (Eds.): COCOON 2006, LNCS 4112, pp. 340–349, 2006.

a maximum number of unit-sized bins so that the level of each bin is at least one [9,10]. Different from the classical bin covering problem, the LBC problem tries to pack items into a minimum number of covered bins so that removing any item from a covered bin turns it into an uncovered one (i.e., its level becomes smaller than one). Formally, the LBC problem can be stated as follows: Given a list $L = \{a_1, a_2, \cdots, a_n\}$ of items, $a_i \in (0, 1]$, pack them into a minimum number m of unit-sized bins $B_1, B_2, \cdots, B_m$ such that $\sum_{a_i \in B_j} a_i \geq 1$ for $1 \leq j \leq m - 1$ and $\sum_{a_i \in B_j} a_i - \min\{a_i | a_i \in B_j\} < 1$ for all $1 \leq j \leq m$. That is, except probably B_m, all other bins are covered and for each bin no item is redundant. In certain sense, the LBC problem can be regarded as finding the worst but reasonable packing for the bin covering problem.

Two related problems have been considered in the past, the plain open-end bin packing problem (POBP) [11,12] and the ordered open-end bin packing problem (OOBP)[13]. The goal of POBP is to pack items into a minimum number of bins, where the level of a bin can exceed one as long as removing the last item brings its level back to less than one. The OOBP further requires that the designated last item in a bin of level exceeding one has to be the largest-indexed item in that bin. LBC can also be viewed as the OOBP with the additional requirement that the item with larger size has lower index.

In this paper, we consider both the offline (Section 3) and online (Section 4) versions of the LBC problem, and present a number of results for each version, such as the complexity analysis and approximation ratio analysis for several classical bin packing strategies. For the online version, each bin is required to be non-redundant but may not be covered.

Cardinality Constrained Maximum Resource Bin Packing

The CCMRBP problem is a generalization of MRBP [1] and can be stated as follows: Given a sequence $L = \{a_1, a_2, \cdots, a_n\}$ of items, $a_i \in (0, 1]$, pack them into a maximum number m of unit-sized bins $B_1, B_2, \cdots, B_m$ such that each bin contains at most C items and the packing satisfies the following constraint.

Constraint 1. *No item $a \in B_j$ can fit into any B_k, $1 \leq k < j, 2 \leq j \leq m$, i.e. either B_k already contains c items or the level of B_k will exceed one after placing a into it.*

In Section 5, we prove that the offline CCMRBP problem is no harder to approximate than the offline MRBP.

Due to space limit, we omit several proofs and many details from this extended abstract.

2   Preliminaries

For an input sequence $L = \{a_1, a_2, \cdots, a_n\}$ of items, let $A(L)$ denote the packing returned by an algorithm A and $OPT(L)$ denote the packing generated by an optimal algorithm OPT. For a set of bins s and any bin $B \in s$, let $c(B)$ be the number of items packed in B, $l(B) = \sum_{a_i \in B} a_i$ be the *level* of B, and

$l(s) = \sum_{B \in s} l(B)$ be the level of s. Denote the size of the smallest (or largest) item in B by $\min(B)$ (or $\max(B)$). Similarly, denote the size of the smallest (or largest) item in s by $\min(s)$ (or $\max(s)$). For a weighting function $w : L \to \Re$, we define $w(B) = \sum_{a_i \in B} w(a_i)$ and $w(s) = \sum_{B \in s} w(B)$ as the total weight of items in B and in s respectively. Let $p = s_1 | s_2$ denote a packing p which consists of a set of bins s_1, followed by a set of bins s_2.

In this paper, we consider the following classical packing algorithms/strategies.

- Next-Fit (NF) maintains only one open bin at any time point and places an item into the bin only if it fits. Otherwise a new bin is opened.
- First-Fit (FF) places an item into the first bin that can accommodate it.
- Worst-Fit (WF) places an item into the lowest level bin among all the bins that can accommodate it. Break tie arbitrarily.
- First-Fit-Increasing (FFI) packs items in a non-decreasing order with respect to their sizes, and places them using First-Fit.
- First-Fit-Decreasing (FFD) packs items in a non-increasing order with respect to their sizes, and places them using First-Fit.

We also presents a modified HARMONIC$_M$ (MH(M)) algorithm [14] for the online LBC problem. Details of the algorithm will be given in Section 4.

An approximation algorithm A is a c-approximation algorithm for $c \geq 1$, if there is a constant b such that for all possible input sequences L, $A(L) \leq cOPT(L) + b$ for minimization problems (or $OPT(L) \leq cA(L) + b$ for maximization problems). The infimum of all such c is called the approximation ratio of the algorithm, R_A. For a parameterized bin packing/covering problem, we consider the parameterized approximation ratio, $R_A(k)$, which is the approximation ratio of A in the case where all items have sizes no more than $\frac{1}{k}$ for some integer k. For an online algorithm A, the performance (or competitive ratio C_A) of A is measured similarly by comparing with the optimal offline algorithm OPT.

3   Offline Lazy Bin Covering Problem

In this section, we first prove that offline LBC is NP-hard, then analyze the approximation ratios of FFD and FFI, and finally show that there exists an asymptotic PTAS for the offline LBC problem.

Theorem 1. *The offline LBC is NP-hard.*

Proof. To prove the NP-hardness, we reduce the partition problem to the decision version of LBC. The partition problem can be stated as follows: Given a set of positive integers $I = \{a_1, a_2, \cdots, a_n\}$, $\sum_{i=1}^{n} a_i = 2b$, decide whether I can be partitioned into two sets, S and $I - S$, such that $\sum_{a_i \in S} a_i = b$. Without lost of generality, we assume that $b \geq 3$ and $a_i \geq 2$ for $1 \leq i \leq n$.

The reduction is constructed as follows. Given an instance I of the partition problem, we construct an instance of LBC $I' = \{p_1, p_2, \cdots, p_n, p_{n+1}, \cdots, p_{n+m}\}$, $m = 2b + 4$, where $p_i = \frac{a_i}{2b}$ (big items) for $1 \leq i \leq n$ and $p_j = \frac{b-1}{2b^2}$ (small items) for $n + 1 \leq j \leq n + m$. Note that $\frac{b-1}{2b^2}$ is the smallest item. Below we show that I

can be partitioned into two equal-sized sets if and only if I' can be packed into two unit-sized bins in the LBC problem.

If I can be partitioned into two equal-sized sets S and $I - S$, we can pack the items in I' as follows. First, for each $a_i \in S$, place the corresponding p_i into the first bin. Then place all p_j for $n + 1 \leq i \leq n + \frac{m}{2}$ into the first bin. The remaining items in I' are packed into the second bin. It is easy to see that the level of each bin is $1 + \frac{b-2}{2b^2}$, which is strictly smaller than $1 + \frac{b-1}{2b^2}$ (i.e., 1 plus the size of the smallest item). Thus removing any item from each bin will make it uncovered, which means the resulting packing is valid.

If I' can be packed into two unit size bins, since $(m - 1)\frac{b-1}{2b^2} = 1 + \frac{b-3}{2b^2} \geq 1$, neither bin contains m small items. Let A be the number of the small items in the first bin, and $\frac{X}{2b}$ be the total size of the big items in the first bin. Similarly, let B be the number of small items in the second bin, and $\frac{Y}{2b}$ be the total size of the big items in the second bin. We have $1 \leq A \leq m - 1$, $1 \leq B \leq m - 1$ and $A + B = m$. From definitions, we also have

$$\begin{cases} A\frac{b-1}{2b^2} + \frac{X}{2b} < 1 + \frac{b-1}{2b^2} \\ B\frac{b-1}{2b^2} + \frac{Y}{2b} < 1 + \frac{b-1}{2b^2}, \end{cases} \text{ That is, } \begin{cases} X < 2b - (A - 1) + \frac{A-1}{b} \\ Y < 2b - (B - 1) + \frac{B-1}{b} \end{cases}$$

Next we want to show that $A = B = \frac{m}{2}$ is the only possible solution to the above set of inequalities. If $A = B = \frac{m}{2}$, we have $X < b + \frac{1}{b}$ and $Y < b + \frac{1}{b}$. Since X and Y are integers and $X + Y = 2b$, we have $X = Y = b$. Therefor I can be partitioned into two equal-sized sets. If A and B are not equal, without loss of generality, assume $A > B$. We then have $b + 3 \leq A \leq 2b + 3$. Consider the following three cases.

Case 1. $A = 2b + 3$. Then $B = 1$. From the above inequalities we have $X < \frac{2}{b}$ which implies $X = 0$ (since $b \geq 3$), and $Y < 2b$. A contradiction to the fact that $X + Y = 2b$.

Case 2. $A = 2b+2$. Then $B = 2$. From the above inequalities we have $X < 1+\frac{1}{b}$ which implies $X = 0$ since $a_i \geq 2$ for $1 \leq i \leq n$, and $Y < 2b - 1 + \frac{1}{b} < 2b$. A contradiction to the fact that $X + Y = 2b$.

Case 3. $b+3 \leq A \leq 2b+1$. Then $3 \leq B \leq b+1$. From the above inequalities we have $Y \leq 2b - (B - 1)$, hence $X + Y < 2b + \frac{A-1}{b} - 2 \leq 2b$. A contradiction to the fact that $X + Y = 2b$.

Thus if I' can be packed into two bins, I can be equally partitioned, and the theorem follows.   □

Next we consider the FFD algorithm. First, note that FFD behaves like the NF-Decreasing algorithm in the offline LBC problem.

Theorem 2. *The parameterized approximation ratio of FFD is*

$$R_{FFD}(k) = \begin{cases} \frac{71}{60}, & \text{if } k = 1, 2, \\ 1 + \frac{1}{k+2} - \frac{1}{\alpha}, & \text{if } k \geq 3, \end{cases}$$

where $\alpha = k(k + 1)(k + 2)$ when k is odd, and $\frac{k(k+1)(k+2)}{2}$ when k is even.

Note that $R_{FFD}(1) = R_{FFD}(2) = R_{FFD}(3)$.

Proof. We first prove the lower bound of the parameterized approximation ratio. We start with the case $k \geq 3$. For this case, we further distinguish two sub-cases: (a) k is odd and (b) k is even. For case (a), let n be an integer divisible by $k(k+1)(k+2)$, L be a sequence of items consisting of $\frac{k-1}{2}n$ items of size $\frac{1}{k}$, n items of size $\frac{1}{k+1}$, and $\frac{k+3}{2}n$ items of size $\frac{1}{k+2}$. For this sequence of items, FFD packs $\frac{k-1}{2k}n$ bins with each containing k items of size $\frac{1}{k}$, $\frac{1}{k+1}n$ bins with each containing $k+1$ items of size $\frac{1}{k+1}$, and $\frac{k+3}{2(k+2)}n$ bins with each containing $k+2$ items of size $\frac{1}{k+2}$. Thus FFD uses in total $(1 + \frac{1}{k+2} - \frac{1}{k(k+1)(k+2)})n$ bins. The optimal algorithm OPT packs n bins with each containing $\frac{k-1}{2}$ items of size $\frac{1}{k}$, one item of size $\frac{1}{k+1}$, and $\frac{k+3}{2}$ items of size $\frac{1}{k+2}$. Thus, the approximation ratio of FFD for this special instance is $(1 + \frac{1}{k+2} - \frac{1}{k(k+1)(k+2)})$, and the lower bound follows. In this instance, the level of each bin in $FFD(L)$ is exactly one, while the level of each bin in $OPT(L)$ is $1 + \frac{1}{k+2} - \frac{1}{k(k+1)(k+2)} < 1 + \frac{1}{k+2}$.

For case (b) (i.e., k is even), let n be an integer divisible by $\frac{k(k+1)(k+2)}{2}$, and L be a sequence of items consisting of $\frac{k-2}{2}n$ items of size $\frac{1}{k}$, $2n$ items of size $\frac{1}{k+1}$, and $\frac{k+2}{2}n$ items of size $\frac{1}{k+2}$. For this instance, FFD uses a total of $(1 + \frac{1}{k+2} - \frac{2}{k(k+1)(k+2)})n$ bins, while OPT packs n bins with each containing $\frac{k-2}{2}$ items of size $\frac{1}{k}$, two items of size $\frac{1}{k+1}$, and $\frac{k+2}{2}$ items of size $\frac{1}{k+2}$. Thus the approximation ratio of FFD for this instance is $(1 + \frac{1}{k+2} - \frac{2}{k(k+1)(k+2)})$ and the lower bound follows. In this instance, the level of each bin in $FFD(L)$ is exactly one, while the level of each bin in $OPT(L)$ is $1 + \frac{1}{k+2} - \frac{2}{k(k+1)(k+2)} < 1 + \frac{1}{k+2}$.

When $k \leq 2$, it is easy to see that $R_{FD}(1) \geq R_{FD}(2) \geq R_{FD}(3) \geq 1 + \frac{1}{3+2} - \frac{1}{3 \times 4 \times 5} = \frac{71}{60}$.

Secondly, we upper bound the approximation ratio of FFD. Let L be any sequence of items. We first prove the upper bound for the case of $k \leq 2$. Here we assume $\frac{1}{0} = +\infty$. Let interval $I_j = [\frac{1}{j}, \frac{1}{j-1})$ for $1 \leq j \leq 6$ and $I_7 = (0, \frac{1}{6})$. We define a weighting function $w : L \to \Re$ as follows:

$$w(a_i) = \begin{cases} \frac{1}{j}, & \text{if } a_i \in I_j, 1 \leq j \leq 6; \\ a_i, & \text{if } a_i \in I_7. \end{cases}$$

From this function, it is clear that the weight of each item is no more than its size. By the problem description, we have the following fact:

Fact 1. *For any bin B, $w(B) < 1 + \min\{w(a_i) \mid a_i \in B\}$.*

Let W be the total weight of all the items in L. Consider a bin B in the packing $FFD(L)$ which is not the last bin and contains only items from a single interval I_j for some j. If $1 \leq j \leq 6$, B will contain exactly j items and $w(B) = 1$. If $j = 7, w(B) = l(B) \geq 1$. Thus $w(B) < 1$ only if B contains items from more than one interval or B is the last bin in $FFD(L)$. Let C be the set of bins in $FFD(L)$ with weight less than one. Obviously, $|C| \leq 6$. Therefore we have $|FFD(L)| - 6 < W$.

Next we show that $W \leq \frac{71}{60}|OPT(L)|$. Clearly, it is sufficient to show that for any bin B in $OPT(L)$, its weight $w(B) \leq \frac{71}{60}$. Suppose this is not true. Then by Fact 1, the weight of the smallest item in B must be larger than $\frac{11}{60}$. Thus B can only contain items with weight in $\{1, \frac{1}{2}, \frac{1}{3}, \frac{1}{4}, \frac{1}{5}\}$. Suppose B contains a items of weight 1, b items of weight $\frac{1}{2}$, c items of weight $\frac{1}{3}$, d items of weight $\frac{1}{4}$ and e items of weight $\frac{1}{5}$. If $e \neq 0$, we have $\frac{71}{60} < a + \frac{b}{2} + \frac{c}{3} + \frac{d}{4} + \frac{e}{5} < \frac{6}{5}$, and $71 < 60a + 30b + 20c + 15d + 12e < 72$. But this is not possible since a, b, c, d, e are all integers. If $e = 0$ and $d \neq 0$, we have $\frac{71}{60} < a + \frac{b}{2} + \frac{c}{3} + \frac{d}{4} < \frac{5}{4}$, and $71 < 60a + 30b + 20c + 15d < 75$. This is also not possible since $60a + 30b + 20c + 15d$ is divisible by 5. If $e = d = 0$ and $c \neq 0$, we have $\frac{71}{60} < a + \frac{b}{2} + \frac{c}{3} < \frac{4}{3}$, and $71 < 60a + 30b + 20c < 80$. This is also not possible since $60a + 30b + 20c$ is divisible by 10. Similarly, if $e = d = c = 0$ and $b \neq 0$, we have $\frac{71}{60} < a + \frac{b}{2} < \frac{3}{2}$, and $71 < 60a + 30b < 90$. This is not possible since $60a + 30b$ is divisible by 30. If $e = d = c = b = 0$ and $a \neq 0$, we have $\frac{71}{60} < a < 2$, and $71 < 60a < 120$, this is not possible since $60a$ is divisible by 60.

Thus $|FFD(L)| < \frac{71}{60}|OPT(L)| + 6$.

We now prove for the case of $k \geq 3$. We define a weighting function $w : L \to \Re$ as following:

$$w(a_i) = \begin{cases} \frac{1}{k+1}, & \text{if } a_i \in [\frac{1}{k+1}, \frac{1}{k}); \\ \frac{1}{k+2}, & \text{if } a_i \in [\frac{1}{k+2}, \frac{1}{k+1}); \\ \frac{1}{k+3}, & \text{if } a_i \in [\frac{1}{k+3}, \frac{1}{k+2}); \\ a_i, & \text{otherwise.} \end{cases}$$

It is clear that the weight of each item is at most its size, and Fact 1 is still true for this weighting function. By similar arguments as in the proof for $k \leq 2$, each bin in $FFD(L)$ has weight at least one except for no more than five bins. Next we claim that each bin B in $OPT(L)$ has total weight at most $1 + \frac{1}{k+2} - \frac{1}{\alpha}$. To prove it, we consider the following two cases.

Case 1. B contains an item with weight at most $\frac{1}{k+3}$. Then we have $w(B) < 1 + \frac{1}{k+3}$ by Fact 1. It is easy to verify that $\frac{1}{k+3} \leq \frac{1}{k+2} - \frac{2}{k(k+1)(k+2)} \leq \frac{1}{k+2} - \frac{1}{\alpha}$ when $k \geq 3$.

Case 2. Each item in B has weight at least $\frac{1}{k+2}$. Then the weight of each item is in $\{\frac{1}{k}, \frac{1}{k+1}, \frac{1}{k+2}\}$. Two subcases occur: (a) B contains at least one item of weight $\frac{1}{k+2}$ and (b) B contains no item of weight $\frac{1}{k+2}$. For case (a), by Fact 1 we have $w(B)\alpha < \alpha + \frac{\alpha}{k+2}$, where α is the least common multiplier of $k, k+1$ and $k+2$. Hence $w(B)\alpha$ and $\frac{\alpha}{k+2}$ are integers, and thus $w(B)\alpha \leq \alpha + \frac{\alpha}{k+2} - 1$, which means $w(B) \leq 1 + \frac{1}{k+2} - \frac{1}{\alpha}$. For case (b), by Fact 1 we have $w(B) < 1 + \frac{1}{k+1}$, and $w(B)k(k+1) < k(k+1) + k$. Since $w(B)k(k+1)$ is an integer, we have $w(B)k(k+1) \leq k(k+1) + k - 1$, i.e. $w(B) \leq 1 + \frac{1}{k+1} - \frac{1}{k(k+1)}$. It is easy to verify that $\frac{1}{k+1} - \frac{1}{k(k+1)} \leq \frac{1}{k+2} - \frac{1}{\alpha}$.

Let W be the total weight of all items in L. We have $|FFD(L)| - 5 \leq W \leq (1 + \frac{1}{k+2} - \frac{1}{\alpha})|OPT(L)|$, and the theorem follows. $\qquad\Box$

Next we consider the FFI algorithm. Different from the FFD algorithm, FFI does not always generate a "legal" packing for an arbitrary instance of the offline LBC problem. This is because it could yield a packing with more than one partially-filled bin, thus violating the constraint of the offline LBC problem. We call the problem without the at-most-one-partially-filled-bin constraint as the *relaxed offline LBC* problem. It is easy to see that this problem is also NP-hard and FFD has the same approximation ratio as the one stated in Theorem 2. In addition, we have the following theorem.

Theorem 3. *For the relaxed offline LBC problem, the parameterized approximation ratio of FFI is*

$$R_{FFI}(k) = \begin{cases} \frac{71}{60}, & k = 1, 2; \\ 1 + \frac{1}{k+2} - \frac{1}{\alpha}, & k \geq 3. \end{cases}$$

where $\alpha = k(k+1)(k+2)$ if k is odd, and $\frac{k(k+1)(k+2)}{2}$ if k is even.

In [15], an asymptotic PTAS is presented for the classical bin packing problem. Using similar techniques, we are able to obtain an APTAS for the offline LBC problem. Details are left for the full paper.

Theorem 4. *There exists an asymptotic PTAS (APTAS) for the offline LBC Problem.*

4 Online Lazy Bin Covering Problem

In this section, we study the online LBC problem. We first analyze the competitive ratios of three classical bin packing algorithms, Next-Fit, Worst-Fit and First-Fit, and then analyze the competitive ratio of a modified HARMONIC$_K$ (MH$_K$) algorithm.

Theorem 5. *The parameterized competitive ratio of Next-Fit is*

$$C_{NF}(k) = \begin{cases} 4, & k = 1; \\ \frac{k+1}{k-1}, & k \geq 2. \end{cases}$$

Proof. First we prove the lower bound of the competitive ratio. For $k = 1$, let the input sequence $L = \langle \epsilon, \epsilon, 1 - \epsilon, \epsilon \rangle^n$, where $\epsilon \leq \frac{1}{3n}$ and n is an even integer. Next-Fit packs n bins with each containing two items of size ϵ and n bins with each containing an item of size $1 - \epsilon$ and an item of size ϵ. The optimal offline algorithm OPT puts $3n$ items of size ϵ into one bin and two $1 - \epsilon$ into each of the other bins, using $\frac{n}{2} + 1$ bins in total. The lower bound follows.

For $k \geq 2$, let the input sequence $L = \langle \langle \frac{1}{k} - \epsilon \rangle^{k-1}, k\epsilon, \epsilon \rangle^{(k+1)n}$, where $\epsilon \leq \frac{1}{(k+1)^2 n}$ and n is an integer. Next-Fit puts $k - 1$ items of size $\frac{1}{k} - \epsilon$, one item of size $k\epsilon$ and one item of size ϵ into one bin, using a total of $(k + 1)n$ bins. OPT puts all items of size $k\epsilon$ and ϵ into one bin and $k + 1$ items of size $\frac{1}{k} - \epsilon$ into each of the other bins, using a total of $(k - 1)n + 1$ bins. Thus the lower bound follows for the case $k \geq 2$.

For the upper bound, let L be any sequence of items and S be the total size of all the items. For $k = 1$, first note that the total level of any two consecutive bins in $NF(L)$ is greater than one. Second, the level of any bin in $OPT(L)$ is less than 2 by definition. Therefore, we have $\frac{1}{2}(|NF(L)| - 1) < S < 2|OPT(L)|$, and immediately, $|NF(L)| \leq 4|OPT(L)|$.

For $k \geq 2$, since all items have size at most $\frac{1}{k}$, we have that every bin in $NF(L)$ has level greater than $1-\frac{1}{k}$ except possibly the last bin. Every bin in $OPT(L)$ has level less than $1 + \frac{1}{k}$ by definition. Thus $\frac{k-1}{k}(|NF(L)| - 1) < S < \frac{1+k}{k}|OPT(L)|$, and $|NF(L)| \leq \frac{k+1}{k-1}|OPT(L)|$. □

Theorem 6. *The parameterized competitive ratio of Worst-Fit is*

$$C_{WF}(k) = \begin{cases} 3, & k = 1; \\ \frac{k+1}{k-1}, & k \geq 2. \end{cases}$$

Note that $C_{WF}(1) = C_{WF}(2)$.

Theorem 7. *The parameterized competitive ratio of First-Fit is* $C_{FF}(k) = \frac{k+1}{k}$.

Proof. First we prove the lower bound. Let the input sequence $L = \langle\langle\frac{1}{k} - \epsilon\rangle^k, k\epsilon\rangle^{(k+1)n}$, where $\epsilon \leq \frac{1}{k(k+1)n}$ and n is an integer. First-Fit packs $(k+1)n$ bins with each bin containing k items of size $\frac{1}{k} - \epsilon$ and one item of size $k\epsilon$. The optimal offline algorithm OPT puts all the items of size $k\epsilon$ into one bin, and $k+1$ items of size $\frac{1}{k} - \epsilon$ into each of the other bins, using a total of $kn+1$ bins. Thus the lower bound follows.

To show the upper bound, for any input sequence L, we let $FF(L) = \{B_1, B_2, \cdots, B_m\}$ be the packing generated by First-Fit, and $\{B'_1, B'_2, \cdots, B'_n\}$ be the bins in $FF(L)$ with level less than one. By the property of First-Fit, $\min(B'_i) + l(B'_{i-1}) - \min\{\min(B'_i), \min(B'_{i-1})\} \geq 1, 2 \leq i \leq n$. Since $l(B'_{i-1}) < 1$, we have $\min\{\min(B'_i), \min(B'_{i-1})\} = \min(B'_{i-1})$, and hence $\min(B'_i) + l(B'_{i-1}) - \min(B'_{i-1}) \geq 1, 2 \leq i \leq n$. Thus we have $\min(B'_n) \geq \min(B'_1) + \sum_{i=1}^{n-1} 1 - l(B'_i)$. Let $S = \sum_{i=1}^{m} l(B_i)$. We have $S \geq (m - n) + \sum_{i=1}^{n} l(B'_i) \geq (m - n) + (n - 1) + l(B'_n) - \min(B'_n) + \min(B'_1) > m - 1$. Note that for any bin in $OPT(L)$, its level is less than $1 + \frac{1}{k}$ by definition. Thus we have $S < (1 + \frac{1}{k})|OPT(L)|$. Therefore $|FF(L)| < (1 + \frac{1}{k})|OPT(L)| + 1$. □

Motivated by the HARMONIC(M) algorithm in [14] for the online bin packing problem, below we present a modified HARMONIC ($MH(M)$) for the online LBC problem, where $M \geq 3$ is the maximum number of open bins at any time. Due to the different problem settings, we modify the "harmonic partitioning" as follows: Let $I_j = [\frac{1}{k+j-1}, \frac{1}{k+j-2}), 1 \leq j < M$, and $I_M = (0, \frac{1}{k+M-2})$. Here we assume $\frac{1}{0} = +\infty$. A bin B is of type j if it only accepts items of size in I_j. Below are the main steps of our algorithm.

Theorem 8. *The parameterized competitive ratio of MH(M) ($M \geq 3$) is*

$$C_{MH(M)}(k) = \max\{\beta, \frac{k+M-1}{k+M-3}\}, \text{ where } \beta = \begin{cases} \frac{71}{60}, & k \leq 2 \\ 1 + \frac{1}{k+2} - \frac{1}{\alpha}, & k \geq 3 \end{cases}$$

and $\alpha = k(k+1)(k+2)$ *when k is odd, and* $\frac{k(k+1)(k+2)}{2}$ *when k is even.*

Algorithm 1. $MH(M)$

1: Open M bins of types $1, 2, \cdots, M$ respectively.
2: For each current item a_i, if $a_i \in I_j$, let B_j be the open bin of type j.
3: **if** $a_i + l(B_j) - \min\{a_i, \min(B_j)\} < 1$ **then**
4:    put a_i into B_j.
5: **else**
6:    close B_j, open a new bin of type j and put a_i into it.
7: **end if**
8: Move to the next item.
9: Repeat steps 2 to 8 until all items are packed.

---

5 Cardinality Constrained Maximum Resource Bin Packing Problem

In this section, we study the offline CCMRBP problem and prove that it is no harder to approximate than the offline MRBP.

Let B be a bin in any feasible packing. B is cardinality-saturated if $c(B) = C$. Otherwise B is load-saturated.

Lemma 1. *There exists an optimal packing* $p = s_1|s_2$ *such that* $s_1 = \{B_1, \cdots, B_j\}$ *is a set of cardinality-saturated bins and* $s_2 = \{B_{j+1}, \cdots, B_m\}$ *is a set of load-saturated bins, for some* $j \in [0, m]$.

Lemma 2. *There exists an optimal packing* $p = s_1|s_2$ *satisfying Lemma 1 and* $\max(s_1) \leq \min(s_2)$.

Proof. Given an optimal packing $p = s_1|s_2$ satisfying Lemma 1. Let $a = \max(s_1)$ and $b = \min(s_2)$. Let B be the bin containing a, and B' be the bin containing b. If $a > b$, we swap a and b. After the swapping B is still cardinality-saturated and the level of B' will increase. Therefore, no item in the bins with index larger than B' can fit into B'. If the new level of B' exceeds one, we can swap the position (index) of B' with the bin right after B', B''. After the swapping, we adjust the items in B' if necessary, i.e. if any item in B' can fit into B'', move it from B' to B''. The level of B' after the adjustment will still be larger than the level of B'' before the adjustment. Thus, no items in the bins positioned after B' can fit into B'. If the new level of B' is no greater than one, the packing is valid and we are done. Otherwise, we keep swapping the position of B' with the bin right after it and adjust the items in B' if necessary. Since p is an optimal packing, eventually the level of B' will be at most one by the moment when it becomes the last bin in the packing. Repeat the above procedure until $\max(s_1) \leq \min(s_2)$.

Lemma 3. *There exists an optimal packing* $p = s_1|s_2$ *satisfying Lemma 2 and the bins in* s_1 *are packed by* FFI.

Theorem 9. *Given an algorithm* A_1 *for the offline MRBP, there exists an algorithm* A_2 *for the offline CCMRBP such that* $R_{A_2} = R_{A_1}$.

Proof. Since the offline MRBP can be regarded as the offline CCMRBP with $C = \lfloor \frac{1}{\min\{a_i \mid a_i \in L\}} \rfloor$, we have $R_{A_2} \geq R_{A_1}$. To show the other direction, for an input sequence L, let A_2 take the following actions. Sort the items in L in a non-decreasing order of their sizes. Let the sorted set be $L = \{a_1, a_2, \cdots, a_n\}$. Try all possible partitions $L_1 = \{a_1, a_2, \cdots, a_k\}$, $L_2 = \{a_{k+1}, a_{k+2}, \cdots, a_n\}$, $0 \leq k \leq n$. For each partition $\langle L_1, L_2 \rangle$, pack L_1 by FFI and pack L_2 by A_1. Then output a valid packing with maximum number of bins. By Lemma 3, $R_{A_2} \leq R_{A_1}$. Thus the theorem follows. □

Corollary 1. *For the offline CCMRBP Problem,* $R_{FFI}(k) = \frac{6}{5}$ *if* $k = 1$ *and* $R_{FFI}(k) = \frac{k^2+k}{k^2+1}$ *if* $k \geq 2$.

Proof. Follow from Theorem 9 and the results in [1] on the offline MRBP. □

References

1. Boyar, J., Epstein, L., Favrholdt, L.M., Kohrt, J.S., Larsen, K.S., Pedersen, M.M., Wøhlk, S.: The maximum resource bin packing problem. In Liskiewicz, M., Reischuk, R., eds.: FCT. Volume 3623 of Lecture Notes in Computer Science., Springer (2005) 397–408
2. Garey, M.R., Graham, R.L., Johnson, D.S.: Resource constrained scheduling as generalized bin packing. J. Comb. Theory, Ser. A **21**(3) (1976) 257–298
3. Johnson, D.S., Garey, M.R.: A 71/60 theorem for bin packing. J. Complexity **1**(1) (1985) 65–106
4. Friesen, D.K., Langston, M.A.: Analysis of a compound bin packing algorithm. SIAM J. Discrete Math. **4**(1) (1991) 61–79
5. Csirik, J.: The parametric behavior of the first-fit decreasing bin packing algorithm. J. Algorithms **15**(1) (1993) 1–28
6. Woeginger, G.J.: Improved space for bounded-space, on-line bin-packing. SIAM J. Discrete Math. **6**(4) (1993) 575–581
7. Csirik, J., Johnson, D.S.: Bounded space on-line bin packing: Best is better than first. Algorithmica **31**(2) (2001) 115–138
8. Shachnai, H., Tamir, T., Yehezkely, O.: Approximation schemes for packing with item fragmentation. (Proc. of WAOA'05)
9. Csirik, J., Kenyon, C., Johnson, D.S.: Better approximation algorithms for bin covering. In: SODA. (2001) 557–566
10. Assmann, S.F., Johnson, D.S., Kleitman, D.J., Leung, J.Y.T.: On a dual version of the one-dimensional bin packing problem. J. Algorithms **5**(4) (1984) 502–525
11. Zhang, G.: Parameterized on-line open-end bin packing. Computing **60**(3) (1998) 267–274
12. Leung, J.Y.T., Dror, M., Young, G.H.: A note on an open-end bin packing problem. Journal of Scheduling **4**(4) (2001) 201–207
13. Yang, J., Leung, J.Y.T.: The ordered open-end bin packing problem. Operations Research **51**(5) (2003) 759–770
14. Lee, C.C., Lee, D.T.: A simple on-line bin-packing algorithm. J. ACM **32**(3) (1985) 562–572
15. de la Vega, W.F., Lueker, G.S.: Bin packing can be solved within $1 + \epsilon$ in linear time. Combinatorica **1**(4) (1981) 349–355

Creation and Growth of Components in a Random Hypergraph Process

Vlady Ravelomanana and Alphonse Laza Rijamamy

LIPN, UMR CNRS 7030
Université Paris XIII, 93430 Villetaneuse, France
and Université de Madagascar, Ankatso – Tana 101, Madagascar
vlad@lipn.univ-paris13.fr, rilazako@yahoo.fr

Abstract. Denote by an ℓ-component a connected b-uniform hypergraph with k edges and $k(b-1)-\ell$ vertices. We prove that the expected number of creations of ℓ-component during a random hypergraph process tends to 1 as ℓ and b tend to ∞ with the total number of vertices n such that $\ell = o\left(\sqrt[3]{\frac{n}{b}}\right)$. Under the same conditions, we also show that the expected number of vertices that ever belong to an ℓ-component is approximately $12^{1/3}(b-1)^{1/3}\ell^{1/3}n^{2/3}$. As an immediate consequence, it follows that with high probability the largest ℓ-component during the process is of size $O((b-1)^{1/3}\ell^{1/3}n^{2/3})$. Our results give insight about the size of giant components inside the phase transition of random hypergraphs.

1 Introduction

A *hypergraph* $\mathcal{H}$ is a pair $(\mathcal{V}, \mathcal{E})$ where $\mathcal{V} = \{1, 2, \cdots, n\}$ denotes the set of vertices of $\mathcal{H}$ and $\mathcal{E}$ is a family of subsets of $\mathcal{V}$ called edges (or hyperedges). For a general treatise on hypergraphs, we refer to Berge [2]. We say that $\mathcal{H}$ is *b-uniform* (or simply *uniform*) if for every edge $e \in \mathcal{E}$, $|e| = b$. In this paper, all considered hypergraphs are b-uniform. We will study the growth of size and complexity of connected components of a random hypergraph process $\{\mathbb{H}(n,t)\}_{0 \leq t \leq 1}$ defined as follows. Let K_n be the complete hypergraph built with n vertices and $\binom{n}{b}$ edges (self-loops and multiple edges are not allowed). $\{\mathbb{H}(n,t)\}_{0 \leq t \leq 1}$ may be constructed by letting each edge e of K_n (amongst the $\binom{n}{b}$ possible edges) appear at random time T_e, with T_e independent and uniformly distributed on $(0, 1)$ and letting $\{\mathbb{H}(n,t)\}_{0 \leq t \leq 1}$ contain the edges such that $T_e \leq t$ (for the random graph counterpart of this model, we refer the reader to [11,17]). This model is closely related to $\{\mathbb{H}(n, M)\}$ where $M \in [1, \binom{n}{b}]$ represents the number of edges picked uniformly at random amongst the $\binom{n}{b}$ possible edges and which are present in the random hypergraph. The main difference between $\{\mathbb{H}(n,M)\}_{0 \leq M \leq \binom{n}{b}}$ and $\{\mathbb{H}(n,t)\}_{0 \leq t \leq 1}$ is that in $\{\mathbb{H}(n,M)\}_{0 \leq M \leq \binom{n}{b}}$, edges are added at fixed (slotted) times $1, 2, \ldots, \binom{n}{b}$ so at any time M we obtain a random graph with n vertices and M edges, whereas in $\{\mathbb{H}(n,t)\}_{0 \leq t \leq 1}$ the edges are added at random times. At time $t = 0$, we have a hypergraph with n vertices and 0 edge, and as the time advances all edges e with r.v. T_e such that $T_e \leq t$ (where t is the current

D.Z. Chen and D.T. Lee (Eds.): COCOON 2006, LNCS 4112, pp. 350–359, 2006.
© Springer-Verlag Berlin Heidelberg 2006

time), are added to the hypergraph until t reaches 1 in which case, one obtains the complete hypergraph K_n.

We define the *excess* (or the *complexity*) of a connected b-uniform hypergraph as (see also [14]):

$$\text{excess}(\mathcal{H}) = \sum_{e \in \mathcal{E}} (|e| - 1) - |\mathcal{V}| = |\mathcal{E}| \times (b - 1) - |\mathcal{V}|. \tag{1}$$

Namely, the complexity (or excess) of connected components ranges from -1 (hypertrees) to $\binom{n}{b}(b - 1) - n$ (complete hypergraph). A connected component with excess ℓ ($\ell \geq -1$) is called an ℓ-component. The notion of excess was first used in [19] where the author obtained substantial enumerative results in the study of connected graphs according to the two parameters number of vertices and number of edges. It was also used in enumerative combinatorics and as well as in various study of random hypergraphs processes[14,15,1].

Numerous results have been obtained for random graphs as witnessed by the books [4,13] and the references therein. In comparison, there are very few works about random hypergraphs. One of the most significant results was obtained by Schmidt-Pruznan and Shamir [18] who studied the component structure for random hypergraphs. In particular, they proved that if $b \geq 2$, $M = cn$ with $c < 1/b(b - 1)$ then asymptotically almost surely (a.a.s. for short) the largest component of $\mathbb{H}(n, M)$ is of order $\log n$ and for $c = 1/b(b - 1)$ it has $\theta(n^{2/3})$ vertices and as $c > 1/b(b-1)$ a.a.s. $\mathbb{H}(n, M)$ has a unique geant component with $\theta(n)$ vertices. This result generalizes the seminal papers of Erdös and Rényi who discovered the abrupt change in the structure of the random graph $\mathbb{G}(n, M)$ when $M = cn$ with $c \sim 1/2$. In [15], Karoński and Łuczak proved limit theorems for the distribution of the size of the largest component of $\mathbb{H}(n, M)$ at the phase transition, i.e., $M = n/b(b - 1) + O(n^{2/3})$.

In this paper, we consider the *continuous time* random hypergraph process described above and will study the creation and growth of components of excess ℓ (or ℓ-components). A connected component which is not a hypertree is said *multicyclic* (following the terms used by our predecessors in [10,11,12]).

1.1 Definitions

We can observe that there are two manners to create a new $(\ell + 1)$ component during the $\{\mathbb{H}(n, t)\}_{0 \leq t \leq 1}$ process :

• either by adding an edge between an existing p-component (with $p \leq \ell$) and $(b - q)$ hypertrees (with $0 \leq q \leq b$) such that the edge encloses q distinct vertices in the p-component,

• or by joining with the last added edge many connected components such that the number of multicyclic components diminishes.

Observe that in the first case, to create an ℓ-component, we must have $p + q - 1 = \ell$. In this case, it is also important to note that the number of multicyclic components remains the same after the addition of the last edge.

The first transition described above will be denoted $p \to \ell$ and the second $\oplus_i p_i \to \ell$. We say that an ℓ-component is *created* by a transition $p \to \ell$ with

$p < \ell$ or by a transition $\oplus_i p_i \to \ell$. For $\ell \geq 0$, we say that an ℓ-component *grows* when it swallows some hypertrees (transition $\ell \to \ell$).

Following Janson in [11], we have two points of view :

- *The static view.* Let $\mathcal{C}_\ell(m)$ denote the collection of all ℓ-components in $\{\mathbb{H}(n,t)\}_{0 \leq t \leq 1}$. Consider the family $\mathcal{C}_\ell^\star = \bigcup_m \mathcal{C}_\ell(m)$ for every ℓ-component that appears at some stage of the continuous process, ignoring when it appears : the elements of $\mathcal{C}_\ell^\star$ are called *static ℓ-components*.
- *The dynamic view.* A connected component can be viewed as "the same" according to its excess even after it has grown by swallowing some hypertrees (transition $\ell \to \ell$). Such component whose excess remains the same can be viewed as a *dynamic ℓ-component* as its size evolves.

We define $V_\ell = |\mathcal{V}_\ell|$ as the number of vertices that at some stage of the process belong to an ℓ-component and $V_\ell^{max} = \max\{|V(C)| : C \in \mathcal{C}_\ell^\star\}$ to be the size of the largest ℓ-component that ever appears.

Let $\alpha(\ell; k)$ be the expected number of times a new edge is added by means of the first type of transition $p \to \ell$ in order to create an ℓ-component with k edges (or with $k \times (b-1) - \ell$ vertices). Note again that in this case, the number of multicyclic components of the $\{\mathbb{H}(n,t)\}_{0 \leq t \leq 1}$ process remains the same after the addition of this edge.

Similarly, let $\beta(\ell; k)$ be the expected number of times an edge is added joining at least two multicyclic components in order to form a newly ℓ-component with a total of k edges. In other terms, $\beta(\ell; k)$ is the expected number of times at least two multicyclic components and some hypertrees merge to form an ℓ-component.

1.2   Our Results and Outline of the Paper

We combine analytic combinatorics [7] and probabilistic theory [13] to study the extremal characteristics of the components of a random hypergraph process inside its phase transition [15] and find that the size of the largest component with k (hyper)edges and $k(b-1) - \ell$ vertices is of order $O((b-1)^{1/3}\ell^{1/3}n^{2/3})$.

This extended abstract is organized as follows. In the next section, we introduce the general expression of the expectations of several random variables of our interest. In section 3, the computations of the expectations are developed focusing on the particular and instructive case of unicyclic components. The last paragraph provides several technical lemmas useful in order to study the extremal case, i.e. whenever the excess ℓ of the component is large. We give there methods on how to investigate the number of creations of ℓ-components as well as the expectation of their size.

2   Connected Components and Expectation of Transitions

2.1   Expected Number of Transitions

In this paragraph, we give a general formal expression of the expectation of the number of the first (resp. second) type of transitions $\alpha(\ell; k)$ (resp. $\beta(\ell; k)$).

We have the following lemma which computes the expected number of transitions $\alpha(\ell; k)$:

Lemma 1. Let $a = k(b - 1) - \ell$. Denote by $\rho(a, k)$ the number of manners to label an ℓ-component with a vertices such that one edge – whose deletion will not increase the number of multicyclic components but will suppress the newly created ℓ-component – is distinguished among the others. Then,

$$\alpha(\ell; k) = \binom{n}{a} \rho(a, k) \int_0^1 t^{k-1}(1 - t)^{\binom{n}{b} - \binom{n-a}{b} - k}\, dt. \tag{2}$$

Proof. There are $\binom{n}{a}$ choices of the $a = k(b - 1) - \ell$ vertices of the newly created ℓ-component. By the definition of $\rho(a; k)$, there are $\binom{n}{a} \times \rho(a; k)$ possible ℓ-components. The probability that the previous component (the one before obtaining the current ℓ-component) belongs to $\{\mathbb{H}(n, t)\}_{0 \leq t \leq 1}$ is given by

$$t^{k-1}(1 - t)^{\sum_{i=1}^{b-1} \binom{n-a}{i}\binom{a}{b-i}+\binom{a}{b}-k+1} \tag{3}$$

where the summation in the exponent represents the number of edges not present between the considered component and the rest of the hypergraph. The conditional probability that the last edge is added during the time interval $(t, t + dt)$ and not earlier is $dt/(1 - t)$. Using the identity

$$\sum_{i=1}^{b-1} \binom{n-a}{i}\binom{a}{b-i} = \binom{n}{b} - \binom{n-a}{b} - \binom{a}{b} \tag{4}$$

and integrating over all times after some algebra, we obtain (2).

Similarly, if we let $\tau(a; k)$ to be the number of ways to label an ℓ-component with $a = (k - 1) - \ell$ vertices and k edges such that one edge – whose suppression augments the number of multicyclic connected components – is distinguished among the others. Then, $\beta(\ell; k)$ can be computed as for $\alpha(\ell; k)$ using exactly $\tau(a; k)$ instead of $\rho(a; k)$.

Next, the following lemma gives some asymptotic values needed when using formula (2).

Lemma 2. Let $a = (b - 1)k - \ell$. We have

$$\binom{n}{a} \int_0^1 t^{k-1}(1 - t)^{\binom{n}{b} - \binom{n-a}{b} - k}\, dt \sim \frac{1}{\sqrt{(b - 1)}\, n^\ell} \frac{k^{(k-1)}\, [(b - 1)!]^k}{\left(k(b - 1) - \ell\right)^{kb - \ell}}$$

$$\times \exp\left(k(b - 2) - \ell - \frac{(b - 1)^4\, k^3}{24\, n^2}\right). \tag{5}$$

Proof. First, using Stirling formula for factorial we get

$$\binom{n}{a} = \frac{1}{\sqrt{2\pi a}} \frac{n^a\, e^a}{a^a} \exp\left(-\frac{a^2}{2n} - \frac{a^3}{6n^2} + O\left(\frac{a^4}{n^3} + \frac{1}{a}\right)\right). \tag{6}$$

Setting $N = \binom{n}{b} - \binom{n-a}{b}$, using standard calculus we then obtain

$$N = \frac{n^{(b-1)}a}{(b-1)!} \left(1 - \frac{a(b-1)}{2n} + \frac{a^2(b-1)(b-2)}{6n^2} + O\left(\frac{b}{n}\right) + O\left(\frac{ab^3}{n^2} + \frac{b^4}{n^2}\right)\right).$$
(7)

Now, using the above formulas we easily find that the integral equals

$$\frac{1}{N\binom{N}{k-1}} = \sqrt{\frac{2\pi}{k}} \frac{k^k}{e^k} \frac{[(b-1)!]^k}{n^{k(b-1)}a^k} \left(1 + O\left(\frac{k}{n^{b-1}b} + \frac{1}{k}\right)\right) \times$$

$$\exp\left(-k\log\left(1 - \frac{a(b-1)}{2n} + \frac{a^2(b-1)(b-2)}{6n^2} + O\left(\frac{b}{n}\right) + O\left(\frac{ab^3}{n^2} + \frac{b^4}{n^2}\right)\right)\right).$$
(8)

Therefore by replacing a with $k(b-1) - \ell$ and using (6), it yields (7)

$$\frac{\binom{n}{a}}{N\binom{N}{k-1}} \sim \frac{1}{\sqrt{(b-1)}\,n^\ell} \frac{k^{(k-1)}\,[(b-1)!]^k}{\left(k(b-1)-\ell\right)^{kb-\ell}} \exp\left(k(b-2)-\ell\right) \exp\left(-\frac{(b-1)^4\,k^3}{24\,n^2}\right).$$

Lemma 2 tells us that the expectations the random variables of interest rely on the asymptotic number of the considered connected components.

2.2   Asymptotic Enumeration of Connected Hypergraphs

As far as we know there are not so many results about the asymptotic enumeration of connected uniform hypergraphs. In this paragraph, we recall some of the results established independently in [14,6,1] (the three papers actually use three different methods). In [1], the authors use the generating functions approach [9,12,7,19,20] to count exactly and asymptotically connected labeled b-uniform hypergraphs. If $A(z) = \sum_n a_n z^n$ and $B(z) = \sum_n b_n z^n$ are two formal power series, $A \preceq B$ means that $\forall n \in \mathbb{N}$, $a_n \leq b_n$. Among other results, the authors of [1] established the following:

Lemma 3. *Let $H_\ell(z)$ be the exponential generating function (EGF for short) of b-uniform connected hypergraphs with excess ℓ. Define by $T(z)$ the EGF of labeled rooted hypertrees. Then,*

$$H_{-1}(z) = T(z) - \frac{(b-1)\,T(z)^b}{b!} \quad with \;\; T(z) = z\exp\left(\frac{T(z)^{(b-1)}}{(b-1)!}\right) = z\frac{\partial H_{-1}(z)}{\partial z}.$$
(9)

For any $\ell \geq 1$, H_ℓ satisfies

$$\frac{\lambda_\ell(b-1)^{2\ell}}{3\,\ell\,T(z)^\ell\,\theta(z)^{3\ell}} - \frac{(\nu_\ell(b-2))(b-1)^{2\ell-1}}{(3\,\ell-1)\,T(z)^\ell\,\theta(z)^{3\ell-1}} \preceq H_\ell(z) \preceq \frac{\lambda_\ell(b-1)^{2\ell}}{3\,\ell\,T(z)^\ell\,\theta(z)^{3\ell}},$$
(10)

where $\lambda_\ell = 3\left(\frac{3}{2}\right)^\ell \frac{\ell!}{2\pi}\left(1 + O\left(\frac{1}{\ell}\right)\right)$ and $\nu_\ell = O(\ell\lambda_\ell)$. Furthermore, λ_ℓ is defined recursively by $\lambda_0 = \frac{1}{2}$ and

$$\lambda_\ell = \frac{1}{2}\lambda_{\ell-1}(3\ell-1) + \frac{1}{2}\sum_{t=0}^{\ell-1}\lambda_t\lambda_{\ell-1-t}, \qquad (\ell \geq 1).$$
(11)

We also need the following result which has been proved independently by Karoński and Łuczak in [14] and Andriamampianiana and Ravelomanana in [1]:

Lemma 4. *For* $\ell \equiv \ell(n)$ *such that* $\ell = o\left(\sqrt[3]{\frac{n}{b}}\right)$ *as* $n \to \infty$, *the number of connected b-uniform hypergraphs built with n vertices and having excess* ℓ *satisfies*

$$\sqrt{\frac{3}{2\pi}} \quad \frac{(b-1)^{\frac{\ell}{2}} \quad e^{\frac{\ell}{2}} \quad n^{n+\frac{3\ell}{2}-\frac{1}{2}}}{12^{\frac{\ell}{2}} \ell^{\frac{\ell}{2}} \left((b-2)!\right)^{\frac{n+\ell}{b-1}}} \exp\left(\frac{n}{b-1} - n\right) \left(1 + O\left(\frac{1}{\sqrt{\ell}}\right) + O\left(\sqrt{\frac{b\,\ell^3}{n}}\right)\right).$$

$$(12)$$

Observe that setting $b = 2$ in (12), we retrieve the asymptotical results of Sir E. M. Wright for connected graphs in his fundamental paper [20].

3 Hypertrees and Unicyclic Components

As typical examples, let us work with unicyclic components. We will compute the expected number of transitions $-1 \to 0$. That is the number of times unicyclic connected components (i.e. 0-components) are created. We will also investigate the number of times unicyclic components merge with hypertrees growing in size but staying with the same complexity (excess 0). In these directions, we have the following result :

Theorem 1. *As* $n \to \infty$, *on the average a b-uniform random hypergraph has about* $\frac{1}{3}\log n$ *dynamic unicyclic components. The expected number of static 0-components is* $\sim \frac{\sqrt{2}\pi^{3/2}24^{1/6}}{6\,\Gamma(\frac{5}{6})} (b-1)^{1/3} n^{1/3} \approx 1.975\,(b-1)^{1/3} n^{1/3}$.

Proof. The creation of unicyclic components can be obtained only by adding an edge joining 2 distinct vertices inside the same hypertree with $(b-2)$ other vertices from $(b-2)$ distinct hypertrees (to complete the edge).

The number of such constructions is therefore given by the coefficients of the following EGF :

$$C'_0(z) = \frac{\left(\vartheta_z H_{-1}(z)\right)^{(b-2)}}{(b-2)!} \times \left(\frac{\vartheta_z^2 - \vartheta_z}{2}\left(H_{-1}(z)\right)\right), \qquad (13)$$

where the combinatorial operator $\vartheta_z = z\frac{\partial}{\partial z}$ corresponds to marking a vertice of the hypergraph in order to distinguish it from the others. For instance, we refer the reader to Bergeron, Labelle and Leroux [3] for the use of distinguishing/marking and pointing in combinatorial species. Recall that the EGFs are as described briefly in Lemma 3 (see also [1]), using $\vartheta_z H_{-1}(z) = T(z)$ and $\vartheta_z T(z) = \frac{T(z)}{1-T(z)^{(b-1)}/(b-2)!}$ we find

$$C'_0(z) = \frac{T(z)^{b-2}}{2\,(b-2)!} \left(\frac{T(z)}{1 - \frac{T(z)^{(b-1)}}{(b-2)!}} - T(z)\right) = \frac{1}{2}\left(\frac{1}{1 - \frac{T(z)^{(b-1)}}{(b-2)!}} - T(z) - 1\right).$$

$$(14)$$

We also have (such expansions are similar to those in [16])

$$\frac{1}{1 - \frac{T(z)^{(b-1)}}{(b-2)!}} = \sum_{k=0}^{\infty} \frac{k^k}{k!\,[(b-2)!]^k}\, z^{(b-1)k}. \qquad (15)$$

Denoting by $\rho'((b-1)k, k)$ the number of manners to label a unicyclic component with $(b-1)k$ vertices and with a distinguished edge such that its deletion will leave a forest of hypertrees, we thus have

$$\rho'\Big((b-1)k, k\Big) = ((b-1)k)!\,\Big[z^{(b-1)k}\Big]\,C'_0(z) \sim \frac{((b-1)k)!\,k^k}{2\,k!\,[(b-2)!]^k} \qquad (16)$$

(where if $A(z) = \sum_n a_n z^n$ then $[z^n]\,A(z) = a_n$).

Next, using Lemma 2 with the above equation, after nice cancellations and summing other all possible values of k, we get

$$\sum_{k=1}^{\frac{n}{(b-1)}} \rho'\Big((b-1)k, k\Big) \binom{n}{(b-1)k} \int_0^1 t^{k-1}(1-t)^{\binom{n}{b}-\binom{n-\frac{(b-1)k}{b}}{b}-k}\,dt$$

$$\sim \sum_{k=1}^{\frac{n}{(b-1)}} \frac{1}{2k} \times \exp\left(-\frac{(b-1)^4\,k^3}{24\,n^2}\right) \sim \frac{1}{2}\int_1^{n/(b-1)} \frac{1}{x}\,e^{-(b-1)^4\,x^3/24n^2}\,dx. \qquad (17)$$

The value of the integral above is $\sim \log n^{2/3} + O(1)$. Thus, the expected number of creations of unicyclic components is $\sim \frac{1}{3}\log n$. which completes the proof of the first part of the theorem. To prove the second part, we have to investigate the number of static 0-components, that is the number of times 0-components merge with hypertrees by the transition $0 \to 0$. The EGF of unicyclic components with a distinguished edge such that its suppression will leave a unicyclic component and a set of $(b-1)$ rooted hypertrees is given by

$$C''_0(z) = \frac{T(z)^{b-1}}{(b-1)!}\,\vartheta_z\Big(H_0(z)\Big) = \frac{T(z)^{b-1}}{(b-1)!}\left(\frac{(b-1)\,T(z)^{b-1}}{2\,(b-2)!\,\theta^2} - \frac{(b-1)\,T(z)^{b-1}}{2\,(b-2)!\,\theta}\right) \qquad (18)$$

where $\theta = 1 - T(z)^{b-1}/(b-2)!$. Denote by $\rho''((b-1)k, k)$ the number of manners to label a unicyclic component with $(b-1)k$ vertices and with a distinguished edge such that its deletion will leave a 0-component with a forest of rooted hypertrees, we claim that

$$\rho''\Big((b-1)k, k\Big) = ((b-1)k)!\,\Big[z^{(b-1)k}\Big]\,C''_0(z)$$

$$\sim \sqrt{\frac{\pi\,(b-1)^3}{8}}\,\frac{k^{k(b-1)+1/2}}{e^{k(b-2)}}\left(\frac{(b-1)^{k(b-1)}}{[(b-2)!]^k}\right). \qquad (19)$$

(We omit the details, since the full proof involves singularity analysis [7] of the EGF C''_0 described above.)

Now, using Lemma 2 and summing over k after some cancellations, the computed expectation is about

$$
\sum_{k=1}^{n/(b-1)} \sqrt{\frac{\pi}{8}}(b-1)\frac{1}{k^{1/2}}\,e^{-(b-1)^4\,k^3/24n^2} \sim \sqrt{\frac{\pi}{8}}(b-1)\int_1^{n/(b-1)} \frac{e^{-(b-1)^4\,x^3/24n^2}}{\sqrt{x}}\,dx
$$

$$
\sim 1/6\,\frac{\sqrt{2}\pi^{3/2}24^{1/6}}{\Gamma(5/6)}\,(b-1)^{1/3}\,n^{1/3} \approx 1.974748319\cdots(b-1)^{1/3}\,n^{1/3}. \tag{20}
$$

Note here that the result stated in Theorem 1 (humbly) generalizes the ones of Janson in [11] since by setting $b = 2$, we retrieve his results concerning unicyclic (graph) components.

Next, we can investigate the number of vertices that ever belong to 0-components. According to the above computations, the expected number of vertices added to $\mathcal{V}_0$ for the creation of such unicyclic components (transition $-1 \to 0$) is about

$$
\frac{1}{2}\sum_{k=1}^{n/(b-1)} (b-1)\,e^{-(b-1)^4\,k^3/24n^2} \sim \frac{1}{6}\,\frac{24^{1/3}\,\Gamma(1/3)}{(b-1)^{1/3}}\,n^{2/3}. \tag{21}
$$

Whenever the excess ℓ is fixed, that is $\ell = O(1)$, the methods developped here for unicyclic components can be generalized, using analytical tools such those in [7]. Thus, we now turn on components with higher complexities.

4   Multicyclic Components with Extremal Complexities

In this section, we focus on the creation and growth of components of higher complexity. First, we will compute the expectations of the number of creations of ℓ-components for $\ell \geq 1$. To this purpose, we need several intermediate lemmas.

Define $h_n(\xi, \beta)$ as follows

$$
\frac{1}{T(z)^\xi\left(1 - \frac{T(z)^{b-1}}{(b-2)!}\right)^{3\xi+\beta}} = \sum_{n\geq 0} h_n(\xi,\beta)\frac{z^n}{n!}. \tag{22}
$$

The following lemma is an application of the saddle point method [5,7] which is well suited to cope with our analysis :

Lemma 5. *Let $\xi \equiv \xi(n)$ be such that $\xi(b-1) \to 0$ but $\frac{\xi(b-1)n}{\ln n^2} \to \infty$ and let β be a fixed number. Then $h_n(\xi n, \beta)$ defined in (22) satisfies*

$$
h_n(\xi n, \beta) = \frac{n!}{\sqrt{2\pi n\Big(b-1\Big)}\,\big((b-1)!\big)^{\frac{\xi n+n}{b-1}}}\left(1 - (b-1)u_0\right)^{(1-\beta)}
$$

$$
\times \exp\left(n\Phi(u_0)\right)\left(1 + O\left(\sqrt{\xi(b-1)}\right) + O\left(\frac{1}{\sqrt{\xi(b-1)n}}\right)\right), \tag{23}
$$

where

$$\Phi(u) = u - \left(\frac{\xi+1}{b-1}\right) \ln u - 3\,\xi \ln\left(1 - (b-1)u\right)$$

$$u_0 = \frac{3/2\,\xi b - \xi + 1 - 1/2\sqrt{\Delta}}{b-1} \quad \text{with } \Delta = 9\,\xi^2 b^2 - 12\,\xi^2 b + 12\,\xi b + 4\,\xi^2 - 12\,\xi. \tag{24}$$

Proof. Omitted in this extended abstract.

Lemma 6. *Let $a = k(b-1) - \ell$. Denote by $c_\ell(a,k)$ the number of manners to label an ℓ-component with a vertices such that one edge – whose deletion will suppress the occurrence of the created ℓ-component – is distinguished among the others. As ℓ tends to ∞ with the number of vertices a such that $\ell = o\left(\sqrt[3]{\frac{a}{b}}\right)$ then*

$$c_\ell(a,k) = a!\,[z^a]\left(\frac{1}{2}\frac{(3\ell)\,(b-1)^{2\ell}\,\lambda_{\ell-1}}{T(z)^\ell\theta^{3\ell+1}}\right) \times \left(1 + O\left(\frac{1}{\sqrt{\ell}} + O\left(\sqrt{\frac{b\,\ell^3}{a}}\right)\right)\right), \tag{25}$$

where $\theta = 1 - T(z)^{b-1}/(b-2)!$ and the sequence (λ_ℓ) is defined with (11).

Sketch of proof. The proof given in this extended abstract is sketched. The main ideas are as follows. The inequalities given by equation (10) in Lemma 3 tell us that when ℓ is large, the main constructions that lead to the creation of a new ℓ-component arises a.a.s. from picking two distinct vertices in an $(\ell-1)$-component and joining them by an edge with $(b-2)$ set of rooted hypertrees. Such constructions are counted by

$$\left(\frac{\vartheta_z^2 - \vartheta_z}{2}H_{\ell-1}(z)\right) \times \frac{T(z)^{b-2}}{(b-2)!}.$$

Using again (10) with (23), one can show that the coefficient of the latter EGF has the same asymptotical behaviour as the following one

$$\frac{3\ell\,(b-1)^{2\ell}\lambda_{\ell-1}}{2\,T(z)^\ell\theta^{3\ell+1}}.$$

(The error terms being the same as those given by the saddle-point Lemma 5 above.)   □

We then have the following result :

Theorem 2. *As $\ell, b \to \infty$ with n but such that $\ell = o\left(\sqrt[3]{\frac{n}{b}}\right)$, the expected number of creations of ℓ-component is ~ 1 and the expected number of vertices that ever belong to an ℓ-component is about $(12\ell(b-1))^{1/3}\,n^{2/3}$. Thus, $\mathbb{E}\left[V_\ell^{max}\right] = O\left(\ell^{1/3}(b-1)^{1/3}n^{2/3}\right)$.*

Sketch of the proof. The proof of this Theorem is a combination of Lemmas 5,6 and 2 together with the asymptotic value of λ_ℓ given in Lemma 3 and with the fact that

$$\sum_{k=0}^{n/(b-1)} k^u \exp\left(-\frac{(b-1)^4 k^3}{24\,n^2}\right) \sim \frac{1}{3}\frac{24^{\frac{u+1}{3}}\,n^{\frac{2(u+1)}{3}}}{(b-1)^{\frac{4(u+1)}{3}}}\,\Gamma\left(\frac{u+1}{3}\right).$$

References

1. Andriamampianina T. and Ravelomanana V. (2005). Enumeration of connected uniform hypergraphs. *In Proc. of the 17-th Formal Power Series and Algebraic Combinatorics.*
2. Berge, C. (1976). *Graphs and hypergraphs* North-Holland Mathematical Library.
3. Bergeron, F., Labelle, G. and Leroux, P. (1998). *Combinatorial Species and Tree-like Structures* Encyclopedia of Mathematics and its Applications, Vol. 67, Cambridge Univ. Press.
4. Bollobás, B. (1985) *Random Graphs* Academic Press, London.
5. De Bruijn, N. G. (1981). *Asymptotic Methods in Analysis.* Dover, New-York.
6. Coja-Oghlan, A., Moore, C. and Sanwalani, V. (2004). Counting Connected Graphs and Hypergraphs via the Probabilistic Method. In Proc. of APPROX-RANDOM 2004.
7. Flajolet, P. and Sedgewick, R. *Analytic Combinatorics.* To appear (chapters are avalaible as Inria research reports). See *http : //algo.inria.fr/flajolet/ Publications/books.html.*
8. Harary, F. and Palmer, E. (1973). *Graphical Enumeration.* Academic Press, New-York and London.
9. Herbert S. Wilf *Generatingfunctionology*
10. Janson, S. (1993). Multicyclic components in random graphs process. *Random Structures and Algorithms,* 4:71–84.
11. Janson, S. (2000). Growth of components in random graphs. *Random Structures and Algorithms,* 17:343-356.
12. Janson, S., Knuth, D. E., Łuczak, T. and Pittel B. (1993). The birth of the giant component. *Random Structures and Algorithms,* 4:233–358.
13. Janson, S., Łuczak, T. and Rucinski A. (2000). *Random Graphs.* John Wiley, New York.
14. Karoński, M. and Łuczak, T. (1997). The number of sparsely edges connected uniform hypergraphs. *Discrete Math.* 171:153–168.
15. Karoński, M. and Łuczak, T. and (2002). The phase transition in a random hypergraph. *J. Comput. Appl. Math.* 142:125–135.
16. Knuth, D. E. and Pittel, B. (1989). A recurrence related to trees. *Proc. Am. Math. Soc.,* 105:335–349.
17. Ravelomanana, V. (to appear). The average size of giant components between the double-jump. *Algorithmica.*
18. Schmidt-Pruznan, J. and Shamir S. (1985). Component structure in the evolution of random hypergraphs *Combinatorica* 5: 81–94.
19. Wright, E. M. (1977). The Number of Connected Sparsely Edged Graphs. *Journal of Graph Theory,* 1:317–330.
20. Wright, E. M. (1980). The Number of Connected Sparsely Edged Graphs. III. Asymptotic results *Journal of Graph Theory,* 4:393–407.

Optimal Acyclic Edge Colouring
of Grid Like Graphs

Rahul Muthu, N. Narayanan, and C.R. Subramanian

The Institute of Mathematical Sciences, Chennai-600113, India
{rahulm, narayan, crs}@imsc.res.in

Abstract. We determine the values of the acyclic chromatic index of a class of graphs referred to as *d-dimensional partial tori*. These are graphs which can be expressed as the Cartesian product of d graphs each of which is an induced path or cycle. This class includes some known classes of graphs like d-dimensional meshes (hypergrids), hypercubes, tori, etc. Our estimates are exact except when the graph is a product of a path and a number of odd cycles, in which case the estimates differ by an additive factor of at most 1. Our results are also constructive and provide an optimal (or almost optimal) acyclic edge colouring in polynomial time.

Keywords: Acyclic Edge Colouring, Graph, Acyclic Chromatic Index, Mesh, Hypercube, Tori.

1 Introduction

All graphs we consider are simple and finite. Throughout the paper, we use $\Delta = \Delta(G)$ to denote the maximum degree of a graph G. A colouring of the edges of a graph is *proper* if no pair of incident edges receive the same colour. A proper colouring $\mathcal{C}$ of the edges of a graph G is *acyclic* if there is no two-coloured (bichromatic) cycle in G with respect to $\mathcal{C}$. In other words, the subgraph induced by the union of any two colour classes in $\mathcal{C}$ is a forest. The minimum number of colours required to edge-colour a graph G acyclically is termed the *acyclic chromatic index* of G and is denoted by $a'(G)$. The notion of acyclic colouring was introduced by Grünbaum in [7].

The acyclic chromatic index and its vertex analogue are closely related to *oriented chromatic number* and *star chromatic number* of a graph G, which have many practical applications (such as in wavelength routing in optical networks [4,9]).

Determining $a'(G)$ either theoretically or algorithmically has been a very difficult problem. Even for the highly structured and simple class of complete graphs, the value of $a'(G)$ is not yet determined. Determining exact values of $a'(G)$ even for very special classes of graphs is still open.

It is easy to see that $a'(G) \geq \chi'(G) \geq \Delta$ for any graph G. Here, $\chi'(G)$ denotes the chromatic index of G (the minimum number of colours used in any proper edge colouring of G). Using probabilistic arguments, Alon, McDiarmid and Reed [1] obtained an upper bound of 64Δ on $a'(G)$. Molloy and Reed [10] refined their analysis to obtain an improvement of $a'(G) \leq 16\Delta$. Recently, Muthu, Narayanan

D.Z. Chen and D.T. Lee (Eds.): COCOON 2006, LNCS 4112, pp. 360–367, 2006.

and Subramanian [12] obtained a better bound of $a'(G) \leq 4.52\Delta$ for graphs G with girth (the length of the shortest cycle) at least 220.

It follows from the work of Burnstein [6] that $a'(G) \leq \Delta + 2$ for all graphs with $\Delta \leq 3$. It was conjectured by Alon, Sudakov and Zaks [2] that $a'(G) \leq \Delta + 2$ for every G and this has been proved for graphs with large girth and for random d-regular (d fixed) graphs. For random d-regular graphs (fixed d), the bound has been improved to $d + 1$ by Nesetril and Wormald [13].

The proofs of the above mentioned bounds are existential in nature and are not constructive. In this work, we look at the class of those graphs which can be expressed as a finite Cartesian product of graphs each of which is an induced path or cycle. We show that (see Theorem 2) $a'(G) \in \{\Delta, \Delta + 1\}$ for each member of this class and also obtain the exact values of $a'(G)$ for all G except when G is a product of a path and a number of odd cycles. Thus we verify the above conjecture for this class of graphs which we refer to as *partial tori*. As special cases, it includes other well-known classes like hypercubes, d-dimensional meshes, etc. All these definitions are given below. Our results are constructive and are proved by an explicit colouring scheme and this can be realized in time polynomial in the size of the graph (see Theorem 3). As a result, we obtain efficient algorithms for optimal acyclic edge colouring of these graphs.

1.1 Definitions and Notation

We use P_k to denote a simple path on k vertices. Without loss of generality (w.l.o.g.), we assume that $V(P_k) = \{0, \ldots, k-1\}$ and $E(P_k) = \{(i, j) : |i - j| = 1\}$. Similarly, we use C_k to denote a cycle $(0, \ldots, k-1, 0)$ on k vertices. We use PATHS to denote the set $\{P_3, P_4, \ldots\}$ of all paths on 3 or more vertices. Similarly, we use CYCLES to denote the set $\{C_3, C_4, \ldots\}$ of all cycles. The standard notation $[n]$ is used to denote the set $\{1, 2, \ldots, n\}$.

Our definition of the class of partial tori is based on the so-called *Cartesian product* of graphs defined below.

Definition 1. *Given two graphs $G_1 = (V_1, E_1)$ and $G_2 = (V_2, E_2)$, the* Cartesian product *of G_1 and G_2, denoted by $G_1 \,\square\, G_2$, is defined to be the graph $G = (V, E)$ where $V = V_1 \times V_2$ and E contains the edge joining (u_1, u_2) and (v_1, v_2) if and only if either $u_1 = v_1$ and $(u_2, v_2) \in E_2$ or $u_2 = v_2$ and $(u_1, v_1) \in E_1$.*

Note that $\square$ is a binary operation on graphs which is commutative in the sense that $G_1 \,\square\, G_2$ and $G_2 \,\square\, G_1$ are isomorphic. Similarly, it is also associative. Hence the graph $G_0 \,\square\, G_1 \,\square\, \cdots \,\square\, G_d$ is unambiguously defined for any d. We use G^d to denote the d-fold Cartesian product of G with itself. It was shown by Sabidussi [14] and Vizing [16] (see also [8]) that any connected graph G can be expressed as a product $G = G_1 \,\square\, \cdots \,\square\, G_k$ of prime factors G_i. Here, a graph is said to be *prime* with respect to the $\square$ operation if it has at least two vertices and if it is not isomorphic to the product of two non-trivial graphs (those having at least two vertices). Also, this factorisation (or decomposition) is unique except for a re-ordering of the factors and will be referred to as the *Unique Prime Factorisation (UPF)* of the graph. Since $a'(G)$ is a graph invariant, we assume, without loss of generality, that any G_i is from $\{K_2\} \cup$ PATHS $\cup$ CYCLES *if* it is either an induced path or an induced cycle.

Definition 2. *A d-dimensional partial torus is a connected graph G whose unique prime factorisation is of the form $G = G_1 \square \cdots \square G_d$ where $G_i \in \{K_2\} \cup$ PATHS $\cup$ CYCLES for each $i \le d$. We denote the class of such graphs by $\mathcal{P}_d$.*

Definition 3. *If each prime factor of a graph $G \in \mathcal{P}_d$ is a K_2, then G is called the d-dimensional hypercube. This graph is denoted by K_2^d.*

Definition 4. *If each prime factor of a graph $G \in \mathcal{P}_d$ is from PATHS, then G is called a d-dimensional mesh. The class of all such graphs is denoted by $\mathcal{M}_d$.*

Definition 5. *If each prime factor of a graph $G \in \mathcal{P}_d$ is from CYCLES, then G is called a d-dimensional torus. The class of all such graphs is denoted by $\mathcal{T}_d$.*

1.2 Results

The proof of the results mentioned in the abstract are based on the following useful theorem whose proof is given later.

Theorem 1. *Let G be a simple graph with $a'(G) = \eta$. Then,*

1. $a'(G \square P_2) \le \eta + 1$,   *if $\eta \ge 2$.*
2. $a'(G \square P_l) \le \eta + 2$,   *if $\eta \ge 2$ and $l \ge 3$.*
3. $a'(G \square C_l) \le \eta + 2$,   *if $\eta > 2$ and $l \ge 3$.*

As a corollary, we obtain the following results.

Theorem 2. *The following is true for each $d \ge 1$.*

- $a'(K_2^d) = \Delta(K_2^d) + 1 = d + 1$ *if $d \ge 2$;* $a'(K_2) = 1$.
- $a'(G) = \Delta(G) = 2d$ *for each $G \in \mathcal{M}_d$.*
- $a'(G) = \Delta(G) + 1 = 2d + 1$ *for each $G \in \mathcal{T}_d$.*
- *Let $G \in \mathcal{P}_d$ be any graph. Let e (respectively p and c) denote the number of prime factors of G which are K_2s (respectively from PATHS and CYCLES). Then,*
  - $a'(G) = \Delta(G) + 1 = e + 2c + 1$ *if $p = 0$.*
  - $a'(G) = \Delta(G) = e + 2p + 2c$ *if either $p \ge 2$, or $p = 1$ and $e \ge 1$.*
  - $a'(G) = \Delta(G) = 2 + 2c$ *if $p = 1$, $e = 0$ and if at least one prime factor of G is an even cycle.*
  - $a'(G) \in \{\Delta = 2 + 2c, \Delta + 1 = 2 + 2c + 1\}$ *if $p = 1$, $e = 0$ and if all prime factors of G (except the one path) are odd cycles. There are examples for both values of $a'(G)$.*

2 Proofs

The following fact about acyclic edge colouring can be easily verified and would be used often in our proofs.

Fact 1. *If a graph G is regular with $\Delta(G) \geq 2$, then $a'(G) \geq \Delta(G) + 1$.*

This is because in any proper edge-colouring of G with $\Delta(G)$ colours, each colour is used on an edge incident incident at every vertex. Hence, for each pair of distinct colours a and b and for each vertex u, there is a unique cycle in G going through u and which is coloured with a and b.

We first present the proof of Theorem 2.

Proof. (of Theorem 2)
Case $G = K_2^d$: Clearly, $a'(K_2) = 1$ and $a'(K_2^2) = a'(C_4) = 3$. For $d > 2$, we start with $G = K_2^2$ and repeatedly and inductively apply Statement (1) of Theorem 1 to deduce that $a'(K_2^d) \leq d + 1$. Combining this with Fact 1, we get $a'(K_2^d) = d + 1$ for $d \geq 2$.

Case $G \in \mathcal{M}_d$: Again, we prove by induction on d. If $d = 1$, then $G \in$ PATHS and hence $a'(G) = 2 = \Delta(G)$. For $d > 1$, repeatedly and inductively apply Statement (2) of Theorem 1 to deduce that $a'(G) \leq 2(d-1)+2 = 2d$. Combining this with the trivial lower bound $a'(G) \geq \Delta(G)$, we get $a'(G) = 2d$ for each $G \in \mathcal{M}_d$ and each $d \geq 1$.

Case $G \in \mathcal{T}_d$: We prove by induction on d. If $d = 1$, then $G \in$ CYCLES and hence $a'(G) = 3 = \Delta(G) + 1$. For $d > 1$, repeatedly and inductively apply Statement (3) of Theorem 1 to deduce that $a'(G) \leq 2(d-1) + 1 + 2 = 2d + 1$. Combining this with Fact 1, we get $a'(G) = 2d + 1$ for each $G \in \mathcal{T}_d$ and each $d \geq 1$.

Case $G \in \mathcal{P}_d$: Let e, p and c be as defined in the statement of the theorem. If $p = 0$, then G is the product of edges and cycles and hence G is regular and $a'(G) \geq \Delta(G)+1$ by Fact 1. Also, we can assume that $c > 0$. Otherwise, $G = K_2^d$ and this case has already been established. Again, without loss of generality, we can assume that the first factor G_1 of G is from CYCLES and $a'(G_1) = 3$. Now, as in the previous cases, we apply induction on d and also repeatedly apply one of the Statements (1) and (3) of Theorem 1 to deduce that $a'(G) \leq \Delta(G) + 1$. This settles the case $p = 0$.

Now, suppose *either $p \geq 2$, or $p = 1$ and $e \geq 1$*. Order the d prime factors of G so that $G = G_1 \,\square\, \cdots \,\square\, G_d$ and the first p factors are from PATHS and the next e factors are copies of K_2. By the previously established cases and from Theorem 1, it follows that

$$a'(G_1 \,\square\, \cdots \,\square\, G_{p+e}) = \Delta(G_1 \,\square\, \cdots \,\square\, G_{p+e}) = 2p + e \geq 3.$$

As before, applying (3) of Theorem 1 inductively, it follows that

$$a'(G) = a'(G_1 \,\square\, \cdots \,\square\, G_{p+e+c}) \leq \Delta(G) = 2p + e + 2c.$$

Combining this with the trivial lower bound establishes this case also.

Suppose $p = 1$, $e = 0$ and at least one prime factor of G is an even cycle. Let $G_1 = P_k$ for some $k \geq 3$ and $G_2 = C_{2l}$ for some $l \geq 2$. We note that it is enough to show that $G' = G_2 \square G_1$ is acyclically colourable with $\Delta(G') = 4$ colours. Extending this colouring to an optimal colouring of G can be achieved by repeated applications of Statement (3) of Theorem 1 as before. Hence we focus on showing $a'(G') = 4$.

Firstly, colour the cycle $G_2 = C_{2l} = \langle 0, 1, \ldots, 2l - 1, 0 \rangle$ acyclically as follows. For each $i, 0 \leq i \leq 2l - 2$, colour the edge $(i, i + 1)$ with 1 if i is even and with 2 if i is odd. Colour the edge $(2l - 1, 0)$ with 3. Now, use the same colouring on each of the k isomorphic copies (numbered with $0, \ldots k - 1$) of G_2. For each $j, 0 \leq j < k - 1$, the jth and $(j+1)^{th}$ copies of G_2 are joined by cross-edges which constitute a perfect matching between similar vertices in the two copies. These cross-edges are coloured as follows. For every i and j, the cross edge joining (i, j) and $(i, j + 1)$ is coloured with 4 **if** $(i + j)$ is even and is coloured with the unique colour from $\{1, 2, 3\}$ which is missing at this vertex i in both copies **if** $(i + j)$ is odd. It can be shown that this colouring is proper and acyclic (refer [11] for details). This shows that $a'(G') = 4$, as desired.

Suppose $p = 1$, $e = 0$ and all prime factors of G (except the one path) are odd cycles. In this case, $a'(G)$ can take both values as the following examples show. If $G = P_3 \square C_3$, then it can be easily verified that $a'(G) = 5 = \Delta + 1$. Also, if $G = P_3 \square C_5$, then $a'(G) = 4 = \Delta$ as shown by the colouring in Figure 1.

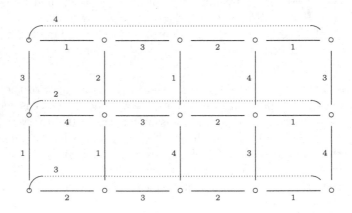

Fig. 1. Colouring of $P_3 \square C_5$

We now present the proof of Theorem 1.

A restricted class of bijections (defined below) would play an important role in this proof.

Definition 6. *A bijection σ from a set $\mathcal{A}$ to an equivalent set $\mathcal{B}$ is a non-fixing bijection if $\sigma(i) \neq i$ for each i.*

Proof. (of Theorem 1)

Since $a'(G) = \eta$, we can edge-colour G acyclically using colours from $[\eta]$. Fix one such colouring $\mathcal{C}_0 : E(G) \to [\eta]$.

Define $\mathcal{C}_1$ to be the colouring defined by $\mathcal{C}_1(e) = \sigma(\mathcal{C}_0(e))$ where $\sigma : [\eta] \to [\eta]$ is any bijection which is *non-fixing*. For concreteness, define $\sigma(i) = (i \bmod \eta) + 1$.

Case 1 $(a'(G \square P_2))$. Let G_0, G_1 be the two isomorphic copies of G induced respectively by the sets $\{(u, 0) : u \in V(G)\}$ and $\{(u, 1) : u \in V(G)\}$. Let G_0 and G_1 be edge coloured respectively by $\mathcal{C}_0$ and $\mathcal{C}_1$. For each of the remaining edges (termed *cross-edges* and which constitute a perfect matching between G_0 and G_1) of the form $((u, 0), (u, 1))$, give a new colour α. Denote by $\mathcal{C}$, the resultant colouring of $G \square P_2$. We claim that $\mathcal{C}$ is proper and acyclic.

It is easy to see that $\mathcal{C}$ is proper. Also note that any bichromatic cycle in $\mathcal{C}$ should necessarily use the colour α (since the colourings of G_0 and G_1 are acyclic).

Suppose that $G \square P_2$ has a bichromatic cycle C using the colours α and some other colour, say i, from the set $[\eta]$. In $\mathcal{C}$, G_0 and G_1 are both coloured α-free and hence any proper α, i-coloured cycle should contain the α-coloured edges an even number of times. Hence we have $|C| \equiv 0 \bmod 4$. Fix a vertex $(u_1, 0)$ as the starting point of C. Then C will look like $C = \langle (u_1, 0) \xrightarrow{\alpha} (u_1, 1) \xrightarrow{i} (u_2, 1) \xrightarrow{\alpha} (u_2, 0) \cdots (u_k, 0) \xrightarrow{i} (u_1, 0) \rangle$.

Notice that k is of even parity (since $|C| \equiv 0 \bmod 4$). For each i-coloured edge $(u_{2\ell+1}, 1) \to (u_{2\ell+2}, 1)$ of G_1 in C, its isomorphic copy in G_0, namely, the edge $(u_{2\ell+1}, 0) \to (u_{2\ell+2}, 0)$ is coloured with a colour $j = \sigma^{-1}(i) \neq i$ (since σ is a non-fixing bijection of $[\eta]$). Now it can be seen that the cycle $\langle (u_1, 0) \xrightarrow{j} (u_2, 0) \xrightarrow{i} (u_3, 0) \ldots \xrightarrow{j} (u_k, 0) \xrightarrow{i} (u_1, 0) \rangle$ is an i, j-coloured cycle in G_0. This is a contradiction to the fact that G_0 is acyclically coloured. Hence the colouring $\mathcal{C}$ is acyclic.

Case 2 $(a'(G \square P_k))$. This case is proved using similar arguments and we alternately use two new colours α_1 and α_2 between G_i and G_{i+1}, where G_i coloured with $\mathcal{C}_{i \bmod 2}$. We omit proof details which is given in [11].

Case 3 $(a'(G \square C_k))$. Consider $G \square C_k$, $k \geq 3$. Here we have k isomorphic copies of G numbered $G_0, G_1, \ldots, G_{k-2}, G_{k-1}$ such that there is a perfect matching between successive copies G_i and $G_{(i+1) \bmod k}$ (see Figure 2). Our colouring is as follows.

For each i, $1 \leq i \leq k - 2$, colour the edges of G_i with $\mathcal{C}_{(i+1) \bmod 2}$.

As before, let α_0, α_1 be two new colours which are not in $[\eta]$. Let $\mathcal{D}_0$ be a colouring of G_0 defined by $\mathcal{D}_0(e) = \tau(\mathcal{C}_0(e))$ where $\tau(i) = i + 1, i < \eta, \tau(\eta) = \alpha_1$.

In order to colour G_{k-1}, define a colouring $\mathcal{D}_1(e) = \mu(\mathcal{C}_0(e))$ where $\mu(i) = i + 2$, $i < \eta - 1$ and $\mu(\eta - 1) = \alpha_{(k+1) \bmod 2}$, $\mu(\eta) = 2$.

Now, colour any edge of the form $((u, i), (u, i+1))$, $0 \leq i < k - 1$ with the new colour $\alpha_{i \bmod 2}$. Colour the edges of the form $((u, k - 1), (u, 0))$ with the colour 1. Denote this colouring of $G \square C_k$ by $\mathcal{C}$.

It can be shown that the colouring $\mathcal{C}$ is proper and acyclic. Again we omit the proof details (refer [11] for details).

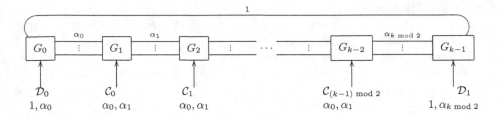

Fig. 2. Colouring of $G \,\square\, C_k$

3   Algorithmic Aspects

There is very little study of algorithmic aspects of acyclic edge colouring. In [3], Alon and Zaks prove that it is NP-complete to determine if $a'(G) \leq 3$ for an arbitrary graph G. They also describe a deterministic polynomial time algorithm which obtains an acyclic $(\Delta + 2)$-edge-colouring for any graph G whose girth g is at least $c\Delta^3$ for some large absolute constant c. Skulrattanakulchai [15] presents a linear time algorithm to acyclically edge colour any graph with $\Delta \leq 3$ using at most 5 colours.

All of our proofs given in the previous section are constructive and readily translate to efficient (polynomial-time) algorithms which find optimal (or almost optimal) acyclic edge colourings of the partial tori. Formally we have,

Theorem 3. *Let $G \in \mathcal{P}_d$ be a graph (on n vertices and m edges) specified by its Unique Prime Factorisation. Then, an acyclic edge colouring of G using Δ or $\Delta + 1$ colours can be obtained in $O(n + m)$ time. Also, the colouring is optimal except when G is a product of a path and a number of odd cycles.*

A brief and formal description of the algorithm appear in the full version of the manuscript [11].

Before we finish, we need to say a few words about how the input is presented to the algorithm. It is known from the work of Aurenhammer, Hagauer and Imrich [5] that the UPF of a connected graph G (on n vertices and m edges) can be obtained in $O(m \log n)$ time. Hence we assume that our connected input $G \in \mathcal{P}_d$ is given by the list of its prime factors $G_1, \ldots, G_d$.

4   Conclusions

If G is isomorphic to the product of a path and a number of odd cycles, $a'(G)$ can take either of the values in $\{\Delta, \Delta + 1\}$. It would be interesting to classify such graphs for which $a'(G) = \Delta$ and also to construct an optimal colouring efficiently for such graphs.

Using similar arguments, a general bound on the acyclic chromatic index of the Cartesian product of two graphs has been obtained. The results are being written up.

References

1. N. Alon, C. J. H. McDiarmid, and B. Reed. Acyclic coloring of graphs. *Random Structures and Algorithms*, 2:277–288, 1991.
2. N. Alon, B. Sudakov, and A. Zaks. Acyclic edge colorings of graphs. *Journal of Graph Theory*, 37:157–167, 2001.
3. N. Alon and A. Zaks. Algorithmic aspects of acyclic edge colorings. *Algorithmica*, 32:611–614, 2002.
4. D. Amar, A. Raspaud, and O. Togni. All to all wavelength routing in all-optical compounded networks. *Discrete Mathematics 235*, pages 353–363, 2001.
5. F. Aurenhammer, J. Hagauer, and W. Imrich. Cartesian graph factorization at logarithmic cost per edge. *Computational Complexity*, 2:331–349, 1992.
6. M. I. Burnstein. Every 4-valent graph has an acyclic 5-coloring. *Soobsc Akad. Nauk Grucin*, 93:21–24 (in Russian), 1979.
7. B. Grünbaum. Acyclic colorings of planar graphs. *Israel J Math*, 14:390–408, 1973.
8. W. Imrich and S. Klavzar. Product graphs : Structure and recognition. *John Wiley & Sons, Inc.*, 2000.
9. A. V. Kostochka, E. Sopena, and X. Zhu. Acyclic and oriented chromatic numbers of graphs. *J. Graph Theory 24(4)*, pages 331–340, 1997.
10. M. Molloy and B. Reed. Further algorithmic aspects of lovaz local lemma. *30th Annual ACM Symposium on Theorey of Computing*, pages 524–529, 1998.
11. Rahul Muthu, N. Narayanan, and C R Subramanian. Optimal acyclic edge colouring of grid like graphs. *Full Manuscript, submitted to a journal. (For a copy, email the authors).*
12. Rahul Muthu, N. Narayanan, and C R Subramanian. Improved bounds on acyclic edge colouring. *Electronic Notes in Discrete Mathematics*, 19:171–177, 2005.
13. J. Nešetřil and N. C. Wormald. The acyclic edge chromatic number of a random d-regular graph is $d + 1$. *Journal of Graph Theory*, 49(1):69–74, 2005.
14. G. Sabidussi. Graph multiplication. *Math. Z.*, 72:446–457, 1960.
15. San Skulrattanakulchai. Acyclic colorings of subcubic graphs. *Information Processing Letters*, 92:161–167, 2004.
16. V. G. Vizing. The cartesian product of graphs (russian). *Vycisl. Sistemy*, 9:30–43 (English translation in *Comp. El. Syst.* 2 (1966) 352–365.), 1963.

An Edge Ordering Problem of Regular Hypergraphs

Hongbing Fan* and Robert Kalbfleisch

Wilfrid Laurier University, Waterloo, ON. Canada N2L 3C5
hfan@wlu.ca

Abstract. Given a pair of integers $2 \leq s \leq k$, define $g_s(k)$ to be the minimum integer such that, for any regular multiple hypergraph $H = (\{1, \ldots, k\}, \{e_1, \ldots, e_m\})$ with edge size at most s, there is a permutation π on $\{1, \ldots, m\}$ (or edge ordering $e_{\pi(1)}, \ldots, e_{\pi(m)}$) such that $g(H, \pi) = \max\{\max\{|d_{H_j}(u) - d_{H_j}(v)| : u, v \in e_{\pi(j+1)}\} : j = 0, \ldots, m-1\} \leq g_s(k)$, where $H_j = (\{1, \ldots, k\}, \{e_{\pi(1)}, \ldots, e_{\pi(j)}\})$. The so-called edge ordering problem is to determine the value of $g_s(k)$ and to find a permutation π such that $g(H, \pi) \leq g_s(k)$. This problem was raised from a switch box design problem, where the value of $g_s(k)$ can be used to design hyper-universal switch boxes and an edge ordering algorithm leads to a routing algorithm. In this paper, we show that (1) $g_2(k) = 1$ for all $k \geq 3$, (2) $g_s(k) = 1$ for $3 \leq s \leq k \leq 6$, and (3) $g_s(k) \leq 2k$ for all $k \geq 7$. We give a heuristic algorithm for the edge ordering and conjecture that there is a constant C such that $g_s(k) \leq C$ for all k and s.

1 Introduction

All hypergraphs considered in this paper are multiple hypergraphs. Let $H = (V, E)$ be a hypergraph, where V is a vertex set and E is an edge set, i.e., $e \subset V$ for every $e \in E$. If edge size is at most s, i.e. $|e| \leq s$ for every $e \in E$, we say H is an s-hypergraph. The degree of a vertex u in H, denoted by $d_H(u)$, is the number of edges which contain (or cover) u, i.e., $d_H(u) = |\{e \in E : u \in e\}|$. H is said to be w-regular/regular if all its vertices have the same degree w. A regular hypergraph is said to be *minimal* if it does not contain a proper spanning (with the same vertex set) regular subhypergraph.

For a fixed integer $k > 0$, up to isomorphism, there are an infinite number of regular hypergraphs on k vertices because multiple edges are allowed (i.e., two or more edges can be the same subset of vertices). But the number of minimal regular hypergraphs on k vertices is finite. To see this, we assume $V = \{1, \ldots, k\}$. Then V has $2^k - 1$ different non-empty subsets, say $S_1, \ldots, S_{2^k-1}$. Subset S_j can be represented by a column vector $(a_{1,j}, \ldots, a_{k,j})^T$, where $a_{i,j} = 1$ if $i \in S_j$ or otherwise $a_{i,j} = 0$. Let A_k denote the matrix $[a_{i,j}]_{k \times (2^k-1)}$. Then a w-regular hypergraph H corresponds to a nonnegative integer vector $X = (x_1, \ldots, x_{2^k-1})$ satisfying $A_k X^T - w(1, \ldots, 1)^T = \mathbf{0}$, where x_j denotes the multiplicity of S_j in H,

---

* This research was partially supported by the Natural Sciences and Engineering Research Council of Canada.

D.Z. Chen and D.T. Lee (Eds.): COCOON 2006, LNCS 4112, pp. 368–377, 2006.

and vice versa. The system of equations $A_k X^T - w(1, \ldots, 1)^T = \mathbf{0}$ is a system of homogeneous linear Diophantine equations as only nonnegative integer solutions are considered. A solution X of a linear Diophantine equation system is said to be minimal if there is no other solution Y such that every component of Y is less than or equal to the corresponding component of X. It is known that the set of all minimal solutions of a homogeneous linear Diophantine equation system is a finite set, called a Hilbert basis [4]. A minimal solution of $A_k X^T - w(1, \ldots, 1)^T = \mathbf{0}$ corresponds to a minimal regular hypergraph on vertex set $\{1, \ldots, k\}$, therefore there are a finite number of minimal regular hypergraphs on k vertices.

Given a regular hypergraph $H = (V, E)$ with $V = \{1, \ldots, k\}$ and $E = \{e_1, \ldots, e_m\}$. Let π be a permutation on $\{1, \ldots, m\}$, then $(e_{\pi(1)}, \ldots, e_{\pi(m)})$ is an ordering of the edges, and it determines a sequence of $m + 1$ subhypergraphs

$$H_j = (\{1, \ldots, k\}, \{e_{\pi(1)}, \ldots, e_{\pi(j)}\}), j = 0, \ldots, m,$$

or $H_0 = (\{1, \ldots, k\}, \emptyset), H_{j+1} = H_j + e_{i_{j+1}}, j = 0, \ldots, m - 1$. We define

$$g(H, \pi) = \max\{\max\{|d_{H_j}(u) - d_{H_j}(v)| : u, v \in e_{\pi(j+1)}\} : j = 0, \ldots, m - 1\}, \quad (1)$$

$$g(H) = \min\{g(H, \pi) : \text{over all permumations } \pi\}. \quad (2)$$

For any pair of integers s, k with $2 \leq s \leq k$, we define

$$g_s(k) = \max\{g(H) : \text{over all regular } s\text{-hypergraphs } H \text{ on } k \text{ vertices}\}. \quad (3)$$

In particular, we denote $g_k(k)$ by $g(k)$. We see that $g_s(k)$ is well-defined. Let H be any regular s-hypergraph on k vertices, then H can be decomposed into a union of edge disjoint minimal spanning regular s-hypergraphs $R_1, \ldots, R_t$, $H = R_1 + \cdots + R_t$. Since we can obtain an ordering of edges of H by ordering edges of $R_1, \ldots, R_t$ respectively and then put them together one following another, we have the following relations $g(H) \leq \max\{g(R_1), \ldots, g(R_t)\}$,

$$g(H) \leq \max\{g(R) : \text{over all minimal regular } s\text{-hypergraphs } R \text{ on } k \text{ vertices}\},$$

$$g_s(k) \leq \max\{g(R) : \text{over all minimal regular } s\text{-hypergraphs } R \text{ on } k \text{ vertices}\}.$$

On the other hand, since a minimal regular s-hypergraph on k vertices is also a regular s-hypergraph, by (3) we have

$$g_s(k) \geq \max\{g(R) : \text{over all minimal regular } s\text{-hypergraphs } R \text{ on } k \text{ vertices}\}$$

$$g_s(k) = \max\{g(R) : \text{over all minimal regular } s\text{-hypergraphs } R \text{ on } k \text{ vertices}\} \quad (4)$$

Since there are a finite number of minimal regular s-hypergraphs on k vertices, $g_s(k)$ is well-defined by (4). Furthermore, $g_2(k) \leq g_3(k) \leq \cdots \leq g_k(k) = g(k)$.

Edge Ordering Problem: 1. Given a pair of integers s, k with $2 \leq s \leq k$, determine the value of $g_s(k)$.

2. Given a regular s-hypergraph H on k vertices, find an edge permutation π_0 such that $g(H, \pi_0) \leq g_s(k)$.

The above edge ordering problem was motivated by the design of switch boxes for interconnection networks, in which both the values of $g_s(k)$ and edge ordering algorithm are crucial. The purpose of this paper is to present the problem and some initial results, and to call for a general solution of the problem.

The rest of this paper is organized as follows. In Section 2, we prove that $g_2(k) = 1$ for all $k \geq 3$. In Section 3, we show that $g_s(k) = 1$ for $3 \leq s \leq k \leq 6$, and that $g_s(k) \leq 2k$ for all $3 \leq s \leq k$. Section 4 presents a heuristic edge ordering algorithm and experimental results. The experimental results show that the algorithm can find an ordering π_0 with $g(H, \pi_0) \leq 1$ for 99% of w-regular hypergraphs on k vertices with $k \leq 6, w \leq 9$. In Section 5, we describe the application of the edge ordering problem in switch box design, and the result of using $g_s(k) \leq 2k$ to design a three-stage interconnection network rearrangeable for group communications of n ports with $O(n^{5/3})$ switches, which is the first three-stage interconnection network design with less than $O(n^2)$ switches.

2   Proof for $g_2(k) = 1$

The following lemma was proved in [7] using the well-known Peterson's theorem in graph theory. Note that a 2-hypergraph becomes a graph if it does not contain singleton edges. We can see if a graph G can be decomposed into a union of 1-factors (complete matching), then $g_2(G) = 0$.

Lemma 2.1. *Every even regular 2-hypergraph can be decomposed into a union of 2-regular 2-hypergraphs.*

Theorem 2.2. $g_2(k) = 1$ *for* $k \geq 3$.

Proof. First we show $g_2(k) \geq 1$. Let H_0 be a 2-regular 2-hypergraph containing a cycle with edges $\{1, 2\}, \{2, 3\}, \{1, 3\}$. For this odd cycle, whatever ordering of the three edges, we have $g(H_0) = 1$, so that $g_2(k) \geq 1$. Further more we see for an 2-regular 2-hypergraph H, $g(H) \leq 1$ and $g(H) = 1$ if and only if H contains an odd cycle.

Next we show $g_2(k) \leq 1$. Let H be any r-regular 2-hypergraph on k vertices, we show that $g(H) \leq 1$. If r is even, then by Lemma 2.1, H can be decomposed into the union of 2-regular 2-hypergraphs, $H = R_1 + \cdots + R_t$, where each R_i is a 2-regular 2-hypergraph. Then $g(H) \leq \max\{g(R_i) : i = 1, \ldots, t\} \leq \max\{1 : i = 1, \ldots, t\} = 1$. Otherwise $r \geq 3$ is an odd number. Let H' be the $(r+1)$-regular 2-hypergraph obtained by adding k singletons $\{1\}, \ldots, \{k\}$. Then by Lemma 2.1, H' can be decomposed into a disjoint union of $m = (r + 1)/2$ 2-regular 2-hypergraphs, $H' = R'_1 + \ldots R'_m$, where each R'_i is a 2-regular 2-hypergraph. Removing the k singletons $\{1\}, \ldots, \{k\}$ from $R'_1, \ldots, R'_m$, we obtain $R_1, \ldots, R_m$. Then H is a disjoint union of $R_1, \ldots, R_m$, i.e., $H = R_1 + \cdots + R_m$. We next show by induction on j that, there is an edge ordering π_j on $R_1 + \cdots + R_j$ such that $g(R_1 + \cdots + R_j, \pi_j) \leq 1$.

Clearly, any ordering of the edges in R_1 has the above property. Assume that the elements of $R_1 + \cdots + R_{n-1}$ have an ordering π_{n-1} with the property. We show that the elements of R_n can be added to the end of π_{n-1} so that the resulting sequence also has the property. Let $G = R_1 + \cdots + R_{n-1}$. Then the degree of a vertex of G is either $2(n-1)$ or $2(n-1) - 1$. The degree of 2-hypergraph R_n is either 1 or 2 because all vertices of 2-hypergraph R'_n have degree 2 and R_n is resulted from R'_n by removing some singletons of $\{1\}, \ldots, \{k\}$. We order the edges of R_n by components as follows. Let C be any connected component of R_n. If all vertices of C have the same degree in G, then any ordering of edges of C has the property. If C contains vertices of both degrees in G, a vertex of C with degree $2(n-1) - 1$ in G must have degree 2 in C. For each maximal path in C with $2(n-1)$ and $2(n-1) - 1$ vertices appearing alternatively in G, order edges at odd positions in the path. Add the remaining edges of C in any order. Applying the same to all connected components of R_n, we obtain an ordering of R_n. Adding this ordering of R_n to the end of π_{n-1}, we obtain an edge ordering, π_n, of $R_1 + \cdots + R_n$. From the induction hypothesis and the construction, we see that the obtained ordering of elements of $R_1 + \cdots + R_n$ has the property. By the induction, we eventually obtain an ordering of R having the property, so that $g(H) \leq 1$. Thus we proved $g_2(H) \leq 1$ for any 2-hypergraph H. Therefore, we have $g_2(k) = 1$.

It is known that the decomposition of an even regular 2-hypergraph into 2-factors can be done in time polynomial in terms of the number of edges. By the proof of the above theorem, we know finding an ordering π such that $g(H, \pi) \leq 1$ for a given w-regular 2-hypergraph H on k vertices can be done in polynomial time in terms of k and w.

3   Proof for $g_s(k) = 1, 3 \leq k \leq 6$ and $g_s(k) \leq 2k$

We calculate the exact value of $g_s(k)$ for $3 \leq s \leq k \leq 6$ using relation (4). That is, we first find the set of all minimal regular s-hypergraphs H on k vertices, determine $g(H)$, and then $g_s(k)$. For example, when $k = 2$, there are only two minimal regular hypergraphs, $H_1 = (\{1, 2\}, \{\{1\}, \{2\}\}), H_2 = (\{1, 2\}, \{\{1, 2\}\})$. Clearly $g(H_1) = 0, g(H_2) = 0$. Therefore, $g_2(2) = g(2) = 0$.

Theorem 3.1. $g_s(k) = 1$ for $3 \leq s \leq k \leq 6$.

Proof. By Theorem 2.2 and the definition of $g_s(k)$, we have $1 = g_2(k) \leq \cdots \leq g_k(k) = g(k)$ for $3 \leq k \leq 6$. Therefore it suffices to prove $g(k) = 1$ for $3 \leq k \leq 6$.

To simplify the proof, we add some reduction rules to eliminate the number of minimal regular hypergraphs to be considered.

Rule 1. Only non-isomorphic minimal regular hypergraphs on k vertices need to be considered. Here, two hypergraph graphs $H_1 = (V_1, E_1)$ and $H_2 = (V_2, E_2)$ are isomorphic if there are two bijections $\phi : V_1 \to V_2, \varphi : E_1 \to E_2$ such that $e = \{v_{i_1}, \ldots, v_{i_t}\} \in E_1$ if and only if $\varphi(e) = \{\phi(v_{i_1}), \ldots, \phi(v_{i_t})\}$. It can be seen that if two regular hypergraphs H_1 and H_2 are isomorphic, then $g(H_1) = g(H_2)$.

Rule 2. We only need to consider minimal regular hypergraphs such that every two edges have intersection. This is because if there are two edges in a minimal regular hypergraph H without intersection, we can merge them into one edge, the resulting minimal regular hypergraph H' satisfies $g(H) \leq g(H')$.

Rule 3. We only need to consider minimal regular hypergraphs containing no singleton. This is because if a minimal regular hypergraph H contains a singleton u, then all other edges of H must contain the singleton u according to Rule 2. Then the degree of u would be bigger than the degree of other vertices. This contradicts H being regular.

Now we consider the case when $k = 3$. Applying the above reduction rules, there are only two non-isomorphic minimal regular hypergraphs with edge intersections. $H_1 = (\{1, 2, 3\}, \{\{1, 2\}, \{2, 3\}, \{1, 3\}\})$ and $H_2 = (\{1, 2, 3\}, \{\{1, 2, 3\}\})$. Clearly $g(H_1) = 1, g(H_2) = 0$, therefore, $g(3) = 1$.

For $k = 4$. First we see that the following minimal regular hypergraphs satisfy the reduction rules.

$$H_1 = (\{1, \ldots, 4\}, \{\{1, 2, 3, 4\}\}),$$
$$H_2 = (\{1, \ldots, 4\}, \{\{1, 2, 3\}, \{1, 2, 4\}, \{3, 4\}\}),$$
$$H_3 = (\{1, \ldots, 4\}, \{\{1, 2, 3\}, \{1, 2, 4\}, \{3, 4, 1\}, \{2, 3, 4\}\}),$$
$$H_4 = (\{1, \ldots, 4\}, \{\{1, 2, 3\}, \{1, 4\}, \{2, 4\}, \{1, 2, 3\}, \{3, 4\}\}).$$

Secondly, we show that these are all of the non-isomorphic minimal regular hypergraphs on $\{1, \ldots, 4\}$ that satisfy the reduction rules. Let H be a regular hypergraph on V satisfying the reduction Rules 2 and 3. We show H must be isomorphic to one of $H_i, i = 1, \ldots, 4$ by six cases.

Case 1. H contains edge $\{1, 2, 3, 4\}$. Then H must be H_1.

Case 2. H contains all types of edges $\{1, 2, 3\}, \{2, 3, 4\}, \{3, 4, 1\}, \{4, 1, 2\}$. Then H is isomorphic to H_3.

Case 3. H contains three types of edges $\{1, 2, 3\}, \{2, 3, 4\}, \{3, 4, 1\}, \{4, 1, 2\}$. Without loss of generality, assume that H contains $\{1, 2, 3\}, \{2, 3, 4\}, \{3, 4, 1\}$. Then H does not contain any one of $\{4\}, \{1\}, \{2\}$. H does not contain $\{1, 4\}$ because $\{\{1, 2, 3\}, \{2, 3, 4\}, \{1, 4\}\}$ forms proper regular sub-hypergraph. Similarly, H does not contain $\{1, 2\}$ or $\{2, 4\}$. Therefore, H only contains the types of edges of $\{1, 2, 3\}, \{2, 3, 4\}, \{3, 4, 1\}, \{3, 4\}, \{2, 3\}, \{1, 3\}, \{3\}$. All of these edges contain 3, therefore H can not be regular, and a contradiction follows.

Case 4. H contains two types of edges $\{1, 2, 3\}, \{2, 3, 4\}, \{3, 4, 1\}, \{4, 1, 2\}$. Without loss of generality, assume that H contains $\{1, 2, 3\}$ and $\{2, 3, 4\}$. Then H must contain edges of size 2. If H does contain $\{1, 4\}$, a contradiction will follow (details are omitted). Therefore, H contains $\{1, 4\}$, then H is isomorphic to H_2.

Case 5. H contains one types of edges $\{1, 2, 3\}, \{2, 3, 4\}, \{3, 4, 1\}, \{4, 1, 2\}$. We may assume that H contains $\{1, 2, 3\}$. Then H must contain at least one of $\{3, 4\}, \{1, 4\}$ and $\{2, 4\}$ to cover vertex 4. If H contains all of $\{3, 4\}, \{1, 4\}$ and $\{2, 4\}$, then H does not contain $\{1, 2\}, \{2, 3\}$ or $\{1, 3\}$. Now H only contains the types of edges in $\{1, 2, 3\}, \{3, 4\}, \{1, 4\}, \{2, 4\}$. Let $x_i, i = 1, \ldots, 4$ denote the number of occurrences of $\{1, 2, 3\}, \{3, 4\}, \{1, 4\}$ and $\{2, 4\}$ in H, respectively. Then the following equations must hold: $x_1 + x_3 = x_1 + x_4 = x_1 + x_2 = x_2 + x_3 + x_4 = r$. Therefore, $x_2 = x_3 = x_4 = \frac{r}{3} \geq 1$, and $x_1 = \frac{2r}{3} \geq 2$. Now H

contains $\{1,2,3\},\{1,2,3\},\{3,4\},\{1,4\},\{2,4\}$, which must be isomorphic to H_4. Otherwise, H either contains two of $\{3,4\},\{1,4\}$ and $\{2,4\}$, or contains only one of $\{3,4\},\{1,4\}$ and $\{2,4\}$. Both cases lead to contradictions.

Case 6. H does not contain any one of $\{1,2,3\}$, $\{2,3,4\},\{3,4,1\}$ and $\{4,1,2\}$. Then H is a graph, we see no regular graph on four vertices would satisfy Rules.

In summary, we know H_1, H_2, H_3, H_4 are all non-isomorphic minimal regular hypergraphs subject to the reduction rules. For each $H_i, i = 1,\ldots,4$, we see $g(H_i) = 1$ with the given edge ordering. Therefore by (4), we have $g_3(4) = g_4(4) = 1$.

For $k = 5$ and 6, we first use the known Hemmecke, Hemmecke, Malkin's 4ti2 package [8] to calculate the Hilbert basis of $A_k X^T - w(1,\ldots,1)^T = 0$, and then implement a brute force algorithm to minimize $g(H,\pi)$ for permutations π. The computation results show that for every minimal regular hypergraph H of 5 or 6 vertices, we found an ordering π_0 such $g(H,\pi_0) \le 1$. Therefore, $g_3(5) = g_4(5) = g_5(5) = 1, f_3(6) = g_4(6) = g_5(6) = g_6(6) = 1$. The details are omitted.

For $3 \le s \le k \ge 7$, no value of $g_s(k)$ is known. But we can prove an upper bound for $g_s(k)$ using the following vector sum result in [9,1].

Lemma 3.2 ([9]). *Let $\mathbf{v}_1,\ldots,\mathbf{v}_t$ be a sequence of k-dimensional vectors with $\sum_{i=1}^{t} \mathbf{v}_i = 0, ||\mathbf{v}_i|| \le 1 (i = 1,\ldots,t)$, there is a permutation $i_1,\ldots,i_t$, such that $\max ||\sum_{h=1}^{j} \mathbf{v}_{i_h}|| \le k$ for every $1 \le j \le t$, where the super norm $||\mathbf{v}||$ of a vector $\mathbf{v}$ is defined to be the maximum of the absolute values of the components of $\mathbf{v}$.*

Theorem 3.3. $g(k) \le 2k$ *for any* $k \ge 2$.

Proof. Let $H = (V, E)$ be any w-regular hypergraph on k vertices. Suppose that $V = \{1,\ldots,k\}$ and $E = \{e_1,\ldots,e_m\}$. Let $\mathbf{v_i} = (n_{i,1},\ldots,n_{i,k}) \in R^k$ be the vector representation of $e_i, i = 1,\ldots,m$.

For $i = 1,\ldots,m$, let $\mathbf{v}_i' = \mathbf{v_i}$, and for $i = m+1,\ldots,m+w$, let $\mathbf{v}_i' = (-1,\ldots,-1) \in R^k$. Then $\sum_{i=1}^{m+w} \mathbf{v}_i' = 0$. By Lemma 3.2, there is a permutation $i_1',\ldots,i_{m+w}'$ of $1,\ldots,m+w$ such that $\mathbf{v}_{i_1'}',\ldots,\mathbf{v}_{i_{m+w}'}'$ satisfy $||\sum_{h=1}^{j} \mathbf{v}_{i_h'}'|| \le k, 1 \le s \le m+w$.

Removing the vectors equal to $(-1,\ldots,-1)$ from the sequence $\mathbf{v}_{i_1'}',\ldots,\mathbf{v}_{i_{m+w}'}'$, we obtain a permutation $\mathbf{v}_{i_1},\ldots,\mathbf{v}_{i_m}$ of $\mathbf{v}_1,\ldots,\mathbf{v}_m$.

For any $1 \le j \le m-1$, let $\mathbf{v}_{i_{j'}'}'$ correspond to $\mathbf{v}_{i_j}$. Then there are $j' - j$ vectors in $\mathbf{v}_{i_1'}',\ldots,\mathbf{v}_{i_{j'}'}'$ which are equal to $(-1,\ldots,-1)$. Then, for every $1 \le j \le m-1$, we have

$$\max\{|d_{H_j}(u) - d_{H_j}(v)| : u, v \in e_{i_{j+1}}\}$$
$$= \max\{d_{H_j}(u) : u \in e_{i_{j+1}}\} - \min\{d_{H_j}(v) : v \in e_{i_{j+1}}\}$$
$$= \max\{\textstyle\sum_{h=1}^{j} n_{i,h} : i \in e_{i_{j+1}}\} - \min\{\textstyle\sum_{h=1}^{j} n_{i,h} : i \in e_{i_{j+1}}\}$$
$$\le \max\{\textstyle\sum_{h=1}^{j} n_{i,h} : 1 \le i \le k\} - \min\{\textstyle\sum_{h=1}^{j} n_{i,h} : 1 \le i \le k\}$$
$$= \max\{\textstyle\sum_{h=1}^{j} n_{i,h} : 1 \le i \le k\} + (j' - j) - (\min\{\textstyle\sum_{h=1}^{j} n_{i,h} : 1 \le i \le k\} + (j' - j))$$
$$= \max\{\textstyle\sum_{h=1}^{j} n_{i,h} + (j - j') : 1 \le i \le k\} - \min\{\textstyle\sum_{h=1}^{j} n_{i,h} + (j' - j) : 1 \le i \le k\}$$
$$\le 2||\textstyle\sum_{h=1}^{j'} \mathbf{v}_{i_h'}'|| \le 2k.$$

By the definitions (1)-(3) for $g_s(k)$, we have $g(k) = g_k(k) \leq 2k$.

In [1], a polynomial time algorithm was given to find a permutation of the vectors with $\max \| \sum_{h=1}^{j} \mathbf{v_{i_h}} \| \leq 3k/2$ for every $1 \leq j \leq t$. Applying that algorithm, we can find an edge ordering π for regular hypergraph H in polynomial time with $g(H, \pi) \leq 3k$.

4   Heuristic Algorithm for Edge Ordering

The brute force algorithm for finding an edge ordering is far from efficient when the number of edges is large. Efforts have been made on finding time efficient heuristic algorithm for edge ordering. This section presents an advanced edge ordering algorithm and primary experimental results. The algorithm determines an ordering of edges by adding edges to construct the hypergraph one edge at a time. The main idea of the algorithm is to define a rank function on the unused edges. To determine which edge is to be added next, the algorithm calculates the ranks of all unused edges and chooses the highest ranking one. The pseudo-code for calculating the rank of an edge is as follows:

```
SET rank to 0
SET number_of_elements to 0
SET number_of_max_elements to 0
SET number_of_min_elements to 0
SET max_degree to smallest integer
SET min_degree to largest integer
FOR each vertex in the hypergraph
    IF this vertex is in the edge THEN
        INCREMENT number_of_elements
        IF the degree of this vertex is greater than max_degree THEN
            SET max_degree to degree of this vertex
        END IF
        IF the degree of this vertex is less than min_degree THEN
            SET min_degree to degree of this vertex
        END IF
        DECREMENT rank
    END IF
END FOR
FOR each vertex in the hypergraph
    IF this vertex is in the edge THEN
        INCREMENT rank by the difference of the degree of the
            lowest priority vertex and the degree of this vertex
        DECREMENT rank by the difference of the degree of this
            vertex and the degree of the highest priority vertex
        IF the degree of this vertex equals min_degree THEN
            INCREMENT number_of_min_elements
        ELSE
            IF the degree of this vertex equals max_degree THEN
                INCREMENT number_of_max_elements
```

```
        END IF
     END IF
  END FOR
  IF number_of_elements equals number_of_min_elements and min_degree
     equals degree of highest priority vertex THEN
        INCREMENT rank by w_max
  ELSE IF number_of_elements equals number_of_max_elements and
     max_degree equals degree of lowest priority vertex THEN
        DECREMENT rank by w_max
  END IF
  IF the difference of max_degree and min_degree is greater than 1 THEN
     RETURN bad edge flag
     ELSE RETURN rank
  END IF
```

The running time of the above heuristic edge ordering algorithm is $O(|V|^2|E|^2)$. We implemented the algorithm and tested on minimal regular hypergraphs on 4, 5 and 6 vertices respectively. The computational results show that all minimal hypergraphs on 4 or 5 vertices passed the test. Table 1 gives a summary of the computation on 6 vertices, in which the first column gives the degree of the minimal regular hypergraphs, the second column gives the number of minimal regular hypergraphs passed the test, and third column gives the number of tested minimal regular hypergraphs.

Table 1. Testing results for minimal regular hypergraphs on 6 vertices

w	passed	tested	passed/tested (%)
1	62	62	100.0
2	1328	1328	100.0
3	12280	12292	99.9
4	27589	27624	99.9
5	20631	20777	99.3
6	8684	8844	98.2
7	3011	3035	99.2
8	753	762	98.8
9	201	201	100.0
Total	74539	74925	99.48

5   Applications in Interconnection Network Designs

This section describes the application of the edge ordering problem in designing switch boxes for interconnection networks. As in [5], a switch box of k sides and w terminals on each sides, written (k, w)-SB, can be represented as a k-partite graph G with the j-th terminal on side i being denoted by a vertex $t_{i,j}, i = 1, \ldots, k, j = 1, \ldots, w$, and a switch joining terminals $t_{i,j}$ and $t_{i',j'}$ ($i \neq i'$) being denoted by an edge $t_{i,j}t_{i',j'}$. Let $V_i = \{t_{i,j} : j = 1, \ldots, w\}$,

$i = 1, \ldots, k$. Then $(V_1, \ldots, V_k)$ forms the partition of the vertices. A routing requirement H for the (k, w)-SB can be represented as a w-regular hypergraph $H = (\{1, \ldots, k\}, \{e_1, \ldots, e_m\})$, where each edge e_i represents a connection requirement which needs to connect $|e_i|$ terminals on sides specified by the elements of e_i. H is said to be routable in G (or G is routable for H) if G contains a spanning subhypergraph with m vertex disjoint components $T_1, \ldots, T_m$ such that each T_i is connected and contains a vertex on side j if and only if $j \in e_i$. $(T_1, \ldots, T_m)$ is called a routing of H in G and T_i a routing of e_i. A (k, w)-SB is said to be s-universal if it is routable for every w-regular s-hypergraph on $\{1, \ldots, k\}$. In particular, a 2-universal (k, w)-SB is also called a universal (k, w)-SB (or simply (k, w)-USB)[3], and a k-universal (k, w)-SB is called a hyper-universal (k, w)-SB (or simply (k, w)-HUSB)[5]. The complete (k, w)-SB, i.e., $K_{k,w} = (V_1 \cup \ldots \cup V_k, \{t_{i,j}t_{i',j'} : i, i' = 1, \ldots, k, i \neq i', j, j' = 1, \ldots, w\})$, is s-universal for all $2 \le s \le k$. The number of switches in $K_{k,w}$ is $k(k-1)w^2/2$.

The optimal s-universal (k, w)-SB design problem is to design an s-universal (k, w)-SB with the minimum number of switches. This problem has been studied extensively in recent years for FPGA routing networks [2,3,5,6,7]. However, this problem is not solved in general except for $k \le 4$ or $s = 2$ and w is even. Recently, we found that using $g_s(k)$ we can design a much better approximation for the optimal s-universal (k, w)-SB as follows. Let $H(k, W, b)$ denote the (k, w)-SB with terminals $t_{i,j}, 1 \le i \le k, 1 \le j \le w$ and switches $t_{i,j}t_{i',j'}, i \neq i', |j - j'| \le b$.

Lemma 5.1. $H(k, w, g_s(k))$ is P_s-universal (k, w)-SB with at most $(2g_s(k) + 1)k(k-1)w/2$ switches.

Proof. For any w-regular s-hypergraph $H = (\{1, \ldots, k\}, \{e_1, \ldots, e_m\})$, we first find an edge ordering $(e_{\pi(1)}, \ldots, e_{\pi(m)})$ such that $g(H, \pi) \le g_s(k)$. Then we find a routing of H in $H(k, w, g_s(k))$ according to this ordering as follows. For $e_{\pi(i)}$, we find the routing of $e_{\pi(i)}$ by choosing the first available terminals on the sides specified by $e_{\pi(i)}$ and switches joining the terminals. By the definition of $H(k, w, g_s(k))$ and the edge ordering property, there are always switches joining the first available terminals. Since the degree of a vertex in $H(k, w, g_s(k))$ is at most $(2g_s(k) + 1)(k - 1)$ and it has kw vertices, the number of switches is at most $(2g_s(k) + 1)k(k - 1)W/2$.

By Lemma 5.1, Theorem 2.2, 3.1, and 3.3, we have $H(k, w, 1)$ is a (k, w)-USB for all $k \ge 2$ and $w \ge 1$, $H(k, w, 1)$ is a (k, w)-HUSB for $3 \le k \le 6$, and $H(k, w, 2k)$ is a (k, w)-HUSB for all $k \ge 2$. Moreover, by the proof of Lemma 5.1, we see an edge ordering algorithm leads to a routing algorithm for $H(k, w, g_s(k))$.

The switch box $H(k, w, g_s(k))$ can be used in three-stage interconnection network for group communications. Yen et al. [10] proposed the so-called Polygonal Switching Network, denoted by $PSN(r, w, k)$, which consists of k copies of $r \times w$ full crossbar at the first and the third stages and a (k, w)-USB at the middle stage. Here a $r \times w$ full crossbar can be represented as a bipartite graph of r vertices on one part and w vertices on the other part; it has rw edges. $PSN(r, w, k)$ forms a three-stage interconnection network rearrangeable for point-to-point communications of rk ports connected by the rk wires from k crossbars provided $r \le w$.

If the (k, w)-USB is substituted by an s-universal (k, w)-SB $H(k, w, g_s(k))$, the resulting $PSN(r, w, k)$ $(r \leq w)$ will be rearrangeable for group communications (with group size at most s), i.e. for any partition of the rk ports with group size at most s, the switches in the network can be reconfigured so that the ports in each group are connected and different groups are not connected. This is because a partition of the ports will results in a routing requirement on the middle switch box, so we can first find a routing for the routing requirement in the switch box and then route to the corresponding ports through the crossbars.

In particular, $PSN(n^{1/3}, n^{2/3}, n^{1/3})$ with $H(n^{1/3}, n^{2/3}, 2n^{1/3})$ at the middle stage will have $O(n^{5/3})$ switches and it is rearrangeable for group communications of n ports. This is the first three-stage interconnection network design rearrangeable for group communications of n ports with less than $O(n^2)$ switches.

Moreover, we see that if there is a constant C such that $g_s(k) \leq C$ for all $k \geq C$, then $PSN(n^{1/2}, n^{1/2}, n^{1/2})$ with $H(n^{1/2}, n^{1/2}, C)$ at the middle stage will have $O(n^{3/2})$ switches, this would be the best possible polygonal switching network design. Therefore, to conclude this paper, we first conjecture the existence of such a constant, and secondly we call for an efficient solution to compute $g_s(k)$ and an efficient algorithm for the edge ordering.

References

1. I. Barany. "A Vector-Sum Theorem and Its Application to Improving Flow Guarantees". *Mathematics of Operations Research*, 6:445–455, Aug. 1981.
2. V. Betz, J. Rose, and A. Marquardt. *Architecure and CAD for Deep-Submicron FPGAs*. Kluwer-Academic Publisher, Boston MA, 1999.
3. Y. W. Chang, D. F. Wong, and C. K. Wong. "Universal Switch Modules for FPGA Design". *ACM Transactions on Design Automation of Electronic Systems.*, 1(1):80–101, Jan. 1996.
4. E. Contejean and H. Devie. "An Efficient Incremental Algorithm for Solving Systems of Linear Diophantine Equations". *Inform. and Comput.*, 113(1):143–172, 1994.
5. H. Fan, J. Liu, and Y. L. Wu. "General Models and a Reduction Design Technique for FPGA Switch Box Designs". *IEEE Transactions on Computers*, 52(1):21–30, Jan. 2003.
6. H. Fan, J. Liu, Y. L. Wu, and C. C. Cheung. "On Optimal Hyper Universal and Rearrageable Switch Box Designs". *IEEE Transactions on Computer Aided Designs*, 22(12):1637–1649, Dec. 2003.
7. H. Fan, J. Liu, Y. L. Wu, and C. K. Wong. "Reduction Design for Generic Universal Switch Blocks". *ACM Transactions on Design Automation of Electronic Systems*, 7(4):526–546, Oct. 2002.
8. R. Hemmecke, R. Hemmecke, and P. Malkin. 4ti2 Version 1.2—Computation of Hilbert bases, Graver bases, toric Gröbner bases, and more. Available at www.4ti2.de, Sep. 2005.
9. S. V. Sevast'yanov. "On Approximate Solutions of Scheduling Problems". *Metody Discretnogo Analiza*, 32:66–75, 1978 (in Russian).
10. M. Yen, S. Chen, and S. Lan. "A Three-Stage One-Sided Rearrangeable Polygonal Switching Network". *IEEE Trans. on Computers*, 50(11):1291–1294, Nov. 2001.

Efficient Partially Blind Signature Scheme with Provable Security⋆

Zheng Gong, Xiangxue Li, and Kefei Chen

Department of Computer Science and Engineering
Shanghai Jiaotong University, Shanghai, 200030, China
{neoyan, xxli, kfchen}@sjtu.edu.cn

Abstract. Partially blind signature was first introduced by Abe and Fujisaki. Subsequently, Abe and Okamoto proposed a provably secure construction for partially blind signature schemes with a formalized definition in their work. In this paper, based on discrete logarithm problem and the Schnorr's blind signature scheme, we propose a new efficient partially blind signature scheme. Follow the construction proposed by Abe and Okamoto, we prove its security in random oracle model. The computation and communication costs are both reduced in our scheme. It will make privacy-oriented applications which based on partially blind signatures more efficient and suitable for hardware-limited environment, such as smart phones and PDAs.

1 Introduction

Blind signature schemes, first introduced by Chaum in [1], allow a user to get a signature without giving the signer any information about the actual message. The signer also can't have a link between the users and the signatures. It's a useful property in privacy oriented e-services such as electronic cash and electronic voting system. However, it may not a good idea to blind everything in the e-cash system[2]. As to prevent a customer's double-spending, the bank has to keep a spent database which stores all spent e-cash to check whether a specified e-cash has been spent or not. Certainly, the spent database kept by the bank may grow unlimitedly. The other problem is to believe the face value of e-cash in the withdraw phase, the signer must assure that the message contains accurate information without seeing it.

Partially blind signature scheme proposed in [2] helps to solve the problems stated above. The scheme allows each of signatures contains an explicit information which both the signer and the user have agreed on. For example, the signer can attach the expiry date and denomination to his blind signatures as an attribute. Accordingly, The attribute of the signatures can be verified independently through those of the certified public key.

Based on different hard problem assumptions, many partially blind signature schemes have been given. The schemes proposed in [2,3,4] are based on RSA

⋆ This work is partially supported by NSFC under the grants 90104005 and 60573030.

D.Z. Chen and D.T. Lee (Eds.): COCOON 2006, LNCS 4112, pp. 378–386, 2006.

algorithm, but the scheme in [2] does not have randomization property, which is important to withstand the chosen plaintext attack[5], and the scheme in [3] was also showed vulnerability on the chosen plaintext attack by [6]. The schemes proposed in [7,8,9] are based on discrete logarithm problem and [9] costs lower computation than [2,10]. The proposed partially blind signature schemes in [10,11] are based on the theories of quadratic residues, and the scheme [11] makes better performance than [10], but the signing protocol in [11] will give two valid signatures corresponding to the same message. The schemes proposed in [12,13] are based on bilinear pairings, but their verification of the signature require pairing operation, which is several times slower than modular exponentiation computation and not suitable for hardware-limited situations in client side, such as smart phones and PDAs.

Our Contribution. Considering both security and efficiency, based on discrete logarithm problem and the blind signature scheme in [14], we propose a new efficient partially blind signature scheme. Follow the construction that given in [7], we prove its security in random oracle model (ROM)[15]. Compared to the schemes in [4,7,9], the computation and communication costs for the user and the signer are both reduced in our scheme.

Organization. The rest of the paper is organized as follows. Section 2 describes the basic definitions associated with partially blind signatures. In section 3, we describe our efficient partially blind signature scheme, and then prove its security in section 4 and compare the performance of the proposed scheme with others related schemes in section 5. Section 6 concludes the paper.

2 Definitions

Abe and Okamoto introduced the notion of partially blind signatures in [7]. For the following provable security, We give the definitions proposed in [8] which provided a compact definitions based on [7]. In the phase of partially blind signatures, the signer and the user are assumed to have agreed on a piece of common information, denoted by **info**. An **info** may be sent from the user to the signer. The paper [7] formalized this notion by providing a function Ag. Function Ag is defined as a polynomial-time deterministic algorithm that completes the negotiation of **info** between the signer and the user correctly. In our scheme, this negotiation is considered to be done outside of the scheme.

Definition 1. *(Partially Blind Signature Scheme) A partially blind signature scheme is a four-tuple$(\mathcal{G}, \mathcal{S}, \mathcal{U}, \mathcal{V})$.*

- *$\mathcal{G}$ is a probabilistic polynomial-time algorithm, that takes security parameter k and outputs a public and secret key pair(pk, sk).*
- *$\mathcal{S}$ and $\mathcal{U}$ are pair of probabilistic interactive Turing machines each of which has a public input tape, a private input tape, a private random tape, a private word tape, a private output tape, a public output tape, and input and output communication tapes. The random tape and the input tapes are read-only,*

and the output tapes are write-only. The private work tape is read-write. The public input tape of $\mathcal{U}$ *contains pk generated by* $\mathcal{G}(1^k)$, *the description of Ag, and* **info**$_u$. *The public input tape of* $\mathcal{S}$ *contains the description of Ag and* **info**$_s$. *The private input type of* $\mathcal{S}$ *contains sk, and that for* $\mathcal{U}$ *contains a message msg. The lengths of* **info**$_s$, **info**$_u$, *and* **msg** *are polynomial in k.* $\mathcal{S}$ *and* $\mathcal{U}$ *engage in the signature issuing protocol and stop in polynomial-time. When they stop, the public output tape of* $\mathcal{S}$ *contains either completed or not-completed. If it is completed, then its private output tape contains common information* **info**. *Similarly, the private output tape of* $\mathcal{U}$ *contains either* $\perp$ *or* (**info**, **msg**, **sig**).

- *V is a polynomial-time algorithm.* $\mathcal{V}$ *takes* (*pk*, **info**, **msg**, **sig**) *and outputs either accept or reject.*

Definition 2. *(Partial Blindness)Let* $\mathcal{U}_0$ *and* $\mathcal{U}_1$ *be two honest users that follow the signature issuing protocol. Let* $\mathcal{S}^*$ *play the following Game A in the presence of an independent umpire.*

1. $(pk, sk) \leftarrow \mathcal{G}(1^k)$.
2. $(\mathbf{msg}_0, \mathbf{msg}_1, \mathbf{info}_{u_0}, \mathbf{info}_{u_1}, Ag) \leftarrow \mathcal{S}^*(1^k, pk, sk)$.
3. *The umpire sets up the input tapes of* $\mathcal{U}_0$, $\mathcal{U}_1$ *as follows:*
   - *The umpire selects* $b \in_R \{0,1\}$ *and places* $\mathbf{msg}_b$ *and* $\mathbf{msg}_{1-b}$ *on the private input tapes of* $\mathcal{U}_0$ *and* $\mathcal{U}_1$, *respectively. b is not disclosed to* $\mathcal{S}^*$.
   - *Place* $\mathbf{info}_{u_0}$ *and* $\mathbf{info}_{u_1}$ *on the public input tapes of* $\mathcal{U}_0$ *and* $\mathcal{U}_1$ *respectively. Also place pk and Ag on their public input tapes.*
   - *Randomly select the contents of the private random tapes.*
4. $\mathcal{S}^*$ *engages in the signature issuing protocol with* $\mathcal{U}_0$ *and* $\mathcal{U}_1$ *in a parallel and arbitrarily interleaved fashion. If either signature issuing protocol fails to complete, the game is aborted.*
5. *Let* $\mathcal{U}_0$ *and* $\mathcal{U}_1$ *output* $(\mathbf{msg}_b, \mathbf{info}_0, \mathbf{sig}_b)$ *and* $(\mathbf{msg}_{1-b}, \mathbf{info}_1, \mathbf{sig}_{1-b})$, *respectively, on their private tapes. If* $\mathbf{info}_0 \neq \mathbf{info}_1$ *holds, then the umpire provides* $\mathcal{S}^*$ *with the no additional information. That is, the umpire gives* $\perp$ *to* $\mathcal{S}^*$. *If* $\mathbf{info}_0 = \mathbf{info}_1$ *holds, then the umpire provides* $\mathcal{S}^*$ *with the additional inputs* $\mathbf{sig}_b, \mathbf{sig}_{1-b}$ *ordered according to the corresponding messages* $\mathbf{msg}_0, \mathbf{msg}_1$.
6. $\mathcal{S}^*$ *outputs* $b' \in_R \{0,1\}$. *The signer S. wins the game if* $b' = b$.

A signature scheme is partially blind if, for every constant $c > 0$, *there exists a bound* k_0 *such that for all probabilistic polynomial-time algorithm* $\mathcal{S}^*$, $\mathcal{S}^*$ *outputs* $b' = b$ *with probability at most* $1/2 + 1/k^c$ *for* $k > k_0$. *The probability is taken over the coin flips of* $\mathcal{G}$, $\mathcal{U}_0$, $\mathcal{U}_1$, *and* $\mathcal{S}^*$.

Definition 3. *(Unforgeability)Let* $\mathcal{S}$ *be an honest signer that follow the signature issuing protocol. Let* $\mathcal{U}^*$ *play the following Game B in the presence of an independent umpire.*

1. $(pk, sk) \leftarrow \mathcal{G}(1^k)$.
2. $Ag \leftarrow \mathcal{U}^*(pk)$.

3. *The umpire places sk, Ag and a randomly taken* **info**$_s$ *on the proper input tapes of* $\mathcal{S}$.
4. $\mathcal{U}^*$ *engages in the signature issuing protocol with* $\mathcal{S}$ *in a concurrent and interleaving way. For each* **info***, let* ℓ_{info} *be the number of executions of the signature issuing protocol where* $\mathcal{S}$ *outputs completed and* **info** *is on its output tapes. (For* **info** *that has never appeared on the private output tape of* $\mathcal{S}$*, define* $\ell_{\text{info}} = 0$.)
5. $\mathcal{U}^*$ *outputs a single piece of common information,* **info***, and* $\ell_{\text{info}} + 1$ *signatures* $(\mathbf{msg}_1, \mathbf{sig}_1), \cdots, (\mathbf{msg}_{\ell_{\text{info}}+1}, \mathbf{sig}_{\ell_{\text{info}}+1})$.

A partially blind signature scheme is unforgeable if, for any probabilistic polynomial-time algorithm $\mathcal{U}^*$ *that plays the above game, the probability that the output of* $\mathcal{U}^*$ *satisfies*

$$\mathcal{V}(pk, \mathbf{info}, \mathbf{msg}_j, \mathbf{sig}_j) = accept$$

for all $j = 1, \cdots, \ell_{\text{info}} + 1$ *is at most* $1/k^c$ *where* $k > k_0$ *for some bound* k_0 *and some constant* $c > 0$. *The probability is taken over the coin flips of* $\mathcal{G}, \mathcal{U}^*$, *and* $\mathcal{S}$.

Definition 4. (DLP (Discrete Logarithm Problem)): *For* $x, g \in_R \mathbb{Z}_p$, *given* $y = g^x (\text{mod } p)$, *compute* $x = \log_g y$. *We assume that DLP is hard, which mean there is no polynomial time algorithm to solve it with non-negligible probability.*

3   The Proposed Partially Blind Signature Scheme

The proposed efficient partially blind signature scheme is based on the theories of DLP. Our scheme consists of five phases: **Initialization, Requesting, Signing, Extraction** and **Verifying**, as described below.

1. **Initialization.** Signer $\mathcal{S}$ selects two large prime numbers p and q (typical length: $|p| = 1024$, $|q| = 160$), which satisfied $q|p - 1$. Then chooses a generator $g \in \mathbb{Z}_p, g^q \equiv 1 (\text{mod } p)$. $\mathcal{S}$ picks up a random number $x \in \mathbb{Z}_q$, computes corresponding $y = g^x (\text{mod } p)$. $a \parallel b$ denotes a concatenates b. $\mathcal{H}, \mathcal{F}, : \{0,1\}^* \mapsto \mathbb{Z}_q$ defined as two public hash functions. M is an arbitrary message space. The public key of $\mathcal{S}$ is the tuple (y, p, q, g), x is the private key.
2. **Requesting.** Assume that User $\mathcal{U}$ wants to get a partially blind signature on message $\mathbf{msg} \in M$, and then prepares a string $\mathbf{info} \in M$ that will be sent to $\mathcal{S}$ for his agreement, this negotiation is considered to be done outside of the scheme. Then $\mathcal{S}$ selects two random numbers $r, d \in_R \mathbb{Z}_q$, computes $z = \mathcal{F}(\mathbf{info})$, then submits $u = g^r z^d (\text{mod } p)$ to $\mathcal{U}$.

    After receiving $u, \mathcal{U}$ also selects three random numbers $v, w, e \in_R \mathbb{Z}_q$, computes $z = \mathcal{F}(\mathbf{info})$ and $b = z^e (\text{mod } p)$. Then computes $C' = \mathcal{H}(\mathbf{msg}\|\mathbf{info}\|t)$ while $t = ubg^v y^w (\text{mod } p)$, sends $C = w - C'$ to $\mathcal{S}$.

3. **Signing.** After receiving C, $\mathcal{S}$ Signs C with the randomizing factor r and his private key x, computes $S = r + (C - z)x (\bmod\ q)$. Then $\mathcal{S}$ sends the other randomizing number d and S to $\mathcal{U}$.
4. **Extraction.** After receiving S and d, $\mathcal{U}$ computes $S' = S + v (\bmod\ q)$ and $N = d + e (\bmod\ q)$. Hence, the resulting signature on the message **msg** and the common information **info** is a tuple $(\mathbf{msg}, \mathbf{info}, S', C', N)$.
5. **Verifying.** For the signature $(\mathbf{msg}, \mathbf{info}, S', C', N)$, because

$$S' = S + v, S = r + (C - z)x$$

and
$$C = w - C', C' = \mathcal{H}(\mathbf{msg} \| \mathbf{info} \| t),$$

we can easily get

$$
\begin{aligned}
g^{S'} y^{z+C'} z^N &= z^{e+d} g^v g^{r+(C-z)x} y^{z+C'} (\bmod\ p) \\
&= ub g^v y^{C-z} y^{z+C'} (\bmod\ p) \\
&= ub g^v y^{C+C'} = ub g^v y^w = t (\bmod\ p).
\end{aligned}
$$

Hence, we have the equation

$$\mathcal{H}(\mathbf{msg} \| \mathbf{info} \| g^{S'} y^{z+C'} z^N (\bmod\ p)) = \mathcal{H}(\mathbf{msg} \| \mathbf{info} \| t) = C'.$$

The partially blind signature is accepted as valid if it satisfies the above equation.

4   Security

In this section, we discuss some security properties of our partially blind signature scheme based on assuming the intractability of the DLP.

4.1   Randomization

Theorem 1. *Given a response S produced by Signer $\mathcal{S}$, user $\mathcal{U}$ cannot remove the random factor r from S in polynomial time.*

Proof. In the scheme, $\mathcal{S}$ selects a large integers r and computes $u = g^r (\bmod\ p)$, and submits u to $\mathcal{U}$. Then $\mathcal{U}$ sends C to $\mathcal{S}$, and $\mathcal{S}$ returns $S = r + (C - z)x$. If $\mathcal{U}$ wants to remove r from the corresponding signature S, he must derive the unique pair (x, r) from (y, u). However, it is difficult for $\mathcal{U}$ to determine (x, r) because the derivation is DLP. Hence, in the proposed scheme, $\mathcal{U}$ cannot remove the random large integer r from the corresponding signature S of **msg**.   □

4.2   Partial Blindness

Due to the **Definition 2**, for each instance numbered i of the proposed scheme, signer $\mathcal{S}^*$ can record C_i received from $\mathcal{U}$ who communicates with $\mathcal{S}^*$ during the instance i of the scheme. The tuple (S_i, C_i, r_i, d_i) is usually referred to as the view of $\mathcal{S}^*$ to the instance i of the scheme. Thus, we have the following theorem.

Theorem 2. *The proposed scheme is partially blind.*

Proof. Since the tuple $(\mathbf{msg}, \mathbf{info}, S', C', N)$ is produced, we have $S' = S_i + v, C' = w - C_i, N = d_i + e$ and $S_i = r_i + (C_i - z)x$. From the view of $\mathcal{S}^*$, Since v, w, e are three random numbers selected by $\mathcal{U}$ from $\mathbb{Z}_q$ and $\mathcal{S}^*$ cannot know v, w, e. The existence of a random triplet (v, w, e) that protects (S', C', N). Hence $\mathcal{S}^*$ can derive (v, w, e) from each view(S_i, C_i, r_i, N_i) such that $C_i = w - C', C' = \mathcal{H}(\mathbf{msg}\|\mathbf{info}\|g^{S'}y^{z+C'}z^N(\mathrm{mod}\ p))$ is satisfied where (S_i, C_i, r_i, N_i) regard as (S, C, r, N). When the instance $i \mapsto \{0, 1\}$, therefore, even an infinitely powerful $\mathcal{S}^*$ can succeed in determining i with probability $1/2$. $\square$

From the proof of **Theorem 2**, we can know the importance of random factors v, w, e. $\mathcal{U}$ must reselect v, w, e in a new instance of the proposed scheme and protect factors v, e as a secret during the proceeding of the scheme. The random factors v, w, e must be destroyed after the signature $(\mathbf{msg}, \mathbf{info}, S', C', N)$ is created.

4.3   Unforgeability

From **Definition 3**, we analyze the successful forgery with following the same security argument given by Abe and Okamoto in [7]. Let us consider two types of forgery against the partially blind signature.

1. A user $\mathcal{U}^*$ can generate a valid partially blind signature while $\ell_{\mathbf{info}} = 0$.
2. Given a large number of valid partially blind signatures$(0 < \ell_{\mathbf{info}} < poly(\log n))$, $\mathcal{U}^*$ can extract a new valid signature.

Theorem 3. *The proposed scheme is unforgeable in the situation of type 1.*

Proof. We assume a successful forger $\mathcal{U}^*$ who plays Game B and produces a valid signature $(\mathbf{msg}, \mathbf{info}, S', C', N)$ with probability $\mu > 1/k^c$, such that $\ell_{\mathbf{info}} = 0$. By exploiting $\mathcal{U}^*$, we construct a machine $\mathcal{M}$ that forges the non-blind signature of the proposed scheme in a passive environment. $\mathcal{M}$ simulates random oracles $\mathcal{F}$ and $\mathcal{H}$.

Let q_F and q_H be the maximum number of queries that $\mathcal{U}^*$ asked from $\mathcal{F}$ and $\mathcal{H}$, respectively. Let q_S be the maximum number of queries of signer $\mathcal{S}$. Selects $i \in \{1, 2, \cdots, q_H + q_S\}$, $\mathcal{U}^*$ sends the tuple $(\mathbf{msg}_i, \mathbf{info}_i, t_i)$ to the oracle $\mathcal{H}$ for computing its hash value $\mathcal{H}(\mathbf{msg}_i\|\mathbf{info}_i\|t_i)$. Simultaneously, $\mathcal{U}^*$ asks $\mathcal{F}$ to get $z = \mathcal{F}(\mathbf{info})$. $\mathcal{F}$ returns $z_i = g^{\omega_i}(\mathrm{mod}\ p)$, where $w_i \in_R \mathbb{Z}_q$. $\mathcal{M}$ knows ω_i from each pair of (z_i, ω_i) in $\mathcal{F}$. All of the parameters are limited by a polynomial in k. As the same proof construction in [7], we can easily know the success probability of $\mathcal{M}$ which is denoted by μ'.

$$\mu' = \frac{\mu}{(q_H + q_S)(q_F + q_S)}.$$

Then we use $\mathcal{M}$ to solve DLP. From the above construction. $\mathcal{M}$ can get a valid signature tuple (t_1, S'_1, C'_1, N_1) in polynomial running time after $1/\mu'$

trials, with probability at least $1 - e^{-1}$ (here, e is base of natural logarithms). Because $\mathcal{U}^*$ only can get hash value from $\mathcal{H}$. Next, we use the standard replay technique [16,17]. That is, we repeat with the same random tape and a different choice of $\mathcal{H}$, we can get another valid signature (t_2, S_2', C_2', N_2) after $2/\mu'$ trials, with probability at least $(1 - e^{-1})/2$, and we have $t_1 = t_2$. From the equation $C' = \mathcal{H}(\mathbf{msg}\|\mathbf{info}\|g^{S'}y^{z+C'}z^N(\bmod\ p))$, we have

$$S_1' + (C_1' + z_1) \cdot x + \omega_1 \cdot N_1 = S_2' + (C_2' + z_2) \cdot x + \omega_2 \cdot N_2.$$

Since $\mathcal{H}$ was changed choice in the second time run, both $S_1' \neq S_2'$ and $C_1' \neq C_2'$ have a overwhelming probability in $1 - 2^{-k}$, $\mathcal{M}$ can get x from

$$x = \frac{(S_2' - S_1') + (\omega_2 \cdot N_2 - \omega_1 \cdot N_1)}{(C_1' - C_2') + (z_1 - z_2)}(\bmod\ q).$$

It means $\mathcal{M}$ can solve DLP in polynomial running time.   □

Next we consider the forgery attempts in situation of type 2. We prove the security of our scheme where the common information is not all the same in Game B.

Theorem 4. *The proposed scheme is unforgeable in the situation of type 2.*

Proof. We assume a successful forger $\mathcal{U}_f^*$ who wins Game B with a probability η, which is a non-negligible in polynomial running time. Then we construct an machine $\mathcal{M}$ that simulates the signer in Game B. Let $\hat{S}$ denote the signer simulated by $\mathcal{M}$. $\mathcal{M}$ simulates two random oracles $\mathcal{F}$ and $\mathcal{H}$. $\mathcal{F}$ returns $z_i = g^{\omega_i}(\bmod\ p)$, where $\omega_i \in_R \mathbb{Z}_q$. We assume $\mathcal{M}$ don't know ω_i this time. $\mathcal{M}$ uses $\mathcal{U}_f^*$ as a black-box and breaks the intractability assumption of DLP to compute ω such that $z = g^{\omega}(\bmod\ p)$.

After $\ell_{\mathbf{info}}$ times execution with $\hat{S}$, $\mathcal{U}_f^*$ has got a set of successful challenge tuple $(\mathbf{msg}_1, \mathbf{info}_1, t_1), (\mathbf{msg}_2, \mathbf{info}_2, t_2), \cdots, (\mathbf{msg}_{\ell_{\mathbf{info}}}, \mathbf{info}_{\ell_{\mathbf{info}}}, t_{\ell_{\mathbf{info}}})$. $\mathcal{U}_f^*$ sends the tuple $(\mathbf{msg}_i, \mathbf{info}_i, t_i)$ to the random oracle $\mathcal{H}$ for computing its hash value $\mathcal{H}(\mathbf{msg}_i\|\mathbf{info}_i\|t_i)$.

From the above construction, $\mathcal{U}_f^*$ can win Game B and forge a valid signature with a successful challenge tuple $(\mathbf{msg}_{\ell_{\mathbf{info}}+1}, \mathbf{info}_{\ell_{\mathbf{info}}+1}, t_{\ell_{\mathbf{info}}+1})$ after $1/\eta$ trails, with probability at least $1 - e^{-1}$. First we consider the situation that there exists $i \in \{1, 2, \cdots, \ell_{\mathbf{info}}\}$, $\mathbf{msg}_i = \mathbf{msg}_{\ell_{\mathbf{info}}+1}$, $\mathbf{info}_i = \mathbf{info}_{\ell_{\mathbf{info}}+1}$ and $t_i = t_{\ell_{\mathbf{info}}+1}$. Because $\mathcal{U}^*$ only can get hash value from $\mathcal{H}$, We have

$$g^{S'_{\ell_{\mathbf{info}}+1}}y^{C'_{\ell_{\mathbf{info}}+1}}z^{N_{\ell_{\mathbf{info}}+1}} = g^{S_i'}y^{C_i'}z^{N_i}$$

such that

$$S'_{\ell_{\mathbf{info}}+1} + N_{\ell_{\mathbf{info}}+1} \cdot \omega = S_i' + N_i \cdot \omega.$$

Hence, we can compute ω from

$$\omega = \frac{S'_{\ell_{\mathbf{info}}+1} - S_i'}{N_{\ell_{\mathbf{info}}+1} - N_i}(\bmod\ q).$$

Then we consider the situation that there does not exist $i \in \{1, 2, \cdots, \ell_{\mathsf{info}}\}$, $\mathbf{msg}_i = \mathbf{msg}_{\ell_{\mathsf{info}}+1}$, $\mathbf{info}_i = \mathbf{info}_{\ell_{\mathsf{info}}+1}$ and $t_i = t_{\ell_{\mathsf{info}}+1}$. This derives to the same forgery attempts in the situation of type 1. $\qquad\square$

From **Theorem 3** and **Theorem 4**, we have the following theorem.

Theorem 5. *The proposed scheme is unforgeable if $\ell_{\mathsf{info}} < poly(\log n)$ for all info.*

5   Performance Concerns

We will discuss the performance of the proposed scheme from the costs of communication and computation. Table 1 gives us a detail costs comparison amongst related partially blind signature schemes[4,7,9]. The techniques to perform the modular exponentiation computation are not used because they need additional storage, which is limited in some application environments.

Table 1. The Comparisons of the partially blind signature schemes

	Our Scheme	Abe00 [7]	Huang03 [9]	Cao05 [4]
Mathematical foundation	DLP	DLP	DLP/CRT	RSA
Signer's computation	$2T_e + 1T_m$	$3T_e + 2T_m$	$2T_e + 4T_m$	$2T_e + 2T_m + T_i$
User's computation	$3T_e + 3T_m$	$4T_e + 4T_m$	$4T_e + 2T_m$	$3T_e + 5T_m$
Verifier's computation	$3T_e + 2T_m$	$4T_e + 2T_m$	$5T_e + 3T_m$	$3T_e + 2T_m$
Signature size	$2|m| + 3|q|$	$2|m| + 4|q|$	$2|m| + 3|n|$	$2|m| + 2|n|$

$^*T_e$: time for one exponentiation computation; T_m: time for one multiplication computation; T_i: time for one inverse computation; Typical length: $|q| = 160bit$, $|n| = 1024bit$.

With regard to estimate the computational costs, we count only modular exponentiation and multiplication. An inverse computation demands the same amount of computation as a modular exponentiation. We also do not calculate the computational costs on hash operations because it is much more faster than modular exponentiation computation, and each schemes takes nearly same times of hash operation. By Table 1, from the computational costs and signature sizes, our scheme all shows more efficient than the schemes in [4,7,9].

6   Conclusion

In this paper, we proposed an efficient partially blind signature scheme based on DLP and the Schnorr's blind signature scheme, and we proved its security in ROM. The computation and communication costs are both reduced in our scheme. It will makes privacy oriented applications which based on partially blind signatures more efficient and suitable for hardware-limited environment, such as smart phones and PDAs.

References

1. D.Chaum. Blind signature for untraceable payments. *Advances in Cryptology-CRYPTO'82*, pp.199-203, 1983.
2. M.Abe and E.Fujisaki. How to date blind signatures. *Advances in Cryptology-ASIACRYPT'96*, LNCS 1163, pp.244-251, 1996.
3. H.Y. Chien, J.K. Jan, Y.M. Tseng. RSA-based partially blind signature with low computation. *In: Proceedings of the eighth international conference on parallel and distributed systems*, pp.385-389, 2001.
4. T.J. Cao et al. A randomized RSA-based partially blind signature scheme for electronic cash, *Computers & Security*. Vol:24, Issue:1, pp.44-49, 2005.
5. A. Shamir and C.P. Schnorr. Cryptanalysis of certain variants of Rabin's signature scheme. *Information proceeding Letters*, vol.19, pp.113-115, 1984.
6. M.S Kwon and Y.K Cho. Randomization enhanced blind signature schemes based on RSA. *IEICE A Fundam 2003*; E86-A(3); 730-3, 2003.
7. M.Abe and T.Okamoto. Provably secure partially blind signatures, *Advance in Cryptology-CRYPTO'00*, LNCS 1880, pp.271-286, 2000.
8. G.Maitland and C.Boyd. A provably secure restrictive partially blind signature scheme, *PKC 2002*, LNCS 2274, pp.99-114, 2002.
9. H.F. Huang and C.C Chang. A new design of efficient partially blind signature scheme, *Journal of Systems and Software*, Vol: 73, Issue: 3, pp.397-403, 2003.
10. C.I. Fan and C.L. Lei. Low-computation partially blind signatures for electronic cash. *IEICE Transaction on Fundamentals of Electronics, Communications and Computer Sciences*, E81-A(5)818-824, 1998.
11. C.I. Fan. Improved low-computation partially blind signatures . *Applied Mathematics and Computation*, Vol: 145, Issue:2-3, pp. 853-867, 2003.
12. Fangguo Zhang et al. Efficient Verifiably Encrypted Signature and Partially Blind Signature from Bilinear Pairings. *Cryptology - INDOCRYPT 2003*, LNCS 2904, pp. 191-204, 2003.
13. S.M. Sherman et al. Two Improved Partially Blind Signature Schemes from Bilinear Pairings. *ACISP 2005*, LNCS 3574, pp. 316-328, 2005.
14. C.P. Schnorr. Discrete log signatures against interactive attacks. *Information and Communications Security*, LNCS 2299, pp. 1-12, 2001.
15. M.Bellare and P.Rogaway. Random oracles are practical: A paradigm for designing efficient protocols. In *ACM CCS*, pp.62-73, 1993.
16. U. Feige, A. Fiat and A. Shamir. Zero-knowledge proofs of identity. *Journal of Cryptography*. 1:77-94. 1988.
17. K. Ohta and T. Okamoto. On concrete security treatment of signatures derived from identification. In H. Krawczyk, editor, *Advances in Cryptology-Proceedings of CRYPTO'98*, LNCS 1462, pp.345-370. 1998.

A Rigorous Analysis for Set-Up Time Models – A Metric Perspective

Eitan Bachmat[1], Tao Kai Lam[2], and Avner Magen[3]

[1] Dept. of Computer Science
Ben Gurion University
[2] EMC cooperation
Hopkinton, MA
[3] Dept. of Computer Science
University of Toronto

Abstract. We consider model based estimates for set-up time. The general setting we are interested in is the following: given a disk and a sequence of read/write requests to certain locations, we would like to know the total time of transitions (set-up time) when these requests are served in an orderly fashion. The problem becomes nontrivial when we have, as is typically the case, only the counts of requests to each location rather then the whole input, in which case we can only hope to estimate the required time. Models that estimate the set-up time have been suggested and heavily used as far back as the sixties. However, not much theory exists to enable a qualitative understanding of such models. To this end we introduce several properties through which we can study different models such as (i) *super-additivity* which means that the set-up time estimate decreases as the input data is refined (ii) *monotonicity* which means that more activity produces more set-up time, and (iii) an approximation guarantee for the estimate with respect to the worst possible time.

We provide criteria for super-additivity and monotonicity to hold for popular models such as the independent reference model (IRM). The criteria show that the estimate produced by these models will be monotone for any reasonable system. We also show that the IRM based estimate functions, upto a factor of 2, as a worst case estimate to the actual set-up time.

To establish our theoretical results we use the theory of finite metric spaces, and en route show a result of independent interest in that theory, which is a strengthening of a theorem of Kelly [4] about the properties of metrics that are formed by concave functions on the line.

1 Introduction

Set-up times which are associated with moving a system from one state to another play a major role in the performance analysis of systems. Perhaps the most glaring example is provided by disk based storage systems in which the states correspond to locations on the disk. In this case the total duration of the movements of the disk's head (from one location to another or from one disk track to another), aka the *set-up time* is the dominant feature in the total service time, and hence a lot of effort is put in order to minimize it by means of reordering

D.Z. Chen and D.T. Lee (Eds.): COCOON 2006, LNCS 4112, pp. 387–397, 2006.
© Springer-Verlag Berlin Heidelberg 2006

the disk's content. Interestingly enough, in this application as well as in other real world applications, the above task becomes a problem with *partial input*. The reason is simple: to collect all transition information will be too costly and will render the original optimization useless as the set-up time will be second to the input collection time. Instead, the only information typically available is the state counts, ie the number of times that each state was requested. In graph terminology we want to know the length of a path in a weighted graph where we only know the number of times that each node was visited.

In order to estimate the set-up time, researchers have used stochastic *models*, ie stochastic processes with parameters that are inherited from the observed count. The simplest of these models, the *Independent Reference Model* (IRM) is very intuitive: the requests at any time are drawn (independently of the previous state) from a distribution proportional to the count vector. This simple model is the most popular model for the analysis of storage system performance; see for for example [1,3,6,7,8,9] among many.

In this paper we consider new and basic properties of set-up time estimates and check whether they hold for the IRM model. In a full version of the paper we will consider other models such as the so called the *Partial Markov Model* (PMM). These properties relate the set-up time estimates to the worst case case and examine the changes in the estimate due to a different way of collecting the data. The applicability of these properties to various models is an evidence to their quality, and moreover they allow for a rigorous study of models that are heavily used, often with not enough underlying rationale. It is interesting to note that while the IRM is one of the oldest models of user access patterns, dating back to the sixties, the basic properties considered above have never been explored. What follows is a brief description of these properties.

Given time intervals $I \subset J$ it is obvious that a system suffers at least as much set-up time during J as it does during I. The *monotonicity* property simply says that the set-up time estimate of the model reflects that fact, ie it gives an estimate for J which is at least as big as the one for I. A model is said to be *super-additive* if the addition of input information (by means of higher resolution of measurements) does not increase the set-up time estimate. It is almost immediate that super-additivity implies monotonicity and that it applies to the worst case time which provides the largest possible set-up time consistent with a given input data. The last property compares the set-up time estimate with the worst case estimate (which is NP hard to compute). Showing that the estimate of a model does not deviate much from the worst case estimate is tantamount to showing that is not over optimistic.

Our Results: We show that monotonicity applies to the IRM , *regardless of the metric involved*. We further show that IRM set-up time estimate is a 1/2 approximation to the worst case. Our results concerning super-additivity have the following curious feature: Super additivity holds in the IRM model *provided* that the "time-metric", ie the times associated with the transition times between pair of states, belongs to the well studied class of metric spaces known as *negative type metrics.* Not all metric spaces belong to this class, but as we show, the time metrics that come from motion of disk drives in fact do, owing to the physical features of the system. Therefore IRM is indeed super additive with respect to these I/O systems. These results show that the IRM can be used to produce reliably conservative estimates which are easy to calculate and that easily lend themselves to compactness-of-input/accuracy tradeoff. Following these observations the first and second authors used the IRM set-up time estimate as a central ingredient in a commercially available application which dynamically reconfigures data in a disk array. Details of the application and successful results from real production environments are to be presented elsewhere.

Techniques: Naturally, much of the notions and proofs come from and use the theory of metric spaces. The classes of interest in this discussion are ℓ_1-metrics and negative type metrics, as well as the general class of metrics. In the process of establishing our results we extend a result of Kelly on the properties of invariant metrics on the real line coming from concave functions.

Organization: The rest of the paper is organized as follows. Section 2 Introduces set-up times and discusses some basic definitions and facts from the theory of metric spaces relevant to our discussion. Section 3 describes the basic models which we will study and introduces the concepts of monotonicity, super additivity, dominance and approximation. In section 4 we prove criteria for monotonicity and super additivity of the IRM estimate in terms of metric properties of the set-up time function. Finally, Section 5 discusses properties of metric arising from the seek times in disk drives.

2   Preliminaries

2.1   Set-Up Time

Throughout the paper we let X represent the states of a system. In this section we let n denote the number of states in X. Following [1] section 6.2, we let the function $d : X \times X \longrightarrow \mathbf{R}^+$, be the set-up time function; namely, for $i, j \in X$, $d(x_i, x_j)$ represents the amount of time which is required to switch the system from state i to state j.

The abstract notion of a state can acquire many different meanings in different applications. For example, the states can refer to different tasks that the system needs to accomplish as in production systems and processors, or, to physical locations where tasks should be conducted as in storage systems. We assume that there is some process which generates a sequence of requests for the states of X.

Given a time interval I let $\mathbf{x}^I = \mathbf{x} = x_1, ..., x_m$ be the sequence of requests for states of X during I. The *Total set-up time* during time interval I is simply the sum of the set-up times between consecutive requests

$$T(\mathbf{x}) = \sum_{j=1}^{m-1} d(x_j, x_{j+1})$$

In some cases we are not given the sequence of requests (a trace) but rather some partial information about the sequence $\mathbf{x}$. We wish to estimate the total set-up time of the sequence using the information available to us. In this paper we shall assume that the partial information available to us is the activity vector $\mathbf{a} = \mathbf{a}_I = (a_1, ..., a_n)$, where a_i is the number of requests for state i during time interval I. We will assume that in general $\mathbf{a}$ can be any vector with integer nonnegative entries. We let $a = \sum_i a_i$ be the total number of requests.

2.2   Metric Spaces

The theory of finite metric spaces will be used in the statements and proofs of our results. The following section provides some basic definitions and facts about metric spaces which will be needed later on.

We continue with a few standard definitions. A pair (X, d) where X is a set and d is a function $d : X \times X \longrightarrow \mathbf{R}^+$ is called a *metric-space* if (i) $d(x, x) = 0$ for all $x \in X$ and $d(x, y) > 0$ for $x \neq y$, (ii) $d(x, y) = d(y, x)$ for all $x, y \in X$ and (iii) $d(x, y) + d(y, z) \geq d(x, z)$ for all $x, y, z \in X$. If instead of property (i) we only require that $d(x, x) = 0$ we say that (X, d) form a *semi-metric*. If we do not require the symmetry property, we say that (X, d) form a *Pseudometric*. One can "symmetrize" such an object by taking $d^*(x, y) = (d(x, y) + d(y, x))/2$. It can be easily seen that d^* satisfies 1' and 3' if d does. Set-up time functions can be reasonably assumed to satisfy the triangle inequality since one way to switch from state x to state z is to first switch from x to y and then from y to z. Set-up time functions cannot always be assumed to be symmetric as can be seen from rotational latency in disk drives.

Certain metric spaces are induced by norms. The ℓ_p norm on $\mathbf{R}^n$ is $\|\mathbf{x}\|_p = (\sum_{i=1}^n |x_i|^p)^{\frac{1}{p}}$ where $\mathbf{x} = (x_1, \ldots, x_n)$. A metric space (X, d) is called an ℓ_p-metric if there exists a mapping $\phi : X \longrightarrow \mathbf{R}^n$ such that $d(x, y) = \|\phi(x) - \phi(y)\|_p$ for all $x, y \in X$. We sometimes say *Euclidean metric* instead of ℓ_2-metric. A space (X, d) is *negative type* if $(X, \sqrt{d})$ is Euclidean.

Some Basic Facts About Metric Spaces. Assume (X, d) is a finite metric space, $X = \{x_1, \ldots, x_n\}$. There are two classical criteria for it to be Euclidean.

- Schoenberg's criterion: (X, d) is Euclidean if and only if for all n real numbers $v_1, \ldots, v_n$ with $\sum_i v_i = 0$ we have $\sum_{i,j} v_i v_j d^2(x_i, x_j) \leq 0$. (This criterion is the reason for the name *negative type*, as by definition, d is Euclidean iff d^2 is negative type.)
- Cayley's criterion: Consider the order $n - 1$ matrix M with entries $M_{i,j} = d^2(x_i, x_n) + d^2(x_j, x_n) - d^2(x_i, x_j)$, $i, j = 1, \ldots, n-1$. Then (X, d) is Euclidean if and only if the matrix M is positive semi definite, ie, all of its eigenvalues are nonnegative.

We say that a metric (X, d) is L_1 if there exist functions f_x, $x \in X$ such that $d(x, y) = \int_{\mathbf{R}} |f_x(t) - f_y(t)| dt$. It is known that a finite metric space is L_1 iff it is ℓ_1. Another well known fact we later use is that every ℓ_1-metric is negative type [5]. Negative type distances do not necessarily satisfy the triangle inequality.

A distance function can be defined on the line given a real positive function F with certain properties. We define the distance d_F between i and j as $d_F(i,j) = F(|i - j|)$. utilized in this paper. We note that if F is convex then d_F satisfies the triangle inequality and thus provides a metric.

3 Models and Their Properties

Recall that our input is an activity vector, that is the count of requests to the different states; however, in order to know the total set-up time we need to know the actual sequence of requests. In the absence of the actual sequence we use models for estimating set-up time. A model for estimating set-up time is an interpretation of an activity vector as a distribution over sequences, and the resulting estimate is then the expected set-up time for a random sequence drawn from this distribution. For example some models will interpret an activity vector $(100, 100)$ as a uniform distribution of sequences that visit either location 1 or 2, while other will consider the distribution in which either all first 100 requests are for the first location or all of them were for the other; clearly the two different models in the above example will produce very different time estimates.

3.1 Examples of Models and Estimates

We now describe a few models M and their associated set-up time estimates.

The IRM (Independent Reference Model). The IRM models independent random requests to states in X, taking into account that the different states are not uniformly popular. The model is parameterized by a probability distribution $p = p_i$ on the set of states X. The model itself is then given by the product measure on X^a. The product measure reflects an underlying assumption that requests are generated independently of each other. To be compatible with the observed activity vector we set the request probability for state i to be $p_i = a_i/a$ and the length of the generated sequence to be a. For this model the expected total set-up time is

$$T(\mathbf{a}, d; IRM) = a \sum_{i,j} p_i p_j d(x_i, x_j) = \frac{1}{a} \sum_{i,j} a_i a_j d(x_i, x_j)$$

We will refer to $T(\mathbf{a}, d; IRM)$ as the IRM estimate. For ease of notation we will sometimes use $T(\mathbf{a}, d)$ instead of $T(\mathbf{a}, d; IRM)$.

The next model is not discussed in details in this extended abstract, and our results about it will be presented in the full version of the paper.

The PMM (Partial Markov Models) r_i of not moving to another state, and in the event of a move, the next state is j with probability q_j, independent of the current requested state. Consequently, the transition probabilities of moving from i to j are $p_{i,j} = (1 - r_i)q_j$ for $i \neq j$ and $p_{i,i} = r_i + (1 - r_i)q_i$. Here $0 \leq r_i, q_i \leq 1$. We call the vector $\mathbf{r} = (r_i)$ the *locality* vector of the model. Given a locality vector $\mathbf{r}$ and an observed activity vector $\mathbf{a}$ for some time interval I there exists a unique partial Markov model P which is compatible with $\mathbf{r}$ and $\mathbf{a}$. By compatibility we mean that $\mathbf{r}$ is the locality vector of P and $\mathbf{a}/a$ is the stationary distribution of P which expresses the expected reference probabilities in the model P. Fix the vector $\mathbf{r} = (r_i)$. We let $P^{\mathbf{r}}$ denote the partial Markov model which for each interval I uses the model P compatible with $\mathbf{r}$ and $\mathbf{a}_I$ to model the request stream during I (note that P^0 is simply the IRM). The $P^{\mathbf{r}}$ estimate is

$$T(\mathbf{a}, d; P^{\mathbf{r}}) = a\left(\sum_{i,j}(a_i/a)P^{\mathbf{r}}_{i,j}d(x_i, x_j)\right)$$

Partial Markov models are useful in capturing locality of reference phenomenon, [1,2], which means that a request to state i is likely to be followed by another request to state i within a short time span. Many applications naturally exhibit this type of behavior. The larger the entries of the locality vector $\mathbf{r}$, the more likely states are to repeat in succession. In the partial Markov model the number of repetitive successions is distributed geometrically.

The Worst Case (Supremum) Model. In the worst case model W we assume that the sequence of states during time interval I was the sequence which maximizes the total set-up time among all sequences which are consistent with the vector $\mathbf{a}$. The measure is thus a δ measure on the worst case sequence. Consequently,

$$T(\mathbf{a}, d; W) = \max \sum_{i=1}^{a} d(x_i, x_{i+1})$$

where the maximum is over all sequences of states in X, of length a that agree with the frequency vector $\mathbf{a}$ and $x_1 = x_{a+1}$. We refer to $T(\mathbf{a}, d; W)$ as the worst case estimate.

3.2   Properties of Models

We introduce notions which will allow us to examine the behavior of model based estimates with regards to changes in the input data and to compare estimates for different models.

Super Additivity. Let I be a time interval and let $I_1, ..., I_k$ be a subdivision of I into subintervals. Accordingly, we have $\mathbf{a}_I = \sum j = 1^k \mathbf{a}_{I_j}$. A model M is said to be *super additive* with respect to a set-up time function d if the inequality

$$T(\mathbf{a}_I, d; M) \geq \sum_{j=1}^{k} T(\mathbf{a}_{I_j}, d; M) \tag{1}$$

always holds. Super additivity may be interpreted as stating that the addition of input information, namely, $\mathbf{a}_{I_j}$ instead of $\mathbf{a}_I$, never increases the estimate.

Monotonicity. We say that a vector $\mathbf{a} = (a_i)$ dominates a vector $\mathbf{b} = (b_i)$ if for all i, $a_i \geq b_i$. We use the notation $\mathbf{a} \geq \mathbf{b}$ to denote dominance. A model M

is said to be *monotone* with respect to d if for any pair of time intervals $I \subset J$ we have $T(a_I, d; M) \leq T(a_J, d; M)$, or stated otherwise, for any pair of vectors $\mathbf{a}, \mathbf{b}$ with nonnegative entries and such that $\mathbf{a} \geq \mathbf{b}$ we have

$$T(\mathbf{a}, d; M) \geq T(\mathbf{b}, d; M) \tag{2}$$

Approximation. Let $0 < \alpha < 1$. Given a set up function d, a model M_1 is said to be provide an α *approximation* to a model M_2 (and vice versa) if for any activity vector $\mathbf{a}$ we have

$$\alpha \leq \frac{T(\mathbf{a}, d; M_1)}{T(\mathbf{a}, d; M_2)} \leq \frac{1}{\alpha} \tag{3}$$

We say that a model M is *conservative* if it α approximates the worst case model W for some $\alpha > 0$.

4 Metric Space Criteria for Properties of Models

In this section we establish criteria for monotonicity and super additivity of the IRM estimates in terms of metric properties of the set-up time function d. We also establish a criterion for the IRM estimate to be a $1/2$ approximation to the worst case estimate.

Theorem 1. *(A criterion for Super additivity) The IRM estimate is super additive with respect to d if and only if d is negative type.*

Proof. It is enough to establish super additivity for a subdivision of I into two subintervals, that is to show that for all nonnegative vectors $\mathbf{a} = (a_i), \mathbf{b} = (b_i)$,

$$T(\mathbf{a} + \mathbf{b}, d) \geq T(\mathbf{a}, d) + T(\mathbf{b}, d) \tag{4}$$

Let $a = \sum_i a_i$ and $b = \sum_i b_i$. Then

$$
\begin{aligned}
&T(\mathbf{a} + \mathbf{b}, d) - T(\mathbf{a}, d) - T(\mathbf{b}, d) \\
&= \sum_{i \neq j} \frac{(a_i + b_i)(a_j + b_j)d(x_i, x_j)}{a + b} - \sum_{i \neq j} \frac{a_i a_j d(x_i, x_j)}{a} - \sum_{i \neq j} \frac{b_i b_j d(x_i, x_j)}{b} \\
&= \frac{1}{ab(a + b)} \sum_{i \neq j} d(x_i, x_j)(a_i b_j ab + a_j b_i ab - a_i a_j b^2 - b_i b_j a^2) \\
&= \frac{1}{ab(a + b)} \sum_{i \neq j} d(x_i, x_j)(a_i b - b_i a)(b_j a - a_j b) \\
&= -\frac{ab}{a + b} \sum_{i \neq j} d(x_i, x_j) \left(\frac{a_i}{a} - \frac{b_i}{b} \right) \left(\frac{a_j}{a} - \frac{b_j}{b} \right).
\end{aligned}
$$

Setting $v_i = \dfrac{a_i}{a} - \dfrac{b_i}{b}$, we get

$$T(\mathbf{a}+\mathbf{b},d) - T(\mathbf{a},d) - T(\mathbf{b},d) = -\frac{ab}{a+b}\sum_{i\neq j} v_i v_j d(x_i,x_j). \tag{5}$$

We note that $\sum_i v_i = 0$, hence by Schoenberg's criterion the IRM estimate is super additive if d is negative type. Conversely if the IRM estimate is super additive then

$$\sum_{i,j} v_i v_j d(x_i,x_j) \leq 0$$

for all $\mathbf{v}$ of the form $\mathbf{a}/a - \mathbf{b}/b$ where $\mathbf{a}, \mathbf{b}$ are vectors with integer non negative entries. After scaling we may deduce that the property holds whenever $\mathbf{a}, \mathbf{b}$ have rational non negative entries and by density of the rationals for all $\mathbf{a}, \mathbf{b}$ with non negative entries. Every vector $\mathbf{v} = (v_1,\ldots,v_h)$ such that $\sum_i v_i = 0$ has a multiple of the form $\frac{1}{a}\mathbf{a} - \frac{1}{b}\mathbf{b}$, where $\mathbf{a}, \mathbf{b}$ have non negative entries. Indeed if $a_i = \max\{v_i,0\}$ and $b_i = \max\{-v_i,0\}$, then $a = b$ and $\frac{1}{a}\mathbf{a} - \frac{1}{b}\mathbf{b} = \frac{1}{a}\mathbf{v}$, hence Schoenberg's criterion holds and d is negative type.

Theorem 2. *(criteria for monotonicity) The IRM estimate is monotone with respect to d if and only for every choice of k, the matrix $B(k,d)_{i,j} = d(x_i,x_k) + d(x_k,x_j) - d(x_i,x_j)$ defines a nonnegative quadratic form when restricted to vectors with nonnegative entries. In particular, if d is a pseudo metric or negative type then the IRM estimate is monotone with respect to d.*

Proof. We check the sign of the partial derivatives of $T(\mathbf{a},d)$ with respect to a_k (where $k \in \{1,\ldots,n\}$ is an arbitrary element).

$$\frac{\partial}{\partial a_k}T(\mathbf{a},d)$$

$$= \frac{a(\sum_i a_i d(x_i,x_k) + \sum_j a_j d(x_j,x_k)) - \sum_{i,j} a_i a_j d(x_i,x_j)}{a^2}$$

$$= \frac{1}{a^2}\sum_{i,j} a_i a_j (d(x_i,x_k) + d(x_j,x_k) - d(x_i,x_j)) = \frac{1}{a^2}\mathbf{a}B\mathbf{a}^t$$

where $B = B(k,d)$ is the matrix with ij entry $d(x_i,x_k) + d(x_j,x_k) - d(x_i,x_j)$. Assume that for all k, $B(k,d)$ is positive semi definite on vectors with nonnegative entries then $\frac{\partial}{\partial a_k}T(\mathbf{a},d) \geq 0$ for all k and all activity vectors $\mathbf{a}$. It follows from the Mean-value Theorem that if $\mathbf{a} \geq \mathbf{b}$ then $T(\mathbf{a},d) \geq T(\mathbf{b},d)$. Conversely if there are $\mathbf{a} \geq 0$ and k such that $\mathbf{a}B(k,d)\mathbf{a}^t < 0$ then taking $\mathbf{b}$ which is identical to $\mathbf{a}$ except that b_k is slightly smaller than a_k we get $T(\mathbf{a},d) < T(\mathbf{b},d)$, which proves the first statement of part 3.

If d is a semi-metric then B has nonnegative entries and so $\mathbf{a}B(k,d)\mathbf{a}^t \geq 0$ and if d is negative type then by Cayley's criterion $\mathbf{a}B(k,d)\mathbf{a}^t \geq 0$ which completes the proof.

Theorem 3. *(Comparison of the IRM estimate and worst case estimate) If d satisfies the triangle inequality then for all activity vectors* **a** *we have*

$$2T(\mathbf{a}, d; IRM) \geq T(\mathbf{a}, d; W) \tag{6}$$

where W is the worst case model.

Proof. Assume first that the activity vector is the vector $(1, 1, \ldots, 1)$. The IRM estimate here is $\frac{1}{n} \sum_{i,j} d(x_i, x_j)$, while the worst case estimate is the length of the longest Hamiltonian cycle in the complete graph on X with edge weights given by d. Assume without loss of generality that the longest Hamiltonian path in X is $1, 2, \ldots, n$. Since d satisfies the triangle inequality we have for $1 \leq i < n$ and for $j \in X$ $d(x_i, x_{i+1}) \leq d(x_i, x_j) + d(x_j, x_{i+1})$ (the $n + 1$ point coincides with the first point). Summing over all i, j we get

$$n \sum_{i=1}^{n} d(x_i, x_{i+1}) \leq 2 \sum_{i,j} d(x_i, x_j).$$

Therefore $2T(\mathbf{a}, d; IRM) \geq T(\mathbf{a}, d; W)$. To complete the proof we need to consider a general activity vector $(a_1, \ldots, a_n)$. Let X' be the metric space with a points that is composed of groups of a_j points of type j. Given d on X we induce a metric on X' by letting the distance between a point of type i and a point of type j be $d(x_i, x_j)$. Clearly X' also satisfies the triangle inequality. We have thus reduced the problem to the case of the activity vector $(1, 1, \ldots, 1)$ and are done.

5   Set-Up Time Functions of a Disk

In this section we show that the radial seek time function of a disk drive, which is the standard set-up function in storage system research is an ℓ_1-metric and in particular is negative type. From this we conclude that the IRM estimates are super additive when applied to disk seek times. Data on disk drives resides on *tracks* which form concentric circles of varying radii r around the center of a platter. To get from a track at radius r_1 to another track at radius r_2 the head of the device performs a radial motion. The time it takes the disk head to perform this radial motion is known as (radial) seek time. Seek time is translation invariant Furthermore, the acceleration and deceleration of the head dictate that the seek time from r_1 to r_2 has the form

$$d_F(r_1, r_2) = F(|r_1 - r_2|)$$

where F is a concave non decreasing function.

If we let X be the set of data locations on the disk then a theorem of Kelly proved in [4] can be interpreted as stating that (X, d_F) is negative type. We prove a stronger result of independent interest using a much simpler proof.

Theorem 4. *Let F be a concave nondecreasing function with $F(0) = 0$ and let $X \subset \mathbf{R}$. Then (X, d_F) is an ℓ_1 metric space.*

Proof. Let $X = \{x_1, \ldots, x_n\}$. Consider

$$Y = \{|x_i - x_j| : 1 \le i, j \le n\}$$

the set of possible distances in X, and order the elements of Y as $0 = y_0 < y_1 < y_2 < \ldots < y_m$. let G be the piecewise linear function which

(i) coincides with F on Y
(ii) is linear on all intervals $[y_i, y_{i+1}]$ and
(iii) is constant on $[y_m, \infty)$ (that is, gets the value $F(y_m)$ there).

Obviously $(X, d_F) = (X, d_G)$ since $F = G$ on the set of all relevant values Y, so it is enough to prove the claim for G, which is also non decreasing and concave. We now define functions $H_{s,t}$ as follows.

$$H_{s,t}(x) = sx \text{ if } x < t \text{ and } st \text{ otherwise.}$$

We also let $s_i = \frac{G(y_i) - G(y_{i-1})}{y_i - y_{i-1}}$ be the sequence of slopes of G. We now claim that G is a convex combination of functions of the form $H_{s,t}$.

The proof proceeds by induction on m. If $m = 0$ then $G = H_{1,0} = 0$. For $m > 0$, look at the function $\tilde{G} = G - H_{s_m, y_m}$. It is not hard to see that $\tilde{G}(0) = 0$, $\tilde{G}$ is constant beyond y_{m-1} and is piecewise linear with breakpoints $y_1, \ldots, y_{m-1}$. A piecewise linear function is concave and nondecreasing if and only if its slopes are decreasing and nonnegative, and so $s_1 \ge s_2 \ge \ldots \ge s_m \ge 0$, and similarly $s_1 - s_m \ge s_2 - s_m \ge \ldots \ge s_{m-1} - s_m \ge 0$. But, these are the slopes of G' and it is therefore concave and nondecreasing. We may now apply the induction hypothesis to $\tilde{G}$ and this proves the claim.

Since a sum of ℓ_1-metrics is also an ℓ_1-metric, we are left with the task of showing that for a function $F = H_{s,y}$, the resulting metric d_F is an ℓ_1-metric. Notice that $d_F(i, j) = s \cdot \min\{|i - j|, y\}$. Let $f_i = \frac{1}{2s}\chi_{[x_i, x_i + y]}$ be the function whose value is $\frac{1}{2s}$ on the interval $[x_i, x_i + y]$ and zero otherwise. It is easy to see that for any $i, j \in \mathbf{R}$

$$d_F(i, j) = s \cdot \min\{|i - j|, y\} = \int_{\mathbf{R}} |f_i(x) - f_j(x)| dx$$

This shows that d_F is an L_1 metric and hence l_1.

Combining theorem 1 with theorem 4 we get

Theorem 5. *The IRM estimate is super additive with respect to the seek time function d_F for any physical disk drive.*

6   Conclusions and Future Work

We have introduced several natural properties of set-up time estimates and studied them for the IRM. We have shown that the IRM estimate satisfies monotonicity which is a "sanity check" for set-up time estimates, and further that the IRM is an easily computable approximation to the worst case estimate. In the specific but important context of seek functions in disk drives we showed that the IRM shares another formal property that holds for worst case estimates namely super additivity. It would be interesting to explore monotonicity, super additivity and various approximation relations among other models. One interesting class of examples are the renewal models which were suggested by Opderbeck and Chu in [6]. The IRM is a special case of such models where the renewal model is based on exponential inter-arrival times. It would be interesting to investigate other cases such as hyperexponential, gamma or Pareto bounded heavy tail distributions. Such an investigation will likely require refined definitions for properties such as monotonicity and super additivity since the associated models are not Markovian.

Acknowledgments. We would like to thank Timothy Chow for helpful discussions regarding a preliminary version of this paper.

References

1. Aven O.I., Coffman E.G. and Kogan Y.A. *Stochastic analysis of computer storage*, D.Reidel publishing, 1987.
2. Coffman E.G. and Denning P.J. *Operating systems theory*, Prentice Hall, 1973.
3. Grossman D.D. and Silverman H.F. Placement of records on a secondary storage device to minimize access time *J. of the ACM 20*, 429-438, 1973.
4. Kelly J.B. Hypermetric spaces and metric transforms, in *Inequalities III*, edited by O.Shisha, 149-158, 1972.
5. Lovasz L., Pyber L., Welsh D.J.A. and Ziegler G.M. Combinatorics in pure mathematics, in *Handbook of combinatorics* , *MIT press* ,*North Holland* 2039-2082, 1995.
6. Opderbeck H. and Chu W.W., The renewal model for program behavior, *SIAM J. of computing 4*, 356–374, 1975.
7. Vanichpun S. and Makowski A., The output of a cache under the independent reference model - where did all the locality of reference go?, *Proceedings of SIGMETRICS*, 2004.
8. Wong C. K. *Algorithmic Studies in mass storage systems*, computer science press, 1983.
9. Yue P.C. and Wong C.K., On the optimality of the probability ranking scheme in storage applications, *J. of the ACM 20*, 624-633, 1973.

Geometric Representation of Graphs in Low Dimension

L. Sunil Chandran[1] and Naveen Sivadasan[2]

[1] Indian Institute of Science, Dept. of Computer Science and Automation,
Bangalore 560012, India
sunil@csa.iisc.ernet.in
[2] Strand Life Sciences, 237, Sir. C.V. Raman Avenue, Rajmahal Vilas,
Bangalore 560080, India
naveen@strandls.com

Abstract. An axis-parallel k–dimensional box is a Cartesian product $R_1 \times R_2 \times \cdots \times R_k$ where R_i (for $1 \leq i \leq k$) is a closed interval of the form $[a_i, b_i]$ on the real line. For a graph G, its *boxicity* box(G) is the minimum dimension k, such that G is representable as the intersection graph of (axis–parallel) boxes in k–dimensional space. The concept of boxicity finds applications in various areas such as ecology, operation research etc.

A number of NP-hard problems are either polynomial time solvable or have much better approximation ratio on low boxicity graphs. For example, the max-clique problem is polynomial time solvable on bounded boxicity graphs and the maximum independent set problem has $\log n$ approximation ratio for boxicity 2 graphs. In most cases, the first step usually is computing a low dimensional box representation of the given graph. Deciding whether the boxicity of a graph is at most 2 itself is NP-hard.

We give an efficient randomized algorithm to construct a box representation of any graph G on n vertices in $1.5(\Delta + 2) \ln n$ dimensions, where Δ is the maximum degree of G. We also show that box$(G) \leq (\Delta + 2) \ln n$ for any graph G. Our bound is tight up to a factor of $\ln n$. The only previously known general upper bound for boxicity was given by Roberts, namely box$(G) \leq n/2$. Our result gives an exponentially better upper bound for bounded degree graphs.

We also show that our randomized algorithm can be derandomized to get a polynomial time deterministic algorithm.

Though our general upper bound is in terms of maximum degree Δ, we show that for almost all graphs on n vertices, its boxicity is upper bound by $c \cdot (d_{av} + 1) \ln n$ where d_{av} is the average degree and c is a small constant. Also, we show that for any graph G, box$(G) \leq \sqrt{8 n d_{av} \ln n}$, which is tight up to a factor of $b\sqrt{\ln n}$ for a constant b.

1 Introduction

Let $\mathcal{F} = \{S_x \subseteq U : x \in V\}$ be a family of subsets of a universe U, where V is an index set. The intersection graph $\Lambda(\mathcal{F})$ of $\mathcal{F}$ has V as vertex set, and two distinct

D.Z. Chen and D.T. Lee (Eds.): COCOON 2006, LNCS 4112, pp. 398–407, 2006.

vertices x and y are adjacent if and only if $S_x \cap S_y \neq \emptyset$. Representations of graphs as the intersection graphs of various geometrical objects is a well studied topic in graph theory. Probably the most well studied class of intersection graphs are the *interval graphs*, where each S_x is a closed interval on the real line.

A well known concept in this area of graph theory is the *boxicity*, which was introduced by F. S. Roberts in 1969 [17]. This concept generalizes the concept of interval graphs. A k–dimensional box is a Cartesian product $R_1 \times R_2 \times \cdots \times R_k$ where R_i (for $1 \leq i \leq k$) is a closed interval of the form $[a_i, b_i]$ on the real line. For a graph G, its *boxicity* is the minimum dimension k, such that G is representable as the intersection graph of (axis–parallel) boxes in k–dimensional space. We denote the boxicity of a graph G by box(G). The graphs of boxicity 1 are exactly the class of interval graphs. This concept finds applications in niche overlap in ecology and to problems of fleet maintenance in operations research. (See [12].)

In many algorithmic problems related to graphs, the availability of certain convenient representations turn out to be extremely useful. Probably, the most well-known and important examples are the tree decompositions and path decompositions [5]. Many NP-hard problems are known to be polynomial time solvable given a tree(path) decomposition of the input graph that has bounded width. Similarly, the representation of graphs as intersections of "disks" or "spheres" lies at the core of solving problems related to frequency assignments in radio networks, computing molecular conformations etc. For the maximum independent set problem which is hard to approximate within a factor of $n^{(1/2)-\epsilon}$ for general graphs, a PTAS is known for disk graphs given the disk representation [13,1] and an FPTAS is known for unit disk graphs [22]. In a similar way, the availability of box representation in low dimension make some well known NP hard problems like the max-clique problem, polynomial time solvable since there are only $O((2n)^k)$ maximal cliques in boxicity k graphs. Though the complexity of finding the maximum independent set is hard to approximate within a factor $n^{(1/2)-\epsilon}$ for general graphs, it is approximable to a $\log n$ factor for boxicity 2 graphs (the problem is NP-hard even for boxicity 2 graphs) given a box representation [2,4].

It was shown by Cozzens [11] that computing the boxicity of a graph is NP–hard. This was later improved by Yannakakis [23], and finally by Kratochvil [16] who showed that deciding whether the boxicity of a graph is at most 2 itself is NP–complete. Therefore it is interesting to design efficient algorithms to represent small boxicity graphs in low dimensions. To the best of our knowledge, the only known strategy till date for computing a box representation for general graphs is by Roberts [17], but it guarantees only a box representation in $n/2$ dimensions for any graph G on n vertices and m edges. In this paper, we give a randomized algorithm that guarantees an exponentially better bound ($O(\ln n)$ instead of $n/2$) for the dimension in case of bounded degree graphs. To be precise, our approach yields a box representation for any graph G on n vertices and maximum degree Δ in $1.5(\Delta + 2) \ln n$ dimensions in $O(\Delta m \ln n)$ time with high probability. We also derandomize our algorithm to obtain a deterministic polynomial time algorithm to do the same.

In a recent manuscript [8] the authors showed that for any graph G, $box(G) \leq tw(G) + 2$, where $tw(G)$ is the treewidth of G. This result implies that the class of 'low boxicity' graphs properly contains the class of 'low treewidth graphs'. It is well known that almost all graphs on n vertices and $m = c \cdot n$ edges (for a sufficiently large constant c) have $\Omega(n)$ treewidth [15]. In this paper we show that almost all graphs on n vertices and m edges have boxicity at most $c' \frac{m}{n} \ln n$ for a small constant c'. An implication of this result is that for almost all graphs, there is an exponential gap between its boxicity and treewidth. Hence it is interesting to take a relook at those NP-hard problems that are polynomial time solvable in bounded treewidth graphs and see whether they are also polynomial time solvable for bounded boxicity graphs.

Researchers have also tried to bound the boxicity of graph classes with special structure. Scheinerman [18] showed that the boxicity of outer planar graphs is at most 2. Thomassen [20] proved that the boxicity of planar graphs is bounded above by 3. Upper bounds for the boxicity of many other graph classes such as chordal graphs, AT-free graphs, permutation graphs etc. were shown in [8] by relating the boxicity of a graph with its treewidth. Researchers have also tried to generalize or extend the concept of boxicity in various ways. The poset boxicity [21], the rectangle number [9], grid dimension [3], circular dimension [14,19] and the boxicity of digraphs [10] are some examples.

1.1   Our Results

We summarize below the results of this paper.

1. We show that for any graph G on n vertices, $box(G) \leq (\Delta + 2) \ln n$. This bound is tight up to a factor of $\ln n$.
2. In fact, we show a randomized algorithm to construct a box representation of G in $1.5(\Delta + 2) \ln n$ dimensions, that runs in $O(\Delta m \ln n)$ time with high probability, where m is the number of edges in G.
3. Next we show a polynomial time deterministic algorithm to construct a box representation in $(\Delta + 2) \ln n$ dimensions by derandomizing the above randomized algorithm.
4. Though the general upper bound that we show is in terms of the maximum degree Δ, we also investigate the relation between boxicity and average degree. We show that for almost all graphs on n vertices and m edges, the boxicity is $O((d_{av} + 1) \ln n)$, where d_{av} is the average degree.
5. We also derive a upper bound for boxicity of general graphs in terms of average degree. We show that for any graph G, $box(G) \leq \sqrt{8 n d_{av} \ln n}$, which is tight up to a factor of $b\sqrt{\ln n}$ for a constant b.

We refer the reader to the complete version [7] for the missing proofs.

1.2   Definitions and Notations

Let G be a undirected simple graph on n vertices. The vertex set of G is denoted as $V(G) = \{1, \cdots, n\}$ (or V in short). Let $E(G)$ denote the edge set of G.

We denote by $\overline{G}$, the complement of G. We say the edge e is missing in G, if $e \in E(\overline{G})$. A graph G' is said to be a super graph of G where $V(G) = V(G')$, if $E(G) \subseteq E(G')$. For a vertex $u \in V$, let $N(u)$ denote the set of neighbors of u in G and let $d(u)$ denote the degree of u in G, i.e. $d(u) = |N(u)|$. Let Δ denote the maximum degree of G.

Definition 1 (Projection). *Let π be a permutation of the set $\{1, \cdots, n\}$. Let $X \subseteq \{1, \cdots, n\}$. The projection of π onto X denoted as π_X is defined as follows. Let $X = \{u_1, \ldots, u_r\}$ such that $\pi(u_1) < \pi(u_2) < \ldots < \pi(u_r)$. Then $\pi_X(u_1) = 1, \pi_X(u_2) = 2, \cdots, \pi_X(u_r) = r$.*

Definition 2 (Interval Representation). *An interval graph can be represented as the intersection graph of closed intervals on real line. To define an interval representation of an interval graph G, we define the two functions $l : V \to R$ and $r : V \to R$. The interval corresponding to a vertex v denoted as $I(v)$ is given by $[l(v), r(v)]$, where $l(v)$ and $r(v)$ are the left and right end points of the interval corresponding to v.*

Definition 3. *We define a map $\mathcal{M}(G, \pi)$ which associates a permutation π of the vertices $\{1, 2, \cdots, n\}$ to an interval super graph G' of G, as follows: Consider any vertex $u \in V(G)$. Let $n_u \in N(u) \cup \{u\}$ be the vertex such that $\pi(n_u) = \min_{w \in N(u) \cup \{u\}} \pi(w)$. Then associate the interval $[\pi(n_u), \pi(u)]$ to the vertex u, and let G' be the resulting interval graph. It is easy to verify that G' is a super graph of G. We define $\mathcal{M}(G, \pi) = G'$.*

1.3   Box Representation and Interval Graph Representation

Let $G = (V, E(G))$ be a graph and let $I_1, \ldots, I_k$ be k interval graphs such that each $I_j = (V, E(I_j))$ is defined on the same set of vertices V. If

$$E(G) = E(I_1) \cap \cdots \cap E(I_k),$$

then we say that $I_1, \ldots, I_k$ is an *interval graph representation* of G. The following equivalence is well-known.

Theorem 1 (Roberts [17]). *The minimum k such that there exists an interval graph representation of G using k interval graphs $I_1, \ldots, I_k$ is the same as* box(G).

Recall that a k–dimensional box representation of G is a mapping of each vertex $u \in V$ to $R_1(u) \times \cdots \times R_k(u)$, where each $R_i(u)$ is a closed interval of the form $[\ell_i(u), r_i(u)]$ on the real line. It is straightforward to see that an interval graph representation of G using k interval graphs $I_1, \ldots, I_k$, is equivalent to a k–dimensional box representation in the following sense. Let $R_i(u) = [\ell_i(u), r_i(u)]$ denote the closed interval corresponding to vertex u in an interval realization of I_i. Then the k–dimensional box corresponding to u is simply $R_1(u) \times \cdots \times R_k(u)$. Conversely, given a k–dimensional box representation of G, the set of intervals $\{R_i(u) : u \in V\}$ forms the ith interval graph I_i in the corresponding interval graph representation.

When we say that a box representation in t dimensions is output by an algorithm, the algorithm actually outputs the interval graph representation: that is, the interval representation of the constituent interval graphs.

2   The Randomized Construction

Consider the following randomized procedure **RAND** which outputs an interval super graph of G. Let Δ be the maximum degree of G.

RAND
    Input: G.
    Output: G' which is an interval super graph of G.
begin
    step1. Generate a permutation π of $\{1, \ldots, n\}$ uniformly at random.
    step2. Return $G' = \mathcal{M}(G, \pi)$.
end.

Lemma 1. *Let* $e = (u, v) \in E(\overline{G})$. *Let* G' *be the output of* **RAND**(G). *Then,*

$$\mathbf{Pr}\left[e \notin E(\overline{G'})\right] = \frac{1}{2}\left(\frac{d(u)}{d(u)+2} + \frac{d(v)}{d(v)+2}\right) \leq \frac{\Delta}{\Delta+2}.$$

Proof. We have to estimate the probability that u and v are adjacent in G'. That is, $I(u) \cap I(v) \neq \emptyset$.

Let $n_u \in N(u)$ be a vertex such that it minimizes $\min_{w \in N(u)} \pi(w)$. Similarly, let $n_v \in N(v)$ be a vertex such that it minimizes $\min_{w \in N(v)} \pi(w)$.

It is easy to see that $I(u) \cap I(v) \neq \emptyset$ if (a) $\pi(n_u) < \pi(v) < \pi(u)$. This is because, if the above condition holds, then, recalling the definition of $\mathcal{M}(G, \pi)$, it follows that $l(u) < r(v) < r(u)$, which implies that $r(v) \in I(u) \cap I(v)$. Similarly, if (b) $\pi(n_v) < \pi(u) < \pi(v)$ then also $I(u) \cap I(v) \neq \emptyset$. On the other hand, it is easy to see that $I(u) \cap I(v) \neq \emptyset$ *only if* either (a) or (b) hold. Again, the above two events ((a) and (b)) are mutually exclusive. Hence

$$\mathbf{Pr}\left[e \notin E(\overline{G'})\right] = \mathbf{Pr}[\pi(n_u) < \pi(v) < \pi(u)] + \mathbf{Pr}[\pi(n_v) < \pi(u) < \pi(v)].$$

We bound $\mathbf{Pr}[\pi(n_u) < \pi(v) < \pi(u)]$ as follows. Let $X = \{u\} \cup N(u) \cup \{v\}$. Let π_X be the projection of π onto X. Clearly, the event $\pi(n_u) < \pi(v) < \pi(u)$ translates to saying that $\pi_X(v) < \pi_X(u)$ and $\pi_X(v) \neq 1$. Note that π_X can be any permutation of $|X|$ elements with equal probability, which is $\frac{1}{(d(u)+2)!}$. The number of permutations where $\pi_X(v) < \pi_X(u)$ equals $(d(u)+2)!/2$. Now the number of permutations where $\pi_X(v) = 1$ equals $(d(u)+1)!$. Note that the set of permutations with $\pi_X(v) = 1$ is a subset of the set of permutations with $\pi_X(v) < \pi_X(u)$. It follows that

$$\mathbf{Pr}[\pi_X(v) < \pi_X(u) \text{ and } \pi_X(v) \neq 1] = \frac{(d(u)+2)!/2 - (d(u)+1)!}{(d(u)+2)!}$$

which is $\frac{d(u)}{2(d(u)+2)}$. Using similar arguments, it follows that $\mathbf{Pr}[\pi(n_v) < \pi(u) < \pi(v)] = \frac{d(v)}{2(d(v)+2)}$. Combing the two bounds, the result follows. ∎

Lemma 2. *Let $I_1, I_2, \cdots, I_t$ be the output generated by t invocations of* **RAND**(*G*)*. If $t \geq \frac{3}{2}(\Delta + 2)\ln n$ then $E(G) = E(I_1) \cap E(I_2) \cap \cdots \cap E(I_t)$ with high probability .*

As mentioned in the proof of Lemma 2, if we fix $t = (\Delta + 2)\ln n$, the resulting intersection graph is G with probability at least $1/2$. Hence we have the following Corollary.

Corollary 1. *Let G be a graph on n vertices and with maximum degree Δ. Then* box$(G) \leq (\Delta + 2)\ln n$.

The following Lemma is straightforward.

Lemma 3. *The* **RAND** *procedure can be implemented in $O(m + n)$ time assuming that a permutation of $\{1, \ldots, n\}$ can be generated uniformly at random in $O(n)$ time.*

The following theorem is a direct consequence of Lemma 2 and Lemma 3.

Theorem 2. *Given a graph G on n vertices and m edges, with high probability, a box representation of G in $(\Delta + 2)\ln n$ dimensions can be constructed in $O(\Delta m \ln n)$ time, where Δ is the maximum degree of G.*

Tight example: We remark that for any given Δ and $n > \Delta + 1$, we can construct a graph G on n vertices and with maximum degree Δ such that box$(G) \geq \lfloor (\Delta + 2)/2 \rfloor$. We assume that Δ is even for the ease of explanation. Roberts [17] has shown that for any even number k, there exists a graph on k vertices with degree $k - 2$ and boxicity $k/2$. We call such graphs as *Roberts graph*. The Roberts graph on n vertices is obtained by removing the edges of a perfect matching from a complete graph on n vertices. We take such a graph by fixing $k = \Delta + 2$ and we let the remaining $n - (\Delta + 2)$ vertices to be isolated vertices. Clearly, the boxicity of such a graph is also $k/2 = (\Delta + 2)/2$, where as the maximum degree is Δ. Thus our upper bound is tight up to a factor of $2\ln n$.

3   Derandomization

In this section we derandomize the above randomized algorithm to obtain a deterministic polynomial time algorithm to output the box representation in $(\Delta + 2)\ln n$ dimensional space for a given graph G on n vertices with maximum degree Δ.

Lemma 4. *Let $G = (V, E)$ be the graph. Let $E(\overline{G})$ be the edge set of the complement of G. Let $H \subseteq E(\overline{G})$. Then we can construct an interval super graph G' of G in polynomial time such that $|E(\overline{G'}) \cap H| \geq \frac{2}{\Delta + 2}|H|$.*

404   L.S. Chandran and N. Sivadasan

Theorem 3. *Let G be a graph on n vertices with maximum degree Δ. The box representation of G in $(\Delta+2)\ln n$ dimensions can be constructed in polynomial time,*

Proof. Let $h = |E(\overline{G})|$. It follows from Lemma 4 that we can construct t interval graphs such that the number of edges of $E(\overline{G})$ which is not missing in any of these t interval graphs is at most $\left(\frac{\Delta}{\Delta+2}\right)^t h$. If $\left(\frac{\Delta}{\Delta+2}\right)^t h < 1$, then we are done. That is, we are done if $t\ln\left(\frac{\Delta}{\Delta+2}\right) + \ln h < 0$ is true. Clearly this is true, if $t > \frac{\ln h}{\ln\left(\frac{\Delta+2}{\Delta}\right)}$. Using the fact that $\ln\frac{\Delta+2}{\Delta} \geq \frac{2}{\Delta} - \frac{1}{2}(\frac{2}{\Delta})^2$, we obtain box$(G) \leq \frac{\Delta^2}{2(\Delta-1)}\ln h \leq (\Delta+2)\ln n$. By Lemma 4, each interval graph is constructed in polynomial time. Hence the total running time is still polynomial. Thus the theorem follows.   ■

Proof (Lemma 4). We derandomize the **RAND** algorithm to devise a deterministic algorithm to construct G'.

Our deterministic strategy defines a permutation π on the vertices $\{1,\cdots,n\}$ of G. The desired G' is then obtained as $\mathcal{M}(G,\pi)$. Let the ordered set $V_n =<v_1,\cdots,v_n>$ denote the final permutation given by π. We construct V_n in a step by step fashion. At the end of step i, we have already defined the first i elements of the permutation, namely the ordered set $V_i =<v_1,\cdots,v_i>$, where each v_j is distinct. Let V_0 denote the empty set. Having obtained V_i for $i \geq 0$, we compute V_{i+1} in the next step as follows.

Given an ordered set V_i of i vertices $<v_1,v_2,\cdots,v_i>$, let $V_i \diamond u$ denote the ordered set of the $i+1$ vertices $<v_1,v_2,\cdots,v_i,u>$. (We will abuse notation and use V_i to denote the underlying unordered set also, when there is no chance of confusion.) Let $V_0 \diamond u$ denote $<u>$.

Consider the **RAND** algorithm whose output is denoted as G''. For each $e \in H$, let x_e denote the indicator random variable which is 1 if $e \in E(\overline{G''})$, and 0 otherwise. Let $X_H = \sum_{e\in H} x_e$.

Let $\mathcal{Z}(V_i)$ for $i \geq 0$ denote the event that the first i elements of the random permutation generated by **RAND** is given by the ordered set $V_i =<v_1,\cdots,v_i>$. Note that $\mathbf{Pr}[\mathcal{Z}(V_0)] = 1$ since the first 0 elements of any permutation is the empty set V_0.

Let $x_e|\mathcal{Z}(V_i)$ denote the indicator random variable corresponding to x_e conditioned on the event $\mathcal{Z}(V_i)$.

Similarly, let the random variable $X_H|\mathcal{Z}(V_i)$ denote the number of missing edges in G'' conditioned on the event $\mathcal{Z}(V_i)$.

For $i \geq 0$, Let $f_e(V_i)$ denote $\mathbf{Pr}[x_e = 1 \mid \mathcal{Z}(V_i)]$ and let $F(V_i)$ denote $\mathbf{E}[X_H \mid \mathcal{Z}(V_i)]$

Note that $f_e(V_0)$ denote $\mathbf{Pr}[x_e = 1]$ and $F(V_0)$ denote $\mathbf{E}[X_H]$.

Clearly
$$F(V_i) = \sum_{e\in H} f_e(V_i).$$

By Lemma 1, we know that for any $e \in H$, $f_e(V_0) \geq \frac{2}{\Delta+2}$. Thus $F(V_0) \geq \frac{2|H|}{\Delta+2}$. Clearly,

$$\mathbf{E}[X_H|\mathcal{Z}(V_i)] = \frac{1}{|V - V_i|} \sum_{u \in V - V_i} \mathbf{E}[X_H|\mathcal{Z}(V_i \diamond u)].$$

Let $u \in V - V_i$ be such that

$$\mathbf{E}[X_H|\mathcal{Z}(V_i \diamond u)] = \max_{w \in V - V_i} \mathbf{E}[X_H|\mathcal{Z}(V_i \diamond w)].$$

Define $V_{i+1} = V_i \diamond u$. It follows that

$$F(V_{i+1}) = \mathbf{E}[X_H|\mathcal{Z}(V_{i+1})] \geq \mathbf{E}[X_H|\mathcal{Z}(V_i)] = F(V_i).$$

In particular, it is also true that $F(V_1) \geq F(V_0)$.

After n steps, we obtain the final permutation V_n. Applying the above inequality n times, it follows that

$$F(V_n) = \mathbf{E}[X_H|\mathcal{Z}(V_n)] \geq \mathbf{E}[Z_H] = F(V_0).$$

Recalling that $F(V_0) \geq \frac{2|H|}{\Delta+2}$, we have $F(V_n) \geq \frac{2|H|}{\Delta+2}$.

Let π be the permutation which maps $< 1, \cdots, n >$ to V_n. The final interval super graph G' output by our deterministic strategy is $\mathcal{M}(G, \pi)$. By definition, $F(V_n)$ is the total number of edges from H that are missing in G'. We have shown that $F(V_n) \geq \frac{2|H|}{\Delta+2}$ as claimed.

It remains to show that the above deterministic strategy takes only polynomial time. For that we need the following lemma.

Lemma 5. *For any ordered set $U_j =< u_1, \cdots, u_j >$ and any $e \in H$, $f_e(U_j)$ can be computed exactly in polynomial time.*

Given a vertex $w \in V - V_i$, $F(V_i \diamond w)$ is simply $\sum_{e \in H} f_e(V_i \diamond w)$. It follows from Lemma 5 that $F(V_i \diamond w)$ can be computed in polynomial time. Recall that given V_i, V_{i+1} is $V_i \diamond u$ where u maximizes $F(V_i \diamond w)$ among the vertices from $w \in V - V_i$. Clearly such a u can also be found in polynomial time. Since there are only n steps before computing V_n, the overall running time is still polynomial. ∎

4   In Terms of Average Degree

It is natural to ask whether our upper bound of $(\Delta + 2) \ln n$ still holds even if we replace Δ by the average degree d_{av}. Unfortunately this is not the case as illustrated by the following example. Consider the following graph $G = (V, E)$ on n vertices. We take a Roberts graph on n_1 vertices such that $n_1(n_1 - 2)/n = d_{av}$ and we let the remaining $n - n_1$ vertices to be isolated vertices. The average degree of this graph is clearly d_{av} (recall the definition of Roberts graph) and its boxicity is at least $n_1/2 \geq \frac{1}{2}\sqrt{nd_{av}}$. If we substitute Δ by d_{av} in our upper bound, we obtain that the boxicity of this graph is at most $(d_{av} + 2) \ln n$, which is far below the actual boxicity. Still, we can prove the following general upper bound in terms of the average degree.

Theorem 4. *For a graph $G = (V, E)$ on n vertices and average degree d_{av},* $\text{box}(G) \leq \sqrt{8nd_{av}\ln(n)}$. *Moreover, there exists a graph G with n vertices and average degree d_{av} such that* $\text{box}(G) \geq \frac{1}{2}\sqrt{nd_{av}}$.

Proof. We show the upper bound as follows. Let $x = \sqrt{\frac{nd_{av}}{2\ln(n)}}$. Let V' denote the set of vertices in G whose degree is greater than or equal to x. It is straightforward to verify that $|V'| \leq \frac{nd_{av}}{x}$. Let G'' be the induced sub graph of G induced on $V - V'$. That is, each vertex in G'' has degree at most x. By Theorem 2, we obtain that $\text{box}(G'') \leq 2x\ln(n)$. Since $\text{box}(G'') + |V'|$ is a trivial upper bound for $\text{box}(G)$, it follows that $\text{box}(G) \leq 2x\ln(n) + \frac{nd_{av}}{x} = 2\sqrt{2nd_{av}\ln(n)}$. The example graph discussed in the beginning of this section serves as the example that illustrate the lower bound. ∎

4.1 Boxicity of Random Graphs

Though in general boxicity of a graph is not upper bound by $(d_{av} + 2)\ln n$, where d_{av} is the average degree, we now show that for almost all graphs, the boxicity is at most $c(d_{av} + 1)\ln n$, for a small positive constant c. We show the following. Let G be a random graph drawn according to the $\mathcal{G}(n, m)$ model [6], where n is the number of vertices and m is the number of edges. Then $\mathbf{Pr}\left[\text{box}(G) \leq 8(\frac{2m}{n} + 1)\ln n\right] \geq 1 - \frac{2}{n^2}$. (Note that $d_{av} = 2m/n$). It follows immediately that for almost all graphs on n vertices and m edges, the boxicity is upper bound by $8(d_{av} + 1)\ln n$.

Theorem 5. *For a random graph G on n vertices and m edges drawn according to $\mathcal{G}(n, m)$ model,*

$$\mathbf{Pr}\left[\text{box}(G) \leq 8\left(\frac{2m}{n} + 1\right)\ln n\right] \geq 1 - \frac{2}{n^2}.$$

References

1. P. AFSHANI AND T. CHAN, *Approximation algorithms for maximum cliques in 3d unit-disk graphs*, in Proc. 17th Canadian Conference on Computational Geometry (CCCG), 2005, pp. 6–9.
2. P. K. AGARWAL, M. VAN KREVELD, AND S. SURI, *Label placement by maximum independent set in rectangles*, Comput. Geom. Theory Appl., 11 (1998), pp. 209–218.
3. S. BELLANTONI, I. B.-A. HARTMAN, T. PRZYTYCKA, AND S. WHITESIDES, *Grid intersection graphs and boxicity*, Discrete mathematics, 114 (1993), pp. 41–49.
4. P. BERMAN, B. DASGUPTA, S. MUTHUKRISHNAN, AND S. RAMASWAMI, *Efficient approximation algorithms for tiling and packing problems with rectangles*, J. Algorithms, 41 (2001), pp. 443–470.
5. H. L. BODLAENDER, *A tourist guide through treewidth*, Acta Cybernetica, 11 (1993), pp. 1–21.
6. B. BOLLOBÁS, *Random Graphs*, Cambridge University Press, 2 ed., 2001.

7. L. S. CHANDRAN AND N. SIVADASAN, *Geometric representation of graphs in low dimension.* http://arxiv.org/abs/cs.DM/0605013.

8. L. S. CHANDRAN AND N. SIVADASAN, *Treewidth and boxicity.* Submitted, Available at http://arxiv.org/abs/math.CO/0505544.

9. Y. W. CHANG AND D. B. WEST, *Rectangle number for hyper cubes and complete multipartite graphs*, in 29th SE conf. Comb., Graph Th. and Comp., Congr. Numer. 132(1998), 19–28.

10. ———, *Interval number and boxicity of digraphs*, in Proceedings of the 8th International Graph Theory Conf., 1998.

11. M. B. COZZENS, *Higher and multidimensional analogues of interval graphs.* Ph. D thesis, Rutgers University, New Brunswick, NJ, 1981.

12. M. B. COZZENS AND F. S. ROBERTS, *Computing the boxicity of a graph by covering its complement by cointerval graphs*, Discrete Applied Mathematics, 6 (1983), pp. 217–228.

13. T. ERLEBACH, K. JANSEN, AND E. SEIDEL, *Polynomial-time approximation schemes for geometric intersection graphs.* To appear in SIAM Journal of Computing.

14. R. B. FEINBERG, *The circular dimension of a graph*, Discrete mathematics, 25 (1979), pp. 27–31.

15. T. KLOKS, *Treewidth: Computations And Approximations*, vol. 842 of Lecture Notes In Computer Science, Springer Verlag, Berlin, 1994.

16. J. KRATOCHVIL, *A special planar satisfiability problem and a consequence of its NP–completeness*, Discrete Applied Mathematics, 52 (1994), pp. 233–252.

17. F. S. ROBERTS, *Recent Progresses in Combinatorics*, Academic Press, New York, 1969, ch. On the boxicity and Cubicity of a graph, pp. 301–310.

18. E. R. SCHEINERMAN, *Intersectin classes and multiple intersection parameters.* Ph. D thesis, Princeton University, 1984.

19. J. B. SHEARER, *A note on circular dimension*, Discrete mathematics, 29 (1980), pp. 103–103.

20. C. THOMASSEN, *Interval representations of planar graphs*, Journal of combinatorial theory, Ser B, 40 (1986), pp. 9–20.

21. W. T. TROTTER AND J. D. B. WEST, *Poset boxicity of graphs*, Discrete Mathematics, 64 (1987), pp. 105–107.

22. E. J. VAN LEEUWEN, *Approximation algorithms for unit disk graphs*, in Proceedings of the 31st International Workshop on Graph-Theoretic Concepts in Computer Science (WG 2005), LNCS 3787, 2005, pp. 351–361.

23. M. YANNAKAKIS, *The complexity of the partial order dimension problem*, SIAM Journal on Algebraic Discrete Methods, 3 (1982), pp. 351–358.

The On-Line Heilbronn's Triangle Problem in d Dimensions[*]

Gill Barequet and Alina Shaikhet

Dept. of Computer Science,
Technion—Israel Institute of Technology,
Haifa 32000, Israel
{barequet, dalina}@cs.technion.ac.il

Abstract. In this paper we show a lower bound for the on-line version of Heilbronn's triangle problem in d dimensions. Specifically, we provide an incremental construction for positioning n points in the d-dimensional unit cube, for which every simplex defined by $d + 1$ of these points has volume $\Omega(1/n^{(d+1)\ln(d-2)+2})$.

1 Introduction

The *off-line* version of the famous *triangle problem* was posed by Heilbronn [Ro51] more than 50 years ago. It is formulated as follows:

> Given n points in the unit square, what is $\mathcal{H}_2(n)$, the maximum possible area of the *smallest* triangle defined by some three of these points?

There is a large gap between the best currently-known lower and upper bounds on $\mathcal{H}_2(n)$, $\Omega(\log n/n^2)$ [KPS82] and $O(1/n^{8/7-\varepsilon})$ (for any $\varepsilon > 0$) [KPS81]. Jiang et al. [JLV02] showed that the expected area of the smallest triangle, when the n points are put uniformly at random in the unit square, is $\Theta(1/n^3)$. Barequet [Ba01] generalized the off-line problem to d dimensions:

> Given n points in the d-dimensional unit cube, what is $\mathcal{H}_d(n)$, the maximum possible volume of the *smallest* simplex defined by some $d + 1$ of these points?

The best currently-known lower bound on $\mathcal{H}_d(n)$ is $\Omega(\log n/n^d)$ [Le03]. Other versions, in which the dimension of the optimized simplex is lower than that of the cube, were investigated in [Le04, BN05, Le05].

The *on-line* version of the triangle problem is harder than the off-line version because the value of n is not specified in advance. In other words, the points are positioned one after the other in a d-dimensional unit cube, while n is incremented by one after every point-positioning step. The procedure can be stopped at any time, and the already-positioned points must have the property that every

---

[*] Work on this paper by the first author has been supported in part by the European FP6 Network of Excellence Grant 506766 (AIM@SHAPE).

D.Z. Chen and D.T. Lee (Eds.): COCOON 2006, LNCS 4112, pp. 408–417, 2006.
© Springer-Verlag Berlin Heidelberg 2006

subset of $d+1$ points defines a polytope whose volume is at least some quantity $\mathcal{H}_d^{\text{on-line}}(n)$, where the goal is to maximize this quantity. Schmidt [Sc71] showed that $\mathcal{H}_2^{\text{on-line}}(n) = \Omega(1/n^2)$. Barequet [Ba04] used nested packing arguments to demonstrate that $\mathcal{H}_3^{\text{on-line}}(n) = \Omega(1/n^{10/3}) = \Omega(1/n^{3.333\cdots})$ and $\mathcal{H}_4^{\text{on-line}}(n) = \Omega(1/n^{127/24}) = \Omega(1/n^{5.292\cdots})$.

In this paper we present a nontrivial generalization of the latter method to d dimensions, showing that for a fixed value of d we have $\mathcal{H}_d^{\text{on-line}}(n) = \Omega(\frac{1}{n^{(d+1)\ln(d-2)+2}})$. Specifically, we provide an incremental procedure for positioning n points (one by one) in a d-dimensional unit cube so that no subset of up to $d+1$ points is "too dense." Specifically, the distance between any two points is at least $a_1/n^{1/d}$ (for some constant $a_1 > 0$), no three points define a triangle whose area is less than $a_2/n^{2/(d-1)}$ (for some constant $a_2 > 0$), and so on. The values of the constants are tuned at the end of the construction. It is then proven that all the d-dimensional simplices defined by $(d+1)$-tuples of the points have volume $\Omega(1/n^{(d+1)\ln(d-2)+2})$.

2 The Construction

2.1 Notation and Plan

We use the following notation. Let $p_{i_1}, p_{i_2}, ..., p_{i_q}$ be any q points in $\Re^d$. Then, $|p_{i_1} p_{i_2}|$ denotes the distance between two points p_{i_1}, p_{i_2}; $|p_{i_1} p_{i_2} p_{i_3}|$ denotes the area of the triangle $p_{i_1} p_{i_2} p_{i_3}$; $|p_{i_1} p_{i_2} p_{i_3} p_{i_4}|$ denotes the 3-dimensional volume of the tetrahedron $p_{i_1} p_{i_2} p_{i_3} p_{i_4}$; and, in general, $|p_{i_1} p_{i_2} \ldots p_{i_q}|$ denotes the volume of the $(q-1)$-dimensional simplex $p_{i_1} p_{i_2} \ldots p_{i_q}$. We denote by C^d the d-dimensional unit cube, and by B_r^d a d-dimensional ball of radius r. The line defined by the pair of points p_{i_1}, p_{i_2} is denoted by $\ell_{i_1 i_2}$.

Throughout the construction we refer to d as a fixed constant. Therefore, we omit factors that depend solely on d, except when they appear in powers of n.

We want to construct a set S of n points in C^d such that

[1] $|p_{i_1} p_{i_2}| \geq V_2 = a_1/n^{1/d}$, for any pair of distinct points $p_{i_1}, p_{i_2} \in S$ and for some constant $a_1 > 0$.

[2] $|p_{i_1} p_{i_2} p_{i_3}| \geq V_3 = a_2/n^{2/(d-1)}$, for any triple of distinct points $p_{i_1}, p_{i_2}, p_{i_3} \in S$ and for some constant $a_2 > 0$.

[3] $|p_{i_1} p_{i_2} p_{i_3} p_{i_4}| \geq V_4 = a_3/n^{\frac{4d^2-5d-1}{d(d-1)(d-2)}}$, for any quadruple of distinct points $p_{i_1}, p_{i_2}, p_{i_3}, p_{i_4} \in S$ and for some constant $a_3 > 0$.

$$\vdots$$

[q − 1] $|p_{i_1} p_{i_2} \ldots p_{i_q}| \geq V_q = a_{q-1} V_{q-1}/(a_{q-2} n^{\frac{d(q-2)+q-3}{d(d-q+2)}})$, for any q-tuple $(4 \leq q \leq d+1)$ of distinct points $p_{i_1}, p_{i_2}, \ldots, p_{i_q} \in S$ and for some constant $a_{q-1} > 0$.

The goal is to construct S incrementally. That is, assume that we have already constructed a subset S_v of v points, for $v < n$, which satisfies the above conditions [1]–[q − 1]. We want to show that there exists a new point $p \in C^d$ that satisfies

[1'] $|pp_{i_1}| \geq V_2 = a_1/n^{1/d}$, for each point $p_{i_1} \in S$.

[2'] $|pp_{i_1}p_{i_2}| \geq V_3 = a_2/n^{2/(d-1)}$, for any pair of distinct points $p_{i_1}, p_{i_2} \in S$.

[3'] $|pp_{i_1}p_{i_2}p_{i_3}| \geq V_4 = a_3/n^{\frac{4d^2-5d-1}{d(d-1)(d-2)}}$, for any triple of distinct points $p_{i_1}, p_{i_2}, p_{i_3}$
$\in S$.

$$\vdots$$

[(q-1)'] $|pp_{i_1}p_{i_2}\cdots p_{i_{q-1}}| \geq V_q = a_{q-1}V_{q-1}/(a_{q-2}n^{\frac{d(q-2)+q-3}{d(d-q+2)}})$, for any q-tuple
$(4 \leq q \leq d+1)$ of distinct points $p_{i_1}, p_{i_2}, \ldots, p_{i_q} \in S$.

We will show this by summing up the volumes of the "forbidden" portions of C^d where one of the inequalities [1']–[(q − 1)'] is violated, and by showing that the sum of these volumes is less than 1. This implies the existence of the desired point p, which we then add to S_v to form S_{v+1}. We continue in this manner until the entire set S is constructed.

2.2 Forbidden Balls

The forbidden regions where one of the inequalities [1'] is violated are v d-dimensional balls of radius $r_1 = a_1/n^{1/d}$.[1] Their total volume is at most

$$v|B_{r_1}^d| = O\left(\frac{v}{n}\right) = O(1).$$

2.3 Forbidden Cylinders

The forbidden regions where one of the inequalities [2'] is violated are $\binom{v}{2}$ d-dimensional "cylinders" G_{ij}, for $1 \leq i < j \leq v$. The cylinder G_{ij} is centered at ℓ_{ij}, its length is at most $\sqrt{d}$, and its cross-section perpendicular to ℓ_{ij} is a $(d-1)$-dimensional sphere of radius $r_2 = \frac{2V_3}{V_2} = \frac{2a_2}{n^{2/(d-1)} \cdot |p_i p_j|} = \Theta\left(\frac{1}{n^{2/(d-1)}|p_i p_j|}\right)$ (see Figure 1).

The overall volume of the "cylinders" (within C^d) is at most

$$\sum_{1 \leq i < j \leq v} (|B_{r_2}^{d-1}|\sqrt{d}) = \sum_{1 \leq i < j \leq v} O\left(\frac{1}{n^2|p_i p_j|^{d-1}}\right). \tag{1}$$

To bound this sum, we fix p_i and sum over p_j. We use a d-dimensional spherical packing argument that exploits the fact that S_v satisfies [1]. Specifically, we have

$$\sum_{j \neq i} \frac{1}{|p_i p_j|^{d-1}} \leq \sum_{t=1}^{O(n^{1/d})} \frac{M_t n^{\frac{d-1}{d}}}{a_1^{d-1}t^{d-1}}, \tag{2}$$

where M_t is the number of points of S_v that lie in the d-dimensional spherical shell centered at p_i with inner radius $a_1 t/n^{1/d}$ and outer radius $a_1(t + 1)/n^{1/d}$;

---

[1] Recall that $|B_r^d| = \pi^{d/2}r^d/\Gamma(d/2+1) = \Theta(r^d)$, where $\Gamma(\cdot)$ is the continuous generalization of the factorial function.

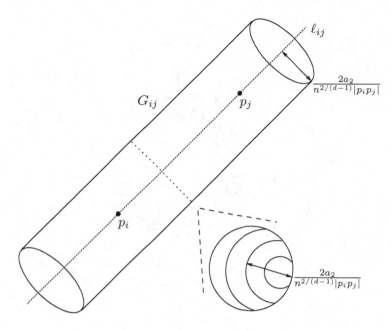

Fig. 1. A cylinder in $\Re^d$

see Figure 2. There are $O(n^{1/d})$ such spherical shells (within C^d). Because of [1], the number of such points is $M_t = O(t^{d-1})$. This follows by an argument of packing spheres of volume $\Theta(1/n)$ within a shell whose volume is $\Theta\left(\frac{t^{d-1}}{n}\right)$.

Hence, the sum in Equation (2) is $O(n)$. Summing this over all p_i, we obtain a final bound of $O(vn)$. Substituting this in Equation (1), we see that the total volume of the forbidden cylinders is $O(v/n) = O(1)$.

2.4 Forbidden Prisms

The forbidden regions where one of the inequalities [3'] is violated are $\binom{v}{3}$ d-dimensional "prisms" ϕ_{ijk}, for $1 \le i < j < k \le v$. The base area (a portion of a 2-dimensional flat) of ϕ_{ijk} is at most d, and its "height" is a $(d-2)$-dimensional sphere of radius $r_3 = \frac{3V_4}{V_3} = O\left(\dfrac{1}{n^{\frac{4d^2-5d-1}{d(d-1)(d-2)}}|p_ip_jp_k|}\right)$. The overall volume of the prisms (within C^d) is at most

$$\sum_{1 \le i < j < k \le v} (|B_{r_3}^{d-2}| \cdot d) = \sum_{1 \le i < j < k \le v} O\left(\dfrac{1}{n^{\frac{4d^2-5d-1}{d(d-1)}}|p_ip_jp_k|^{d-2}}\right). \tag{3}$$

To bound this sum, we fix p_i, p_j and sum over p_k. We use a d-dimensional cylindrical packing argument that exploits the fact that S_v satisfies [1] and [2].

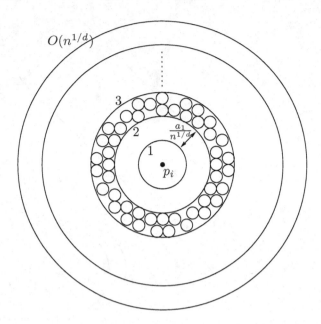

Fig. 2. A spherical packing of balls in $\Re^d$

The cylinders are centered at ℓ_{ij}; see Figure 3, where the line ℓ_{ij} emanates from p_i toward p_j through the dth dimension. Specifically, we have

$$\sum_{k \neq i,j} \frac{1}{|p_i p_j p_k|^{d-2}} \leq \frac{N_0 n^{\frac{2(d-2)}{d-1}}}{a_2^{d-2}} + \sum_{t=1}^{O(n^{1/d})} \frac{2^{d-2} N_t n^{\frac{d-2}{d}}}{a_1^{d-2} t^{d-2} |p_i p_j|^{d-2}}, \tag{4}$$

where N_0 is the number of points of S_v that lie in the innermost d-dimensional cylinder of the packing (centered at ℓ_{ij} and of radius $a_1/n^{1/d}$), and N_t is the number of points of S_v that lie in the cylindrical shell centered at ℓ_{ij} with inner radius $a_1 t/n^{1/d}$ and outer radius $a_1(t+1)/n^{1/d}$.

Obviously, $N_0 = O(n^{1/d})$, since the volume of the $(d-1)$-dimensional cross-sectional sphere of the innermost cylinder is $O(1/n^{\frac{d-1}{d}})$ and because of [1]. Also, we have $N_t = O(t^{d-2} n^{1/d})$. This follows by an argument of packing spheres of volume $\Theta(1/n)$ within a shell whose volume is $\Theta\left(\frac{t^{d-2}}{n^{\frac{d-1}{d}}}\right)$.

Hence, the quantity in Equation (4) is

$$O\left(n^{\frac{2d^2-3d-1}{d(d-1)}} + \frac{n}{|p_i p_j|^{d-2}}\right).$$

Substituting this in Equation (3), we obtain the upper bound on the total volume of the forbidden prisms

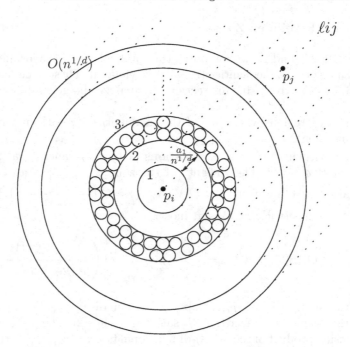

Fig. 3. A d-dimensional cylindrical packing (an extruded $(d-1)$-dimensional spherical packing) of balls in $\Re^d$

$$O\left(\sum_{1 \le i < j \le v} \left(\frac{1}{n^2} + \frac{1}{n^{\frac{3d^2 - 4d - 1}{d(d-1)}} |p_i p_j|^{d-2}} \right) \right)$$

$$= O\left(\frac{v^2}{n^2} + \frac{1}{n^{\frac{3d^2 - 4d - 1}{d(d-1)}}} \sum_{1 \le i < j \le v} \frac{1}{|p_i p_j|^{d-2}} \right). \tag{5}$$

We bound the sum in the second summand similarly to our bounding of the term in Equation (2) (in Section 2.3). We fix p_i and use a d-dimensional spherical packing argument within spherical shells centered at p_i. Arguing as above, we obtain

$$\sum_{j \ne i} \frac{1}{|p_i p_j|^{d-2}} \le \sum_{t=1}^{O(n^{1/d})} \frac{M_t n^{\frac{d-2}{d}}}{a_1^{d-2} t^{d-2}} = \sum_{t=1}^{O(n^{1/d})} \frac{O(t^{d-1}) n^{\frac{d-2}{d}}}{a_1^{d-2} t^{d-2}} = O(n).$$

Summing this over all p_i, we obtain a final bound of $O(vn)$. Substituting this in Equation (5), we see that the total volume of the forbidden prisms is

$$O\left(\frac{v^2}{n^2} + \frac{v}{n^{\frac{2d^2 - 3d - 1}{d(d-1)}}} \right) = O(1).$$

2.5   General Forbidden Zones

In Sections 2.2, 2.3, and 2.4 we computed the total volume of the forbidden zones in which the respective inequalities [1']–[3'] are violated. These zones correspond to $q = 2, 3, 4$, respectively. In this section we analyze the general case $4 < q \leq d + 1$.

The forbidden regions where one of the inequalities $[(q - 1)']$ is violated are $\binom{v}{q-1}$ d-dimensional zones $\psi_{i_1 i_2 \ldots i_{q-1}}$ (for $1 \leq i_1 < i_2 < \ldots < i_{q-1} \leq v$ and $4 < q \leq d + 1$), whose "bases" are portions of $(q - 2)$-dimensional flats with volume at most $d^{(q-2)/2}$. The "height" of the zone $\psi_{i_1 i_2 \ldots i_{q-1}}$ is a $(d - q + 2)$-dimensional sphere of radius $r_{q-1} = O(V_q/V_{q-1}) = O\left(\frac{V_q}{|p_{i_1} p_{i_2} \cdots p_{i_{q-1}}|}\right)$. The total volume of the zones (within C^d) is at most

$$\sum_{1 \leq i_1 < \ldots < i_{q-1} \leq v} O(|B_{r_{q-1}}^{d-q+2}| d^{\frac{q-2}{2}}) = \sum_{1 \leq i_1 < \ldots < i_{q-1} \leq v} O\left(\frac{V_q^{d-q+2}}{|p_{i_1} p_{i_2} \cdots p_{i_{q-1}}|^{d-q+2}}\right).$$

$$(6)$$

To bound this sum, we fix $p_{i_1}, p_{i_2}, \ldots, p_{i_{q-2}}$ and sum over $p_{i_{q-1}}$. We use a packing argument that exploits the fact that S_v satisfies $[1]$–$[q - 2]$. The packing consists of the Cartesian product of the $(q - 3)$-dimensional flat $\pi = \pi_{i_1 i_2 \ldots i_{q-2}}$ that passes through $p_{i_1}, p_{i_2}, \ldots, p_{i_{q-2}}$, and spheres whose centers belong to π and extend to the $(d - q + 3)$-dimensional space orthogonal to π. Specifically, we have

$$\sum_{i_{q-1} \neq i_1, \ldots, i_{q-2}} \frac{1}{|p_{i_1} p_{i_2} \cdots p_{i_{q-1}}|^{d-q+2}}$$

$$\leq \frac{Z_0}{V_{q-1}^{d-q+2}} + \sum_{t=1}^{O(n^{1/d})} \left(Z_t \cdot O\left(\frac{n^{\frac{1}{d}}}{a_1 t |p_{i_1} p_{i_2} \cdots p_{i_{q-2}}|}\right)^{d-q+2}\right), \tag{7}$$

where Z_0 is the number of points of S_v that lie in the innermost shape of the packing (centered at the flat π and of radius $a_1/n^{1/d}$), and Z_t is the number of points of S_v that lie in the shell centered at π with inner radius $a_1 t/n^{1/d}$ and outer radius $a_1(t + 1)/n^{1/d}$.

Obviously, $Z_0 = O(n^{\frac{q-3}{d}})$, since the volume of the innermost shape is $O(1/n^{\frac{d-q+3}{d}})$ and because of [1]. Also, we have $Z_t = O(t^{d-q+2} n^{\frac{q-3}{d}})$. This follows by an argument of packing spheres of volume $\Theta(1/n)$ within a shell whose volume is $\Theta(t^{d-q+2}/n^{\frac{d-q+3}{d}})$.

Hence, the sum in Equation (7) is

$$O\left(\frac{n^{\frac{q-3}{d}}}{V_{q-1}^{d-q+2}} + \frac{n}{|p_{i_1} p_{i_2} \cdots p_{i_{q-2}}|^{d-q+2}}\right). \tag{8}$$

Substituting this in Equation (6), we obtain the upper bound on the total volume of the forbidden zones

$$O\left(V_q^{d-q+2}\sum_{1\le i_1<\cdots<i_{q-2}\le v}\left(\frac{n^{\frac{q-3}{d}}}{V_{q-1}^{d-q+2}}+\frac{n}{|p_{i_1}p_{i_2}\cdots p_{i_{q-2}}|^{d-q+2}}\right)\right)$$

$$=O\left(n^{\frac{q-3}{d}}v^{q-2}\left(\frac{V_q}{V_{q-1}}\right)^{d-q+2}+\sum_{1\le i_1<\cdots<i_{q-2}\le v}n\left(\frac{V_q}{V_{q-2}}\right)^{d-q+2}\right).$$

Combining this with the equality $V_q=\dfrac{a_{q-1}V_{q-1}}{a_{q-2}n^{\frac{d(q-2)+q-3}{d(d-q+2)}}}$, we see that the total forbidden volume is

$$O\left(1+\sum_{1\le i_1<\cdots<i_{q-2}\le v}n\left(\frac{V_q}{V_{q-2}}\right)^{d-q+2}\right). \tag{9}$$

In order to show that the bound in Equation (9) is $O(1)$, it remains to prove that the second summand in it is smaller than 1. This amounts to proving that the second summand in Equation (8) is smaller than the first summand in it. From $[q-2]$ we know that $V_{q-1}=\dfrac{a_{q-2}V_{q-2}}{a_{q-3}n^{\frac{d(q-3)+q-4}{d(d-q+3)}}}$ for $4<q\le d+1$. By substituting this in Equation (8), we obtain the equal quantity

$$O\left(n^{\frac{q-3}{d}}\left(\frac{a_{q-3}n^{\frac{d(q-3)+q-4}{d(d-q+3)}}}{a_{q-2}V_{q-2}}\right)^{d-q+2}+\frac{n}{V_{q-2}^{d-q+2}}\right)$$

$$=O\left(\frac{n^{\frac{(q-3)d^2+(-q^2+7q-13)d-2q^2+12q-17}{d(d-q+3)}}}{V_{q-2}^{d-q+2}}+\frac{n}{V_{q-2}^{d-q+2}}\right). \tag{10}$$

The second summand in Equation (8) is smaller than the first summand in it if and only if the second summand in Equation (10) is smaller than the first summand in it. That is, we have to prove that

$$n^{\frac{d^2(q-3)+d(-q^2+7q-13)-2q^2+12q-17}{d(d-q+3)}}>n,$$

i.e., the inequality

$$\frac{(q-3)d^2+(-q^2+7q-13)d-2q^2+12q-17}{d(d-q+3)}>1,$$

which, after simple manipulations, is

$$(q-4)d^2-(q-4)^2d-2q^2+12q-17>0.$$

However, it is easily verified that

$$(q-4)d^2 - (q-4)^2 d - 2q^2 + 12q - 17 = (q-4)(d-q+2)(d+2) - 1 \geq 5 > 0,$$

using the facts that $q \geq 5$, $d - q \geq -1$, and $d \geq 4$.

2.6 Epilogue

We are now ready to bound $\mathcal{H}_d^{\mathrm{on-line}}(n)$, the maximum possible volume of the smallest simplex defined by some $d+1$ of n points in the d-dimensional unit cube. In other words, we want to lower bound V_{d+1}. For this purpose we use its recursive definition and write

$$V_{d+1} = \prod_{q=4}^{d+1} \left(\frac{a_{q-1}}{a_{q-2} n^{\frac{d(q-2)+q-3}{d(d-q+2)}}} \right) \cdot V_3 = \frac{a_d}{n^{\left(\sum_{q=4}^{d+1} \frac{d(q-2)+q-3}{d(d-q+2)}\right) + \frac{2}{d-1}}}. \tag{11}$$

Let us upper bound the power of n in Equation 11:

$$\sum_{q=4}^{d+1} \frac{d(q-2)+q-3}{d(d-q+2)} + \frac{2}{d-1}$$

$$= \sum_{q=4}^{d+1} \left(\frac{q-1-1/d}{d-q+2} \right) - \frac{1}{d} \sum_{q=4}^{d+1} \left(\frac{d-q+2}{d-q+2} \right) + \frac{2}{d-1}$$

$$= \sum_{t=1}^{d-2} \frac{d+1-1/d-t}{t} - (1 - 2/d) + \frac{2}{d-1}$$

$$< (d+1-1/d)(\ln(d-2)+1) - (d-2) - (1-2/d) + 2/(d-1)$$

$$= (d+1)\ln(d-2) + 2 - (\ln(d-2)-1)/d + 2/(d-1)$$

$$< (d+1)\ln(d-2) + 2,$$

where we use the facts that $\sum_{t=1}^{k} 1/t < \ln k + 1$ and $2/(d-1) - (\ln(d-2)-1)/d < 0$ for d sufficiently large ($d \geq 24$). We see that $V_{d+1} > \frac{a_d}{n^{(d+1)\ln(d-2)+2}}$.

It remains to show that the constants $a_1, a_2, \ldots, a_d$ can be fixed so that the total volume of the forbidden zones is strictly less than 1. To this aim note that among these constants, the total volume of the forbidden balls depends only on a_1, the total volume of the forbidden prisms depends only on a_1, a_2, and so on. This allows us to fix the values of the constants sequentially so that the total volume of any type of forbidden shapes is strictly less than $1/d$. (See [Ba04] for the implementation of this technique for $d = 3, 4$.)

This completes the proof of the main theorem:

Theorem 1. $\mathcal{H}_d^{\mathrm{on-line}}(n) = \Omega(1/n^{(d+1)\ln(d-2)+2})$. $\square$

3 Conclusion

In this paper we show by using nested packing arguments that $\mathcal{H}_d^{\mathrm{on-line}}(n) = \Omega(1/n^{(d+1)\ln(d-2)+2})$. This compares favorably with the best-known lower bound [Le03] in the off-line case $\mathcal{H}_d^{\mathrm{off-line}}(n) = \Omega(\log n/n^d)$.

References

[Ba01]   G. BAREQUET, A lower bound for Heilbronn's triangle problem in d dimensions, *SIAM J. on Discrete Mathematics*, 14 (2001), 230–236.

[Ba04]   G. BAREQUET, The on-line Heilbronn's triangle problem, *Discrete Mathematics*, 283 (2004), 7–14.

[BN05]   G. BAREQUET AND J. NAOR, Large k-D simplices in the d-dimensional unit cube, *Proc. 17th Canadian Conf. on Computational Geometry*, Windsor, Ontario, Canada, 30–33, August 2005.

[JLV02] T. JIANG, M. LI, AND P. VITÁNYI, The average-case area of Heilbronn-type triangles, *Random Structures and Algorithms*, 20 (2002), 206–219.

[KPS81] J. KOMLÓS, J. PINTZ, AND E. SZEMERÉDI, On Heilbronn's triangle problem, *J. London Math. Soc. (2)*, 24 (1981), 385–396.

[KPS82] J. KOMLÓS, J. PINTZ, AND E. SZEMERÉDI, A lower bound for Heilbronn's problem, *J. London Math. Soc. (2)*, 25 (1982), 13–24.

[Le03]   H. LEFMANN, On Heilbronn's problem in higher dimension, *Combinatorica*, 23 (2003), 669–680.

[Le04]   H. LEFMANN, Large triangles in the d-dimensional unit-cube, *Proc. 10th Ann. Int. Computing and Combinatorics Conf.*, Jeju Island, South Korea, *Lecture Notes in Computer Science*, 3106, Springer-Verlag, 43–52, August 2004.

[Le05]   H. LEFMANN, Distributions of points in d dimensions and large k-point simplices, *Proc. 11th Ann. Int. Computing and Combinatorics Conf.*, Kunming, China, *Lecture Notes in Computer Science*, 3595, Springer-Verlag, 514–523, August 2005.

[Ro51]   K.F. ROTH, On a problem of Heilbronn, *Proc. London Math. Soc.*, 26 (1951), 198–204.

[Sc71]   W.M. SCHMIDT, On a problem of Heilbronn, *J. London Math. Soc. (2)*, 4 (1971), 545–550.

Counting d-Dimensional Polycubes and Nonrectangular Planar Polyominoes*

Gadi Aleksandrowicz and Gill Barequet

Dept. of Computer Science,
Technion—Israel Institute of Technology,
Haifa 32000, Israel
{gadial@tx, barequet@cs}.technion.ac.il

Abstract. A planar polyomino of size n is an edge-connected set of n squares on a rectangular 2-D lattice. Similarly, a d-dimensional polycube (for $d \geq 2$) of size n is a connected set of n hypercubes on an orthogonal d-dimensional lattice, where two hypercubes are neighbors if they share a $(d - 1)$-dimensional face. There are also two-dimensional polyominoes that lie on a triangular or hexagonal lattice. In this paper we describe a generalization of Redelmeier's algorithm for counting two-dimensional rectangular polyominoes [Re81], which counts all the above types of polyominoes. For example, our program computed the number of distinct 3-D polycubes of size 18. To the best of our knowledge, this is the first tabulation of this value.

Keywords: Polycubes, lattice animals, subgraph counting.

1 Introduction

A polyomino of size (or order) n is an edge-connected set of n squares on the regular square lattice $\mathbb{Z}^2$. *Fixed* polyominoes are considered distinct if they have different shapes *or* orientations. In the literature, the symbol $A(n)$ usually denotes the number of fixed polyominoes of size n. We have $A(1) = 1$, $A(2) = 2$, $A(3) = 6$, $A(4) = 19$, and so on. Two main open problems related to polyominoes are the number of polyominoes of order n, and the limit of their growth rate as n tends to infinity.

Redelmeier [Re81] introduced the first efficient algorithm for counting polyominoes, in the sense that it generates all the polyominoes sequentially and without repetitions. Thus, it only has to count the number of generated polyominoes but does not have to compare every generated polyomino to all the previously-generated polyominoes. Since the algorithm generates each polyomino in constant time, its total running time is $O(A(n))$. Redelmeier implemented his algorithm in Algol W (and for efficiency also in the PDP assembly language). The program required about 10 months of CPU time on a PDP-11/70 to compute the number of all fixed polyominoes of up to order 24.

* Work on this paper by the second author has been supported in part by the European FP6 Network of Excellence Grant 506766 (AIM@SHAPE).

D.Z. Chen and D.T. Lee (Eds.): COCOON 2006, LNCS 4112, pp. 418–427, 2006.

The best currently-known algorithm (in terms of running time) for counting fixed polyominoes is that of Jensen [Je01]. This is a so-called transfer-matrix algorithm, which does not generate all the polyominoes. Instead, it generates classes of polyominoes with identical "boundaries," while being able to compute efficiently the number of polyominoes in each such class. Jensen was able to compute $A(n)$ up to $n = 56$ [Je03].

3-D polyominoes are called *polycubes* [Lu71]. A poly-cube of size n is a face-connected set of n cubes in Euclidean three-dimensional space. We denote by $A_3(n)$ the number of distinct fixed polycubes of size n. Figure 1(a,b) shows the three (resp., 15) polycubes of size 2 (resp., 3). Therefore, $A_3(2) = 3$, $A_3(3) = 15$, and so on. Polyominoes and polycubes have triggered the imagination of not only mathematicians. Extensive studies of them can also be found in the statistical-physics literature, where fixed polyominoes and polycubes are usually referred to as *lattice animals* (of dimension 2 or 3). Animals play an important role in computing the mean cluster density in percolation processes, in particular those of fluid flow in random media [BH57].

(a) 3-D dominoes

(b) 3-D trominoes

Fig. 1. Fixed 3-D dominoes and trominoes

Similarly, d-dimensional polycubes of size n are connected sets of n hypercubes on the orthogonal d-D lattice $\mathbb{Z}^d$, where connections are through $(d-1)$-dimensional faces. We denote by $A_d(n)$ the number of distinct fixed d-dimensional polycubes of size n.

We are aware of only three previous attempts to count fixed polycubes:

- Lunnon [Lu72b] analyzed three-dimensional polycubes by considering symmetry groups, and computed (manually!) $A_3(n)$ up to $n = 6$.
- Lunnon [Lu75] computed multi-dimensional polyominoes by considering polycubes that could fit into restricted boxes. For example, he computed $A_3(n)$ up to $n = 12$.
- Sykes et al. [SGG76] used a method proposed by Martin [Ma74] in order to derive and analyze series expansions on a three-dimensional lattice (but did not compute new values of $A_3(n)$).

In the plane we also consider regular hexagonal and triangular lattices. Consistent with the previous definition, fixed hexagonal and triangular polyominoes are edge-connected sets of cells of the respective lattices. (Again, "fixed" polyominoes are considered distinct if they have different shapes *or* orientations.) We denote by $H(n)$ and $T(n)$ the respective numbers of such polyominoes. Lunnon [Lu72a] computed the values of these series up to $H(12)$ and $T(16)$. To the best of our knowledge, values of $T(n)$ were computed up to $n = 28$ (see sequence A001420 in [EIS], J. Myers credited). Vöge and Guttmann [VG03] used a transfer-matrix approach to tabulate $H(n)$ up to $n = 35$.

Redelmeier's algorithm is based on counting connected subgraphs in the underlying graph of the two-dimensional orthogonal lattice. However, it does not depend on any property of the graph. Therefore, in order to generalize the algorithm to higher dimensions and to other types of lattices, we basically need to compute the respective underlying graphs and apply the same subgraph-counting algorithm. The intricate issues in this task, for us, were to define precisely which subgraphs need to be counted, and to handle the exponential growth in the size of the graphs in higher dimensions efficiently. The subgraph-counting method is inferior to the transfer-matrix mathod, but the latter cannot be adapted easily to higher dimensions.

In this paper we describe a generalization of Redelmeier's algorithm for counting fixed polyominoes that includes all the types mentioned above. For example, we used our implementation of the algorithm to compute the values of $A_3(18)$ and $T(29)$–$T(31)$. To the best of our knowledge, this is the first tabulation of these values, as well as a few terms of $A_d(n)$, for $d > 3$.

2   The Original Algorithm

In this section we briefly describe Redelmeier's algorithm for counting two-dimensional polyominoes. The reader is referred to the original paper [Re81] for the full details.

Redelmeier's algorithm is a procedure for connected-subgraph counting, where the underlying graph is induced by the square lattice. Since translated copies of a fixed polyomino are considered identical, one must decide upon a canonical form. Redelmeier's choice was to fix the leftmost square of the bottom row of a polyomino at the origin, that is, at the square $(0,0)$. (Note that coordinates are associated with squares and not with their corners.) Thus, he needed to count the number of edge-connected sets of squares (that contain the origin) in

$$\{(x, y) \mid (y > 0) \text{ or } (y = 0 \text{ and } x \geq 0)\}.$$

The squares in this set are located above the thick line in Figure 2(a). The shaded area in this figure consists of all the *reachable* cells, that is, possible locations of squares of polyominoes of order 5. Counting these sets of squares amounts to counting all the connected subgraphs of the graph shown in Figure 2(b), which contain the vertex a_1. The algorithm [Re81] is shown in Figure 3. Line 4(a)

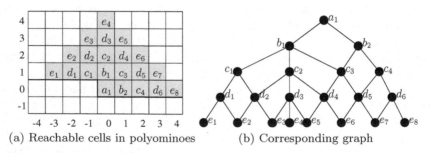

(a) Reachable cells in polyominoes   (b) Corresponding graph

Fig. 2. Polyominoes as subgraphs of a specific base graph

```
Initialize the parent to be the empty polyomino, and the untried set
to contain only the origin. The following steps are repeated until the
untried set is exhausted.
1. Remove an arbitrary element from the untried set.
2. Place a cell at this point.
3. Count this new polyomino.
4. If the size is less than n:
   (a) Add new neighbors to the untried set.
   (b) Call this algorithm recursively with the new parent being the
       current polyomino, and the new untried set being a copy of the
       current one.
   (c) Remove the new neighbors from the untried set.
5. Remove newest cell.
```

Fig. 3. Redelmeier's algorithm [Re81, p. 196]

deserves some attention. By "new neighbors" we mean only neighbors of the
new cell c that was placed in line 2, which *were not neighbors* of any cells of the
polyomino before c was placed. This ensures that we will not count the same
polyomino more than once.

This sequential subgraph-counting algorithm can be applied to any graph
(which, indeed, is exactly what we do in the next sections), and it has the
property that it never produces the same subgraph twice.

3   Polycubes in Higher Dimensions

In this section we describe the extension of Redelmeier's algorithm to counting
polycubes in orthogonal lattices in higher dimensions.

3.1   Three Dimensions

We first demonstrate the extension of the algorithm in three dimensions. To
this aim we need only modify the underlying graph so that it will represent a
three-dimensional cubic lattice. Then we must decide upon a canonical form of

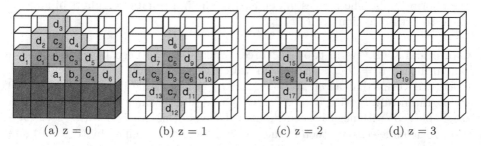

(a) z = 0   (b) z = 1   (c) z = 2   (d) z = 3

Fig. 4. Reachable cubes in a 3D lattice

a polycube. We fix the origin at the leftmost cube in the "closest" row in the bottom layer. This way polycubes are built only at

$$\{(x, y, z) \mid (z > 0) \text{ or } ((z=0) \text{ and } (y > 0)) \text{ or } ((z = 0) \text{ and } (y = 0) \text{ and } x \geq 0)\}.$$

The origin cube is shown in light grey in Figure 4(a). The colored areas in Figure 4 are the possible locations of cells of polycubes of order 4. The corresponding graph in which the polycubes are counted is drawn in Figure 5.

3.2   Higher Dimensions

We now describe how to generalize the algorithm to d dimensions.

Cell Labeling. Our first version of the program, which counted only 3-D polycubes, built the underlying graph by using three nested loops and mapping the 3-D lattice cells to a linear array. It computed the position in the array by the formula $z_0 \cdot X \cdot Y + y_0 \cdot X + x_0$, where X and Y were the ranges of the x and y coordinates, respectively, and (x_0, y_0, z_0) was the location of the three-dimensional cell.

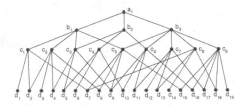

Fig. 5. The underlying graph in three dimensions

It is not possible to generalize this method to the d-dimensional case, since we cannot code an unknown number d of nested loops. Instead, we use a single loop as follows. Each point in the d-dimensional space is specified as a vector $x = (x_1, x_2, \ldots, x_d)$. In order to convert this vector into a single integer, we regard it as a number in some base $t \in \mathbb{N}$ with d digits. Consequently, x translates to the number $\Gamma(x) = \sum_{k=1}^{d} x_k t^{k-1}$.

We choose t large enough so that no two cells are mapped to the same number. The minimum possible choice of t is simply the size of the range of coordinates attainable by reachable cells of polycubes. This number is the generalization of X and Y in three dimensions. Obviously, $t = 2n-2$, where n is the polycube size. A clear benefit of this representation is the ability to move efficiently between neighboring cells. The images under $\Gamma(\cdot)$ of the immediate neighbors of a cell x along the kth direction are $\Gamma(x) \pm t^{k-1}$.

Canonical Cell. As in two and three dimensions, we need to decide upon a canonical cell $\bar{0} = (c_1, \ldots, c_d)$, from which the algorithm starts executing. It is easy to verify that a cell $x = (x_1, \ldots, x_d)$ is reachable if and only if $x \geq \bar{0}$ lexicographically. Setting $\bar{0} = (0, \ldots, 0)$ is undesirable since this would result in reachable cells with negative coordinates. To avoid this, we fix $\bar{0} = (n-2, \ldots, n-2, 0)$. Reachable cells are then in the d-dimensional box defined by $(0, \ldots, 0)$ and $x_M = (2n-3, \ldots, 2n-3, n-1)$. The function $\Gamma(\cdot)$ subsequently maps reachable

ALGORITHM GraphOrthodD(int n, int d)
  begin
    1. $o := \Gamma((n-2,\ldots,n-2,0))$; $M := (2n-2)^{d-1}n - 1$;
    2. for $i = o,\ldots,M$ do
      2.1 $b := 1$; *counter* := 0;
      2.2 for $j = 1,\ldots,d$ do
        2.2.1 if $i + b \geq o$ then do
          2.2.1.1 neighbors$[i][counter] := i + b$;
          2.2.1.2 *counter* := *counter*+1;
        end if
        2.2.2 if $i - b \geq o$ then do
          2.2.2.1 neighbors$[i][counter] := i - b$;
          2.2.2.2 *counter* := *counter*+1;
        end if
        2.2.3 $b := b(2n - 2)$;
      end for
      2.3 neighbors_num$[i] := emphcounter$;
    end for
  end GraphOrthodD

Fig. 6. Computing the graph for a d-dimensional orthogonal lattice

cells to numbers in the range 0 to $M = \Gamma(x_M) = \sum_{k=1}^{d-1}(2n-3)(2n-2)^{k-1} + (n-1)(2n-2)^{d-1} = (2n-2)^{d-1}n - 1$.[1]

Building the Graph. Figure 6 shows the algorithm for building the d-D graph. The cell-neighborhood relations are kept in the array neighbors$[x][y]$, where x is the identity number of the current cell and y is a serial number in the range $(1,\ldots,2d)$ (all possible directions in d dimensions).[2] The array neighbors_num holds the the numbers of actual neighbors of each cell, and the variable b is used to traverse the lattice in all possible directions.

4   Nonrectangular Planar Lattices

In this section we describe our extensions of the algorithm to counting planar triangular and hexagonal polyominoes. For both cases Lunnon [Lu72a] considered cells with three coordinates (albeit in the plane) while we invoke a preprocessing step that computes a neighborhood graph of cells with only two coordinates.

4.1   Triangular Polyominoes

Triangular polyominoes are also called *polyiamonds*.

---

[1] In fact, x_M is not reachable. The reachable cell with the largest image under Γ is
   $y = (n-2,\ldots,n-2,n-1)$ (the top cell of a "stick" aligned with the dth direction).
   We have $\Gamma(y) = (t^{d+1} - 2t^{d-1} - t + 2)/(2(t-1))$, where $t = 2n - 2$.
[2] The actual implementation used the range $(0,\ldots,2d-1)$.

Cell representation. For a triangular lattice we use equilateral triangles in two inverse orientations. We model a triangular lattice by a rectangular lattice with a restriction on the neighborhood relations: A cell whose x coordinate is even (resp., odd) has only an upper (resp., lower) neighboring cell but not a lower (resp., upper) neighbor. All cells have right and left neighbors; see Figure 7.

Fig. 7. Representing a triangular lattice by a rectangular lattice (bold segments indicate removed adjacency relations)

Canonical cell. As in the orthogonal case, we choose the canonical cell to be the leftmost triangle in the lowest row. Since we have two types of triangles ("up" and "down," e.g., $(0, 1)$ and $(0, 0)$, respectively, in Figure 7), there are two possible types of canonical cells. Nevertheless, it turns out that it suffices to count only triangular polyominoes whose canonical cell is of one type, say, "down." Recall that we denote by $T(n)$ the number of triangular polyominoes of size n. Let $T'(n)$ mark the number of triangular polyominoes whose canonical call is "down." Observe that $T(n) = T'(n) + T'(n-1)$. To see this, note that when the canonical cell is "up," its only possible neighbor is its right "down" neighbor. Therefore, the number of triangular polyominoes of size n whose canonical cell is "up" is the same as the number of polyominoes of size $(n-1)$ whose canonical cell is "down," the right neighbor of the original canonical cell.

Building the graph. We build the graph as in the orthogonal case. We set the left and right neighbors of a cell in exactly the same way, and choose whether to set an upper or lower neighbor by comparing the parity of the sum of coordinates of the current cell to that of the canonical cell (where equality implies that the neighbor is "up"). The entire algorithm is given in the full version of the paper.

4.2 Hexagonal Polyominoes

In [VG03], hexagonal polyominoes have already been counted by a transfer-matrix method, which is much faster than our subgraph-counting approach. For completeness, we also describe this case here.

Cell representation. We can model a hexagonal lattice by using a rectangular lattice with a special type of adjacency relationship. Namely, consider a hexagonal lattice as a rectangular lattice in which every second column is "slid down" by half a square. Every two adjacent squares (even along half of a cell boundary edge) in the modified lattice are considered neighbors (see Figure 8). Thus, we allow the cell to have six neighbors in the hexagonal lattice instead of four as in the rectangular lattice. The two additional neighbors are diagonal—either the upper or the lower diagonal cells, depending on the parity of the x coordinate of the cell. In the example shown in Figure 8, the two additional neighbors of $(1, 1)$ (odd x coordinate) are $(0, 0)$ and $(2, 0)$ (the two lower diagonal cells), while the neighbors of $(2, 1)$ (even x coordinate) are $(1, 2)$ and $(3, 2)$ (the two upper diagonal cells).

Canonical cell. In the hexagonal case we need a sharper definition for the canonical cell. While it is still defined as the leftmost cell in the lowest row, the meaning of a "row" is now restricted only to hexagons that are aligned horizontally. In other words, in our example, $(1,0)$ is not in the same row as $(2,0)$. Consequently, every row in the rectangular lattice becomes two rows in the hexagonal lattice— an upper and a lower hexagonal row. We

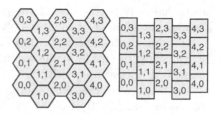

Fig. 8. Representing a hexagonal lattice by a rectangular lattice

fix the canonical cell to always be in the upper hexagonal row. While building the graph, we do not include cells that are below that upper row. Thus, if $(2,0)$ is the canonical cell, then $(3,0)$ is not reachable and is, therefore, not included in the graph.

Building the graph. The algorithm for building the graph is quite similar to that of the rectangular case. While processing the current cell, we also verify that it is not in a level lower than that of the canonical cell (the "lowest hexagonal row"). We also take care to to add the diagonal neighbors. The entire algorithm is given in the full version of the paper.

5   Results

Time and space complexity. The algorithm for building the graph in the d-dimensional orthogonal lattice (see Figure 6) consists of one main loop (2) that handles every cell once. For each cell it computes its up to $2d$ neighbors, each neighbor in constant time. Since there are $M = (2n-2)^{d-1}n - 1 = \Theta(n^d)$ cells, the time needed to create the graph is $\Theta(dn^d)$. However,

Table 1. Numbers of fixed three-dimensional polycubes

n	$A_3(n)$	Ref.
1	1	
2	3	
3	15	
4	86	
5	534	
6	3,481	[Lu72b]
7	23,502	
8	162,913	
9	1,152,870	
10	8,294,738	
11	60,494,549	
12	446,205,905	[Lu75]
13	3,322,769,321	
14	24,946,773,111	
15	188,625,900,446	
16	1,435,074,454,755	
17	10,977,812,452,428	[EIS][@]
18	84,384,157,287,999	Here

[@]Attributed to A. Flammenkamp in 1999.

the running time of the algorithm is dominated by the time needed to count the subgraphs, i.e., $\Theta(A_d(n))$. It is widely believed that this term is roughly exponential in n, where the base of the exponent is a constant that depends only on d (e.g., about 4.06 for $d = 2$ [Je03]). The space complexity of the algorithm is the size of the graph, that is, $\Theta(dn^d)$. Similarly, for the planar triangular and hexagonal lattices, the size of the graphs is $\Theta(n^2)$, and the running times are linear in the number of counted polyominoes.

We implemented the algorithm in C and ran the program (with the various lattice types) in an MS Windows environment. Values of $A_3(n)$ were computed on an IBM workstation with one 1.7GHz Pentium4 processor and 1GB of RAM. All the other values were computed on an IBM X500 with four 2.4GHz XEON processors and 3.5GB of RAM.

Table 2. Numbers of fixed d-dimensional polycubes

n	$A_4(n)$	$A_5(n)$	$A_6(n)$	$A_7(n)$	$A_8(n)$	$A_9(n)$
1	1	1	1	1	1	1
2	4	5	6	7	8	9
3	28	45	66	91	120	153
4	234	495	901	1,484	2,276	3,309
5	2,162	6,095	13,881	27,468	49,204	
6	21,272	80,617	231,008	551,313		
7	218,740	1,121,075	4,057,660			
8	2,323,730	16,177,405	74,174,927			
9	25,314,097	240,196,280	1,398,295,989			
10	281,345,096	3,648,115,531				
11	3,178,474,308	56,440,473,990				
12	36,400,646,766	886,696,345,225				
13	421,693,622,520	14,111,836,458,890				
14	4,933,625,049,464					
15	58,216,226,287,844					

On the 3-D orthogonal lattice the program yielded, after 51.37 days, the values reported in Table 1. The values of $A_3(n)$ for all $1 \leq n \leq 17$ agree with previous works.

Table 2 shows found values of $A_d(n)$ in higher dimensions. The running times in 4, 5, and 6-D were 47.44 days, 12.03 hours, and 2:00 minutes, resp. In higher dimensions the running times were negligible since memory constraints did not allow high values of n.

Table 3. Numbers of fixed triangular polyominoes

n	$T(n)$	n	$T(n)$	n	$T(n)$
1	2	12	39,169	23	4,236,446,214
2	3	13	110,194	24	12,341,035,217
3	6	14	311,751	25	36,009,329,450
4	14	15	886,160	26	105,229,462,401
5	36	16	2,529,260	27	307,942,754,342
6	94	17	7,244,862	28	902,338,712,971
7	250	18	20,818,498	29	2,647,263,986,022
8	675	19	59,994,514	30	7,775,314,024,683
9	1,838	20	173,338,962	31	22,861,250,676,074
10	5,053	21	501,994,070		
11	14,016	22	1,456,891,547		

Table 3 lists the numbers of triangular polyominoes computed in 19.9 days.

References

[BH57]   S.R. BROADBENT AND J.M. HAMMERSLEY, Percolation processes: I. Crystals and mazes, *Proc. Cambridge Philosophical Society*, 53 (1957), 629–641.

[EIS]   http://www.research.att.com/~njas/sequences (the on-line encyclopedia of integer sequences).

[Je01]   I. JENSEN, Enumerations of lattice animals and trees, *J. of Statistical Physics*, 102 (2001), 865–881.

[Je03]   I. JENSEN, Counting polyominoes: A parallel implementation for cluster computing, *Proc. Int. Conf. on Computational Science*, part III, Melbourne, Australia and St. Petersburg, Russia, *LNCS*, 2659, Springer, 203–212, June 2003.

[Lu71]   W.F. LUNNON, Counting polyominoes, in: *Comp. in Number Theory* (A.O.L. Atkin and B.J. Birch, eds.), Academic Press, London, 1971, 347–372.

[Lu72a]   W.F. LUNNON, Counting hexagonal and triangular polyominoes, in: *Graph Theory and Comp.* (R.C. Read, ed.), Academic Press, New York, 1972, 87–100.

[Lu72b]   W.F. LUNNON, Symmetry of cubical and general polyominoes, in: *Graph Theory and Computing* (R.C. Read, ed.), Academic Press, New York, 1972, 101–108.

[Lu75]   W.F. LUNNON, Counting multidimensional polyominoes, *The Computer Journal*, 18 (1975), 366–367.

[Ma74]   J.L. MARTIN, Computer techniques for evaluating lattice constants, in: *Phase Transitions and Critical Phenomena*, vol. 3 (C. Domb and M.S. Green, eds.), Academic Press, London, 97–112, 1974.

[Re81]   D.H. REDELMEIER, Counting polyominoes: Yet another attack, *Discrete Mathematics*, 36 (1981), 191–203.

[SGG76]   M.F. SYKES, D.S. GAUNT, AND M. GLEN, Percolation processes in three dimensions, *J. of Physics, A: Math. and General*, 10 (1976), 1705–1712.

[VG03]   M. VÖGE AND A.J. GUTTMANN, On the number of hexagonal polyominoes, *Theoretical Computer Science*, 307 (2003), 433–453.

Approximating Min-Max (Regret) Versions of Some Polynomial Problems

Hassene Aissi, Cristina Bazgan, and Daniel Vanderpooten

LAMSADE, Université Paris-Dauphine, France
{aissi, bazgan, vdp}@lamsade.dauphine.fr

Abstract. While the complexity of min-max and min-max regret versions of most classical combinatorial optimization problems has been thoroughly investigated, there are very few studies about their approximation. For a bounded number of scenarios, we establish a general approximation scheme which can be used for min-max and min-max regret versions of some polynomial problems. Applying this scheme to shortest path and minimum spanning tree, we obtain fully polynomial-time approximation schemes with much better running times than the ones previously presented in the literature.

Keywords: min-max, min-max regret, approximation, fptas, shortest path, minimum spanning tree.

1 Introduction

The definition of an instance of a combinatorial optimization problem requires to specify parameters, in particular objective function coefficients, which may be uncertain or imprecise. Uncertainty/imprecision can be structured through the concept of *scenario* which corresponds to an assignment of plausible values to parameters. There exist two natural ways of describing the set of all possible scenarios. In the *interval data case*, each numerical parameter can take any value between a lower and an upper bound. In the *discrete scenario case*, which is considered here, the scenario set is described explicitly. Kouvelis and Yu [6] proposed the min-max and min-max regret criteria, stemming from decision theory, to construct solutions hedging against parameters variations. The min-max criterion aims at constructing solutions having a good performance in the worst case. The min-max regret criterion, less conservative, aims at obtaining a solution minimizing the maximum deviation, over all possible scenarios, of the value of the solution from the optimal value of the corresponding scenario.

Complexity of the min-max and min-max regret versions has been studied extensively during the last decade. In [6], for the discrete scenario case, the complexity of min-max (regret) versions of several combinatorial optimization

---

* This work has been partially funded by grant CNRS/CGRI-FNRS number 18227. The second author was partially supported by the ACI Sécurité Informatique grant-TADORNE project 2004.

D.Z. Chen and D.T. Lee (Eds.): COCOON 2006, LNCS 4112, pp. 428–438, 2006.

problems was studied, including shortest path and minimum spanning tree. In general, these versions are shown to be harder than the classical versions. More precisely, if the number of scenarios is not constant, these problems become strongly NP-hard, even when the classical problems are solvable in polynomial time. On the other hand, for a constant number of scenarios, min-max (regret) versions of these polynomial problems usually become weakly NP-hard.

While the complexity of these problems was studied thoroughly, their approximation was not studied until now, except in [2]. That paper investigated the relationships between min-max (regret) and multi-objective versions, and showed the existence, in the case of a bounded number of scenarios, of fully polynomial-time approximation schemes (fptas) for min-max versions of several classical optimization problems (shortest path, minimum spanning tree, knapsack). The interest of studying these relationships is that, unlike for min-max (regret) versions, fptas, which determine an approximation of the non-dominated set (or Pareto set), have been proposed for the multi-objective version (see, e.g., [9, 11]). Approximation algorithms for the min-max version, which basically consist of selecting one min-max solution from an approximation of the non-dominated set, are then easy to derive but critically depend on the running time of the approximation scheme for the multi-objective version.

In this paper, we adopt an alternative perspective and develop a general approximation scheme, using the scaling technique, which can be applied to min-max (regret) versions of some problems, provided that some conditions are satisfied. The advantage of this approach is that the resulting fptas usually have a much better running time than those derived using multi-objective fptas.

After presenting some background concepts in section 2, we introduce in section 3 the general approximation scheme. In section 4 we present applications of this general scheme to shortest path and minimum spanning tree, giving in each case fptas with better running times than previously known fptas based on multi-objective versions.

2 Preliminaries

We consider in this paper the class $\mathcal{C}$ of 0-1 problems with a linear objective function defined as:

$$\begin{cases} \min \sum_{i=1}^{m} c_i x_i & c_i \in \mathbb{N} \\ x \in X \subset \{0,1\}^m \end{cases}$$

This class encompasses a large variety of classical combinatorial problems, some of which are polynomial-time solvable (shortest path problem, minimum spanning tree, ...) and others are NP-hard (knapsack, set covering, ...). The size of a solution $x \in X$ is the number of variables x_i which are set to 1.

2.1 Min-Max, Min-Max Regret Versions

Given a problem $\mathcal{P} \in \mathcal{C}$, the min-max (regret) version associated to $\mathcal{P}$ has for input a finite set of scenarios S where each scenario $s \in S$ is represented by a vector $(c_1^s, \ldots, c_m^s)$. We denote by $val(x, s) = \sum_{i=1}^m c_i^s x_i$ the value of solution $x \in X$ under scenario $s \in S$ and by val_s^* the optimal value in scenario s.

The min-max optimization problem corresponding to $\mathcal{P}$, denoted by MIN-MAX $\mathcal{P}$, consists of finding a solution x having the best worst case value across all scenarios, which can be stated as: $\min_{x \in X} \max_{s \in S} val(x, s)$.

Given a solution $x \in X$, its *regret*, $R(x, s)$, under scenario $s \in S$ is defined as $R(x, s) = val(x, s) - val_s^*$. The *maximum regret* $R_{max}(x)$ of solution x is then defined as $R_{max}(x) = \max_{s \in S} R(x, s)$.

The min-max regret optimization problem corresponding to $\mathcal{P}$, denoted by MIN-MAX REGRET $\mathcal{P}$, consists of finding a solution x minimizing the maximum regret $R_{max}(x)$ which can be stated as: $\min_{x \in X} \max_{s \in S} \{val(x, s) - val_s^*\}$.

When $\mathcal{P}$ is a maximization problem, the max-min and min-max regret versions associated to $\mathcal{P}$ are defined similarly.

2.2 Approximation

Let us consider an instance I, of size $|I|$, of an optimization problem and a solution x of I. We denote by $opt(I)$ the optimum value of instance I. The *performance ratio* of x is $r(x) = \max\left\{\frac{val(x)}{opt(I)}, \frac{opt(I)}{val(x)}\right\}$, and its *error* is $\varepsilon(x) = r(x) - 1$. For a function f, an algorithm is an $f(n)$-*approximation algorithm* if, for any instance I of the problem, it returns a solution x such that $r(x) \le f(|I|)$. An optimization problem has a *fully polynomial-time approximation scheme* (an *fptas*, for short) if, for every constant $\varepsilon > 0$, it admits an $(1 + \varepsilon)$-approximation algorithm which is polynomial both in the size of the input and in $1/\varepsilon$. The class of problems having an fptas is denoted by *FPTAS*.

2.3 Matrix Tree Theorem

In this section we describe classical results concerning the *matrix tree theorem* that will enable us to derive approximation schemes for min-max and min-max regret versions of spanning tree.

The matrix tree theorem provides a way of counting all the spanning trees in a graph (see, e.g., [10]). Consider a graph $G = (V, E)$ with $|V| = n$, $|E| = m$ and let c_{ij} denote the cost of edge $(i, j) \in E$.

Define an $n \times n$ matrix A whose entries are given as follows:

$$a_{ij} = \begin{cases} -c_{ij} \text{ if } i \ne j \text{ and } (i, j) \in E \\ \sum_{(i, \ell) \in E} c_{i\ell} \text{ if } i = j \\ 0 \text{ otherwise} \end{cases}$$

Define A_r as the submatrix of A obtained by deleting the rth row and column and $\mathcal{D}(A_r)$ as its determinant. The matrix tree theorem [10] states the following equality:

$$\mathcal{D}(A_r) = \sum_{T \in \mathcal{T}} \prod_{(i,j) \in T} c_{ij} \tag{1}$$

where $\mathcal{T}$ is the set of all spanning trees of G.

As indicated in [3], this theorem can be extended to count the number of spanning trees of value v for each possible value v using a matrix depending on one variable. Following this idea, we can extend the matrix tree theorem to the multiple scenarios case as in [5]. Define the $n \times n$ matrix $A(y_1, \ldots, y_k)$ as follows:

$$
a_{ij}(y_1, \ldots, y_k) = \begin{cases} -\prod_{s=1}^{k} y_s^{c_{ij}^s} & \text{if } i \neq j \text{ and } (i,j) \in E \\ \sum_{(i,\ell) \in E} \prod_{s=1}^{k} y_s^{c_{i\ell}^s} & \text{if } i = j \\ 0 & \text{otherwise} \end{cases}
$$

Then, the determinant of the submatrix $A_r(y_1, \ldots, y_k)$ obtained by deleting any r^{th} row and column is given by

$$
\mathcal{D}(A_r(y_1, \ldots, y_k)) = \sum_{v_1, \ldots, v_k \in V^{\mathcal{T}}} a_{v_1, \ldots, v_k} \prod_{s=1}^{k} y_s^{v_s} \tag{2}
$$

where $a_{v_1, \ldots, v_k}$ is the number of spanning trees with value v_s in scenario s, for all $s \in S$ and $V^{\mathcal{T}}$ is the set of values reached on all scenarios, for all spanning trees of G.

Equality (2) is obtained by replacing each c_{ij} in (1) by $\prod_{s=1}^{k} y_s^{c_{ij}^s}$. Then each product term in (1) corresponding to tree T becomes $\prod_{s=1}^{k} y_s^{\sum_{(i,j) \in T} c_{ij}^s}$.

3   A General Approximation Scheme

We establish now a general result giving a sufficient condition for the existence of fptas for min-max (regret) versions of problems $\mathcal{P}$ in $\mathcal{C}$.

Theorem 1. *Given a problem* MIN-MAX (REGRET) $\mathcal{P}$*, if*

1. *for any instance I, a lower and an upper bound L and U of opt can be computed in time $p(|I|)$, such that $U \leq q(|I|)L$, where p and q are two polynomials with q non decreasing and $q(|I|) \geq 1$,*
2. *and there exists an algorithm that finds for any instance I an optimal solution in time $r(|I|, U)$ where r is a non decreasing polynomial,*

   then MIN-MAX (REGRET) $\mathcal{P}$ *is in FPTAS.*

Proof. Let I be an instance of MIN-MAX $\mathcal{P}$ or MIN-MAX REGRET $\mathcal{P}$ defined on a scenario set S where each scenario $s \in S$ is represented by a vector $(c_1^s, \ldots, c_m^s)$. We use the technique of *scaling* in order to provide an fptas. In order to obtain a solution with an error bounded by ε, we need a lower bound of $opt(I)$. Moreover, for obtaining a polynomial algorithm, we have to use a lower bound that is polynomially related to an upper bound.

When I is an instance of MIN-MAX $\mathcal{P}$, consider $\bar{I}$ the instance of MIN-MAX $\mathcal{P}$ derived from I where each scenario $s \in S$ is represented by a vector $(\bar{c}_1^s, \ldots, \bar{c}_m^s)$, with $\bar{c}_i^s = \lfloor \frac{tc_i^s}{\varepsilon L} \rfloor$ and t is an upper bound of the size of any feasible solution of I.

Let x^* and $\overline{x}^*$ denote respectively an optimal solution of instance I and $\overline{I}$. Let $val(x, s)$ denote the value of a solution x in scenario s for $\overline{I}$. We have

$$c_i^s < \frac{\varepsilon L}{t}(\overline{c}_i^s + 1), \text{ for all } s \in S,$$

and thus,
$$val(\overline{x}^*, s) < \tfrac{\varepsilon L}{t}\overline{val}(\overline{x}^*, s) + \varepsilon L, \text{ for all } s \in S,$$

which implies
$$\max_{s \in S} val(\overline{x}^*, s) < \tfrac{\varepsilon L}{t} \max_{s \in S} \overline{val}(\overline{x}^*, s) + \varepsilon L.$$

Since $\overline{x}^*$ is an optimal solution in $\overline{I}$, we have

$$opt(\overline{I}) = \max_{s \in S} \overline{val}(\overline{x}^*, s) \leq \max_{s \in S} \overline{val}(x^*, s)$$

and thus, the value of an optimal solution of $\overline{I}$ has, in I, the value

$$\max_{s \in S} val(\overline{x}^*, s) < \frac{\varepsilon L}{t} \max_{s \in S} \overline{val}(x^*, s) + \varepsilon L \leq opt(I) + \varepsilon L \leq opt(I)(1 + \varepsilon).$$

A similar result can be obtained for MIN-MAX REGRET $\mathcal{P}$. Let I be an instance of MIN-MAX REGRET $\mathcal{P}$ and let $\overline{I}$ denote the instance derived from I, by scaling each entry c_i^s as follows: $\overline{c}_i^s = \lfloor \frac{2tc_i^s}{\varepsilon L} \rfloor$, where t is an upper bound of the size of any feasible solution of I. Let x^* and $\overline{x}^*$ denote respectively an optimal solution of instance I and $\overline{I}$ and let x_s^*, $\overline{x}_s^*$ denote respectively, an optimal solution of instance I and $\overline{I}$ in scenario s.

Then, we have, for all $s \in S$,

$$val(\overline{x}^*, s) - val(x_s^*, s) < \frac{\varepsilon L}{2t}\overline{val}(\overline{x}^*, s) - val(x_s^*, s) + \frac{\varepsilon}{2}L$$

$$\leq \frac{\varepsilon L}{2t}(\overline{val}(\overline{x}^*, s) - \overline{val}(x_s^*, s)) + \frac{\varepsilon}{2}L$$

$$\leq \frac{\varepsilon L}{2t}(\overline{val}(\overline{x}^*, s) - \overline{val}(\overline{x}_s^*, s)) + \frac{\varepsilon}{2}L$$

and thus

$$\max_{s \in S}\{val(\overline{x}^*, s) - val(x_s^*, s)\} < \max_{s \in S}\left\{\frac{\varepsilon L}{2t}(\overline{val}(\overline{x}^*, s) - \overline{val}(\overline{x}_s^*, s))\right\} + \frac{\varepsilon}{2}L$$

$$\leq \max_{s \in S}\left\{\frac{\varepsilon L}{2t}(\overline{val}(x^*, s) - \overline{val}(\overline{x}_s^*, s))\right\} + \frac{\varepsilon}{2}L$$

$$\leq \max_{s \in S}\{val(x^*, s) - val(x_s^*, s) + val(x_s^*, s) - \frac{\varepsilon L}{2t}\overline{val}(\overline{x}_s^*, s)\} + \frac{\varepsilon}{2}L$$

$$\leq \max_{s \in S}\{val(x^*, s) - val(x_s^*, s) + val(\overline{x}_s^*, s) - \frac{\varepsilon L}{2t}\overline{val}(\overline{x}_s^*, s)\} + \frac{\varepsilon}{2}L$$

$$\leq \max_{s \in S}\{val(x^*, s) - val(x_s^*, s)\} + \varepsilon L \leq opt(I)(1 + \varepsilon)$$

We show in the following that such a solution $\overline{x}^*$ of instance $\overline{I}$ for MIN-MAX $\mathcal{P}$ or MIN-MAX REGRET $\mathcal{P}$ can be obtained in polynomial time in $|I|$ and $\frac{1}{\varepsilon}$.

The bounds L and U can be computed in time $p(|I|)$ by hypothesis. In order to compute an optimal solution for $\overline{I}$, we apply the algorithm (that exists by hypothesis) that runs in time $r(|\overline{I}|, U(\overline{I}))$.

In the case where $\overline{I}$ is an instance of MIN-MAX $\mathcal{P}$, since $opt(\overline{I}) \leq \frac{topt(I)}{\varepsilon L} \leq \frac{tU}{\varepsilon L} \leq \frac{tq(|I|)}{\varepsilon}$, and q, r are non decreasing, the total time for computing the $(1+\varepsilon)$-approximation is $p(|I|) + r(|\overline{I}|, U(\overline{I})) \leq p(|I|) + r(|\overline{I}|, q(|\overline{I}|)L(\overline{I})) \leq p(|I|) + r(|I|, q(|I|)\frac{tq(|I|)}{\varepsilon})$.

In the case where $\overline{I}$ is an instance of MIN-MAX REGRET $\mathcal{P}$, since $opt(\overline{I}) \leq \frac{2topt(I)}{\varepsilon L} + t \leq \frac{2tU}{\varepsilon L} + t \leq \frac{2tq(|I|)}{\varepsilon} + t$, and q, r are non decreasing, the total time for computing the $(1 + \varepsilon)$-approximation is $(k + 1)p(|I|) + r(|\overline{I}|, U(\overline{I})) \leq (k + 1)p(|I|) + r(|\overline{I}|, q(|\overline{I}|)L(\overline{I})) \leq (k + 1)p(|I|) + r(|I|, q(|I|)(\frac{tq(|I|)}{\varepsilon} + t))$. □

We discuss now the two conditions of the previous theorem. The following result shows that the first condition can be satisfied easily if the underlying problem $\mathcal{P}$ is solvable in polynomial time.

Proposition 1. *If a minimization problem $\mathcal{P}$ is solvable in polynomial time, then for any instance on a set of k scenarios of MIN-MAX $\mathcal{P}$ and MIN-MAX REGRET $\mathcal{P}$, there exist a lower and an upper bound L and U of opt computable in polynomial time, such that $U \leq kL$.*

Proof. Consider an instance I of MIN-MAX $\mathcal{P}$ defined on a set S of k scenarios where each scenario $s \in S$ is represented by $(c_1^s, \ldots, c_m^s)$ and let X be the set of feasible solutions of I. We define the following instance I' of a single scenario problem $\min_{x \in X} \sum_{s \in S} \frac{1}{k} val(x, s)$ obtained by taking objective function coefficients $c_i' = \sum_{s=1}^{k} \frac{c_i^s}{k}$, $i = 1, \ldots, m$. Let x^* be an optimal solution of I'. We take as lower and upper bounds $L = \sum_{s \in S} \frac{1}{k} val(x^*, s)$ and $U = \max_{s \in S} val(x^*, s)$. Clearly, we have

$$L = \min_{x \in X} \sum_{s \in S} \frac{1}{k} val(x, s) \leq \min_{x \in X} \sum_{s \in S} \frac{1}{k} (\max_{s \in S} val(x, s)) = \min_{x \in X} \max_{s \in S} val(x, s) = opt$$

and

$$\min_{x \in X} \max_{s \in S} val(x, s) \leq \max_{s \in S} val(x^*, s) \leq \sum_{s \in S} val(x^*, s) = k \sum_{s \in S} \frac{1}{k} val(x^*, s) = kL$$

Consider now an instance I of MIN-MAX REGRET $\mathcal{P}$ defined on a set S of k scenarios and let X be the set of feasible solutions of I. Let $x^* \in X$ be an optimal solution of the single scenario instance I' derived from I as for the min-max case. We take as lower and upper bounds $L = \sum_{s \in S} \frac{1}{k}(val(x^*, s) - val_s^*)$ and $U = \max_{s \in S}(val(x^*, s) - val_s^*)$. Clearly, we have

$$L = \min_{x \in X} \frac{1}{k} \sum_{s \in S} (val(x, s) - val_s^*) \leq \min_{x \in X} \frac{1}{k} k \max_{s \in S} (val(x, s) - val_s^*) = opt$$

and

$$\min_{x \in X} \max_{s \in S}(val(x,s) - val_s^*) \le \max_{s \in S}(val(x^*,s) - val_s^*) \le \sum_{s \in S}(val(x^*,s) - val_s^*) = kL$$

If any instance of $\mathcal{P}$ of size n is solvable in time $p(n)$, where p is a polynomial, then bounds L and U are computable in $O(p(|I|/k))$. $\quad\square$

If $\mathcal{P}$ is polynomially approximable, then the first condition of Theorem 1 can be satisfied for MIN-MAX $\mathcal{P}$. More precisely, if $\mathcal{P}$ is $f(n)$-approximable where $f(n)$ is a polynomial, given an instance I of MIN-MAX $\mathcal{P}$, let $\widetilde{x}$ be an $f(|I|/k)$-approximate solution in I' (defined as in the proof of Proposition 1), then we have $L = \frac{1}{f(|I|/k)} \sum_{s \in S} \frac{1}{k} val(\widetilde{x},s)$ and $U = \max_{s \in S} val(\widetilde{x},s)$, and thus $U \le kf(|I|/k)L$.

The second condition of Theorem 1 can be weakened for MIN-MAX $\mathcal{P}$ by requiring only a pseudo-polynomial algorithm, that is an algorithm polynomial in $|I|$ and $max(I) = \max_{i,s} c_i^s$. Indeed, knowing an upper bound U, we can eliminate any variable x_i such that $c_i^s > U$ on at least one scenario $s \in S$. Condition 2 is then satisfied applying the pseudo-polynomial algorithm on this modified instance.

MIN-MAX and MIN-MAX REGRET versions of some problems, like shortest path, knapsack, admit pseudo-polynomial time algorithms based on dynamic programming [6]. For some dynamic programming formulations, we can easily obtain algorithms satisfying condition 2, by discarding partial solutions with value more than U on at least one scenario. We illustrate this approach in section 4.1 for the shortest path problem.

For other problems, which are not known to admit pseudo-polynomial algorithms based on dynamic programming, specific algorithms are required. We present an algorithm verifying condition 2 for MIN-MAX SPANNING TREE (section 4.2).

Unfortunately, these algorithms cannot be adapted directly in order to obtain algorithms satisfying condition 2 for min-max regret versions. The basic difficulty here is that, if we can find an algorithm in $r(|I|, U(I))$ for any instance I of MIN-MAX $\mathcal{P}$, the direct extension of this algorithm for the corresponding instance I' of MIN-MAX REGRET $\mathcal{P}$ will be in $r(|I'|, U(I') + opt_{max})$ where $opt_{max} = \max_{s \in S} val_s^*$ is a value which is not necessarily polynomially related to $U(I')$.

However, for problems whose feasible solutions have a fixed size such as spanning tree, we reduced the min-max regret version to a min-max version in [2]. In this context, we need to consider instances where some coefficients are negative and possibly non integral but any feasible solution has a non-negative integral value. For an optimization problem $\mathcal{P}$, we denote by $\mathcal{P}'$ the extension of $\mathcal{P}$ to these instances. More precisely, we proved the following theorem.

Theorem 2. ([2]) *For any polynomial-time solvable minimization problem $\mathcal{P}$ whose feasible solutions have a fixed size and for any function $f : \mathbb{N} \to (1, \infty)$, if MIN-MAX $\mathcal{P}'$ has a polynomial-time $f(n)$-approximation algorithm, then MIN-MAX REGRET $\mathcal{P}$ has a polynomial-time $f(n)$-approximation algorithm.*

4 Applications

In this section, we apply the previous results to min-max (regret) shortest path, and minimum spanning tree. We also compare the running time for our algorithms and for the fptas obtained using an approximation of the non-dominated set, and show a significant improvement.

4.1 Shortest Path

In [6], Kouvelis and Yu proved the NP-hardness of min-max and min-max regret versions of shortest path, even for two scenarios.

Consider an instance I defined by a directed graph $G = (V, A)$, with $V = \{1, \ldots, n\}$ and $|A| = m$, and a set S of k scenarios giving for each arc $(i, j) \in A$ its cost c_{ij}^s under scenario s. Denote by c_{ij} the vector of size k formed by c_{ij}^s, $s \in S$. We are interested in optimal paths from 1 to any other vertex.

We give now pseudo-polynomial algorithms satisfying condition 2 of Theorem 1 for MIN-MAX (REGRET) SHORTEST PATH.

Proposition 2. *Given U an upper bound on the optimal value, then* MIN-MAX SHORTEST PATH *and* MIN-MAX REGRET SHORTEST PATH *can be solved in time* $O(nmU^k)$.

Proof. We propose for each problem, an enumeration algorithm based on a dynamic programming formulation, that produces the set of all vectors of values (or regrets), for which all coordinates are less than or equal to U, and selects from this set an optimal vector. Let $u = (U, \ldots, U)$ denote the vector of size k.

Considering first MIN-MAX SHORTEST PATH, we describe an algorithm that computes at each stage ℓ, the set V_j^ℓ of all possible vectors of values at most U corresponding to paths from 1 to j of length at most ℓ, $\ell = 1, \ldots, n-1$, $j = 2, \ldots, n$. The algorithm starts by initializing $V_1^0 = \{(0, \ldots, 0)\}$, where $(0, \ldots, 0)$ is a vector of size k and computes V_j^ℓ at each stage ℓ for each vertex j, $\ell = 1, \ldots, n-1$, $j = 2, \ldots, n$ as follows:

$$V_j^\ell = \cup_{i \in \Gamma^{-1}(j)} \{v^j = v^i + c_{ij} : v^i \in V_i^{\ell-1} \text{ and } v^j \leq u\} \tag{3}$$

Finally, the algorithm selects, as an optimal vector, a vector in V_j^{n-1} such that its largest coordinate is minimum, for $j = 2, \ldots, n$.

Consider now MIN-MAX REGRET SHORTEST PATH. Let $(val_s^*)^i$, $s \in S$, $i = 1, \ldots, n$, be the value of a shortest path in graph G from 1 to i under scenario s and let $(val^*)^i$ be the vector of size k of these values $(val_s^*)^i$, $s \in S$.

We describe an algorithm that computes at each stage ℓ, the set R_j^ℓ of all possible vectors of regrets at most U corresponding to paths from 1 to j of length at most ℓ, $\ell = 1, \ldots, n-1$, $j = 2, \ldots, n$. Consider arc $(i, j) \in A$ and let P_i be a path in G from 1 to i of regret $r_s^i = val(P_i, s) - (val_s^*)^i$, $s \in S$. Denote by P_j the path constructed from P_i by adding arc (i, j). The regret of P_j is $r_s^j = val(P_i, s) + c_{ij}^s - (val_s^*)^j = r_s^i + (val_s^*)^i + c_{ij}^s - (val_s^*)^j$, $s \in S$. The algorithm starts by initializing $R_1^0 = \{(0, \ldots, 0)\}$ and for $1 \leq \ell \leq n-1$ and $2 \leq j \leq n$ let

$$R_j^\ell = \cup_{i \in \Gamma^{-1}(j)} \{r^j = r^i + (val^*)^i + c_{ij} - (val^*)^j : r^i \in R_i^{\ell-1} \text{ and } r^j \leq u\} \tag{4}$$

Finally, the algorithm selects, as an optimal vector, a vector in R_j^{n-1} such that its largest coordinate is minimum, for $j = 2, \ldots, n$.

We point out that, for both algorithms, any path of interest can be obtained using standard bookkeeping techniques that do not affect the complexity of these algorithms.

In order to prove the correctness of these algorithms, we show that V_j^{n-1}, resp. R_j^{n-1}, contains all vectors of values, resp. regrets, at most U corresponding to paths from 1 to j, $j = 2, \ldots, n$. For this, we need to justify that we can eliminate, at any stage, any vector which violates the upper bound U, without losing any vector at the end.

Indeed, for the min-max version, if such a solution v^i is obtained then any of its extensions computed in (3) would also violate U due to the non-negativity of vectors c_{ij}.

Similarly, for the min-max regret version, if such a solution r^i is obtained then any of its extensions computed in (4) would also violate U since vectors of the form $(val^*)^i + c_{ij} - (val^*)^j$ are non-negative.

Both algorithms can be implemented in time $O(nmU^k)$.   □

Corollary 1. MIN-MAX (REGRET) SHORTEST PATH *are in FPTAS.*

Proof. Using Theorem 1, Propositions 1 and 2, we derive an fptas whose running time is $O(\frac{mn^{k+1}}{\varepsilon^k})$.   □

Warburton describes in [11] an fptas for approximating the non-dominated set for the multi-objective version of the shortest path problem. From this fptas, Warburton derives an fptas for MIN-MAX SHORTEST PATH in acyclic graphs with running time $O(\frac{n^{2k+1}}{\varepsilon^{2k-2}})$, whereas our running time, for general graphs, is better.

4.2   Minimum Spanning Tree

In [6], Kouvelis and Yu proved the NP-hardness of min-max and min-max regret versions of minimum spanning tree, even for two scenarios. We first describe algorithms for MIN-MAX SPANNING TREE with running time polynomial in a suitably chosen upper bound on the optimal value.

Consider an instance of MIN-MAX SPANNING TREE represented by a graph $G = (V, E)$ where $|V| = n$, $|E| = m$, c_{ij}^s is the cost of edge (i, j) in scenario $s \in S$ and $|S| = k$.

Proposition 3. *Given U an upper bound on the optimal value, then* MIN-MAX SPANNING TREE *can be solved in time $O(mn^4U^k \log U)$.*

Proof. We can solve MIN-MAX SPANNING TREE using an extension of the matrix tree theorem to the multiple scenarios case as presented in section 2.3.

The optimal value opt of MIN-MAX SPANNING TREE can be computed by considering, for each monomial in (2), the largest power $v_{max} = \max_{s=1,\ldots,k} v_s$. The minimum value of v_{max} over all monomials corresponds to opt.

Actually, instead of computing all monomials, we can use, as suggested in [5], the algorithm presented in [7]. When applied to matrix $A_r(y_1, \ldots, y_k)$, this algorithm can compute the determinant polynomial up to a specified degree in each variable in opposition to the classical method of Edmonds [4]. In this case, it is sufficient to compute the polynomial determinant up to degree U in each variable y_s for $s = 1, \ldots, k$. The algorithm in [7] requires $O(n^4)$ multiplications and additions of polynomials. The time needed to multiply two multivariate polynomials of maximum degree d_s in variable y_s for $s = 1, \ldots, k$ is $\prod_{s=1}^{k} d_s \log \prod_{s=1}^{k} d_s$ [1]. Thus, the running time to compute the polynomial determinant is $O(n^4 U^k \log U)$.

Once an optimal vector is identified, a corresponding spanning tree can be constructed using self reducibility [8]. It consists of testing iteratively, for each edge if the graph obtained by contracting this edge admits a spanning tree of the required vector of adjusted values on all scenarios (subtracting iteratively the vector of costs c_{ij}^s, $s \in S$, for each edge (i, j) being tested to the required vector of values). In at most $m - (n-1)$ iterations such a spanning tree is obtained. Hence, the self reducibility requires $O(m)$ computations of determinant polynomial. $\square$

Corollary 2. MIN-MAX SPANNING TREE *is in FPTAS.*

Proof. Using Theorem 1, Propositions 1 and 3, we derive an fptas whose running time is $O(\frac{mn^{k+4}}{\varepsilon^k} \log \frac{n}{\varepsilon})$. $\square$

Corollary 3. MIN-MAX REGRET SPANNING TREE *is in FPTAS.*

Proof. Notice that Theorem 1 and Proposition 3 remain true even for the instances of spanning tree where some coefficients are negative but any feasible solution has a non-negative value. Thus, MIN-MAX SPANNING TREE′ is in FPTAS. The result follows from Theorem 2. The running time of the fptas is $O(\frac{mn^{k+4}}{\varepsilon^k} \log \frac{n}{\varepsilon})$. $\square$

In this case, we obtain fptas with better running times for MIN-MAX (REGRET) SPANNING TREE. Indeed, the running time of the fptas obtained in [2] using the general multi-objective approximation scheme presented in [9] is $O(\frac{n^{k+4}}{\varepsilon^{2k}} (\log U)^k \log \frac{n}{\varepsilon})$.

References

1. A. V. Aho, J. E. Hopcroft, and J. D. Ullmann. *The design and analysis of computer algorithms.* Addison-Wesley Reading, 1976.
2. H. Aissi, C. Bazgan, and D. Vanderpooten. Approximation complexity of min-max (regret) versions of shortest path, spanning tree, and knapsack. In *Proceedings of the 13th Annual European Symposium on Algorithms (ESA 2005), Mallorca, Spain,* LNCS 3669, pages 862–873, 2005.
3. F. Barahona and R. Pulleyblank. Exact arborescences, matching and cycles. *Discrete Applied Mathematics,* 16:91–99, 1987.
4. J. Edmonds. System of distinct representatives and linear algebra. *Journal of Research of the National Bureau of Standards,* 71:241–245, 1967.

5. S. P. Hong, S. J. Chung, and B. H. Park. A fully polynomial bicriteria approximation scheme for the constrained spanning tree problem. *Operations Research Letters*, 32(3):233–239, 2004.

6. P. Kouvelis and G. Yu. *Robust discrete optimization and its applications*. Kluwer Academic Publishers, Boston, 1997.

7. M. Mahajan and V. Vinay. A combinatorial algorithm for the determinant. In *Proceedings of the 8th Annual ACM-SIAM Symposium on Discrete Algorithms (SODA 1997), New Orleans, USA*, pages 730–738, 1997.

8. C. H. Papadimitriou. *Computational complexity*. Addison Wesley, 1994.

9. C. H. Papadimitriou and M. Yannakakis. On the approximability of trade-offs and optimal access of web sources. In *IEEE Symposium on Foundations of Computer Science (FOCS 2000), Redondo Beach, California, USA*, pages 86–92, 2000.

10. W.T. Tutte. *Graph Theory*, volume 21 of *Encyclopedia of Mathematics and its Applications*. Addison-Wesley, 1984.

11. A. Warburton. Approximation of Pareto optima in multiple-objective, shortest-path problems. *Operations Research*, 35:70–79, 1987.

The Class Constrained Bin Packing Problem with
· Applications to Video-on-Demand*

E.C. Xavier and F.K. Miyazawa

Instituto de Computação, Universidade Estadual de Campinas
Caixa Postal 6176, 13084-971, Campinas–SP, Brazil
eduardo.xavier@ic.unicamp.br, fkm@ic.unicmp.br

Abstract. In this paper we present approximation results for the class constrained bin packing problem that has applications to Video-on-Demand Systems. In this problem we are given bins of capacity B with C compartments, and n items of Q different classes. The problem is to pack the items into the minimum number of bins, where each bin contains items of at most C different classes and has total items size at most B. We present several approximation algorithms for off-line and online versions of the problem.

1 Introduction

In this paper we study the class constrained version of the well known bin packing problem, which we denote by CCBP (Class Constrained Bin Packing). In this problem we are given a tuple $I = (L, s, c, C, Q)$ where $L = (a_1, \ldots, a_n)$ is a list of items, each item $a_i \in L$ with size $0 < s_{a_i} \leq B$ and class $c_{a_i} \in \{1, \ldots, Q\}$, and a set of bins, each one with capacity B and C compartments. A packing $\mathcal{P}$ of L is a partition of the items into bins, where each part has total items size at most B and the number of different classes in each part is at most C. The problem is to find a packing of L into the minimum number of bins. In the online version of the CCBP problem the items must be packed in the order $(a_1, \ldots, a_n)$, where each item a_i must be packed without knowledge of further items. We consider that $1 < C < Q$, otherwise the CCBP problem can be solved as the original bin packing problem. That is, if $C = 1$ then items of different classes must be packed into different bins and if $C \geq Q$ then the class constraints are irrelevant. We also consider the version of this problem with bins of different sizes. In this case we have T different bins size. The input instance is a tuple $I = (L, s, c, w, C, Q)$ where $w : \{1, \ldots, T\} \to \mathbb{R}^+$ is a function of bins size. We assume w.l.o.g. that for each $i \in \{1, \ldots, T\}$, $w(i) \leq B$. In this case, the problem is to pack all items into bins such that the total size of used bins is minimized. This problem is denoted by VCCBP (Variable Class Constrained Bin Packing). Packing problems with class constraints have many applications in multimedia storage systems, resource allocation and in operations research like manufacturing systems [9,8,10,11,5,7,3].

---

* This work has been partially supported by CAPES, CNPq (Proc. 306526/04–2, 471460/04–4, 490333/04–4) and ProNEx–FAPESP/CNPq (Proc. 2003/09925-5).

D.Z. Chen and D.T. Lee (Eds.): COCOON 2006, LNCS 4112, pp. 439–448, 2006.
© Springer-Verlag Berlin Heidelberg 2006

Notation: In the online case, the bins used to pack the items are classified as *open* or *closed*. An empty bin is declared open when it receives its first item, and remains so until it is declared closed. Only open bins may receive items. Once a bin is closed, it cannot be declared open again. We consider the bounded and unbounded space versions for the online CCBP problem. In the l-bounded space problem an algorithm must keep at any time during its execution at most l open bins. In the unbounded version, an algorithm may keep an unbounded number of open bins. We define $s(I) = s(L) = \sum_{a \in L} s(a)$. Given an integer M, we denote by $[M]$ the set $\{1, \ldots, M\}$. Given two sequences $L_a = (a_1, \ldots, a_n)$ and $L_b = (b_1, \ldots, b_m)$, we denote the concatenation of these two lists by $L_a \| L_b$. Given a packing $\mathcal{P}$ we denote by $|\mathcal{P}|$ the number of bins in $\mathcal{P}$. Throughout this paper, we use the terms color and class with the same meaning. We say that a bin is *colored* if it contains items of C different classes

Related Work: A special case of the CCBP problem is the Bin Packing problem, which is a well known problem in the literature. We refer the reader to Coffman et al. [1] for a survey on approximation algorithms for bin packing problems. Recently the class-constrained versions of packing problems have obtained attention. In [3,2], Dawande et. al claimed to present an approximation scheme for the off-line VCCBP problem when the number of different classes Q in the input instance is bounded by a constant. In [9], Shachnai and Tamir presented a dual polynomial time approximation scheme for the off-line class constrained bin packing problem (CCBP) when the number of different classes in the input instance is bounded by a constant. In [8], Shachnai and Tamir presented theoretical results for a Multiple Knapsack problem with class constraints where all items have unit size. They introduced this problem with applications to video-on-demand servers. Subsequently to this work, Golubchik et al. [5] presented an approximation scheme to the problem. Later, Kashyap and Khuller [7], and Shachnai and Tamir [10] also presented approximation schemes to the problem, but they consider that the class requirement of items are not equal to all classes. Shachnai and Tamir in [11], presented algorithms for the online CCBP problem when all items have equal size. In this case they provide a lower bound of 2 to the problem and also algorithms that have a competitive ratio of 2.

Results: In this paper we show that the bounded space online CCBP problem does not admit a constant competitive ratio algorithm. Moreover if any item of the instance have size at least $\varepsilon < B$ we show that there is no algorithm with competitive ratio better than $O(1/C\varepsilon)$. For the unbounded space problem, we show that the First-Fit algorithm has competitive ratio in $[2.7, 3]$. We also present another online algorithm with competitive ratio in $[2.666, 2.75]$. We also present some results for the off-line problem. When all items have equal size, we present an $(1 + 1/C)$-approximation algorithm. When items have size at most B/m, for some integer m, we show an algorithm with approximation factor $(1 + 1/C + 1/\min\{C, m\})$. In all these cases, we consider that the number of different classes Q is part of the input. The VCCBP problem was first considered by Dawande et al. [3,2] where a tentative of an APTAS was considered when Q is bounded by a constant. We observed that their algorithm does not lead to an APTAS as claimed. In this paper we show the points where their algorithm fails and present an APTAS for the VCCBP problem for fixed Q.

Organization: In Section 2 we present the application of the CCBP problem to data placement of videos. In Section 3, we present practical approximation algorithms for the CCBP problem. In Section 4, we present lower and upper bounds for the online CCBP problem and in Section 5 we present an APTAS for the VCCBP problem when Q is bounded by a constant.

2 Applications of the CCBP Problem to Data Placement on Video-on-Demand Servers

The first work to consider packing problems with class constraints as a data placement problem is the one presented by Shachnai and Tamir [8]. They considered the knapsack version of the CCBP problem. In this case N bins are given, and the objective is to pack the maximum number of items satisfying the class constraints in each bin. Suppose we have a server of videos with N hard disks, each disk $j \in \{1, \ldots, N\}$ with capacity C_j and load capacity B_j. That is, each disk j can store C_j movies and can attend at most B_j simultaneous requests for videos. The problem is to construct a server such that, based on expected requests for movies (computed by movies popularity), the number of attended requests is maximized. The total load capacity of the server is $B_T = \sum_{j=1}^{N} B_j$. The movies considered to be stored in the server are $F_1, F_2, \ldots, F_f$ with popularity parameters $p_1, p_2, \ldots, p_f$, where $\sum_{i=1}^{f} p_i = 1$. Given these popularity parameters we compute expected requests for each movie at any time. These expected requests are, for each i, defined as $r_i = B_T p_i$. Notice that $\sum_{i=1}^{f} r_i = B_T$ (we assume that each r_i is an integer).

Consider for example that we have a server with two hard disks. Disk 1 has $C_1 = 2$ and $B_1 = 4$ and disk 2 has $C_2 = 2$ and $B_2 = 8$. There are three movies F_1, F_2 and F_3, with popularity parameters $p_1 = 1/4$, $p_2 = 1/4$ and $p_3 = 1/2$. Computing the expected requests one obtain $r_1 = 3$, $r_2 = 3$ and $r_3 = 6$. One optimal solution can be constructed as follows: one copy of movie F_1 is done in disk 1 (with 2 loads), a copy of movie F_2 is done in disk 1 (with 2 loads) and disk 2 (1 load) , and a copy of movie F_3 is done in disk 2 (with 6 loads). Notice that not all load capacity of the disks can be used. We call a perfect placement when all load capacity is used, i.e, all requests are allocated.

This problem was shown to be NP-hard by Shachnai and Tamir [8], and Golubchik et al. [5] show that even if all disks are equal, i.e, have same load and store capacities, the problem remains NP-hard. We can also consider the following problem: given a set of requests for a set of movies, construct a server using the minimum number of disks. This problem is also NP-hard since, given an instance for the data placement with N disks, a perfect placement exists, if and only if we can find a packing for all requests using at most N disks. When all disks are equal, we can see this data placement problem as a special case of the CCBP problem. In this case we have an instance $I = (L, s, c, C, Q)$, where each item $i \in L$ is a request for a load of class $c_i \in Q$ (the movie type). All items have the same size and C is the capacity of the disks, i.e, the number of different movies that the disk can store. That is, we want to construct a video server storing the videos and distributing all the requests minimizing the number of used disks.

3 Practical Approximation Algorithms

In this section we consider the problem where all items have unit size. As we saw, this problem is NP-hard and has applications in the data placement problem for video-on-demand. In this case, we can consider that items are given as a list of sets $U_1, \ldots, U_Q$, where each set U_i has n_i items of unit size with class i. Each bin packs at most B items of at most C different sets. The problem is to pack all sets of items in the minimum number of bins.

We adapt here, an algorithm known as Moving-Window (MW) first presented by Shachnai and Tamir [8] and also used later by Golubchik et al. [5] and Kashyap and Khuller [7]. All these papers are devoted to the knapsack version of the problem, where one must pack the maximum number of items in a given number of bins.

Moving-Window (MW): The algorithm keeps a vector $R = (R[1], R[2], \ldots, R[Q])$ representing non-packed items in such a way that $R[i]$ is the number of remaining items to be packed of some set U_j. The vector is maintained in non-decreasing order of the values $R[i]$ during all the execution of the algorithm. If at any given moment, it is packed part of the items represented by $R[i]$, then the vector must be reordered. In any iteration of the algorithm, it tries to pack C different sets to obtain a packing of a new bin. For that, the algorithm keeps a window of C sets. At first, the window goes from $R[1]$ to $R[C]$. If $\sum_{i=1}^{C} R[i] \geq B$ then the algorithm packs the corresponding sets of $R[1], R[2], \ldots, R[j]$, where $j \leq C$ is the first index such that $\sum_{i=1}^{j} R[i] \geq B$. Notice that $R[j]$ may be partially packed. The totally packed sets are removed from the vector. If $\sum_{i=1}^{C} R[i] < B$ then the algorithm moves the window to the right, until that for the first time the window has C sets such that their sizes are greater or equal than B. If this is the case, the C sets are packed and the vector R is reordered (if the last considered set was partially packed). Then the algorithm restarts. If in some iteration, the window reaches the end of the vector R, i.e, the C largest sets have total size smaller than B, then the algorithm generates bins by packing entirely C sets in each bin, except perhaps in the last bin that can have less than C sets.

Let $B_1, \ldots, B_N$ be the bins generated by the algorithm MW in the order they were generated. Let N_F be the number of full bins and N_C be the number of bins that are not full which we call colored. Let $N = N_F + N_C$. Notice that bins $B_1, \ldots, B_{N_F}$, are the full bins. When the algorithm creates the first non-full bin, when the window reaches the end of R and the C largest sets have total size smaller than B, all other generated bins becomes non-full having C different sets each, except perhaps the last.

Lemma 1. *If any of the first N_F bins produced by the algorithm MW packs less than C different sets (classes), then the algorithm produces an optimal solution.*

In this way, we consider that for each of the N_F first bins, the algorithm packs in each iteration, exactly C different sets and at most one of these sets is partially packed. Clearly, for the remaining N_C bins, all of them packs totally C different sets except perhaps in the last bin. Let OPT(I) be the number of bins used by an optimal solution to pack instance I. Notice that we must have $N_F \leq \text{OPT}(I) - 1$ otherwise the algorithm generated an optimal solution. We have the following result.

Lemma 2. *After the* MW *algorithm has generated the first* OPT(I) *bins, there exists at most* N_F *sets to be packed.*

With this result we can give the approximation factor of the MW algorithm. We can also show that the bound of the next theorem is tight.

Theorem 1. *The* MW *algorithm has an asymptotic approximation factor of* $(1 + \frac{1}{C})$ *for the* CCBP *problem when all items are equal sized.*

Notice that the MW algorithm is based on a heuristic that tries to pack C different sets in each bin. But the way the algorithm works, it tends to pack small and large sets in different bins. A good heuristic is to pack large and small sets together, in such a way that each generated bin has a good use of its capacity, while trying to pack C different sets in each bin. For that, we propose a new algorithm that we call Modified-Moving-Window (MW'). Experimental tests shows that this algorithm produce better solutions than the MW algorithm.

Modified-Moving-Window (MW'): This algorithm is similar to the MW algorithm since it also keeps a window of size C over a vector $R = (R[1], R[2], \ldots, R[Q])$ that is maintained in non-decreasing order of the values $R[i]$. The algorithm also moves a window of size C until the total size of the sets in the window contains B or more items. In the MW' algorithm, we consider that the vector R is a circular list. At first, the window consists of the sets $R[1], \ldots, R[C]$. If the total size of these sets is greater than or equal to B, then the algorithm packs the sets $R[1], \ldots, R[j]$, where $j \leq C$ is the first index such that $\sum_{i=1}^{j} R[i] \geq B$, with the last set $R[j]$ probably partially packed. If the total size of these sets is smaller than B then instead of doing a move to the right, as in the original MW algorithm, the algorithm performs a move to the left and considers the sets $R[Q], R[1], \ldots, R[C-1]$. The algorithm performs moves to the left until the total size of the C sets are greater than or equal to B. In this case it packs the C sets and restarts. If the algorithm performs C moves to the left, and then considers the C largest sets, and these sets have total size less than B, then the algorithm generates a packing as the original MW algorithm, by packing entirely C sets in each bin.

It is not hard to prove similar results to Lemma 1 and Lemma 2 to the MW' algorithm. We can prove the following result which is also a tight bound.

Theorem 2. *The* MW' *algorithm has an asymptotic approximation factor of* $(1 + \frac{1}{C})$ *for the* CCBP *problem where all items have the same size.*

Now we consider the case where items in each set may have different sizes. This case is also interesting for applications of the data-placement problem to video-on-demand servers. Suppose that users have different network access speeds. In this case, requests for load resources may have different sizes. Suppose that the maximum size of an item is bounded by B/m for some $m \geq 1$. Problems with this restriction are also called parametric packing problems. Given an integer m, we denote this version of the problem as Parametric Class Constrained Bin Packing (PCCBP$_m$) problem.

Let I be an instance of the PCCBP$_m$ problem, $m \geq 1$. Consider that the input instance I consists of sets $U_1, \ldots, U_Q$. We now present an algorithm to pack this instance. Although items may have different sizes, consider that each item with size s, $s > 1$, is broken into s unit sized pieces. Now apply the MW algorithm for this modified instance. Now consider this packing for the original items. For each full bin it may happen that the last item is fractionally packed. For each bin where this happens, remove the item from the bin. Notice that there are at most N_F items removed from the generated packing. For these remaining items, generate new bins packing at least $\min\{m, C\}$ items in each bin except perhaps in the last bin.

Theorem 3. *There exists an algorithm for the* PCCBP_m *problem, where each item has size at most* B/m, *for some* $m \geq 1$, *with asymptotic approximation factor bounded by* $(1 + 1/\min\{m, C\} + 1/C)$.

4   The Online CCBP Problem

From now on, we consider that the capacity of the bin is $B = 1$, and each item e has size $0 < s_e \leq 1$. In this section we consider the online class constrained bin packing problem. In this case each item in the list of items $L = (a_1, \ldots, a_n)$, is packed without knowledge of subsequent items in the list.

First we present inapproximability results for the bounded space online CCBP problem. In this case, the basic strategy is to compare the result obtained by any on-line algorithm with the optimum off-line packing. The idea to obtain the lower bounds is to construct an instance where the list of items consists of n sublists. All sublists are identical and consists of very small items all of them of different classes. Since any bounded space algorithm cannot keep too many opened bins, each one of the lists are almost packed separately to each other. In an optimal off-line solution we can group items of the same class of all sublists and pack them together.

Theorem 4. *Let l be a constant, then the l-bounded space online* CCBP *problem does not admits an algorithm with constant competitive ratio. Moreover, if each item has size at least $\varepsilon < 1$ then the online* CCBP *problem does not admits an algorithm with competitive ratio better than* $O(1/C\varepsilon)$.

Given these results, for the remaining of this section we only consider the unbounded space online CCBP problem. Given an online algorithm $\mathcal{A}$ for the bin-packing problem, we can obtain an online algorithm $\mathcal{A}^*$ for the online CCBP problem in a straightforward manner. To pack the next item e, the algorithm $\mathcal{A}^*$ works as follows: Let c_e be the class of the item e, $\mathcal{B}$ be the list of bins in the order they were opened. Let $\mathcal{B}_e$ be the list of bins of $\mathcal{B}$, in the same order of $\mathcal{B}$, where each bin has at least one item of class c_e or has items of at most $C - 1$ different classes. The item e is packed with algorithm $\mathcal{A}$ into the bins of $\mathcal{B}_e$.

One of the most famous algorithm for the bin-packing problem is the First-Fit (FF) algorithm. This algorithm packs the next item into the first bin, in the order they were opened, that has sufficient space for the item. Notice that the algorithm FF* is online, since it only looks for the item it is packing and it is unbounded since it keeps all bins opened. In fact it closes a bin only if the bin is full.

We can show the following result for the algorithm FF*. We note that the upper bound was previously shown by Dawande et al. [2].

Theorem 5. *The algorithm* FF* *has a competitive ratio in* $[2.7, 3]$ *for the online* CCBP *problem.*

The idea to prove the upper bound of this theorem is to consider separately bins that are filled by at least half of its capacity and bins that are not. In the first case the number of bins is bounded by $2\mathrm{OPT}(I)$. In the later case we can prove that all bins are colored, except perhaps the last, and then using the fact that $\lceil Q/C \rceil \leq \mathrm{OPT}(I)$, we can bound the number of used bins by $\mathrm{OPT}(I) + 1$. The idea to prove the lower bound is to use an intricate instance presented by Johnson et al. [6] that provides a lower bound of 1.7 for the FF algorithm in the bin packing problem and add at the beginning of the input list very small items all of them of different classes.

Now we present another online algorithm, which we denote by $\mathcal{A}_C$, with competitive ratio in the interval $(2.666, 2.75]$. **Algorithm** $\mathcal{A}_C$: Let $\mathcal{P}_i$ be lists of empty bins, for $i = 1, 2, 3$. For each item e given in the input instance let $i = 1$ if $1 \geq s_e > 1/2$; $i = 2$ if $1/2 \geq s_e > 1/3$; and $i = 3$ if $1/3 \geq s_e$. The algorithm packs item e using the algorithm FF^* in the bins of $\mathcal{P}_i$. We can show the following result for this algorithm.

Theorem 6. *Algorithm* $\mathcal{A}_C$ *has a competitive ratio in* $[2.666, 2.75]$.

5 An APTAS for Bounded Number of Classes

In this section we present an APTAS for the off-line VCCBP problem. The input instance for this problem is a tuple $I = (L, s, c, w, C, Q)$. The problem is to find a packing of all items minimizing the total size of used bins. In this section we consider that the maximum size of a bin is 1 and that the number of different classes Q in the input instance, is bounded by a constant. As was noticed by Dawande et al. [3,2], we only use bins such that their size are at least ε, since this condition does not affect too much the cost of the solution, i.e, the algorithm remains an APTAS.

We give a brief description of the algorithm of Dawande et al. [3,2] and present the points where their algorithm fails. Let I be an instance for the VCCBP problem and let L_b be the items in L with size at least ε^2 (big items) and let L_s be the remaining items in L (small items). Let $n = |L_b|$. The algorithm sorts the list L_b in non-increasing order of size and partition this list into groups (lists) $L_1, \ldots, L_M$, each one with $\lceil n\varepsilon^2 \rceil$ items except perhaps the last list that can have less than $\lceil n\varepsilon^2 \rceil$ items. Call the first item in each group as the group-leader. Let L'_i be the list having $|L'_i| = |L_i|$ items, where each item has size equal to the size of the group-leader of L_i. Let $L' = L'_1 \| \ldots \| L'_M$.

For the list L' it is possible to generate all configurations of bins in constant time. Given an item size and an item color, denote by d_i the number of items of this type $i \in [t]$. Let N be the total number of bin configurations. Let x_j be a variable that represents the number of times a configuration $j \in [N]$ is used in a solution, a_{ij} be the coefficient that represents the number of times an item type $i \in [t]$ is used in configuration j and w_j the size of the bin used in configuration j. The next step of the algorithm is to solve the following linear program:

$$\min \sum_{j=1}^{N} w_j x_j$$

$$\sum_{j=1}^{N} a_{ij} x_j \geq d_i \qquad \forall\, i \in [t] \quad (1) \qquad \text{(LP)}$$

$$x_j \geq 0 \qquad \forall\, j \in [N], (2)$$

The algorithm solves this linear program and generates an integer solution by rounding up the variables x. The solution is a packing for the list L' that is used to generate a packing for the list L_b.

The next step of the algorithm is to pack the small items into the packing obtained from the linear program step. To do this, it uses a first-fit strategy: Pack an item in the first bin that has enough space to accommodate it and that satisfy the color constraints. Dawande et al. [3,2] claimed that this algorithm is an APTAS for the VCCBP problem. The list L_b was partitioned into lists $L_1 \| \ldots \| L_M$. Let L_i'' be a list having $|L_i''| = |L_i|$ items, where each item has size equal to the group-leader of the list L_{i+1}, for $i = 1, \ldots, M-1$, and L_M'' be an empty list. Let $L'' = L_1'' \| \ldots \| L_M''$. Clearly OPT$(L'') \leq$ OPT(L_b).

Dawande et al. claimed that the following relation is valid

$$OPT(L') \leq OPT(L'') + \lceil n\varepsilon^2 \rceil \leq OPT(L_b) + \lceil n\varepsilon^2 \rceil,$$

given the argument that L' and L'' differ only in their first and last groups. Although it seems to be true, notice that the color of items of L_i' and L_{i-1}'' may be different. Then, it is not clear how to construct a packing for $L_2' \| \ldots \| L_M'$ given a packing for L''. Let X be the number of bins used by their algorithm. After packing the small items using the first-fit strategy, they claimed that at least $X - \lceil \frac{Q}{C} \rceil$ bins have residual capacity at most ε. This is also not true. Suppose all small items have different colors from the big items. It is easy to construct examples where optimal packings for the big items given by the linear program have all bins with C different colors and the residual space is larger than a given ε. In this way, no small item will be packed in the bins obtained by the linear program step and then, all these bins will have residual capacity greater than ϵ.

Now we present an APTAS for the VCCBP problem. We show how to pack big items doing a linear rounding for each different color. The algorithm to pack the big items generates a polynomial number of packings for the big items, and also provide information of how to pack small items. The algorithm to pack the small items is based in the solution of a linear program. The algorithm generates a polynomial number of packings such that at least one is very close to the optimal.

Let L_b be the items in L with size at least ε^2 (big items) and let L_s be the remaining items in L (small items). The algorithm that packs the list L_b, which we denote by A_{LR}, uses the linear rounding technique, presented by Fernandez de la Vega and Lueker [4], and considers only items with size at least ε^2. The algorithm A_{LR} returns a pair $(\mathcal{P}_B, \mathbb{P})$, where $\mathcal{P}_B$ is a packing for a list of very big items and $\mathbb{P}$ is a set of packings for the remaining items of L_b.

Let X and Y be two lists of items. We say that $X \preceq Y$ if there is an injection $f_c : X_c \to Y_c$ for each $c \in [Q]$ such that $s(e) \leq s(f(e))$ for all $e \in X_c$. For a list of items X, denote by $\overline{X}$ the list with precisely $|X|$ items with size equal to the size of the smallest item in X.

The algorithm A_{LR} consists in the following: Let $L_1, \ldots, L_Q$ be the partition of the input list L_b into colors $1, \ldots, Q$ and let $n_c = |L_c|$ for each color c. The algorithm A_{LR} sorts each list L_c in non-increasing order of size and then partition the list L_c into at most $M = \lceil 1/\varepsilon^3 \rceil$ groups $L_c^1, L_c^2, \ldots, L_c^M$, where $L_c = L_c^1 \| \ldots \| L_c^M$. Each group has $\lfloor n_c \varepsilon^3 \rfloor$ items except perhaps the last list (with the smallest items) that can have less than $\lfloor n_c \varepsilon^3 \rfloor$ items. Let $L_B = \cup_{c=1}^{Q} L_c^1$. The algorithm generates a packing $\mathcal{P}_B$ of L_B with cost at most $O(\varepsilon)\mathrm{OPT}(I)$ and a set $\mathbb{P}$ with a polynomial number of packings for the items in $L_b \setminus L_B$. The packing $\mathcal{P}_B$ is generated by the algorithm FF^* (presented in Section 4) with bins of size 1. The algorithm generates a set of packings $\mathbb{P}$, of polynomial size, for the list $(\overline{L_1^1}\| \dots \|\overline{L_1^{M-1}} \| \dots \|\overline{L_Q^1}\| \dots \|\overline{L_Q^{M-1}})$. This can be done in polynomial time as the next lemma guarantees.

Lemma 3. *Given an instance $I = (L_b, s, c, w, C, Q)$, where the number of distinct item sizes of each color is at most a constant M, the number of different colors is bounded by a constant Q and each item $e \in L_b$ has size $s_e \geq \varepsilon^2$, then there exists a polynomial time algorithm that generates all possible packings of L_b. Moreover, each bin of each generated packing has an indication of the possible colors that may be used by further small items.*

Since $\overline{L_c^i} \succeq L_c^{i+1}$, for $i = 1, \dots, M-1$ and each color c, it is easy to construct a packing for the list $L_1^2\| \dots \|L_1^M\| \dots \| L_Q^2\| \dots \|L_Q^M$, given a packing for the list $(\overline{L_1^1}\| \dots \|\overline{L_1^{M-1}}\| \dots \|\overline{L_Q^1}\| \dots \|\overline{L_Q^{M-1}})$.

Let $\mathcal{P} = \{B_1, \dots, B_k\}$ be a packing of L_b and suppose we have to pack a list L_s of small items, with size at most ε^2, into $\mathcal{P}$. The packing of the small items is obtained from a solution of a linear program. Let $N_i \subseteq [Q]$ be the set of possible colors that may be used to pack the small items into the bin B_i of the packing $\mathcal{P}$. For each color $c \in N_i$, define a non-negative variable x_c^i. The variable x_c^i indicates the total size of small items of color c to be packed in the bin B_i. Denote by $s(B_i)$ the total size of items already packed in the bin B_i and by $w(B_i)$ the capacity of bin B_i. Consider the following linear program denoted by LPS:

$$\max \sum_{i=1}^{k} \sum_{c \in N_i} x_c^i$$

$$s(B_i) + \sum_{c \in N_i} x_c^i \leq w(B_i) \qquad \forall\, i \in [k] \quad (1) \quad \text{(LPS)}$$

$$\sum_{i=1}^{k} x_c^i \leq s(S_c) \qquad \forall\, c \in [C], (2)$$

where S_c is the set of small items of color c in L_s.

Given a packing $\mathcal{P}$, and a list L_s of small items, the algorithm first solves the linear program LPS, and then packs small items in the following way: For each variable x_c^i it packs, while possible, the small items of color c into the bin B_i, so that the total size of the packed small items is at most x_c^i. The possible remaining small items are packed using the algorithm FF^* into new bins of size 1. The algorithm to pack small items has polynomial time, since the linear program *LPS* can be solved in polynomial time.

The small items that are packed into new bins use at most

$$\left\lceil \frac{(s(L_s) - \sum_{i=1}^{k}\sum_{c \in N_i} x_c^i)}{(1 - \varepsilon^2)} + \frac{|\mathcal{P}|\varepsilon^2 Q}{(1 - \varepsilon^2)} \right\rceil + \lceil Q/C \rceil$$

new bins, since each bin is filled by at least $(1 - \varepsilon^2)$ except perhaps by at most $\lceil Q/C \rceil$ bins. The algorithm packs the small items in each packing $\mathcal{P} \in \mathbb{P}$. In the end, the algorithm generates another set of packings $\mathbb{P}'$ for all items. At least one of the generated packings has size at most $(1+O(\varepsilon))\mathrm{OPT}(I)+\beta$, for a constant β. The algorithm returns the packing with smallest cost. We can prove that the presented algorithm is an APTAS for the VCCBP.

Theorem 7. *Let* $I = (L, s, c, w, C, Q)$, *be an instance for the* VCCBP *problem. The packing* $\mathcal{P}$ *returned by the algorithm satisfy* $w(\mathcal{P}) \leq (1 + O(\varepsilon))\mathrm{OPT}(I) + \beta$, *where* β *is a constant.*

References

1. E. G. Coffman, Jr., M. R. Garey, and D. S. Johnson. Approximation algorithms for bin packing: a survey. In D. Hochbaum, editor, *Approximation Algorithms for NP-hard Problems*, chapter 2, pages 46–93. PWS, 1997.
2. M. Dawande, J. Kalagnanam, and J. Sethuraman. Variable sized bin packing with color constraints. Technical report, IBM, T.J. Watson Research Center, NY, 1998.
3. M. Dawande, J. Kalagnanam, and J. Sethuraman. Variable sized bin packing with color constraints. In *Proceedings of Graco 2001*, volume 7 of *Electronic Notes in Dicrete Mathematics*, 2001.
4. W. Fernandez de la Vega and G. S. Lueker. Bin packing can be solved within $1 + \epsilon$ in linear time. *Combinatorica*, 1(4):349–355, 1981.
5. L. Golubchik, S. Khanna, S. Khuller, R. Thurimella, and A. Zhu. Approximation algorithms for data placement on parallel disks. In *Proceedings of SODA*, pages 223–232, 2000.
6. D. S. Johnson, A. Demers, J. D. Ullman, M. R. Garey, and R. L. Graham. Worst-case performance bounds for simple one-dimensional packing algorithms. *SIAM Journal on Computing*, 3:299–325, 1974.
7. S. R. Kashyap and S. Khuller. Algorithms for non-uniform size data placement on parallel disks. In *Proceedings of FSTTCS*, volume 2914 of *Lecture Notes in Computer Science*, pages 265–276, 2003.
8. H. Shachnai and T. Tamir. On two class-constrained versions of the multiple knapsack problem. *Algorithmica*, 29:442–467, 2001.
9. H. Shachnai and T. Tamir. Polynomial time approximation schemes for class-constrained packing problems. *Journal of Scheduling*, 4(6):313–338, 2001.
10. H. Shachnai and T. Tamir. Approximation schemes for generalized 2-dimensional vector packing with application to data placement. In *Proceedings of 6th RANDOM-APPROX*, volume 2764 of *Lecture Notes in Computer Science*, pages 165–177, 2003.
11. H. Shachnai and T. Tamir. Tight bounds for online class-constrained packing. *Theoretical Computer Science*, 321(1):103–123, 2004.

MAX-SNP Hardness and Approximation of Selected-Internal Steiner Trees

Sun-Yuan Hsieh[*] and Shih-Cheng Yang

Department of Computer Science and Information Engineering,
National Cheng Kung University,
No. 1, University Road, Tainan 701, Taiwan
hsiehsy@mail.ncku.edu.tw

Abstract. In this paper, we consider an interesting variant of the well-known Steiner tree problem: Given a complete graph $G = (V, E)$ with a cost function $c : E \rightarrow \mathbf{R}^+$ and two subsets R and R' satisfying $R' \subset R \subseteq V$, a *selected-internal Steiner tree* is a Steiner tree which contains (or spans) all the vertices in R such that each vertex in R' cannot be a leaf. The *selected-internal Steiner tree problem* is to find a selected-internal Steiner tree with the minimum cost. In this paper, we show that the problem is MAX SNP-hard even when the costs of all edges in the input graph are restricted to either 1 or 2. We also present an approximation algorithm for the problem.

1 Introduction

The Steiner tree problem is succinctly a minimum interconnection problem. The most basic version is in a graph: Given a weighted graph $G = (V, E)$ with a cost function $c : E \rightarrow \mathbf{R}^+$ on the edges and a subset $R \subseteq V$ of vertices, a *Steiner tree* is a connected and acyclic subgraph of G which spans all the vertices in R. The vertices in R are usually referred to as *terminals* and the others in $V \setminus R$ are *Steiner* (or *optional*) vertices. Note that a Steiner tree may contain some Steiner vertices. The *total cost* (*cost* for short) of a Steiner tree is defined to be the sum of the costs of all its edges. The so-called *Steiner tree problem* (STP for short) is to find a *Steiner minimum tree* (or an *optimal Steiner tree*), i.e., a Steiner tree with the minimum cost in G [4]. The decision version of STP was shown to be NP-complete [6].

There are practical applications such that some terminal vertices are required to be internal vertices in a Steiner tree and the others may not. For example, in a network resource allocation, some specified servers (terminals) are allowed to act as transmitters and the others need not have this restriction. Consequently, in a solution tree, some terminals are restricted to be internal vertices and the others can be leaves or internal vertices. In this paper, we study an interesting variant of the Steiner tree problem: Given a complete graph $G = (V, E)$ with a cost function $c : E \rightarrow \mathbf{R}^+$ and two subsets R and R' satisfying $R' \subset R \subseteq V$, the *selected-internal Steiner tree problem* (SISTP for short) is to find a Steiner

---

[*] Correspondence author.

D.Z. Chen and D.T. Lee (Eds.): COCOON 2006, LNCS 4112, pp. 449–458, 2006.
© Springer-Verlag Berlin Heidelberg 2006

minimum tree which spans all the vertices in R such that each vertex in R' cannot be a leaf. For convenience, we call such a tree as *optimal selected-internal Steiner tree*, and call the vertices in R' the *demanded terminals*. We first show that SISTP is MAX SNP-hard. We then present an approximation algorithm for the problem.

2   Preliminaries

This paper considers finite, simple, and loopless graph $G = (V, E)$, where V and E are the vertex and edge sets of G, respectively. We also use the notations $V(G)$ and $E(G)$ to denote the vertex and edge sets of G, respectively. For a graph G, the *degree* of v in G, denoted by $\deg_G(v)$, is the number of edges incident to v in G. A *path* of *length* k from a vertex v_0 to a vertex v_k in a graph $G = (V, E)$ is a sequence $\langle v_0, v_1, v_2, \ldots, v_k \rangle$ of vertices such that $(v_{i-1}, v_i) \in E$ for $i = 1, 2, \ldots, k$. We use $P_G[u, u']$ to denote a path from u to u' in G. A path is *simple* if all vertices in the path are distinct.

A *tree* T is a connected, acyclic, and undirected graph. A vertex in T with degree 1 is called a *leaf*. A nonleaf vertex in T is an *internal vertex*. For graph-theoretic terminologies and notations not mentioned here, please refer to [9].

Definition 1. A function $c : V \times V \to \mathbf{R}^+$ is called a *metric* if it satisfies the following three conditions: (1) $c(x, y) \geq 0$ for any $x, y \in V$, where equality holds if and only if $x = y$; (2) $c(x, y) = c(y, x)$ for any $x, y \in V$; (3) The triangle inequality $c(x, y) \leq c(x, z) + c(z, y)$ holds for any $x, y, z \in V$.

The cost function c used throughout this paper is *metric*. For an edge e in a tree T, $c(e)$ is the *cost* of e, and $c(T)$ is the sum of all the edge costs of T. If we reduce the co-domain of c from $\mathbf{R}^+$ to $\{1, 2\}$, then the restricted SISTP is called the *(1,2)-Selected-Internal Steiner Tree Problem* (SISTP(1,2) for short). The problem STP(1,2) can be defined similarly.

Definition 2. Given two optimization problem Π_1 and Π_2, we say that Π_1 *L-reduces* to Π_2 if there exist polynomial-time algorithms A_1 and A_2, and positive constants α and β such that for any instance I_1 of Π_1, the following conditions are satisfied:

1. The algorithm A_1 produces an instance $A_1(I_1)$ of Π_2 such that $\mathsf{OPT}(A_1(I_1)) \leq \alpha \cdot \mathsf{OPT}(I_1)$, where $\mathsf{OPT}(I_1)$ and $\mathsf{OPT}(A_1(I_1))$ represent the costs of optimal solutions of I_1 and $A_1(I_1)$, respectively.
2. Given any solution of $A_1(I_1)$ with cost $cost_2$, the algorithm A_2 produces a solution of I_1 with cost $cost_1$ in polynomial time such that $|cost_1 - \mathsf{OPT}(I_1)| \leq \beta \cdot |cost_2 - \mathsf{OPT}(A_1(I_1))|$.

A problem is said to be MAX SNP-hard if all MAX SNP problems can be L-reduced to this problem. Note that the problem itself may not be MAX SNP. It was shown that if any MAX SNP-hard problem has a *polynomial-time approximation scheme* (PTAS), then P=NP [1,2].[1] In other words, it is very unlikely

---

[1] A problem has a PTAS if the problem can be approximated within a factor, $1 + \epsilon$, in polynomial time for any $\epsilon > 0$.

that for a MAX SNP-hard problem to have a PTAS. On the other hand, if Π_1 L-*reduces* to Π_2 and Π_2 has a PTAS, then Π_1 has a PTAS [7].

3   MAX SNP-Hardness Results

In this section, we will show that SISTP(1,2) is MAX SNP-hard.

Algorithm $A_1(I_1)$

Input: The input instance $I_1 = (G_1, R_1, c_1)$ of STP(1,2), where $G_1 = (V_1, E_1)$ is a
    complete graph, $R_1 \subseteq V_1$, and the cost function $c_1 : E_1 \to \{1, 2\}$.
Output: The input instance $I_2 = (G_2, R_2, R_2', c_2)$ of SISTP(1,2).

1: Create one *auxiliary vertex* a and *auxiliary edge set* $E' = \{(a, v_i)| \ v_i \in V_1\}$.
2: Construct $G_2 = (V_2, E_2)$, where $V_2 = V_1 \cup \{a\}$ and $E_2 = E_1 \cup E'$. Let $R_2' = \{a\}$
    and $R_2 = R_1 \cup \{a\}$.
3: Define the cost function $c_2 : E_2 \to \{1, 2\}$ such that for each $e \in E_2$,

$$c_2(e) = \begin{cases} c_1(e) & \text{if } e \in E_1, \\ 2 & \text{otherwise.} \end{cases}$$

Fig. 1. An algorithm for the input instance transformation from STP(1,2) to SISTP(1,2)

Lemma 1.   [3] *STP(1,2) is MAX SNP-hard.*

Next, we go on to show the MAX SNP-hardness for the problem by providing an
L-reduction from STP(1,2) to SISTP(1,2). An algorithm for the input instance
transformation from STP(1,2) to SISTP(1,2) is presented in Figure 1.

Lemma 2. $OPT(A_1(I_1)) \leq 5OPT(I_1).$

Proof. Let S^* be an optimal Steiner tree of I_1. Let $e = (u, v)$ be any edge of S^*.
By deleting the edge e from S^* and adding two auxiliary edges (a, u) and (a, v),
we obtain another tree $S^* - (u, v) + (a, u) + (a, v)$, which is clearly a selected-
internal Steiner tree for $A_1(I_1)$. Note that a becomes an internal vertex of the
resulting tree. Then,

$$\begin{aligned} \text{OPT}(A_1(I_1)) &\leq c_2(S^*) - c_2(e) + c_2(a, u) + c_2(a, v) \\ &= c_1(S^*) - c_1(e) + 4 \\ &\leq \text{OPT}(I_1) + 4 \\ &\leq \text{OPT}(I_1) + 4\text{OPT}(I_1) \\ &\leq 5\text{OPT}(I_1). \end{aligned} \tag{1}$$

Q.E.D.

We now present a polynomial-time algorithm A_2 shown in Figure 2 to obtain
a solution of I_1 from a solution of $A_1(I_1)$. Let $T_1, T_2, \ldots, T_k$ be the resulting trees

Algorithm $A_2(T, A_1(I_1))$

Input: A selected-internal Steiner tree T of $A_1(I_1)$.
Output: A Steiner tree of I_1.

1: Delete the auxiliary vertex a in T. Since a is an internal vertex of T, at least two subtrees resulted. Let $T_1, T_2, \ldots, T_k$ be the resulting subtrees after deleting a such that each tree contains at least one terminal, i.e., $V(T_i) \cap R \neq \emptyset$ for all $1 \leq i \leq k$.
2: Merge $T_1, T_2, \ldots, T_k$ into one tree using $k - 1$ edges in $E(G_1) \setminus (\bigcup_{i=1}^{k} E(T_i))$ in which the cost of these $k - 1$ edges is as smallest as possible.

Fig. 2. An algorithm generates a solution of I_1 from a solution of $A_1(I_1)$

obtained after executing Step 1 of Algorithm A_2. A path $\langle v_1, v_2, \ldots, v_l \rangle$ is said to be a (p, q)-*bridge*, where $p, q \in \{1, 2, \ldots, k\}$ and $p \neq q$, iff $v_1 \in V(T_p)$, $v_l \in V(T_q)$, and $v_2, v_3, \ldots, v_{l-1} \in V(G_1) - \bigcup_{i=1}^{k} V(T_i)$. The following two propositions are useful to show Lemma 3.

Proposition 1. *If T is an optimal selected-internal Steiner tree of $A_1(I_1)$, then each subtree obtained after deleting the auxiliary vertex a from T contains at least one terminal.*

Proof. Assume, by contradiction, that there exists some subtree T_i with no terminal. Then, $T - T_i$ is another selected-internal Steiner tree whose cost is smaller than that of T. This is a contradiction. **Q.E.D.**

Proposition 2. *Let T be an optimal selected-internal Steiner tree of $A_1(I_1)$, and let $T_1, T_2, \ldots, T_k$ be the resulting trees obtained after executing Step 1 of Algorithm A_2. If $k > 2$, then the cost of any (p, q)-bridge is at least 2.*

Proof. Let r_p and r_q be the roots of T_p and T_q, respectively. Assume, by contradiction, that there is a (p, q)-bridge whose cost is 1. Clearly, such a bridge is exactly an edge, denoted by e. By the fact that $k > 2$, $T - (a, r_q) + e$ remains a selected-internal Steiner tree whose cost is smaller then that of T. This is contrary to the assumption that T is optimal. **Q.E.D.**

Lemma 3. *If T is an optimal selected-internal Steiner tree of the instance $A_1(I_1)$, then an output tree returned by Algorithm A_2 is an optimal Steiner tree of the instance I_1.*

Proof. Let $T_1, T_2, \ldots, T_k$ be the resulting trees obtained after executing Step 1 of Algorithm A_2. Let $r_1, r_2, \ldots, r_k$ be the roots of $T_1, T_2, \ldots, T_k$, respectively. Also let T' be an output tree returned by Algorithm A_2. Assume, by contradiction, that T' is not optimal, i.e., there exists another Steiner tree of I_1 whose cost is smaller than that of T'. Let T'' be such a tree with the smallest Steiner vertices.

CASE 1. $k > 2$. According to Algorithm A_2 and Proposition 2, the $k - 1$ edges used to merge $T_1, T_2, \ldots, T_k$ into one tree are all cost 2. That is, T' is obtained from T by deleting k cost-2 edges, and adding $k - 1$ cost-2 edges. Then, we have that

$$c(T) = c(T') + 2 \tag{2}$$

According to Proposition 2 and the assumption that T'' contains the smallest Steiner vertices, there is a cost-2 edge (x, y) in T''. By the execution of Algorithm A_1, each vertex v of T'' is adjacent to the auxiliary vertex a, through the auxiliary edge (v, a). Then, we can obtain another selected-internal Steiner tree $Q = T'' - (x, y) + (x, a) + (y, a)$ is another selected-internal Steiner tree such that

$$
\begin{aligned}
c(Q) &= c(T'') - 2 + 4 \\
&= c(T'') + 2.
\end{aligned} \tag{3}
$$

Since $c(T'') < c(T')$, $c(Q) < c(T)$ by Equations 2 and 3. This contradicts to the assumption that T is optimal.

CASE 2. $k = 2$. Assume that e is the edge used in Step 2 of Algorithm A_2 to merge T_1 and T_2.

  CASE 2.1.   The cost of e equals 1. By the execution of Algorithm A_2, we have that

$$c(T) = c(T') + 3. \tag{4}$$

  If there is a cost-2 edge (x, y) in T'', then $T'' - (x, y) + (x, a) + (y, a)$ is a selected-internal Steiner tree with cost $c(T'') + 2$, which is obviously smaller than $c(T) = c(T') + 3$ by the fact that $c(T'') < c(T')$. This is a contradiction. Otherwise, if all the edges of T'' have cost 1, then we can select an arbitrary edge (x', y') and obtain a selected-internal Steiner tree $T'' - (x', y') + (x', a) + (y', a)$ with cost $c(T'') + 3$. According to Equation 4, the cost of this selected-internal Steiner tree is smaller than that of T, which is a contradiction.

  CASE 2.2.   The cost of e equals 2. It is not difficult to show that T'' contains a cost-2 edge. As with a proof similar to that of CASE 2.1, the result holds.   **Q.E.D.**

Lemma 4. *Let $cost_2$ denote the cost of any solution S of $I_2 = A_1(I_1)$ and let $cost_1$ denote the cost of a solution of I_1 returned by $A_2(S, I_2)$. Then, $|cost_1 - OPT(I_1)| \leq 3|cost_2 - OPT(A_1(I_1))|$.*

Proof. If $|cost_2 - OPT(A_1(I_1))| = 0$, then S is an optimal solution of $A_1(I_1)$. By Lemma 3, $cost_1$ is the optimal cost of I_1. Therefore, $0 = |cost_1 - OPT(I_1)| = |cost_2 - OPT(A_1(I_1))| \leq 3|cost_2 - OPT(A_1(I_1))|$. Now we consider the case of $|cost_2 - OPT(A_1(I_1))| > 0$, i.e., the solution of I_2 with cost $cost_2$ is not optimal. By the execution of Algorithm A_2 (Step 2), we know that $cost_1 \leq cost_2 - 2$.

Hence,

$$|cost_1 - \mathsf{OPT}(I_1)| \leq |cost_2 - 2 - \mathsf{OPT}(I_1)|$$
$$\leq |cost_2 - 2 - (\mathsf{OPT}(A_1(I_1)) - 4)| \quad /* \text{ by Equation 1 } */$$
$$\leq |cost_2 - \mathsf{OPT}(A_1(I_1)) + 2|$$
$$\leq 3|cost_2 - \mathsf{OPT}(A_1(I_1))|. \qquad \textbf{Q.E.D.}$$

Note that Algorithms A_1 and A_2 can be implemented to run in polynomial time. By Lemmas 2 and 4, we have the following result.

Theorem 1. *SISTP(1,2) is MAX SNP-hard.*

4   An Approximation Algorithm for SISTP

In this section, we present an approximation Algorithm A_{SISTP} for SISTP. Let A_{STP} denote the best-known approximation algorithm for STP with ratio $\rho = 1 + \frac{\ln 3}{2} \approx 1.55$ [8], and also let $S_A = (V_A, E_A)$ be the Steiner tree returned by A_{STP}. To make sure that the solution of SISTP exists, in what follows, we assume that $|R \setminus R'| \geq 2$ if $R' \neq \emptyset$.

Let T be a Steiner tree of the instance $I = (G, R, R', c)$ of the problem SISTP. A vertex $v \in V(T)$ is said to be a *demand-leaf* iff v is a leaf of T and $v \in R'$. The following property is useful to our algorithm.

Lemma 5. *Let T be a Steiner tree of the instance $I = (G, R, R', c)$ of the problem SISTP such that $|R \setminus R'| \geq 2$. If v is a demand-leaf of T, then there is an internal vertex $m_v \in V(T)$ satisfying one of the following two conditions:*

1. *$\deg_T(m_v) = 2$ and $m_v \notin R'$;*
2. *$\deg_T(m_v) \geq 3$.*

Proof. If neither Condition (1) nor Condition (2) holds, then the resulting tree T is a path $\langle v, v_1, v_2, \dots, v_{n-1} \rangle$ such that $v, v_1, v_2, \dots, v_{n-2}$ are all in R'. Then, $|R \setminus R'| \leq 1 < 2$, which contradicts to the assumption that $|R \setminus R'| \geq 2$.   **Q.E.D.**

The skeleton of our algorithm is first to apply A_{STP} to obtain a Steiner tree $S_A = (V_A, E_A)$ spanning R, and then transform it to a selected-internal Steiner tree by using Lemma 5 to make each demand-leaf of S_A to be an interval vertex. We now present our approximation algorithm in Figure 3. We call the two vertices m_v and t_v selected by Algorithm A_{SISTP} for each demand-leaf v as the *medium vertex* and the *target vertex* of v, respectively. The following properties are obtained from the algorithm.

Lemma 6. *During the execution of Algorithm A_{SISTP}, assume that v is a demand-leaf of the current tree T handled by the algorithm in some iteration of the for-loop. Then, we have the following observations:*

1. $\deg_T(m_v)$ will be decreased by 1 in the next iteration.
2. $\deg_T(t_v)$ will be unchanged in the next iteration.
3. $\deg_T(v)$ will be increased by 1, and fixed as 2 until the algorithm terminates.

Proof. Straightforward.   **Q.E.D.**

Algorithm $A_{SISTP}(G, R, R', c)$

Input: A complete graph $G = (V, E)$ with a metric cost function $c : E \to \mathbf{R}^+$, and
two subsets R and R' satisfying $R' \subset R \subseteq V$ and $|R \setminus R'| \geq 2$ if $R' \neq \emptyset$.
Output: A selected-internal Steiner tree T_s.

1: Use A_{STP} to find a Steiner tree $S_A = (V_A, E_A)$ spanning R in G.
2: $S'_A \leftarrow S_A$.
3: **if** there is a demand-leaf in S'_A **then**
4:     **for** each demand-leaf v in the current tree S'_A **do**
5:         Select the nearest vertex $m_v \in V(S'_A)$ satisfying one of the following condition:
6:         (1) $\deg_{S'_A}(m_v) = 2$ and $m_v \notin R'$
7:         (2) $\deg_{S'_A}(m_v) \geq 3$. $\triangleright$ The existence of m_v is ensured by Lemma 5
8:         Choose a vertex $t_v \in V(S'_A)$ which is adjacent to m_v but does not belong
to the path $P_{S'_A}[v, m_v]$.
9:         $E(S'_A) \leftarrow E(S'_A) \cup \{(v, t_v)\}$.
10:         $E(S'_A) \leftarrow E(S'_A) \setminus \{(m_v, t_v)\}$.
11: **Return** $T_s \leftarrow S'_A$.

Fig. 3. An algorithm to construct a selected-internal Steiner tree

Lemma 7. *Suppose that v_0 is a demand-leaf of S_A and v_k is the medium ver-*
tex of v_0 selected by Algorithm A_{SISTP}, i.e., $m_{v_0} = v_k$. Let $P_{S_A}[v_0, v_k] =$
$\langle v_0, v_1, \ldots, v_k \rangle$ be a path of S_A. Then, during the executing of Algorithm A_{SISTP}
before v_0 being handled, there always exists a path $P_T[v_0, v_k]$ in the current tree T,
which is extended from $P_{S_A}[v_0, v_k]$ such that the following three properties hold:

1. *The length of $P_T[v_0, v_k]$ is at least k.*
2. *The path $P_T[v_0, v_k]$ contains all the vertices of $P_{S_A}[v_0, v_k]$ and also retains*
their relative order as $v_0 \to v_1 \to v_2 \to \cdots \to v_k$.
3. *If $P_T[v_0, v_k] \setminus P_{S_A}[v_0, v_k] \neq \emptyset$, then the vertices in $P_T[v_0, v_k] \setminus P_{S_A}[v_0, v_k]$ are*
all in R', and each has a fixed degree of 2 until the algorithm terminates.

Proof. The lemma can be shown by induction on the number h of demand-leaves
handled before v_0.   **Q.E.D.**

Lemma 8. *Suppose that u and v are two different demand-leaves of S_A such*
that u is handled by the algorithm before v. Then, the two paths $P_{S_A}[u, m_u]$ and
$P_{S_A}[v, m_v]$ are edge-disjoint.

We can generalize Lemma 8 to obtain the following Lemma.

Lemma 9. *Let* $v_1, v_2, \ldots, v_l$ *be an order of the demand-leaves of* S_A *handled by Algorithm* A_{SISTP}. *Then, the paths* $P_{S_A}[v_1, m_{v_1}], P_{S_A}[v_2, m_{v_2}], \ldots, P_{S_A}[v_l, m_{v_l}]$ *are pairwise edge-disjoint.*

The following lemma is used to analyze the approximation ratio of our algorithm in Theorem 2.

Lemma 10. *Suppose that* v *is a demand-leaf of the current tree* T, *which is being handled by the algorithm* A_{SISTP}. *Then, the target vertex* t_v *does not belong to* $P_{S_A}[v, m_v]$.

Proof. According to Line 8 of the algorithm, it is clear that t_v does not belong to $P_T[v, m_v]$, i.e., t_v is not a vertex in $P_T[v, m_v]$. Assume that $P_T[v, m_v] = \langle v_0(= v), v_1, v_2, \ldots, v_k(= m_v) \rangle$. By Lemma 7(2), the vertices of $P_{S_A}[v, m_v]$ are contained in $P_T[v, m_v]$ and the relative order of the vertices in $P_{S_A}[v, m_v]$ are retained in $P_T[v, m_v]$, i.e., $P_{S_A}[v, m_v] = \langle v_0, v_{i_1}, v_{i_2}, \ldots, v_{i_j}, v_k \rangle$, where $0 < i_1 < i_2 < \cdots < i_j < k$. Therefore, t_v is not a vertex in $P_{S_A}[v, m_v]$.   **Q.E.D.**

The following lemma can be obtained using triangle inequalities.

Lemma 11. *Let* $P = \langle v_1, v_2, \cdots, v_{k-1}, v_k \rangle$ *be a path of a graph* $G = (V, E)$ *with a metric cost function* $c: E \to \mathbf{R}^+$, *and let* $P' = \langle v_1, v_2, \cdots, v_{k-2}, v_{k-1} \rangle$. *Then,*

$$c(v_1, v_k) - c(v_{k-1}, v_k) \leq c(P'), \text{ where } c(P') = \sum_{j=1}^{k-2} c(v_j, v_{j+1}).$$

Let $L_R = \{v|\ v \in R \text{ is a leaf of } S_A\}$ and $L_{R'} = \{v|\ v \text{ is a demand-leaf of } S_A\}$. Note that $L_{R'} \subseteq L_R$. Define $\phi = \min\{|L_R \setminus L_{R'}|, |R \setminus R'| - 2\}$. We now show our main result.

Theorem 2. *Let* $e_1, e_2, \ldots, e_i$ *denote the first* i *smallest-cost edges of* $S_A = (V_A, E_A)$. *Algorithm* A_{SISTP} *is a* $(2 - \frac{\sigma}{c(S_A)})\rho$-*approximation algorithm for SISTP, where* ρ *is the best-known approximation ratio of the Steiner tree problem and* σ *is defined as follows:*

$$\sigma = \begin{cases} c(S_A) & \text{if } R' = \emptyset, \\ \sum_{i=1}^{\phi} c(e_i) & \text{otherwise.} \end{cases}$$

Moreover, $0 \leq \frac{\sigma}{c(S_A)} \leq 1$.

Proof. It is clear that Algorithm A_{SISTP} correctly constructs a selected-internal Steiner tree. We now analyze the approximation ratio. Let T_s, T^* and S^* be the output of A_{SISTP}, the optimal solution of SISTP and the optimal solution of STP, respectively. Since S_A is the output of Algorithm A_{STP}, we have that

$$c(S_A) \leq \rho c(S^*). \tag{5}$$

Since T^* is a feasible solution of STP,

$$c(S^*) \leq c(T^*). \tag{6}$$

By Equations 5 and 6, we have that

$$c(S_A) \leq \rho c(T^*).$$

Next, we consider the following two cases according to the demanded-terminal set R'.

CASE 1: $R' = \emptyset$. Then, $c(T_s) = c(S_A) \leq \rho c(T^*) = (2 - \frac{c(S_A)}{c(S_A)})\rho c(T^*)$. The theorem holds.

CASE 2: $R' \neq \emptyset$ and $|R \setminus R'| \geq 2$. According to Algorithm A_{SISTP},

$$c(T_s) = c(S_A) + \sum_{v \in L_{R'}} (c(v, t_v) - c(m_v, t_v)).$$

By Lemmas 10 and 11, we know that

$$\sum_{v \in L_{R'}} (c(v, t_v) - c(m_v, t_v)) \leq \sum_{v \in L_{R'}} c(P_{S_A}[v, m_v]).$$

Define Q to be the set obtained by selecting arbitrary $\phi(= \min\{|L_R \setminus L_{R'}|, |R \setminus R'| - 2\})$ elements from $L_R \setminus L_{R'}$. Note that

$$Q = \begin{cases} \emptyset & \text{if } \phi = 0, \\ L_R \setminus L_{R'} & \text{if } \phi = |L_R \setminus L_{R'}|, \\ \text{a proper subset of } L_R \setminus L_{R'} & \text{otherwise.} \end{cases}$$

If we transform $L_{R'} \cup Q$ into internal vertices using Algorithm A_{SISTP} $(G, R, Q \cup R', c)$, then the resulting tree remains a selected-internal Steiner tree. (Note that the algorithm actually transform only $L_{R'}$ into internal vertices.) By above observation together with Lemma 9, we have

$$\sum_{v \in L_{R'}} c(P_{S_A}[v, m_v]) + \sum_{v \in Q} c(P_{S_A}[v, m_v]) \leq c(S_A).$$

Therefore, we can obtain the following result: $c(T_s) = c(S_A) + \sum_{v \in L_{R'}} (c(v, t_v) -$

$c(m_v, t_v)) \leq c(S_A) + \sum_{v \in L_{R'}} c(P_{S_A}[v, m_v]) \leq c(S_A) + \left(c(S_A) - \sum_{v \in Q} c(P_{S_A}[v, m_v])\right)$

$\leq 2c(S_A) - \sum_{i=1}^{|Q|} c(e_i) = 2c(S_A) - \sum_{i=1}^{\phi} c(e_i) = \left(2 - \frac{\sum_{i=1}^{\phi} c(e_i)}{c(S_A)}\right) c(S_A) \leq$

$\left(2 - \frac{\sum_{i=1}^{\phi} c(e_i)}{c(S_A)}\right) \rho c(T^*)$. Therefore, $\frac{c(T_s)}{c(T^*)} \leq \left(2 - \frac{\sum_{i=1}^{\phi} c(e_i)}{c(S_A)}\right) \rho$. It is clear that

$0 \leq \frac{\sigma}{c(S_A)} \leq 1$. Therefore, the result holds.   Q.E.D.

References

1. S. Arora, C. Lund, R. Motwani, M. Sudan, and M. Szegedy, "Proof verification and the hardness of approximation problems," *Journal of the Association for Computing Machinery*, vol. 45, pp. 501–555, 1998.
2. G. Ausiello, P. Crescenzi, G. Gambosi, V. Kann, A. Marchetti-Spaccamelai, and M. Protasi, *Complexity and Approximation-Combinatorial Optimization Problems and Their Approximability Properties*, Springer Verlag, Berlin, 1999.
3. M. Bern and P. Plassmann, "The Steiner problem with edge lengths 1 and 2," *Information Processing Letters*, vol. 32(4), pp. 171–176, 1989. 5Zelikovsky, "On wire-length estimations for row-based placement,"
4. X. Cheng and D. Z. Du, *Steiner Trees in Industry*, Kluwer Academic Publishers, Dordrecht, Netherlands, 2001.
5. T. H. Cormen, C. E. Leiserson, R. L. Rivest, and C. Stein, *Introduction to Algorithms* (Second Edition), MIT Press, Cambridge, 2001.
6. R. Karp, "Reducibility among combinatorial problems," in R. E. Miller, J. W. Thatcher eds.: *Complexity of Computer Computations*, Plenum Press, New York, pp. 85–103, 1972.
   5*Information Processing Letters*, vol. 84(2), pp. 103–107,
7. C. Papadimitriou and M. Yannakakis, "Optimization, approximation and complexity classes," *Journal of Computer and System Science*, vol. 43, pp. 770–779, 1991.
8. G. Robins and A. Zelikovsky, "Improved Steiner tree approximation in graphs," in *Proceedings of the 11th Annual ACM-SIAM Symposium on Discrete Algorithms (SODA)*, pp. 770–779, 2000.
9. D. B. West, *Introduction to Graph Theory* (Second Edition), Prentice Hall, Upper Saddle River, 2001.

Minimum Clique Partition Problem with Constrained Weight for Interval Graphs*

Jianbo Li[1], Mingxia Chen[2], Jianping Li[3,**], and Weidong Li[3]

[1] School of Management and Economics,
Kunming University of Science and Technology, Kunming, P.R. China
[2] Department of Science and Technology, Yunnan University, Kunming, P.R. China
[3] Department of Mathematics, Yunnan University, Kunming, P.R. China
{mxchen, jianping}@ynu.edu.cn

Abstract. Interval graphs play important roles in analysis of *DNA* chains in Benzer [1], restriction maps of *DNA* in Waterman and Griggs [11] and other related areas. In this paper, we study a new combinatorial optimization problem, named as the minimum clique partition problem with constrained weight, for interval graphs. For a weighted interval graph G and a bound B, partition the weighted intervals of this graph G into the smallest number of cliques, where each clique, consisting of some intervals whose intersection on a real line is not empty, has its weight not beyond B. We obtain the following results: (1) This problem is NP-hard in the strong sense, and it cannot be approximated within a ratio $\frac{3}{2} - \varepsilon$ in polynomial-time for any $\varepsilon > 0$; (2) We design some approximation algorithms with different constant ratios to this problem; (3) For the case where all intervals have the same weight, we also design an optimal algorithm to solve the problem in linear time.

Keywords: Interval graph, cliques, approximation algorithm.

1 Introduction

An undirected graph $G = (V, E)$ is an interval graph if, for each vertex $v \in V$, v can be associated an open interval I_v of the real line, such that any pair of distinct vertices $u, v \in V$ are connected by an edge in E if and only if $I_u \cap I_v \neq \emptyset$. The family $\{I_v\}_{v \in V}$ is an interval representation of G.

Interval graphs have many applications in the molecular biology, scheduling of tasks executed, timing of traffic lights, and so on. Benzer [1] invented interval graphs to study analysis of *DNA* chains, i.e., the linearity of the chain for higher organisms, and interval graph aids in locating genes along the *DNA* sequence; Waterman and Griggs [11] utilized interval graphs to study an important representation of *DNA* called restriction maps; Papadimitriou and Yannakakis [7]

---

* The work is fully supported by the National Natural Science Foundation of China [Project No. 10561109, 10271103] and Natural Science Foundation of Yunnan Province [Project No. 2003F0015M]. The partial work of this author was done while visiting Department of Computer Science, City University of Hong Kong.
** Correspondence author.

D.Z. Chen and D.T. Lee (Eds.): COCOON 2006, LNCS 4112, pp. 459–468, 2006.
© Springer-Verlag Berlin Heidelberg 2006

utilized interval graphs to study the scheduling interval-order tasks; Roberts [8] utilized interval graphs to consider the problem of timing of traffic lights to optimize some criterion such as average waiting time. Other applications of interval graphs shall be found in [3,5,9,12].

Since such an interval graph is a special one of perfect graphs [5], there many research papers have studied some combinatorial optimization problems in interval graphs, such as computing the maximum coloring number, a maximum independent set and a maximum clique in such an interval graph. In this paper, we study the graph partition problem with some bounds in weighted interval graphs, where each clique consists of some intervals in G whose intersection on a real line is not empty.

Our problem, named as *Minimum Clique Partition Problem with Constrained Weight for Interval Graphs* (MCPCW), is stated in details as follows:

INSTANCE: a weighted interval graph $G = (V, E; w)$ with intervals $I_1, \ldots, I_n$, having weights $w_1, \ldots, w_n$, and a bound B;

QUESTION: Find a partition of these n intervals into the smallest number of cliques, each clique having its weight not beyond B.

We knew that Kaplan and Shamir [6] studied a problem related to ours, they studied some pathwidth, bandwidth and completion problems to proper interval graphs with small cliques. We noticed Bodlaender and Jansen [2] studied the restrictions of graph partition problems in several classes of graphs without weights, and they obtained some results: the problem to partition a cograph into bounded cliques (independent sets, respectively) remains NP-hard, and the problem to partition an unweighted interval graph into bounded cliques is solvable in linear-time by using an algorithm of Papadimitriou and Yannakakis [7]. But, when we study the clique partition problem with a bound B for weighted interval graphs, it becomes NP-hard to compute such a minimum number of cliques, each having weight not beyond B. To our knowledge by now, there is no approximation algorithms to this new combinatorial optimization problem, then we design some approximation algorithms with constant ratios to the problem, and we also redesign a new linear-time algorithm to solve the problem in the version where all intervals have the same weights. Our new linear-time algorithm is completely different from the one of Bodlaender and Jansen in [2], and the technique they used heavily depends on the one of scheduling interval-order tasks in [7], but our linear-time algorithm in the section 4 only depends on the new sorting technique defined in the section 3 and the GREEDY method.

This paper is structured in the following sections. In the section 2, we prove that the MCPCW problem is NP-hard in a strong sense, by transformed from the 3-PARTITION problem, and that it cannot be approximated within a ratio $\frac{3}{2} - \varepsilon$ for any $\varepsilon > 0$; Some approximation algorithms with constant ratios are designed to this problem in the section 3; For the case where each interval has the same weight, we design an optimal algorithm to solve the problem in linear-time in the section 4; We give the conclusions and remarks in the last section.

2   Hardness of the MCPCW Problem

In this section, we study the hardness of the MCPCW problem, and then we also prove that the MCPCW problem cannot be approximated within a ratio $\frac{3}{2} - \varepsilon$ in polynomial time for any $\varepsilon > 0$.

Theorem 1. *The* MCPCW *problem is NP-hard in a strong sense.*

Proof. We prove the NP-hardness of the MCPCW problem by transforming any instance of 3-PARTITION problem to an instance of the MCPCW problem. The 3-PARTITION problem is one of the earliest known natural NP-hard problems in a strong sense [4].

Consider an instance $\mathcal{I}$ of the 3-PARTITION problem: Given the set $S = \{a_1, a_2, \ldots, a_{3k}\}$ of $3k$ integers to satisfy $\frac{B}{4} < a_j < \frac{B}{2}$ for each $1 \leq j \leq 3k$ and $\sum_{j=1}^{3k} a_j = kB$, ask whether S can be partitioned into k subsets $S_1, S_2, \ldots, S_k$ such that, for each $i = 1, 2, \ldots, k$, S_i exactly contains three elements of S and $\sum_{a \in S_i} a = B$.

We construct a reduction τ from $\mathcal{I}$ of the 3-PARTITION problem to an instance $\tau(\mathcal{I})$ of the MCPCW problem: a weighted interval graph G with intervals $I_1, \ldots, I_{3k}$, for $j = 1, 2, \ldots, 3k$, each interval I_j having its left endpoint $o(I_j) = j-1$, right endpoint $d(I_j) = 3k$ in the real line and possessing its weight a_j, and the bound $B = \frac{\sum_{j=1}^{3k} a_j}{k}$.

Now, we prove the following claim: There exists a feasible solution to an instance $\mathcal{I}$ of 3-PARTITION problem if and only if the instance $\tau(\mathcal{I})$ of the MCPCW problem has its optimal solution with value k.

In fact, for any feasible solution of an instance $\mathcal{I}$ of 3-PARTITION problem, the set S is partitioned into k subsets $S_1, S_2, \ldots, S_k$ such that, for $i = 1, 2, \ldots, k$, S_i exactly contains three elements of S and $\sum_{a \in S_i} a = B$, then we construct a partition to the instance $\tau(\mathcal{I})$ of the MCPCW problem: for each $S_i = \{a_{i_1}, a_{i_2}, a_{i_3}\}$, keep the clique $C_i = \{I_{i_1}, I_{i_2}, I_{i_3}\}$, and then we obtain the partition of these $3k$ intervals into k cliques, each clique exactly having its weight B.

Conversely, if the instance $\tau(\mathcal{I})$ of the MCPCW problem has an optimal partition $\{C_1, C_2, \ldots, C_k\}$ with the smallest integer k, having $\sum_{a \in C_i} a \leq B$ for each $i = 1, 2, \ldots, k$. By the facts $\sum_{j=1}^{3k} a_j = kB$ and $\frac{B}{4} < a_j < \frac{B}{2}$ for each $1 \leq j \leq 3k$, we obtain $\sum_{a \in C_i} a = B$ for each $1 \leq i \leq k$ and then each clique C_i exactly contains three elements from S, i.e., $S_i = \{a_{i_1}, a_{i_2}, a_{i_3}\}$ and $\sum_{a \in C_i} a = B$ for each $1 \leq i \leq k$. So the instance $\mathcal{I}$ of the 3-PARTITION problem has the partition $S_1, S_2, \ldots, S_k$.

Hence, the NP-hardness in a strong sense of the MCPCW problem follows the fact that the 3-PARTITION problem is one of the earliest known natural NP-hard problems in a strong sense. This reaches at the conclusion of the theorem.   ∎

We know that the MCPCW problem is NP-hard in a strong sense from the theorem 1; moreover, we obtain the following strong result.

Theorem 2. *For any $\varepsilon >$, there is no approximation algorithm of a ratio $\frac{3}{2} - \varepsilon$ for the* MCPCW *problem.*

Proof. Suppose that there were such an approximation algorithm $\mathcal{A}$, then we show how to solve the PARTITION problem by the algorithm $\mathcal{A}$, i.e., deciding if there is a way to partition n nonnegative numbers $a_1, a_2, \ldots, a_n$ into two sets, each adding up to $\frac{1}{2}\Sigma_{i=1}^{n}a_i$.

For an instance $\mathcal{I}$ of the PARTITION problem consisting of n nonnegative numbers $a_1, a_2, \ldots, a_n$, we construct a reduction τ from $\mathcal{I}$ of the PARTITION problem to an instance $\tau(\mathcal{I})$ of the MCPCW problem: an interval graph G with intervals $I_1, \ldots, I_n$, for $j = 1, 2, \ldots, n$, each interval I_j having its left endpoint $o(I_j) = j - 1$, right endpoint $d(I_j) = n$ in the real line and possessing its weight a_j, and the bound $B = \frac{1}{2}\Sigma_{i=1}^{n}a_i$.

Clearly, the answer to the PARTITION problem is 'yes' if and only if the MCPCW problem exactly has two cliques of weight $\frac{1}{2}\Sigma_{i=1}^{n}a_i$.

When we use the algorithm $\mathcal{A}$ on the instance $\tau(\mathcal{I})$, it produces an output to satisfy $m \leq (\frac{3}{2} - \varepsilon)OPT$, where OPT is the optimal value to the instance $\tau(\mathcal{I})$. If $OPT = 2$, then the preceding formula implies $m = 2$, showing that the PARTITION problem has a feasible solution; If $OPT \geq 3$, then we get $m \geq OPT \geq 3$, implying that the PARTITION problem has no feasible solution. So the algorithm $\mathcal{A}$ solves the PARTITION problem in polynomial-time. But the PARTITION problem remains NP-hard [4], a contradiction.

Hence, the theorem holds.   ∎

3   Some Approximation Algorithms for MCPCW

Since the MCPCW problem is NP-hard and there is no polynomial-time algorithm to optimally solve it, we design some approximation algorithms for it in this section.

In order to simply describe our approximation algorithms, we shall utilize an optimal algorithm from Tarjan [10] to compute a maximum independent set in such an interval graph, by utilizing some technique of minimum-cost flow with value 1, and this optimal algorithm runs in time $\mathcal{O}(n)$. For convention, we denote such an algorithm as MAX-SET.

Before we design approximation algorithms for the MCPCW problem, we give the rules of the sequel index sorting. For a weighted interval graph $G = (V, E; w)$ with intervals $I_1, \ldots, I_n$, denote $o(i)$ and $d(i)$ respectively as the left endpoint and the right endpoint, located on the real line from left to right, of the interval I_i for each $1 \leq i \leq n$, each interval I_i having its weight w_i.

We denote a linear order '$\leq$' on G: for any two intervals I_i, I_j of G, denote $I_i \leq I_j$ if and only if (1) either $d(i) < d(j)$, or (2) $d(i) = d(j)$ and $o(i) < o(j)$, or (3) $d(i) = d(j)$, $o(i) = o(j)$ and $w_i \leq w_j$. It takes $\mathcal{O}(n \log n)$ steps to sort these n intervals, heavily depending on this linear order '$\leq$' on the rule (3). When all intervals have the same weight, i.e., when we do not care the weights of these n intervals, the precede order '$\leq$' is also linear, but the sorting time in this case runs in $\mathcal{O}(n)$. We shall changeably to utilize these two linear orders in the sequel, but their different running times depend on the rule choices. For any subgraph

G' of the interval graph G, it is known that the partial order '$\leq$' on G' is a linear order [9], too.

We design first approximation algorithm for the MCPCW problem:

Algorithm Clique-Partition I

INPUT: a weighted interval graph $G = (V, E; w)$ with intervals $I_1, \ldots, I_n$, having weights $w_1, \ldots, w_n$, and a bound B;

OUTPUT: m disjoint cliques consisting of these n intervals, each clique having its weight not beyond B.

Begin

Step 1: Use the algorithm MAX-SET to compute a maximum independent set $\mathcal{I} = \{I_{i_1}, I_{i_2}, \ldots, I_{i_r}\}$ in G;

Step 2: Use the GREEDY method to obtain r disjoint cliques $C_1, C_2, \ldots, C_r$, where C_t contains the interval I_{i_t}, for each $1 \leq t \leq r$, such that $\{C_1, C_2, \ldots, C_r\}$ is a partition of these n intervals;

Step 3: For each clique $C_j = \{I_{j_1}, I_{j_2}, \ldots, I_{j_{m_j}}\}$, where $1 \leq j \leq r$ and $I_{j_1} = I_{i_j}$, choose some suitable cliques as follows: for $1 \leq t \leq m_j$ and the 'alive cliques' $C_j^1, \ldots, C_j^k$ containing the intervals $I_{j_1}, I_{j_2}, \ldots, I_{j_{t-1}}$ (for convention, the 'alive clique' is empty when $t = 1$ for the initiation), add the current interval I_{j_t} into some 'alive clique' $C_j^{k'}$ if the total weight sum of I_{j_t} and the intervals in the original clique $C_j^{k'}$ is not greater than B, where $1 \leq k' \leq k$, otherwise open a new 'alive clique' as C_j^{k+1} to contain the current interval I_{j_t} as the first interval, until $t > m_j$;

Step 4: Output all cliques obtained from the step 3.

End of Clique-Partition I

Theorem 3. *The algorithm Clique-Partition I is an approximation algorithm with ratio 3 for the MCPCW problem, it runs in the time $\mathcal{O}(n^2)$.*

Proof. For $1 \leq j \leq r$, let Out_j be the set of cliques produced by the step 3 of the algorithm Clique-Partition I on each clique C_j and denote $OUT = \bigcup_{j=1}^{r} Out_j$. Then, such $|Out_j|$ cliques must contain at least $|Out_j| - 1$ cliques whose weight is greater than $\frac{B}{2}$, otherwise $|Out_j|$ will be decreased. Thus

$$\sum_{i=1}^{m} w_i > \frac{B}{2}(|Out_1| - 1) + \frac{B}{2}(|Out_2| - 1) + \cdots + \frac{B}{2}(|Out_r| - 1)$$

$$= \frac{B}{2}(|Out_1| + |Out_2| + \cdots + |Out_r| - r)$$

$$= \frac{B}{2}(|OUT| - r)$$

implying

$$|OUT| < \frac{2\sum_{i=1}^{m} w_i}{B} + r \leq 2|OPT| + |OPT| = 3|OPT|$$

where the second inequality depends on the two facts that the optimal solution has two lower bounds $\frac{\sum_{i=1}^{m} w_i}{B}$ and r, i.e., $|OPT| \geq \frac{\sum_{i=1}^{m} w_i}{B}$ and $|OPT| \geq r$. So the algorithm Clique-Partition I has a ratio 3.

Now, we analyze the complexity of the algorithm Clique-Partition I: (1) the step 1 needs $\mathcal{O}(n)$ steps to compute a maximum independent set in such an interval graph in Tarjan [10]; (2) by using GREADY method, the step 2 needs $\mathcal{O}(n)$ steps to to find such a clique partition of these n intervals; (3) since each interval must be chosen in a clique and the index sorting only depends the rules (1) and (2), the steps in the step 3 totally needs time in $\mathcal{O}(m_1^2) + \mathcal{O}(m_2^2) + \cdots + \mathcal{O}(m_r^2)$, i.e., at most $\mathcal{O}(n^2)$. Hence, the whole algorithm needs running time in $\mathcal{O}(n^2)$.

This establishes the conclusion of the theorem. ∎

Now, we design another new approximation algorithm for the MCPCW problem that has a better ratio than before. For convention, the interval is called as a *large* interval if this interval has its weight greater than $\frac{B}{2}$, otherwise the interval is called as a *small* interval. Our second approximation algorithm for the MCPCW problem is designed:

Algorithm Clique-Partition II
INPUT: a weighted interval graph $G = (V, E; w)$ with intervals $I_1, \ldots, I_n$, having
    weights $w_1, \ldots, w_n$, and a bound B;
OUTPUT: m disjoint cliques consisting of these n intervals, each clique having
    its weight not beyond B.
Begin
Step 1: For each large interval I_i, let a clique C_i only contain such a large
    interval I_i; and after remove all large intervals from G, the current interval
    graph G only contains small intervals;
        (/*We only consider all small intervals below/*)
Step 2: Sort all small intervals according to the precede rules (1) and (2),
    without loss of generality, the small intervals in G are sorted as $I_1, \ldots, I_{n'}$
    depending on the rules (1) and (2);
Step 3: Depending on the linear order '$\leq$' on G, choose a smallest element,
    I_{min}, in the current graph G as the alive interval; and find the *maximal*
    clique C from the current graph G to contain such an alive interval I_{min},
    without loss of generality, all intervals in such maximal clique C are sorted
    as $I_i, I_{i+1}, \ldots, I_j$, where $I_i = I_{min}$;
Step 4: Use the similar method at the step 3 in the algorithm Clique-Partition I,
    and then obtain m_i disjoint cliques $C_{i_1}, C_{i_2}, \ldots, C_{i_{m_i}}$ from the current alive
    clique C, simultaneously, the cliques $C_{i_1}, C_{i_2}, \ldots, C_{i_{m_i-1}}$ must have weights
    greater than $\frac{B}{2}$;
    – Step 4.1: If the clique $C_{i_{m_i}}$ has its weight greater than $\frac{B}{2}$, then put $G :=$
      $G - \bigcup_{t=1}^{m_i} C_{i_t}$ and produce the m_i cliques $C_{i_1}, C_{i_2}, \ldots, C_{i_{m_i}}$;
    – Step 4.2: If $m_i \geq 2$ and the clique $C_{i_{m_i}}$ has its weight not greater than $\frac{B}{2}$,
      then put $G := G - \bigcup_{t=1}^{m_i-1} C_{i_t}$ and produce the $m_i - 1$ cliques $C_{i_1}, C_{i_2}, \ldots,$
      $C_{i_{m_i-1}}$;
    – Step 4.3: If $m_i = 1$ and the clique $C_{i_{m_i}}$ ($=C$) has its weight not greater
      than $\frac{B}{2}$, then put $G := G - C$ and produce the clique C;
Step 5: Continue to execute the step 3 until $G = \emptyset$;

Step 6: Output all cliques obtained from the steps 1 and 4.

End of Clique-Partition II

We shall notice the facts: (1) when the step 4.1 or 4.3 executes for a time, the step 4 will exactly produce the m_i cliques $C_{i_1}, C_{i_2}, \ldots, C_{i_{m_i}}$ for this time; (2) but when the step 4.2 executes for a time, the step 4 will only produce the $m_i - 1$ cliques $C_{i_1}, C_{i_2}, \ldots, C_{i_{m_i-1}}$ for this time.

Now, we provide a proof of the correctness for the algorithm Clique-Partition II and its running complexity.

Theorem 4. *The algorithm Clique-Partition II is an approximation algorithm with ratio 2 for the MCPCW problem, it runs in the time $\mathcal{O}(n^2)$.*

Proof. For convention, we may assume that the output cliques are ordered as $C_1^0, \ldots, C_{j_0}^0, C_1^1, \ldots, C_{j_1}^1, C_1^2, \ldots, C_{j_2}^2, \ldots, C_1^t, \ldots, C_{j_t}^t$, where $C_1^0, \ldots, C_{j_0}^0$ are sequentially produced at the step 1, each clique having its weight greater than $\frac{B}{2}$, and $C_{j_1}^1, C_{j_2}^2, \ldots, C_{j_t}^t$ are sequentially produced at the step 4.3, each clique having its weight not greater than $\frac{B}{2}$ except the last clique $C_{j_t}^t$ (we note that the clique $C_{j_t}^t$ has its weight greater than $\frac{B}{2}$ when $C_{j_t}^t$ is the last clique produced before the algorithm stops at the step 4.1, otherwise the clique $C_{j_t}^t$ has its weight not greater than $\frac{B}{2}$), and the other cliques are sequentially produced at the steps 4.1 or 4.2, each clique having its weight greater than $\frac{B}{2}$. So the number of output cliques is $m = j_0 + j_1 + j_2 + \cdots + j_t$. And these m cliques are disjoint by the choices in the algorithm.

By the choice at the step 4.3 of our algorithm, any two cliques $C_{j_k}^k, C_{j_{k'}}^{k'}$ of $C_{j_1}^1$, $C_{j_2}^2, \ldots, C_{j_t}^t$ can not be covered simultaneously by a clique in any optimal solution OPT, otherwise at least one of $C_{j_k}^k, C_{j_{k'}}^{k'}$ is not maximal clique, contradicting the choice at the step 4.3. Since the other cliques obtained at the step 4.1 or 4.2 or the step 1 have the weight greater than $\frac{B}{2}$, then the cliques $C_1^1, \ldots, C_{j_1}^1$ must be covered by at least $\lceil \frac{j_1}{2} \rceil$ cliques from any optimal solution OPT, and the cliques $C_1^2, \ldots, C_{j_2}^2$ must be covered by at least $\lceil \frac{j_2}{2} \rceil$ cliques from any optimal solution OPT, and so on.

For each $1 \leq k \leq t$, denote $\varepsilon(j_k) = 1$ if j_k is odd and $\varepsilon(j_k) = 0$ otherwise. When $\varepsilon(j_k) = 0$, i.e., j_i is even, the j_i cliques $C_1^k, \ldots, C_{j_k}^k$ must be covered by at least $\frac{j_k + \varepsilon(j_k)}{2}$ cliques in any optimal solution OPT; but when $\varepsilon(j_k) = 1$, i.e., j_i is odd, the $j_i - 1$ cliques $C_1^k, \ldots, C_{j_k-1}^k$ must be covered by at least $\frac{j_k-1}{2}$ cliques in any optimal solution OPT, and both of the clique $C_{j_k}^k$ and any clique from $C_1^0, \ldots, C_{j_0}^0$ must be covered by at least one clique in any optimal solution OPT. This shows that

$$|OPT| \geq \max\{\frac{j_0 - \sum_{k=1}^t \varepsilon(j_k)}{2}, 0\} + \sum_{k=1}^t \frac{j_k + \varepsilon(j_k)}{2}$$

$$\geq \frac{j_0 + j_1 + \cdots + j_t}{2} = \frac{m}{2}$$

which implies $m \leq 2|OPT|$. So the algorithm Clique-Partition II has a ratio 2 for the MCPCW problem.

Now, we analyze the complexity of the algorithm Clique-Partition II: (1) the step 1 needs at most $2n$ steps to find all large intervals to construct the cliques, each having weight greater than $\frac{B}{2}$; (2) the step 2 needs at most $2n$ steps to sort the smaller intervals, depending on the rules 1-2; (3) for the current interval graph G at each time executed, it needs a constant time to choose the minimum element I_{min} from the current interval graph G, so it totally needs time at most in $\mathcal{O}(n)$; (4) since each interval must be chosen in a clique, so the steps during the steps 3-4, except the steps to find minimum element from the current G, totally needs time in $\mathcal{O}(n^2)$ similarly to the step 3 in the algorithm Clique-Partition I. Hence, the whole algorithm needs running time in $\mathcal{O}(n^2)$.

This establishes the conclusion of the theorem.   ∎

4   Linear Algorithm for the Special Case of MCPCW

In this section, we study the MCPCW problem in the version where all intervals have the same weight 1. When we utilize the algorithms Clique-Partition I or II on this special interval graph, we obtain a feasible solution whose value is not greater than three or two times that of optimal solution. But when we slightly modify the algorithm Clique-Partition II in some ways, we design an optimal algorithm in the linear-time for the the MCPCW problem in this special version.

Our method to design an optimal algorithm in linear-time depend on the following ideas: (1) sort all intervals depending on the rules 1-2; (2) choose a suitable maximal clique C; (3) choose some cliques, each having its weight not greater than B, from the intervals of C by distinguishing the cardinality of such suitable maximal clique C; (4) repeatedly execute steps (2)-(3) until $G = \emptyset$.

Since the precede partial order '$\leq$' is a linear order on the original interval graph G, then this partial order '$\leq$' is also a linear order on the current interval subgraph of G. We can sort all intervals in the time $\mathcal{O}(n)$, and choose the smallest element, I_{min}, in the constant time at each choice in the current interval subgraph of G; Again, execute this process repeatedly until this subgraph becomes empty. Our linear optimal algorithm is described in detail as follows:

Algorithm: Clique-Partition III
INPUT: a interval graph $G = (V, E; w)$ with intervals $I_1, \ldots, I_n$, and a bound B;
OUTPUT: m disjoint cliques consisting of these n intervals, each clique containing intervals not beyond B.
Begin
Step 1: Sort all intervals of G according to the precede rules 1-2 of the linear order '$\leq$' on G;
Step 2: Choose the smallest element, I_{min}, in G as the alive interval; and find the maximal clique C from G to contain such an alive interval I_{min}, and then sort all intervals in C according to the precede rules 1-2, without loss of generality, all intervals in C are sorted as $I_{i_1}, I_{i_2}, \ldots, I_{i_r}$, where $I_{i_1} = I_{min}$;

Step 3: For the current alive clique $C = \{I_{i_1}, I_{i_2}, \ldots, I_{i_r}\}$, choose the new cliques from C, depending on the following choice regulations:
 – Step 3.1: If $r < B$, i.e., $|C| < B$, then output the alive clique C only containing these r intervals; and put $G := G - \{I_{i_1}, I_{i_2}, \ldots, I_{i_r}\}$;
 – Step 3.2: If $r \geq B$, set $r = sB + r_0$, where $s = \lfloor \frac{r}{B} \rfloor$ and $0 \leq r_0 < B$, then output the s cliques $C_1 = \{I_{i_1}, \ldots, I_{i_B}\}$, $C_2 = \{I_{i_{B+1}}, \ldots, I_{i_{2B}}\}$, $\ldots$, $C_s = \{I_{i_{(s-1)B+1}}, \ldots, I_{i_{sB}}\}$; and $G := G - \{I_{i_1}, \ldots, I_{i_B}, \ldots, I_{i_{(s-1)B+1}}, \ldots, I_{i_{sB}}\}$;
Step 4: Continue to execute the step 2 until $G = \emptyset$;
Step 5: Output all cliques at the step 3.
End of Clique-Partition III

Theorem 5. *The algorithm Clique-Partition III is a linear optimal algorithm for the special version of the MCPCW problem, where all intervals have the same weight 1.*

Proof. For convention, we may assume that the output cliques are ordered as C_1^1, $\ldots$, $C_{j_1}^1$, C_1^2, $\ldots$, $C_{j_2}^2$, $\ldots$, C_1^t, $\ldots$, $C_{j_t}^t$, where $C_{j_1}^1, C_{j_2}^2, \ldots, C_{j_t}^t$ are sequentially produced at the step 3.1 for the case $r < B$, each clique having its weight less than B except the last clique $C_{j_t}^t$ (we note that the clique $C_{j_t}^t$ exactly has its weight B when $C_{j_t}^t$ is the last clique produced before the algorithm stops at the step 3.2), and the other cliques are sequentially produced at the step 3.2 for the case $r \geq B$ in the sequential times, each clique exactly having its weight B. So the number of output cliques is $m = j_1 + j_2 + \cdots + j_t$. And these m cliques are disjoint by the choices in the algorithm.

By the choice at the step 3.1 of our algorithm, any two cliques $C_{j_k}^k$, $C_{j_{k'}}^{k'}$ of $C_{j_1}^1, C_{j_2}^2, \ldots, C_{j_t}^t$ can not be covered simultaneously by a clique in any optimal solution OPT, otherwise at least one of $C_{j_k}^k$ and $C_{j_{k'}}^{k'}$ is not maximal clique, contradicting the choice at the step 3.1.

Since the other cliques obtained at the step 3.2 have the same weight B, then the cliques $C_1^1, \ldots, C_{j_1}^1$ must be covered by at least j_1 cliques from any optimal solution OPT, and the cliques $C_{j_1}^1, C_1^2, \ldots, C_{j_2}^2$ must be covered by at least $j_2 + 1$ cliques from any optimal solution OPT, and so on. Then, for any optimal solution OPT, we must have $|OPT| \geq j_1 + j_2 + \cdots + j_t = m$, implying $|OPT| = m$.

Hence, the output cliques $C_1^1, \ldots, C_{j_1}^1, C_1^2, \ldots, C_{j_2}^2, \ldots, C_1^t, \ldots, C_{j_t}^t$ are the elements of an optimal solution OPT to the special version of the MCPCW problem.

Now, we analyze the complexity of the algorithm Clique-Partition III: (1) the step 1 needs $2n$ steps to sort the n intervals; (2) for the current interval graph G at each time, it needs a constant time to choose the minimum element I_{min} from the current interval graph G, so it totally needs at most time in $\mathcal{O}(n)$ to find such minimum elements; (3) since each interval must be chosen in a clique, so the steps during the steps 2-3, except the steps to find minimum element from the current G, totally needs time in $\mathcal{O}(n)$. Hence, the whole algorithm needs running time in $\mathcal{O}(n)$, i.e., the whole algorithm runs in linear-time.

This establishes the conclusion of the theorem. ∎

5   Conclusion

In this paper, we study the minimum clique partition problem with constrained weight for interval graphs, and we have proved that this new problem is *NP*-hard and it cannot be approximated within a ratio $\frac{3}{2} - \varepsilon$ in polynomial-time for any $\varepsilon > 0$, and then we have designed some approximation algorithms with different constant ratios to this problem and an optimal algorithm in linear-time to solve the problem for the version where all intervals have the same weights.

For the future work, we shall design an approximation algorithm for the MCPCW problem with a ratio $\frac{3}{2}$, which shall show the tight ratio $\frac{3}{2}$, by adding result of the theorem 2. On the other way, we shall design some approximation algorithms within a ratio 2 to possess lower complexity.

References

1. S. Benzer, *On the topology of the genetic fine structure*, Proc. Nat. Acad. Sci. USA 45 (1959), 1607-1620.
2. Hans L. Bodlaender and K. Jansen, *Restrictions of graph partition problems: Part I*, Theoretical Computer Science 148(1995), 93-109.
3. M. Carlisle and E. Lloyd, *On the k-coloring of intervals*, Discrete Applied Mathematics 59 (1995), 225-235.
4. M.R. Garey and D.S. Johnson, *Computers and Intractability: A Guide to the Theory of NP-Completeness*, W.H. Freeman, San Francisco (1979).
5. M.C. Golumbic, *Algorithmic Graph Theory and Perfect Graphs*, Academic Press, New York-London-Toronto, 1980.
6. H. Kaplan and R. Shamir, *Pathwidth, bandwidth and completion problems to proper interval graphs with small cliques*, Siam Journal on Computing Vol.25, No.3 (1996), 540-561.
7. C.H. Papadimitriou and M. Yannakakis, *Scheduling interval-order tasks*, Siam Journal on Computing, Vol.8 No.3 (1979), 405-409.
8. F.S. Roberts, Graph Theory and its Applications to the Problem of Society (CBMS-NSF Monograph 29), SIAM Publications, 1978.
9. A. Schrijver, *Combinatorial Optimization: Polyhedra and Efficiency*, Springer 2002.
10. R. Tarjan, *Data Structures and Network Algorithms*, SIAM, Philadelphia, PA, 1983.
11. M.S. Waterman and J.R. Griggs, *Interval graphs and maps of DNA*, Bulletin of Mothematical Biology Vol.48, No.2(1986), 189-195.
12. D.B. West, *Introduction to Graph Theory* (second edition), Prntice-Hall, Inc. 2001.

Overlap-Free Regular Languages

Yo-Sub Han[1,*] and Derick Wood[2]

[1] System Technology Division, Korea Institute of Science and Technology,
P.O. BOX 131, Cheongryang, Seoul, Korea
emmous@kist.re.kr
[2] Department of Computer Science, The Hong Kong University of Science
and Technology, Clear Water Bay, Kowloon, Hong Kong SAR
dwood@cs.ust.hk

Abstract. We define a language to be overlap-free if any two distinct strings in the language do not overlap with each other. We observe that overlap-free languages are a proper subfamily of infix-free languages and also a proper subfamily of comma-free languages. Based on these observations, we design a polynomial-time algorithm that determines overlap-freeness of a regular language. We consider two cases: A language is specified by a nondeterministic finite-state automaton and a language is described by a regular expression. Furthermore, we examine the prime overlap-free decomposition of overlap-free regular languages and show that the prime overlap-free decomposition is not unique.

1 Introduction

Regular languages are popular in many applications such as editors, programming languages and software systems in general. People often use regular expressions for searching in text editors or for UNIX command; for example, vi, emacs and grep. Moreover, regular expression searching is also used in pattern matching.

The pattern matching problem is to find all matching substrings of a text T with respect to a pattern L. If L is a regular language given by a regular expression, then the problem becomes the regular-expression matching problem. Many researchers have investigated various regular-expression matching problems [1, 3, 7, 18]. One question in regular-expression matching is how many matching substrings are in T. Given a regular expression E and a text T, there can be at most n^2 matching substrings in T with respect to $L(E)$, where n is the size of T. For example, $E = (a + b)^*$ and $T = aababababa \cdots abaa$ over the alphabet $\{a, b\}$. These matching substrings often overlap and nest with each other. To avoid this situation, researchers restrict the search to find and report only a linear subset of the matching substrings. We call it *linearizing restriction*. There are two well-known linearizing restrictions in the literature: The *longest match* rule, which is a generalization of the *leftmost longest match* rule of IEEE POSIX [14] and the *shortest-match substring search* rule of Clarke and Cormack [3]. These two rules have different semantics and, therefore, identify different matching

---

* The author was supported by KIST Tangible Space Initiative Grant 2E19020.

D.Z. Chen and D.T. Lee (Eds.): COCOON 2006, LNCS 4112, pp. 469–478, 2006.

substrings for same pattern and text in general. On the other hand, Han and Wood [10] showed that if the pattern language is infix-free, then both rules give the same output. Furthermore, they proposed another linearizing restriction, *leftmost non-overlapping match* rule that only reports non-overlapping matching substrings of T. This new rule leads us to define a new subfamily of regular languages, *overlap-free regular languages*. We define a language L to be overlap-free if any two strings in L do not overlap with each other. (We give a formal definition in Section 3.) If we use an overlap-free regular language as pattern, it guarantees that all matching substrings of a text do not overlap with each other and, therefore, ensures a linear number of matching substrings.

As a continuation of our investigations of subfamilies of regular languages, it is natural to examine overlap-free regular languages and the prime overlap-free decomposition problem since overlap-free regular languages are a proper subfamily of regular languages. Our goal is to design an efficient algorithm that determines overlap-freeness of a given regular language and to study the prime overlap-free decomposition and its uniqueness.

We define some basic notions in Section 2. In Section 3, we define overlap-free languages and design an efficient algorithm that determines overlap-freeness of a given regular language L based on the structural properties of L. Then, in Section 4, we demonstrate that an overlap-free regular language does not have a unique prime overlap-free decomposition. We also develop an algorithm for computing a prime overlap-free decomposition from a minimal deterministic finite-state automaton (DFA) of an overlap-free regular language.

2   Preliminaries

Let Σ denote a finite alphabet of characters and Σ^* denote the set of all strings over Σ. A language over Σ is any subset of Σ^*. The character $\emptyset$ denotes the empty language and the character λ denotes the null string. A finite-state automaton (FA) A is specified by a tuple $(Q, \Sigma, \delta, s, F)$, where Q is a finite set of states, Σ is an input alphabet, $\delta \subseteq Q \times \Sigma \times Q$ is a (finite) set of transitions, $s \in Q$ is the start state and $F \subseteq Q$ is a set of final states. Let $|Q|$ be the number of states in Q and $|\delta|$ be the number of transitions in δ. Then, the size $|A|$ of A is $|Q| + |\delta|$. Given a transition (p, a, q) in δ, where $p, q \in Q$ and $a \in \Sigma$, we say that p has an *out-transition* and q has an *in-transition*. Furthermore, p is a *source state* of q and q is a *target state* of p. A string x over Σ is accepted by A if there is a labeled path from s to a state in F such that this path spells out the string x. Thus, the language $L(A)$ of an FA A is the set of all strings that are spelled out by paths from s to a final state in F. We say that A is *non-returning* if the start state of A does not have any in-transitions and A is *non-exiting* if the final state of A does not have any out-transitions. We assume that A has only *useful* states; that is, each state of A appears on some path from the start state to some final state.

Given two strings x and y over Σ, x is a *prefix* of y if there exists $z \in \Sigma^*$ such that $xz = y$ and x is a *suffix* of y if there exists $z \in \Sigma^*$ such that $zx = y$.

Furthermore, x is said to be a *substring* or an *infix* of y if there are two strings u and v such that $uxv = y$. Given a set X of strings over, X is *infix-free* if no string in X is an infix of any other string in X. Similarly, X is *prefix-free* if no string in X is a prefix of any other string in X.

3   Overlap-Free Regular Languages

Given two strings x and y, we say that x and y overlap with each other if either a suffix of x is a prefix of y or a suffix of y is a prefix of x. For example, $x = abcd$ and $y = cdee$ overlap.

Definition 1. *Given a (regular) language L, we define L to be* overlap-free *if any two distinct strings in L do not overlap with each other.*

Since we examine overlap of strings, we can think of the derivative operation [2]. The *derivative* $x \backslash L$ of a language L with respect to a string x is the language $\{y \mid xy \in L\}$.

Proposition 1. *If a language L is overlap-free, then $x \backslash L \cup L$ is prefix-free for any string x.*

Let us examine the relationship with other families of languages. By Definition 1, overlap-free languages are a proper subfamily of infix-free languages. Golomb et al. [6] introduced comma-free languages: A language L is comma-free if $LL \cap \Sigma^+ L \Sigma^+ = \emptyset$. Comma-free languages are also a proper subfamily of infix-free languages [15]. We compare these two subfamilies of infix-free languages and establish the following result:

Proposition 2. *Overlap-free languages are a proper subfamily of comma-free languages.*

A regular language is represented by an FA or described by a regular expression. Thus, we define a regular expression E to be overlap-free if $L(E)$ is overlap-free and an FA A to be overlap-free if $L(A)$ is overlap-free.

We now investigate the decision problem of overlap-freeness of a regular language. Given a language L, L is prefix-free if and only if $L \cap L\Sigma^+ \neq \emptyset$ [15]. If L is a regular language, then we can check the emptiness of $L \cap L\Sigma^+$ in polynomial time. Thus, if we can find a proper string x, then we can use Proposition 1 for deciding overlap-freeness of L. However, we do not know which string is proper unless we check the emptiness of $(x \backslash L \cup L) \cap (x \backslash L \cup L)\Sigma^+$ and certainly it is undesirable to try all possible strings over Σ. Recently, Han et al. [8] introduced state-pair graphs and proposed an algorithm for determining infix-freeness of a regular language L based on the structural properties of L. Based on state-pair graphs, we design algorithms that determine overlap-freeness of a regular language. Since an overlap-free language must be infix-free, we assume that a given language L is infix-free. Note that we can check infix-freeness of L in quadratic time in the size of the representation of L [8]; if L is not infix-free, then L is not overlap-free.

First, we consider when a language is given by an FA. Given an FA $A = (Q, \Sigma, \delta, s, F)$, we assign a unique number for each state in A from 1 to m, where m is the number of states in A.

Definition 2. *Given an FA $A = (Q, \Sigma, \delta, s, F)$, we define the state-pair graph $G_A = (V_G, E_G)$ of A, where V_G is a set of nodes and E_G is a set of edges, as follows:*

$$V_G = \{(i,j) \mid i \text{ and } j \in Q\} \text{ and}$$
$$E_G = \{((i,j), a, (x,y)) \mid (i,a,x) \text{ and } (j,a,y) \in \delta \text{ and } a \in \Sigma\}.$$

The crucial property of state-pair graphs is that if there is a string w spelled out by two distinct paths in A, for example, one path is from i to x and the other path is from j to y, then, there is a path from (i,j) to (x,y) in G_A that spells out the same string w. Note that state-pair graphs do not require given FAs to be deterministic. The complexity of the state-pair graph $G_A = (V_G, E_G)$ for an FA $A = (Q, \Sigma, \delta, s, F)$ is as follows:

Proposition 3. *Given an FA $A = (Q, \Sigma, \delta, s, F)$ and its state-pair graph G_A, $|G_A| \leq |Q|^2 + |\delta|^2$.*

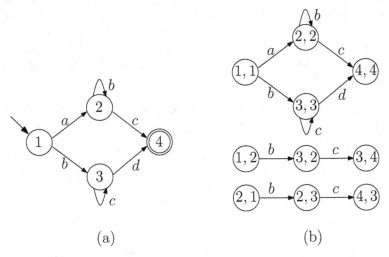

(a)   (b)

Fig. 1. (a) is an FA A for $L(ab^*c + bc^*d)$ and (b) is the corresponding state-pair graph G_A. We omit all nodes without transitions in G_A. Note that $L(A)$ is not overlap-free.

Fig. 1 illustrates the state-pair graph for a given FA A. Note that the language $L(A) = L(ab^*c + bc^*d)$ in Fig. 1 is not overlap-free since abc and bcd overlap, and the overlapped string bc appears on the path from $(1,2)$ to $(3,4)$ in G_A.

Since we assume that $L(A)$ is infix-free, a final state of A has no out-transitions and the start state has no in-transitions. Namely, A is non-returning and non-exiting. Therefore, if A has more than one final state, then all final states can be merged into a single final state since they are equivalent. From now on, we assume that a given FA is non-returning and non-exiting and has only one final state.

Theorem 1. *Given an FA $A = (Q, \Sigma, \delta, s, f)$, $L(A)$ is overlap-free if and only if the state-pair graph G_A for A has no path from $(1, i)$ to (j, m), where $i \neq m$ and $j \neq 1$, and 1 denotes the start state and m denotes the final state.*

We can identify such a path in Theorem 1 in linear time in the size of G_A using Depth-First Search (DFS) [4]. Thus, we obtain the following result from Proposition 3 and Theorem 1:

Theorem 2. *Given an FA $A = (Q, \Sigma, \delta, s, f)$, we can determine whether or not $L(A)$ is overlap-free in $O(|Q|^2 + |\delta|^2)$ worst-case time.*

Since $O(|\delta|) = O(|Q|^2)$ in the worst-case for NFAs, the runtime is $O(|Q|^4)$ in the worst-case. On the other hand, if a regular language is given by a regular expression E, then we can construct an FA for E that improves the worst-case running time. Since the complexity of state-pair graphs is closely related to the number of states and the number of transitions of input FAs, we use an FA construction that gives fewer states and transitions. One possibility is the Thompson construction [18].

Given a regular expression E, the Thompson construction takes $O(|E|)$ time and the resulting Thompson automaton has $O(|E|)$ states and $O(|E|)$ transitions [13]; namely, $O(|Q|) = O(|\delta|) = O(|E|)$. Even though Thompson automata are a subfamily of NFAs, they define all regular languages. Therefore, we can use Thompson automata to determine overlap-freeness of a regular language. Since Thompson automata allow null-transitions, we include the null-transition case to construct the edges for state-pair graphs as follows:

$$V_G = \{(i,j) \mid i \text{ and } j \in Q\} \text{ and}$$
$$E_G = \{((i,j), a, (x,y)) \mid (i, a, x) \text{ and } (j, a, y) \in \delta \text{ and } a \in \Sigma \cup \{\lambda\}\}.$$

The complexity of the state-pair graph based on this new construction is the same as before; namely, $O(|Q|^2 + |\delta|^2)$. Therefore, we establish the following result for checking regular expression overlap-freeness.

Theorem 3. *Given a regular expression E, we can determine whether or not $L(E)$ is overlap-free in $O(|E|^2)$ worst-case time.*

Furthermore, we can use state-pair graphs for determining comma-freeness of regular languages. A regular language L is comma-free if and only if $LL \cap \Sigma^+ L \Sigma^+ = \emptyset$. Because of the assumption that a given FA A is infix-free, (otherwise, $L(A)$ is not comma-free.) A has a single final state that has no out-transitions. Using this structural property, we construct an FA A' for LL by catenating two As; see Fig. 2 for an example.

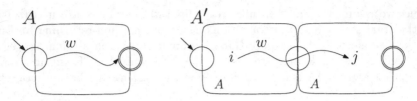

Fig. 2. Given an FA $A = (Q, \Sigma, \delta, s, f)$, we construct A' by merging the final state of one A and the start state of the other A. If $L(A)$ is not comma-free, then there exist two paths, one is from $A' = AA$ and the other is from A, and both path spell out the same string w.

Now we construct the state-pair graph for $L(A)$. The construction of state-pair graph for the comma-free case is slightly different from the state-pair graph in Definition 2. Given an FA $A = (Q, \Sigma, \delta, s, f)$, let $A' = (Q', \Sigma, \delta', s', f')$ be the catenation of two As; namely, $L(A') = L(A)L(A)$. The state-pair graph $G_A = (V_G, E_G)$ for the comma-free case is defined as follows:

$$V_G = \{(i, j) \mid i \in Q \text{ and } j \in Q'\} \text{ and}$$
$$E_G = \{((i, j), a, (x, y)) \mid (i, a, x) \in \delta, (j, a, y) \in \delta' \text{ and } a \in \Sigma\}.$$

Theorem 4. *Given an FA $A = (Q, \Sigma, \delta, s, f)$, $L(A)$ is comma-free if and only if there is no path from $(1, i)$ to (m, j), for $i \neq 1$ and $j \neq m$, in the state-pair graph G_A for A. Moreover, we can determine comma-freeness in $O(|Q|^2 + |\delta|^2)$ worst-case time.*

A subfamily of languages with certain properties is often closed under catenation. For example, prefix-free languages, bifix-free languages, infix-free languages and outfix-free languages are all closed under catenation, respectively [8, 9, 11]. Now we characterize the family of overlap-free (regular) languages in terms of closure properties.

Theorem 5. *The family of overlap-free (regular) languages is closed under intersection but not under catenation, union, complement or star.*

4   Prime Overlap-Free Regular Languages and Decomposition

Decomposition is the reverse operation of catenation. If $L = L_1 \cdot L_2$, then L is the catenation of L_1 and L_2 and $L_1 \cdot L_2$ is a decomposition of L. We call L_1 and L_2 *factors* of L. Note that every language L has a decomposition, $L = \{\lambda\} \cdot L$, where L is a factor of itself. We call $\{\lambda\}$ a *trivial* language. We define a language L to be *prime* if $L \neq L_1 \cdot L_2$, for any non-trivial languages L_1 and L_2. Then, the prime decomposition of L is to decompose L into $L_1 L_2 \cdots L_k$, where $L_1, L_2, \cdots, L_k$ are prime languages and $k \geq 1$ is a constant.

Mateescu et al. [16, 17] showed that the primality of regular languages is decidable and the prime decomposition of a regular language is not unique. Czyzowicz et al. [5] showed that for a given prefix-free regular language L, the prime prefix-free decomposition is unique and the decomposition can be computed in $O(m)$ worst-case time, where m is the size of the minimal DFA for L. Han et al. [8] investigated the prime infix-free decomposition of infix-free regular languages and demonstrated that the prime infix-free decomposition is not unique. On the other hand, the prime outfix-free decomposition of outfix-free regular languages is unique [11]. We investigate prime overlap-free regular languages and decomposition.

4.1   Prime Overlap-Free Regular Languages

Definition 3. *We define a regular language L to be a prime overlap-free language if $L \neq L_1 \cdot L_2$, for any overlap-free regular languages L_1 and L_2.*

From now on, when we say prime, we mean prime overlap-free.

Definition 4. *We define a state b in a DFA A to be a bridge state if the following conditions hold:*

1. *State b is neither a start nor a final state.*
2. *For any string $w \in L(A)$, its path in A must pass through b only once.*
3. *State b is not in any cycles in A.*
4. *$L(A_1)$ and $L(A_2)$ are overlap-free.*

Given an overlap-free DFA $A = (Q, \Sigma, \delta, s, f)$ with a bridge state $b \in Q$, we can partition A into two subautomata A_1 and A_2 as follows: $A_1 = (Q_1, \Sigma, \delta_1, s, b)$ and $A_2 = (Q_2, \Sigma, \delta_2, b, f)$, where Q_1 is a set of states that appear on some path from s and b in A, δ_1 is a set of transitions that appear on some path from s and b in A, $Q_2 = Q \setminus Q_1 \cup \{b\}$ and $\delta_2 = \delta \setminus \delta_1$. See Fig. 3 for an example.

Note that the second requirement in Definition 4 ensures that the decomposition of $L(A)$ is $L(A_1) \cdot L(A_2)$ and the third requirement is from the property that overlap-free FAs must be non-returning and non-exiting.

Theorem 6. *An overlap-free regular language L is prime if and only if the minimal DFA A for L does not have any bridge states.*

We tackle the decomposition problem based on FA partitioning using bridge states. Note that Czyzowicz et al. [5] demonstrated the use of FA partitioning for the prefix-free decomposition and Han and Wood [12] proposed an efficient algorithm that computes shorter regular expressions from FAs based on FA partitioning. In many applications, FAs become more and more complicated and the size of FAs is too large to fit into main memory. Therefore, FA decomposition is necessary and FA partitioning is one approach for solving this problem.

4.2   Prime Decomposition of Overlap-Free Regular Languages

The prime decomposition for an overlap-free regular language L is to represent L as a catenation of prime overlap-free regular languages. If L is prime, then L

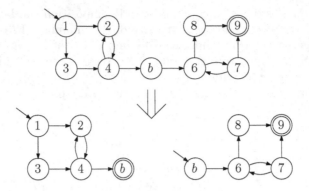

Fig. 3. An example of the partitioning of an FA at a bridge state b

itself is a prime decomposition. Thus, given an overlap-free regular language L, we, first, determine whether or not L is prime. If L is not prime, then there should be some bridge state(s) and we decompose L using the bridge state(s). Let A_1 and A_2 be two subautomata partitioned at a bridge state for L. If both $L(A_1)$ and $L(A_2)$ are prime, then a prime decomposition of L is $L(A_1) \cdot L(A_2)$. Otherwise, we repeat the preceding procedure for a non-prime language.

Let B denote a set of bridge states for a given minimal DFA A. The number of states in B is at most m, where m is the number of states in A. Note that once we partition A at $b \in B$ into A_1 and A_2, then only the states in $B \setminus \{b\}$ can be bridge states in A_1 and A_2. (It is not necessary for all remaining states to be bridge states as demonstrated in Fig. 4.) Therefore, we can determine the primality of $L(A)$ by checking whether or not A has bridge states. Moreover, we can compute a prime decomposition of $L(A)$ using these bridge states. Since there are at most m bridge states in A, we can compute a prime decomposition of $L(A)$ after a finite number of decompositions at bridge states.

Note that the first three requirements in Definition 4 are based on the structural properties of A. We call a state that satisfies the first three requirements a *candidate bridge state*. We first compute all candidate bridge states and, then we determine whether or not each candidate bridge state satisfies the fourth requirement in Definition 4.

Proposition 4 (Han et al. [8]). *Given a minimal DFA $A = (Q, \Sigma, \delta, s, f)$, we can identify all candidate bridge states in $O(|Q| + |\delta|)$ worst-case time.*

Let C_B denote a set of candidate bridge states that we compute from an overlap-free DFA A based on Proposition 4. Then, for each state $b_i \in C_B$, we check whether or not two subautomata A_1 and A_2 partitioned at b_i are overlap-free. If both A_1 and A_2 are overlap-free, then L is not prime and, thus, we decompose L into $L(A_1) \cdot L(A_2)$ and continue to check and decompose for each A_1 and A_2, respectively, using the remaining states in $C_B \setminus \{b_i\}$.

Theorem 7. *Given a minimal DFA $A = (Q, \Sigma, \delta, s, f)$ for an overlap-free regular language, we can determine primality of $L(A)$ in $O(m^3)$ worst-case time* *and compute a prime decomposition for $L(A)$ in $O(m^4)$ worst-case time, where $m = |Q|$.*

The algorithm for computing a prime decomposition for $L(A)$ in Theorem 7 looks similar to the algorithm for the infix-free regular language case studied by Han et al. [8]. However, there is one big difference between these two algorithms because of the different closure properties of two families: In fact, Han et al. [8] speeded up their algorithm by linear factor based on the fact that infix-free languages are closed under catenation whereas overlap-free languages are not closed as shown in Theorem 5.

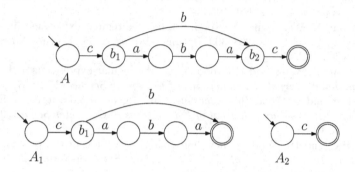

Fig. 4. States b_1 and b_2 are bridge states for A. However, once we decompose A at b_2, then b_1 is no longer a bridge state in A_1 since b_1 now violates the fourth requirement in Definition 4. Similarly, if we decompose A at b_1, then b_2 is not a bridge state.

We observe that a bridge state b_i of a minimal DFA A may not be a bridge state anymore if A is partitioned at a different bridge state b_j. See Fig. 4 for an example: It hints that the prime overlap-free decomposition might not be unique. Note that the prime prefix-free decomposition for a prefix-free regular language is unique [5] whereas the prime infix-free decomposition for an infix-free regular language is not unique [8]. Since overlap-free languages are a proper subfamily of prefix-free languages and a proper subfamily of infix-free languages, it is natural to examine the uniqueness of prime overlap-free decomposition. The following example demonstrates that the prime overlap-free decomposition is not unique.

$$L(c(aba + b)c) = \begin{cases} L_1(c(aba + b)) \cdot L_2(c). \\ L_2(c) \cdot L_3((aba + b)c). \end{cases}$$

The language L is overlap-free but not prime and it has two different prime decompositions, where L_1, L_2 and L_3 are prime overlap-free languages.

References

[1] A. Aho. Algorithms for finding patterns in strings. In J. van Leeuwen, editor, *Algorithms and Complexity*, volume A of *Handbook of Theoretical Computer Science*, 255–300. The MIT Press, Cambridge, MA, 1990.

[2] J. Brzozowski. Derivatives of regular expressions. *Journal of the ACM*, 11:481–494, 1964.

[3] C. L. A. Clarke and G. V. Cormack. On the use of regular expressions for searching text. *ACM Transactions on Programming Languages and Systems*, 19(3):413–426, 1997.

[4] T. H. Cormen, C. E. Leiserson, R. L. Rivest, and C. Stein. *Introduction to Algorithms*. McGraw-Hill Higher Education, 2001.

[5] J. Czyzowicz, W. Fraczak, A. Pelc, and W. Rytter. Linear-time prime decomposition of regular prefix codes. *International Journal of Foundations of Computer Science*, 14:1019–1032, 2003.

[6] S. Golomb, B. Gordon, and L. Welch. Comma-free codes. *The Canadian Journal of Mathematics*, 10:202–209, 1958.

[7] Y.-S. Han, Y. Wang, and D. Wood. Prefix-free regular-expression matching. In *Proceedings of CPM'05*, 298–309. Springer-Verlag, 2005. Lecture Notes in Computer Science 3537.

[8] Y.-S. Han, Y. Wang, and D. Wood. Infix-free regular expressions and languages. *International Journal of Foundations of Computer Science*, 17(2):379–393, 2006.

[9] Y.-S. Han and D. Wood. The generalization of generalized automata: Expression automata. *International Journal of Foundations of Computer Science*, 16(3):499–510, 2005.

[10] Y.-S. Han and D. Wood. A new linearizing restriction in the pattern matching problem. In *Proceedings of FCT'05*, 552–562. Springer-Verlag, 2005. Lecture Notes in Computer Science 3623.

[11] Y.-S. Han and D. Wood. Outfix-free regular languages and prime outfix-free decomposition. In *Proceedings of ICTAC'05*, 96–109. Springer-Verlag, 2005. Lecture Notes in Computer Science 3722.

[12] Y.-S. Han and D. Wood. Shorter regular expressions from finite-state automata. In *Proceedings of CIAA'05*, 141–152. Springer-Verlag, 2005. Lecture Notes in Computer Science 3845.

[13] J. Hopcroft and J. Ullman. *Formal Languages and Their Relationship to Automata*. Addison-Wesley, Reading, MA, 1969.

[14] IEEE. *IEEE standard for information technology: Portable Operating System Interface (POSIX) : part 2, shell and utilities*. IEEE Computer Society Press, Sept. 1993.

[15] H. Jürgensen and S. Konstantinidis. Codes. In G. Rozenberg and A. Salomaa, editors, *Word, Language, Grammar*, volume 1 of *Handbook of Formal Languages*, 511–607. Springer-Verlag, 1997.

[16] A. Mateescu, A. Salomaa, and S. Yu. On the decomposition of finite languages. Technical Report 222, TUCS, 1998.

[17] A. Mateescu, A. Salomaa, and S. Yu. Factorizations of languages and commutativity conditions. *Acta Cybernetica*, 15(3):339–351, 2002.

[18] K. Thompson. Regular expression search algorithm. *Communications of the ACM*, 11:419–422, 1968.

On the Combinatorial Representation of Information

Joel Ratsaby

Ben Gurion University of the Negev, ISRAEL
ratsaby@bgu.ac.il

Abstract. Kolmogorov introduced a combinatorial measure of the information $I(x : y)$ about the unknown value of a variable y conveyed by an input variable x taking a given value x. The paper extends this definition of information to a more general setting where 'x = x' may provide a vaguer description of the possible value of y. As an application, the space $\mathcal{P}(\{0,1\}^n)$ of classes of binary functions $f : [n] \rightarrow \{0,1\}$, $[n] = \{1, \ldots, n\}$, is considered where y represents an unknown function $t \in \{0,1\}^{[n]}$ and as input, two extreme cases are considered: x = $x_{\mathcal{M}_d}$ and x = $x_{\mathcal{M}'_d}$ which indicate that t is an element of a set $G \subseteq \{0,1\}^n$ that satisfies a property $\mathcal{M}_d$ or $\mathcal{M}'_d$ respectively. Property $\mathcal{M}_d$ (or $\mathcal{M}'_d$) means that there exists an $E \subseteq [n]$, $|E| = d$, such that $|\text{tr}_E(G)| = 1$ (or 2^d) where $\text{tr}_E(G)$ denotes the trace of G on E. Estimates of the information value $I(x_{\mathcal{M}_d} : t)$ and $I(x_{\mathcal{M}'_d} : t)$ are obtained. When d is fixed, it is shown that $I(x_{\mathcal{M}_d} : t) \approx d$ and $I(x_{\mathcal{M}'_d} : t) \approx 1$ as $n \rightarrow \infty$.

Keywords: Information theory, combinatorial complexity, VC-dimension.

1 Introduction

Kolmogorov [5] sought for a measure of information of 'finite objects'. He considered three approaches, the so-called combinatorial, probabilistic and algorithmic. The probabilistic approach corresponds to the well-established definition of the Shannon entropy which applies to stochastic settings where an 'object' is represented by a random variable. In this setting, the entropy of an object and the information conveyed by one object about another are well defined. Kolmogorov's algorithmic-notion of the information contained in a finite binary string s is the length of the minimal-size program that can compute it and is denoted by $K(s)$. This notion of the information contained in s, which is fundamentally different from the Shannon information since s is non-stochastic, has been developed into the so-called *Kolmogorov Complexity* field [6].

In the combinatorial approach, Kolmogorov investigated another non stochastic measure of information for an object x. Here x is taken to be a variable with a range of possible values in some finite set $E = \{x_1, \ldots, x_n\} \subset X$ where X is any set of objects. To distinguish between a variable and its possible values we use sans serif fonts such as x to denote variables and normal fonts x to denote fixed elements of sets. We write x $\subset E$ to denote that the unknown value of the

D.Z. Chen and D.T. Lee (Eds.): COCOON 2006, LNCS 4112, pp. 479–488, 2006.

variable x is contained in E while $x \in E$ refers to a specific x as being an element of E. Kolmogorov [5] defined the 'entropy' of x as $H(\mathsf{x}) = \log|E|$ where $|E|$ denotes the cardinality of E and all logarithms henceforth are taken with respect to 2. If the value of x is known to be x then this much entropy is 'eliminated' by providing $\log|E|$ bits of 'information'.

The conditional entropy between two variables x and y is defined based on a set $A \subseteq X \times Y$ that consists of all 'allowed' values of pairs $(x, y) \in X \times Y$. The entropy of y is defined as $H(\mathsf{y}) = \log|\Pi_Y(A)|$ where $\Pi_Y(A) \equiv \{y \in Y : (x, y) \in A\}$ denotes the projection of A on Y. Let

$$A_x = \{y \in Y : (x, y) \in A\} \tag{1}$$

then the conditional combinatorial entropy of y given 'x $= x$' is defined as

$$H(\mathsf{y}|x) = \log|A_x|. \tag{2}$$

Kolmogorov defines the information conveyed by 'x $= x$' about y by the quantity

$$I(x : \mathsf{y}) = H(\mathsf{y}) - H(\mathsf{y}|x) \tag{3}$$

where in both definitions one of the variables, in this case x, takes a known fixed value x while the second variable y is left unknown.

In many applications, knowing 'x $= x$' conveys only vague information about y. For instance, in problems which involve the analysis of discrete classes of structures, e.g., sets of Boolean functions on a finite domain, an algorithmic search is made for some optimal element in this set based only on partial information. Formally, let n be a positive integer and consider the domain $[n] = \{1, \ldots, n\}$. Let $F = \{0, 1\}^{[n]}$ be the set of all binary functions $f : [n] \to \{0, 1\}$. The power set $\mathcal{P}(F)$ represents the family of all sets $G \subseteq F$. Let us denote by $\mathcal{M}$ a property of a set G and write $G \models \mathcal{M}$. Suppose that we seek to know some unknown target function $t \in F$. Any partial information about t which may be expressed by $t \in G \models \mathcal{M}$ can effectively reduce the search space. Typically, one is interested in some estimate of the value of such partial information. Kolmogorov's framework may be applied here by letting the variable x take as values x the possible descriptions of properties $\mathcal{M}$ of subsets $G \subseteq F$. The variable y represents the unknown 'object', i.e., the target t, which may be any element in F. The input 'x $= x$' conveys that t is in some subset G that has some particular property $\mathcal{M}_x$. Therefore as a measure of information, one option is to compute the value $I(x : \mathsf{y})$ of the information in x about y.

Kolmogorov's combinatorial representation of information (3) is not sufficient in this setting since it requires that the target y be restricted to a fixed set A_x on knowledge of 'x $= x$'. To see this, suppose it is given that x $= x$, i.e., that t is in a set that satisfies property $\mathcal{M}_x$. Consider the collection $\{G_z\}_{z \in Z_x}$ of all subsets $G_z \subseteq F$ that have this property. Clearly, $t \in \bigcup_{z \in Z_x} G_z$ hence we may at first consider $A_x = \bigcup_{z \in Z_x} G_z$. However, this ignores some useful information implicit in this collection as we now show: consider two properties $\mathcal{M}_0$ and $\mathcal{M}_1$ with corresponding index sets Z_{x_0} and Z_{x_1} such that $\bigcup_{z \in Z_{x_0}} G_z = \bigcup_{z \in Z_{x_1}}$

$G_z \equiv F' \subseteq F$. Suppose that most of the sets G_z, $z \in Z_{x_0}$ are small while the sets G_z, $z \in Z_{x_1}$ are large. Clearly, property $\mathcal{M}_0$ is more informative than $\mathcal{M}_1$ since starting with knowledge that t is in a set that satisfies property $\mathcal{M}_0$ should take less additional information (once it becomes available) to completely specify t. If, as above, we let $A_{x_0} = \bigcup_{z \in Z_{x_0}} G_z$ and $A_{x_1} = \bigcup_{z \in Z_{x_1}} G_z$ then we have $I(x_0 : y) = I(x_1 : y)$ which wrongly implies that both properties are equally informative. A more general definition of information which applies also to such setting is needed and is proposed in this paper.

The remaining sections are organized as follows: in Section 2 we state a new definition of combinatorial information. In Section 3 we apply this to the setting of binary function classes and state two results. Section 4 contains the technical work.

2   Combinatorial Formulation of Information

Our aim is to extend Kolmogorov's information measure (3) to a more general setting (as discussed in Section 1) where the knowledge of 'x = x' may leave some vagueness about the possible set of values of y. As in [5], we seek a non-probabilistic representation of the information conveyed by x about y and the set $A \subseteq X \times Y$ represents the 'degree of freedom' of x and y. As a first attempt let us try to extend (3) by using one extension of the combinatorial conditional entropy (see [7]) which treats both x and y as unknown variables and is defined as follows:

$$H(y|x) = \max_{x \in X} \log |A_x|$$

where A_x is defined in (1). Substituting this for $H(y|x)$ in (3) gives

$$I(x : y) = \log |\Pi_Y(A)| - \max_{x \in X} \log |A_x|. \tag{4}$$

There are two immediate difficulties with this definition: first, here x is unknown and therefore the definition departs from Kolmogorov's definitions of (2) and (3) where the value of x is known to be x. The second problem can perhaps be best seen from the following example:

Example 1. Let $X = Y = Z = \mathbb{N}$, where $\mathbb{N}$ denotes the natural numbers. Let $z_0 \in Z$ and let $E \subset \mathbb{N}$ be a set with $z_0 \notin E$. Let $A \subset Z \times Y$ satisfy the following: $A_z = E$ if $z = z_0$, $A_z = \{z\}$ if $z \in E$ where A_z is defined as in (1). Suppose 'x = x' means that the unknown value of y is an element of at least one set A_z, $z \in Z_x$ and $Z_x = \{z_0\} \bigcup E$. How much information is conveyed about the unknown value of the variable y by the statement 'x = x'?

This is an example of partial-information where knowing 'x = x' still leaves some uncertainty about the set of possible values of y. If (4) is used then the information in x conveyed about y is zero since $\log |\Pi_Y(A)| = \log |E| = \max_{z \in Z_x} \log |A_z|$. A zero value is clearly not representative of the amount of information since knowing 'x = x' means that for half the number of possible pairs $\{(z, y) \in A :$

$z \in Z_x$} the value of y can be exactly determined. Hence the information value should be greater than zero.

Consider extending Kolmogorov's representation of uncertainty by letting A and B be two sets which consist of all permissible pairs $(z, y) \in Z \times Y$ and $(x, z) \in X \times Z$, respectively. We view the set B as defining the allowed pairs (x, z) of property descriptions x and class-index values $z \in Z_x$. The sets $A_z \subseteq Y$, by definition, satisfy the property described by x. We propose the following combinatorial measure for the information conveyed by 'x $= x$' about y:

Definition 1. *Let* X, Y, Z *be general sets of objects. Let* $A \subset Z \times Y$, $B \subset X \times Z$. *For any* $x \in \Pi_X(B)$ *denote by* $Z_x = \{z \in Z : (x, z) \in B\} \subset \Pi_Z(B)$ *and for any* $z \in Z_x$ *let* $A_z = \{y \in Y : (z, y) \in A\} \subset \Pi_Y(A)$. *Let* 'x $= x$' *mean that the unknown value of* y *is contained in at least one set* A_z *where* $z \in Z_x$. *Then the information conveyed by* 'x $= x$' *about* y *is defined as*

$$I(x : y) \equiv \log |\Pi_Y(A)| - \sum_{k \geq 2} \omega_x(k) \log k \qquad (5)$$

where $\omega_x(k) = \frac{|\{z : |A_z| = k, z \in Z_x\}|}{|Z_x|}$.

In words, (5) is a sequence of values $\{I(x : y)\}_{\{x : x \in \Pi_X(B)\}}$ that correspond to inputs x each describing some property common to all sets $A_z \subset \Pi_Y(A)$, $z \in Z_x \subset \Pi_Z(B)$ whose union covers all the possibilities for the unknown value of y. Henceforth, the sets A, B are assumed fixed and known.

In order to understand the motivation behind Definition 1, first note that it is consistent with Kolmogorov's definition (3) in that x appears as taking a known value x and the representation of uncertainty is done as in [5] via a set-theoretic approach since all expressions in (5) involve set-quantities, e.g., cardinality, projections. Note that (3) is a special case of Definition 1 with Z_x being a singleton set. The factor of $\log k$ comes from $\log |A_z|$ which from (2) is the combinatorial conditional-entropy $H(y|z)$.

The knowledge conveyed by 'x $= x$' still results in some uncertainty which is represented by a set Z_x of possible values for z. This induces an uncertainty in the value of y which is now manifested through several sets $\{A_z\}_{z \in Z_x}$ each satisfying the property described by x and whose union covers the range of possible values of y. A detailed application of Definition 1 is considered in Section 3.

There is a straightforward analogy between this combinatorial measure of information and Shannon's information formula. Let $\mathcal{Z}_x$ and $\mathcal{Y}$ be two random variables with $\mathcal{Y}$ having a uniform probability distribution given $\mathcal{Z}_x$. Then

$$I(\mathcal{Y} : \mathcal{Z}_x) = H(\mathcal{Y}) - H(\mathcal{Y}|\mathcal{Z}_x)$$

with $H(\mathcal{Y}|\mathcal{Z}_x) = \sum_z P_{\mathcal{Z}_x}(z) H(\mathcal{Y}|\mathcal{Z}_x = z) = \sum_{k \geq 2} P_{\mathcal{Z}_x}(H(\mathcal{Y}|z) = \log k) \log k$. The factor $\omega_x(k)$ in the sum of (5) is analogous to the probability $P_{\mathcal{Z}_x}(H(\mathcal{Y}|z) = \log k)$.

Let us now evaluate this information measure for Example 1. We have $|Z_x| = |E| + 1$ and the sum in (5) has only two terms, $k = 1$ which applies for all $z \in E$ and $k = |E|$ for $z = z_0$. Hence

$$\sum_{k \geq 2} \frac{|\{z : |A_z| = k, z \in Z_x\}|}{|Z_x|} \log k = \frac{|E|}{|E| + 1} \log 1 + \frac{1}{|E| + 1} \log |E| = \frac{1}{|E| + 1} \log |E|.$$

Since $\Pi_Y(A) = E$ then $I(x : y) = (1 - \alpha) \log |E|$ where $\alpha = 1/(|E| + 1)$. As $H(y) = \log |\Pi_Y(A)| = \log |E|$, then $I(x : y)$ equals $(1 - \alpha)$ times the combinatorial entropy of y. It thus reflects the fact that for a fraction $(1 - \alpha)$ of the set Z_x the knowledge of 'x = x' identifies the value of y exactly (zero uncertainty) and for the remaining $\alpha|Z_x|$ elements, this knowledge leaves the uncertainty about y unchanged at $\log |E|$ bits. In the next section, we apply this information measure to binary function classes.

3   Binary Function Classes

As in Section 1, let n be a positive integer, denote by $[n] = \{1, \ldots, n\}$, $F = \{0, 1\}^n$ and write $\mathcal{P}(F)$ for the power set which consists of all subsets $G \subseteq F$. An element f of F is referred to as a binary function $f : [n] \to \{0, 1\}$. Let $G \models \mathcal{M}$ represent the statement "G satisfies property $\mathcal{M}$". In order to apply Kolmogorov's combinatorial representation of information we let the variable y represent the unknown target t in F and the input variable x describe the possible properties $\mathcal{M}$ of sets $G \subseteq F$ which may contain t. The aim is to compute the value of information $I(x_{\mathcal{M}} : t)$ for various inputs x = $x_{\mathcal{M}}$.

Before we proceed, let us recall a few basic definitions from set theory. For any fixed subset $E \subseteq [n]$ of cardinality d and any $f \in F$ denote by $f_{|E} \in \{0, 1\}^d$ the restriction of f on E. For a set $G \subseteq F$ of functions, the set $tr_E(G) = \{f_{|E} : f \in G\}$ is called the *trace* of G on E. The properties considered next are based on the trace of a class and are defined in terms of an integer variable d in the following general form: $d = \max\{|E| : E \subseteq [n]$, condition on $tr_E(G)$ holds$\}$. The first definition taking such form is the so-called Vapnik-Chervonenkis dimension.

Definition 2. *The Vapnik-Chervonenkis dimension of a set $G \subseteq F$, denoted $VC(G)$, is defined as $VC(G) \equiv \max\{|E| : E \subseteq [n], |tr_E(G)| = 2^{|E|}\}$.*

The next definition considers the other possible extreme for the size of the trace.

Definition 3. *Let $L(G)$ be defined as $L(G) \equiv \max\{|E| : E \subseteq [n], |tr_E(G)| = 1\}$.*

For any class $G \subseteq F$ define the following two properties: $\mathcal{M}_d \equiv$ '$L(G)$ is at least d', $\mathcal{M}'_d \equiv$ '$VC(G)$ is at least d'.

As an application of the information-measure of Definition 1 we state the following results (for clarity, we defer the proof sketches to the next section). Henceforth, for two sequences a_n, b_n, we write $a_n \approx b_n$ to denote that $\lim_{n \to \infty} \frac{a_n}{b_n} = 1$. Denote the standard normal probability distribution and cumulative distribution by $\phi(x) = (1\sqrt{2\pi}) \exp(-x^2/2)$ and $\Phi(x) = \int_{-\infty}^{x} \phi(z)dz$, respectively.

Theorem 1. *Let $1 \leq d \leq n$ and t be an unknown element of F. Then the information value in knowing that $t \in G$, where $G \models \mathcal{M}_d$, is*

$$I(x_{\mathcal{M}_d} : t) = \log |F| - \sum_{k \geq 2} \omega_{x_{\mathcal{M}_d}}(k) \log k$$

$$\approx n - \frac{\Phi(-a) \log \left(\frac{2^n}{1+2^d} \right) + 2^{-(n-d)/2} \phi(a) + O(2^{-(n-d)})}{1 - \left(\frac{2^d}{1+2^d} \right)^{2^n}}$$

with n increasing and where $a = 2^{(n-d)/2} - 2(1 + 2^d) 2^{-(n+d)/2}$.

Remark 1. For large n, the above is approximately

$$I(x_{\mathcal{M}_d} : t) \simeq n - \log \left(\frac{2^n}{1 + 2^d} \right) \simeq n - (n - d) = d.$$

A rough explanation to this result is as follows: given that it has a cardinality k, the chance that a random class satisfies property $\mathcal{M}_d$ decreases exponentially with respect to k. As shown in the proof, this implies that the majority of classes that satisfy $\mathcal{M}_d$ have cardinality $k = 2^n/(1 + 2^d)$.

The next result is for the property $\mathcal{M}'_d$.

Theorem 2. *Let d be any fixed integer satisfying $1 \leq d \leq n - 1$ and t be an unknown element of F. Then the information value in knowing that $t \in G$, where $G \models \mathcal{M}'_d$, is*

$$I(x_{\mathcal{M}'_d} : t) = \log |F| - \sum_{k \geq 2} \omega_{x_{\mathcal{M}'_d}}(k) \log k$$

$$\approx n - \frac{(n-1) \left(2^n \Phi(a) + 2^{n/2} \phi(a) \left(1 + \frac{a^2}{(n-1)2^n} \right) \right)}{2^n \Phi(a) + 2^{n/2} \phi(a)}$$

as n increases, where $a = \frac{2^n - 2^{d+1}}{2^{n/2}}$.

Remark 2. For large n, the information value is approximately $I(x_{\mathcal{M}'_d} : t) \simeq 1$. A rough explanation is as follows: for all $k \geq 2^d$, the chance that a random class of cardinality k has property $\mathcal{M}'_d$ tends to 1 (Lemma 3 below). When d is insignificant compared to n this implies that the property holds for almost every class. The majority of classes in $\mathcal{P}(F)$ have cardinality $k = 2^{n-1}$ which is one half the total number of functions on $[n]$. Thus there is approximately 1 bit of information in knowing that t is an element of a class that has this property.

4   Technical Results

In this section we provide the sketch of proofs of Theorems 1 and 2. Our approach is to estimate the number of sets $G \subseteq F$ that satisfy a property $\mathcal{M}$. We employ a probabilistic method by which a random class is generated and the probability that it satisfies $\mathcal{M}$ is computed. As we use the uniform probability distribution on elements of the power set $\mathcal{P}(F)$ then probabilities yield cardinalities of the corresponding sets. The computation of $\omega_x(k)$ and hence of (5) follows directly. It is worth noting that, as in [4], the notion of probability is only used here for simplifying some of the counting arguments and thus, unlike Shannon's information, it plays no role in the actual definition of information. Before proceeding with the proofs, in the next section we describe the probability model for generating a random class.

4.1   Random Class Generation

In this subsection we describe the underlying probabilistic processes with which a random class is generated. A random class $\mathcal{F}$ is constructed through 2^n independent coin tossings, one for each function in F, with a probability of success (i.e., selecting a function into $\mathcal{F}$) equal to p. The probability distribution $P_{n,p}$ is formally defined on $\mathcal{P}(F)$ as $P_{n,p}(\mathcal{F} = G) = p^{|G|}(1-p)^{2^n-|G|}$. In our application, we choose $p = 1/2$ and denote the probability distribution as $P_n \equiv P_{n,\frac{1}{2}}$. Hence for any element $G \in \mathcal{P}(F)$, the probability that the random class $\mathcal{F}$ equals G is

$$\alpha_n \equiv P_n(\mathcal{F} = G) = \left(\frac{1}{2}\right)^{2^n} \tag{6}$$

and the probability of $\mathcal{F}$ having a cardinality k is $P_n(|\mathcal{F}| = k) = \binom{2^n}{k}\alpha_n$, $1 \leq k \leq 2^n$. The following fact easily follows from the definition of the conditional probability: for any set $B \subseteq \mathcal{P}(F)$,

$$P_n(\mathcal{F} \in B|\,|\mathcal{F}| = k) = \frac{\sum_{G \in B} \alpha_n}{\binom{2^n}{k}\alpha_n} = \frac{|B|}{\binom{2^n}{k}}. \tag{7}$$

Denote by $F^{(k)} = \{G \in \mathcal{P}(F) : |G| = k\}$ the collection of binary-function classes of cardinality k, $1 \leq k \leq 2^n$. Consider the uniform probability distribution on $F^{(k)}$ which is defined as follows: given parameters n and $1 \leq k \leq 2^n$ then for any $G \in \mathcal{P}(F)$,

$$P^*_{n,k}(G) = \frac{1}{\binom{2^n}{k}}, \quad \text{if } G \in F^{(k)}, \tag{8}$$

and $P^*_{n,k}(G) = 0$ otherwise. Hence from (7) and (8) it follows that for any $B \subseteq \mathcal{P}(F)$,

$$P_n(\mathcal{F} \in B|\,|\mathcal{F}| = k) = P^*_{n,k}(\mathcal{F} \in B). \tag{9}$$

It will be convenient to use another probability distribution which estimates $P^*_{n,k}$ and is defined by the following process of random-class construction. First, construct a random $n \times k$ binary matrix by fair-coin tossings with the nk elements taking values 0 or 1 independently with probability $1/2$. Denoting by $Q^*_{n,k}$ the probability measure corresponding to this process, then for any matrix $U \in \mathcal{U}_{n \times k}(\{0,1\})$,

$$Q^*_{n,k}(U) = \frac{1}{2^{nk}}. \tag{10}$$

Denote by S a *simple* binary matrix as one all of whose columns are distinct [1]. It is easy to verify that the conditional distribution of the set of columns of a random binary matrix, knowing that the matrix is simple, is the uniform distribution $P_{n,k}^*$. As it turns out, the distribution $Q_{n,k}^*$ leads to simpler computations of the asymptotic probability of several types of events that are associated with the properties of Theorems 1 and 2. The following result will enable us to replace $P_{n,k}^*$ by $Q_{n,k}^*$ (due to space limitation we omit the proof which can be found in [9]).

Lemma 1. *Assume $k_n \ll 2^{n/2}$ and let $B \subseteq \mathcal{P}(F)$. If $P_{n,k_n}^*(B)$ and $Q_{n,k_n}^*(B)$ converge with increasing n then they converge to the same limit.*

We now proceed to sketch the proofs of the Theorems in Section 3.

4.2   Proofs

Note that for any property $\mathcal{M}$, the quantity $\omega_x(k)$ in (5) is the ratio of the number of classes $G \in F^{(k)}$ that satisfy $\mathcal{M}$ to the total number of classes that satisfy $\mathcal{M}_x$. It is therefore equal to $P_n(|\mathcal{F}| = k \mid \mathcal{F} \models \mathcal{M}_x)$. Our approach starts by computing the probability $P_n(\mathcal{F} \models \mathcal{M}_x \mid |\mathcal{F}| = k)$ from which $P_n(|\mathcal{F}| = k \mid \mathcal{F} \models \mathcal{M}_x)$ and then $\omega_x(k)$ may be obtained.

4.3   Proof Sketch of Theorem 1

We start with an auxiliary lemma which states that the probability $P_n(\mathcal{F} \models \mathcal{M}_d \mid |\mathcal{F}| = k)$ possesses a zero-one behavior.

Lemma 2. *Let $\mathcal{F}$ be a class of cardinality k_n and randomly drawn according to the uniform distribution P_{n,k_n}^* on $F^{(k_n)}$. Then as n increases, the probability $P_{n,k_n}^*(\mathcal{F} \models \mathcal{M}_d)$ that $\mathcal{F}$ satisfies property $\mathcal{M}_d$ tends to 0 or 1 if $k_n \gg \log(2n/d)$ or $k_n \ll (\log(n))/d$, respectively.*

Proof sketch: For brevity, we sometimes write k for k_n. Using Lemma 1 it suffices to show that $Q_{n,k}^*(\mathcal{F} \models \mathcal{M}_d)$ tends to 1 or 0 under the stated conditions. For any set $S \subset [n]$, $|S| = d$ and any fixed $v \in \{0,1\}^d$, under the probability distribution $Q_{n,k}^*$, the event E_v that every function $f \in \mathcal{F}$ satisfies $f_{|S} = v$ has a probability $(1/2)^{kd}$. Denote by E_S the event that all functions in the random class $\mathcal{F}$ have the same restriction on S. It is easy to show that $Q_{n,k}^*(E_S) = 2^d(1/2)^{kd} = 2^{-(k-1)d}$. The event that $\mathcal{F}$ has property $\mathcal{M}_d$, i.e., that $L(\mathcal{F}) \geq d$, equals the union of E_S, over all $S \subseteq [n]$ of cardinality d. It follows that $Q_{n,k}^*(\mathcal{F} \models \mathcal{M}_d) \leq 2^{-(k-1)d}n^d(1 - o(1))/d!$. For $k = k_n \gg \log(2n/d)$ the right hand side tends to zero which proves the first statement. Let the mutually disjoint sets $S_i = \{id + 1, id + 2, \ldots, d(i + 1)\} \subseteq [n]$, $0 \leq i \leq m - 1$ where $m = \lfloor n/d \rfloor$. The event that $\mathcal{M}_d$ is not true equals $\bigcap_{S:|S|=d} \overline{E}_S$. It is easy to show that its probability is no larger than $1 - mQ_{n,k}^*(E_{[d]})$ which equals $1 - \lfloor \frac{n}{d} \rfloor 2^{-(k-1)d}$ and tends to zero when $k = k_n \ll (\log(n))/d$. $\square$

By the premise of Theorem 1, the input 'x $= x$' describes the target t as an element of a class that satisfies property $\mathcal{M}_d$. In this case the quantity $\omega_x(k)$ is the ratio of the number of classes of cardinality k that satisfy $\mathcal{M}_d$ to the total number of classes that satisfy $\mathcal{M}_d$. Since by (6) the probability distribution P_n is uniform over the space $\mathcal{P}(F)$ whose size is 2^{2^n} then using (9) it follows that the sum in (5) equals

$$\sum_{k=2}^{2^n} \omega_x(k) \log(k) = \sum_{k=2}^{2^n} \frac{P_{n,k}^*(\mathcal{M}_d)P_n(k)}{\sum_{j=1}^{2^n} P_{n,j}^*(\mathcal{M}_d)P_n(j)} \log k. \tag{11}$$

Let $N = 2^n$, then by Lemma 1 and from the proof of Lemma 2, as n (hence N) increases, it follows that

$$P_{n,k}^*(\mathcal{M}_d) \approx Q_{n,k}^*(\mathcal{M}_d) = \left(\frac{1}{2}\right)^{d(k-1)} A(N,d) \tag{12}$$

where $A(N,d)$ satisfies $\frac{\log N}{d} \le A(N,d) \le \frac{\log^d N}{d!}$. Let $p = 1/(1+2^d)$ then using (12) the ratio in (11) is

$$\frac{\sum_{k=2}^N \binom{N}{k} p^k (1-p)^{N-k} \log k}{\sum_{j=1}^N \binom{N}{j} p^j (1-p)^{N-j}}.$$

Using the DeMoivre-Laplace limit theorem [2], the binomial distribution $P_{N,p}(k)$ with parameters N and p satisfies $P_{N,p}(k) \approx \frac{1}{\sigma}\phi\left(\frac{k-\mu}{\sigma}\right)$, $N \to \infty$ where $\phi(x)$ is the standard normal probability density function and $\mu = Np, \sigma = \sqrt{Np(1-p)}$. Simple algebra then yields that (11) is asymptotically equal to

$$\sum_{k=2}^{2^n} \omega_x(k) \log k \approx \frac{\Phi(-a) \log\left(\frac{2^n}{1+2^d}\right) + 2^{-(n-d)/2}\phi(a) + O(2^{-(n-d)})}{1 - \left(\frac{2^d}{1+2^d}\right)^{2^n}} \tag{13}$$

where $a = 2^{(n-d)/2} - 2(1+2^d)2^{-(n+d)/2}$. In Theorem 1, the set $\Pi_Y(A)$ is the class F (see Definition 1) hence $\log |F| = n$ and $I(x:t) = n - \sum_{k\ge 2} \omega_x(k) \log k$. Combining with (13) the statement of Theorem 1 follows. $\blacksquare$

4.4   Proof Sketch of Theorem 2

We start with an auxiliary lemma that states a threshold value for the cardinality of a random element of $F^{(k)}$ that satisfies property $\mathcal{M}_d'$.

Lemma 3. *For any integer $d > 0$ let k be an integer satisfying $k \ge 2^d$. Let $\mathcal{F}$ be a class of cardinality k and randomly drawn according to the uniform distribution $P_{n,k}^*$ on $F^{(k)}$. Then $\lim_{n\to\infty} P_{n,k}^*(\mathcal{M}_d') = 1$.*

Proof sketch: It suffices to prove the result for $k = 2^d$ since $P_{n,k}^*(\mathcal{F} \models \mathcal{M}_d') \ge P_{n,2^d}^*(\mathcal{F} \models \mathcal{M}_d')$. As in the proof of Lemma 2, by Lemma 1 it suffices to show that $Q_{n,2^d}^*(\mathcal{F} \models \mathcal{M}_d')$ tends to 1. Denote by U_d the 'complete' matrix with d rows and 2^d columns formed by all 2^d binary vectors of length d, ranked for instance in alphabetical order. The event "$\mathcal{F} \models \mathcal{M}'_d$" occurs if there exists a subset $S = \{i_1, \ldots, i_d\} \subseteq [n]$ such that the submatrix whose rows are indexed by S and columns by $[2^d]$, is equal to U_d. Let S_i, $0 \le i \le m - 1$, be the sets defined in the proof of Lemma 2 and consider the m corresponding events, the ith event defined as having a submatrix whose rows are indexed by S_i and is equal to U_d. The probability that at least one of the events occurs is $1 - (1 - 2^{-d2^d})^{\lfloor n/d \rfloor}$ which tends to 1 as n increases. $\qquad\square$

When $k < 2^d$ there does not exist a set $E \subseteq [n]$ of cardinality d such that $|\text{tr}_E(\mathcal{F})| = 2^d$ and hence $P^*_{n,k}(\mathcal{M}'_d) = 0$. Hence with Lemma 3, it follows that the sum in (5) is

$$\sum_{k=2^d}^{2^n} \frac{P^*_{n,k}(\mathcal{M}'_d) P_n(k) \log k}{\sum_{j=2^d}^{2^n} P^*_{n,j}(\mathcal{M}'_d) P_n(j)}. \tag{14}$$

From the proof of Lemma 3, it follows that for all $k \ge 2^d$, $P^*_{n,k}(\mathcal{M}'_d) \approx 1 - (1 - \beta)^{rk}$, $\beta = 2^{-d2^d}, r = \frac{n}{d2^d}$. Since β is an exponentially small positive real we approximate $(1 - \beta)^{rk}$ by $1 - rk\beta$ and take $P^*_{n,k}(\mathcal{M}'_d) \approx rk\beta$. As in the proof of Theorem 1, resorting to a normal-approximation of the binomial we obtain that (14) tends to

$$\frac{\log(N/2) \left(\Phi(a)N/2 + \left(1 + \frac{a^2}{N \log(N/2)} \right) \phi(a)\sqrt{N}/2 \right)}{\Phi(a)N/2 + \phi(a)\sqrt{N}/2}$$

where $a = (N/2 - 2^d)/\sqrt{N/4}$. Substituting back for a and assuming that $d+1 \le n$ then the above tends to $\log(N/2) = \log N - 1$. With $N = 2^n$, and by Definition 1, the statement of the Theorem follows. $\qquad\blacksquare$

Bibliography

[1] R. Anstee, B. Fleming, Z. Furedi, and A. Sali. Color critical hypergraphs and forbidden configurations. In *Proc. EuroComb'2005*, pages 117–122. DMTCS, 2005.

[2] William Feller. *An Introduction to Probability Theory and Its Applications*, volume 1. Wiley, New York, third edition, 1968.

[3] S. Janson, T. Luczak, and A. Ruciński. *Random Graphs*. Wiley, New York, 2000.

[4] A. N. Kolmogorov. On tables of random numbers. *Sankhyaa, The Indian J. Stat.*, A(25):369–376, 1963.

[5] A. N. Kolmogorov. Three approaches to the quantitative definition of information. *Problems of Information Transmission*, 1:1–17, 1965.

[6] M. Li and P. Vitanyi. *An introduction to Kolmogorov complexity and its applications*. Springer-Verlag, New York, 1997.

[7] A. Romashchenko, A. Shen, and N. Vereshchagin. Combinatorial interpretation of Kolmogorov complexity. *Theoretical Computer Science*, 271:111–123, 2002.

[8] V. V. Vyugin. Algorithmic complexity and stochastic properties of finite binary sequences. *The Computer Journal*, 42:294–317, 1999.

[9] Bernard Ycart and Joel Ratsaby. VC and related dimensions of random function classes. submitted, 2006.

Finding Small OBDDs for Incompletely Specified Truth Tables Is Hard

Jesper Torp Kristensen and Peter Bro Miltersen

Department of Computer Science, University of Aarhus, Denmark

Abstract. We present an efficient reduction mapping undirected graphs G with $n = 2^k$ vertices for integers k to tables of partially specified Boolean functions $g : \{0,1\}^{4k+1} \rightarrow \{0,1,\perp\}$ so that for any integer m, G has a vertex colouring using m colours if and only if g has a consistent ordered binary decision diagram with at most $(2m+2)n^2 + 4n$ decision nodes. From this it follows that the problem of finding a minimum-sized consistent OBDD for an incompletely specified truth table is **NP**-hard and also hard to approximate.

1 Introduction

In this paper we consider the following problem: *Given a partially defined Boolean function $f : \{0,1\}^k \rightarrow \{0,1,\perp\}$ (with $\perp$ being interpreted as "don't care"), find or approximate the minimum representation of f as an Ordered Binary Decision Diagram* (OBDD). For details about OBDDs, see the comprehensive monograph by Wegener [15]. Throughout the paper, we consider OBDDs with a *fixed* variable ordering. For concreteness and simplicity, we assume the ordering to be $x_1 < x_2 < \ldots < x_k$ for Boolean functions on k variables and always define the functions we use with its arguments in the same order, i.e., the i'th argument of a function $g : \{0,1\}^k \rightarrow \{0,1\}$ is assigned to the variable x_i. The *size* of an OBDD is the number of its decision nodes. We say that an OBDD D *represents* or *is consistent with* $f : \{0,1\}^k \rightarrow \{0,1,\perp\}$ when the fully defined Boolean function $g_D : \{0,1\}^k \rightarrow \{0,1\}$ defined by the diagram is consistent with f, i.e., satisfies $g_D(x) = f(x)$ whenever $f(x) \neq \perp$.

The corresponding minimization problem for *fully* defined Boolean functions was shown to be in **P** in the original papers introducing OBDDs by Bryant [2,3]. Indeed, his efficient algorithm for minimizing OBDD size is one of the main attractions of using OBDD representation for Boolean functions. The minimum-size OBDD problem for partially defined Boolean functions was considered previously in two almost simultaneous papers [12,7], both showing **NP**-hardness for versions of the problem. The hardness results of the two papers differ mainly by the way the partially defined Boolean function is to be represented.

More precisely, Sauerhoff and Wegener [12] showed the following decision problem D_1 to be **NP**-complete.

D_1: Given two OBDDs representing two Boolean functions $g_1, g_2 : \{0,1\}^k \rightarrow \{0,1\}$ and an integer s, does the partially defined Boolean function f given by

D.Z. Chen and D.T. Lee (Eds.): COCOON 2006, LNCS 4112, pp. 489–496, 2006.
© Springer-Verlag Berlin Heidelberg 2006

$f(x) = \perp$ for those x for which $g_1(x) = 0$ and $f(x) = g_2(x)$ for those x for which $g_1(x) = 1$ have an OBDD of size less than s?

Hirata, Shimozono and Shonohara [7,13] showed the following decision problem D_2 to be **NP**-complete[1].

D_2: Given two explicitly listed sets $S_0, S_1 \subseteq \{0,1\}^k$ and an integer s, does the partially defined Boolean function f given by $f(x) = 0$ for $x \in S_0$, $f(x) = 1$ for $x \in S_1$ and $f(x) = \perp$ otherwise have an OBDD of size less than s?

To compare the strengths of the two results, we observe that it is immediate that the problem D_2 polynomial-time many-one reduces to D_1: Given two sets S_0 and S_1 we can easily construct small OBDDs representing functions g_1 and g_2 so that $g_1(x) = 1$ if and only if $x \in S_0 \cup S_1$ and $g_2(x) = 1$ if and only if $x \in S_1$. On the other hand, conversion from representation of the input as two OBDDs to representation as two explicitly given sets in general incurs an exponential blowup in size and is hence not a polynomial-time reduction. Hence, the **NP**-hardness result of Hirata, Shimozono and Shonohara is stronger than the one of Sauerhoff and Wegener.

In this paper we look at a *third* input representation and consider the following decision problem.

D_3: Given a *table* of $f : \{0,1\}^k \to \{0,1,\perp\}$ as a string of length 2^k over $\{0,1,\perp\}$ and an integer s, does f have an OBDD of size less than s?

The main result of the present paper is that D_3 is **NP**-complete. To be precise, we establish the following reduction.

Theorem 1. *There is a polynomial time computable reduction mapping undirected graphs G with $n = 2^k$ vertices for integers k to tables of partially specified Boolean functions $g : \{0,1\}^{4k+1} \to \{0,1,\perp\}$ so that for any integer K, G has a vertex colouring using K colours if and only if g has a consistent ordered binary decision diagram with at most $(2K + 2)n^2 + 4n$ decision nodes.*

Then, **NP**-hardness of D_3 follows from the **NP**-hardness of graph colouring (see, e.g., Garey and Johnson [6]).

To compare the strength of our result to the result of Hirata, Shimozono and Shonohara, we observe that it is immediate that the problem D_3 polynomial-time many-one reduces to D_2: Given a table of f, we can certainly efficiently list the sets $S_0 := \{x | f(x) = 0\}$ and $S_1 := \{x | f(x) = 1\}$. On the other hand, conversion from representation as two sets S_1, S_2 to a full table on the domain $\{0,1\}^n$ may incur an exponential blowup in size. This happens when the sets S_0 and S_1 are small (i.e., when f is undefined on most of the domain $\{0,1\}^n$). Hence, our **NP**-hardness result is stronger than the **NP**-hardness result of Hirata, Shimozono and Shonohara. Also, the proof of Hirata, Shimozono and Shonohara uses functions undefined everywhere on $\{0,1\}^k$ except on a subset of size $k^{O(1)}$, so their proof does not tell us anything about the hardness of the problem in a

---

[1] Very similar results were obtained by Pitt and Warmuth [11] and Simon [14] for deterministic finite automata, a model closely related to OBDDs.

situation where the functions considered *are* defined on a non-negligible fraction of the domain $\{0, 1\}^k$ and it does not yield our hardness result.

We find our stronger result well-motivated, as we'll explain next: A practical relevance of concrete **NP**-hardness results are their *redirection of attention* from the construction of efficient algorithms towards the construction of good heuristics for the problems at hand. This point is made explicitly by Sauerhoff and Wegener who cite several studies in the VLSI verification domain where the problem of finding minimum size OBDDs for given partial Boolean functions arise. For these applications, the input mode of Sauerhoff and Wegener is indeed the relevant one: The Boolean functions arising when formally verifying correctness of VLSI chips have truth tables so huge that representing them explicitly is out of the question, so typically, they are defined by OBDDs to begin with, as assumed by Sauerhoff and Wegener. Thus, for these applications our result provides no new "redirection signal".

However, there are other natural applications of using OBDDs for partially defined functions where the function to be encoded *is* given explicitly as a table. An application studied in the master's thesis [10] of the first author is the compression of *endgame tables* for chess. Such an endgame table may provide, for any chess position with a given set of pieces (say, a King and a Queen for White and a King and a Rook for Black) a Boolean value indicating whether the player with material advantage has a winning strategy. Given an encoding of chess positions as Boolean vectors, we may think of the table as a Boolean function $f : \{0, 1\}^n \rightarrow \{0, 1\}$ where $f(x)$ is the value of the chess position with Boolean encoding x. One may vary the way chess positions are represented as Boolean vectors, but any natural and efficiently computable encoding will have many Boolean vectors not representing any position. The values assigned to such vectors are inconsequential, so we may think of them as *undefined* values and hence of the table as defining a partially defined Boolean function. The potential usefulness of endgame tables for chess playing software is obvious. However, to be actually useful for such applications, an endgame table must support fast lookup and thus it should, preferably, reside in fast memory. For most endgame tables, this means that some *compression scheme* has to be applied on the table. Unfortunately, most state-of-the-art lossless compression schemes do *not* support efficient retrieval of individual bits of the compressed table (i.e., efficient table lookup). Here, representing the table by an OBDD seems to be an attractive alternative. From a theoretical point of view, Kiefer, Flajolet and Yang [9] showed that representation by OBDDs has the important *universality* property: The compression rate achieved asymptotically (i.e., for long inputs, and up to a low-order additive term) matches the *block entropy* of the string to be compressed for any constant block size. At the same time, by construction, a table represented by an OBDD supports fairly fast lookups (we may lookup an entry in the table by following a path from the root to a leaf in the OBDD). In his master's thesis [10], the first author obtained encouraging practical results on using OBDDs to compress endgame tables for chess while preserving efficient lookup. To achieve this, heuristics had to be used to minimize the OBDDs. The hardness result of the present paper indicates that such heuristics cannot be replaced with efficient algorithms.

We finally note that we may combine our reduction with known results concerning hardness of approximation for graph colouring to show that the minimum consistent OBDD problem is also hard to approximate. In particular, Feige and Killian [5] showed the following theorem[2]. Recall that **ZPP** is the class of decision problems which can be solved in expected polynomial time by a randomized algorithm.

Theorem 2 (Feige and Killian). *For any $\epsilon > 0$, if* **NP** $\neq$ **ZPP**, *no polynomial time algorithm distinguishes between the following two classes of graphs:*

- *Graphs $G = (V, E)$ with chromatic number less than $|V|^\epsilon$.*
- *Graphs $G = (V, E)$ with chromatic number bigger than $|V|^{1-\epsilon}$.*

Combining Theorem 1 with Theorem 2, noticing that we in Theorem 2 without loss of generality can assume that the graphs considered have $n = 2^k$ vertices for an integer k, we immediately obtain:

Corollary 1. *Let $\epsilon > 0$ be an arbitrary constant. If* **NP** $\neq$ **ZPP**, *no polynomial time algorithm distinguishes between the following two classes of incompletely specified truth tables $f : \{0,1\}^k \to \{0,1,\perp\}$:*

- *Truth tables for which a consistent OBDD of size less than $2^{(0.5+\epsilon)k}$ exists.*
- *Truth tables for which all consistent OBDDs have size more than $2^{(0.75-\epsilon)k}$*

In particular, unless **NP** equals **ZPP**, no efficient approximation algorithm for the minimum consistent OBDD problem has an approximation factor of $2^{(0.25-\epsilon)k}$, for any constant $\epsilon > 0$. Somewhat weaker non-approximability results for chromatic number assuming only **NP** $\neq$ **P** are known [1]; these may be combined with our reduction to show similarly weaker non-approximability results for our minimum consistent OBDD problem. We omit the details.

2   The Reduction

We consider an auxiliary problem. Given a family (s_i) of truth tables $s_i : \{0,1\}^k \to \{0,1,\perp\}$ of *partially* defined Boolean functions and a family (g_i) of truth tables $g_j : \{0,1\}^k \to \{0,1\}$ of *fully* defined Boolean functions, we say that the family (g_j) *covers* the family (s_i) if for every s_i there is some g_j consistent with s_i. The *minimum truth table cover* problem is the following optimization problem: Given a family (s_i) of $n = 2^k$ truth tables of partially defined Boolean functions (represented as a collection of n strings of length 2^k over $\{0,1,\perp\}$), find the smallest family (g_j) that covers (s_i).

We present a reduction from the graph colouring problem to the minimum truth table cover problem:

---

[2] Subsequently, the theorem was refined by Khot [8] and Engebretsen and Holmerin [4] who replaced the constant ϵ in Theorem 2 with specific subconstant functions. However, when combining inapproximability results for chromatic number with our reduction, such improvements are more or less irrelevant.

Lemma 1. *There is a polynomial time computable reduction mapping undirected graphs G with $n = 2^k$ vertices for integers k to a collection of n tables of partially specified Boolean functions $s_i : \{0,1\}^k \to \{0,1,\bot\}, i = 1,\ldots,n$ so that for any integer K, G has a vertex colouring using K colours if and only if (s_i) has a truth table cover of size K.*

Proof. Given a graph $G = (V, E)$ with $V = \{0, \ldots, n-1\}$, we define

$$s_i(j) = \begin{cases} 0 & \text{if } i \neq j \wedge (i, j) \in E; \\ 1 & \text{if } i = j; \\ \bot & \text{otherwise.} \end{cases}$$

Note that we in the definition of s_i identify an integer j with its binary representation. We shall do so in the following as well. It is an easy observation that the reduction has the desired property.

In the rest of the section, we reduce the minimum truth table cover problem to the minimum consistent OBDD problem, thus completing the proof of Theorem 1.

We need in our reduction an auxiliary family of functions $g_j^{p,m} : \{0,1\}^p \to \{0,1\}$ where p is an arbitrary non-negative integer, $1 \leq m \leq 2^{2^p}$ and $0 \leq j \leq m - 1$. The family must have the following properties.

1. For fixed p, m, the functions $g_j^{p,m}, j \in \{0, \ldots, m-1\}$ are all different.
2. The truth table for $g_j^{p,m}$ can be generated in time polynomial in 2^p (given the parameters p, m, j),
3. For fixed p, m, the family $(g_j^{p,m}), j \in \{0, \ldots, m-1\}$ is computed by a multi-source OBDD (an OBDD with m sources, one for each member of the family) of size at most $m + 2\sqrt{m} + 3p$.

Note that the third property makes the construction of the family a bit tricky: The sources of the desired multi-source OBDD use almost its entire "node budget". We give an inductively defined construction. For $p = 0$, the construction is trivial as we must have $m = 1$ or $m = 2$. For $p > 0$ we let $q = \lceil \sqrt{m} \rceil$. Note that $q \leq 2^{2^{p-1}}$ since $\sqrt{2^{2^p}} = 2^{2^{p-1}}$ is an integer. We define for integers $i, j \in \{0, \ldots, q-1\}$:

$$g_{jq+i}^{p,m}(x_1 x_2 \ldots x_p) = \begin{cases} g_i^{p-1,q}(x_2 \ldots x_p) & \text{if } x_1 = 0; \\ g_j^{p-1,q}(x_2 \ldots x_p) & \text{if } x_1 = 1. \end{cases}$$

The construction clearly satisfies properties 1 and 2. Also, if we let $B^{p,m}$ be the size of a multi-source OBDD computing the family $(g_j^{p,m})$ we have by induction that $B^{p,m} = B^{p-1,q} + m \leq q + 2\sqrt{q} + 3(p-1) + m \leq m + 2\sqrt{m} + 3p$, so it also satisfies property 3.

We consider the values $k \geq 5$ and $n = 2^k$ fixed in the discussion to follow. For $j \in \{0, \ldots, n^2 - 1\}$ we let $b_j = g_j^{k,n^2}$. By property 3 of the family (g_j^{k,n^2}), the family (b_j) is computed by a multi-source OBDD of size at most $n^2 + 3n$.

Our reduction from minimum truth table cover to the minimum consistent OBDD problem is then defined as follows. It maps the minimum truth table cover instance $\{s_i\}_{i=1,\ldots,n}$, $s_i : \{0,1\}^k \rightarrow \{0,1,\perp\}$ to the truth table of the partial function $g : \{0,1\}^k \times \{0,1\}^{2k} \times \{0,1\} \times \{0,1\}^k \rightarrow \{0,1,\perp\}$ defined by:

$$g(i,j,t,z) = \begin{cases} b_j(z) & \text{if } t = 0; \\ s_i(z) & \text{if } t = 1. \end{cases} \tag{1}$$

(where we again identify integers with their binary notation). By property 2 of the family (g_j^{k,n^2}) the reduction is polynomial time computable. In the remainder of this section, we show that the composition of the reduction with the reduction of Lemma 1 has the property claimed in Theorem 1.

Lemma 2. *For any integer K, if (s_i) has a cover of size K, then g has a consistent OBDD of size at most $(2K + 2)n^2 + 4n$.*

Proof. We can assume $K \leq n$. Let T be the cover. Let s_i' be a total function in T consistent with s_i. Then, a total function h consistent with g is

$$h(i,j,t,z) = \begin{cases} b_j(z) & \text{if } t = 0; \\ s_i'(z) & \text{if } t = 1. \end{cases}$$

Let us give an upper bound for the size of an OBDD computing h. For each truth table $s \in T$, there is an OBDD of size at most $n - 1$ computing s (as $n - 1$ is the number of decision nodes in a complete decision tree on $k = \log n$ Boolean variables). There is a multi-source OBDD computing all functions b_j of size $n^2 + 3n$ by construction. The number of different subfunctions of h of the form $(j,t,z) \rightarrow h(i_0,j,t,z)$ (for some i_0) is K, the size of the cover. Each of these subfunctions can be computed by an OBDD with an additional $2^{2k+1} - 1 = 2n^2 - 1$ nodes above the OBDDs for (s_i') and (b_j). Having constructed OBDDs for all these subfunctions, an OBDD for h needs at most an additional $n - 1$ nodes to read the first k input bits to decide which subfunction to use. Thus, h can be computed by an OBDD of size at most $(n-1)+K(2n^2-1)+K(n-1)+n^2+3n \leq (2K + 2)n^2 + 4n$.

Lemma 3. *Let a minimum-sized OBDD G consistent with g be given. Viewing G as a graph, the subgraph of G induced by nodes reading variables $x_{k+1}, \ldots, x_{3k}$, x_{3k+1} (i.e. nodes reading the Boolean variables defining arguments j and t in equation (1)) forms a forest of disjoint complete binary trees (each tree containing $2^{2k+1} - 1 = 2n^2 - 1$ nodes).*

Proof. Let a minimum-sized OBDD G consistent with g be given, computing a function h. First note that since all the functions b_j are different, any OBDD consistent with g must read all variables $x_{k+1}, \ldots, x_{3k}$ on all paths through the OBDD. For the same reason, the left and right son of any node reading any variable $x_{k+1}, \ldots, x_{3k}$ must be different. Thus, the subgraph of G induced by nodes reading variables $x_{k+1}, \ldots, x_{3k}, x_{3k+1}$ is a union of complete binary trees. To prove the lemma, we just have to prove that they are disjoint. This follows if we show that any two nodes v and v' both reading a variable x_{k+m}, $m \in \{1, \ldots, 2k\}$ cannot share a son u. Assume to the contrary that they do and without loss of generality that u is a left son of v (corresponding to reading $x_{k+m} = 0$ in v).

The node v corresponds to a subfunction of h of the form

$$(x, t, z) \to h(a_1, c_1 \cdot x, t, z)$$

for constants $a_1 \in \{0, 1\}^k$ and $c_1 \in \{0, 1\}^{m-1}$ and variables $x \in \{0, 1\}^{2k-m+1}$, $t \in \{0, 1\}$, $z \in \{0, 1\}^k$. Here $c_1 \cdot x$ denotes concatenation of the bit-strings c_1 and x.

The node v' corresponds to a subfunction of h of the form

$$(x, t, z) \to h(a_2, c_2 \cdot x, t, z)$$

for constants $a_2 \in \{0, 1\}^k$ and $c_2 \in \{0, 1\}^{m-1}$ and variables $x \in \{0, 1\}^{2k-m+1}$, $t \in \{0, 1\}$, $z \in \{0, 1\}^k$.

Since u is a son of v as well as v' and all the b_j's are different we must have that $c_1 = c_2$ and that u is a *left* son of v'. Also, we must have the partial truth tables s_{a_1} and s_{a_2} are consistent, i.e., that they agree on inputs where neither has value $\perp$. Thus, we can get a smaller OBDD than G also consistent with g by removing the node v' and redirecting to v any incoming arc to v'. This contradicts G being minimum-sized.

Lemma 4. *Assume $n > 3$. For any integer K, if g has a consistent OBDD of size at most $(2K + 2)n^2 + 4n$, then (s_i) has a cover of size at most K.*

Proof. We can assume $K \leq n$. Let a minimum-sized OBDD consistent with g of size at most $(2K + 2)n^2 + 4n$ be given, computing a function h. According to Lemma 3, the nodes reading variables $x_{k+1}, \ldots x_{3k+1}$ induces a collection of disjoint complete binary trees. There must be at most K trees in this collection: Otherwise the contribution of nodes from the trees would amount to at least $(2n^2 - 1)(K + 1)$ nodes. Also, all members of the family (b_i) are subfunctions of g and since they are distinct and fully defined, each must be computed at a distinct node in the diagram, yielding n^2 additional nodes. In total, there would be at least $(2n^2-1)(K+1)+n^2$ nodes which is strictly more than $(2K+2)n^2+4n$ nodes.

Let (v_i) be the roots of the trees. The corresponding subfunctions of h are $(x, t, y) \to h(a_i, x, t, y)$ for constants a_i. The functions $j \to h(a_i, 0, 1, j)$ then form a cover for the family (s_i) of size at most K.

Combining Lemma 1, Lemma 2 and Lemma 4, we have proved Theorem 1 and are done.

Acknowledgements

We would like to thank Martin Sauerhoff for detailed comments and corrections and Hans Ulrich Simon for pointers to the literature. The research of Peter Bro Miltersen was supported by BRICS, a center of the Danish National Research Foundation and by a grant from the Danish Science Research Council.

References

1. Mihir Bellare, Oded Goldreich, and MadhuSudan. Free bits, PCPs, and non-approximability—towards tight results. *SIAM J. Comput.*, 27(3): 804–915, 1998.
2. R. E. Bryant. Symbolic manipulation of boolean functions using a graphical representation. In *Proceedings of the 22nd ACM/IEEE Design Automation Conference*, Los Alamitos, Ca., USA, 1985. IEEE Computer Society Press.
3. Randal E. Bryant. Graph-based algorithms for boolean function manipulation. *IEEE Transactions on Computers*, 35(8): 677–691, 1986.
4. Lars Engebretsen and Jonas Holmerin. Towards optimal lower bounds for clique and chromatic number. *Theoret. Comput. Sci.*, 299(1-3): 537–584, 2003.
5. Uriel Feige and Joe Kilian. Zero knowledge and the chromatic number. *J. Comput. System Sci.*, 57(2): 187–199, 1998.
6. Michael R. Gareyand David S. Johnson. *Computers and intractability*. W. H. Freeman and Co., San Francisco, Calif., 1979.
7. Kouichi Hirata, Shinichi Shimozono, and Ayumi Shinohara. On the hardness of approximating the minimum consistent OBDD problem. In *Algorithm theory—SWAT'96 (Reykjavík, 1996)*, volume 1097 of Lecture Notes in Comput. Sci., pages 112–123. Springer, Berlin, 1996.
8. Subhash Khot. Improved in approximability results for MaxClique, chromatic number and approximate graph coloring. In *42nd IEEE Symposiumon Foundations of Computer Science (Las Vegas, NV, 2001)*, pages 600–609. IEEE Computer Soc., 2001.
9. J.C. Kiefer, P. Flajolet, and E.-H Yang. Data compression via binary decision diagrams. In *Proc. of the 2000 IEEE Intern. Symp. Inform. Theory, Sorrento, Italy, June 25–30*, page 296. 2000.
10. Jesper Torp Kristensen. Generation and compression of endgame tables in chess with fast random access using OBDDs. Master's thesis, University of Aarhus, Department of Computer Science, 2005.
11. Leonard Pittand Manfred K. Warmuth. The minimum consistent DFA problem cannot be approximated within any polynomial. *J. Assoc. Comput. Mach.*, 40(1): 95–142, 1993.
12. Martin Sauerhoff and Ingo Wegener. On the complexity of minimizing the OBDD size for in completely specified functions. *IEEE Transactions on Computer-Aided Design*, 15:435–1437, 1996.
13. Shinichi Shimozono, Kouichi Hirata, and Ayumi Shinohara. On the hardness of approximating the minimum consistent a cyclic DFA and decision diagram. *Inform. Process. Lett.*, 66(4):165–170, 1998.
14. Hans-Ulrich Simon. On approximate solutions for combinatorial optimization problems. *SIAM J. Discrete Math.*, 3(2):294–310, 1990.
15. Ingo Wegener. *Branching programs and binary decision diagrams*. SIAM Monographs on Discrete Mathematics and Applications. Society for Industrial and Applied Mathematics (SIAM), 2000.

Bimodal Crossing Minimization*

Christoph Buchheim, Michael Jünger, Annette Menze, and Merijam Percan

Universität zu Köln, Institut für Informatik,
Pohligstraße 1, 50969 Köln, Germany
{buchheim, mjuenger, menze, percan}@informatik.uni-koeln.de

Abstract. We consider the problem of drawing a directed graph in two dimensions with a minimum number of crossings such that for every node the incoming edges appear consecutively in the cyclic adjacency lists. We show how to adapt the planarization method and the recently devised exact crossing minimization approach in a simple way. We report experimental results on the increase in the number of crossings involved by this additional restriction on the set of feasible drawings. It turns out that this increase is negligible for most practical instances.

1 Introduction

The importance of automatic graph drawing stems from the fact that many different types of data can be modeled by graphs. In most applications, the interpretation of an edge is asymmetric, so that the graph is intrinsically directed. This is the case, e.g., for metabolic networks. Here, the incoming edges of a reaction node correspond to reactants, while the outgoing edges correspond to products of the modeled reaction. Consequently, a good layout of such a network should separate incoming from outgoing edges, e.g., by letting the incoming edges enter on one side of the node and letting the outgoing edges leave on the opposite side. By this, the human viewer is able to distinguish reactants from products much more easily; see Figure 1.

In spite of its practical relevance, the direction of edges is ignored by many graph drawing algorithms. The graph is processed as an undirected graph first; only after the positions of nodes and edges have been determined the direction is visualized by replacing lines by arrows. An important exception is given by hierarchical drawings, in which incoming and outgoing edges are separated by definition. Furthermore, a polynomial time algorithm for hierarchical drawings of digraphs that allows directed cycles and produces the minimum number of bends is given by Bertolazzi, Di Battista and Didimo in [1]. However, the restriction to this special type of drawing might lead to many more crossings than necessary.

In this paper, our aim is to adapt the planarization method in order to obtain the desired separation of incoming and outgoing edges. We focus on the planarization step itself, i.e., the computation of a planar embedding of the graph after eventually adding virtual nodes representing edge crossings. The objective

---

* Partially supported by the Marie Curie RTN ADONET 504438 funded by the EU.

D.Z. Chen and D.T. Lee (Eds.): COCOON 2006, LNCS 4112, pp. 497–506, 2006.
© Springer-Verlag Berlin Heidelberg 2006

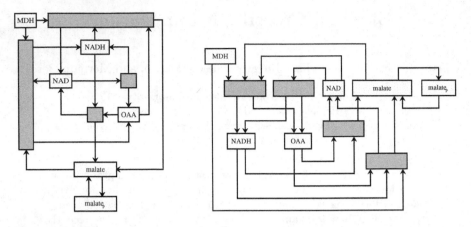

Fig. 1. Two drawings of the same graph, both have three crossings, with unsorted (left) and sorted (right) incoming and outgoing edges; gray nodes represent reactions

is to add as few such nodes as possible. For a comprehensive survey over the planarization approach, see [11].

In order to obtain the separation of edges, we consider the additional bimodal restriction that all incoming edges appear consecutively in all cyclic adjacency lists. We show how to adapt the well-known approach based on finding a planar subgraph first and then reinserting the missing edges one after the other in a very efficient way. We use an experimental evaluation to investigate the question of how many additional crossings have to be expected from restricting the class of feasible embeddings in this way. The results show that—for practical instances— this increase is usually negligible.

We do not address the question of how to realize the resulting embedding by an actual drawing of the graph. Notice however that once we have such an embedding at hand, it is easily possible to adapt, e.g., the orthogonal layout algorithm such that incoming and outgoing edges lie on opposite sides [12].

In Section 2 we recall the concept of bimodality and describe the basic transformation used by the evaluated algorithms. Next we propose a postprocessing technique that can be combined with any crossing reduction approach, see Section 3. Then we look into the planarization method; the problem of finding a planar subgraph is considered in Section 4, while edge reinsertion is dealt with in Section 5. In Section 6, we discuss a recently developed exact approach for crossing minimization. In Section 7, we present an experimental evaluation showing that the number of crossings computed by different methods does not grow much by our additional requirement. Section 8 summarizes the results.

2   Bimodal Embeddings

An embedding of a graph $G = (V, E)$ is called *bimodal* if and only if for every vertex v of G the circular list of the edges around v is partitioned into two (possibly empty) linear lists of edges, one consisting of the incoming edges and the other consisting if the outgoing edges. A planar digraph is *bimodally planar* if and only if it has a bimodal embedding that is planar. This structure was first investigated by Bertolazzi, Di Battista, and Didimo in [1]. Bimodal planarity of a graph G can be decided by testing planarity of a simple transformation of G in $\mathcal{O}(|V|)$ time [1]. The transformation is applied in the following way: for every node v of G expand v by an expansion edge e and add all incoming edges of v to one end-node v_- of e and all outgoing edges to the other end-node v_+ of e. The resulting graph is denoted by $G_d = (V_d, E_d)$ in the following. We call G_d the *d-graph* of G. An illustration of this construction is given in Figure 2.

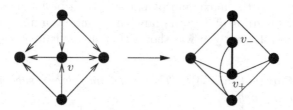

Fig. 2. A directed graph G and its d-graph G_d. The bold edge is an expansion edge. Note that G_d is equal to $K_{3,3}$.

Throughout this paper, we will denote the set of all expansion edges by E'. We use this simple transformation for adapting techniques for undirected crossing minimization to the directed variant. Planar directed graphs are not necessarily bimodally planar. By Kuratowski's theorem, this can only happen if a $K_{3,3}$ or K_5 subdivision is created by the transformation into a d-graph. Note that for graphs with all nodes of degree at most three the transformation of G to G_d is trivial, as no nodes are split in this case. In particular, this holds for cubic graphs that are defined by the property that all nodes have degree three. Therefore, a directed cubic graph is bimodally planar if and only if it is planar. This is also true for graphs in which each node has at most one incoming edge or at most one outgoing edge.

3 Naive Post-processing Approach

We first discuss a post-processing procedure that can be used after applying any crossing reduction algorithm or heuristic to the d-graph G_d. Our aim is to embed G_d such that no expansion edge crosses any other edge; contracting expansion edges then yields an embedding of G with the desired separation of incoming and outgoing edges. So assume that any embedding of G_d is given. We first delete all edges crossing any expansion edge. If two expansion edges cross each other, we delete one of them. Next, we reinsert all deleted edges one after another, starting with the deleted expansion edges. As explained in Section 5.1 below, we can insert a single edge with a minimal number of crossings for the fixed embedding computed so far such that crossings with expansion edges are prevented. If reinserting an expansion edge produces any crossings, the crossed (non-expansion) edges have to be deleted and put to the end of the queue of edges to be reinserted. At the end of this reinsertion process, no expansion edge will cross any other edge. However, the number of crossings of the remaining edges might grow significantly in this approach. In the following sections, we explain how to get better results by adapting well-known crossing minimization approaches for our purposes, especially the planarization approach.

4   Maximum Bimodally Planar Subgraphs

It is well-known that the maximum planar subgraph problem—the problem of finding a planar subgraph of a given graph that contains a maximum number of edges—is NP-hard. Recently, it was shown that this remains true even for cubic graphs:

Theorem 1 (Faria et al. [4]). *The maximum planar subgraph problem is NP-hard for cubic graphs.*

As a cubic graph is equal to its d-graph, we derive that this also holds for the maximum bimodally planar subgraph problem:

Corollary 1. *It is an NP-hard problem to compute a maximum bimodally planar subgraph of a directed graph, even for a cubic graph.*

For computing *maximal* bimodally planar subgraphs, i.e., bimodally planar subgraphs such that adding any further edge of G destroys bimodally planarity, we do the following: it is easy to see that the bimodally planar subgraphs of G are in one-to-one correspondence to the planar subgraphs of G_d containing all expansion edges. Thus we have to modify a given maximal planar subgraph algorithm such that it never deletes any expansion edge. Methods for finding maximal planar subgraphs have been studied intensively [8,10,3]; here we only discuss the incremental method; see Section 4.1. We also have a look at the exact approach; see Section 4.2.

4.1   Incremental Method

Starting with the empty subgraph $(V_H, \emptyset)$ of some graph $H = (V_H, E_H)$, the incremental method tries to add one edge from E_H after the other. Whenever adding an edge would destroy planarity, it is discarded, otherwise it is added permanently to the subgraph being constructed. The result is a maximal planar subgraph of H, which however is not a maximum planar subgraph in general. To find a maximal bimodally planar subgraph of G, we have to compute a maximal planar subgraph of its d-graph G_d. However, this subgraph must always contain all expansion edges, so that the latter can be contracted at the end. We thus have to start with the subgraph (V_d, E')—which is obviously planar—instead of the empty subgraph $(V_d, \emptyset)$. Then we try to add the remaining edges $E_d \setminus E'$ as before. The resulting subgraph of G_d corresponds to a maximal bimodally planar subgraph H of G.

4.2   Exact Method

An exact approach for finding a maximum planar subgraph of $H = (V_H, E_H)$ based on polyhedral techniques was devised in [9]. The problem is modeled by an integer linear program (ILP) as follows: for every edge $e \in E_H$, a binary variable x_e is introduced, having value one if and only if e belongs to the chosen subgraph. To enforce that the modeled subgraph is planar, one has to make sure that it contains no Kuratowski subgraph of H, i.e., no subdivision of K_5 or $K_{3,3}$. In terms of the model, this is equivalent to the constraint $\sum_{e \in K} x_e \leq |K| - 1$ for every (edge set of a) Kuratowski graph K in G. As we search for a planar subgraph containing the maximal number of edges, the number of variables set to one should be maximized. The integer linear program is thus

$$\max \ \sum_{e \in E_H} x_e$$

$$\text{s.t.} \ \sum_{e \in K} x_e \leq |K| - 1 \ \text{ for all Kuratowski subgraphs } K \text{ of } G$$

$$x_e \in \{0, 1\} \quad \text{for all } e \in E_H .$$

This ILP can now be solved by branch-and-cut. However, in order to improve the runtime of such algorithms and hence obtain a practical solution method, one has to further investigate this formulation and exhibit other classes of valid inequalities as well as fast techniques for finding violated constraints for a given fractional solution. For details, the reader is referred to [9]. This solution approach can easily be adapted to our situation: we have to ensure that the edges in E' always belong to the chosen subgraph, i.e., we have to add the constraint $x_e = 1$ to the ILP, for each expansion edge $e \in E'$. Observe that this type of constraint is harmless with respect to the complexity of the problem, as it cuts out a face from the polytope spanned by the feasible solutions of the ILP.

5   Edge Reinsertion

After calculating a maximal (resp., maximum) bimodally planar subgraph, the deleted edges have to be reinserted. Our objective is to reinsert them one by one so that the minimum number of crossings are produced for each edge. This can be done in two different ways: either by inserting an edge into a fixed bimodally planar embedding of the bimodally planar subgraph, see Section 5.1, or by inserting an edge optimally over all bimodally planar embeddings of the bimodally planar subgraph, see Section 5.2. Again, we have to treat expansion edges differently, as they may not be involved in any edge crossings.

5.1   Fixed Embedding

Given a fixed embedding $\Gamma(G_d)$ of G_d, it is easy to insert an edge $e(v, w)$ into $\Gamma(G_d)$ such that a minimal number of crossings is produced. For this, one can use the *extended dual graph* D of $\Gamma(G_d)$, the nodes of which are the faces of $\Gamma(G_d)$ plus two nodes v_D and w_D corresponding to v and w. For each edge in $E_d \setminus E'$, we have the dual edge in D. Additionally, we connect v (resp., w) with all nodes in D corresponding to faces that are adjacent to v (resp., w) in $\Gamma(G_d)$. Then we calculate the shortest path from v to w in the extended dual graph and insert the edge e into $\Gamma(G_d)$ along this path, replacing crossings by dummy nodes. Clearly, the shortest path does not cross any edge of E' as its dual edge is not included in D. This can be done in $\mathcal{O}(|V|)$ time.

5.2 All Embeddings

In the previous section we have considered reinserting an edge into a fixed embedding. For getting fewer edge crossings, a powerful method is to calculate the shortest path between two nodes v and w over all embeddings. In [6] a linear time algorithm is presented for finding an optimal embedding which allows to insert e with the minimum number of crossings. It uses the SPQR-tree and BC-tree data-structures for representing all planar embeddings of a connected graph. In the same straightforward way as explained in the previous section, this approach can be adapted such that no expansion edge is crossed by any reinserted edge. The resulting algorithm runs in $\mathcal{O}(|V|)$ time.

6   Exact Bimodal Crossing Minimization

It is a well-known fact that the general crossing minimization problem for undirected graphs is NP-hard [5]. More recent results show that this is even true for graphs with all nodes of degree three:

Theorem 2 (Hliněný [7], Pelsmajer, Schaefer, Štefankovič [13]). *The crossing minimization problem is NP-hard for cubic graphs.*

Corollary 2. *It is an NP-hard problem to compute a drawing of G separating incoming and outgoing edges such that the number of crossings is minimal. This even holds for cubic graphs.*

Despite the NP-hardness of undirected crossing minimization, an exact approach has been devised recently [2]; a branch-and-cut algorithm is proposed for minimizing the number of crossings over all possible drawings. The first step in this approach is to replace every edge of the graph by a path of length (at most) $|E|$. After this, one may assume that every edge has a crossing with at most one other edge. The ILP model used in this approach contains a variable x_{ef} for all pairs of edges $(e, f) \in E \times E$, having value one if and only if there is a crossing between e and f in the drawing to be computed. By appropriate linear constraints, one can ensure that the given solution is realizable, i.e., corresponds to some drawing of G. Again, it is easy to adjust this method to our problem, i.e., the problem of computing a crossing-minimal drawing with incoming and outgoing edges separated. For this, we can apply the above algorithm to the graph G_d. Then we only have to make sure that the expansion edges do not have any crossings in the computed solution. We can thus do the adjustment as follows: first observe that the edges in E' do not have to be replaced by a path at all, as they are not allowed to produce crossings. Now we can just omit the variable x_{ef} whenever $e \in E'$ or $f \in E'$, and thereby set this variable to zero implicitly. The resulting ILP will thus have exactly the same number of variables as the original ILP for the non-transformed graph. It will not become harder structurally, as it arises from setting variables to zero.

7 Experimental Comparison

In the previous sections, we showed how to adapt several crossing minimization algorithms and heuristics in a simple way such that for directed graphs the sets of incoming and outgoing edges are separated in the adjacency lists. This is obtained by transforming the original directed graph into a new undirected graph where certain edges do not allow any crossings. From the nature of this transformation and the described modifications, it is obvious that the runtime is not affected negatively. We also observed this in our experiments. For this reason, we focused on the number of crossings in the evaluation reported in the following: we are interested in comparing the number of crossings when (a) the direction of edges is ignored, i.e., crossing minimization is done as usual, and (b) we apply the transformation in order to separate incoming from outgoing edges. Theoretically, the crossing number cannot decrease by our modification, but it is possible that it grows considerably. However, our experiments show that for practical graphs the number of crossings is not increased significantly. In fact, the increase in the number of crossings is marginal compared with the variance due to the randomness of the heuristics, such that for many instances the number of crossings after the transformation even decreases. Combining this observation with the simpleness of implementation and the fact that runtime does not increase, our claim is that these techniques should always be applied when dealing with (meaningfully) directed edges. For the experiments, we used the instances of the Rome library of directed graphs [14], consisting of two sets of graphs called **north** and **random**. The former contains 1277 directed acyclic graphs on 10 to 100 nodes derived from real-world instances. The latter contains 909 directed acyclic graphs randomly generated in a specific way, they are much denser in general. We first applied the simple incremental method (Section 4.1) combined with the optimal edge reinsertion over all embeddings (Section 5.2). As mentioned above, it turned out that the increase in the number of crossings when separating incoming and outgoing edges is very small in general. This is shown in Figure 3 (a) and (b), where each instance is given by a plus sign. Its x-coordinate is the number of crossings before the transformation and the y-coordinate is the number of crossings afterwards. In particular, each cross on the diagonal line represents an instance with the same number of crossings before and afterwards. A cross above the diagonal represents an instance for which the number of crossings increases. Due to the randomness of the heuristics, there are also crosses below the diagonal, in particular for the **random** instances.

Another interesting finding is the negligible increase in the number of crossings for planar graphs: if G is planar, then G_d is not necessarily planar. Anyway, if

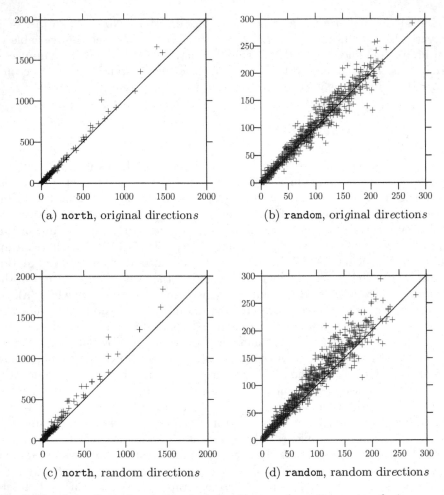

Fig. 3. Numbers of crossings before and after the transformation, using the incremental planar subgraph heuristic. For non-planar graphs, the average increase is 0.36 % (a), 0.59 % (b), 0.95 % (c), and 0.87 % (d), respectively.

we consider all 854 planar `north` instances, then the average number of crossings after the transformation is only 0.04, i.e., in most cases the graph remains planar. The set of `random` instances does not contain any planar graph. We next applied the optimal planar subgraph method (Section 4.2), again in combination with the optimal edge reinsertion over all embeddings (Section 5.2). As many instances could not be solved within a reasonable running time, we had to set a time limit of five CPU minutes (on an Athlon processor with 2.0 GHz). Within this time limit, 89 % of the `north` instances and 33 % of the `random` instances could be solved. The results are shown in Figure 4; the general picture is similar to the one for the incremental method.

The directed graphs contained in the libraries `north` and `random` are all acyclic. This fact might favor a small number of additional crossings. For this

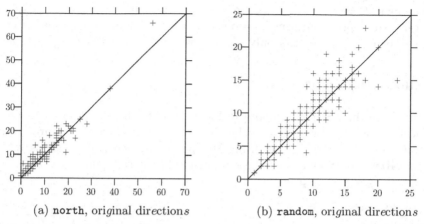

(a) **north**, *original directions*   (b) **random**, *original directions*

Fig. 4. Numbers of crossings before and after the transformation, using the optimal planar subgraph method. For non-planar graphs, the average increase is 3.30 % (a) and 4.21 % (b), respectively.

reason, we also examined graphs with a random direction for each edge. To allow us to compare the corresponding results to the results presented so far, we used the **north** and **random** instances again, this time with the direction of each edge reversed with a probability of 1/2. The results obtained with the incremental heuristic are displayed in Figure 3 (c) and (d). In fact, the increase in the number of crossings induced by sorting adjacency lists is more obvious now compared to Figure 3 (a) and (b), but it is still very small. Nevertheless, we conjecture that in theory the requirement of separating incoming and outgoing edges may induce a quadratic number of edge crossings even for planar graphs. We have constructed a family of directed planar graphs G_k such that G_k has $O(k)$ edges and such that the planarization heuristic has always produced $\Omega(k^2)$ crossings when separating incoming from outgoing edges. The graph G_k is defined as follows: it consists of two wheel graphs W_{2k} sharing their rim; one of them has all spokes directed from the rim to the hub, the other one has spokes with alternating direction. We applied the planarization method to the graphs G_k many times, with enforced separation of incoming and outgoing edges. For all k, the smallest number of crossings we could find was $\sum_{i=1}^{k} \lfloor i/2 \rfloor = \Theta(k^2)$. We conjecture that this is the minimum number of crossings for all bimodal drawings of G_k. This would mean that a quadratic number of crossings is unavoidable even for planar graphs.

8 Conclusion

We can summarize the statement of this paper as follows: whenever the direction of edges in a graph carries significant information, this should be stressed by separating incoming and outgoing edges in the adjacency lists. We have shown how crossing reduction algorithms can be adapted in order to comply with this requirement. The necessary changes are not only easy to implement but also neutral with respect to runtime. As our experiments show, the number of crossings can be expected to grow only slightly for practical instances.

Acknowledgement

We thank Dietmar Schomburg and Ralph Schunk for helpful discussions on the application of our approach to biochemical networks. Moreover, we are grateful to Maria Kandyba for implementing the presented heuristic methods and to Katrina Riehl and Stefan Hachul for proof-reading this paper.

References

1. P. Bertolazzi, G. Di Battista, and W. Didimo. Quasi-upward planarity. *Algorithmica*, 32:474–506, 2002.
2. C. Buchheim, D. Ebner, M. Jünger, G. W. Klau, P. Mutzel, and R. Weiskircher. Exact Crossing Minimization, in: P. Healy, N.S. Nikolov (eds.), *Graph Drawing 2005*. LNCS 3843, pp. 37–48, 2006.
3. H. N. Djidjev. A linear algorithm for the maximal planar subgraph problem. In *WADS '95*, volume 955 of *LNCS*, pages 369–380. Springer-Verlag, 1995.
4. L. Faria, C. M. H. de Figueiredo, and C. F. X. de Mendonça N. Splitting number is NP-complete. *Discrete Applied Mathematics*, 108(1–2):65–83, 2001.
5. M. R. Garey and D. S. Johnson. Crossing number is NP-complete. *SIAM Journal on Algebraic and Discrete Methods*, 4(3):312–316, 1983.
6. C. Gutwenger, P. Mutzel, and R. Weiskircher. Inserting an edge into a planar graph. *Algorithmica*, 41(4):289–308, 2005.
7. P. Hliněný. Crossing number is hard for cubic graphs. In *MCFS 2004*, pages 772–782, 2003.
8. R. Jayakumar, K. Thulasiraman, and M. N. S. Swamy. $O(n^2)$ algorithms for graph planarization. *IEEE Transactions on Computer-Aided Design*, 8:257–267, 1989.
9. M. Jünger and P. Mutzel. Maximum planar subgraphs and nice embeddings: Practical layout tools. *Algorithmica*, 16(1):33–59, 1996.
10. J. A. La Poutré. Alpha-algorithms for incremental planarity testing. In *STOC '94*, pages 706–715, 1994.
11. A. Liebers. Planarizing graphs – a survey and annotated bibliography. *Journal of Graph Algorithms and Applications*, 5(1):1–74, 2001.
12. A. Menze. Darstellung von Nebenmetaboliten in automatisch erzeugten Zeichnungen metabolischer Netzwerke. Master's thesis, Institute of Biochemistry, University of Cologne, June 2004.
13. M. J. Pelsmajer, M. Schaefer, and D. Štefankovič. Crossing number of graphs with rotation systems. Technical report, Department of Computer Science, DePaul University, 2005.
14. Rome library of directed graphs.
    http://www.inf.uniroma3.it/people/gdb/wp12/directed-acyclic-1.tar.gz.

Fixed Linear Crossing Minimization by Reduction to the Maximum Cut Problem*

Christoph Buchheim[1] and Lanbo Zheng[2,3]

[1] Computer Science Department, University of Cologne, Germany
buchheim@informatik.uni-koeln.de
[2] School of Information Technologies, University of Sydney, Australia
[3] IMAGEN program, National ICT Australia
lzheng@it.usyd.edu.au

Abstract. Many real-life scheduling, routing and location problems can be formulated as combinatorial optimization problems whose goal is to find a linear layout of an input graph in such a way that the number of edge crossings is minimized. In this paper, we study a restricted version of the linear layout problem where the order of vertices on the line is fixed, the so-called fixed linear crossing number problem (FLCNP). We show that this $\mathcal{NP}$-hard problem can be reduced to the well-known maximum cut problem. The latter problem was intensively studied in the literature; efficient exact algorithms based on the branch-and-cut technique have been developed. By an experimental evaluation on a variety of graphs, we show that using this reduction for solving FLCNP compares favorably to earlier branch-and-bound algorithms.

1 Introduction

For a given simple graph $G = (V, E)$ with vertex set V and edge set E, a linear embedding of G is a special type of embedding in which vertices of V are placed on a horizontal line L and edges are drawn as semicircles above or below L; see Fig. 1. This type of drawing was first introduced by Nicholson [12] in order to develop a heuristic algorithm for the general $\mathcal{NP}$-complete crossing minimization problem [4]. However, Masuda et al. proved that it is still $\mathcal{NP}$-hard to find a linear embedding of a given graph with a minimum number of crossings, even if the ordering of vertices on L is predetermined [10]. The latter problem is called the *fixed linear crossing number problem* (FLCNP).

Crossing minimization for linear embeddings has important applications in different areas such as sorting permutations [6], fault tolerant VLSI design [13], complexity theory [3], and compact graph encodings [11]. Moreover, the problem FLCNP is of general interest in graph drawing and information visualization, where the number of edge crossings has a big effect on the readability of graph layout [2]. It was also shown to be a subproblem in communications network management graphics facilities such as CNMgraf [5]. Sorting with parallel stacks

---

* Partially supported by the Marie Curie RTN ADONET 504438 funded by the EU.

D.Z. Chen and D.T. Lee (Eds.): COCOON 2006, LNCS 4112, pp. 507–516, 2006.
© Springer-Verlag Berlin Heidelberg 2006

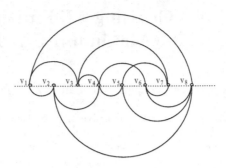

Fig. 1. A linear embedding

is similar to FLCNP where the layout of vertices is fixed, although the objective is to find a layout with no crossings at all.

Recently, heuristic methods, as well as exact algorithms, have been proposed to find optimal or near-optimal solutions of linear layout problems. Cimikowski [1] presented different powerful heuristics as well as an exact branch-and-bound algorithm for FLCNP. In the worst case, the latter enumerates all possible assignments of edges to the two sides of L (up to symmetry). However, the idea of branch-and-bound is to use known bounds on the objective function in order to skip most feasible solutions during the enumeration. Cimikowski's algorithm is able to find exact solutions for graphs with up to 50 edges.

In this paper, we introduce a new exact algorithm for the problem FLCNP that is based on a reduction to the *maximum cut problem* (MAXCUT). The same reduction yields a simple test for fixed linear planarity. Computational results for our approach are compared to those obtained with the exact algorithm of [1], on exactly the same test data and equipment. Our approach yields a remarkable improvement in terms of computational efficiency.

This paper is organized as follows. In Sect. 2, the problem under consideration is formalized and necessary notation is introduced. In Sect. 3, we describe the reduction from FLCNP to MAXCUT and present a corresponding optimization algorithm. Experimental results are analyzed in Sect. 4, and Sect. 5 concludes.

2   Preliminaries

Throughout this paper, we consider an undirected, simple graph $G = (V, E)$ with vertex set V and edge set E. A *vertex ordering* (or *vertex permutation*) of G is a bijection $\delta: V \rightarrow \{1, 2, \ldots, |V|\}$. For a pair of vertices (v, w), we will shortly write $v < w$ instead of $\delta(v) < \delta(w)$.

In a *fixed linear embedding* of G, we assume that the vertices of G are placed on a straight horizontal line L according to a fixed vertex ordering. Moreover, each edge is drawn as a semicircle; see Fig. 1. Consequently, edges may be routed above or below L but never cross L. Notice that three edges cannot intersect in one point unless it is a common endpoint.

For a given graph G and vertex ordering δ, a pair of edges $e_1 = (v_1, w_1)$ and $e_2 = (v_2, w_2)$ is *potentially crossing* if e_1 and e_2 cross each other when routed on the same side of L. Clearly, e_1 and e_2 are potentially crossing if and only if $v_1 < v_2 < w_1 < w_2$ or $v_2 < v_1 < w_2 < w_1$.

In this paper, we are interested in the number of crossings in fixed linear embeddings of G. There is a crossing between e_1 and e_2 if, and only if:

− e_1 and e_2 are potentially crossing, and
− e_1 and e_2 are embedded on the same side of L.

We are going to address the following optimization problem:

Fixed linear crossing number problem (FLCNP): Given a graph $G = (V, E)$ with a fixed vertex ordering, find a corresponding linear embedding of G with a minimum number of edge crossings.

It is easy to see that the number of edge crossings in a linear embedding only depends on the order of vertices and the sides to which the edges are assigned, but not on the exact positions of the vertices. In particular, as the vertex ordering is fixed as part of the input of FLCNP, the only remaining choice is whether edges are drawn above or below the line L. Thus, with respect to the crossing number, there are essentially $2^{|E|}$ different fixed linear embeddings of G. Nevertheless, the problem FLCNP was shown to be $\mathcal{NP}$-hard by Masuda et al. [10].

3   A New Algorithm

The exact algorithm used by Cimikowski [1] to solve the problem FLCNP is based on the branch-and-bound technique: basically, all possible solutions of the problem are enumerated. The set of solutions is given by a binary enumeration tree, where each inner node corresponds to a decision whether a chosen edge is drawn above or below the horizontal line. In the worst case, an exponential number of solutions has to be enumerated. However, the basic idea of branch-and-bound is the pruning of branches in this tree: at some node of the tree, a certain set of edges is already fixed. According to this information, one can derive a lower bound on the number of crossings subject to these fixed edges. If this lower bound is at most as good as a feasible solution that has already been found, e.g., by some heuristics, it is clear that the considered subtree cannot contain a better solution, so it does not have to be explored.

In the following, we describe a different approach for solving FLCNP exactly. We show that the problem is, in fact, a special case of the well-known MAXCUT problem; see Sect. 3.1. The latter has been studied intensively in the literature. In particular, branch-and-cut algorithms have been developed; see Sect. 3.2, which we use for our experimental evaluation presented in Sect. 4.

3.1   Reduction to MAXCUT

In this section, we show that the problem FLCNP can easily be reduced to the maximum cut problem given as follows:

Maximum Cut Problem (MAXCUT): Given an undirected graph $G' = (V', E')$, find a partition of V' into disjoint sets V_1 and V_2 such that the number of edges from E' that have one endpoint in V_1 and one endpoint in V_2 is maximal.

For an instance of FLCNP, i.e., a given graph $G = (V, E)$ with a fixed vertex permutation, we construct the associated *conflict graph* $G' = (V', E')$ as follows: the vertices of G' are in one-to-one correspondence to the edges of G, i.e., $V' = E$. Two vertices of G' corresponding to edges $e_1, e_2 \in E$ are adjacent if, and only if, e_1 and e_2 are potentially crossing. See Fig. 2 for an illustration.

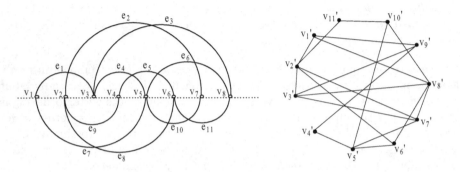

Fig. 2. The graph G and its associated conflict graph G'

Definition 1. *Let G be a graph with a fixed vertex permutation. Given a vertex partition (V_1, V_2) of its conflict graph G', the associated* cut embedding *is the fixed linear embedding of G where edges corresponding to V_1 and V_2 are embedded to the half spaces above and below the vertex line, respectively.*

Theorem 1. *Consider a partition (V_1, V_2) of V'. Then the corresponding cut embedding is a fixed linear embedding of G with a minimum number of crossings if, and only if, (V_1, V_2) is a maximum cut in G'.*

Proof. Let F' be the set of edges in G' with one endpoint in V_1 and one endpoint in V_2, i.e., the cut given by (V_1, V_2). By definition of G', we know that every crossing in the cut embedding associated to (V_1, V_2) corresponds to an edge in G' such that either both its endpoints belong to V_1, or both belong to V_2, i.e., to an edge in $E' \setminus F'$. Thus, the number of crossings is $|E'| - |F'|$. As $|E'|$ is constant for a fixed vertex permutation, the result follows. □

Theorem 2. *For a graph $G = (V, E)$ with a fixed vertex permutation, there is a planar fixed linear embedding of G if, and only if, the associated conflict graph G' of G is bipartite.*

Proof. Suppose H is a planar fixed linear embedding of G. Let E_1 and E_2 represent the two edge sets above and below the horizontal vertex line, respectively. Then the vertices of G' consist of two vertex sets V_1 corresponding to E_1 and V_2

corresponding to E_2. Since H is planar, there is no edge connecting vertices from the same set. So G' is bipartite. On the other hand, if G' is bipartite, the resulting cut embedding of G is obviously planar.   □

Observe that testing whether the graph G' is bipartite can be done in linear time (with respect to G') by two-coloring a DFS-tree.

3.2 Solving MAXCUT by Branch-and-Cut

By Theorem 1, we can use any algorithm for MAXCUT in order to solve FLCNP. One of the most successful approaches for solving MAXCUT to optimality in practice is branch-and-cut.

It would go beyond the scope of this paper to explain this approach in detail. Roughly, the problem is modelled as an integer linear program (ILP). This ILP is first solved as a linear program (LP), i.e., the integrality constraints are relaxed. LPs are solved very quickly in practice. If the LP-solution is integer, we can stop. Otherwise, one tries to add cutting planes that are valid for all integer solutions of the ILP but not necessary for (fractional) solutions of the LP. If such cutting planes are found, they are added to the LP and the process is reiterated.

We have to resort to the branching part only if no more cutting planes are found. In general, only a small portion of the enumeration tree has to be explored, as many branches can be pruned. Compared to a pure branch-and-bound approach as presented in [1], the number of subproblems to be considered is very small in general. This, however, depends on the quality of the cutting planes being added. The latter in turn depend on the specific problem; finding good cutting planes is a sophisticated task. Fortunately, the MAXCUT problem has been investigated intensively, so that many classes of cutting planes are known.

More detailed information on algorithms for MAXCUT using cutting plane techniques can be found in [7,9]. Observe that MAXCUT can also be adressed by semidefinite programming methods; see e.g. [8]. These methods perform well on very dense instances, while being outperformed by ILP approaches on sparse or large graphs. For this reason, we chose the latter method for our experiments.

4 Experimental Results

In order to evaluate the practical performance of our new exact approach to FLCNP presented in the previous section, we performed extensive experiments. In this section, we report the results and compare them to the results obtained with the branch-and-bound algorithm proposed by Cimikowski [1]. The set of test instances is exactly the same as used in [1].

These instances mainly arise from network models of computer architectures; in general they are hamiltonian. The fixed order of nodes, as part of the input of FLCNP, is then determined by a hamiltonian cycle in the graph, as an ordering of the vertices along a hamiltonian cycle tends to yield a smaller number of crossings in general. In our experiments, we always used the same ordering as chosen in [1] for ensuring comparability.

More specifically, the networks considered are the following, see also [1]:

- *complete graphs* K_n for $n = 5, \ldots, 13$
- *hypercubic networks*: this class of graphs includes the *hypercubes* Q_d and several derivatives of hypercubes such as the *cube-connected-cycles* CCC_d, the *twisted cubes* TQ_d, the *crossed cubes* CQ_d, the *folded cubes* FLQ_d, the *hamming cubes* HQ_d, the *binary de Bruijn graphs* DB_d and the *undirected de Bruijn graphs* UDB_d, the *wrapped butterfly graphs* WBF_d and the *shuffle-exchange graphs* SX_d
- other interconnection networks, including the $d \times d$ *tori* $T_{d,d}$, the *star graphs* ST_d, the *pancake graphs* PK_d, and the *pyramid graphs* PM_d
- *circular graphs*: the circular graph $C_n(a_1, \ldots, a_k)$ is regular and hamiltonian.

In Table 1, we contrast our runtime results with those of the branch-and-bound algorithm presented in [1]; we list all instances for which runtimes are reported in [1]. For a better comparison, we ran both algorithms on the same machine, a Pentium 4 with 2.8 GHz. The running times for the branch-and-bound algorithm were obtained with the original implementation used in [1]. In the remainder of this section, all running times are given in CPU seconds.

Table 1. Running times for exact approaches

instance	B & B [1]	MAXCUT
Q_4	0.01	0.00
CCC_3	0.02	0.00
SX_4	0.01	0.00
FLQ_4	0.13	0.42
UDB_5	0.43	0.07
$C_{26}(1,3)$	0.46	0.00
$T_{6,6}$	1.27	0.04
CCC_4	2.59	0.01
K_{10}	2.27	3.21
SX_5	2.16	1.84
$C_{20}(1,2,3)$	16.69	0.39
$T_{7,7}$	64.89	0.15
$C_{22}(1,2,3)$	73.16	0.39
K_{11}	148.21	24.56
Q_5	612.35	1.67
K_{12}	1925.51	79.15
K_{13}	> 86400.00	2119.12

Notice that in our approach we did not use any initial heuristics, in order to give clearer and independent runtime figures. Nevertheless, as obvious from Table 1, our approach is much faster than the branch-and-bound algorithm. This is particularly true for sparse instances, e.g., Q_5. However, our approach outperforms [1] also on the larger complete graphs.

For all other instances, only heuristic results are given in [1]. Tables 2 and 3 state the results of our approach, sorted as in [1]. In all tables, the columns

Table 2. Experimental results, part I

instance	best heuristic [1]	worst heuristic [1]	exact solution	MAXCUT runtime
K_5	1	1	1	0.00
K_6	3	4	3	0.00
K_7	9	11	9	0.01
K_8	18	24	18	0.06
K_9	36	46	36	0.83
K_{10}	60	80	60	3.21
K_{11}	100	130	100	24.56
K_{12}	150	200	150	79.15
K_{13}	225	295	225	2119.12
Q_4	8	8	8	0.00
Q_5	62	80	**60**	1.67
Q_6	370	512	**368**	17924.24
Q_7	1874	2688	[1894]	> 86400.00
CCC_3	0	4	0	0.00
CCC_4	16	24	16	0.01
CCC_5	104	148	104	3.99
SX_4	7	8	7	0.00
SX_5	60	74	60	1.84
SX_6	281	333	[285]	> 86400.00
SX_7	1315	1554	[1319]	> 86400.00
UDB_4	5	5	5	0.00
UDB_5	28	34	28	0.07
UDB_6	149	183	**148**	19.88
UDB_7	629	815	[646]	> 86400.00
UDB_8	2384	3065	[2387]	> 86400.00
FLQ_3	4	6	4	0.00
FLQ_4	36	44	36	0.42
FLQ_5	208	256	208	5981.78
FLQ_6	1036	1320	[1061]	> 86400.00
FLQ_7	4712	6144	[4804]	> 86400.00
TQ_3	1	1	1	0.00
TQ_4	8	10	8	0.00
TQ_5	65	83	**63**	1.42
TQ_6	372	516	372	28694.06
TQ_7	1866	2693	[1916]	> 86400.00
CQ_3	1	1	1	0.00
CQ_4	12	12	12	0.01
CQ_5	88	106	88	6.47
CQ_6	494	588	[508]	> 86400.00
CQ_7	2475	3056	[2481]	> 86400.00
HQ_3	5	6	5	0.00
HQ_4	50	57	50	1.86
HQ_5	303	361	[303]	> 86400.00
HQ_6	1523	1885	[1531]	> 86400.00
HQ_7	6913	8734	[7057]	> 86400.00
WBF_3	22	30	22	0.02
WBF_4	164	205	**158**	22.54
WBF_5	904	1066	[948]	> 86400.00

Table 3. Experimental results, part I

instance	best heuristic [1]	worst heuristic [1]	exact solution	MAXCUT runtime
$T_{3,3}$	3	4	3	0.00
$T_{4,4}$	8	8	8	0.00
$T_{5,5}$	20	30	20	0.02
$T_{6,6}$	24	38	24	0.04
$T_{7,7}$	48	70	48	0.15
$T_{8,8}$	48	80	48	0.16
$T_{9,9}$	88	142	88	0.71
$T_{10,10}$	80	190	80	0.69
ST_4	11	13	11	0.01
ST_5	570	699	[572]	> 86400.00
PK_4	10	11	10	0.01
PK_5	500	564	[514]	> 86400.00
PM_3	4	26	4	0.00
PM_4	439	796	439	28964.87
$C_{20}(1,2)$	0	4	0	0.00
$C_{20}(1,2,3)$	22	28	22	0.39
$C_{20}(1,2,3,4)$	70	98	70	1.49
$C_{22}(1,2)$	0	2	0	0.00
$C_{22}(1,2,3)$	24	32	24	0.39
$C_{22}(1,3,5,7)$	200	254	200	191.70
$C_{24}(1,3)$	12	16	12	0.00
$C_{24}(1,3,5)$	72	92	72	1.68
$C_{24}(1,3,5,7)$	216	282	216	266.11
$C_{26}(1,3)$	14	18	14	0.00
$C_{26}(1,3,5)$	82	102	82	22.79
$C_{26}(1,4,7,9)$	364	446	364	19392.85
$C_{28}(1,3)$	16	20	**14**	0.00
$C_{28}(1,3,5)$	86	110	86	3.38
$C_{28}(1,2,3,4)$	98	138	98	3.90
$C_{28}(1,3,5,7,9)$	560	714	[560]	> 86400.00
$C_{30}(1,3,5)$	96	120	**90**	2.77
$C_{30}(1,3,5,8)$	302	348	**[298]**	> 86400.00
$C_{30}(1,2,4,5,7)$	392	470	[396]	> 86400.00
$C_{32}(1,2,4,6)$	160	202	160	21.65
$C_{34}(1,3,5)$	110	132	**104**	5.83
$C_{34}(1,4,8,12)$	574	670	**[572]**	> 86400.00
$C_{36}(1,2,4)$	36	60	36	0.03
$C_{36}(1,3,5,7)$	328	422	328	5624.53
$C_{38}(1,7)$	84	98	84	14.70
$C_{38}(1,4,7)$	190	236	190	149.04
$C_{40}(1,5)$	56	64	56	5.04
$C_{42}(1,4)$	42	46	42	0.07
$C_{42}(1,3,6)$	158	170	**150**	651.18
$C_{42}(1,2,4,6)$	210	284	210	115.16
$C_{44}(1,4,5)$	180	200	180	53.72
$C_{44}(1,4,7,10)$	632	830	[648]	> 86400.00
$C_{46}(1,4)$	46	50	46	0.07
$C_{46}(1,5,8)$	296	374	**294**	1104.35

show the following data: the name of the instance, the number of crossings produced by the best and worst heuristics of [1], respectively, the optimal number of crossings (when successfully computed by our approach), and the runtime of our algorithm. However, as some instance are far too large for exact solution, we had to set a general time limit of 24 hours. Whenever this limit was reached, we report the best crossing number found instead of the optimal solution; the figures are then put into brackets. Where an optimal solution was found for an instance that was not solved to proven optimality before, we use italics. Bold figures indicate that our algorithm could improve the best heuristic solution.

It is remarkable that many instances can be solved very quickly by our approach while others cannot even be solved in one CPU day. In other words, the border line between easy instances (those solvable within 25 seconds, say) and hard ones (those unsolved even in one day) is very sharp, few instances do not fall into one of these categories.

Our results can also help to evaluate the quality of the heuristic methods. In fact, it turns out that many heuristics proposed by [1] are able to find optimal or near-optimal solutions even for larger instances. In summary, we think that small to medium sized instances should be solved to optimality in general, whereas for larger instances one can at least be confident that the heuristic solution is not too far away from the optimum.

The algorithm we used for solving the MAXCUT problem is generally better adapted to sparse graphs. This is reflected in the runtime figures presented in this section. Therefore, practical instances tend to be easy for our approach.

5   Conclusion and Future Work

We have presented a new exact algorithm for the fixed linear crossing number problem, running significantly faster than earlier exact algorithms. The essential part of our approach is the reduction to the maximum cut problem. After this transformation, the problem can be solved with a sophisticated mathematical programming algorithm, based on the extensive knowledge that has been gathered for the maximum cut problem by intensive research. Moreover, testing the existence of a planar fixed linear embedding of a given graph can be done in an easy way using this transformation. We believe that this principle can also be applied to other linear embedding problems with different objective functions.

Our experimental results show that many medium sized instances can be solved very quickly by our approach. However, for many large instances we cannot find optimal solutions. For these instances, the heuristics proposed by Cimikowski [1] are a good compromise between running time and quality. In fact, our evaluation shows that in most cases at least one of these heuristics is able to find the optimal solution.

In consequence, we plan to integrate good heuristics into our branch-and-cut algorithm in order to further improve running times. In general, this can be done in the same way as in the branch-and-bound approach. We are convinced that this will considerably increase the performance of our approach.

Since the generation of edge crossings largely depends on the original vertex ordering, it is crucial to study the general version of the linear crossing number problem. We plan to develop heuristic or exact algorithms finding vertex orderings leading to a minimal number of potential edge crossings. Having done this, we will be able to evaluate our approach on instances without a predetermined order of vertices. In particular, we plan to test its performance on the graphs in the well-known Rome library. As these graphs are usually very sparse, we are convinced that we will be able to solve most of these instances to optimality.

Acknowledgement. We would like to thank Frauke Liers for providing us her implementation of a branch-and-cut algorithm for the maximum cut problem. Moreover, we are grateful to Robert Cimikowski for making his implementation and experimental data available to us.

References

1. R. Cimikowski. Algorithms for the fixed linear crossing number problem. *Disc. Appl. Math.*, 122:93–115, 2002.
2. G. Di Battista, P. Eades, R. Tamassia, and I. G. Tollis. Algorithms for drawing graphs: an annotated bibliography. *Computational Geometry: Theory and Applications*, 4:435–282, 1994.
3. Z. Galil, R. Kannan, and E. Szemerédi. On nontrivial separators for k-page graphs and simulations by nondeterministic one-tape Turing machines. *J. Comput. System Sci.*, 38(1):134–149, 1989.
4. M. R. Garey and D. S. Johnson. Crossing number is NP-complete. *SIAM J. Alg. Disc. Meth.*, 4:312–316, 1983.
5. R. S. Gilbert and W. K. Kleinöder. CNMgraf – graphic presentation services for network management. In *Proc. 9th Symposium on Data Communication*, pages 199–206, 1985.
6. T. Harju and L. Ilie. Forbidden subsequences and permutations sortable on two parallel stacks. In *Where mathematics, computer science, linguistics and biology meet*, pages 267–275. Kluwer, 2001.
7. M. Laurent. The max-cut problem. In M. Dell'Amico, F. Maffioli, and S. Martello, editors, *Annotated Bibliography in Combinatorial Optimization*. Wiley, 1997.
8. M. Laurent and F. Rendl. Semidefinite programming and integer programming. In *Discrete Optimization*, pages 393–514. Elsevier, 2005.
9. F. Liers, M. Jünger, G. Reinelt, and G. Rinaldi. Computing exact ground states of hard Ising spin glass problems by branch-and-cut. In *New Optimization Algorithms in Physics*, pages 47–69. Wiley-VCH, 2004.
10. S. Masuda, K. Nakajima, T. Kashiwabara, and T. Fujisawa. Crossing minimization in linear embeddings of graphs. *IEEE Trans. Comput.*, 39(1):124–127, 1990.
11. J. I. Munro and V. Raman. Succinct representation of balanced parentheses and static trees. *SIAM J. Comput.*, 31(3):762–776, 2001.
12. T. A. J. Nicholson. Permutation procedure for minimizing the number of crossings in a network. In *Proc. IEEE*, volume 115, pages 21–26, 1968.
13. A. L. Rosenberg. DIOGENES, circa 1986. In *Proc. VLSI Algorithms and Architectures*, volume 227 of *LNCS*, pages 96–107. Springer, 1986.

On the Effectiveness of the Linear Programming Relaxation of the 0-1 Multi-commodity Minimum Cost Network Flow Problem

Dae-Sik Choi and In-Chan Choi[*]

Department of Industrial Systems and Information Engineering, Korea University
Anamdong, Seongbookku, Seoul 136, South Korea

Abstract. Several studies have reported that the linear program relaxation of integer multi-commodity network flow problems often provides integer optimal solutions. We explore this phenomenon with a 0-1 multi-commodity network with mutual arc capacity constraints. Characteristics of basic solutions in the linear programming relaxation problem of the 0-1 multi-commodity problem are identified. Specifically, necessary conditions for a linear programming relaxation to have a non-integer solution are presented. Based on the observed characteristics, a simple illustrative example problem is constructed to show that its LP relaxation problem has integer optimal solutions with a relatively high probability. Furthermore, to investigate whether or not and under what conditions this tendency applies to large-sized problems, we have carried out computational experiments by using randomly generated problem instances. The results of our computational experiment indicate that there exists a narrow band of arc density in which the 0-1 multi-commodity problems possess no integer optimal solutions.

1 Introduction

The integer multi-commodity minimum cost network flow problem (IMNFP), which has been applied in various fields such as transportation, production, and communication systems, involves finding optimal integral flows that satisfy arc capacity constraints on an underlying network. The problem is known to be NP-hard even in its simplest form, viz. in a planar graph with unit arc capacities [1]. Moreover, coupled by various side constraints, many IMNFP problems in practice usually take further complication. Subsequently, several studies have developed heuristic procedures or efficient branch-and-bound based procedures for IMNFP problems with side constraints ([2],[3],[5],[7],[9],[11], [12]).

Some of these studies have reported that the linear program (LP) relaxation of instances of the IMNFP with or without side constraints often provides integer optimal solutions or excellent bounds ([5], [9], [12]). Löbel [9] considered the IMNFP with arc cover constraints as coupling constraints for vehicle scheduling in public transit

---

[*] Corresponding author: Prof. I.C. Choi, Fax: +82-2-929-5888, E-mail:ichoi@korea.ac.kr

D.Z. Chen and D.T. Lee (Eds.): COCOON 2006, LNCS 4112, pp. 517–526, 2006.
© Springer-Verlag Berlin Heidelberg 2006

and suggested the column generation technique for solving large-scale linear problems. In his computational experiments using instances based on real-world data, the LP relaxation gave tight bounds in several problem instances. In fact, the LP relaxation was observed to yield integer optimal solutions for a few problem instances. Faneyte, Spieksma, and Woeginger[5] studied the 0-1 MNFP with node cover constraints as coupling constraints for the crew-scheduling problem and presented a branch-and-price algorithm for arc-chain formulation. Their computational experiments also showed that the LP relaxation for most of the instances based on practical data for a crane rental company gave an integer optimal solution. In addition, they indicated the short length of feasible paths in their instances as one possible explanation for this phenomenon. For some instances with longer possible paths, however, the LP relaxation provided an integer optimal solution. For the IMNFP on a ring network, Ozdaglar and Bertsekas [12] reported that the LP relaxation gave an integer optimal solution for almost all instances. Moreover, similar findings for some instances of the IMNFP with side constraints, viz. the LP relaxation often provides an integer optimal solution, are observed in [2], [3], [7], and [11].

In some special classes of the IMNFP, it has been shown that the LP relaxation gives an integer optimal solution. Evans [4] described a sufficient condition under which the IMNFP can be transformed into an equivalent single-commodity problem, and Kleitmann, Martin-Lof, Rothschild, and Whinston [8] showed that if each node in a network were a source or sink for at least $(k-1)$ of the k commodities, the optimal solution would be integral. Along this line of research, we investigate the 0-1 MNFP, a sub-class problem of IMNFP, to explore the effectiveness of the LP relaxation. The 0-1 MNFP has been applied to several practical problems, such as the crew-scheduling problem and telecommunications ([2], [5], [11]).

We have identified some characteristics of basic feasible solutions of the LP relaxation. Also, by using them, we have constructed an example to show that its LP relaxation problem has integer optimal solutions with a relatively high probability. As the construction of the example is pathological, we have conducted computational experiments by using randomly generated problem instances in order to investigate whether or not and under what conditions the LP relaxation provides integer optimal solutions.

This paper is organized as follows. Section 2 below describes the problem under consideration, along with some definitions of notation. It also includes the characteristics of basic feasible solutions of the LP relaxation and a simple illustrative example. Section 3 discusses the results of computational experiments and Section 4 contains concluding remarks.

2 The 0-1 MNFP and Characteristics of the Problem

We consider the 0-1 MNFP on a digraph $G(V, E)$ with a node set V and an arc set E. Given a set of commodities K, each commodity k is assumed to have a single origin and single destination. Let i and j be an arc and path index, respectively. Let Ψ^k and P_j denote the set of origin-destination paths of commodity k and the set of arcs in path j, respectively. Also, let c_j ($j \in \Psi^k$) and u_i represent the cost of shipping commodity k along path j and the bundle (mutual arc) capacity of arc i, respectively. Without loss of generality, we will assume that all arcs have capacities. Moreover, δ_{ij} denotes the Kronecker delta to indicate whether an arc i belongs to P_j ; i.e. δ_{ij} equals 1 if $i \in P_j$ and equals zero otherwise. The decision variable y_j is a binary variable to indicate whether or not commodity is shipped along path j. Then, the arc-chain formulation of the 0-1 MNFP is expressed as

$$Min \quad \sum_k \sum_{j \in \psi^k} c_j y_j$$

$$s.t \quad \sum_{j \in \psi^k} y_j = 1 \ for \ k \in K \tag{1}$$

$$\sum_k \sum_{j \in \psi^k} \delta_{ij} y_j + z_i = u_i \ for \ i \in E \tag{2}$$

$$y_j \in \{0,1\} \ for \ j \in \psi^k, k \in K \tag{3}$$

In the above formulation, z_i is a slack variable. Constraints (1) represent that each commodity must be shipped on a unique path and Constraints (2) represent that total flow on an arc can not exceed its capacity. An LP relaxation problem is obtained by relaxing integrality constraints (3) and the rank of the constraint matrix in the relaxed problem is $(|K|+|E|)$, where $|K|$ denotes the cardinality of set K. Given a basic feasible solution of the LP relaxation problem, let B_y and B_z denote index sets of basic path variables and basic slack variables, respectively. Moreover, we use B_y^k as an index set of basic path variables for commodity k. Also, let N_z denote a set of arcs of which slack variables are nonbasic. As the arcs that belong to N_z are saturated, we will call them *saturated nonbasic arcs*. Note that there can be saturated basic arcs because of degeneracy.

In a basic feasible solution of the LP relaxation problem, at least one path variable should be basic for each commodity, i.e. $|B_y^k| \geq 1$ for all k. Thus, a basic matrix B of the LP relaxation problem can be expressed as

$$B = \begin{bmatrix} \overbrace{I}^{|K|} & \overbrace{D_1}^{|N_z|} & \overbrace{0}^{|B_z|} \\ D_2 & D_3 & 0 \\ D_4 & D_5 & I \end{bmatrix} \begin{matrix} \}|K| \\ \}|N_z| \\ \}|B_z| \end{matrix} \tag{4}$$

Each of the first $|K|$ columns corresponds to one path variable for each commodity, which we shall call a *primary path variable*. The next $|N_z|$ columns correspond to non-primary path variables in the basis and the last $|B_z|$ columns correspond to slack variables of the arcs in B_z. The first $|K|$ rows are flow constraints and the next $|N_z|$ and last $|B_z|$ rows correspond to the capacity constraints of saturated arcs in N_z and basic arcs in B_z, respectively. Now, let $D = D_3 - D_2 D_1$. Then, the inverse of the basic matrix is expressed as

$$B^{-1} = \begin{bmatrix} I + D_1 D^{-1} D_2 & -D_1 D^{-1} & 0 \\ -D^{-1} D_2 & D^{-1} & 0 \\ -D_4 + (D_5 - D_4 D_1) D^{-1} D_2 & -(D_5 - D_4 D_1) D^{-1} & I \end{bmatrix}$$

The integrality of basic solutions is closely related to the characteristics of matrix D. The matrix D is square and its rank is $|N_z|$. Moreover, it can be described by using the relationship of paths and saturated arcs. We assume that, in the matrix B, the $(i+|K|)^{th}$ row corresponds to the capacity constraints of arc i and the jth column corresponds to path j. Also, we assume the rth column (for $r=1,2,\ldots,|K|$) corresponds to the primary path of commodity r. Let $i=1,2,\ldots, |N_z|$ and $j=|K|+1, \ldots, |K|+|N_z|$ be the arc and path index, respectively. Also, let k_j be an commodity index of path j. In addition, $(D)_{ij}$, $(D)_{i.}$, and $(D)_{.j}$ denote an element, row, and column of matrix D, respectively. Element $(D_1)_{r,j-|K|} = 1$ if $r= k_j$; otherwise it is zero. It notes that $(D_1)_{r,j-|K|} = 1$ if path j is a path of commodity r. Moreover, element $(D_2)_{ir}=1$ if $i\in P_r$, otherwise $(D_2)_{ir}=0$. Therefore,

$$(D_2 D_1)_{i,j-|K|} = (D_2)_{i.}(D_1)_{.j-|K|} = \begin{cases} 1 \text{ if } i\in P_{k_j} \\ 0 \text{ otherwise} \end{cases}$$

Element $(D_3)_{ij}$ is given as

$$(D_3)_{i,j-|K|} == \begin{cases} 1 \text{ if } i\in P_j \\ 0 \text{ otherwise} \end{cases}$$

Then, the element of matrix D is given as, for $i=1,2,\ldots, |N_z|$ and $j=|K|+1, \ldots, |K|+|N_z|$

$$(D)_{i,j-|K|} = (D_3)_{i,j-|K|} - (D_2)_{i.}(D_1)_{.j-|K|} = \begin{cases} 1 & \text{if } i\in P_j, i\notin P_{k_j} \\ -1 & \text{if } i\notin P_j, i\in P_{k_j} \\ 0 & \text{otherwise} \end{cases} \qquad (5)$$

The above observation clearly indicates that that matrix D can be obtained from the relationship of the basic path variables and saturated nonbasic arcs and that the elements in matrix D are 0, 1, or -1. Similar observation applies to the term $D_5 - D_4 D_1$, which is a term in the inverse of a basic matrix. As a parenthetical note, if D is unimodular, then the corresponding basic feasible solution is integral.

Now, we state some properties of basic solutions.

Proposition 1. Every integer feasible solution of the 0-1 MNFP is a basic feasible solution of its LP relaxation problem.

Proof) For a given integer feasible solution, each commodity has exactly one path with its flow equal to 1. Let the index set of these variables be B_y and the arc set E be B_z. Then, from a constraint matrix, a matrix B consisting of columns corresponding to the path variables in B_y and slack variables of arcs in B_z is given as

$$B = \begin{bmatrix} I_{|K|\bowtie|K|} & 0 \\ D_4 & I_{|E|\bowtie|E|} \end{bmatrix}$$

Matrix B is nonsingular and its rank is $(|K|+|E|)$. Therefore, the matrix B is a basis of an LP relaxation. ∎

As one of the properties for non-integer basic solutions, a relationship between $|K|$, $|N_z|$, and $|B_y|$ is established in [6] and [10].

Proposition 2. (By Maurras and Vaxès [10] and Farvoleden *et al* [6]) Every non-integer basic feasible solution of the LP relaxation problem satisfies $|B_y| = |N_z|+|K|$.

Proof. From equation (5), the result immediately follows. ∎

In the above proposition, if $|N_z|=0$, then $|B_y|=|K|$ and the corresponding solution is integral.

Proposition 3. There should be at least two saturated nonbasic arcs in a non-integer basic feasible solution of the LP relaxation problem; viz. $|N_z| \geq 2$.

Proof. Suppose that $|N_z| < 2$. If $|N_z|=0$, the solution is integral. Moreover, if $|N_z|=1$, matrix D is [1] and the solution is integral. ∎

Proposition 3 above can be strengthened by the main proposition of the paper below. It states that there should be at least two saturated nonbasic arcs on the same path for a non-integer basic solution.

Proposition 4. For a non-integer basic feasible solution of the LP relaxation problem with a basic matrix B of the form given in (4), $| N_z \cap P_j | \geq 2$ for some $j \in B_y$.

Proof. Suppose that $| N_z \cap P_j | \leq 1$ for all $j \in B_y$. By (5), each column of matrix D has at most two nonzero elements because $| N_z \cap P_j | \leq 1$ for all $j \in B_y$. Since in this case the matrix D takes the form of node-arc incidence matrix obtained by removing a node in a general network, it is unimodular. This is a contradiction because the solution is assumed to be nonintegral. ∎

The above two propositions describe necessary conditions for the LP relaxation problem to have non-integer vertices and they are closely related to the form of basic matrices.

It is possible to approximate the probability that LP relaxations have integer optimal solutions for simple problems, in which we relax the assumption that all arcs have capacities. As an example, we consider an instance with $|K|=2$, $|E|=2$, and four feasible paths for each commodity, in which E is a set of arcs with capacities only. Fig. 1(a) shows the constraint and basic matrices corresponding to a non-integer basic solution and one of their extended forms. The four paths include (i) a path including none of the two capacity-constrained arcs, (ii) a path including both of the two capacity-constrained arcs, and (iii) a couple of paths including exactly one of the two capacity-constrained arcs. Each of the paths may include some uncapacitated arcs that are not included in E, but they will not appear in the constraint matrix. There are nine feasible integer solutions and, by Proposition 1, all of them are basic solutions of the LP relaxation. By Propositions 2 and 3, $|N_z| = 2$, $|B_y| = 4$, $|B_y^k| = 2$ (for $k=1,2$). Then, two non-integer basic feasible solutions can be drawn to satisfy the necessary condition in Proposition 4. The extended form of a basic matrix corresponds to the case in which one path of commodity 1 includes all saturated nonbasic arcs and each path of commodity 2 includes exclusively one saturated nonbasic arc.

Suppose that the arc capacity is set equal to 1 and the cost for each path is selected randomly among integer values between 1 and C_{max} in the above example. Consider the pseudo-probability P_u that all optimal solutions are integral and the pseudo-probability P_o that at least one of the optimal solutions is integral. To get P_u and P_o for a given C_{max}, we obtained optimal solutions for all possible cases (C_{max}^8 cases) of path

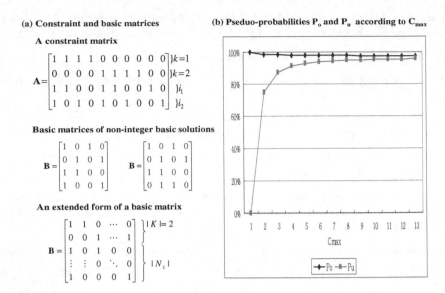

Fig. 1. The constraint and basic matrices of a non-integer basic solution and the pseudo-probabilities P_o and P_u according to C_{max} for an instance with $|K|=2$, $|E|=2$

costs and counted the number of corresponding cases. The term "pseudo" is used to indicate that the sample space generated as such provides "pseudo" elementary events, which may not be equally likely. Fig. 1(b) shows the results. When all path costs are identically one, i.e. $C_{max}=1$, $P_u=0$ and $P_o=1$. Moreover, $P_u=95.7\%$ and $P_o=97.5\%$ at $C_{max}=13$. Also, pseudo-probability P_u increases while pseudo-probability P_o decreases as C_{max} increases and that they converges as C_{max} increases. Because P_o is always greater than or equals to P_u for a specific value of C_{max} the pseudo-probability P_o is greater than 95.7%. Considering the pseudo-probability and the simple ratio of the number of integer solutions to that of basic feasible solutions (9/11), we conclude that the probability that the LP relaxation has an integer optimal solution must be high in our example problem. For simple instances like the example above, it may be possible to identify all basic feasible solutions and to approximate the probabilities P_o and P_u. In general, however, it will be difficult to calculate the probability for a large-sized problem instance.

3 Computational Results

Conceivably, as the number of uncapacitated arcs increases, so does the chance of obtaining integer solution of the LP relaxation problem. Moreover, there might be a specific band of arc density within which the chance of obtaining non-integer optimal solution of the LP relaxation problem is high. If so, then many observations made by earlier studies on the integrality of the LP relaxation solution could be partially explained.

We performed computational experiments using randomly generated instances of the 0-1 MNFP to search for trends that the LP relaxation had an integer optimal solution in large-size instances. For this purpose, we considered four factors: the number of commodities ($|K|$=10, 30, 50, 70), the arc density (d=0.5, 1, 2, 3, 4, 5, 6, 7, 8, 9, 10, 15%), the number of nodes ($|V|$=100, 200, 300), and the maximum arc capacity (U_{max}=1, 0.1*$|K|$). The arc density d, defined as the number of arcs over the number of possible arcs, specified the number of arcs in the random instance, viz. $d*(|V|-1)*|V|$ arcs, where $|V|$ denotes the number of nodes.

To generate instances, we used a path-based generation scheme. First, for each commodity, source and destination nodes were randomly selected and a path from the source to the destination node was constructed. The length of a path was randomly determined between 1 and the minimum of $|V|-1$ and $d*(|V|-1)*|V|/|K|$. If needed, additional arcs were randomly generated. In addition, the arc capacities were determined randomly between 1 and the maximum arc capacity U_{max}, and each arc was assigned a random cost between 1 and 100. For each combination of the four factors, 100 instances were generated and Cplex 9.0 was used to solve the generated instances. We obtained LP optimal solutions of the generated problem instances and counted the number of instances of which the optimal solution was integral.

Tables 1 and 2 show the results for U_{max} with 1 and 0.1*$|K|$, respectively. In all but three cases, there were at least 50 instances with an integer optimal solution, particularly more than 90 instances for all cases when U_{max}=0.1*$|K|$. Moreover, the minimum number of instances with an integer optimal solution was 29, when U_{max} =1, $|V|$=300, $|K|$=70, and d=1%.

In Table 1, for the given number of nodes and commodities, the number of instances with an integer optimal solution is minimal at a specific arc density and increases when the arc density is far apart from the specific level. There are two explanations for this: in our procedure to generate problem instances, candidate paths likely share few arcs at low arc densities; at high arc densities, there are many arcs and the paths with low cost likely share few arcs.

Table 1. The number of instances in which the LP relaxation yields optimal solution when U_{max}=1

d \ $|K|$	$|V|$=100				$|V|$=200				$|V|$=300			
	10	30	50	70	10	30	50	70	10	30	50	70
0.5%	-	-	-	-	91	100	100	100	92	59	97	99
1%	100	100	100	100	97	64	83	100	100	91	65	28
2%	97	97	100	100	100	91	68	45	100	97	93	81
3%	97	74	71	95	99	97	90	70	100	99	97	93
4%	98	80	58	51	100	99	98	87	100	99	99	100
5%	100	92	70	49	100	98	95	85	100	100	99	98
6%	100	90	78	42	100	98	96	98	100	100	100	99
7%	99	96	88	67	100	100	99	97	100	100	99	100
8%	99	92	83	67	100	100	100	100	100	99	100	99
9%	100	98	85	68	100	100	98	98	100	100	100	99
10%	99	98	90	78	100	99	100	99	100	100	100	100
15%	100	100	97	90	100	100	98	100	100	100	100	100

Table 2. The number of instances in which the LP relaxation yields optimal solution when $U_{max}=0.1*|K|$

d ⟍ $|K|$	$|V|=100$				$|V|=200$				$|V|=300$			
	10	30	50	70	10	30	50	70	10	30	50	70
0.5%		-	-	-	91	100	100	100	92	86	99	100
1%	100	100	100	100	97	89	98	100	100	98	97	98
2%	97	98	100	100	100	100	99	100	100	100	100	100
3%	97	95	98	100	99	100	99	100	100	100	100	99
4%	98	97	98	100	100	100	100	100	100	100	100	100
5%	100	99	100	99	100	100	100	100	100	100	100	100
6%	100	98	99	100	100	100	100	100	100	100	100	100
7%	99	99	100	100	100	100	100	98	100	100	100	100
8%	99	99	99	100	100	100	100	100	100	100	100	100
9%	100	99	100	100	100	100	100	100	100	100	100	100
10%	99	100	99	100	100	99	100	100	100	100	100	100
15%	100	99	100	100	100	100	100	100	100	100	100	100

Although the results in Tables 1 and 2 indicate that instances give integer optimal solutions with a high probability for the given number of nodes and commodities, there may be arc densities at which most instances give non-integer optimal solutions. In our cases indeed, the results reveal that the range of arc densities at which most instances have non-integer optimal solutions is very narrow, if it exists. As an example, consider the case with $U_{max}=1$, $|V|=300$, and $|K|=70$ in Table 1; most instances may have non-integer optimal solutions at some arc densities between 0.5 and 2%; however, the range of arc densities at which most instances have non-integer optimal solutions is very narrow for this case, if exists, as shown in Fig. 2. In Fig. 2, the minimum number of instances with an integer optimal solution was 21, when d=0.8%.

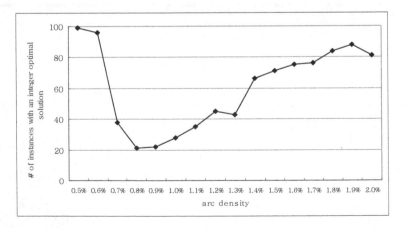

Fig. 2. The result according to the arc density when $|V|=300$ and $|K|=70$

In Table 1, the range of arc densities, at which the number of instances with integer optimal solutions is large, widens as the number of commodities decreases. In the case with $U_{max}=1$, $|V|=100$, and $|K|=70$, there are less than 90 problem instances with integer optimal solutions at arc densities from 4 to 10%. In contrast, for $|K|=30$, the number is less than 90 at arc densities of 3 and 4%. As the number of nodes increases, the range of arc densities at which there are more than 90 instances with an integer optimal widens.

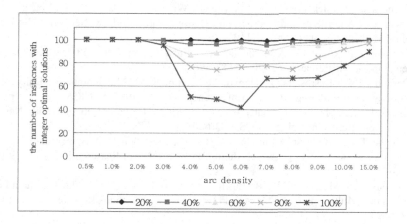

Fig. 3. The results for d_c levels of 20, 40, 60, 80, and 100% when $|V|=100$ and $|K|=70$

When the 0-1 MNFP is applied to crew or vehicle scheduling problems, it would have many arcs without capacity. To see the effects of varying the number of arcs with capacity, we performed additional experiments. For $|V|=100$ and $|K|=70$, five different ratios d_c of arcs with capacity were used to generate instances, *i.e.*, 20, 40, 60, 80, and 100%, where d_c is defined as the ratio of the number of arcs with capacity over the number of arcs. For each case, 100 instances were generated and tested. Fig. 3 shows that the number of instances with integer optimal solutions increases as d_c decreases.

4 Conclusion

Motivated by the observations made by several studies, in which the LP relaxations of instances of the IMNFP often gave integer optimal solutions or excellent bounds, we have examined the 0-1 MNFP with mutual arc capacity constraints to explore this phenomenon. The characteristics of basic feasible solutions in the LP relaxation were examined and the necessary conditions for a basis in the LP relaxation problem to be non-integral were identified. Our computational experiments showed that the LP relaxation frequently provided an integer optimal solution, except when the arc density was within a specific range. Moreover, when the capacities of the arcs were large or the proportion of arcs with capacities was small, the LP relaxation yielded an integer optimal solution.

Our results are applicable to the 0-1 MNFP with side constraints related to a single path, such as hop constraints, because the constraints are considered implicitly in sub-problems used to generate a feasible path. However, further study is needed to address other side constraints such as resource constraints.

References

1. Brandes U., Schlickenrieder, W., Neyer G., Wagner, D, Weihe K.: PlaNet, a Software Package of Algorithms and Heuristics for Disjoint Paths in Planar Networks. Discrete Applied Mathematics, 92 (1999) 91-110
2. Cappanera, P., Gallo, G.: A Multicommodity Flow Approach to the Crew Rostering Problem. Operations Research, 52 (2004) 583-596
3. Desauliers, G., Desrosiers, J., Dumas, Y., Solomon, M. M., Soumis, F.: Daily Aircraft Routing and Scheduling. Management Science, Vol. 43 (1997) 841-855
4. Evans, J. R.: A Single Commodity Transformation for Certain Multicommodity Networks. Operations Research, Vol. 26 (1978) 673-680
5. Faneyte, Diego B.C., Spieksma, Frits C.R., Woeginger, G. J.: A Branch-and-price Algorithm for a Hierarchical Crew Scheduling Problems. Naval Research Logistics, 49 (2002) 743-759
6. Farvoleden, J. M., Powell, W. B., Lustig I. J.: A primal Solution for the Arc-chain Formulation of a Multicommodity Network Flow Problem. Operations Research, Vol. 41 (1993) 669-693
7. Gouveia, L.: Multicommodity Flow Models for Spanning Trees with Hop Constraints. European Journal of Operational Research, 95 (1996) 178-190
8. Kleitmann, D. J., Martin-Lof, A., Rothschild, B., Whinston, A.: A Matching Theorem for Graphs. Journal of Combinatorial Theory, 8 (1970) 104-114
9. Löbel, A.: Vehicle Scheduling in Public Transit and Lagrangean Pricing. Management Science, Vol. 44 (1998) 1637-1649
10. Maurras, J.- F., Vaxès, Y.: Multicommodity Network Flow with Jump Constraints. Discrete Mathematics, 165/166 (1997) 481-486
11. Moz, M., Pato, M. V.: An Integer Multicommodity Flow Model Applied to the Rerostering of Nurse Schedules. Annals of Operations Research, 119 (2003) 285-301
12. Ozdaglar, A. E., Bertsekas D. P.: Optimal Solution of Integer Multicommodity Flow Problems with Application in Optical Networks. Proc. of Symposium on Global Optimization, Santorini, Greece (2003)

Author Index

Aissi, Hassene   428
Aleksandrowicz, Gadi   418
Amano, Kazuyuki   104
Andersson, Mattias   196
Arvind, V.   126
Atallah, Mikhail J.   2
Avis, David   205

Bachmat, Eitan   387
Barequet, Gill   408, 418
Bayouth, John E.   156
Bazgan, Cristina   428
Bengtsson, Fredrik   255
Benkert, Marc   166
Bereg, Sergey   176
Buatti, John M.   156
Buchheim, Christoph   497, 507

Chan, Wun-Tat   309, 320
Chandran, L. Sunil   398
Chang, Wen-Chieh   235
Chen, Jingsen   255
Chen, Kefei   378
Chen, Mingxia   459
Chen, Xi   3, 13
Chen, Zhixiang   245
Chin, Francis Y.L.   320
Choi, Dae-Sik   517
Choi, In-Chan   517

Daescu, Ovidiu   176
Daley, Mark   94
Das, Bireswar   126
Deng, Xiaotie   3, 13
Dou, Xin   156
Dubey, Chandan K.   42

Eulenstein, Oliver   235

Fan, Hongbing   368
Fowler, Richard H.   245
Fu, Bin   245
Fung, Stanley P.Y.   320

Gao, Yong   226
Golovkins, Marats   83
Gong, Zheng   378
Goto, Kazuya   63
Gudmundsson, Joachim   166, 196

Hadjicostis, Christoforos   284
Han, Yo-Sub   469
Harada, Shigeaki   33
Hsieh, Sun-Yuan   449
Huang, Xiuzhen   136
Hung, Regant Y.S.   330

Iliopoulos, Costas S.   146
Ito, Takehiro   63

Jiang, Minghui   176
Jünger, Michael   497

Kalbfleisch, Robert   368
Kanj, Iyad A.   299
Karakostas, George   23
Katoh, Naoki   205
Knauer, Christian   166
Kolliopoulos, Stavros G.   23
Kristensen, Jesper Torp   489

Lam, Tao Kai   387
Lee, Inbok   146
Levcopoulos, Christos   196
Li, Jianbo   459
Li, Jianping   459
Li, Weidong   459
Li, Xiangxue   378
Lin, Min Chih   73
Lin, Mingen   340
Lin, Tzu-Chin   52
Liu, Becky Jie   13
Lotker, Zvi   216

Magen, Avner   387
Majumdar, Debapriyo   216
Maruoka, Akira   33, 104
McQuillan, Ian   94
Mehta, Shashank K.   42

Menze, Annette   497
Miltersen, Peter Bro   489
Miyazawa, F.K.   439
Moet, Esther   166
Mohamed, Manal   146
Mölle, Daniel   265
Mukhopadhyay, Partha   126
Muthu, Rahul   360

Nagamochi, Hiroshi   274
Nakanishi, Masaki   116
Nakhleh, Luay   299
Narayanan, N.   360
Narayanaswamy, N.S.   216
Nishizeki, Takao   63

Ohsaki Makoto   205

Percan, Merijam   497
Pin, Jean-Eric   83
Poon, Chung Keung   320
Poon, Sheung-Hung   186
Preparata, Franco P.   1

Rahman, M. Sohel   146
Ratsaby, Joel   479
Ravelomanana, Vlady   350
Richter, Stefan   265
Rijamamy, Alphonse Laza   350
Rossmanith, Peter   265

Salomaa, Kai   94
Sato, Takayuki   104
Shaikhet, Alina   408
Sivadasan, Naveen   398
Smyth, William F.   146

Streinu, Ileana   205
Subramanian, C.R.   360
Suzuki, Tomoya   116
Szwarcfiter, Jayme L.   73

Takimoto, Eiji   33
Tanigawa, Shin-ichi   205
Thulasiraman, Krishnaiyan   284
Ting, H.F.   330

van Oostrum, René   166
Vanderpooten, Daniel   428

Wang, Biing-Feng   52
Watanabe, Katsumasa   116
Weber, Ingmar   216
Wolff, Alexander   166
Wong, Prudence W.H.   309, 320
Wood, Derick   469
Wu, Xiaodong   156

Xavier, E.C.   439
Xia, Ge   299
Xiao, Ying   284
Xu, Jinhui   340

Yamashita, Shigeru   116
Yang, Shih-Cheng   449
Yang, Yang   340
Yu, Hung-I   52
Yung, Fencol C.C.   309

Zheng, Feifeng   320
Zheng, Lanbo   507
Zhou, Xiao   63
Zhu, Binhai   245

Lecture Notes in Computer Science

For information about Vols. 1–3999

please contact your bookseller or Springer

Vol. 4127: E. Damiani, P. Liu (Eds.), Data and Applications Security XX. X, 319 pages. 2006.

Vol. 4112: D.Z. Chen, D. T. Lee (Eds.), Computing and Combinatorics. XIV, 528 pages. 2006.

Vol. 4099: Q. Yang, G. Webb (Eds.), PRICAI 2006: Trends in Artificial Intelligence. XXVIII, 1263 pages. 2006. (Sublibrary LNAI).

Vol. 4098: F. Pfenning (Ed.), Term Rewriting and Applications. XIII, 415 pages. 2006.

Vol. 4097: X. Zhou, O. Sokolsky, L. Yan, E.-S. Jung, Z. Shao, Y. Mu, D.C. Lee, D. Kim, Y.-S. Jeong, C.-Z. Xu (Eds.), Emerging Directions in Embedded and Ubiquitous Computing. XXVII, 1034 pages. 2006.

Vol. 4096: E. Sha, S.-K. Han, C.-Z. Xu, M.H. Kim, L.T. Yang, B. Xiao (Eds.), Embedded and Ubiquitous Computing. XXIV, 1170 pages. 2006.

Vol. 4090: S. Spaccapietra, K. Aberer, P. Cudré-Mauroux (Eds.), Journal on Data Semantics VI. XI, 211 pages. 2006.

Vol. 4088: Z.-Z. Shi, R. Sadananda (Eds.), Agent Computing and Multi-Agent Systems. XVII, 827 pages. 2006. (Sublibrary LNAI).

Vol. 4079: S. Etalle, M. Truszczyński (Eds.), Logic Programming. XIV, 474 pages. 2006.

Vol. 4077: M.-S. Kim, K. Shimada (Eds.), Advances in Geometric Modeling and Processing. XVI, 696 pages. 2006.

Vol. 4076: F. Hess, S. Pauli, M. Pohst (Eds.), Algorithmic Number Theory. X, 599 pages. 2006.

Vol. 4075: U. Leser, F. Naumann, B. Eckman (Eds.), Data Integration in the Life Sciences. XI, 298 pages. 2006. (Sublibrary LNBI).

Vol. 4074: M. Burmester, A. Yasinsac (Eds.), Secure Mobile Ad-hoc Networks and Sensors. X, 193 pages. 2006.

Vol. 4073: A. Butz, B. Fisher, A. Krüger, P. Olivier (Eds.), Smart Graphics. XI, 263 pages. 2006.

Vol. 4072: M. Harders, G. Székely (Eds.), Biomedical Simulation. XI, 216 pages. 2006.

Vol. 4071: H. Sundaram, M. Naphade, J.R. Smith, Y. Rui (Eds.), Image and Video Retrieval. XII, 547 pages. 2006.

Vol. 4070: C. Priami, X. Hu, Y. Pan, T.Y. Lin (Eds.), Transactions on Computational Systems Biology V. IX, 129 pages. 2006. (Sublibrary LNBI).

Vol. 4069: F.J. Perales, R.B. Fisher (Eds.), Articulated Motion and Deformable Objects. XV, 526 pages. 2006.

Vol. 4068: H. Schärfe, P. Hitzler, P. Øhrstrøm (Eds.), Conceptual Structures: Inspiration and Application. XI, 455 pages. 2006. (Sublibrary LNAI).

Vol. 4067: D. Thomas (Ed.), ECOOP 2006 – Object-Oriented Programming. XIV, 527 pages. 2006.

Vol. 4066: A. Rensink, J. Warmer (Eds.), Model Driven Architecture – Foundations and Applications. XII, 392 pages. 2006.

Vol. 4065: P. Perner (Ed.), Advances in Data Mining. XI, 592 pages. 2006. (Sublibrary LNAI).

Vol. 4064: R. Büschkes, P. Laskov (Eds.), Detection of Intrusions and Malware & Vulnerability Assessment. X, 195 pages. 2006.

Vol. 4063: I. Gorton, G.T. Heineman, I. Crnkovic, H.W. Schmidt, J.A. Stafford, C.A. Szyperski, K. Wallnau (Eds.), Component-Based Software Engineering. XI, 394 pages. 2006.

Vol. 4062: G. Wang, J.F. Peters, A. Skowron, Y. Yao (Eds.), Rough Sets and Knowledge Technology. XX, 810 pages. 2006. (Sublibrary LNAI).

Vol. 4061: K. Miesenberger, J. Klaus, W. Zagler, A. Karshmer (Eds.), Computers Helping People with Special Needs. XXIX, 1356 pages. 2006.

Vol. 4060: K. Futatsugi, J.-P. Jouannaud, J. Meseguer (Eds.), Algebra, Meaning and Computation. XXXVIII, 643 pages. 2006.

Vol. 4059: L. Arge, R. Freivalds (Eds.), Algorithm Theory – SWAT 2006. XII, 436 pages. 2006.

Vol. 4058: L.M. Batten, R. Safavi-Naini (Eds.), Information Security and Privacy. XII, 446 pages. 2006.

Vol. 4057: J.P. W. Pluim, B. Likar, F.A. Gerritsen (Eds.), Biomedical Image Registration. XII, 324 pages. 2006.

Vol. 4056: P. Flocchini, L. Gąsieniec (Eds.), Structural Information and Communication Complexity. X, 357 pages. 2006.

Vol. 4055: J. Lee, J. Shim, S.-g. Lee, C. Bussler, S. Shim (Eds.), Data Engineering Issues in E-Commerce and Services. IX, 290 pages. 2006.

Vol. 4054: A. Horváth, M. Telek (Eds.), Formal Methods and Stochastic Models for Performance Evaluation. VIII, 239 pages. 2006.

Vol. 4053: M. Ikeda, K.D. Ashley, T.-W. Chan (Eds.), Intelligent Tutoring Systems. XXVI, 821 pages. 2006.

Vol. 4052: M. Bugliesi, B. Preneel, V. Sassone, I. Wegener (Eds.), Automata, Languages and Programming, Part II. XXIV, 603 pages. 2006.

Vol. 4051: M. Bugliesi, B. Preneel, V. Sassone, I. Wegener (Eds.), Automata, Languages and Programming, Part I. XXIII, 729 pages. 2006.

Vol. 4049: S. Parsons, N. Maudet, P. Moraitis, I. Rahwan (Eds.), Argumentation in Multi-Agent Systems. XIV, 313 pages. 2006. (Sublibrary LNAI).

Vol. 4048: L. Goble, J.-J.C.. Meyer (Eds.), Deontic Logic and Artificial Normative Systems. X, 273 pages. 2006. (Sublibrary LNAI).

Vol. 4047: M. Robshaw (Ed.), Fast Software Encryption. XI, 434 pages. 2006.

Vol. 4046: S.M. Astley, M. Brady, C. Rose, R. Zwiggelaar (Eds.), Digital Mammography. XVI, 654 pages. 2006.

Vol. 4045: D. Barker-Plummer, R. Cox, N. Swoboda (Eds.), Diagrammatic Representation and Inference. XII, 301 pages. 2006. (Sublibrary LNAI).

Vol. 4044: P. Abrahamsson, M. Marchesi, G. Succi (Eds.), Extreme Programming and Agile Processes in Software Engineering. XII, 230 pages. 2006.

Vol. 4043: A.S. Atzeni, A. Lioy (Eds.), Public Key Infrastructure. XI, 261 pages. 2006.

Vol. 4042: D. Bell, J. Hong (Eds.), Flexible and Efficient Information Handling. XVI, 296 pages. 2006.

Vol. 4041: S.-W. Cheng, C.K. Poon (Eds.), Algorithmic Aspects in Information and Management. XI, 395 pages. 2006.

Vol. 4040: R. Reulke, U. Eckardt, B. Flach, U. Knauer, K. Polthier (Eds.), Combinatorial Image Analysis. XII, 482 pages. 2006.

Vol. 4039: M. Morisio (Ed.), Reuse of Off-the-Shelf Components. XIII, 444 pages. 2006.

Vol. 4038: P. Ciancarini, H. Wiklicky (Eds.), Coordination Models and Languages. VIII, 299 pages. 2006.

Vol. 4037: R. Gorrieri, H. Wehrheim (Eds.), Formal Methods for Open Object-Based Distributed Systems. XVII, 474 pages. 2006.

Vol. 4036: O. H. Ibarra, Z. Dang (Eds.), Developments in Language Theory. XII, 456 pages. 2006.

Vol. 4035: T. Nishita, Q. Peng, H.-P. Seidel (Eds.), Advances in Computer Graphics. XX, 771 pages. 2006.

Vol. 4034: J. Münch, M. Vierimaa (Eds.), Product-Focused Software Process Improvement. XVII, 474 pages. 2006.

Vol. 4033: B. Stiller, P. Reichl, B. Tuffin (Eds.), Performability Has its Price. X, 103 pages. 2006.

Vol. 4032: O. Etzion, T. Kuflik, A. Motro (Eds.), Next Generation Information Technologies and Systems. XIII, 365 pages. 2006.

Vol. 4031: M. Ali, R. Dapoigny (Eds.), Advances in Applied Artificial Intelligence. XXIII, 1353 pages. 2006. (Sublibrary LNAI).

Vol. 4029: L. Rutkowski, R. Tadeusiewicz, L.A. Zadeh, J. Zurada (Eds.), Artificial Intelligence and Soft Computing – ICAISC 2006. XXI, 1235 pages. 2006. (Sublibrary LNAI).

Vol. 4028: J. Kohlas, B. Meyer, A. Schiper (Eds.), Dependable Systems: Software, Computing, Networks. XII, 295 pages. 2006.

Vol. 4027: H.L. Larsen, G. Pasi, D. Ortiz-Arroyo, T. Andreasen, H. Christiansen (Eds.), Flexible Query Answering Systems. XVIII, 714 pages. 2006. (Sublibrary LNAI).

Vol. 4026: P.B. Gibbons, T. Abdelzaher, J. Aspnes, R. Rao (Eds.), Distributed Computing in Sensor Systems. XIV, 566 pages. 2006.

Vol. 4025: F. Eliassen, A. Montresor (Eds.), Distributed Applications and Interoperable Systems. XI, 355 pages. 2006.

Vol. 4024: S. Donatelli, P. S. Thiagarajan (Eds.), Petri Nets and Other Models of Concurrency - ICATPN 2006. XI, 441 pages. 2006.

Vol. 4021: E. André, L. Dybkjær, W. Minker, H. Neumann, M. Weber (Eds.), Perception and Interactive Technologies. XI, 217 pages. 2006. (Sublibrary LNAI).

Vol. 4020: A. Bredenfeld, A. Jacoff, I. Noda, Y. Takahashi (Eds.), RoboCup 2005: Robot Soccer World Cup IX. XVII, 727 pages. 2006. (Sublibrary LNAI).

Vol. 4019: M. Johnson, V. Vene (Eds.), Algebraic Methodology and Software Technology. XI, 389 pages. 2006.

Vol. 4018: V. Wade, H. Ashman, B. Smyth (Eds.), Adaptive Hypermedia and Adaptive Web-Based Systems. XVI, 474 pages. 2006.

Vol. 4017: S. Vassiliadis, S. Wong, T.D. Hämäläinen (Eds.), Embedded Computer Systems: Architectures, Modeling, and Simulation. XV, 492 pages. 2006.

Vol. 4016: J.X. Yu, M. Kitsuregawa, H.V. Leong (Eds.), Advances in Web-Age Information Management. XVII, 606 pages. 2006.

Vol. 4014: T. Uustalu (Ed.), Mathematics of Program Construction. X, 455 pages. 2006.

Vol. 4013: L. Lamontagne, M. Marchand (Eds.), Advances in Artificial Intelligence. XIII, 564 pages. 2006. (Sublibrary LNAI).

Vol. 4012: T. Washio, A. Sakurai, K. Nakajima, H. Takeda, S. Tojo, M. Yokoo (Eds.), New Frontiers in Artificial Intelligence. XIII, 484 pages. 2006. (Sublibrary LNAI).

Vol. 4011: Y. Sure, J. Domingue (Eds.), The Semantic Web: Research and Applications. XIX, 726 pages. 2006.

Vol. 4010: S. Dunne, B. Stoddart (Eds.), Unifying Theories of Programming. VIII, 257 pages. 2006.

Vol. 4009: M. Lewenstein, G. Valiente (Eds.), Combinatorial Pattern Matching. XII, 414 pages. 2006.

Vol. 4008: J.C. Augusto, C.D. Nugent (Eds.), Designing Smart Homes. XI, 183 pages. 2006. (Sublibrary LNAI).

Vol. 4007: C. Àlvarez, M. Serna (Eds.), Experimental Algorithms. XI, 329 pages. 2006.

Vol. 4006: L.M. Pinho, M. González Harbour (Eds.), Reliable Software Technologies – Ada-Europe 2006. XII, 241 pages. 2006.

Vol. 4005: G. Lugosi, H.U. Simon (Eds.), Learning Theory. XI, 656 pages. 2006. (Sublibrary LNAI).

Vol. 4004: S. Vaudenay (Ed.), Advances in Cryptology - EUROCRYPT 2006. XIV, 613 pages. 2006.

Vol. 4003: Y. Koucheryavy, J. Harju, V.B. Iversen (Eds.), Next Generation Teletraffic and Wired/Wireless Advanced Networking. XVI, 582 pages. 2006.

Vol. 4001: E. Dubois, K. Pohl (Eds.), Advanced Information Systems Engineering. XVI, 560 pages. 2006.